GLOBAL SOIL REGIONS

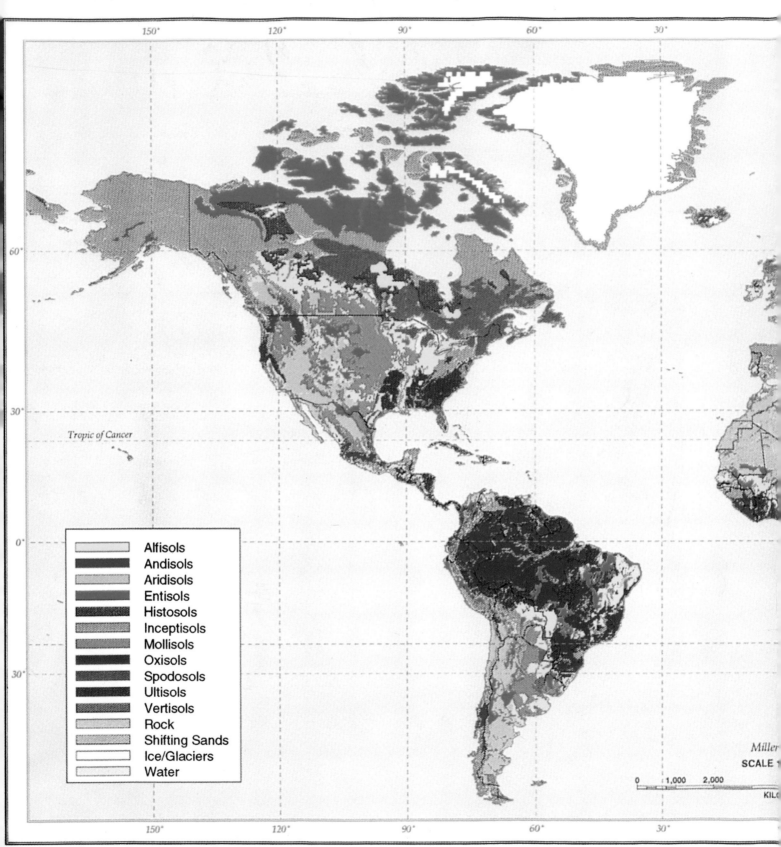

Legend:
- Alfisols
- Andisols
- Aridisols
- Entisols
- Histosols
- Inceptisols
- Mollisols
- Oxisols
- Spodosols
- Ultisols
- Vertisols
- Rock
- Shifting Sands
- Ice/Glaciers
- Water

Tropic of Cancer

Miller

SCALE

0   1,000   2,000

KILO

USDA Natural Resource Conservation Service
Soil Survey Division/World Soil Resources

jection

0,000,000

6,000   8,000

TERS

March 1994
U.S. Agency for International Development

The Nature and Properties of Soils

ELEVENTH EDITION

NYLE C. BRADY
EMERITUS PROFESSOR OF SOIL SCIENCE
CORNELL UNIVERSITY

RAY R. WEIL
PROFESSOR OF SOIL SCIENCE
UNIVERSITY OF MARYLAND AT COLLEGE PARK

PRENTICE-HALL INTERNATIONAL, INC.

Production Editor: Adele Kupchik
Managing Editor: Mary Carnis
Acquisitions Editor: Charles Stewart
Manufacturing Buyer: Ilene Sanford
Director of Manufacturing & Production: Bruce Johnson
Editorial Assistants: Meryl Chertoff & Mollie Pfeiffer
Marketing Manager: Debbie Yarnell
Formatting/page make-up: North Market Street Graphics
Interior illustrations: Mark Ammerman
Printer/Binder: Van Hoffman Press

© 1996 by Prentice-Hall, Inc.
A Simon & Schuster Company
Upper Saddle River, New Jersey 07458

This edition may be sold only in those countries to which it is consigned by Prentice-Hall International. It is not to be re-exported and it is not for sale in the U.S.A., Mexico, or Canada.

Earlier editions by T. Lyttleton Lyon and Harry O. Buckman copyright 1922, 1929, 1937, and 1943 by Macmillan Publishing Co., Inc. Earlier edition by T. Lyttleton, Harry O. Buckman, and Nyle C. Brady copyright 1952 by Macmillan Publishing Co., Inc. Earlier editions by Harry O. Buckman and Nyle C. Brady copyright © 1960 and 1969 by Macmillan Publishing Co., Inc. Copyright renewed 1950 by Bertha C. Lyon and Harry O. Buckman, 1957 and 1965 by Harry O. Buckman, 1961 by Rita S. Buckman. Earlier editions by Nyle C. Brady copyright © 1974, 1984 and 1990 by Macmillan Publishing Company.

Printed in the United States of America

10 9 8 7 6 5 4 3 2 1

ISBN 0-13-243189-0

Prentice-Hall International (UK) Limited, London
Prentice-Hall of Australia Pty. Limited, Sydney
Prentice-Hall Canada Inc., Toronto
Prentice-Hall Hispanoamericana, S.A., Mexico
Prentice-Hall of India Private Limited, New Delhi
Prentice-Hall of Japan, Inc., Tokyo
Simon & Schuster Asia Pte. Ltd., Singapore
Editora Prentice-Hall do Brasil, Ltda., Rio de Janeiro
Prentice-Hall, Upper Saddle River, New Jersey

CONTENTS

PREFACE

A fundamental knowledge of soil science is a prerequisite to meeting the many natural-resource challenges that will face humanity in the 21st century. This new edition of the classic book on soil science emphasizes the soil as a natural resource and highlights the many interactions between the soil and other components of the ecosystem. Throughout the text, soil properties and processes are illustrated with examples from forest, range, agricultural, wetland, and constructed ecosystems. Our priority is to explain the fundamental principles of soil science that have applications relevant to students in many fields of study. We recognize that for some students this book will be their only formal exposure to soil science, while for other students this book represents the initial step in a comprehensive soil science education.

We have therefore sought the advice of professors and students to help us make this book an exciting introduction to the fascinating world of soil science. We also intend for this book to be a "keeper," one that will serve natural-resource and agricultural students as a reliable reference as they pursue their professional activities in the years to come. Therefore, while much is new in substance and style in this edition, we have been careful to maintain the level of rigor and thoroughness that has made previous editions so valuable.

The study of soils can be both fascinating and intellectually satisfying. The soil is an ideal system in which to observe practical applications for basic principles of biology, chemistry, and physics. In turn, these principles can be used to minimize the degradation and destruction of one of our most important natural resources.

We gratefully acknowledge the able research and editorial assistance of Karen Lowell and Joyce Torio. The book has greatly benefited from comments and suggestions contributed by many colleagues, especially those in universities around the country who responded to our questionnaire on the 10th edition. For their advice and counsel we would like to give special thanks to Hari Eswaran, Sharon Waltman, and Diane Shields (USDA/NRCS), George Ley (Tanzania Ministry of Agriculture), Bruce James, Martin Rabenhorst, Paul Shipley, and Richard Weismiller (University of Maryland).

Last, but not least, we must express our special thanks to our wives, Martha and Trish, for their constant encouragement and for their permitting us to utilize almost every free moment during the past 18 months to concentrate on the revision of this textbook.

N. C. B. and R. R. W.

1

THE·SOILS AROUND US

*If we take time to learn the language of the land,
the soil will speak to us.*
—R. WEIL AND W. KROONTJE, LISTENING TO THE LAND

The Earth, our unique home in the vastness of the universe, is in crisis. Depletion of the ozone layer in the upper atmosphere is threatening us with an overload of ultraviolet radiation. Tropical rain forests, and the incredible array of plant and animal species they contain, are disappearing at an unprecedented rate. Groundwater supplies are being contaminated in many areas and depleted in others. In parts of the world, the capacity of soils to produce food is being degraded, even as the number of people needing food is increasing. It will be a great challenge for the current generation to bring the global environment back into balance (Figure 1.1).

FIGURE 1.1 Our planet is unique in that it is covered with life-sustaining water and soil. Great care will be required in preserving the quality of both if our species is to continue to thrive. (Photo courtesy of NASA)

1

Soils are crucial to life on earth. From ozone depletion to rain forest destruction to water pollution, the global ecosystem is impacted in far-reaching ways by the processes carried out in the soil. To a great degree, soil quality determines the nature of plant ecosystems and the capacity of land to support animal life and society. As human societies become increasingly urbanized, fewer people have intimate contact with the soil, and individuals tend to lose sight of the many ways in which they depend upon soils for their prosperity and survival. The degree to which we are dependent on soils is likely to increase, not decrease, in the future. Of course, soils will continue to supply us with nearly all of our food and much of our fiber. On a hot day, would you rather wear a cotton shirt or one made of polyester? In addition, biomass grown on soils is likely to become an increasingly important source of energy and industrial feedstocks, as the world's finite supplies of petroleum are depleted over the coming century. The early signs of this trend can be seen in the soybean oil–based inks, the cornstarch plastics, and the wood alcohol fuels that are becoming increasingly important on the market (Figure 1.2).

The art of soil management is as old as civilization. As we move into the 21st century, new understandings and new technologies will be needed to protect the environment and at the same time produce food and biomass to support society. The study of soil science has never been more important to farmers, engineers, natural resource managers, and ecologists alike.

1.1 FUNCTIONS OF SOILS IN OUR ECOSYSTEM

In any ecosystem, whether your backyard, a farm, a forest, or a regional watershed, soils have five key roles to play (Figure 1.3). First, soil supports the growth of higher plants, mainly by providing a medium for plant roots and supplying nutrient elements that are essential to the entire plant. Properties of the soil often determine the nature of the veg-

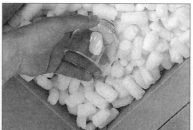

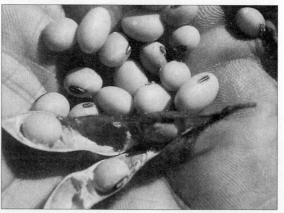

FIGURE 1.2   In the future, we will be increasingly dependent on soils to grow renewable resources that can substitute for dwindling supplies of crude oil. Plastics and inks, for example, can be manufactured from soybean seed oil instead of from petroleum. Soybean oil is edible and far less toxic to the environment. Plastics, such as packaging foam "peanuts," made from cornstarch are readily biodegradable. (Photos courtesy of R. Weil)

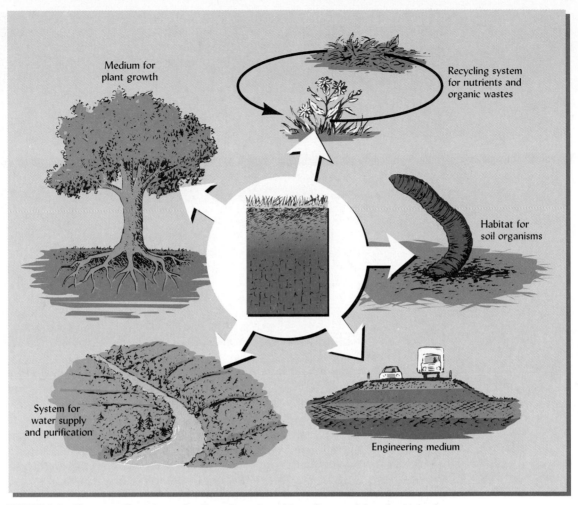

FIGURE 1.3 The many functions of soil can be grouped into five crucial ecological roles.

etation present and, indirectly, the number and types of animals (including people) that the vegetation can support. Second, soil properties are the principal factor controlling the fate of water in the hydrologic system. Water loss, utilization, contamination, and purification are all affected by the soil. Third, the soil functions as nature's recycling system. Within the soil, waste products and dead bodies of plants, animals, and people are assimilated, and their basic elements are made available for reuse by the next generation of life. Fourth, soils provide habitats for a myriad of living organisms, from small mammals and reptiles to tiny insects to microscopic cells of unimaginable numbers and diversity. Finally, in human-built ecosystems, soil plays an important role as an engineering medium. Soil is not only an important building material in the form of earth fill and bricks (baked soil material), but provides the foundation for virtually every road, airport, and house we build.

1.2 MEDIUM FOR PLANT GROWTH

Imagine a growing tree or a corn plant. The aboveground portion of this plant may be most familiar, but the portion of the plant growing below the soil surface, its root system, may be nearly as large as the portion we see above ground. What things do these plants obtain from the soils in which their roots proliferate? It is clear that the soil mass provides physical support, anchoring the root system so that the plant does not fall over. Occasionally, as in Figure 1.4, plants will fall over during windy conditions if shallow soils or restrictive soil layers make them top-heavy.

Plant roots depend on the process of respiration to obtain energy. Since root respiration, like our own respiration, produces carbon dioxide (CO_2) and uses oxygen (O_2), an

FIGURE 1.4 This wet, shallow soil failed to allow sufficiently deep roots to develop to prevent this tree from blowing over when snow-laden branches made it top-heavy during a winter storm. (Photo courtesy of R. Weil)

important function of the soil is "ventilation"—allowing CO_2 to escape and fresh O_2 to enter the root zone. This ventilation is accomplished via the network of soil pore spaces.

An equally important function of the soil pores is to absorb rainwater and hold it where it can be used by plant roots. As long as plant leaves are exposed to sunlight, the plant requires a continuous stream of water to use in cooling, nutrient transport, turgor maintenance, and photosynthesis. Since plants use water continuously, but it usually rains only occasionally, the water-holding capacity of soils is essential for plant survival. A deep soil may store enough water to allow plants to survive long periods without rain (see Figure 1.5).

FIGURE 1.5 The mango tree in this central Africa field is in full leaf and beginning to flower, even though no rain had fallen for the five months prior to when this photograph was taken. The trees are still using water stored from the previous rainy season in this deep clayey soil. It is the local belief that abundant flowering on the mango tree presages a good rainy season to come. Note that the more shallow-rooted grasses growing in thinner soils on the hillside are in a light-colored, dried-up, dormant condition. (Photo courtesy of R. Weil)

As well as moderating moisture changes in the root environment, the soil also moderates temperature fluctuations. Perhaps you can recall digging in garden soil on a summer afternoon and feeling how hot the soil was at the surface and how much cooler just a few centimeters below. The insulating properties of soil protect the deeper portion of the root system from the extremes of hot and cold that often occur at the soil surface. For example, it is not unusual for the temperature at the surface of bare soil to exceed 35 or 40°C in midafternoon, a condition that would be lethal to most plant roots. Just a few centimeters deeper, however, the temperature may be 10°C cooler, allowing roots to function normally.

There are many potential sources of **phytotoxic substances** in soils. These toxins may result from human activity, or they may be produced by plant roots, by microorganisms, or by natural chemical reactions. Many soil managers consider a function of a good soil to be protection of plants from toxic concentrations of such substances by ventilating gases, by decomposing or adsorbing organic toxins, or by suppressing toxin-producing organisms. At the same time, it is true that some microorganisms in soil produce organic, growth-stimulating compounds. These substances, when taken up by plants in small amounts, may improve plant vigor.

Soils supply plants with inorganic, mineral nutrients in the form of dissolved ions. These mineral nutrients include such metallic elements as potassium, calcium, iron, and copper, as well as such nonmetallic elements as nitrogen, sulfur, phosphorus, and boron (Table 1.1). The plant takes these elements out of the soil solution and incorporates most of them into the thousands of different organic compounds that constitute plant tissue. A fundamental role of soil in supporting plant growth is to provide a continuing supply of dissolved mineral nutrients in amounts and relative proportions appropriate for plant growth. While plants may use minute quantities of organic compounds from soils, uptake of these substances is certainly not necessary for normal growth. The organic metabolites, enzymes, and structural compounds making up a plant's dry matter consist mainly of carbon, hydrogen, and oxygen which the plant obtains by photosynthesis from air and water, not from the soil.

It is true that plants can be grown in nutrient solutions without soil (**hydroponics**), but then the plant-support functions of soils must be engineered into the system and maintained at a high cost of time, effort, and management. Although hydroponic production can be feasible on a small scale for a few high-value plants, the production of the world's food and fiber and the maintenance of natural ecosystems will always depend on the use of millions of square kilometers of productive soils.

1.3 REGULATOR OF WATER SUPPLIES

There is much concern about the quality and quantity of the water in our rivers, lakes, and underground aquifers. Governments and citizens everywhere are working to stem

TABLE 1.1 **Elements Essential for Plant Growth and Their Sources[a]**

Used in relatively large amounts (>0.1% of dry plant tissue)		Used in relatively small amounts (<0.1% of dry plant tissue)
Mostly from air and water	*Macronutrients (from soil solids)*	*Micronutrients (from soil solids)*
Carbon (C)	Nitrogen (N)	Iron (Fe)
Hydrogen (H)	Phosphorus (P)	Manganese (Mn)
Oxygen (O)	Potassium (K)	Boron (B)
	Calcium (Ca)	Molybdenum (Mo)
	Magnesium (Mg)	Copper (Cu)
	Sulfur (S)	Zinc (Zn)
		Nickel (Ni)
		Chlorine (Cl)
		Cobalt (Co)

[a]Many other elements are taken up from soils by plants, but are not essential for plant growth. Some of these elements (such as sodium, silicon, iodine, fluorine, barium, and strontium) do enhance the growth of certain plants, but do not appear to be as universally required for normal growth as are the 18 listed in this table.

FIGURE 1.6 The condition of the soils covering these foothills of the Blue Ridge in West Virginia and Maryland will greatly influence the quality and quantity of water flowing down the Potomac River past Washington, D.C., over 100 kilometers downstream. (Photo courtesy of R. Weil)

the pollution that threatens the value of our waters for fishing, swimming, and drinking. For progress to be made in improving water quality, we must recognize that nearly every drop of water in our rivers, lakes, estuaries, and aquifers has either traveled through the soil or flowed over its surface.[1] Imagine, for example, a heavy rain falling on the hills surrounding the river in Figure 1.6. If the soil allows the rain to soak in, some of the water may be stored in the soil and used by the trees and other plants, while some may seep slowly down through the soil layers to the groundwater, eventually entering the river over a period of months or years as base flow. If contaminated, as the water soaks through the upper layers of soil it is purified and cleansed by soil processes that remove many impurities and kill potential disease organisms.

Contrast the preceding scenario with what would occur if the soil were so shallow or impermeable that most of the rain could not penetrate the soil, but ran off the hillsides on the soil surface, scouring surface soil and debris as it picked up speed, and entering the river rapidly and nearly all at once. The result would be a destructive flash flood of muddy water. Clearly, the nature and management of soils in a watershed will have a major influence on the purity and amount of water finding its way to aquatic systems.

1.4 RECYCLER OF RAW MATERIALS

What would a world be like without the recycling functions performed by soils? Without reuse of nutrients, plants and animals would have run out of nourishment long ago. The world would be covered with a layer, possibly hundreds of meters high, of plant and animal wastes and corpses. Obviously, recycling must be a vital process in ecosystems, whether forests, farms, or cities. The soil system plays a pivotal role in the major geochemical cycles. Soils have the capacity to assimilate great quantities of organic waste, turning it into beneficial **humus**, converting the mineral nutrients in the wastes to forms that can be utilized by plants and animals, and returning the carbon to the atmosphere as carbon dioxide, where it again will become a part of living organisms through plant photosynthesis. Some soils can accumulate large amounts of carbon as soil organic mat-

---

[1]This excludes the relatively minor quantity of precipitation that falls directly into bodies of fresh surface water.

ter, thus having a major impact on such global changes as the much-discussed *greenhouse effect* (see Sections 1.12 and 12.2).

1.5   HABITAT FOR SOIL ORGANISMS

When we speak of protecting ecosystems, most people envision a stand of old-growth forest with its abundant wildlife, or perhaps an estuary such as the Chesapeake Bay with its oyster beds and fisheries. (Perhaps, once you have read this book, you will envision a handful of soil when someone speaks of an ecosystem.) Soil is not a mere pile of broken rock and dead debris. A handful of soil may be home to *billions* of organisms, belonging to thousands of species. In even this small quantity of soil, there are likely to exist predators, prey, producers, consumers, and parasites (Figure 1.7).

How is it possible for such a diversity of organisms to live and interact in such a small space? One explanation is the tremendous range of niches and habitats in even a uniform-appearing soil. Some pores of the soil will be filled with water in which swim organisms such as roundworms, diatoms, and rotifers. Tiny insects and mites may be crawling about in other, larger pores filled with moist air. Micro-zones of good aeration may be only millimeters from areas of **anoxic** conditions. Different areas may be enriched with decaying organic materials; some places may be highly acidic, some more basic. Temperature, too, may vary widely.

Hidden from view in the world's soils are communities of living organisms every bit as complex and intrinsically valuable as their counterparts that roam the savannas, forests, and oceans of the earth. Soils harbor much of the earth's genetic diversity. Soils, like air and water, are important components of the larger ecosystem. Yet only now is soil quality taking its place, with air quality and water quality, in discussions of environmental protection.

1.6   ENGINEERING MEDIUM

"*Terra firma*, solid ground." We usually think of the soil as being firm and solid, a good base on which to build roads and all kinds of structures. Indeed, most structures rest on

FIGURE 1.7   The soil is home to a wide variety of organisms, both relatively large and very small. Here, a predatory centipede is about to encounter a detritus-eating sowbug. (Photo courtesy of R. Weil)

the soil, and many construction projects require excavation into the soil. Unfortunately, as can be seen in Figure 1.8, some soils are not as stable as others. Reliable construction on soils, and with soil materials, requires knowledge of the diversity of soil properties, as discussed later in this chapter. Designs for roadbeds or building foundations that work well in one location on one type of soil may be inadequate for another location with different soils.

Working with natural soils or excavated soil materials is not like working with concrete or steel. Properties such as bearing strength, compressibility, shear strength, and stability are much more variable and difficult to predict for soils than for manufactured building materials. Although this book will not go into great detail about engineering properties of soils, many of the physical properties discussed will have direct application to engineering uses of soil. For example, Chapter 8 discusses the swelling properties of certain types of clays in soils. The engineer should be aware that when soils with swelling clays are wetted they expand with sufficient force to crack foundations and buckle pavements. Much of the information on soil properties and soil classification discussed in later chapters will be of great value to people planning land uses that involve construction or excavation.

1.7 SOIL AS A NATURAL BODY

You may have noticed that this book sometimes refers to "the soil," sometimes to "a soil," and sometimes to "soils." *The soil* is often said to cover the land as the peel covers an orange. However, while the peel is relatively uniform around the orange, the soil is highly variable from place to place on Earth. In fact, the soil is a collection of individually different soil bodies. One of these individual bodies, *a soil*, is to *the soil* as an individual tree is to the earth's vegetation. Just as one may find sugar maples, oaks, hemlocks, and many other species of trees in a particular forest, so, too, might one find Christiana clay loams, Sunnyside sandy loams, Elkton silt loams, and other kinds of soils in a particular landscape.

A soil is a three-dimensional natural body in the same sense that a mountain, lake, or valley is. By dipping a bucket into a lake you may sample some of its water. In the same way, by digging or augering a hole into a soil, you may retrieve some soil material. Thus you can take a sample of soil material or water into a laboratory and analyze its contents, but you must go out into the field to study a soil or a lake.

FIGURE 1.8 Better knowledge of the soils on which this road was built may have allowed its engineers to develop a more stable design, thus avoiding this costly and dangerous situation. (Photo courtesy of R. Weil)

In most places, the rock exposed at the earth's surface has crumbled and decayed to produce a layer of unconsolidated debris overlying the hard, unweathered rock. This unconsolidated layer is called the *regolith*, and varies in thickness from virtually nonexistent in some places (i.e., exposed bare rock) to tens of meters in other places. The regolith material, in many instances, has been transported many kilometers from the site of its initial formation and then deposited over the bedrock which it now covers. Thus, all or part of the regolith may or may not be related to the rock now found below it.

The upper part of this regolith, usually the upper 1 to 2 meters, has been affected by the activities of living organisms, and is quite different from the deeper layers. Here, at the interface between the worlds of rock (the lithosphere), air (the atmosphere), water (the hydrosphere), and living things (the biosphere), soil is born. The transformation of inorganic rock and debris into a living soil is one of nature's most fascinating displays. Although generally hidden from everyday view, the soil and regolith can often be seen in road cuts and other excavations (Figure 1.9).

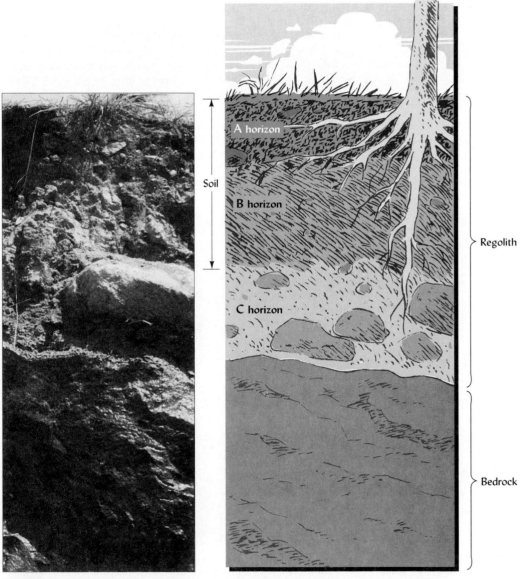

FIGURE 1.9 Relative positions of the regolith, its soil, and the underlying bedrock. Note that the soil is a part of the regolith, and that the A and B horizons are part of the **solum** (from the Latin word *solum*, which means soil or land). The C horizon is the part of the regolith that underlies the solum, but may be slowly changing into soil in its upper parts. Sometimes the regolith is so thin that it has been changed entirely to soil; in such a case, soil rests directly on the bedrock. (Photo courtesy of R. Weil)

A soil is the product of both destructive and creative (synthetic) processes. Weathering of rock and microbial decay of organic residues are examples of destructive processes, whereas the formation of new minerals, such as certain clays, and of new stable organic compounds are examples of synthesis. Perhaps the most striking result of synthetic processes is the formation of horizontal layers called **soil horizons**. The development of these horizons in the upper regolith is a unique characteristic of soil that sets it apart from the deeper regolith materials.

1.8   THE SOIL PROFILE AND ITS LAYERS (HORIZONS)

Soil scientists often dig a large hole, called a *soil pit*, usually several meters deep and about a meter wide, to expose soil horizons for study. The vertical section exposing a set of horizons in the wall of such a pit is termed a ***soil profile***. Road cuts and other ready-made excavations can expose soil profiles and serve as windows on the soil. In an excavation open for some time, horizons are often obscured by soil material that has been washed by rain from upper horizons to cover the exposed face of lower horizons. For this reason, horizons may be more clearly seen if a fresh face is exposed by scraping off a layer of material several centimeters thick from the pit wall. Observing how soils exposed in road cuts vary from place to place can add a fascinating new dimension to travel. Once you have learned to interpret the different horizons (see Chapter 2), soil profiles can warn you about potential problems in using the land, as well as tell you much about the environment and history of a region. For example, soils developed in a dry region will have very different horizons from those developed in a humid region.

Horizons within a soil may vary in thickness and have somewhat irregular boundaries, but generally they parallel the land surface (Figure 1.10). This horizontal alignment is expected since the differentiation of the regolith into distinct horizons is largely the result of influences, such as air, water, solar radiation, and plant material,

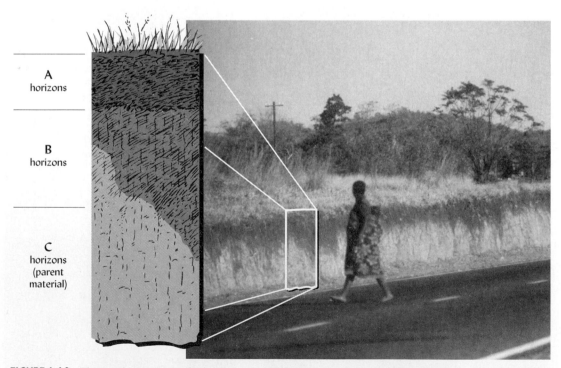

FIGURE 1.10 This road cut in central Africa reveals soil layers or horizons which parallel the land surface. Taken together, these horizons comprise the profile of this soil, as shown in the enlarged diagram. The upper horizons are designated *A horizons*. They are usually higher in organic matter and darker in color than the lower horizons. Some constituents, such as iron oxides and clays, have been moved downward from the A horizons by percolating rainwater. The lower horizon, called a *B horizon*, is sometimes a zone in which clays and iron oxides have accumulated, and in which distinctive structure has formed. The presence and characteristics of the horizons in this profile distinguish this soil from the thousands of other soils in the world. (Photo courtesy of R. Weil)

originating at the soil–atmosphere interface. Since the weathering of the regolith occurs first at the surface and works its way down, the uppermost layers have been changed the most, while the deepest layers are most similar to the original regolith, which is referred to as the soil's *parent material*. In places where the regolith was originally rather uniform in composition, the material below the soil may have a similar composition to the parent material from which the soil formed. In other cases, the regolith material has been transported long distances by wind, water, or glaciers and deposited on top of dissimilar material. In such a case, the regolith material found below a soil may be quite different from the upper layer of regolith in which the soil formed.

Organic matter from decomposed plant leaves and roots tends to accumulate in the uppermost horizons of a soil profile, giving these horizons a darker color than the lower horizons. Also, as weathering tends to be most intense in the upper layers, in many soils these layers have lost some clay or other weathering products by leaching to the horizons below. The organic-matter-enriched horizons nearest the soil surface are called the *A horizons*. In some soils, intensely weathered and leached horizons that have not accumulated organic matter occur in the upper part of the profile, usually just below the A horizon. These horizons may be designated *E horizons* (Figure 1.11).

The layers underlying the A horizon contain comparatively less organic matter than the horizons nearer the surface. Varying amounts of silicate clays, iron and aluminum oxides, gypsum, or calcium carbonate may accumulate in the underlying horizons. The accumulated materials may have been washed down from the horizons above, or they may have been formed in place through the weathering process. These underlying layers are referred to as *B horizons* (Figure 1.11).

In some soil profiles, the component horizons are very distinct in color, with sharp boundaries that can be seen easily by even novice observers. In other soils, the color changes between horizons may be very gradual, and the boundaries more difficult to locate. However, color is only one of many properties by which one horizon may be dis-

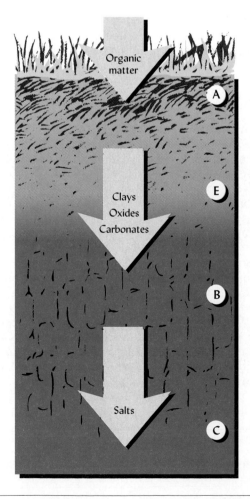

FIGURE 1.11 Horizons begin to differentiate as materials are added to the upper part of the profile and other materials are translocated to deeper zones. Under certain conditions, usually associated with forest vegetation and high rainfall, a leached E horizon forms between organic-matter-rich A and the B horizons. If sufficient rainfall occurs, soluble salts will be carried below the soil profile, perhaps all the way to the groundwater.

FIGURE 1.12 (Left) This soil profile was exposed by digging a pit about 2 meters deep in a well-developed soil (a Hapludalf) in southern Michigan. The top horizon can be easily distinguished because it has a darker color than those below it. However, some of the horizons in this profile are difficult to discern on the basis of color differences, especially in this black-and-white photo. The white string was attached to the profile to clearly demarcate some of the horizon boundaries. Then a trowel full of soil material was removed from each horizon and placed on a horizontal board, at right. It is clear from the way the soil either crumbled or held together that soil material from horizons with very similar colors may have very different physical properties. (Photos courtesy of R. Weil)

tinguished from the horizon above or below it (see Figure 1.12). The study of soils in the field is a sensual as well as an intellectual activity. Delineation of the horizons present in a soil profile often requires a careful examination, using all the senses. In addition to seeing the colors in a profile, a soil scientist may feel, smell, and listen[2] to the soil, as well as conduct chemical tests, in order to distinguish the horizons present.

1.9 TOPSOIL AND SUBSOIL

The organically enriched A horizon at the soil surface is sometimes referred to as *topsoil*. When a soil is plowed and cultivated, the natural state of the upper 12 to 25 centimeters (5 to 10 inches) is modified. In this case, the topsoil may also be called the *plow layer* (or the *furrow slice* in situations where a moldboard plow has turned or "sliced" the upper part of the soil). Even where a plow is no longer used, the plow layer will remain evident for many years. For example, if you stroll through a typical New England forest you will see towering trees some 100 years old, but should you dig a shallow pit in the forest floor, you might still be able to see the smooth boundary between the century-old plowed layer and the lighter-colored, undisturbed soil below.

In cultivated soils, the majority of plant roots can be found in the topsoil (Figure 1.13). The topsoil contains a large part of the nutrient and water supplies needed by plants. The chemical properties and nutrient supply of the topsoil may be easily altered by mixing in organic and inorganic amendments, thereby making it possible to improve or maintain the soil's fertility and, to a lesser degree, its productivity. The physical structure of the topsoil, especially the part nearest the surface, is also readily affected by management operations such as tillage and application of organic materials. Maintaining an open structure at the soil surface is especially critical for providing balanced air and water supplies to plant roots and for avoiding excessive losses of soil and water by runoff. Sometimes the plow layer is removed from a soil and sold as topsoil for use at another site. This use of topsoil is especially common to provide a rooting medium suitable for

---

[2]For example, the grinding sound emitted by wet soil rubbed between one's fingers indicates the sandy nature of the soil.

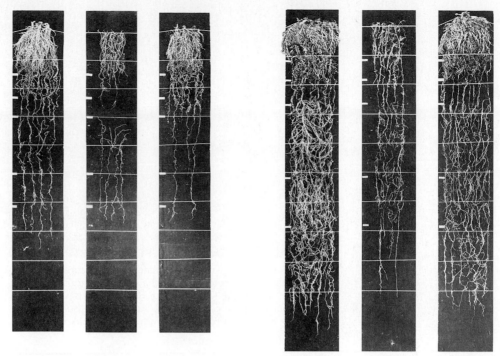

FIGURE 1.13 Plant roots respond to the varying conditions they find in the horizons of a soil profile. Roots tend to proliferate in the better-aerated, more fertile, and looser A horizon than in the horizons below. Improvement of soil fertility in the A horizon not only enhances root growth there, but may also increase the vigor and extent of the root system deeper in the profile as well. The roots shown are from a corn crop grown on an Illinois soil (Cisne series) that received no fertilizers or crop residues (left) or that received both fertilizers and crop residues (right). (Photos courtesy of J. B. Fehrenbacher, University of Illinois)

lawns and shrubs around newly constructed buildings, where the original topsoil was removed or buried, and the underlying soil layers were exposed during grading operations (Figure 1.14).

The soil layers that underlie the topsoil are referred to as *subsoil*. Not normally seen from the surface, the subsoil lies below the usual depth of tillage. However, the characteristics of the subsoil horizons can greatly influence most land uses. Much of the water needed by plants is stored in the subsoil. Many subsoils also supply important quantities of certain plant nutrients. In some soils there is an abrupt change in properties between the topsoil and the subsoil. In other soils, the change is gradual and the upper part of the subsoil may be quite similar to the topsoil. In most soils the properties of the topsoil are far more conducive to plant growth than those in the subsoil. That is why there is often a good correlation between the productivity of a soil and the thickness of the topsoil in a profile.

Impermeable subsoil layers can impede root penetration, as can very acid subsoils. Poor drainage in the subsoil can result in waterlogged conditions in the topsoil. Because of its relative inaccessibility, it is usually much more difficult and expensive to physically or chemically modify the subsoil. On the other hand, good fertilization of the topsoil can produce vigorous plants whose roots are capable of greater exploration in subsoil layers (Figure 1.13).

1.10 SOIL: THE INTERFACE OF AIR, MINERALS, WATER, AND LIFE

We stated that where the regolith meets the atmosphere, the worlds of air, rock, water, and living things are intermingled. In fact, the four major components of soil are air, water, mineral matter, and organic matter. The relative proportions of these four components greatly influence the behavior and productivity of soils. In a soil, the four

FIGURE 1.14 The large mound of material at this construction site consists of topsoil (A horizon material) carefully separated from the lower horizons and pushed aside during initial grading operations. This stockpile was then seeded with grasses to give it a protective cover. After the construction activities are complete, the stockpiled topsoil will be used in landscaping the grounds around the new building. (Photo courtesy of R. Weil)

components are mixed in complex patterns; however, the proportion of soil volume occupied by each component can be represented in a simple pie chart. Figure 1.15 shows the approximate proportions (by volume) of the components found in a loam surface soil in good condition for plant growth. Although a handful of soil may at first seem to be a solid thing, it should be noted that only about half the soil volume consists of solid material (mineral and organic); the other half consists of pore spaces filled with air or water. Of the solid material, typically most is mineral matter derived from the rocks of the earth's crust. Only about 5% of the *volume* in this ideal soil consists of organic matter. However, the influence of the organic component on soil properties is often far greater than its small proportion would suggest. Since it is far less dense than mineral matter, the organic matter accounts for only about 2% of the *weight* of this soil.

The spaces between the particles of solid material are just as important to the nature of a soil as are the solids themselves. It is in these pore spaces that air and water circulate, roots grow, and microscopic creatures live. Plant roots need both air and water. In an

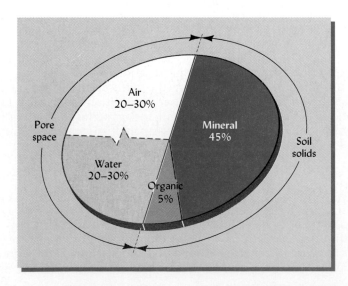

FIGURE 1.15 Volume composition of a loam surface soil when conditions are good for plant growth. The broken line between water and air indicates that the proportions of these two components fluctuate as the soil becomes wetter or drier. Nonetheless, a nearly equal proportion of air and water is generally ideal for plant growth.

optimum condition for most plants, the pore space will be divided roughly equally among the two, with 25% of the soil volume consisting of water and 25% consisting of air. If there is much more water than this, the soil will be waterlogged. If much less water is present, plants will suffer from drought. The relative proportions of water and air in a soil typically fluctuate greatly as water is added or lost. Soils with much more than 50% of their volume in solids are likely to be too compacted for good plant growth. Compared to surface soil layers, subsoils tend to contain less organic matter, less total pore space, and a larger proportion of small pores (*micropores*) which tend to be filled with water rather than with air.

1.11 MINERAL (INORGANIC) CONSTITUENTS OF SOILS

Except in the case of organic soils, most of a soil's solid framework consists of **mineral**[3] particles. The larger soil particles, which include stones, gravel, and coarse sands, are generally rock fragments of various kinds. That is, these larger particles are often aggregates of several different minerals. Most smaller particles tend to be made of a single mineral. Thus, any particular soil is made up of particles that vary greatly in both size and composition.

The mineral particles present in soils are extremely variable in size. Excluding, for the moment, the larger rock fragments such as stones and gravel, soil particles range in size over 4 orders of magnitude: from 2.0 millimeters (mm) to smaller than 0.0002 mm in diameter. **Sand** particles are probably most familiar to us. Individual sand particles are large enough (2.0 to 0.05 mm) to be seen by the naked eye and feel gritty when rubbed between the fingers. Sand particles do not adhere to one another, therefore sands do not feel sticky. **Silt** particles are somewhat smaller (0.05 to 0.002 mm). Silt particles are too small to see without a microscope or to feel individually, so silt feels smooth but not sticky, even when wet. The smallest class of mineral particles are the **clays** (<0.002 mm), which adhere together to form a sticky mass when wet and hard clods when dry. The smaller particles (<0.001 mm) of clay (and similar-sized organic particles) have **colloidal**[4] properties and can be seen only with the aid of an electron microscope. Because of their extremely small size, colloidal particles possess a tremendous amount of surface area per unit of mass. Since the surface of soil colloids (both mineral and organic) exhibit electromagnetic charges that attract positive and negative ions as well as water, this fraction of the soil is the seat of most of the soil's chemical and physical activity.

The proportions of particles in these different size ranges is called **soil texture**. Terms such as *sandy loam*, *silty clay*, and *clay loam* are used to identify the soil texture. Texture has a profound influence on many soil properties, and it affects the suitability of a soil for most uses. To understand the degree to which soil properties can be influenced by texture, imagine dressing in a bathing suit and lying first on a sandy beach, and then in a clayey mud puddle. The difference in these two experiences would be due largely to the properties described in lines 5 and 7 of Table 1.2. Other properties related to particle size are also listed in Table 1.2. Note that clay-sized particles play a dominant role in holding certain inorganic chemicals and supplying nutrients to plants.

To anticipate the effect of clay on the way a soil will behave, it is not enough to know only the *amount* of clay in a soil. It is also necessary to know the *kinds* of clays present. As home builders and highway engineers know all too well, certain clayey soils, such as those high in smectite clays, make very unstable material on which to build because the clays swell when the soil is wet and shrink when the soil dries. This shrink-and-swell action can easily crack foundations and cause even heavy retaining walls to collapse. These clays also become extremely sticky and difficult to work when they are

---

[3]The word *mineral* is used in soil science in three ways: (1) as a general adjective to describe inorganic materials derived from rocks, (2) as a specific noun to refer to distinct minerals found in nature, such as quartz and feldspars (see chapter 2 for detailed discussions of soil-forming minerals and the rocks in which they are found), and (3) as an adjective to describe chemical elements, such as nitrogen and phosphorus, in their inorganic state in contrast to their occurrence as part of organic compounds.

[4]Colloidal systems are two-phase systems in which very small particles of one substance are dispersed in a medium of a different substance. Clay and organic soil particles smaller than about 0.001 mm (1 micrometer, μm) in diameter are generally considered to be colloidal in size. Other examples of colloidal systems in which very small solid particles are dispersed in a liquid medium are milk and blood.

TABLE 1.2 Some General Properties of the Three Major Size Classes of Inorganic Soil Particles

	Property	Sand	Silt	Clay
1.	Range of particle diameters in mm	2.0–0.05	0.05–0.002	Smaller than 0.002
2.	Means of observation	Naked eye	Microscope	Electron microscope
3.	Dominant minerals	Primary	Primary and secondary	Secondary
4.	Attraction of particles for each other	Low	Medium	High
5.	Attraction of particles for water	Low	Medium	High
6.	Ability to hold chemicals and nutrients in plant-available form	Very low	Low	High
7.	Consistency when wet	Loose, gritty	Smooth	Sticky, malleable
8.	Consistency when dry	Very loose, gritty	Powdery, some clods	Hard clods

wet. Other types of clays, formed under different conditions, can be very stable and easy to work with. Learning about the different types of clay minerals will help us understand many of the physical and chemical differences among soils in various parts of the world.

Primary and Secondary Minerals

Minerals that have persisted with little change in composition since they were extruded in molten lava (e.g., quartz, micas, and feldspars) are known as **primary minerals**. They are prominent in the sand and silt fractions of soils. Other minerals, such as silicate clays and iron oxides, were formed by the breakdown and weathering of less resistant minerals as soil formation progressed. These minerals are called **secondary minerals** and tend to dominate the clay and, in some cases, silt fractions.

The inorganic minerals in the soil are the original source of most of the chemical elements essential for plant growth. Although the bulk of these nutrients is held rigidly as components of the basic crystalline structure of the minerals, a small but important portion is in the form of charged ions on the surface of fine colloidal particles (clays and organic matter). Mechanisms of critical importance to growing plants allow plant roots to have access to these surface-held nutrient ions (see Section 1.16).

Soil Structure

Sand, silt, and clay particles can be thought of as the building blocks from which soil is constructed. The manner in which these building blocks are arranged together is called **soil structure**. The particles may remain relatively independent of each other, but more commonly they are associated together in aggregates of different-size particles. These aggregates may take the form of roundish granules, cubelike blocks, flat plates, or other shapes. Soil structure (the way particles are arranged together) is just as important as soil texture (the relative amounts of different sizes of particles) in governing how water and air move in soils. Both structure and texture fundamentally influence the suitability of soils for the growth of plant roots.

1.12 SOIL ORGANIC MATTER

Soil organic matter consists of a wide range of organic (carbonaceous) substances, including living organisms (the soil **biomass**), carbonaceous remains of organisms which once occupied the soil, and organic compounds produced by current and past metabolism in the soil. The remains of plants, animals, and microorganisms are continuously broken down in the soil and new substances are synthesized by other microorganisms. Over periods of time ranging from hours to centuries, organic matter is lost from the soil as carbon dioxide produced by microbial respiration. Because of such loss, repeated additions of new plant and/or animal residues are necessary to maintain soil organic matter.

Under conditions which favor plant production more than microbial decay, large quantities of atmospheric carbon dioxide used by plants in photosynthesis are sequestered in the abundant plant tissues which eventually become part of the soil organic matter. Since carbon dioxide is a major cause of the greenhouse effect which is believed to be warming the earth's climate, the balance between accumulation of soil organic matter and its loss through microbial respiration has global implications. In fact, more carbon is stored in the world's soils than in the world's plant biomass and atmosphere combined.

Even so, organic matter comprises only a small fraction of the mass of a typical soil. By weight, typical well-drained mineral surface soils contain from 1 to 6% organic matter. The organic matter content of subsoils is even smaller. However, the influence of organic matter on soil properties, and consequently on plant growth, is far greater than the low percentage would indicate.

Organic matter binds mineral particles into a granular soil structure that is largely responsible for the loose, easily managed condition of productive soils. Part of the soil organic matter that is especially effective in stabilizing these granules consists of certain gluelike substances produced by various soil organisms, including plant roots (Figure 1.16).

Organic matter also increases the amount of water a soil can hold and the proportion of water available for plant growth (Figure 1.17). In addition, it is a major soil source of the plant nutrients phosphorus and sulfur, and the primary source of nitrogen for most plants. As soil organic matter decays, these nutrient elements, which are present in organic combinations, are released as soluble ions that can be taken up by plant roots. Finally, organic matter, including plant and animal residues, is the main food that supplies carbon and energy to soil organisms. Without it, biochemical activity so essential for ecosystem functioning would come to a near standstill.

Humus, usually black or brown in color, is a collection of very complex organic compounds which accumulate in soil because they are relatively resistant to decay. Just as clay is the colloidal fraction of soil mineral matter, so humus is the colloidal fraction of soil organic matter. Because of their charged surfaces, both humus and clay act as contact bridges between larger soil particles; thus both play an important role in the formation of soil structure. The surface charges of humus, like those of clay, attract and hold both nutrient ions and water molecules. However, gram for gram, the capacity of humus to hold nutrients and water is far greater than that of clay. Unlike clay, humus contains certain components that can have a hormone-like stimulatory effect on plants. All in all, small amounts of humus may remarkably increase the soil's capacity to promote plant growth.

1.13 SOIL WATER: A DYNAMIC SOLUTION

Water is of vital importance in the ecological functioning of soils. The presence of water in soils is essential for the survival and growth of plants and other soil organisms. The soil moisture regime, often reflective of climatic factors, is a major determinant of the productivity of terrestrial ecosystems, including agricultural systems. Movement of water, and substances dissolved in it, through the soil profile is of great consequence to

FIGURE 1.16 Abundant organic matter, including plant roots, helps create physical conditions favorable for the growth of higher plants as well as microbes (left). In contrast, soils low in organic matter, especially if they are high in silt and clay, are often cloddy (right) and not suitable for optimum plant growth.

FIGURE 1.17 Soils higher in organic matter are darker in color and have greater water-holding capacities than soils low in organic matter. The soil in each container has the same texture, but the one on the right has been depleted of much of its organic matter. The same amount of water was applied to each container. As the lower photo shows, the depth of water penetration was less in the high organic matter soil (left) because of its greater water-holding capacity. It required a greater volume of the low organic matter soil (right) to hold the same amount of water.

the quality and quantity of local and regional water resources. Water moving through the regolith is also a major driving force in soil formation.

Two main factors help explain why **soil water** is different from our everyday concept of, say, drinking water in a glass.

1. Water is held within soil pores with varying degrees of tenacity depending on the amount of water present and the size of the pores. The attraction between water and the surfaces of soil particles greatly restricts the ability of water to flow as it would flow in a drinking glass.

2. Because soil water is never pure water, but contains hundreds of dissolved organic and inorganic substances, it may be more accurately called the **soil solution**. An important function of the soil solution is to serve as a constantly replenished, dilute nutrient solution bringing dissolved nutrient elements (e.g., calcium, potassium, nitrogen, and phosphorus) to plant roots.

When the soil moisture content is optimum for plant growth (Figure 1.15), the water in the large and intermediate-sized pores can move in the soil and can be used by plants. As some of the moisture is removed by the growing plants, however, that which remains is in the tiny pores and in thin films around soil particles. The soil solids strongly attract this soil water and consequently compete with plant roots for it. Thus, not all soil water is *available* to plants. Depending on the soil, one-fourth to two-thirds of the moisture remains in the soil after the plants have wilted or died for lack of water.

The soil solution contains small but significant quantities of soluble inorganic compounds, some of which supply elements that are essential for plant growth. Refer to Table 1.1 for a listing of the 18 **essential elements** along with their sources. The soil solids, particularly the fine organic and inorganic colloidal particles, release these elements to the soil solution from which they are taken up by growing plants. Such exchanges, which are critical for higher plants, are dependent on both soil water and the fine soil solids.

One other critical property of the soil solution is its *acidity* or *alkalinity*. Many chemical and biological reactions are dependent on the levels of the H^+ and OH^- ions in the soil. The levels of these ions also influence the solubility, and in turn the availability, of several essential nutrient elements (including iron and manganese) to plants.

The concentrations of H^+ and OH^- ions in the soil solution are commonly ascertained by measuring the *pH* of the soil solution. Technically, the pH is the negative logarithm of the concentration of H^+ ions in the soil solution. Figure 1.18 shows very simply the relationship between pH and the concentration of H^+ and OH^- ions. It should be studied carefully along with Figure 1.19, which shows ranges in pH commonly encountered in soils from different climatic regions. Sometimes referred to as a *master variable*, the pH controls the nature of many chemical and microbial reactions in the soil. It is of great significance in essentially all aspects of soil science.

1.14 SOIL AIR: A CHANGING MIXTURE OF GASES

Approximately half of the volume of the soil consists of pore spaces of varying sizes (refer to Figure 1.15) which are filled with either water or air. When water enters the soil, it displaces air from some of the pores; the air content of a soil is therefore inversely related to its water content. If we think of the network of soil pores as the ventilation system of the soil connecting airspaces to the atmosphere, we can understand that when the smaller pores are filled with water the ventilation system becomes clogged. Think how stuffy the air would become if the ventilation ducts of a classroom became clogged. Because oxygen could not enter the room, nor carbon dioxide leave it, the air in the room would soon become depleted of oxygen and enriched in carbon dioxide and water vapor by the respiration (breathing) of the people in it. In an air-filled soil

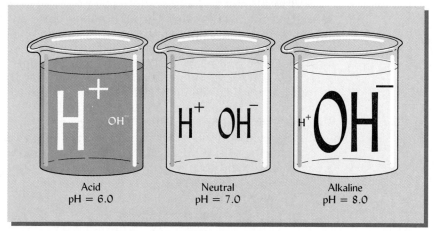

Acid	Neutral	Alkaline
pH = 6.0	pH = 7.0	pH = 8.0

FIGURE 1.18 Diagrammatic representation of acidity, neutrality, and alkalinity. At neutrality the H^+ and OH^- ions of a solution are balanced, their respective numbers being the same (pH 7). At pH 6, the H^+ ions are dominant, being 10 times greater, whereas the OH^- ions have decreased proportionately, being only one-tenth as numerous. The solution therefore is acid at pH 6, there being 100 times more H^+ ions than OH^- ions present. At pH 8, the exact reverse is true; the OH^- ions are 100 times more numerous than the H^+ ions. Hence, the pH 8 solution is alkaline. This mutually inverse relationship must always be kept in mind when pH data are used.

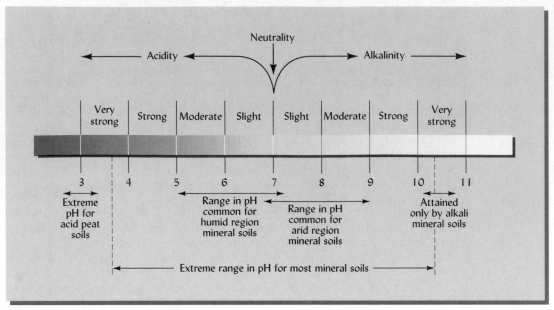

FIGURE 1.19 Extreme range in pH for most mineral soils and the ranges commonly found in humid region and arid region soils. Also indicated are the maximum alkalinity for alkali soils and the minimum pH likely to be encountered in very acid peat soils.

pore surrounded by water-filled smaller pores, the metabolic activities of plant roots and microorganisms have a similar effect.

Therefore, soil air differs from atmospheric air in several respects. First, the composition of soil air varies greatly from place to place in the soil. In local pockets, some gases are consumed by plant roots and by microbial reactions, and others are released, thereby greatly modifying the composition of the soil air. Second, soil air generally has a higher moisture content than the atmosphere; the relative humidity of soil air approaches 100% unless the soil is very dry. Third, the content of carbon dioxide (CO_2) is usually much higher, and that of oxygen (O_2) lower, than contents of these gases found in the atmosphere. Carbon dioxide in soil air is often several hundred times more concentrated than the 0.03% commonly found in the atmosphere. Oxygen decreases accordingly and, in extreme cases, may be 5 to 10%, or even less, compared to about 20% for atmospheric air.

The amount and composition of air in a soil are determined to a large degree by the water content of the soil. The air occupies those soil pores not filled with water. As the soil drains from a heavy rain or irrigation, large pores are the first to be filled with air, followed by medium-sized pores, and finally the small pores, as water is removed by evaporation and plant use. This explains the tendency for soils with a high proportion of tiny pores to be poorly aerated. In such soils, water dominates, and the soil air content is low, as is the rate of diffusion of the air in to and out of the soil from the atmosphere. The result is high levels of CO_2 and low levels of O_2, unsatisfactory conditions for the growth of most plants. In extreme cases, lack of oxygen both in the soil air and dissolved in the soil water may fundamentally alter the chemical reactions that take place in the soil solution. This is of particular importance to understanding the functions of wetland soils.

1.15 INTERACTION OF FOUR COMPONENTS TO SUPPLY PLANT NUTRIENTS

As you read our discussion of each of the four major soil components, you may have noticed that the impact of one component on soil properties is seldom expressed independently from that of the others. Rather, the four components interact with each other to determine the nature of a soil. Thus, soil moisture, which directly meets the needs of plants for water, simultaneously controls much of the air and nutrient supply

TABLE 1.3   Quantities of Six Essential Elements Found in Representative Soils in Temperate Regions

Essential element	Humid region soil			Arid region soil		
	In solid framework (kg/ha)	Exchangeable (kg/ha)	In soil solution (kg/ha)	In solid framework (kg/ha)	Exchangeable (kg/ha)	In soil solution (kg/ha)
Ca	8,000	2,250	60–120	20,000	5,625	140–280
Mg	6,000	450	10–20	14,000	900	25–40
K	38,000	190	10–30	45,000	250	15–40
P	900	—	0.05–0.15	1,600	—	0.1–0.2
S	700	—	2–10	1,800	—	6–30
N	3,500	—	7–25	2,500	—	5–20

to the plant roots. The mineral particles, especially the finest ones, attract soil water, thus determining its movement and availability to plants. Likewise, organic matter, because of its physical binding power, influences the arrangement of the mineral particles into clusters and, in so doing, increases the number of large soil pores, thereby influencing the water and air relationships.

Essential Element Availability

Perhaps the most important interactive process involving the four soil components is the provision of essential nutrient elements to plants. Plants absorb essential nutrients, along with water, directly from one of these components: the soil solution. However, the amount of essential nutrients in the soil solution at any one time is far less than is needed to produce a mature plant. Consequently, the soil solution nutrient levels must be constantly replenished from the inorganic or organic parts of the soil and from fertilizers or manures added to agricultural soils.

Fortunately, relatively large quantities of these nutrients are associated with both inorganic and organic soil solids. By a series of chemical and biochemical processes, nutrients are released from these solid forms to replenish those in the soil solution. For example, through *ion exchange*, essential elements such as Ca^{2+} and K^+ are released from the colloidal surfaces of clay and humus to the soil solution. The intimate contact between soil solution ions and adsorbed[5] ions makes this exchange possible. In the example below, an H^+ ion in the soil solution is shown to exchange readily with an adsorbed K^+ ion on the colloidal surface. The K^+ ion is then available in the soil solution for uptake by crop plants.

$$\boxed{\text{colloid}}\ K^+ + H^+ \text{ ion} \longrightarrow \boxed{\text{colloid}}\ H^+ + K^+ \text{ ion}$$

Adsorbed   Soil solution   Adsorbed   Soil solution

The K^+ ion thus released can be readily taken up (absorbed) by plants. Some scientists consider that this ion exchange process is among the most important of chemical reactions in nature.

Nutrient ions are also released to the soil solution as soil microorganisms decompose organic tissues. Plant roots can readily absorb all of these nutrients from the soil solution provided there is enough O_2 in the soil air to support root metabolism.

Most soils contain large amounts of plant nutrients relative to the annual needs of growing vegetation. However, the bulk of most nutrient elements is held in the structural framework of primary and secondary minerals and organic matter. Only a small fraction of the nutrient content of a soil is present in forms which are readily available to plants. Table 1.3 will give you some idea of the quantities of various essential elements present in different forms in typical soils of humid and arid regions.

---

[5]*Adsorption* refers to the attraction of ions to the surface of particles, in contrast to *absorption*, the process by which ions are taken *into* plant roots.

Figure 1.20 illustrates how the two solid soil components interact with the liquid component (soil solution) to provide essential elements to plants. Plant roots do not ingest soil particles, no matter how fine, but are only able to absorb nutrients that are dissolved in the soil solution. Because elements in the coarser soil framework of the soil are only slowly released into the soil solution over long periods of time, the bulk of most nutrients in a soil is not readily available for plant use. Nutrient elements in the framework of colloid particles are somewhat more readily available to plants, as these particles break down much faster because of their greater surface area. Thus, the structural framework is the major storehouse and, to some extent, a significant source of essential elements in many soils.

The distribution of nutrients among the various components of a fertile soil, as illustrated in Figure 1.20, may be likened to the distribution of financial assets in the portfolio of a wealthy individual. In such an analogy, nutrients readily available for plant use would be analogous to cash in the individual's pocket. A millionaire would likely keep most of his or her assets in long-term investments such as real estate or bonds (the coarse fraction solid framework), while investing a smaller amount in short-term stocks and bonds (colloidal framework). For more immediate use, an even smaller amount might be kept in a checking account (exchangeable nutrients), while a tiny fraction of the overall wealth might be carried to spend as currency and coins (nutrients in the soil solution). As the cash is used up, the supply is replenished by making a withdrawal from the checking account. The checking account, in turn, is replenished occasionally by the sale of long-term investments. It is possible for a wealthy person to run short of cash even though he or she may own a great deal of valuable real estate. In an analogous way, plants may use up the readily available supply of a nutrient even though the total supply of that nutrient in the soil is very large. Luckily, in a fertile soil, the process described

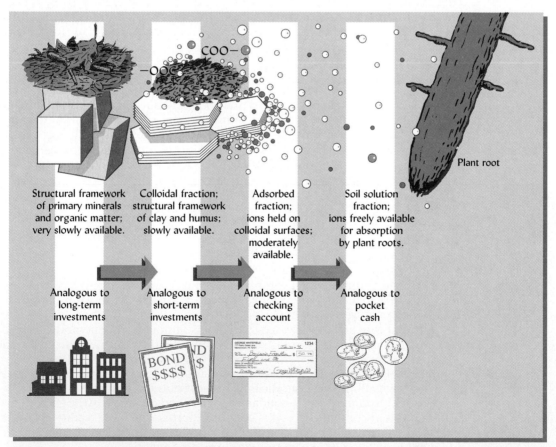

FIGURE 1.20 Nutrient elements exist in soils in various forms characterized by different accessibility to plant roots. The bulk of the nutrients is locked up in the structural framework of primary minerals, organic matter, clay, and humus. A smaller proportion of each nutrient is adsorbed in a swarm of ions near the surfaces of soil colloids (clay and organic matter). From the swarm of adsorbed ions, a still smaller amount is released into the bulk soil solution where uptake by plant roots can take place.

in Figure 1.20 can help replenish the soil solution as quickly as plant roots remove essential elements.

1.16 NUTRIENT UPTAKE BY PLANT ROOTS

To be taken up by a plant, a nutrient element must be in a soluble form and must be located *at the root surface*. Often, parts of a root are in such intimate contact with soil particles (see Figure 1.21) that a direct exchange may take place between nutrient ions adsorbed on the surface of soil colloids and H^+ ions from the surface of root cell walls. In any case, the supply of nutrients in contact with the root will soon be depleted. This fact raises the question of how a root can obtain additional supplies once the nutrient ions at the root surface have all been taken up into the root. There are three basic mechanisms by which the concentration of nutrient ions at the root surface is maintained.

First, **root interception** comes into play as roots continually grow into new, undepleted soil. For the most part, however, nutrient ions must travel some distance in the soil solution to reach the root surface. This movement can take place by **mass flow**, as when dissolved nutrients are carried along with the flow soil water toward a root that is actively drawing water from the soil. In this type of movement, the nutrient ions are somewhat analogous to leaves floating down a stream. On the other hand, plants can continue to take up nutrients even at night, when water is not being drawn into the roots. Nutrient ions continually move by **diffusion** from areas of greater concentration toward the nutrient-depleted areas of lower concentration around the root surface.

In the diffusion process, the random movements of ions in all directions causes a *net* movement from areas of high concentration to areas of lower concentrations, independent of any mass flow of the water in which the ions are dissolved. Factors such as soil compaction, cold temperatures, and low soil moisture content, which reduce root interception, mass flow, or diffusion, can result in poor nutrient uptake by plants even in soils with adequate supplies of soluble nutrients. Furthermore, the availability of nutrients for uptake can also be negatively or positively influenced by the activities of microorganisms that thrive in the immediate vicinity of roots. Maintaining the supply of available nutrients at the plant root surface is thus a process that involves complex interactions among different soil components.

It should be noted that the plant membrane separating the inside of the root cell from the soil solution is permeable to dissolved ions only under special circumstances. Plants do not merely take up, by mass flow, those nutrients that happen to be in the

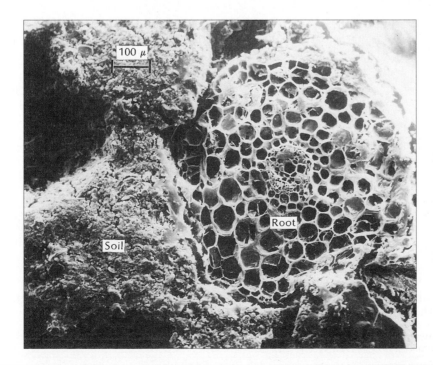

FIGURE 1.21 Scanning electron micrograph of a cross section of a peanut root surrounded by soil. Note the intimacy of contact. [Courtesy Tan and Nopamornbodi (1981)]

water that roots are removing from the soil. Nor do dissolved nutrient ions brought to the root's outer surface by mass flow or diffusion cross the root cell membrane and enter the root passively by diffusion. On the contrary, a nutrient is taken up when a chemical carrier molecule in the root cell membrane forms an activated complex with the nutrient and then travels across the membrane to the interior of the root cell before releasing the nutrient. The carrier mechanism, activated by root metabolic energy, allows the plant to accumulate concentrations of nutrients inside the root cell that far exceed the nutrient concentrations in the soil solution. Because different nutrients are taken up by specific types of carrier molecules, the plant is able to exert some control over how much and in what relative proportions essential elements are taken up.

Since nutrient uptake is an active metabolic process, conditions that inhibit root metabolism may also inhibit nutrient uptake. Examples of such conditions include excessive soil water content or soil compaction resulting in poor soil aeration, excessively hot or cold soil temperatures, and aboveground conditions which result in low translocation of sugars to plant roots. We can see that plant nutrition involves biological, physical, and chemical processes and interactions among many different components of soils and the environment.

1.17 CONCLUSION

Soil, like water and air, is a fundamental natural resource. The earth's soil is comprised of numerous soil individuals, each of which is a three-dimensional natural body in the landscape. Each individual soil is characterized by a unique set of properties and soil horizons as expressed in its profile. The nature of the soil layers seen in a particular profile is closely related to the nature of the environmental conditions at a site.

Soils perform five broad ecological functions: they act as the principal medium for plant growth, regulate water supplies, recycle raw materials and waste products, and serve as a major engineering medium for human-built structures. They are also home to many kinds of living organisms. Soil is thus a major ecosystem in its own right. The soils of the world are extremely diverse, each type of soil being characterized by a unique set of soil horizons. A typical surface soil in good condition for plant growth consists of about half solid material (mostly mineral, but with a crucial organic component, too) and half pore spaces filled with varying proportions of water and air. These components interact to influence a myriad of complex soil functions, a good understanding of which is essential for wise management of our terrestrial resources.

2

FORMATION OF SOILS FROM PARENT MATERIALS

It is a poem of existence . . . not a lyric but a slow-moving epic whose beat has been set by eons of the world's experience . . .
—JAMES MICHENER, CENTENNIAL

A prime reason for our studying soils is to find out why they vary so much from place to place, and what we can do to take advantage of this variation for the good of all earthly creatures. By understanding the processes that lead to the formation of soils we can satisfy our human curiosity as to why the soils in tropical Africa differ so much from those in Europe and North America. For a more local geographic region we can learn why soils developed on limestone are so different from those formed from sandstones. In all cases, knowledge of soil-forming processes helps us to classify soils and to determine how they might best be managed.

We learned in Chapter 1 that *the* soil is a collection of *individual* soils, each with distinctive profile characteristics. This concept of soils as natural bodies derived initially from field studies by a brilliant Russian team of soil scientists led by V. V. Dukochaev. They noted similar profile layering in soils hundreds of miles apart, provided that the climate and vegetation were similar at the two locations. Such observations and much careful subsequent field and laboratory research led to the recognition of five major factors that control the formation of soils.

1. Parent materials (geological or organic precursors to the soil)
2. Climate (primarily precipitation and temperature)
3. Biota (living organisms, especially native vegetation, microbes, soil animals, and human beings)
4. Topography (configuration of the soil surface)
5. Time the parent materials are subject to soil formation

In fact, soils often are defined in terms of these factors as "dynamic natural bodies having properties derived from the combined effect of **climate** and **biotic activities**, as modified by **topography**, acting on **parent materials** over periods of **time.**"

These factors do not exert their influences independently. Indeed, interdependence is the rule. For example, they jointly impact on weathering processes that cause the breakdown of rocks and minerals as well as the associated syntheses of new minerals and new soil layers that characterize the horizons of today's soils. Because of the vital significance of these jointly stimulated weathering activities, we will discuss them before turning to each of the five major soil-forming factors.

2.1 WEATHERING OF ROCKS AND MINERALS

The influence of weathering and especially the physical and chemical breakdown of particles is evident everywhere. Nothing escapes it. It breaks up rocks and minerals, modifies or destroys their physical and chemical characteristics, and carries away the soluble products. Likewise, it synthesizes new minerals of great significance. The rates of both of these processes are determined, however, by the nature of the rocks and minerals being impacted.

Characteristics of Rocks and Minerals

The rocks in the earth's outer surface are commonly classified as **igneous, sedimentary,** and **metamorphic.** Those of igneous origin are formed from molten magma and include such common rocks as **granite** and **diorite** (Figure 2.1).

Igneous rock is composed of **primary** minerals[1] such as quartz, the feldspars, and the dark-colored minerals, including biotite, augite, and hornblende. In general, gabbro and basalt, which are dark colored and high in easily weathered iron- and magnesium-containing minerals, are more easily broken down than are the granites and other lighter-colored rocks.

Sedimentary rocks have resulted from the deposition and recementation of weathering products from other rocks. For example, quartz sand weathered from a granite and deposited on the shore or beach of a prehistoric sea may through geological changes have become cemented into a solid mass called a **sandstone.** Similarly, recemented clays are termed **shale.** Other important sedimentary rocks are listed in Table 2.1 along with their dominant minerals. The resistance of a given sedimentary rock to weathering is determined by its particular dominant minerals and by the cementing agent.

Metamorphic rocks are those that have formed by the metamorphism or change in form of other rocks. Igneous and sedimentary masses subjected to tremendous pressures and high temperature succumb to metamorphism. Igneous rocks are commonly modified to form **gneisses** and **schists;** those of sedimentary origin, such as sandstone and shale, may be changed to **quartzite** and **slate,** respectively. Some of the common metamorphic rocks are shown in Table 2.1. As was the case for igneous and sedimentary rock, the particular mineral or minerals that dominate a given metamorphic rock will influence its resistance to weathering (see Table 2.2 for a listing of the more common minerals).

Weathering: A General Case

Weathering is basically a combination of the processes of destruction and synthesis (see Figure 2.2). Initially, the weathering process breaks down rocks physically into smaller rocks and eventually into the individual minerals of which they are composed. Sand

---

[1]Primary minerals have not been altered chemically since they formed as molten lava solidified. Secondary minerals are recrystallized products of the chemical breakdown and/or alteration of primary minerals.

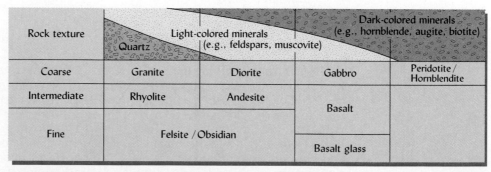

FIGURE 2.1 Classification of some igneous rocks in relation to mineralogical composition and the size of mineral grains in the rock (rock texture). Worldwide, light-colored minerals and quartz are generally more prominent than are the dark-colored minerals.

TABLE 2.1 Some of the More Important Sedimentary and Metamorphic Rocks and the Minerals Commonly Dominant in Them

	Type of rock	
Dominant mineral	Sedimentary	Metamorphic
Calcite ($CaCO_3$)	Limestone	Marble
Dolomite ($CaCO_3 \cdot MgCO_3$)	Dolomite	Marble
Quartz (SiO_2)	Sandstone	Quartzite
Clays	Shale	Slate
Variable	Conglomerate[a]	Gneiss[b]
Variable		Schist[b]

[a] Small stones of various mineralogical makeup are cemented into conglomerate.
[b] The minerals present are determined by the original rock, which has been changed by metamorphism. Primary minerals present in the igneous rocks commonly dominate these rocks, although some secondary minerals are also present.

particles are commonly made up of individual minerals. Simultaneously, rock fragments and the minerals therein are attacked by chemical forces and changed to new minerals, either by minor modifications (alterations) or by complete chemical changes. Chemical changes are accompanied by a continued decrease in particle size and by the release of soluble constituents, which are subject to loss in drainage waters or recombination into new (secondary) minerals.

Three groups of minerals that remain in soils are shown in Figure 2.2: (1) silicate clays, (2) very resistant end products, including iron and aluminum oxide clays, and (3) very resistant primary minerals such as quartz. In highly weathered soils of humid tropical regions, the oxides of iron and aluminum predominate.

Two basic processes, physical **disintegration** and chemical **decomposition**, are involved in the changes indicated in Figure 2.2. Both are operative, moving from left to right in the weathering diagram. Disintegration results in a decrease in size of rocks and minerals without appreciably affecting their composition. During decomposition, however, definite chemical changes take place, soluble materials are released, and new minerals are synthesized, some of which are resistant end products.

TABLE 2.2 The More Important Primary and Secondary Minerals Found in Soils Listed in Order of Decreasing Resistance to Weathering Under Conditions Common in Humid Temperate Regions

The primary minerals are also found abundantly in igneous and metamorphic rocks. Secondary minerals are commonly found in sedimentary rocks.

Primary minerals		Secondary minerals		
		Geothite	FeOOH	Most resistant
		Hematite	Fe_2O_3	
		Gibbsite	$Al_2O_3 \cdot 3H_2O$	
Quartz	SiO_2			
		Clay minerals	Al silicates	
Muscovite	$KAl_3Si_3O_{10}(OH)_2$			
Microcline	$KAlSi_3O_8$			
Orthoclase	$KAlSi_3O_8$			
Biotite	$KAl(Mg,Fe)_3Si_3O_{10}(OH)_2$			
Albite	$NaAlSi_3O_8$			
Hornblende[a]	$Ca_2Al_2Mg_2Fe_3Si_6O_{22}(OH)_2$			
Augite[a]	$Ca_2(Al,Fe)_4(Mg,Fe)_4Si_6O_{24}$			
Anorthite	$CaAl_2Si_2O_8$			
Olivine	$(Mg,Fe)_2SiO_4$			
		Dolomite	$CaCO_3 \cdot MgCO_3$	
		Calcite	$CaCO_3$	
		Gypsum	$CaSO_4 \cdot 2H_2O$	Least resistant

[a] The given formula is only approximate since the mineral is so variable in composition.

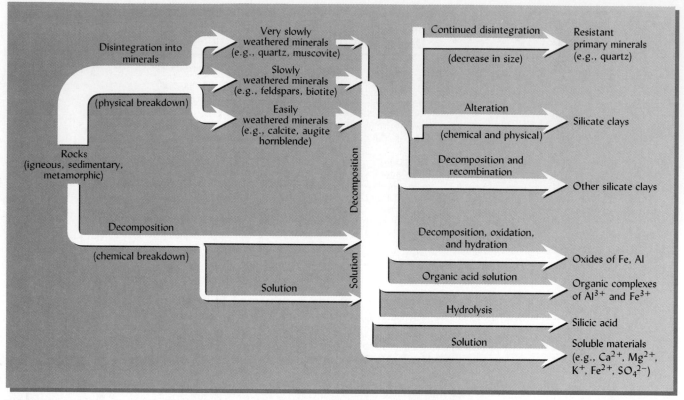

FIGURE 2.2 Pathways of weathering that occur under moderately acid conditions common in humid temperate regions. The disintegration of rocks into minerals is a physical process, whereas decomposition, recombination, and solution are chemical processes. Alteration of minerals involves both physical and chemical processes. Note that resistant primary minerals, new secondary minerals synthesized as weathering occurs, and soluble materials are the products of the weathering. In arid regions the physical processes would predominate, but in humid tropical areas decomposition and recombination would be most prominent.

2.2 PHYSICAL WEATHERING (DISINTEGRATION)

TEMPERATURE. During the day, rocks become heated; at night they often cool much below the temperature of the air. This warming causes some minerals within a given rock to expand more than others. Likewise, when cooled, some minerals contract more than others. With every temperature change, therefore, differential stresses are set up that eventually must produce cracks, thus encouraging mechanical breakdown.

Because heat is conducted slowly, the outer surface of a rock is often warmer or colder than the inner and more protected portions. This differential heating and cooling sets up lateral stresses that, in time, may cause the surface layers to peel away from the parent mass. This phenomenon is referred to as *exfoliation,* which at times is sharply accelerated by the freezing of included water (Figure 2.3). The force[2] developed by the freezing of water is equivalent to about 1465 metric tons (megagrams) per square meter (Mg/m^2) or 150 tons/ft^2, a pressure that widens cracks in huge boulders and dislodges mineral grains from smaller fragments.

WATER, ICE, AND WIND. Rainwater beats down on the land and then travels oceanward, continually shifting, sorting, and reworking the sediments that it carries. When loaded with sediments, water has a tremendous cutting power, as is amply demonstrated by the gorges, ravines, and valleys around the world. The rounding of riverbed rocks and beach sand grains is further evidence of the abrasion that accompanies water movement.

Ice can detach and transport tremendous amounts of material (Figure 2.4). A look at a Greenland or Alaskan glacier illustrates its power. Wind, an important carrying agent, also exerts an abrasive action when armed with fine debris. As dust is transferred and

---

[2]The international system of measurements called the *SI system* is used in this text. A list of the most commonly used units and their equivalents in the old British system is shown inside the back cover. The comparative SI and British units will both be shown in the text in some cases.

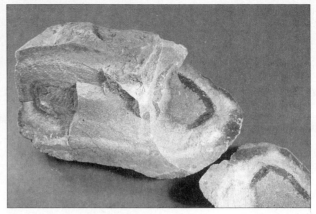

FIGURE 2.3 Two illustrations of rock weathering. (Left) An illustration of concentric weathering called *exfoliation*. A combination of physical and chemical processes stimulate the mechanical breakdown, which produces layers that appear much like the leaves of a cabbage. (Right) Concentric bands of light and dark colors indicate that chemical weathering (oxidation and hydration) has occurred from the outside inward, producing iron compounds that differ in color. (Right photo courtesy of R. Weil; left photo from previous edition)

deposited from one area to another, abrasion of one particle against another occurs. The rounded rock remnants in some arid areas of the west are caused largely by wind action.

PLANTS AND ANIMALS. Roots of higher plants sometimes exert a prying effect on rocks, which results in some disintegration. However, such influences, as well as those exerted by animals, are of little importance in producing parent material when compared to the drastic physical effects of water, ice, wind, and temperature changes.

2.3 CHEMICAL PROCESSES OF WEATHERING (DECOMPOSITION)

Chemical decomposition starts as soon as rock and minerals begin to disintegrate. This is especially noticeable in warm and humid regions, where chemical and physical processes are intense and tend to accelerate each other.

Chemical weathering is accelerated by the presence of water (with its omnipresent solutes), oxygen, and the organic and inorganic acids which result from the microbial breakdown of plant residues. These agents commonly act in concert to convert primary minerals (e.g., feldspars and micas) to secondary minerals (e.g., silicate clays) and to soluble forms that carry essential elements that support plant growth (see Figure 2.5).

WATER AND ITS SOLUTIONS. Water and its dissolved salts and acids are perhaps the most pervasive factors in the weathering of minerals. Through **hydrolysis, hydration,** and **dissolution,** water enhances the degradation, alteration, and resynthesis of minerals. A simple example is the action of water on microcline, a potassium-containing feldspar.

$$KAlSi_3O_8 + H_2O \xrightarrow{\text{hydrolysis}} HAlSi_3O_8 + K^+ + OH^-$$
$$\text{(solid)} \quad\quad \text{(liquid)} \quad\quad\quad\quad\quad \text{(solid)} \quad\quad \text{(solution)}$$

$$2HAlSi_3O_8 + 11H_2O \xrightarrow{\text{hydrolysis}} Al_2O_3 + 6H_4SiO_4$$
$$\text{(solid)} \quad\quad\quad \text{(liquid)} \quad\quad\quad\quad\quad \text{(solid)} \quad\quad \text{(solution)}$$

$$Al_2O_3 + 3H_2O \xrightarrow{\text{hydration}} Al_2O_3 \cdot 3H_2O$$
$$\quad\quad\quad\quad\quad\quad\quad\quad\quad\quad\quad\quad\quad \text{(hydrated solid)}$$

Note that these reactions illustrate dissolution and hydration as well as hydrolysis[3] (see Figure 2.2). The potassium released is soluble and is subject to adsorption by soil colloids, to uptake by plants, and to removal in the drainage water. Likewise, the silicic acid

[3]Hydrolysis involves the splitting of water into its H^+ and OH^- components, while hydration attaches intact water molecules to a compound.

FIGURE 2.4 Effects of water on weathering and the breakdown of rock. (Upper) The V-like notch carved by water in this sandstone cliff in Montana is evidence of the cutting power of water laden with sediment. The cave below the notch was carved by eddies under the ancient waterfall. (Lower) The expansion of water as it freezes has broken up these Appalachian sedimentary rocks into ever smaller fragments. (Photos courtesy of R. Weil)

(H_4SiO_4) is soluble. It can be removed slowly in drainage water, or it can be recombined with other compounds to form secondary minerals such as the silicate clays. Hydration is evident through the formation of the hydrated aluminum oxide ($Al_2O_3 \cdot 3H_2O$). Iron oxides also are commonly hydrated.

ACID SOLUTION WEATHERING. Weathering is accelerated by the presence of the hydrogen ion in water, such as that contained in carbonic and organic acids. For example, the presence of carbonic acid (H_2CO_3) results in the chemical solution of calcite in limestone, as illustrated in the following reaction:

FIGURE 2.5 Scanning electron micrographs illustrating silicate clay formation from weathering of a granite rock in southern California. (a) A potassium feldspar (K-spar) is shown surrounded by smectite (Sm) and a vermiculite (both silicate clays). (b) Mica (M) and quartz (Q) associated with smectite (Sm). (Courtesy of J. R. Glasmann)

$$CaCO_3 + H_2CO_3 \longrightarrow Ca^{2+} + 2HCO_3^-$$

Calcite   (solution)   (solution)
(solid)

Other, much stronger acids, such as nitric acid (HNO_3), sulfuric acid (H_2SO_4), and some organic acids are found in soils. Hydrogen ions are also associated with soil clays. Each of these sources of acidity is available for reaction with other soil minerals.

Soil microorganisms produce organic acids such as oxalic, citric, and tartaric acid. These compounds provide H^+ ions that help solubilize aluminum. They also form organic complexes (chelates) with the Al^{3+} ions held within the structure of silicate minerals. By so doing they remove the Al^{3+} from the mineral, which then is subject to further disintegration (see Figure 2.6).

OXIDATION. Oxidation reactions often occur in conjunction with the action of water, especially in rocks that contain iron, an element that is easily oxidized. In some minerals, iron is present in the reduced or two-valent ferrous (Fe^{2+}) form. If ferrous iron is oxidized to the three-valent ferric (Fe^{3+}) state, other ionic adjustments must be made because a three-valent ion is replacing a two-valent one. These adjustments result in a less stable mineral, which is then subject to both disintegration and decomposition.

In other cases, ferrous iron may be released from the mineral and is almost simultaneously oxidized to the ferric form. An example of this is the hydration of olivine and the release of ferrous oxide, which may be oxidized immediately to ferric oxide (geothite).

$$3MgFeSiO_4 + 2H_2O \longrightarrow H_4Mg_3Si_2O_9 + SiO_2 + 3FeO$$

Olivine   Serpentine   Ferrous oxide

$$4FeO + O_2 + 2H_2O \longrightarrow 4FeOOH$$

Ferrous oxide   Geothite
(valence = 2)   (valence = 3)

When ions such as Fe^{2+} are removed or are oxidized within the minerals, the rigidity of the mineral structure is weakened, and the mechanical breakdown is made easier. This provides a favorable environment for further chemical reactions (see Figure 2.3b).

INTEGRATED WEATHERING PROCESSES. The chemical weathering processes just described occur simultaneously and are interdependent. For example, hydrolysis of a given primary mineral may release ferrous iron that is quickly oxidized to the ferric (three-valent) form, which, in turn, is hydrated to give a hydrous oxide of iron. Hydrolysis also may release soluble cations, silicic acid, and aluminum or iron compounds. These substances can be recombined to form silicate clays and other secondary silicate minerals. In this manner, weathering transforms primary geologic materials into the compounds of which soils are made. These reactions are illustrated in a general way in Figure 2.2.

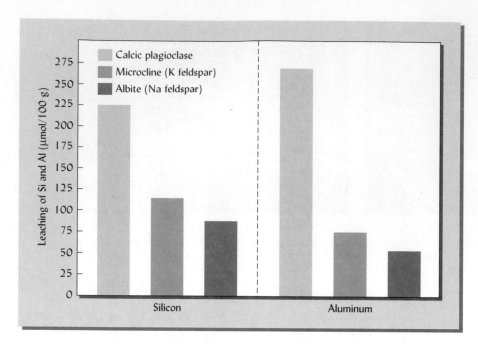

FIGURE 2.6 The leaching of silicon and aluminum by organic acids from three feldspars. Acids such as citric, tartaric, and oxalic form organic complexes with aluminum (called *chelates*) and pull the Al^{3+} from the crystal structure, thereby leading to its breakdown. Note that the calcic plagioclase released more of both elements than did the potassium and sodium feldspars. [Drawn from data (averages of leaching with eight different organic acids) in E. P. Manley and L. J. Evans, "Dissolution of feldspars by low-molecular-weight aliphatic and aromatic acids," *Soil Science*, **141**:106–12, © by Williams & Wilkins (1986)]

2.4 PARENT MATERIALS

Geological processes have brought to the earth's surface numerous **parent materials** in which soils form (Figure 2.7). The nature of the parent material profoundly influences soil characteristics. For example, soil texture or the size of the soil particles (see Section 4.2) is largely influenced by parent materials. In turn, soil texture helps control the downward movement of water, thereby affecting the translocation of fine soil particles and plant nutrients.

The chemical and mineralogical compositions of parent material also can influence weathering directly and, simultaneously, can affect the natural vegetation. For example, the presence of limestone in parent material will delay the development of acidity, a process that moist climates encourage. In addition, the leaves of trees growing in limestone materials are relatively high in calcium and other base-forming metallic cations. As these high-base leaves are incorporated into the soil and are decomposed, they further delay the process of acidification or, in humid temperate areas, the progress of soil development.

Parent material also influences the quantity and type of clay minerals present in the soil profile. The parent material itself may contain varying amounts and types of clay minerals, perhaps from a previous weathering cycle. Also, the nature of parent material greatly influences the kinds of clays that can develop as the soil evolves. In turn, the nature of the clay minerals present markedly affects the kind of soil that develops.

Classification of Parent Materials

Inorganic parent materials can either be formed in place (sedentary) as residual weathered material or they can be transported from one location and deposited at another (see Figure 2.8). The mode of transport may be by water, wind, ice, or gravity, as follows:

1. Sedentary, formed in place   **Residual**
2. Transported
     a. By gravity   **Colluvial**
     b. By water   **Alluvial**
                    Marine
                    Lacustrine
     c. By ice   **Glacial**
     d. By wind   **Eolian**

These terms properly relate only to the placement of the parent materials. However, they are sometimes applied to the soils that form from these deposits—for example,

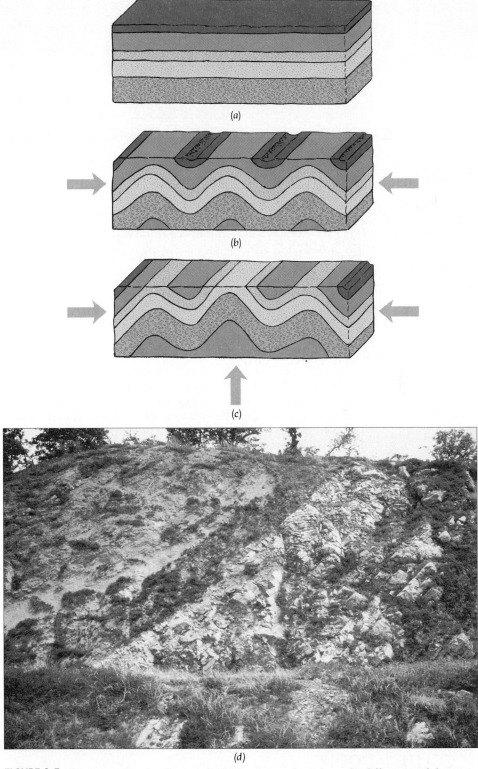

FIGURE 2.7 Diagrams showing how geological processes have brought different rock layers to the surface in a given area. (a) Unaltered layers of sedimentary rock with only the uppermost layer exposed. (b) Lateral geological pressures deform the rock layers through a process called *crustal warping*. At the same time, erosion removes much of the top layer, exposing part of the first underlying layer. (c) Localized upward pressures further re-form the layers, thereby exposing two more underlying layers. As these four rock layers are weathered, they give rise to the parent materials on which different kinds of soils can form. (d) Crustal warping that lifted up the Appalachian Mountains tilted these sedimentary rock formations that were originally laid down horizontally. This deep roadcut in Virginia illustrates the abrupt change in soil parent material as one walks along the ground surface at the top of this photograph. (Photo courtesy of R. Weil)

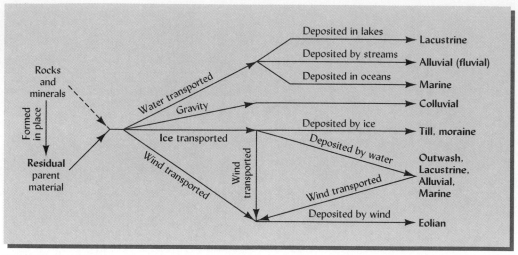

FIGURE 2.8 How various kinds of parent material are formed, transported, and deposited.

glacial soils, alluvial soils, and residual soils. These groupings are very general because a wide diversity occurs within each soil group.

2.5   RESIDUAL PARENT MATERIAL

Residual parent material develops in place from the underlying rock. It may have experienced long and often intense weathering. In a warm, humid climate, it is likely to be thoroughly oxidized and well leached. Residual material is generally comparatively low in calcium and magnesium because these constituents have been largely leached out.

Red and yellow colors due to oxides of iron are characteristic when weathering has been intense, as in hot, humid areas. In cooler and especially drier climates, weathering of residual materials is much less drastic, and the oxidation and hydration of the iron may be hardly noticeable. Also, the calcium content is often higher and the colors of the weathering mass are subdued. Large areas of this type of material are found on the Great Plains and in other regions of the western United States.

Residual materials are widely distributed on all continents. In the United States, the physiographic map (Figure 2.9) shows six great eastern and central provinces where residual materials are prominent: (1) the Piedmont Plateau, (2) the Appalachian Mountains and plateaus, (3) the limestone valleys and ridges near the Appalachians, (4) the limestone uplands, (5) the sandstone uplands, and, (6) the Great Plains region. The first three groups alone encompass about 10% of the area of the United States. In addition, great expanses of these sedentary accumulations are found west of the Rocky Mountains.

A great variety of soils occupy the regions covered by residual debris because of the marked differences in the nature of the rocks from which these materials have evolved. The varied soils are also a reflection of wide differences in other soil-forming factors such as climate and vegetation. As will be seen in Sections 2.12 and 2.13, the profile of a well-developed soil is profoundly influenced by climate and associated vegetation.

2.6   COLLUVIAL DEBRIS

Colluvial debris is made up of poorly sorted fragments of rock detached from the heights above and carried down the slopes mostly by gravity. Frost action has much to do with the development of such deposits. Rock fragment (talus) slopes, cliff rock debris (detritus), and similar heterogeneous materials are good examples. Avalanches are made up largely of such accumulations.

Parent material developed from colluvial accumulation is dependent on the sources of the material. It is frequently coarse and stony because physical rather than chemical weathering has been dominant. Stones, gravel, and fine materials are interspersed and

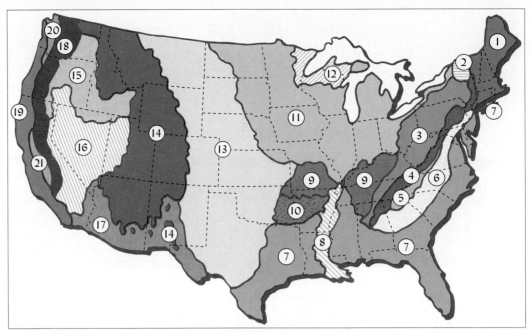

FIGURE 2.9 Generalized physiographic and regolith map of the United States. The regions are as follows (major residual areas italicized).

1. New England: mostly glaciated metamorphic rocks.
2. Adirondacks: glaciated metamorphic and sedimentary rocks.
3. *Appalachian Mountains and plateaus:* shales and sandstones.
4. *Limestone valleys and ridges:* mostly limestone.
5. Blue Ridge mountains: sandstones and shales.
6. *Piedmont Plateau:* metamorphic rocks.
7. Atlantic and Gulf coastal plain: sedimentary rocks with sands, clays, and limestones.
8. Mississippi floodplain and delta: alluvium.
9. *Limestone uplands:* mostly limestone and shale.
10. *Sandstone uplands:* mostly sandstone and shale.
11. Central lowlands: mostly glaciated sedimentary rocks with till and loess, a wind deposit of great agricultural importance (see Figure 2.18).
12. Superior uplands: glaciated metamorphic and sedimentary rocks.
13. *Great Plains region:* sedimentary rocks.
14. *Rocky Mountain region:* sedimentary, metamorphic, and igneous rocks.
15. Northwest intermountain: mostly igneous rocks; loess in Columbia and Snake river basins (see Figure 2.14).
16. Great Basin: gravels, sands, alluvial fans from various rocks; igneous and sedimentary rocks.
17. Southwest arid region: gravel, sand, and other debris of desert and mountain.
18. *Sierra Nevada and Cascade mountains:* igneous and volcanic rocks.
19. *Pacific Coast province:* mostly sedimentary rocks.
20. Puget Sound lowlands: glaciated sedimentary.
21. California central valley: alluvium and outwash.

the coarse fragments are rather angular. Soils developed from colluvial materials are generally not of great agricultural importance because of their small area, inaccessibility, and unfavorable physical and chemical characteristics. However, some useful timber and grazing lands in very mountainous regions have colluvial materials.

2.7 ALLUVIAL STREAM DEPOSITS

There are three general classes of alluvial deposits: floodplains, alluvial fans, and deltas. They will be considered in order.

FLOODPLAINS. Streams commonly overflow their banks and flood the surrounding area. That part of a valley that is inundated during floods is a floodplain. Sediment carried by the swollen stream is deposited during the flood, with the coarser materials being laid down near the river channel and finer materials farther away. Figure 2.10 illustrates the

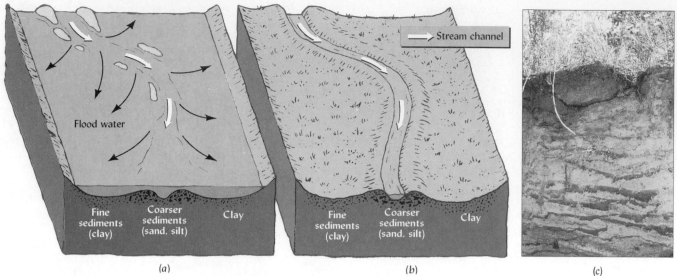

(a)   *(b)*   *(c)*

FIGURE 2.10   Illustration of floodplain development. (a) A stream is at flood stage, has overflowed its banks, and is depositing sediment in the floodplain. The coarser particles are being deposited near the stream channel where water is moving most rapidly, while the finer clay particles are being deposited where water movement is slower. (b) After the flood the sediments are in place and vegetation is growing. (c) Contrasting layers of sand, silt, and clay characterize the alluvial floodplain. Each layer resulted from separate flooding episodes. (Redrawn from *Physical Geology,* 2d ed., by F. R. Flint and B. J. Skinner, copyright © 1977 John Wiley & Sons, Inc. Reprinted by permission of John Wiley & Sons, Inc.) (Photo courtesy of R. Weil)

relationship between a stream and its surrounding floodplain. If, over a period of time, there is a change in grade, a stream may cut down through its already well-formed alluvial deposits. This leaves **terraces** above the floodplain on one or both sides. Often two or more terraces of different heights can be detected. These suggest periods when the stream was at each of these elevations.

The floodplain along the Mississippi River is the largest in the United States (Figure 2.11), varying from 20 to 75 miles in width. But throughout the country, floodplains large and small provide significant parent materials for other important soil areas. The soils derived from these sediments are generally rich in nutrients and productive, although some may require drainage. While alluvial soils are often uniquely suited to forestry and crop production, their use for home sites and urban development should

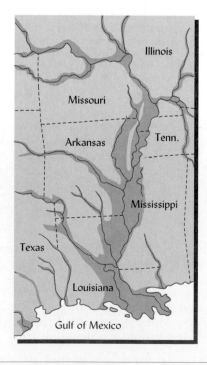

FIGURE 2.11   Floodplain and delta of the Mississippi River. This is the largest continuous area of alluvial soil in the United States.

generally be avoided. Building on a floodplain, no matter how great the investment in flood control measures, all too often leads to tragic loss of life and property during serious floods.

Productive soils are found on the floodplains of many countries. The Nile River Valley in Egypt and the Sudan, the Euphrates, Ganges, Indus, Brahmaputra, and Hwang Ho river valleys of Asia, and the Amazon Basin of Brazil are good examples. Some of these floodplain deposits are used for the production of wetland rice, grown by most of the low-income people of these areas.

ALLUVIAL FANS. Streams that leave a narrow valley in an upland area and suddenly descend to a much broader valley below deposit sediment in the shape of a fan (alluvial fans) (see Figures 2.12 and 19.12). Fan material generally is gravelly and stony, somewhat porous, and well drained.

Alluvial fan debris is found over wide areas in mountainous and hilly regions. The soils derived from this debris often prove very productive, although they may be quite coarse-textured. In certain glaciated sections, such deposits also occur in large enough areas to be of considerable agricultural importance.

DELTA DEPOSITS. Much of the finer sediment carried by streams is not deposited in the floodplain but is discharged into a lake reservoir or ocean into which the stream flows. Some of the suspended material settles near the mouth of the river, forming a delta. Such delta deposits are by no means universal, being found at the mouths of only a few rivers of the world. A delta often is a continuation of a floodplain (its front, so to speak). It is clayey in nature, and is likely to be swampy as well.

Important agricultural areas have been developed where flood control and drainage are applied to delta sediments. The combination deltas and floodplains of the Mississippi, Ganges, Amazon, Hwang Ho, Po, Tigris, and Euphrates rivers are striking examples. The Nile Valley of Egypt illustrates the fertility and productivity of soils originating from such parent material.

Increasing attention has been given to the possibility of returning some of the drained wetland areas alongside streams (either floodplains or deltas) to their natural state. These areas were natural habitats for birds and other wildlife. Furthermore, their drainage and protection from floods are costly and often ineffective. Farmers and the general public pay high costs to keep these areas in agricultural production, and some steps have been taken to reestablish the wetland states of the river-flooded areas.

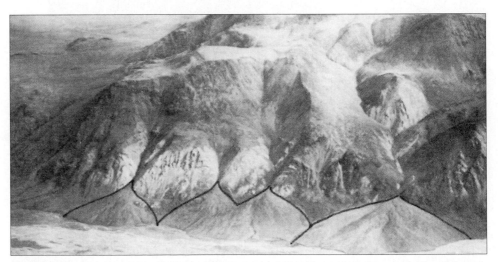

FIGURE 2.12 Characteristically shaped alluvial fans alongside a river valley in Alaska. Although the areas are small and sloping, they can develop into well-drained soils. (Courtesy U.S. Geological Survey)

2.8 MARINE SEDIMENTS

Much of the sediment carried away by streams eventually is deposited in the oceans, lakes, seas, and gulfs. The coarser fragments are deposited near the shore and the finer particles at a distance (Figure 2.13). Over long periods of time, these underwater sediments build up and can become quite deep. In some areas, changes in the elevation of the earth's crust have resulted in these marine deposits being raised above sea level. The deposits are then subject to weathering and to soil formation, giving rise in some cases to valuable agricultural soils. Such regions are found along the Atlantic and Gulf coasts of the United States and make up about 10% of the country's land area.

Marine deposits are quite variable in texture. Some are sandy, as is the case in much of the Atlantic seaboard coastal plain. Others are high in clay, as are deposits found in the Atlantic and Gulf coastal flatwoods and in the interior pinelands of Alabama and Mississippi. All of these sediments came from the erosion of upland areas, some of

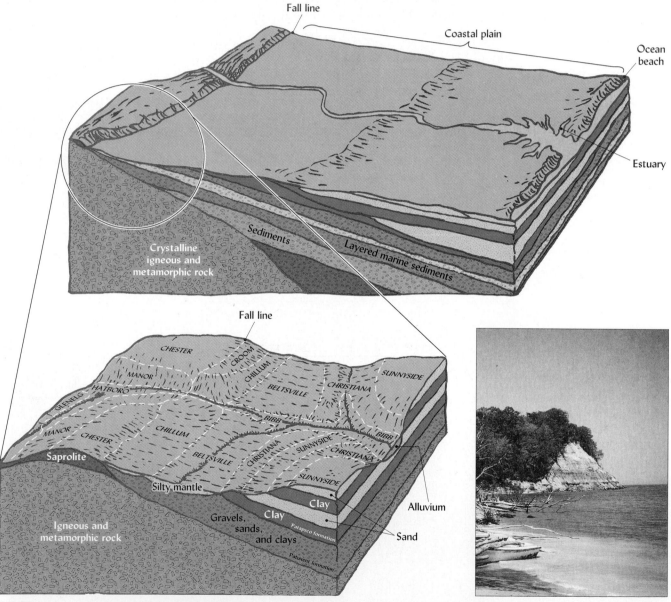

FIGURE 2.13 Diagrams showing sediments laid down by marine waters adjacent to coastal residual igneous and metamorphic rocks. Note that the marine sediments are alternate layers of fine clay and coarse-textured sands and gravels. The photo in the lower right shows such layering on coastal marine sediment. The soils developing from the marine and residual parent materials are also shown. The river had cut through the marine sediments and has deposited some alluvial material in its basin. This diagram and photo illustrate the relationship between marine sediments and residual materials in the Coastal Plain of southeastern United States. [Photo courtesy of R. Weil; soils diagram from Miller (1976)]

which were highly weathered before the transport took place. However, the marine sediments have been subjected to the soil-forming process for a shorter period of time than their upland counterparts. As a consequence, the properties of the soils that form are heavily influenced by those of the marine parent materials. Many marine sediments are high in sulfur and go through a period of acid-forming sulfur oxidation at some stage of soil formation. Fortunately, with proper management and fertilization, soils developed on some marine deposits are quite productive.

2.9   PARENT MATERIALS TRANSPORTED BY ICE AND MELT WATERS

During the Pleistocene epoch (about 10^4–10^7 years ago), northern North America, northern and central Europe, and parts of northern Asia were invaded by a succession of great ice sheets. The Pleistocene ice at its maximum extension covered perhaps 20% of the land area of the world. It is surprising to learn that present-day glaciers, which we consider as mere remnants of the Great Ice Age, occupy an area about one-third that occupied by the Pleistocene glaciers. The present volume of ice, however, is much less, since contemporary glaciers are comparatively thin. Even so, if all present-day glaciers were to melt, the world sea level would increase by about 65 m (210 ft). One of the troublesome consequences of global warming is the probable melting of these present-day glaciers and the associated increase in sea level and flooding of many coastal areas around the world.

In North America, Pleistocene-epoch glaciers covered most of what is now Canada, southern Alaska, and the northern part of the contiguous United States. The southernmost extension was down the Mississippi Valley, where the least resistance was met because of the lower and smoother topography (Figure 2.14).

Europe and central North America apparently sustained four distinct ice invasions over a period of 1–1.5 million years (Table 2.3). These invasions were separated by long, interglacial, ice-free intervals of warm or even semitropical climates. We now may be enjoying the mildness of another interglacial period.

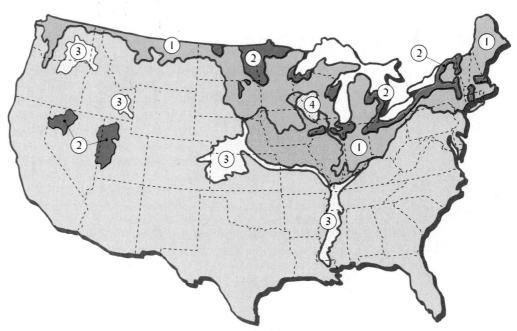

FIGURE 2.14 Areas in the United States covered by the continental ice sheet and the deposits either directly from, or associated with, the glacial ice. (1) Till deposits of various kinds; (2) glacial-lacustrine deposits; (3) the loessial blanket (note that the loess overlies great areas of till in the Midwest); (4) an area, mostly in Wisconsin, that escaped glaciation and is partially loess covered.

TABLE 2.3 Nomenclature Used to Identify Periods of Glaciation and Interglaciation in North America During the Pleistocene Period

Period name	Type of period	Approximate date starting
Holocene	Interglacial	10,000 B.C.
Wisconsin	Glacial	100,000
Sangamonian	Interglacial	225,000
Illinoian	Glacial	325,000
Yarmouthian	Interglacial	600,000
Kansan	Glacial	700,000
Aftonian	Interglacial	900,000
Nebraskan	Glacial	1–1.5 million B.C.

As the glacial ice was pushed forward, the existing regolith with much of its mantle of soil was swept away, hills were rounded, valleys filled, and, in some cases, the underlying rocks were severely ground and gouged. Thus, the glacier became filled with rock and all kinds of unconsolidated materials, and it carried great masses of these materials as it pushed ahead (Figure 2.15). Finally, as the ice melted and the glacier retreated, a mantle of glacial debris or **drift** remained. This provided a new regolith and fresh parent material for soil formation.

The area covered by glaciers in North America is estimated at 10.4 million km^2 (4 million mi^2), and at least 20% of the United States is influenced by the deposits. An examination of Figure 2.14 indicates the magnitude of the ice invasion at maximum glaciation in this country.

Glacial Till and Associated Deposits

The name *drift* is applied to all material of glacial origin, whether deposited by the ice or associated waters. The materials deposited directly by the ice, called *glacial till,* are heterogeneous mixtures of debris, which vary in size from boulders to clay. Glacial till is found mostly as irregular deposits called *moraines.* Figure 2.16 shows how glacial sheets deposited several types of soil parent materials.

Since glacial till and the soils derived from such material are so variable, the term *glacial soil* is of value only in suggesting the mode of deposition of the parent materials. It indicates practically nothing about the characteristics of a soil so designated.

Glacial Outwash and Lacustrine Sediments

Torrents of water were constantly gushing from the ice lobes of glaciers, especially during the summer. The vast loads of sediment carried by such streams were either dumped immediately or carried to other areas for deposition. As long as the water had ready egress, it flowed away rapidly to deposit its load downstream.

An *outwash* plain is formed by streams that are heavily laden with glacial sediment flowing from the ice (Figure 2.16). Since the sediment is sorted by flowing water, sands and gravels are common. Outwash deposits are found in valleys and on plains, where the glacial waters were able to flow away freely. These valley fills are common in the United States. Figure 2.17 shows the sorted layering of coarse and fine materials in glacial outwash overlain by mixed materials of glacial till.

When the ice front came to a standstill where there was no ready escape for the water, ponding began and ultimately very large lakes were formed (Figure 2.16). Particularly prominent were those south of the Great Lakes and in the Red River Valley of Minnesota (see Figure 2.14). The latter lake, called Glacial Lake Agassiz, was about 1200 km (745 mi) long and 400 km (248 mi) wide at its maximum extension. Somewhat smaller lakes also were formed in the intermountain regions west of the Rockies (e.g., Williamette Valley in Oregon), as well as in the Connecticut River Valley in New England and elsewhere.

The *lacustrine deposits* formed in these glacial lakes range from coarse delta materials and beach deposits near the shore to larger areas of fine silts and clay deposited from the deeper, more still waters at the center of the lake. Areas of inherently fertile (though not always well-drained) soils have developed from these materials as the lakes dried.

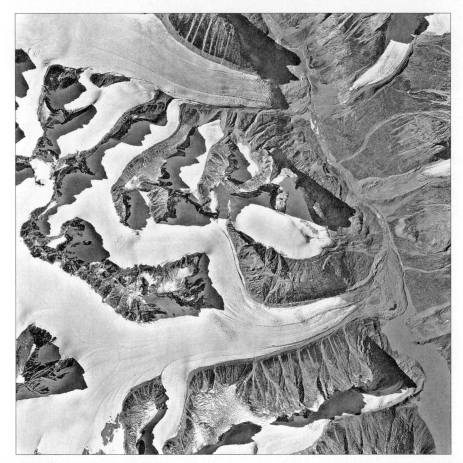

FIGURE 2.15 (Upper) Tongues of a modern-day glacier in Canada. Note the evidence of transport of materials by the ice and the "glowing" appearance of the two major ice lobes. (Lower) This U-shaped valley in the Rocky Mountains illustrates the work of glaciers in carving out land forms. The glacier left the valley floor covered with glacial till. Some of the material gouged out by the glacier was deposited many miles down the valley. (Upper photo A-16817-102 courtesy National Air Photo Library, Surveys and Mapping Branch, Canadian Department of Energy, Mines, and Resources; Lower photo courtesy of R. Weil)

2.10 PARENT MATERIALS TRANSPORTED BY WIND

Wind-transported (**eolian**) parent materials are among the most important in the United States, especially in the central part of the country. These materials are associated mostly with the initially barren, unprotected deposits left by glaciers.

During the winter months in glacial periods, winds picked up fine alluvial materials deposited in previous summers by the ice-fed streams, and moved them southward.

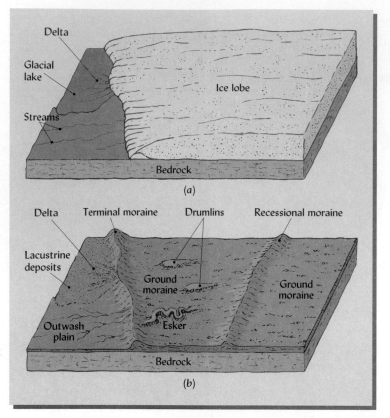

FIGURE 2.16 Illustration of how several glacial materials were deposited. (a) A glacier ice lobe moving to the left, feeding water and sediments into a glacial lake and streams, and building up glacial till near its front. (b) After the ice retreats, terminal, ground, and recessional moraines are uncovered along with cigar-shaped hills (drumlins), the beds of rivers that flowed under the glacier (eskers), and lacustrine, delta, outwash deposits.

Fine particles from glacial till and even residual materials were similarly transported.

These windblown materials, composed primarily of silt with some fine sand and clay, are called **loess** (pronounced "luss"). They covered existing soils and parent materials—in some areas to depths of more than 8 m (26 ft).

Loess and other associated eolian deposits are found over wide areas of the central United States, as shown in Figure 2.18. The deepest deposits are located along the Mississippi and Missouri rivers. Loessial deposits also are found in other countries. For example, deposits reaching 30–100 m (98–328 ft) in depth are found in some 800,000 km^2 (309,000 mi^2) in central and western China. These materials have been windblown from the dryland areas of central Asia and are generally not associated directly with glaciers.

Eolian deposits other than loess exist, such as volcanic ash and sand dunes. Soils from volcanic ash occur in Montana, Oregon, Washington, Idaho, Nebraska, and Kansas. They are light and porous and generally are of less agricultural value than soils developed from loess.

FIGURE 2.17 The stratification of different-size particles can be seen in the glacial outwash in the lower part of this soil profile in North Dakota. The outwash is overlain by a layer of glacial till containing a random assortment of particles, ranging in size from small boulders to clays. Note the rounded edges of the rocks, evidence of the churning action within the glacier. (Photo courtesy of R. Weil)

FIGURE 2.18 Approximate distribution of loess in central United States. The soil that has developed therefrom is generally a silt loam, often somewhat sandy. Note especially the extension down the eastern side of the Mississippi River and the irregularities of the northern extensions. Smaller areas of loess occur in Washington, Oregon, and Idaho (see Figure 12.14).

2.11 ORGANIC DEPOSITS

The organic material accumulates in wet places where plant growth exceeds the rate of residue decomposition. For centuries, residues accumulated from water-loving plants such as pondweeds, cattails, sedges, reeds, mosses, shrubs, and certain trees. These residues sank into the water, where their oxidation was curtailed. As a result, organic deposits up to several meters in depth are common (Figure 2.19). Collectively, these organic deposits are called *peat.*

Distribution and Nature of Peats

Peat deposits are found all over the world, but most extensively in the cool climates and in areas that have been glaciated. About 75% of the 340 million hectares of peat lands in the world are found in Canada and northern Russia. The United States is a distant third, with about 20 million hectares.

The rate of accumulation of organic deposits varies from one area to another. In the warm climate of Florida, peat deposits of the everglades accumulated for 3000–4000 years at the rate of about 0.8 mm/yr. In Wisconsin accumulation rates have been estimated to be 0.2 to 0.4 mm/yr.

Based on the nature of the parent materials, four kinds of peat are recognized:

1. *Moss peat,* the remains of mosses such as sphagnum
2. *Telmatic* or *herbaceous peat,* residues of herbaceous plants such as sedges, reeds, and cattails
3. *Terrestic* or *woody peat,* from the remains of woody plants, including trees and shrubs
4. *Limnic* or *sedimentary peat,* remains of aquatic plants (e.g., algae) and of fecal material of aquatic animals

Organic deposits generally contain two or more of these kinds of peats. Alternating layers of different peats are common, as are mixtures of the peats. This is not surprising, since different species of plants will dominate for periods of time, and mixtures of the plants are usually present. Because the succession of plants, as the residues accumulate, tends to favor trees (see Figure 2.19), Telmatic (woody) peats are often at the surface of the organic material.

In cases where the wetland area has been drained, woody peats tend to make very productive agricultural soils that are highly praised, especially for vegetable production. While moss peats have high water-holding capacities, they tend to be quite acid. Sedi-

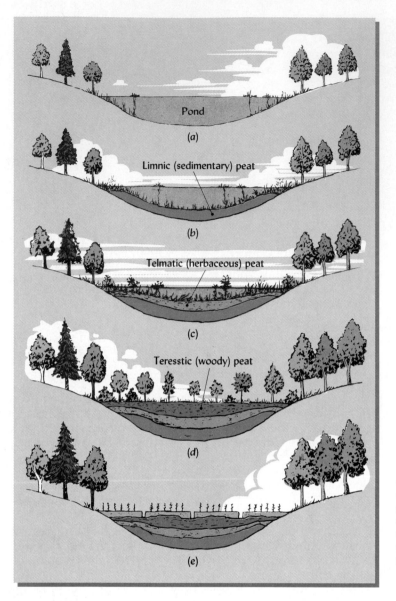

Pond

(a)

Limnic (sedimentary) peat

(b)

Telmatic (herbaceous) peat

(c)

Teresstic (woody) peat

(d)

(e)

FIGURE 2.19 Stages in the development of a typical woody peat bog and the area after clearing and draining. (a) Nutrient runoff from the surrounding uplands encourages aquatic plant growth, especially around the pond edges. (b, c) Organic debris fills in the bottom of the pond. (d) Eventually trees cover the entire area. (e) If the land is cleared and a drainage system installed, the area becomes a most productive muck soil.

mentary peat is generally undesirable as a soil. This material is highly colloidal and compact, and is rubbery when wet. Upon drying, it resists rewetting and remains in a hard, lumpy condition. Fortunately, it occurs mostly deep in the profile and is unnoticed unless it interferes with drainage of the bog area.

The organic material is called *peat* or *fibric* if the organic remains are sufficiently intact to permit the plant forms to be identified. If most of the material has decomposed sufficiently so that identification of the plant parts is difficult, the term *muck* or *sapric* is used. In mucky peats (*hemic*) some of the plant parts can be recognized and some cannot.

The draining and cultivating of wetland areas is subject to considerable controversy. These areas are important natural habitats for wildlife, and their drainage reduces the area of such habitats. While organic material is the foundation for some very productive agricultural soils, environmentalists argue that once drained, the organic deposits will decompose and disappear after a century or so, and therefore these areas should be left in (or returned to) their natural state.

Recognizing that soils are defined as dynamic natural bodies having properties derived from the combined effects of **climate** and **biotic activities** as modified by **topography** acting on **parent materials,** we will now turn to the other four factors of soil formation, starting with climate.

Climate is perhaps the most influential of the four factors acting on parent material because it determines the nature of the weathering that occurs. For example, temperature and precipitation affect the rates of chemical, physical, and biological processes responsible for profile development. For every 10°C rise in temperature, the rates of biochemical reactions double. Furthermore, biochemical changes by soil organisms are sensitive to temperature as well as moisture. Temperature and moisture relations influence the organic matter content of soil (see Figure 12.15). The very modest profile development characteristic of cold areas contrasted with the deep-weathered profiles of the humid tropics is further evidence of climatic control (Figure 2.20).

Water from rain and snowfall is a primary requisite for weathering and soil development. Along with warm temperature, it encourages the deep weathering of tropical soils. Not only are plant nutrients leached from these soils, but the basic framework of the original soil minerals is destroyed and new minerals are created. Likewise, a deficiency of water is a major factor in determining the characteristics of soils of arid regions. Soluble salts are not leached from these soils, and in some cases they build up to levels that curtail plant growth.

The amount of water leaching through a soil is determined not only by the total annual precipitation but by its seasonal distribution. One hundred millimeters of rainfall distributed evenly on a month-by-month basis throughout the year likely causes less soil leaching than if the rainfall were concentrated mostly in a four–five-month period. Likewise, rain falling on a steep slope subject to high runoff losses into nearby streams would likely result in less water infiltration into the soil than the same amount of rain falling on flat or rolling topographies.

Climate also influences the natural vegetation. In humid regions, plentiful rainfall provides an environment favorable for the growth of trees (Figure 2.21). In contrast, grasslands are the dominant native vegetation in semiarid regions, and shrubs and brush of various kinds dominate in arid areas. Thus, climate also exerts its influence through a second soil-forming factor, the living organisms.

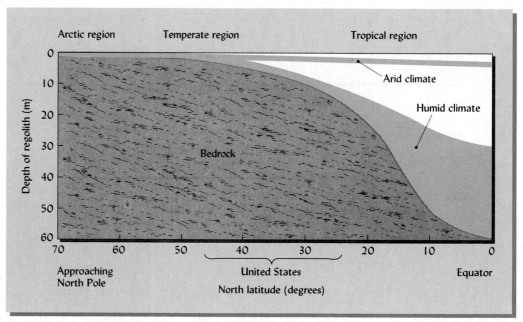

FIGURE 2.20   Illustration of the effects of two climatic variables, temperature and moisture (precipitation), on the ranges of depth of weathering as indicated by ranges of regolith depth. In cold climates (arctic regions) the regolith is shallow under both humid and arid conditions. At lower latitudes (higher temperatures), the depth of the regolith increases sharply in humid area but is little affected in arid regions. Under humid tropical climates the regolith may be 50 or more meters in depth. Note that soil depth may be only a fraction of regolith depth.

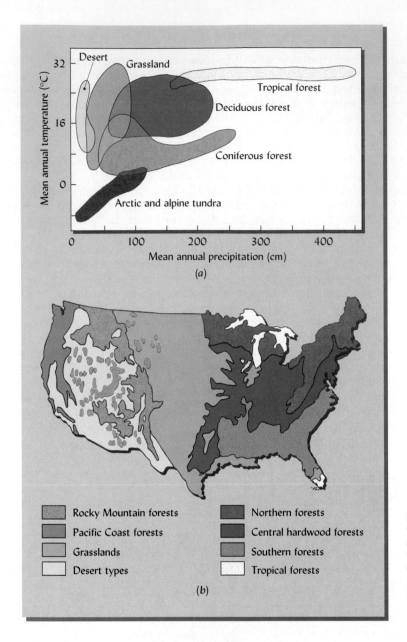

FIGURE 2.21 The effect of climate on vegetation. (a) General relationship among mean annual temperatures and precipitation and types of vegetation. (b) General types of vegetation in the United States. [(a) From NSF (1975); (b) redrawn from a more detailed map of U.S. Geological Survey]

2.13 BIOTA: LIVING ORGANISMS

Soil organisms play a major role in profile differentiation. Organic matter accumulation, profile mixing, nutrient cycling, and structural stability are all enhanced by the activities of organisms in the soil. Vegetative cover reduces natural soil erosion rates, thereby slowing down the rate of mineral surface removal.

The effect of vegetation on soil formation is seen by comparing properties of soils formed under grassland and forest vegetation (Figure 2.22). The organic matter content of the grassland soils is generally higher than that of comparable soils in forested areas, especially in the subsurface horizons. The higher organic content gives the soil a darker color and higher moisture- and cation-holding capacity compared to the forest soil. Also, structural stability of soil aggregates tends to be encouraged by the grassland vegetation.

The mineral element content in the leaves, limbs, and stems of the natural vegetation strongly influences the characteristics of the soils that develop, especially their acidity. Coniferous trees (e.g., pines, firs) tend to be low in metallic cations such as calcium, magnesium, and potassium. The cycling of nutrients from the litter falling from conifers will be low compared to that of some deciduous trees (e.g., beech, oaks, maples) that are much higher in metallic cations (Figure 2.23). Consequently, soil acidity is more likely to develop under pine vegetation with its low content of base-forming

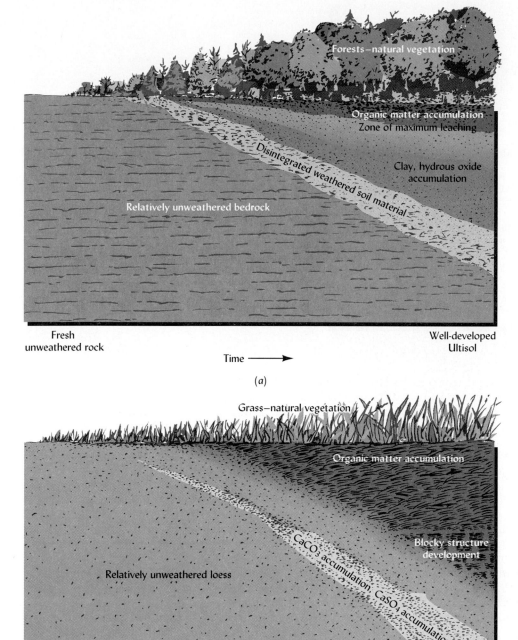

Forests—natural vegetation

Organic matter accumulation
Zone of maximum leaching

Clay, hydrous oxide accumulation

Disintegrated weathered soil material

Relatively unweathered bedrock

Fresh
unweathered rock

Well-developed
Ultisol

Time ⟶

(a)

Grass—natural vegetation

Organic matter accumulation

Blocky structure development

CaCO₃ accumulation, CaSO₄ accumulation

Relatively unweathered loess

Fresh
loess deposit

Well-developed
Mollisol

Time ⟶

(b)

FIGURE 2.22   How two soil profiles may have developed in climatic areas that encouraged as natural vegetation either (a) forests or (b) grasslands. Organic matter accumulation in the upper horizons occurs in time, the amount and distribution depending on the type of natural vegetation present. Clay and iron oxide accumulate and characteristic structures develop in the lower horizons. The end products differ markedly from the soil materials from which they formed.

cations than under oak or maple tree vegetation.[4] Also, the removal of base-forming cations by leaching is more rapid under coniferous vegetation.

In addition, an interaction develops among the natural vegetation, other soil organisms, and the characteristics of the soil. Thus, as soils develop under native grasslands, bacteria (*Azotobacter*) that can fix atmospheric nitrogen into compounds usable by

---

[4]Soil acidity is determined primarily by the relative proportion of H⁺ ions, which come from acids such as H_2CO_3 and HNO_3, and of metallic cations (e.g., Ca^{2+}, Mg^{2+}, and K^+), which are capable of forming bases such as $Ca(OH)_2$ and KOH. Thus, Ca^{2+}, Mg^{2+}, and K^+ are known as **base-forming** cations. See Section 9.3 for further explanation.

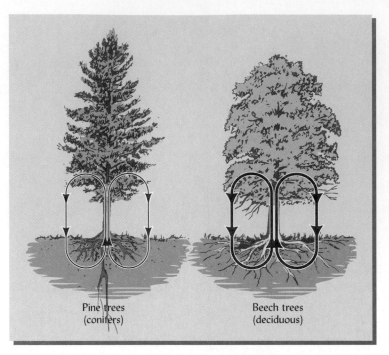

Pine trees
(conifers)

Beech trees
(deciduous)

FIGURE 2.23 Nutrient cycling is an important factor in determining the relationship between vegetation and the soil that develops. If residues from the vegetation are low in bases, as is the case for most conifers such as pine species, (left), acid weathering conditions are favored. High base-containing plant residues, such as many deciduous trees including the European beach (right), tend to neutralize acids, thereby favoring only slightly acid to neutral weathering conditions.

plants tend to flourish. This provides an opportunity for an increase in the nitrogen content of the soils and, as we shall see in Chapter 11, the organic matter content as well. Other microorganisms attack plant and animal residues, producing slimy organic materials. These, along with an abundance of plant roots, help bind soil particles into stable aggregates. Living organisms are a critical factor in determining soil character.

The effects of animals on soil-formation processes must not be overlooked. Large animals such as gophers, moles, and prairie dogs bore into the lower soil horizons, bringing materials to the surface. Their tunnels are often open to movement of water and air into the subsurface layers. In localized areas, they enhance mixing of the lower and upper horizons by creating and later refilling underground tunnels.

Earthworms and other small burrowing animals have significant effects on soil formation. They bring about considerable soil mixing as they burrow through the soil. Earthworms ingest soil particles and organic residues that pass through their bodies and enhance the availability of plant nutrients. They aerate and stir the soil and increase the stability of soil aggregates, thereby assuring ready infiltration of water. Ants and termites as they build mounds also transport soil materials from one horizon to another.

Human activities can also significantly influence soil formation. For example, several large areas of prairie grasslands in Indiana and Michigan are believed to have been maintained in grassland by repeated fires set by Native Americans. In more recent times, destroying of natural vegetation (trees, grass) and subsequently tilling the soil for crop production has abruptly modified the soil-forming factors. Likewise, irrigating a soil in an arid area drastically influences the soil-forming factors, as does adding fertilizer and lime to soils of low fertility. Although human activities are pertinent only in recent geological times, they have had significant influences on soil-forming processes in some areas.

2.14 TOPOGRAPHY

Topography relates to the configuration of the land surface and is described in terms of differences in elevation and slope—in other words, the lay of the land. Topography can hasten or delay the work of climatic forces. In smooth, flat country, excess water is removed less rapidly than in rolling areas. Rolling to hilly topography encourages natural erosion of the surface layers, which reduces the possibility of a deep soil (Figure 2.24). On the other hand, if water stands for part or all of the year on a given area, the climate is less effective in regulating soil development.

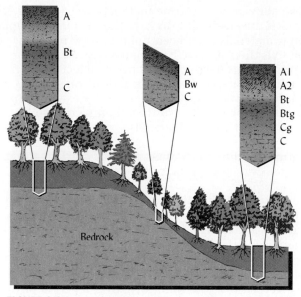

FIGURE 2.24 Topography influences soil properties, including soil depth. The diagram on the left shows the effect of slope on the profile characteristics and the depth of a soil on which forest trees are the natural vegetation. The photo on the right illustrates the same principle under grassland vegetation. Often a relatively small change in slope can have a great effect on soil development. See Section 2.17 for explanation of horizon symbols. (Photo courtesy of R. Weil)

There is a definite interaction among topography, vegetation, and soil formation. In the grassland–forest transition zones, trees commonly occupy the slight depressions in an otherwise prairie vegetation. This is apparently a moisture effect. As would be expected, the nature of the soil in the depressions is quite different from that in the uplands. Topography, therefore, not only modifies climate and vegetative effects, but often has a major direct effect on soil formation and on the type of soil that forms.

The slope of the land with respect to the rays of the sun influences soil temperature and soil moisture and, in turn, soil formation. In the northern hemisphere, south-facing slopes are more perpendicular to the sun's rays and are generally warmer and thereby commonly lower in moisture than their north-facing counterparts (see also Section 7.9). Consequently, soils on the south slopes tend to be lower in organic matter and are not so deep.

Topography also influences the deposition of plant residues that become parent materials for organic soils. Low-lying areas in humid regions receive runoff waters from the uplands, providing a habitat for water-tolerating plants, the residues of which accumulate and give rise to peat bogs and, in turn, organic soils (see Figure 2.19).

In arid and semiarid areas, topography also influences the buildup of soluble salts in low-lying areas. Salts from surrounding upland soils move on the surface and through the underground water table to the lower areas, where they can accumulate to levels that are toxic to plants when the water evaporates.

On steeper slopes in semiarid regions, soil development is commonly retarded compared to more level areas where more of the rain soaks in rather than running off into streams and gullies. The drier soils on the slopes support only sparse vegetation, and the soil-formation process is delayed.

2.15 TIME

The length of **time** that materials have been subjected to weathering influences soil formation. Time effects can be seen by comparing soils of a glaciated region with those in a comparable unglaciated area nearby. The effect of time is also apparent when comparing soils on recent Wisconsin glacial material with those on adjacent older deposits of the Illinoian or Kansan age. The influence of parent material is much more apparent in the soils of glaciated regions, where insufficient time has elapsed since the ice retreated to permit the full development of many soils.

Soils located on alluvial or lacustrine materials (see Sections 2.7 and 2.9) generally have not had as much time to develop as have the surrounding upland soils. Also, some coastal-plain parent materials have been uplifted only in recent geological time, and, consequently, the soils thereon have had relatively short exposure to weathering.

Residual parent materials have generally been subject to weathering for longer periods of time than have materials transported from one site to another. Soils of the uplands in the southeast part of the United States, for example, have been submitted to weathering for a much longer period of time than nearby soils in low-lying areas that are developed on marine or alluvial materials.

Soils develop so slowly that it is usually not possible to measure time-related changes in soil formation directly. Indirect methods such as those using carbon dating must be turned to for evidence on the time required for different aspects of soil development to occur.

The interaction of time with other factors affecting soil formation must be emphasized. The time required for the development of a horizon definitely will be influenced by the parent material, the climate, and the vegetation. These factors are highly interdependent in determining the kind of soil that develops.

It should be emphasized that two or more of the factors influencing soil formation are commonly active simultaneously and interdependently; thus, climate and parent material influence vegetation, which, in turn, over a period of time, has helped influence the nature of the parent material available. Likewise, there is a relation between some parent materials and topography. The interaction and interdependence of these factors introduces some complexity in evaluating just how a given soil was formed.

Now that each of the five major factors of soil formation have been considered, we are ready to give attention to the processes that lead directly to the formation of soils.

2.16   SOIL FORMATION IN ACTION

An examination of a soil pit or a fresh roadcut reveals a soil profile with distinctive horizontal layers, some of which are highly visible. Such layers would not be found if a similar cut were to be made in unconsolidated materials recently moved by a bulldozer or laid down following a recent major volcanic eruption. Obviously, then, significant changes have occurred as soils develop from relatively unconsolidated parent materials. A study of soil formation (genesis) gives some notion as to how these changes occur and why they can stimulate the development of so many different kinds of soil.

FOUR MAJOR PROCESSES.   Soil genesis is brought about by a series of processes, the most significant of which are: (1) **transformations** such as weathering and organic matter breakdown, by which some soil constituents are modified or destroyed, and others are synthesized; (2) **translocation** or movement of inorganic and organic materials from one horizon up or down to another, the materials being moved mostly by water but also by soil organisms; (3) **additions** of materials to the soil profile from outside sources such as organic matter from leaves, dust from the atmosphere, or soluble salts from the groundwater; and (4) **losses** of materials from the soil profile by leaching to groundwater, erosion of surface materials or other forms of removal. Often transformation and translocation results in the accumulation of materials in a particular horizon.

A SIMPLIFIED EXAMPLE.   The role of these major processes can be seen by following the changes that take place as soils develop from relatively uniform parent material. When plants begin to grow and their residues are deposited on the surface of the parent materials, soil formation has truly begun. The plant residues are disintegrated and partly decomposed by soil organisms that also synthesize new organic compounds that make up humus. Earthworms that burrow into and live in the soil along with other small animals such as ants and termites mix these organic materials with the underlying mineral matter near the soil surface. This mixture, which comes into being rather quickly, is commonly the first soil horizon developed; it is different in color (darker) and composition from the original parent material (see Figure 2.25).

As plant residues decay, organic acids are formed. These acids are carried by percolating waters into the soil, where they stimulate the weathering processes. They solubilize some chemicals that are translocated (leached) from upper to lower horizons or are

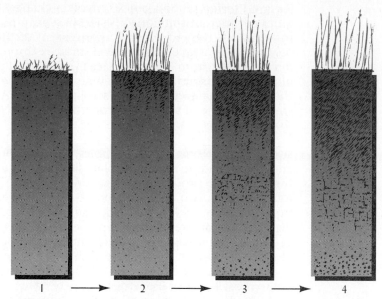

FIGURE 2.25 Illustration of stages of development of a hypothetical soil from relatively uniform parent material such as loess. At stage 1, no layering exists. The accumulation of organic matter near the surface from plant residues has just begun. By stage 2, organic matter has been incorporated by earthworms into the upper few centimeters of soil, soluble ions (e.g., Ca^{2+}, SO_4^{2-}) have moved downward, and clays are beginning to form. By stage 3, more organic matter has accumulated and moved more deeply into the soil. Further leaching of soluble ions has occurred and clays are beginning to be translocated to lower horizons. By stage 4, as we may find the soil today, even higher levels of organic matter are found deeper in the profile. Clays have moved into the subsoil and have stimulated the formation of distinctive blocklike groups of soil particles (structure) in the subsoil. Deeper in the regolith, ions leached from the upper surface may be precipitated as insoluble compounds, forming a layer of calcium carbonate.

completely removed from the emerging soil by percolating waters. On sloping lands erosion may remove some materials from the upper horizon.

As weathering proceeds, some primary minerals are disintegrated and altered to form different kinds of silicate clays. Others are decomposed, and the decomposition products are recombined into new minerals such as other silicate clays and hydrous oxides of iron and aluminum (see Figure 2.25).

These newly formed minerals may accumulate in place where they are formed or may move downward and accumulate in lower soil zones. As materials are removed from one layer and accumulate in another, soil horizons are formed. Upper horizons may be characterized by the removal of specific constituents, while the accumulation of these or other constituents may characterize the lower horizons. In either case, soil horizons are created that are different in character from the original parent material. As the soil matures, the various horizons within the profile become more distinctly different from each other.

TRANSPORT OR ACCUMULATION OF SIMPLE COMPOUNDS. Weathering also produces soluble materials, including positively charged ions (cations, e.g., Ca^{2+}) and negatively charged ions (anions, e.g., SO_4^{2-}). In humid regions, these materials are moved downward in rainwater and are either taken up by plant roots and recycled back into the soil or removed in the drainage water. In areas with lower rainfall, however, once some of these ions are moved to the subsurface horizon, they are combined and precipitate as insoluble compounds such as calcite ($CaCO_3$) and gypsum ($CaSO_4 \cdot 2H_2O$) (see Figure 2.25). Layers containing these compounds are common in soils of semiarid and arid areas.

In some soils in arid areas, soluble salts such as sodium chloride (NaCl) and sodium sulfate (Na_2SO_4) also accumulate at or near the soil surface, having been transported there by upward-moving water that then evaporates into the atmosphere (see Section 10.2).

PHYSICAL PROPERTIES.   Soil genesis involves changes in physical as well as chemical properties of different soil horizons. The best examples of such changes relate to soil structure (that is, the grouping or arrangement of the soil particles into aggregates[5]). The organo-mineral layer near the soil surface commonly has a granular-type structure, differentiating it from the subsurface horizons and the original parent material. Likewise, blocky and prismatic-type structures characterize some subsurface horizons, especially those with clay accumulation (see Figure 2.25). As will be discussed in Section 3.2, soil structure is one of the major properties used to characterize soil horizons and, in turn, to classify soils.

SOIL GENESIS IN NATURE.   The examples we have considered of weathering, translocation, and accumulation illustrate rather simply how soil horizons come into being and, in turn, how the process of soil genesis proceeds. But before studying specific horizons found in soils, two additional points should be emphasized. First, the parent materials from which many soils develop are not uniform, varying considerably according to depth. These differences in the properties of a given parent material commonly existed before soil genesis started. For example, water-deposited materials may vary widely in texture with depth.[6] Consequently, in characterizing soils, consideration must be given not only to the **genetic** horizons and properties that come into being during soil genesis, but to those layers or properties that may have been **inherited** from the parent material.

Second, soils are dynamic bodies whose genetic horizons continue to be developed. Consequently, in some soils the process of horizon differentiation has only begun while in others it is well advanced. The latter are sometimes referred to as *well-developed soils.* This means that the soil profiles we see today are likely quite different from those that existed a few thousand years ago and from those that will exist a thousand years from now.

2.17   THE SOIL PROFILE

The layering or horizon development described in the previous section gradually gives rise to natural bodies called **soils.** Each soil is characterized by a given sequence of these horizons. A vertical exposure of this sequence is termed a ***soil profile.*** Attention now will be given to the major horizons making up soil profiles and the terminology used to describe them.

The Master Horizons and Layers

Five **master** soil horizons are recognized and are designated using the capital letters O, A, E, B, and C. If two different geologic parent materials (e.g., loess over glacial till) are present within the soil profile the numeral 2 is placed in front of the master horizon symbols for horizons developed in the second layer of parent material. Subordinate layers or distinctions within these master horizons are designated by lowercase letters. A common sequence of horizons within a profile is shown in Figure 2.26.

O HORIZONS (ORGANIC).   The O group is comprised of organic horizons that form above the mineral soil. They result from litter derived from dead plants and animals. O horizons usually occur in forested areas and are generally absent in grassland regions. The specific horizons are:

Oi   Organic horizon of the original plant and animal residues, only slightly decomposed (fibric)

Oe   Organic horizon, residues intermediately decomposed (hemic)

Oa   Organic horizon, residues highly decomposed (sapric)

---

[5]Soil structure was discussed briefly in Section 1.11. See Figure 4.11 for a description of the various structural types.

[6]As mentioned in Section 1.11, soil texture is related to the size of the individual soil particles (sand, clay, etc.). See Figure 4.8 for the sand, silt, and clay components of different textural classes.

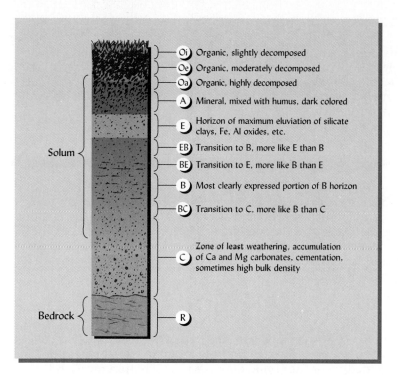

FIGURE 2.26 Hypothetical mineral soil profile showing the major horizons that may be present in a well-drained soil in the temperate humid region. Any particular profile may exhibit only some of these horizons, and the relative depths vary. In addition, however, a soil profile may exhibit more detailed subhorizons than indicated here. The solum includes the A, E, and B horizons plus some cemented layers of the C horizon.

A HORIZONS. These are the topmost mineral horizons. They generally contain enough partially decomposed (humified) organic matter to darken the soil color more than that of the lower horizons. A horizons are often coarser in texture, having lost some of the finer materials to lower horizons.

E HORIZONS. These are zones of maximum leaching or **eluviation** (from Latin *ex* or *e,* out, and *lavere,* to wash) of clay, iron, and aluminum oxides, which leaves a concentration of resistant minerals, such as quartz, in the sand and silt sizes. An E horizon is generally lighter in color than the A horizon and is found underneath the A horizon.

B HORIZONS (ILLUVIAL). These are subsurface horizons in which the accumulation of materials by **illuviation** (from the Latin *il,* in, and *lavere,* to wash) has taken place. In humid regions, the B horizons are the layers of maximum accumulation of materials such as iron and aluminum oxides and silicate clays, some of which may have formed in place. In arid and semiarid regions, calcium carbonate, calcium sulfate, and other salts may accumulate in the B horizon.[7]

The B horizons are sometimes referred to incorrectly as the *subsoil,* a term that lacks precision. In soils with shallow A horizons, part of the B horizon may be incorporated into the plow layer and thus become part of the topsoil. In other soils with deep A horizons, the plow layer or topsoil may include only the upper part of the A horizons, and the subsoil would include the lower part of the A horizon along with the B horizon. This emphasizes the need to differentiate between terms commonly used in soil management (*topsoil, subsoil*) and those used to describe the soil profile.

C HORIZON. The C horizon is the unconsolidated material underlying the solum (A and B horizons). It may or may not be the same as the parent material from which the solum formed. The C horizon is outside the zones of major biological activities and is generally little affected by the processes that formed the horizons above it. Its upper layers may in time become a part of the solum as weathering and erosion continue.

R LAYERS. These are found underlying consolidated rock, with little evidence of weathering.

---

[7]In some soils of arid and semiarid regions the accumulation of these calcium compounds takes place in the C horizon (below).

Transition Horizons

These horizons are transitional between the master horizons (O, A, E, B, and C). They may be dominated by properties of one horizon but have prominent characteristics of another. The two applicable capital letters are used to designate the transition horizons (e.g., AE, EB, BE, and BC), the dominant horizon being listed before the subordinate one. Letter combinations such as E/B are used to designate transition horizons where distinct parts of the horizon have properties of E while other parts have properties of B.

Subordinate Distinctions

Master horizons are further characterized by specific properties such as distinctive color or the accumulation of materials such as clays and salts. These subordinate distinctions are identified by using lowercase letters that designate specific characteristics. A list of these subordinate distinctions and their meanings are shown in Table 2.4. By illustration, a Bt horizon is a B horizon characterized by clay accumulation (t); likewise, in a Bk horizon, carbonates (k) have accumulated.

Horizons in a Given Profile

It is not likely that the profile of any one soil will show all of the horizons that collectively are cited in Figure 2.26. The ones most commonly found in well-drained soils are Oe (or Oa if the land is forested); A or E (or both, depending on circumstances); B or Bw; and C. Conditions of soil genesis will determine which others are present and their clarity of definition.

When a virgin soil is plowed and cultivated, the upper 15–20 cm (6–8 in.) becomes the furrow slice. The cultivation, of course, destroys the original layered condition of this upper portion of the profile, and the furrow slice becomes more or less homogeneous. In some soils, the A and E horizons are deeper than the furrow slice (Figure 2.27). In other cases, where the upper horizons are quite thin, the plowline is just at the top of, or even down in, the B horizon.

In some cultivated land, serious erosion has produced a **truncated** profile. As the surface soil was swept away, the plowline was gradually lowered to maintain a sufficiently thick furrow slice. Hence, the furrow slice in many cases is almost entirely within the B zone, and the C horizon is correspondingly near the surface. Many farmers are today cultivating B horizon soil material. In profile study and description, such a situation requires careful analysis. Comparison to a nearby noneroded site can show how much erosion has occurred.

TABLE 2.4   Lowercase Letter Symbols to Designate Subordinate Distinctions Within Master Horizons

Letter	Distinction	Letter	Distinction
a	Organic matter, highly decomposed	o	Accumulation of Fe and Al oxides
b	Buried soil horizon	p	Plowing or other disturbance
c	Concretions or nodules	q	Accumulation of silica
d	Dense unconsolidated materials	r	Weathered or soft bedrock
e	Organic matter, intermediate decomposition	s	Illuvial accumulation of O.M.[a] and Fe and Al oxides
f	Frozen soil	ss	Slickensides
g	Strong gleying (mottling)	t	Accumulation of silicate clays
h	Illuvial accumulation of organic matter	v	Plinthite (high iron, red material)
i	Organic matter, slightly decomposed	w	Distinctive color or structure
k	Accumulation of carbonates	x	Fragipan (high bulk density, brittle)
m	Cementation or induration	y	Accumulation of gypsum
n	Accumulation of sodium	z	Accumulation of soluble salts

[a]O.M. = organic matter.

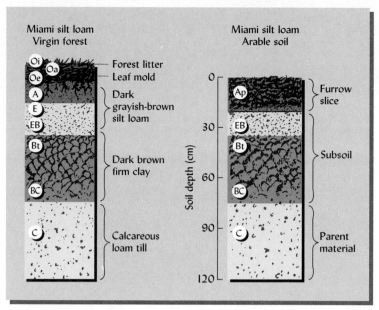

FIGURE 2.27 Generalized profile of the Miami silt loam, one of the Alfisols of the eastern United States, before and after land is plowed and cultivated. The surface layers are mixed by tillage and are termed the *Ap* (plowed) *horizon*. If erosion occurs, they may disappear, at least in part, and some of the B horizon will be included in the furrow slice.

2.18   CONCLUSION

Soil formation is stimulated by **climate** and **living organisms** acting on **parent materials** over periods of **time** and under the modifying influence of **topography**. These five major factors of soil formation determine the kinds of soil that will develop. When all of these factors are the same from one location to another, the kind of soil at the two locations should be the same.

The parent materials from which soils develop vary widely around the world and from one location to another only a few meters away. Knowledge of these materials, their source or origin, mechanisms for their weathering, and means of transport and deposition are essential if we are to gain an understanding of soils, and especially if we are to classify them properly.

Soil genesis starts when layers or horizons not present in the parent material begin to appear in the soil profile. Organic matter accumulation in the upper horizons, the downward movement of soluble ions, the synthesis and downward movement of clays, and the development of specific soil particle groupings (structure) in both the upper and lower horizons are evidences that the process of soil formation is under way.

An understanding of the processes of soil formation and of the five major factors influencing this formation is invaluable in site selection and in predicting the nature of soil bodies likely to be found on a particular site. Conversely, analysis of the horizon properties of a soil profile can tell us much about the nature of the climatic, biological, and geological conditions (past and present) at the site.

Characterization of the horizons in the profile leads to the identity of a soil individual, which is then subject to classification—the topic of our next chapter.

REFERENCES

Daly, R. A. 1934. *The Changing World of the Ice Age* (New Haven: Yale University Press).

Jenny, Hans. 1941. *Factor of Soil Formation: A System of Quantitative Pedology* (originally published by McGraw-Hill), Dover Pub., Mineola, N.Y.

Miller, F. P. 1976. Maryland Soils Bul. 212 (Univ. of Maryland Coop. Ext. Serv., College Park, Md.).

National Science Foundation. 1975. "All that unplowed land." *Mosaic*, **6**:17–21 (Washington, D.C.: National Science Foundation).

Petersen, G. W., R. L. Cummingham, and R. P. Matelski. 1971. "Moisture characteristics of Pennsylvania soils: III. Parent material and drainage relationships," *Soil Sci. Soc. Amer. Proc.*, **35**:115–119.

Soil Survey Division Staff. 1993. *Soil Survey Manual,* Agric. Hndbk. 18 (Washington, D.C.: U.S. Government Printing Office).

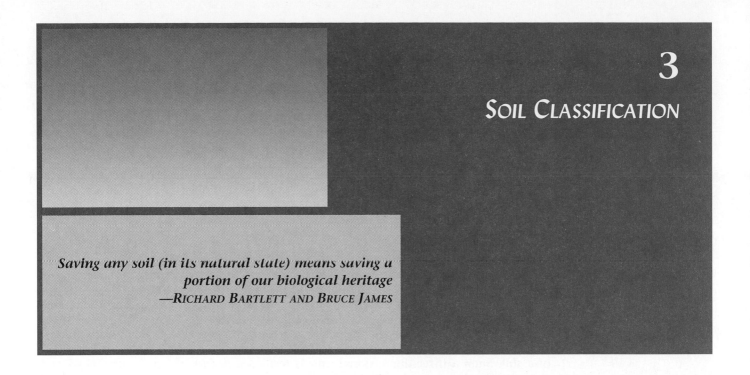

3

SOIL CLASSIFICATION

Saving any soil (in its natural state) means saving a portion of our biological heritage
—RICHARD BARTLETT AND BRUCE JAMES

The sustainable use of land anywhere in the world is largely dependent on soil properties. To make proper land-use decisions we must know the differences in soil properties from one site to another. Furthermore, we must be able to communicate these differences by the names we give the soils and, in turn, through the systems we use to classify them. Through such systems we can take advantage of research and experience at one location to predict the response of similarly classified soils at another location. We can create a universal language of soils that enhances communications among users of soils around the world. When soil names such as *Mollisol* or *Oxisol* are used, a mental picture appears in the mind of soil scientists everywhere, whether they live in the United States, Europe, Japan, developing countries, or elsewhere.

Throughout history humans have used various systems to name and classify soils. From the time crops were first cultivated, humans noticed differences in soils and classified them, if only in terms of "good" and "bad." Beginning with the early Chinese, Egyptian, Greek, and Roman civilizations, people have acknowledged differences in soils and given them descriptive terms such as "black cotton soils," "forest soils," "rice soils," or "olive soils."

Soils also have been classified in terms of the geological parent materials from which they were formed. Terms such as *sandy* or *clayey* soils as well as *limestone* soils and *alluvial* soils have a geological connotation and are still used today. Likewise, the topographic position of the soil on the landscape has been used to identify the soil class, with names such as *piedmont, coastal plain, flatlands,* or *bottom lands.*

A classification system is needed to help us organize our knowledge about the thousands of individual natural bodies we recognize as soils. Furthermore, as natural bodies, soils need to be classified not just on the basis of their suitability for crop production but for all uses, including that of being left in their natural state. Soil classification systems are essential to foster global communications about soils, not only among soil scientists, but among all concerned with the management and conservation of soils.

3.1 CONCEPT OF INDIVIDUAL SOILS

When in the 1880s the Russian soil scientist V. V. Dukochaev and his associates conceived the idea that soils were natural bodies, they opened up an entirely new vision for the study of soils. Unfortunately, however, poor international communications and the reluctance of some scientists to accept such radical ideas delayed the acceptance by scientists in Europe and the United States of the natural bodies concept. Some field scientists recognized the pertinence of the Russian studies, but it was not until the late 1920s that these studies were seriously considered. C. F. Marbut of the U.S. Department of Agriculture was one who grasped the concept of soils as natural bodies, and in 1927 he developed a soil classification scheme based on these principles.

The natural body concept of soils assumes the existence of individual entities, each of which we call *a* soil. Just as human individuals differ from one another, individual soils have characteristics that distinguish them from other individual soils. Likewise, just as human individuals may be grouped according to some characteristic such as height or weight, soil individuals may be grouped with other soil individuals having one or more characteristics in common. Similarly, sets of groups, each having some characteristic in common, can occur as one moves up the classification ladder to *the* soil.

Pedon and Polypedon

There are no sharp demarcations between the properties of one soil individual and those of another. Rather there is a gradation in such properties as one moves from one soil individual to a second. The gradation in soil properties can be compared to the gradation in the wavelengths of light as you move from one color to another. The change is gradual, and yet we identify a boundary that differentiates what we call "green" from what we call "blue." Consequently, it is necessary to characterize a soil individual in terms of the smallest three-dimensional unit that embodies the primary characteristics of a soil. Such a three-dimensional unit is termed a *pedon* (rhymes with "head on," from the Greek *pedon,* ground) (see Figure 3.1). It is the smallest sampling unit that can characterize a soil individual.

Pedons occupy from about 1 to 10 m^2 of land area. Because of this very small size, a pedon obviously cannot be used as the basic unit for a workable field soil classification system. However, a group of pedons, termed a *polypedon,* closely associated in the field and similar in properties, is of sufficient size to serve as a basic classification unit, or a *soil individual.* Such groupings conceptually approximate what in the United States has been called a *soil series.* The nearly 19,000 soil series characterized in the United States are the basic units used to classify the nation's soils.[1]

Groupings of Soil Individuals

Two extremes in the concept of soils now have been identified. One extreme is that of a natural body called *a* soil, characterized by a three-dimensional sampling unit (pedon), related groups of which (polypedons) are included in a soil individual. At the other extreme is *the* soil, a collection of all these natural bodies that is distinct from water, solid rock, and other natural parts of the earth's crust. The two extremes represent opposite ends of elaborate soil classification schemes that are used to organize knowledge of soils.

These schemes generally involve the grouping of different soil individuals with some common characteristics into classes. These classes, in turn, may be further grouped into higher-category classes and the process continued until the uppermost category, **the** soil, is reached. It is the rationale and criteria that one uses to ascertain how the individual soils and, in turn, higher-category classes are grouped that differentiates one classification system from another.

3.2 COMPREHENSIVE CLASSIFICATION SYSTEM: SOIL TAXONOMY[2]

There are a number of soil classification systems in use in different parts of the world.[3] A refined Russian system with heavy emphasis on factors of soil formation continues to be used, not only in countries of the former Soviet Union but in some other European

---

[1]It is well to point out the conceptional soil **classification** units may not coincide strictly with **mapping** units as used in the field. Thus, as applied in the preparation of field maps, the series designation may include aggregates of polypedons along with some inclusions of others. See Section 19.7 for further explanation.

[2]The original concept of *Soil Taxonomy* is from Soil Survey Staff (1975).

[3]See Appendix B for Canadian and FAO systems.

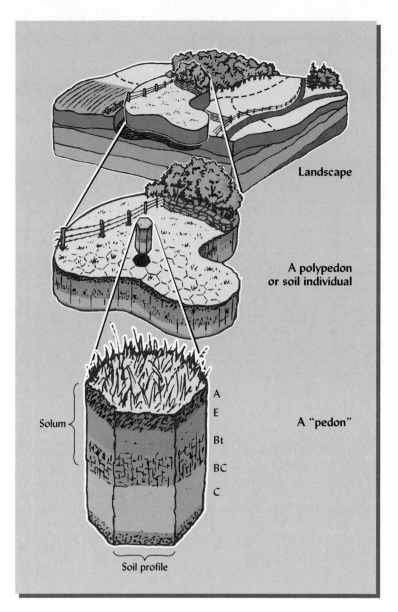

Landscape

A polypedon
or soil individual

A "pedon"

Solum

A
E
Bt
BC
C

Soil profile

FIGURE 3.1 A schematic diagram to illustrate the concept of pedon and of the soil profile that characterizes it. Note that several contiguous pedons with similar characteristics are grouped together in a larger area (outlined by broken lines) called a *polypedon* or *individual soil*. Several individual soils are present in this landscape.

countries as well. Also, classification systems have been developed and are in use in France and its former colonies, Belgium, the United Kingdom, Australia, Canada, and Brazil. In each case the systems meet the needs of the concerned countries. Also, a world soil map produced by two United Nations–associated organizations, Food and Agriculture (FAO), and the United Nations Education, Science and Cultural Organization (UNESCO), involves a partial classification system that is used to inventory and describe the world's soil resources.

In the United States, Marbut's 1927 classification scheme was improved in 1935, 1938, and 1949, the latter revision lasting for about 25 years. But in 1951 a decision was made to begin to develop an entirely new system of classification. Accordingly, the Soil Survey Staff of the U.S. Department of Agriculture, in cooperation with soil scientists in the United States and in other countries, developed a new comprehensive system of soil classification based on soil properties. This system has been in use in the United States since 1965, and it is used, at least to some degree, by scientists in 45 other countries. This system will be used in this text.

The comprehensive soil classification system, called ***Soil Taxonomy***,[4] maintains the natural body concept and has two other major features that make it most useful. First,

---

[4]Taxonomy is the science of principles of classification. For a review of the achievements and challenges of *Soil Taxonomy,* see SSSA (1984). See Soil Survey Staff (1994) for the latest *Keys to Soil Taxonomy.*

the system is based on *soil properties* that are easily verified. This lessens the likelihood of controversy over the classification of a given soil which can occur when scientists deal with systems based on soil genesis or presumed mechanisms of soil formation.

The second significant feature of *Soil Taxonomy* is the *unique nomenclature* employed, which gives a definite connotation of the major characteristics of the soils in question. It is truly international in makeup since it was meant to apply universally. Consideration will be given to the nomenclature used after brief reference is made to the major criteria for the system—soil properties.

Bases of Soil Classification

Soil Taxonomy is based on the properties of soils as they are found today. Although one of the objectives of the system is to group soils similar in genesis, the specific criteria used to place soils in these groups are those of soil properties. In so doing, soil genesis is not ignored. Since soil properties often are related directly to soil genesis, it is difficult to emphasize soil properties without at least indirectly considering soil genesis as well.

All of the chemical, physical, and biological properties presented in this text are used as criteria for *Soil Taxonomy.* A few examples are moisture and temperature status throughout the year, as well as color, texture, and structure of the soil. Chemical and mineralogical properties such as the contents of organic matter, clay, iron and aluminum oxides, silicate clays, salts, the pH, the percentage base saturation,[5] and soil depth are other important criteria for classification. Precise measurements of these properties are made, not just comparative assessments. The presence or absence of certain diagnostic soil horizons also determines the place of a soil in the classification system. These special horizons that are formed in the surface or the subsurface will now receive our attention.

Diagnostic Surface Horizons

The diagnostic horizons that occur at the soil surface are called ***epipedons*** (from the Greek *epi,* over, and *pedon,* soil). The epipedon includes the upper part of the soil darkened by organic matter, the upper eluvial horizons, or both. It may include part of the B horizon (see Section 2.17) if the latter is significantly darkened by organic matter. Seven epipedons are recognized (Table 3.1), but only five occur naturally over wide areas (Figure 3.2). The other two, called ***anthropic*** and ***plaggen,*** are the results of intensive human use of soils. They are common in parts of Europe and Asia where soils have been utilized for many centuries.

---

[5]The percentage base saturation is the percentage of a soil's total cation adsorption (exchange) capacity that is satisfied by base-forming cations such as Ca^{2+}, Mg^{2+}, and K^+ (see Section 8.12).

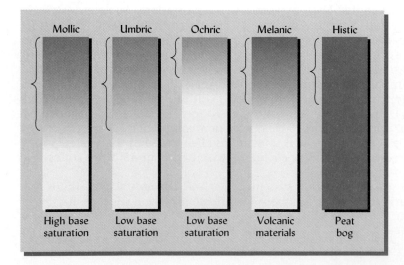

FIGURE 3.2 Representative profile characteristics of five surface diagnostic horizons (epipedons). The comparative organic matter levels and distribution are indicated by the darkness of colors. The Mollic and Umbric epipedons have similar organic matter distribution but the percentage base saturation is higher (greater than 50%) in the Mollic epipedon and lower (less than 50%) in the Umbric epipedon. The Ochric epipedon is much lower in organic matter content; consequently, it is light in color and sometimes hard when dry. Two other epipedons have very high organic matter contents and are very dark in color. The Melanic epipedon is formed on recently deposited volcanic materials, usually in cool wet areas. The Histic epipedon is formed from organic deposits laid down in peat bogs. The approximate depth of each epipedon is shown by the brackets.

TABLE 3.1 Major Features of Diagnostic Horizons in Mineral Soils Used to Differentiate at the Higher Levels of *Soil Taxonomy*

Diagnostic horizon (and designation)	Major features
Surface horizons = epipedons	
Mollic (A)	Thick, dark colored, high base saturation, strong structure
Umbric (A)	Same as mollic except low base saturation
Ochric (A)	Light colored, low organic content, may be hard and massive when dry
Melanic (A)	Thick, black, high in organic matter (>6% organic C) common in volcanic ash soils
Histic (O)	Very high in organic content, wet during some part of year
Anthropic (A)	Human-modified mollic-like horizon, high in available P
Plaggen (A)	Human-made sodlike horizon created by years of manuring
Subsurface horizons	
Argillic (Bt)	Silicate clay accumulation
Natric (Btn)	Argillic, high in sodium, columnar or prismatic structure
Spodic (Bhs)	Organic matter, Fe and Al oxides accumulation
Cambic (B)	Changed or altered by physical movement or by chemical reactions
Agric (A or B)	Organic and clay accumulation just below plow layer resulting from cultivation
Oxic (Bo)	Highly weathered, primarily mixture of Fe, Al oxides and non-sticky-type silicate clays
Duripan (m)	Hardpan, strongly cemented by silica
Fragipan (x)	Brittle pan, usually loamy textured, weakly cemented
Albic (E)	Light colored, clay and Fe and Al oxides mostly removed
Calcic (k)	Accumulation of $CaCO_3$ or $CaCO_3 \cdot MgCO_3$
Gypsic (y)	Accumulation of gypsum
Salic (z)	Accumulation of salts
Kandic	Accumulation of low-activity clays
Petrocalcic	Cemented calcic horizon
Petrogypsic	Cemented gypsic horizon
Placic	Thin pan cemented with iron alone or with manganese or organic matter
Sombric	Organic matter accumulation
Sulfuric	Highly acid with Jarosite mottles

MOLLIC EPIPEDON. (Latin *mollis,* soft) This mineral surface horizon is noted for its dark color associated with its high organic matter content (>0.6% organic C throughout the thickness), for its thickness (generally >25 cm), and for its softness even when dry. It has a high base saturation (greater than 50% base-forming cations). Mollic epipedons are moist at least three months a year and the soil temperature is usually 5°C or higher to a depth of 50 cm. These epipedons are characteristic of soils developed under native prairies. Mollisols are among the world's most productive soils.

UMBRIC EPIPEDON. (Latin *umbra,* shade; hence, dark) Has the same general characteristics as the mollic epipedons except that the percentage base saturation is less than 50%. This mineral horizon commonly develops in areas with somewhat higher rainfall and where the parent material has lower content of base-forming cations compared to the mollic epipedon.

OCHRIC EPIPEDON. (Greek *ochros,* pale) A mineral horizon that is either too thin, too light in color, or too low in organic matter to be either a mollic or umbric horizon. It is usually not as deep as the mollic or umbric epipedons. As a consequence of its low organic matter content it may be hard and massive when dry.

MELANIC EPIPEDON. (Greek *melas, melan,* black) This is very black in color due to its high organic matter content (organic carbon >6%). It is characteristic of soils high in such minerals as allophane, developed from volcanic ash. It is more than 30 cm thick and is light in weight and fluffy.

Histic Epipedon.   (Greek *histos,* tissue) Consists primarily of organic materials (peat and muck). Formed in wet areas, it too is black in color. It is very low in density.

Each epipedon name is associated with a series of properties that can be used to efficiently characterize soils as they exist in the field. The same can be said of the diagnostic subsoil horizons that will now receive our attention.

Diagnostic Subsurface Horizons

Many subsurface horizons are used to characterize different soils in *Soil Taxonomy.* The 18 that are considered diagnostic horizons are shown along with their major features in Table 3.1 and Figure 3.3. Each of these layers is used as a distinctive property to help place a soil in its proper class in the system. We will discuss briefly a few of these subsurface horizons.

The **argillic horizon** is a subsurface layer of accumulation of high-activity silicate clays that have moved downward from the A or E horizon or have formed in place. The clays often are found as coatings on pore walls and surfaces of the structural groupings.

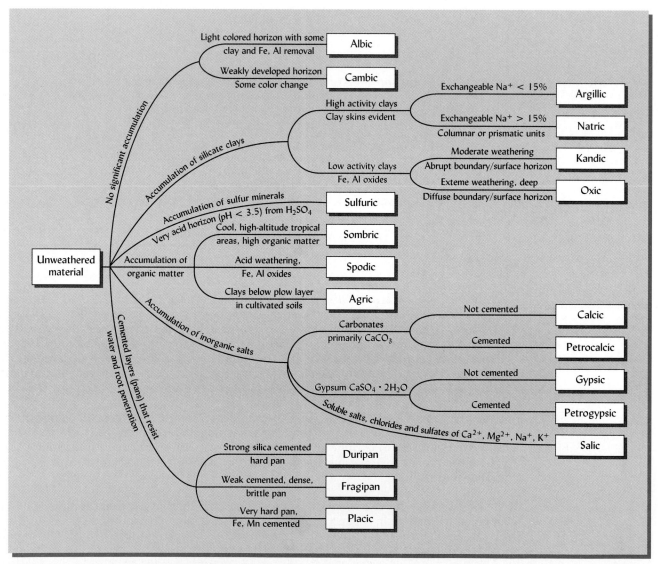

FIGURE 3.3   Names and major distinguishing characteristics of subsurface diagnostic horizons. Note that the focus is primarily on soil properties, not on how the properties have presumably evolved. Among the characteristics emphasized is the accumulation of silicate clays, organic matter, Fe and Al oxides, calcium compounds, and soluble salts, as well as materials that become cemented or highly acidified, thereby constraining root growth. The presence or absence of these horizons plays a major role in determining in which class a soil falls in *Soil Taxonomy.* See Chapter 8 for discussion of low- and high-activity clays.

The coatings usually appear as shiny surfaces or as clay bridges between sandgrains. Termed *cutans* or *clay skins,* they are concentrations of clay translocated from upper horizons (see Plate 18).

The **natric horizon** likewise has silicate clay accumulation (with clay skins), but the clays are accompanied by more than 15% exchangeable sodium on the colloidal complex and by columnar or prismatic soil structural units. The natric horizon is found mostly in arid and semiarid areas.

The **kandic horizon** has an accumulation of Fe and Al oxides as well as silicate clays (e.g., kaolinite), but clay skins need not be evident. The clays are low in activity as shown by their low cation-holding capacities (<16 $cmol_c$/Kg clay). The A horizon that overlies a kandic horizon has commonly lost much of its clay content.

The **oxic horizon** is a highly weathered subsurface horizon that is very high in Fe and Al, oxides, and low-activity silicate clays (e.g., kaolinite). The cation-holding capacity is <16 $cmol_c$/Kg clay. It is at least 30 cm deep and has <10% weatherable minerals in the fine fraction. It is generally physically stable, crumbly, and not very sticky despite its high clay content. It is found mostly in humid tropical and subtropical regions.

The **spodic horizon** is an illuvial horizon that is characterized by the accumulation of aluminum oxide (with or without iron oxide) and of colloidal organic matter. It is commonly found in highly leached forest soils of cool humid climates, typically on sandy-textured parent materials.

The **sombric horizon** is an illuvial horizon that is dark in color because of high organic matter accumulation. It has a low degree of base saturation and is found mostly in the cool, moist soils of high plateaus and mountains in tropical and subtropical regions.

The **albic horizon** is light-colored eluvial horizon that is low in clay and oxides of Fe and Al. These materials have largely been removed from this horizon and have moved downward.

A number of horizons have accumulations of saltlike chemicals that have leached from upper horizons in the profile. **Calcic horizons** contain an accumulation of carbonates (mostly $CaCO_3$) that often appear as white chalklike nodules. **Gypsic horizons** have an accumulation of gypsum ($CaSO_4 \cdot 2H_2O$), and **salic horizons** an accumulation of soluble salts. These are found mostly in soils of arid and semiarid regions.

In some subsurface diagnostic horizons, the materials are cemented, resulting in relatively impermeable layers called *pans* (**duripan, fragipan,** and **placic horizons**). These tend to resist water movement and the penetration of plant roots. Such pans constrain plant growth and may encourage water runoff and erosion because rainwater cannot move readily downward through the soil. These, along with diagnostic subsurface horizons, are referred to in Figure 3.3.

Soil Moisture Regimes (SMR)

A soil moisture regime refers to the presence or absence of either groundwater or of water at a sufficiently high level to be easily available to plants. The water is present or absent at given periods in the year in what is termed the *control section* of the soil. The upper boundary of the control section is the depth that 2.5 cm of water will penetrate within 24 hours when added to a dry soil. The lower boundary is the depth that 7.5 cm of water will penetrate. The control section ranges from 10 to 30 cm for soils high in fine particles (clay) and from 30 to 90 for sandy soils. Several moisture regime classes are used to characterize soils. The percentage distribution of areas with different soil moisture regimes is shown in Table 3.2.

Aquic.   The soil is saturated with water and virtually free of gaseous oxygen for sufficient periods of time for evidence of poor aeration (gleying, mottling) to occur. More than 8% of the world's total land area is classed as Aquic.

Udic.   The soil moisture is sufficiently high most years to meet plant needs. This regime is common for soils in humid climatic regions and characterizes about one-third of the worldwide land area. An extremely wet moisture regime throughout the year is termed *perudic.*

Ustic.   The soil moisture is intermediate between Udic and Aridic regimes—generally high enough to meet plant needs during the growing season, although significant periods of drought may occur. Some 18% of the world's land area is classified as Ustic.

TABLE 3.2   Percent of Global Area Occupied by Soils with Different Soil Moisture and Temperature Regimes

Soil temperature regimes	Soil moisture regimes						
	Aridic	Xeric	Ustic	Udic	Perudic	Aquic	Total
Pergelic	3.7			5.3		1.9	10.9
Cryic	4.3			8.5	0.0	0.5	13.5
Frigid	0.5		0.2	0.4			1.2
Mesic	5.4	0.8	0.6	5.3	0.1	0.2	12.5
Thermic	3.3	2.6	1.5	3.0	0.2	0.6	11.4
Hyperthermic	15.7		1.5	1.0	0.0	0.3	18.5
Isofrigid	0.0			0.0	0.0	0.0	0.1
Isomesic				0.2	0.0	0.0	0.3
Isothermic	0.6		1.0	0.7	0.0	0.0	2.4
Isohyperthermic	2.1		13.0	8.4	0.4	1.9	26.0
Water						1.2	1.2
Ice						1.4	1.4
Total	35.9	3.5	18.0	33.1	1.0	8.3	100.0

From Eswaran (1993).

ARIDIC.   The soil contains very low moisture levels except for short wet periods. This regime is characteristic of soils in arid regions. The term *torric* is also used to indicate the same moisture condition in areas that are both hot and dry. The aridic regime characterizes about 36% of the global land area.

XERIC.   This soil moisture regime is found in typical Mediterranean-type climates, with cool, moist winters and warm, dry summers. Like the Ustic regime, it is characterized by having long periods of drought in the summer.

These terms are used to diagnose the soil moisture regime and are helpful not only in classifying soils but also in suggesting the most rational long-term use of soils to ensure their sustainability.

Soil Temperature Regimes

Several soil temperature regimes are used to define classes of soils in *Soil Taxonomy*. These regimes are based on **mean annual soil temperature, mean summer temperatures**, and the **difference between mean summer and winter temperatures**. The major specific soil temperature regimes and their mean annual temperatures at a depth of 50 cm are shown in Table 3.3.

PERGELIC.   0°C (32°F) and lower. Worldwide, nearly 11% of soil areas are so classified (see Table 3.3). Permafrost is present if the soil is moist. Ice lenses are common in soils in the United States.

TABLE 3.3   Soil Temperature Classes Relating to the Mean Annual Temperature at a Depth of 50 cm and to Differences Between Summer and Winter Temperatures

Percent of world total is in parentheses.

Mean annual temperature	Differences between summer and winter temperatures	
°C	Greater than 5°C	Less than 5°C
<0	Pergelic (10.9)	
0–8	Cryic (13.5)	
<8	Frigid (1.2)	Isofrigid (0.1)
8–15	Mesic (12.5)	Isomesic (0.3)
15–22	Thermic (11.4)	Isothermic (2.4)
>22	Hyperthermic (18.5)	Isohyperthermic (26.0)

CRYIC. 0–8°C (32–47°F). The summer temperature is 13–15°C without an insulating (organic) horizon and 6–8°C with such a horizon. This regime characterizes about 13% of the world's land.

In the following soils the difference between the mean annual summer and winter temperatures is more than 5°C at a depth of 50 cm:

Frigid. 0–8°C (32–47°F). Somewhat warmer in summer than the cryic regime. (1.2% of world total.)

Mesic. 8–15°C (47–59°F). (12.5% of world total.)

Thermic. 15–22°C (59–72°F). (11.4% of world total.)

Hyperthermic. 22°C (72°F) and higher. (18.5% of world total.)

If the prefix *iso* is used to describe a soil temperature regime (**isofrigid**, 0.1%; **isomesic**, 0.3%; **isothermic**, 2.4%; and **isohyperthermic**, 26%), the mean annual temperature is the same as above, but the difference between the mean annual summer and winter temperature is less than 5°C. Note that soils characterized by high temperatures (above 22°C) occupy nearly 45% of the world's land area (hyperthermic and isohyperthermic areas).

The diagnostic horizons, moisture regimes, and temperature regimes just discussed are the principal criteria used to define the various categories in *Soil Taxonomy* which we will now consider.

3.3 CATEGORIES OF SOIL TAXONOMY

There are six categories of classification in *Soil Taxonomy:* (1) *order,* the highest (broadest) category, (2) *suborder,* (3) *great group,* (4) *subgroup,* (5) *family,* and (6) *series,* the most specific category. These categories are hierarchical in that the lower categories fit within the higher categories; thus, each order has several suborders, each suborder has several subgroups, etc. This system may be compared with those used for the classification of plants. The comparison would be as shown in Table 3.4, where white clover (*Trifolium repens*) and Miami silt loam are the examples of plants and soils, respectively.

Just as *Trifolium repens* identifies a specific kind of plant, the Miami series identifies a specific kind of soil. The similarity continues up the classification scale to the highest categories—phylum for plants and order for soils. With this general background, a brief description of the six soil categories and the nomenclature used in identifying them is presented.

ORDER. The **order** category is based largely on soil-forming processes as indicated by the presence or absence of major diagnostic horizons. A given order includes soils whose properties suggest that they are not too dissimilar in their genesis. As an example, many soils that developed under grassland vegetation have the same general sequence of horizons and are characterized by a mollic epipedon—a thick, dark, surface horizon that is high in base-forming cations. Soils with these properties are thought to have been formed by the same general genetic processes and thereby are included in the same order: Mollisols. There are 11 soil orders in *Soil Taxonomy* (Figure 3.5). The names and major characteristics of each soil order are shown in Table 3.5.

TABLE 3.4 **Comparison of the Classification of a Common Cultivated Plant, White Clover (*Trifolium repens*), and a Soil, Miami Series**

Plant classification			Soil classification	
Phylum	Pterophyta		Order	Alfisols
Class	Angiospermae	Increasing specificity	Suborder	Udalfs
Subclass	Dicotyledoneae		Great Group	Hapludalfs
Order	Rosales		Subgroup	Typic Hapludalfs
Family	Leguminosae		Family	Fine loamy, mixed, mesic
Genus	*Trifolium*		Series	Miami
Species	*repens*		Phase[a]	Miami, eroded phase

[a]Technically not a category in *Soil Taxonomy* but used in field surveying.

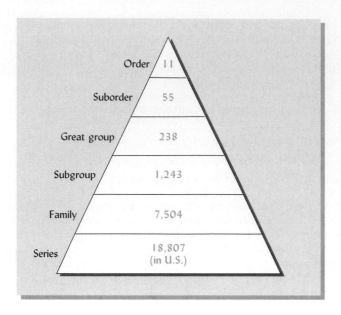

FIGURE 3.4 The categories of *Soil Taxonomy* and approximate number of units in each category. (Data from U.S. Department of Agriculture, personal correspondence)

SUBORDER. The **suborders** are subdivisions of orders that emphasize properties that suggest some common features of soil genesis. Thus, the soil moisture and temperature regimes with associated natural vegetation, which help determine the nature of the genetic processes, also will help determine the suborder in which a given soil is categorized. Some 55 soil suborders are recognized.

GREAT GROUP. Diagnostic horizons (Table 3.1 and Figures 3.2 and 3.3) are the primary bases for differentiating the **great groups** in a given suborder. Soils in a given great group have the same kind and arrangement as these horizons. More than 230 great groups are recognized, some 187 of which are used in the United States.

SUBGROUP. The **subgroups** are subdivisions of the great groups. The central concept of a great group makes up one subgroup (Typic). Other subgroups may have characteristics that are intergrades between those of the central concept and soils of other orders, suborders, or great groups. More than 1240 subgroups are recognized, about 1000 of which are found in the United States.

FAMILY. In the **family** category are found soils with a subgroup having similar physical and chemical properties affecting their response to management and especially to the penetration of plant roots (e.g., soil–water–air relationships). Differences in texture, mineralogy, temperature, and soil depth are primary bases for family differentiation. Some 7500 families have been recognized.

TABLE 3.5 Names of Soil Orders in *Soil Taxonomy* with Their Derivation and Major Characteristics

Name	Formative element Derivation	Pronunciation	Major characteristics
Entisols	Nonsense symbol	Re*cent*	Little profile development, ochric epipedon common
Inceptisols	L. *inceptum,* beginning	In*cept*ion	Embryonic soils with few diagnostic features, ochric or umbric epipedon; cambic horizon
Mollisols	L. *mollis,* soft	*Moll*ify	Mollic epipedon, high base saturation, dark soils, some with argillic or natric horizons
Alfisols	Nonsense symbol	Ped*alf*er	Argillic or natric horizon; high to medium base saturation
Ultisols	L. *ultimus,* last	*Ult*imate	Argillic horizon, low base saturation
Oxisols	Fr. *oxide,* oxide	*Oxide*	Oxic horizon, no argillic horizon, highly weathered
Vertisols	L. *verto,* turn	In*vert*	High in swelling clays; deep cracks when soil dry
Aridisols	L. *aridus,* dry	*Arid*	Dry soil, ochric epipedon, sometimes argillic or natric horizon
Spodosols	Gk. *Spodos,* wood ash	*Podzol*; odd	Spodic horizon commonly with Fe, Al, and humus accumulation
Histosols	Gk. *Histos,* tissue	*Hist*ology	Peat or bog; >30% organic matter
Andisols	Modified from Ando	*And*esite	From volcanic ejecta, dominated by allophane or Al-humic complexes

SERIES. The **series** category is the most specific unit of the classification system. It is a subdivision of the family, and its differentiating characteristics are based primarily on the kind and arrangement of horizons. Conceptually, it comprises a primary polypedon, although in the field, aggregates of polypedons and associated inclusions are included in the soil series mapping units. There are more than 18,800 soil series recognized in the United States (Figure 3.4).

3.4 NOMENCLATURE OF SOIL TAXONOMY

A unique feature of *Soil Taxonomy* is that the nomenclature used to identify different soil classes is very informative about the nature of the soils named. The names of the classification units are combinations of syllables, most of which are derived from Latin or Greek and are root words in several modern languages. Since each part of a soil name conveys a concept of soil character or genesis, the name automatically describes the general kind of soil being classified. For example, soils of the order **Aridisols** (from the Latin *aridus,* dry, and *solum,* soil) are characteristically dry soils. Those of the order **Inceptisols** (from the Latin *inceptum,* beginning, and *solum,* soil) are soils with only the beginnings of profile development. Thus, the names of orders are combinations of (1) formative elements, which generally define the characteristics of the soils, and (2) the ending *sol.*

The names of suborders automatically identify the order of which they are a part. For example, soils of the suborder **Aquolls** are the wetter soils (from the Latin *aqua,* water) of the Mollisols order. Likewise, the name of the great group identifies the suborder and order of which it is a part. **Argiaquolls** are Aquolls with clay or argillic (Latin *argilla,* white clay) horizons.

The nomenclature as it relates to the different categories in the classification system might be illustrated as follows:

M*oll*isols	order
Aqu*oll*s	suborder
Argiaqu*oll*s	great group
Typic Argiaqu*oll*s	subgroup

Note that the three letters *oll* identify each of the lower categories as being in the Mollisols order. Likewise, the suborder name **Aquolls** is included as part of the great group and subgroup name. If one is given only the subgroup name, the great group, suborder, and order to which the soil belongs are automatically known.

Family names in general identify subsets of the subgroup that are similar in texture, mineral composition, and mean soil temperature at a depth of 50 cm. Thus the name **Typic Argiaquolls, fine, mixed, mesic** identifies a family in the **Typic Argiaquolls** subgroup with a fine texture, mixed clay mineral content, and mesic (8–15°C) soil temperature.

Soil series are named after a geographic feature (town, river, etc.) near where they were first recognized. Thus, names such as "Fort Collins," "Cecil," "Miami," "Norfolk," and "Ontario" identify soil series first described near the city or locality named.

In field soil surveying, soil series are sometimes further differentiated on the basis of surface soil texture or other characteristics. These field mapping units are called *soil phases.* Names such as "Fort Collins loam," "Cecil clay," or "Cecil, eroded phase" are used to identify such phases. Note, however, that soil phases, practical as they may be in local situations, are not a category in the *Soil Taxonomy* system.

With this brief explanation of the nomenclature of the new system, we will now consider the general nature of soils in each of the soil orders.

3.5 SOIL ORDERS

Eleven orders are recognized. The names of these orders and their major characteristics are shown in Table 3.5. Note that all order names have a common end, *sol* (from the Latin *solum,* soil).

The general conditions that enhance the formation of mineral soils in the different orders is shown in Figure 3.5. From soil profile characteristics scientists have ascertained

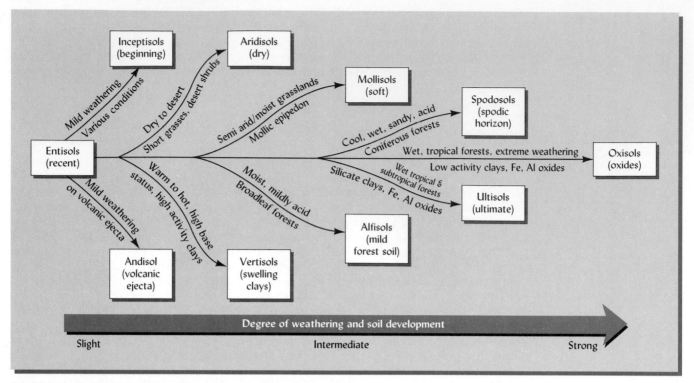

FIGURE 3.5 Diagram showing general degree of weathering and soil development in the different orders of mineral soils classified in *Soil Taxonomy*. Also shown are the general climatic and vegetative conditions under which soils in each order are formed.

the degree of development of soils in the different orders as shown in this figure. Note that soils with essentially no profile layering (Entisols) have the least development, while the deeply weathered soils of the humid tropics (Oxisols and Ultisols) show the greatest soil development. The effect of climate (temperature, moisture) and of vegetation (forests, grasslands) on the kinds of soils that develop is also indicated in Figure 3.5. Study Table 3.5 and Figure 3.5 to better understand the relationship between soil properties and the terminology used in *Soil Taxonomy*.

Figure 3.6 is a simplified general soil map showing the distribution of soil orders in the United States. The major areas of the country dominated by each order are evident. Profiles for each soil order are shown in color on Plates 1–11. More detailed general soil maps (along with the map legends) for the world appear inside the front cover and for the United States in Appendix A. We shall now discuss briefly each soil order and the characteristics, general management, and requirements for each.

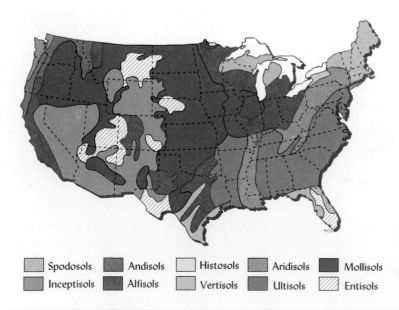

FIGURE 3.6 Simplified general soil map of the United States showing patterns of soil orders based on *Soil Taxonomy*. For more detailed soil maps showing the distribution of both soil orders and suborders for the United States, see Appendix A.

3.6 ENTISOLS (RECENT: LITTLE IF ANY PROFILE DEVELOPMENT)

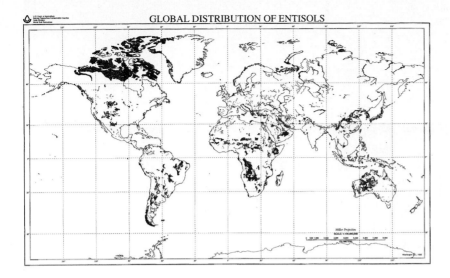

GLOBAL DISTRIBUTION OF ENTISOLS

SUBORDERS

Aquents (wet)	Orthents (typical)
Arents (plow induced)	Psamments (sandy)
Fluvents (alluvial deposits)	

Soils with the least profile development are included in the Entisol order. There is little or no evidence of soil formation. Entisols are either young soils or their parent materials have not reacted to soil-forming factors. They may be formed on parent materials such as fresh lava flows or recent alluvium (Fluvents) for which there has been too little time for soil formation to take place. Some recently bulldozed sites would include Entisols. In extremely dry areas, too little water and vegetation may prevent soil formation. Likewise, soil formation is delayed if the soil is frequently saturated with water (Aquents). Some Entisols are found on steep slopes where the rates of erosion may be greater than the rate of soil formation, thereby preventing horizon development.

Entisols are weakly developed mineral soils without natural genetic (subsurface) horizons or with only the beginnings of such horizons (Plate 1). Most have an ochric epipedon and a few have human-made anthropic or agric epipedons. Some have albic subsurface horizons. The extremes of highly productive soils in recent alluvium and of infertile soils and barren sands as well as shallow soils on bedrock are included.

Distribution and Use

Globally, Entisols occupy nearly 15 million square kilometers of land area, about 11% of the total. In the United States, they comprise 8% of the soil area (see Table 3.6). Entisols are found under a wide variety of environmental conditions in the United States (see Figure 3.6 and endpapers). For example, in the Rocky Mountain region and in southwest Texas, shallow, medium-textured Entisols (Orthents) over hard rock are common. These are used mostly as rangeland. Sandy Entisols (Psamments) (Figure 3.7) are found in Florida, Alabama, and Georgia, and typify the sand hill section of Nebraska. Psamments are used for cropland in humid areas. Some of the citrus-, vegetable-, and peanut-producing areas of the South are typified by Psamments. Poorly drained and flooded areas alongside the Mississippi River (Aquents) are included in this order.

Entisols are probably found under even more widely varied environmental conditions outside the United States (see endpapers). Psamments are typical of the Sahara desert and Saudi Arabia and dominate parts of southern Africa and central and north central Australia. Entisols on recent alluviums (Fluvents) are prominent in the intensively cultivated rice lands of Asia that have produced this major food staple for generations. Entisols having medium to fine textures (Orthents) are found in northern Quebec and parts of Alaska, Siberia, and Tibet. Orthents are typical of some mountain

A

C

FIGURE 3.7 Profile of a Psamment formed on sandy alluvium in Virginia. Note the accumulation of organic matter in the A horizon but no other evidence of profile development. The knife is 24 cm long. (Photo courtesy of R. Weil)

areas such as the Andes in South America and some of the uplands of an area extending from Turkey eastward to Pakistan.

The agricultural productivity of the Entisols varies greatly depending on their location and properties. With adequate fertilization and a controlled water supply, some Entisols are quite productive; in fact, Entisols developed on alluvial flood plains are among the world's most productive soils. On such soils, major civilizations of the world developed. However, restrictions on the depth, clay content, or water balance of most Entisols limit the intensive use of large areas of these soils.

Management

The overall characteristics of Entisols are so variable that it is not possible to generalize on the management needs of these soils. The more productive soils found on recent alluvium can be cropped intensively since they are not subject to soil erosion. The broad areas of Entisols developed on sands (Psamments) should not be used for annual crops but in most cases should be left under their natural vegetation. Since these sandy soils have low water-holding capacities, any practice that conserves water, such as leaving crop residues on the soil surface, should be encouraged. Entisols on steep slopes should not be used for agriculture, but their natural vegetative cover should be maintained.

3.7 INCEPTISOLS (FEW DIAGNOSTIC FEATURES: INCEPTION OF B HORIZON)

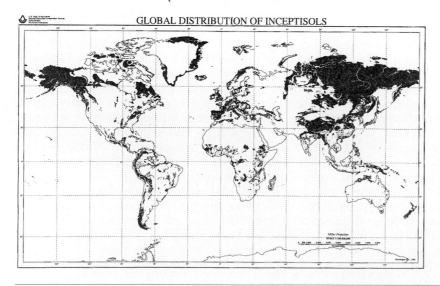

GLOBAL DISTRIBUTION OF INCEPTISOLS

Aquepts (wet)   Tropepts (continually warm)

Ochrepts (light surface)   Umbrepts (acid, dark surface)

Plaggepts (long-term manuring)

The words *beginning, incipient,* and *embryonic* are used to describe Inceptisols. The early stages of profile development are evident and some diagnostic features are present. However, the well-defined profile characteristics of soils thought to be more mature have not yet developed. These soils show more significant profile development than the Entisols but less so than the other mineral soil orders. While all soil-forming processes may be active in these soils, none predominates. The presence of a weakly expressed cambic subsurface horizon is common on steep slopes. Likewise, in cool grassland areas a weakly expressed mollic or umbric epipedon may be present.

Distribution and Use

Inceptisols are widely distributed throughout the world, some 16% of the world's soils being so classified (Table 3.6). As was the case for Entisols, Inceptisols are found in most

TABLE 3.6 Approximate Land Areas of Different Soil Orders and Suborders of the United States and the World (in Parentheses), Their Major Land Use and Natural Fertility Status

Order and suborder	Land area (%)	Major land uses	Fertility
Alfisols	13.4 (13.5)		
Aqualfs	1.0	Cropland	High
Boralfs	3.0	Forest	High
Udalfs	5.9	Cropland	High
Ustalfs	2.6	Cropland	High
Xeralfs	0.9	Rangeland	High
Aridisols	11.5 (23.4)		
Argids	8.6	Rangeland	Low
Cambids, etc.	2.9	Rangeland	Low
Entisols	7.9 (11.0)		
Aquents	0.2	Wetland	Moderate
Orthents	5.2	Rangeland, forest	Moderate
Psamments	2.2	Cropland	Low
Histosols	0.5 (1.2)	Wetland, cropland	Moderate
Inceptisols	16.3 (16.0)		
Aquepts	11.9	Cropland	Moderate
Ochrepts	4.3	Cropland, forest	Moderate to low
Umbrepts	0.7	Forest	Low
Mollisols	24.6 (4.1)		
Aquolls	1.3	Cropland	High
Borolls	4.9	Cropland	High
Udolls	4.7	Cropland	High
Ustolls	8.8	Cropland, rangeland	High
Xerolls	4.8	Cropland, rangeland	High
Spodosols	5.1 (3.6)		
Aquods	0.7	Forest	Low
Orthods	4.4	Forest	Low
Ultisols	12.9 (8.3)		
Aquults	1.1	Forest	Low to moderate
Humults	0.8	Forest	Low
Udults	10.0	Forest, cropland	Low
Vertisols	1.0 (2.4)		
Uderts	0.4	Cropland	High
Usterts	0.6	Cropland	High
Andisols	1.9 (1.8)	Cropland, forest	Moderate
Oxisols	0.01 (8.7)	Cropland, forest	Low
Miscellaneous land	4.5 (5.6)	Barren land	

Data from USDA Natural Resources Conservation Service, Soil Survey Staff (personal communication), and Eswaran (1993)

[a]Includes Andisols.

[b]Andisol world percentage is included in the Inceptisols; United States percentage is an approximation.

climatic and physiographic conditions. They are often prominent in mountainous areas, especially in the tropics. They are probably the most important soil order in the lowland rice-growing areas of Asia.

Inceptisols are found in each of the continents (see endpapers). Some inceptisols, called *Ochrepts* (from the Greek *ochros*, pale), with thin, light-colored surface horizons, extend from southern New York through central and western Pennsylvania, West Virgina, and eastern Ohio. Ochrepts also dominate an area extending from southern Spain through central France to Germany, and are present as well in Chile, North Africa, eastern China, and western Siberia.

Tropepts (Inceptisols formed under continually warm conditions) are found in northwestern Australia, central Africa, southwestern India, and southwestern Brazil. Wet Inceptisols or Aquepts (from the Latin *aqua*, water) are found in areas along the Amazon and Ganges rivers.

There is considerable variability in the natural productivity of Inceptisols. For instance, those found in the Pacific Northwest of the United States are quite fertile and provide some of the world's best wheatland. In contrast, some of the low organic matter Ochrepts in southern New York and northern Pennsylvania are not naturally productive. They have been allowed to reforest following earlier periods of crop production.

Management

Inceptisols are about as variable as Entisols and have equally variable management requirements. Their response to management depends upon the climatic and topographic conditions that characterize their location. Those found on recent alluvium may be highly productive. Such soils can be intensively used over extended periods of time without serious degradation. Inceptisols formed on hillsides, however, are best left in their native vegetation to prevent accelerated soil erosion. Those formed under wet conditions can be drained and used for agriculture, although there is increasing recognition that undisturbed wetlands serve as essential habitats for wildlife. The management practices for all Inceptisols should prevent soil erosion and should sustain their long-term productivity.

3.8   ANDISOLS

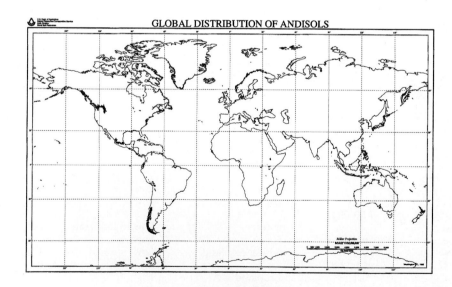

GLOBAL DISTRIBUTION OF ANDISOLS

SUBORDERS

Aquands (wet)	Ustands (moist/dry)
Cryands (cold)	Vitrands (volcanic glass)
Torrands (hot, dry)	Xerands (dry summers)
Udands (humid)	

Andisols are formed on volcanic ash and cinders and are commonly found at high elevations near the volcano source. Since the volcanic materials have been deposited in recent geological time, Andisols have not been highly weathered. Like the Entisols and Inceptisols, Andisols are young soils—having been developed for only 5000–10,000 years. Unlike the previous two orders of immature soils, Andisols have a unique set of properties due to a common type of parent materials.

The colloidal fraction of at least the upper 35 cm of Andisols is dominated by the silicate minerals allophane and imogolite and/or aluminum-humus complexes. The combination of these minerals and the high organic matter results in a high water-holding capacity. The upper layers are characteristically dark in color and have low-bulk densities.

The minerals in the volcanic materials have weathered in place, producing amorphous or poorly crystallized minerals such as allophane, imogolite, and ferrihydrite, but no downward translocation of these colloids has taken place. Andisols are characterized by the melanic epipedon, a surface diagnostic horizon that has a high organic matter content. These soils were previously classified as suborders of the Inceptisols order, a fact that implies their minimum profile development.

Distribution and Use

Andisols are found in areas where significant depths of volcanic ash and other ejecta have accumulated (Figure 3.8). Globally, they make up less than 2% of the soil area (Table 3.6). However, in the Pacific area they are important and productive soils that support intensive agricultural production, especially in cool, high-elevation areas.

Andisols having a Udic (moist) soil moisture regime (Udands) are widely used in Asia, producing enough food to support very high population densities. The somewhat drier Ustands are also used intensively for agriculture. Significant areas of Udands and Ustands occur along the Rift Valley of eastern Africa (see Plate 2). Andisols are found to a minor extent in cold climates (Cryands) in Canada and Russia, and in hot, dry climates (Torrands) in Mexico and Syria. Very recent eruptions such as that of Mount Saint Helens in the northwestern part of the United States are giving rise to Vitrands.

Melanic A horizon

Pumice layer

Weathered layers of volcanic ash and pumice

Buried A horizon

Oldest layers of volcanic pumice

Underlying layer of expanding clay

FIGURE 3.8 An Andisol developed on layers of volcanic ash and pumice in central Africa. (Photo courtesy of R. Weil)

In the United States, the area of Andisols is not extensive since recent volcanic action is not widespread. However, Andisols do occur in some very productive wheatland areas of Washington, Idaho, and Oregon. Likewise, this soil order is found in some Latin American countries such as Ecuador and Colombia.

Management

Andisols appear to be quite easily managed. They are light in density and thereby require less energy for tillage than do other, more dense soils. The most serious problem associated with the management of Andisols is their high capacity to hold phosphorus in forms not readily available to plants (see Figure 14.17). Proper management of crop residues and fertilizers can usually overcome this deficiency. Also, because of their low density, they tend to be loose and fluffy, making them somewhat more prone to wind erosion. Fortunately, Andisols are found mostly in areas with sufficient rainfall to keep the soil moist so that wind erosion is generally not a serious problem.

3.9 MOLLISOLS (DARK SOILS OF GRASSLANDS: *MOLLIS*, SOFT)

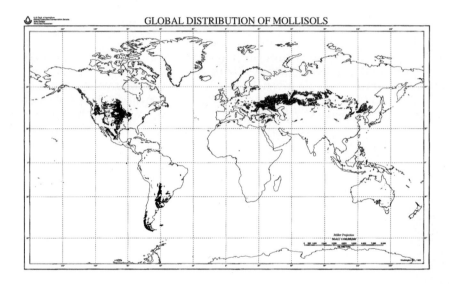

GLOBAL DISTRIBUTION OF MOLLISOLS

SUBORDERS

Albolls (albic horizon)	Udolls (humid)
Aquolls (wet)	Ustolls (moist/dry)
Borolls (cold)	Xerolls (dry summers)
Rendolls (calcareous)	

The Mollisols order includes some of the world's most productive soils. Formed under grassland vegetation, these soils are characterized primarily by the presence of a **mollic epipedon.** This humus-rich surface horizon is thick (often 60–80 cm in depth), and dominated by base-forming cations such as Ca^{2+} and Mg^{2+} (high base saturation) (Table 3.5 and Plate 7). They may have an argillic (clay), natric, albic, or cambic subsurface horizon, but not an oxic or spodic horizon. The surface horizon generally has granular or crumb structures, largely resulting from an abundance of organic matter and swelling-type clays. Mollisols in humid regions generally have higher organic matter and darker, thicker mollic epipedons than their lower-moisture-regime counterparts. The granules are not hard when the soils are dry. This justifies the use of the name *Mollisol,* implying softness (see Table 3.5).

Most of the Mollisols have developed under grass vegetation (Figure 3.9). Grassland soils of the central part of the United States, lying between Andisols on the west and the Alfisols on the east, make up the central core of this order. However, a few soils developed under forest vegetation (primarily in depressions) have a mollic epipedon and are included among the Mollisols. The soil temperature and soil moisture characteristics of Mollisols and other soil orders are shown in Figure 3.10.

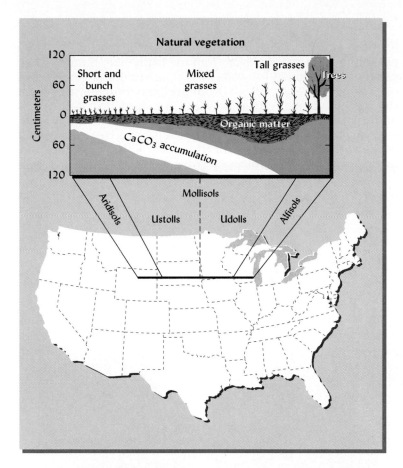

FIGURE 3.9 Correlation between natural grassland vegetation and certain soil orders is graphically shown for a transect across north central United States. The control, of course, is climate. Note the deeper organic matter and deeper zone of calcium accumulation, sometimes underlain by gypsum, as one proceeds from the drier areas in the west toward the more humid region where prairie soils are found. Alfisols may develop under grassland vegetation, but more commonly occur under forests and have lighter colored surface horizons.

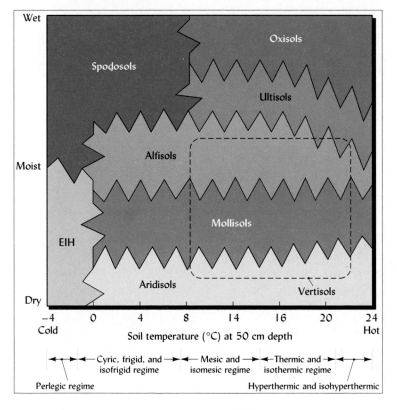

FIGURE 3.10 Diagram showing the general soil moisture and soil temperature regimes that characterize the most extensive soils in each of seven soil orders. Soils of the other four orders (Andisols, Entisols, Inceptisols, and Histosols) may be found under any of the soil moisture and temperature conditions (including the area marked EIH). Major areas of Vertisols are found only where clayey materials are in abundance and are most extensive where the soil moisture and temperature conditions approximate those shown inside the box with broken lines. Note that these relationships are only approximate and that less extensive areas of soils in each order may be found outside the indicated ranges. For example, some Ultisols (Ustults) and Oxisols (Ustox) have soil moisture levels for at least part of the year that are much lower than this graph would indicate. (The terms used at the bottom to describe the soil temperature regimes are those used in helping to identify soil families.)

Distribution and Use

Mollisols are dominant in the Great Plains states and Illinois (see Figure 3.9). Those where soil moisture is not limiting are called *Udolls* (from the Latin *udus,* humid). They are associated with nearby wet soils termed *Aquolls.* A region extending from North Dakota to southern Texas is characterized by Ustolls (from the Latin *ustus,* burnt), which are intermittently dry during the summer. Farther west in parts of Idaho, Utah, Washington, Oregon, and California are found sizable areas of Xerolls (from the Greek *xeros,* dry), which are the driest of the Mollisols. This Mollisol order characterizes a larger land area in the United States than any other soil order (nearly 25%) (Table 3.6). Figure 3.11 shows landscapes common for Udolls and Ustolls.

FIGURE 3.11 Typical landscapes dominated by Ustolls (top) and Udolls (bottom). These productive soils produce much of the food and feed in the United States. (Photos courtesy of R. Weil)

The largest area of Mollisols in the world is that stretching from east to west across the heartlands of Kasikstan, Ukraine, and Russia (cool to cold, mostly Borolls). Other sizable areas are found in Mongolia and northern China and in northern Argentina, Paraguay, and Uruguay. Mollisols occupy only about 4% of the world's total soil area, but they account for a much higher percentage of total crop production.

The high native fertility of Mollisols (especially the Udolls) makes them among the world's more productive soils, and accounts for the fact that most of these soils are cultivated. When they were first cleared for cultivation, much of their native organic matter was oxidized, releasing nitrogen and other nutrients in sufficient quantities to produce high crop yields even without the use of fertilizers. Yields on these soils were unsurpassed by other unirrigated areas.

Even after more than a century of cultivation, Mollisols are among the most productive soils, although some fertilization is generally required. Photographs of two Mollisols are shown in Figure 3.12. Very few areas of undisturbed Mollisols exist in the United States today, and efforts to preserve the few remnants of the native prairies are under way.

Management

Because of their high productivity, many Mollisols are intensively cultivated. Cash crops are grown each year, often without rotation among the crops. For example, planting corn year after year with heavy use of fertilizers is an all-too-common practice in parts of the midwest. Continuous cultivation with row crops leads to serious deterioration of soil structure and soil erosion, especially if the land is rolling, with moderate slopes. Researchers have demonstrated that crop rotations involving forage crops as well as row crops on Mollisols can better maintain soil organic matter and granular structure, protect the land from erosion, and still provide adequate crop productivity.

Intensive cropping of Mollisols has involved the application of heavy rates of farm manure and of chemical fertilizers and pesticides. In recent years, studies have shown that some of these chemicals move into the groundwater and subsequently into streams and rivers and into rural water supplies. Moderation in the use of these chemicals is essential. Environmental regulators and cooperating farmers have taken steps to reduce the potential contamination of the country's water supplies caused by the excess use of agricultural chemicals.

FIGURE 3.12 Monoliths of profiles representing three soil orders. The suborder names are also shown (in parentheses). Note the spodic horizons in the Spodosol characterized by humus (Bh) and iron (Bs) accumulation. In the Alfisol is found the illuvial clay horizon (Bt), and the structural B horizon (Bw) is indicated. The thick dark surface horizon (mollic epipedon) characterizes both Mollisols. Note that the zone of calcium carbonate accumulation (Ck) is higher in the Ustoll, which has developed in a dry climate. The E/B horizon in the Alfisol has characteristics of both E and B horizons.

The conservation of moisture is also a requisite for high yields from Mollisols. Conservation practices that maintain a cover of crop residues on the soil have been found to be worthwhile in protecting Mollisols from erosion and in increasing moisture conservation.

3.10   ALFISOLS (ARGILLIC OR NATRIC HORIZON, MEDIUM-HIGH BASES)

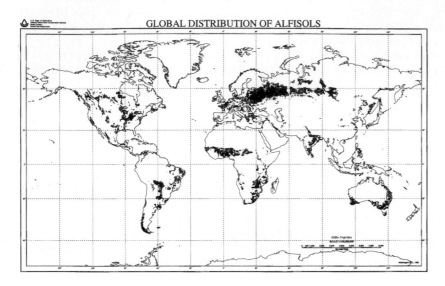

GLOBAL DISTRIBUTION OF ALFISOLS

SUBORDERS

Aqualfs (wet)	Ustalfs (moist/dry)
Boralfs (cold)	Xeralfs (dry summers)
Udalfs (humid)	

Alfisols have developed under forests in cool to warm humid areas, and are characterized by a subsurface horizon in which silicate clay has accumulated by illuviation. They have gray to brown surface horizons (commonly an ochric epipedon) and medium- to high-base status (see Plate 1, for example). Clay skins or other signs of clay movement are present in the B horizon (Plate 18). The cation exchange capacity of the clay horizon is more than 35% saturated with base-forming cations (Ca^{2+}, Mg^{2+}, etc.). This horizon is termed *argillic* because of its accumulation of silicate clays. The horizon is termed *natric* if, in addition to the clay, it is more than 15% saturated with sodium and has prismatic or columnar structure (see Figure 4.11).

The Alfisols appear to be more strongly weathered than the Inceptisols, but less so than Spodosols. They are formed in cool to hot humid areas but also are found in the semiarid tropics. Most often, Alfisols are developed under native deciduous forests, although in some cases grass is the native vegetation.

Distribution and Use

About 13% of the land area in the United States is classified as Alfisols (Table 3.6). Alfisols are typified by **Udalfs** (from the Latin *udus*, humid) in Ohio, Indiana, Michigan, Wisconsin, Minnesota, Pennsylvania, and New York (see Figures 3.6 and 3.12). They are found in sizable areas of **Xeralfs** (from the Greek *xeros*, dry) in central California; some cold **Boralfs** (from the Greek *boreas*, northern) in the Rockies and Minnesota; **Ustalfs** (from the Latin *ustus*, burnt) in areas of hot summers, including Texas and New Mexico; and the wet **Aqualfs** (from the Latin *aqua*, water), in parts of the midwest.

Nearly 14% of the world's land is occupied by Alfisols. They occur in several other countries around the world (see endpapers). A large area dominated by Boralfs is found in northern Europe, extending from the Baltic States through western Russia. A second large area is found in Siberia. Ustalfs are prominent in sub-Saharan Africa, eastern Brazil, eastern India, and southeastern Asia. Large areas of Udalfs are found in central China,

England, France and central Europe, and southeastern Australia. Xeralfs are prominent in southwestern Australia, Italy, and central Spain.

In general, Alfisols are productive soils. Good hardwood forest growth and crop yields are favored by their medium- to high-base status, generally favorable texture, and location (except for some Xeralfs) in regions with sufficient rainfall at least part of the year. In the United States these soils rank favorably with the Mollisols and Ultisols in their productive capacity.

Management

Since most temperate region Alfisols are found in humid climates, their moisture supply is generally sufficient for good crop production. They are not as highly weathered as their counterpart Spodosols and Utisols, so their nutrient-supplying power is also good. The management challenge is to effectively use the moisture and nutrient-supplying capabilities of these soils without subjecting them to erosion and nutrient depletion.

Alfisols generally require two kinds of management concern. First, they are mildly acidic, making it necessary to apply lime to encourage optimum crop production especially of legumes. Second, they are often found on sloping to steep land, making them susceptible to soil erosion. The steeper slopes should remain in forest culture, or if they are already cultivated, they should be returned to the natural vegetative state. Crop rotation and maintenance of land cover with crop residues are essential to ensure sustainable agriculture on these soils.

In recent years chemical fertilizer use has expanded considerably on these soils. As is the case for the Mollisols, in some areas leaching of the applied nutrients into drainage waters has resulted in undesirable contamination of local water supplies. Management practices should monitor the amount and timing of fertilizer applications to enhance crop production without excess use of these chemicals.

3.11 ULTISOLS (ARGILLIC HORIZON, LOW BASES)

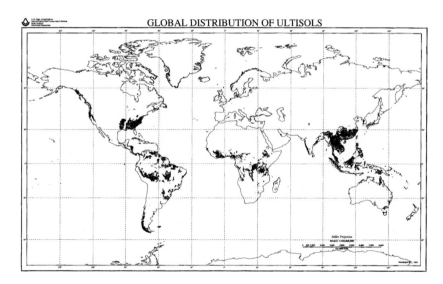

GLOBAL DISTRIBUTION OF ULTISOLS

SUBORDERS

Aquults (wet)	Ustults (moist/dry)
Humults (high humus)	Xerults (dry summers)
Udults (humid)	

Ultisols have developed primarily in forested, humid tropical and subtropical areas. They often have an ochric or umbric epipedon but are characterized by an argillic (clay) horizon with relatively low activity clays. The horizon has a low-base status (less than 35% of the exchange capacity satisfied with base-forming metallic cations). The mean annual temperatures at 50 cm depth in Ultisols are above 8°C. Ultisols commonly have an epipedon that is quite acid and low in plant nutrients.

Ultisols are more highly weathered and acidic than Alfisols, but less acid than Spodosols. Most ultisols have developed under moist conditions in warm to tropical climates. Except for the wetter members of the order, their subsurface horizons are commonly red or yellow in color, evidence of accumulations of oxides of iron (Plate 10). Ultisols still have some weatherable minerals, however, in contrast to the Oxisols. Ultisols are formed on old land surfaces, normally under forest vegetation, although savanna or even swamp vegetation is also common.

Distribution and Use

Nearly 13% of the soil area in the United States is classified in the Ultisols order (Table 3.6). Most of the soils of the southeastern part of the United States fall in this order (Figure 3.6 and Appendix A). Udults, the moist, well-drained Ultisols, extend from the east coast (Maryland to Florida) to and beyond the Mississippi River Valley and are the most extensive of soils in the humid southeast. Humults (high in organic matter) are found in the United States in Hawaii, eastern California, Oregon, Washington. Humults are also present in the highlands of some tropical countries. Xerults (Ultisols in Mediterranean-type climates) occur locally in southern Oregon and northern and western California.

Ultisols occupy about 8% of the land area worldwide. Extensive areas of Ultisols are found in the humid tropics in close association with some Oxisols. Ustults are found in the semiarid areas with a marked dry season. Together with the Ustalfs, the Ustults occupy large areas in Africa and India. Ultisols are prominent on the east and northeast coasts of Australia (see endpapers). Large areas of Udults are located in southeastern Asia and in southern China. Important agricultural areas also are found in southern Brazil and Paraguay.

Although Ultisols are not naturally as fertile as Alfisols or Mollisols, they respond well to good management. They are located in most regions of long growing seasons and of ample moisture for good crop production. The silicate clays of Ultisols are usually of the nonsticky type,[6] which along with oxides of iron and aluminum assures ready workability. Where adequate levels of fertilizers are applied, Ultisols are quite productive. In the United States, the well-managed Ultisols compete well with Mollisols and the Alfisols as first-class agricultural soils. They also support an abundant growth of forest species.

Management

Although not naturally as productive as Mollisols or Alfisols, when properly managed Ultisols can be just as productive. Soil moisture is generally favorable for crop growth, as are the temperatures. The low-activity clays in these soils are generally easier to work with than are the high-activity clays in Mollisols and Aridisols.

Since these soils have been subjected to reasonably intense leaching by rainfall, for optimum crop production supplemental nutrients must be applied in crop residues, manures, or chemical fertilizers. Likewise, lime must be used to combat acidity, although the quantity needed is lower than that required for soils with high-activity clays.

Ultisols have been subjected to significant soil erosion as evidenced by the presence of clayey, reddish soil material at the surface in some cases. The topsoil has been eroded, leaving the red-colored B horizon at the surface. Soil conservation practices are needed to prevent further soil deterioration. In areas with significant slope, the land must be reforested, as has already been the case in large areas of the southeastern United States. Maintaining vegetative cover on the cultivated soils will help prevent further soil erosion. It will also prevent excessive flooding of downstream areas from uncontrolled runoff from upstream watersheds.

---

[6]Silicate clays vary considerably in their chemical and physical properties. Some are sticky when wet and are cloddy and hard when dry. Others are not sticky and are desirable for ease of tillage of the soil. Silicate clay types will be covered in Chapter 8.

3.12 OXISOLS (OXIC HORIZON, HIGHLY WEATHERED)

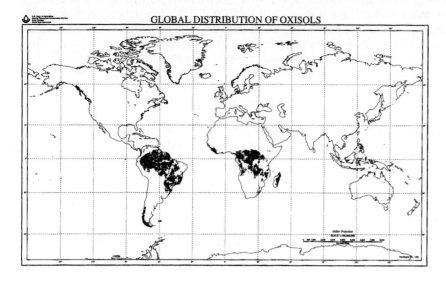

GLOBAL DISTRIBUTION OF OXISOLS

SUBORDERS

Aquox (wet)	Udox (humid)
Perox (very humid)	Ustox (moist/dry)
Torrox (hot, dry)	

The Oxisols are the most highly weathered soils in the classification system (Figure 3.5). They are found mostly in tropical areas. They have ochric or umbric epipedons, but their most important diagnostic feature is a deep **oxic** subsurface horizon. This horizon is generally very high in clay-size particles dominated by hydrous oxides of iron and aluminum. Weathering and intense leaching have removed a large part of the silica from the silicate materials in this horizon. Some quartz and 1:1-type silicate clay minerals remain, but the hydrous oxides are dominant.

The clay content of Oxisols is generally high, but the clays are of the low-activity, nonsticky type. Consequently, when the clay dries out it is not hard and cloddy, but is easily tilled. Also, Oxisols are resistant to compaction so that free water movement through the profile is the norm. The depth of weathering in Oxisols is much greater than for most of the other soils—16 m or more having been observed. Road and building construction on Oxisols have more stable foundations than on less weathered soils that have expansion-type clays.

Distribution and Use

Oxisols are found on nearly 9% of the world's land (Table 3.6). They occupy old land surfaces and occur mostly in the tropics. Relatively less is known about them than most of the other soil orders. They occur in large geographic areas, however, and millions of people in the tropics depend on them for food and fiber production.

The largest known areas of Oxisols occur in South America and Africa (see endpapers). Orthox (Oxisols having a short, dry season or none) occur in northern Brazil and neighboring countries as well as in the Caribbean area (Plate 8). An area of Ustox (hot, dry summers) nearly as large appears in Brazil to the south of the Orthox. In the humid areas of central Africa, Oxisols are prominent and in some cases dominant.

Most Oxisols are formed under rain forest vegetation, although in large areas the natural vegetation is tropical savannas (grasses). Very high rainfall is common in most regions with Oxisols, but some areas with Oxisols have lower rainfall levels today than during earlier periods when the soils were forming.

Management

Most Oxisols are extremely low in plant nutrients, especially after the natural vegetation has been removed. Native farmers ages ago learned to take advantage of the ability of deep-rooted trees to refurbish the nutrients in the upper layers of the soil. The trees act as recyclers, absorbing nutrients from deep subsoil layers, bringing them up into their leaves and stems. When the leaves drop onto the soil and are decomposed, the nutrients are released by soil microbes and the fertility of the upper layers is enhanced.

Unfortunately, increases in human population have placed great stress on such natural recycling systems, making it necessary to complement them with fertilizer chemicals. Continued attempts should be made, however, to maximize the role of these natural systems. Probably the best use of Oxisols other than supporting rain forests is the culture of mixed-canopy perennial crops, especially tree crops. Such cultures can restore the nutrient cycling system that characterized the soil–plant relationships before the rain forest was removed.

By including legumes in the crop rotation, farmers can make use of a natural system of providing nitrogen for uptake by the crops. Microorganisms associated with the roots of legumes capture nitrogen from the atmosphere and incorporate it into nitrogen compounds the plants can use (see Chapter 11). Unfortunately, however, the legume growth is commonly constrained by deficiencies of other nutrients, particularly phosphorus. Even when phosphorus compounds are added, they are tied up in insoluble forms by the iron and aluminum oxides. Even so, research has shown that modest amounts of phosphorus and nitrogen fertilizers encourage crop rotations that markedly increase food production on Oxisols.

3.13   VERTISOLS (DARK, SWELLING CLAYS)

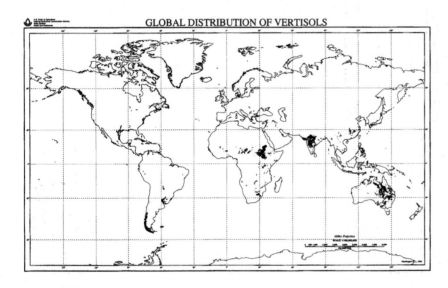

GLOBAL DISTRIBUTION OF VERTISOLS

SUBORDERS

Aquerts (wet)	Uderts (humid)
Cryerts (cold)	Usterts (moist/dry)
Torrerts (hot, dry)	Xererts (very dry)

The Vertisols order of mineral soils is characterized by a high content (>30%) of sticky or swelling-and-shrinking-type clays to a depth of 1 m or more. In dry seasons these clays shrink and thereby cause the soils to develop deep, wide cracks that are diagnostic for this order (Figure 3.13, Plate 11). A significant amount of material from the upper part of the profile may slough off into the cracks, giving rise to a partial inversion of the soil (Figure 3.14). This accounts for the term *invert,* which is used to characterize this order in a general way. As the subsoil swells, blocks of soil rub past each other under pressure, giving rise to another diagnostic feature called *slickensides* (see Figure 3.15).

Vertisols are found mostly in subhumid to semiarid environments and where the average soil temperatures are higher than 8°C (Figure 3.10). They commonly develop from limestone or other basic parent materials. For this reason, they are somewhat more homogeneous than are soils of other orders.

FIGURE 3.13 Wide cracks formed during the dry season in the surface layers of this Vertisol in India. Surface debris can slough off into these cracks and move to subsoil. When the rains come, water can move quickly to the lower horizons, but the cracks are soon sealed, making the soils relatively impervious to the water.

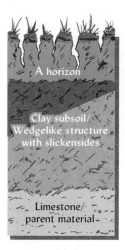

FIGURE 3.14 Vertisols are characterized by their high content of swelling-type clays and angular wedgelike structure in the subsoil. (Left) Sometime after the end of the rainy season the surface soil has dried somewhat, the clay in the soil shrinks, and small cracks begin to appear. (Center) Later, in the dry season the cracks enlarge to a depth of 1 meter or more, permitting granular surface soil to be sloughed off into the cracks through the action of wind and animals. (Right) When the rains begin, the clay swells the cracks closed again, compressing the topsoil debris in the lower profile. The volume occupied by the topsoil results in pressures that force some of the soil upward, yielding an uneven surface with microbasins and knolls, a feature called *gilgai*.

Distribution and Use

There are several small but significant areas of Vertisols in the United States (Figure 3.6). They make up the 1% of land area in this country. Two areas are located in humid areas, one in eastern Mississippi and western Alabama and the other along the southeast coast of Texas. These soils are of the **Uderts** (from the Latin *udus,* humid) suborder because their moist condition prevents cracks from persisting for more than three months of the year.

Two other Vertisol areas are found in east central and southern Texas, where the soils are drier. Since cracks persist for more than three months of the year in these areas, the soils belong to the **Usterts** (from the Latin *ustus,* burnt) suborder, characteristic of areas with hot, dry summers. **Xererts** (Greek *xeros,* dry) are also located in California.

FIGURE 3.15 An example of a slickenside in a Vertisol. Massive rubbing of soil blocks against each other in response to the swelling clays have brought about this shiny appearance. The white concretions of carbonate minerals are evidence of the high base status of these soils. (Photo courtesy of R. Weil)

Globally, Vertisols comprise about 2.5% of the total land area. Large areas of Vertisols are found in India, Ethiopia, the Sudan, and northern and eastern Australia (see endpapers). Smaller areas occur in sub-Saharan Africa and in Mexico, Venezuela, Bolivia, and Paraguay. These latter soils probably are of the Usterts or Xererts suborders, since dry conditions persist long enough for the wide cracks to stay open for periods of three months or longer.

In each of the major Vertisol areas mentioned, some of the soils are used for crop production. Even so, their very fine texture and marked shrinking and swelling characteristics make them less suitable than other soils for crop production and for building foundations and highway bases. They are sticky and plastic when wet and hard when dry. Because they are too sticky when wet and too hard when dry, the period of time during which they can be plowed or otherwise tilled is very short. In areas such as those in India and the Sudan, where slow-moving animals or human power are commonly used to till the soil, the cultivators cannot perform tillage operations on time and are limited to the use of small, near-primitive tillage implements that their animals can pull through the "heavy" soil.

Despite their limitations, Vertisols are widely tilled, especially in India, Ethiopia, and the Sudan. Sorghum, corn, millet, and cotton are crops commonly grown, but the yields are generally low. Recent research shows that these large soil areas can produce greatly increased yields of food crops with improved soil management practices.

Management

Even though Vertisols are generally located on rolling but not steep topography, their high clay content and poor physical condition make them very susceptible to erosion. When one sees on the surface of a tilled Vertisol white limestone that in untilled land is found in the parent material at 1 meter depth, it is obvious that erosion has removed most of the solum, exposing the parent material. Conservation practices that keep the surface covered and protect the soils from the beating action of rain are recommended on these soils. In some cases, the soils should be taken out of agriculture, thereby permitting the natural vegetation to reestablish.

Timing of surface tillage and seedbed preparation is critical for Vertisols. Typically, they are said to be too wet to plow one day and too dry the next. This characteristic is a problem in some developing countries since mechanized tillage is not available and the critical tillage cannot be done on time. Thus, Vertisols are more suited for mechanized farming than for traditional low-input methods. Where irrigation is available, the sealing qualities of the swelling-type clays in Vertisols make some of these soils quite suitable for flooded rice production.

Plate 1 Alfisols—an Aeric Epiaqualf from western New York.

Plate 2 Andisols—a Typic Melanudand from western Tanzania.

Plate 3 Aridisols—a Typic Haplocambid from western Nevada.

Plate 4 Entisols—a Typic Quartzipsamment from eastern Texas.

Plate 5 Histosols—a Limnic Medisaprist from southern Michigan.

Plate 6 Inceptisols—a Typic Dystrochrept from West Virginia.

Plate 7 Mollisols—a Typic Hapludoll from Rio de Janeiro, Brazil.

Plate 8 Oxisols—a Tropeptic Hapludox from central Puerto Rico.

Plate 9 Spodosols—a Typic Haplorthod from northern New York.

Plate 10 Ultisols—a Typic Hapludult from western Arkansas.

Plate 11 Vertisols—a Typic Haplustert from Queensland, Australia.

Plate 12 Redoximorphic mottles from an Aquic Paleudalf Btg horizon.

Plate 13 Green colors from reduced structural iron around roots in forested Aquic Hapludalf A horizon.

Plate 14 Landsat Thematic Mapper image of Palo Verde Valley, California, irrigation scheme using composite of bands 2, 3, and 4. Lush vegetation appears bright red.

Plate 15 Landsat Thematic Mapper image of Potomac River of Washington, D.C., and Potomac River. Composite image with natural colors.

Plate 16 Normal (left) and phosphorus-deficient (right) corn.

Plate 17 Zinc deficiency in sweet corn.

Plate 18 Clay skins in an Ultisol Bt horizon. Bar = 1 cm.

Plate 19 Eroded calcareous soil with iron-deficient sorghum.

Plate 20 Zinc deficiency on peach tree.

Photo 1 courtesy Bill Waltman and D. J. Lathwell, Cornell University; photos 2, 12, 13, and 16–20 courtesy R. Weil, University of Maryland; photos 3, 4, 7, and 11 courtesy R. W. Simonson; photos 5, 6, and 8–10 courtesy Soil Science Society of America; photo 14 courtesy of Earth Satellite Corp, Rockville, Md.; photo 15 courtesy of Chesapeake Bay Foundation, Annapolis, Md.

3.14 ARIDISOLS (DRY SOILS)

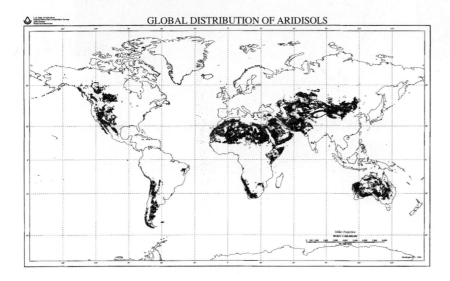

GLOBAL DISTRIBUTION OF ARIDISOLS

Argids (clay)   Cambids (typical)   Durids (duripan)   Salids (salty)

Calcids (carbonate)   Cryids (cold)   Gypsids (gypsum)

Aridisols are dry soils. They occupy a larger area globally than any other soil order (more than 23%). A deficiency of water is a major characteristic of these soils. At no period of 90 consecutive days is the soil moisture level sufficiently high to support plant growth. The natural vegetation is desert shrubs and short grasses.

Aridisols are characterized by an ochric epipedon that is generally light in color and low in organic matter (Plate 3). The process of soil formation for Aridisols brings about a redistribution of soluble materials and their subsequent accumulation at a lower level in the profile. These soils may have a horizon of accumulation of calcium carbonate (calcic), gypsum (gypsic), soluble salts (salic), or exchangeable sodium (natric). They may also have an argillic horizon, possibly indicating a wetter period at some time during their soil formation.

Except where there is groundwater or irrigation, the soil layers are moist only for short periods during the year. These short moist periods may be sufficient for drought-adapted desert shrubs and annual plants, but not for conventional crop production. If groundwater is present near the soil surface, soluble salts often accumulate in the upper horizons to levels that most crop plants cannot tolerate. Unique features that some Aridisols include are a subsurface layer of wind-polished gravel termed **desert pavement** and hard, cemented subsurface layers called **petrocalcic horizons** that interface with both agricultural and urban use of the soils (see Figure 3.16).

Distribution and Use

Nearly 12% of the land area in the United States is classified as Aridisols. They dominate the land in most of the western part of the country. A large area of Aridisols called **Argids** (from the Latin *argilla,* white clay), which have a horizon of clay accumulation, occupies much of the southern parts of California, Nevada, Arizona, and central New Mexico (Figure 3.6). The Argids also extend down into northern Mexico. Smaller areas of **Orthids** (Aridisols without argillic or natric horizons) are found in several western states (see Plate 3).

Vast areas of Aridisols are present in the Sahara desert in Africa, the Gobi and Takla-makan deserts in China, and the Turkestan desert of the former Soviet Union. Most of the soils of southern and central Australia are Aridisols, as are those of southern Argentina, southwestern Africa, Pakistan, and the Middle East countries.

Without irrigation, Aridisols are not suitable for growing cultivated crops. Some areas are used for sheep or goat grazing, but the production per unit area is low. Some

FIGURE 3.16 Two features characteristic of some Aridisols. (Left) Wind-rounded pebbles have given rise to a desert pavement. (Right) A petrocalcic horizon of cemented calcium carbonate. (Photo courtesy of R. Weil)

xerophitic plants such as a jojoba have been cultivated to produce various industrial feedstocks such as oil and rubber. Where irrigation water and fertilizers are available, Aridisols can be made highly productive. Irrigated valleys in arid areas are among the most productive in the world. However, they must be carefully managed to prevent the accumulation of soluble salts (see Chapter 10).

Management

Unirrigated Aridisols can be readily overexploited. The overgrazing of Aridisol areas leads to bare soils and often to wind erosion (Figure 3.17). The desertification of areas of sub-Saharan Africa is evidence of such degradation. Discipline to restrict overgrazing is difficult where the land is publicly or communally owned, but it must be accomplished.

The irrigation of these soils, especially in river valleys, has been highly successful. Because of the low organic matter content of these soils, their ability to supply nitrogen is limited and must be supplemented with manures or fertilizers. In the irrigation process, care must be taken to provide good drainage to prevent the undesirable buildup of soluble salts.

FIGURE 3.17 It is easy to overgraze the vegetation on an Aridisol, and an overgrazed Aridisol is subject to ready wind and water erosion. This gully was formed during infrequent but heavy rainstorms in an area in Arizona that at one time had been well covered with desert plants such as alkali sacaton and vine mesquite, plants that were destroyed by the overgrazing. (Photo courtesy U.S. Natural Resource Service)

3.15 SPODOSOLS (ACID, SANDY, FOREST SOILS, LOW BASES)

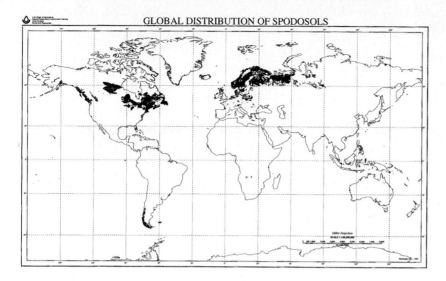

GLOBAL DISTRIBUTION OF SPODOSOLS

SUBORDERS

Aquods (wet)　　　　Humods (humus)

Cryods (icy cold)　　Orthods (typical)

Spodosols are characteristic of cold, moist to wet climates where intensive acid leaching is the norm. They are mineral soils that have a spodic horizon, a subsurface horizon with an accumulation of organic matter and of oxides of aluminum with or without iron oxides (Plate 9 and Figure 3.12). This illuvial horizon, which is not very thick, usually occurs under an eluvial horizon, normally an albic horizon, which is light in color and, therefore, has been described as "wood ash."

Spodosols form mostly on coarse-textured, acid, parent materials subject to ready leaching. They occur only in moist to wet areas and are most common where it is cold or temperate (Figure 3.10). Forests are the natural vegetation under which most of these soils have developed.

Forest species such as pine trees, whose needles are low in metallic cations, seem to encourage the development of Spodosols. As the litter from these low-base species decomposes, strong acidity develops. Percolating water leaches acids down the profile, and iron and aluminum are bound by organic acids and carried down to the spodic horizon where they accumulate. Most minerals except quartz are removed by this acid leaching. In the lower horizons, oxides of aluminum and iron as well as organic matter precipitate, thus yielding the interesting Spodosol profiles.

Distribution and Use

Spodosols are found in about 5% of the land area of the United States. Many of the soils in the Northeast, as well as those of northern Michigan and Wisconsin, belong to this order (Figures 3.6 and 3.12). Most of them are Orthods, the common Spodosols described above. Some, however, are Aquods because they are seasonally saturated with water and possess characteristics associated with this wetness. Important areas of Aquods occur in Florida.

The area of Spodosols in the Northeast extends into Canada. Other large areas of this soil are found in northern Europe and Siberia. Smaller but important areas occur in the southern part of South America and in the cool mountainous areas of temperate regions. Globally, spodosols are found on about 4% of the total land area.

Spodosols are not naturally fertile. When properly fertilized, however, these soils can become quite productive. For example, the productive potato-producing soils of northern Maine are Spodosols, as are some of the vegetable- and fruit-producing soils of Florida, Michigan, and Wisconsin. Even so, the low native fertility of most Spodosols generally makes them uncompetitive for tilled crops unless they receive considerable levels of fertilizers. They are now covered mostly with forests, the vegetation under which they originally developed. Most Spodosols should remain as forest habitats. Because they are already quite acid and poorly buffered, many Spodosols are susceptible to damage from acid rain (see Chapter 9).

Management

Most Spodosols continue to support forest vegetation. Even some that at an earlier date were cultivated have been allowed to revert to forests. This is probably the most sustainable use of these soils. Since their coarse (sandy) texture encourages water infiltration, they are not as subject to erosion as are the more fine-textured soils. At the same time, the presence of forest cover can moderate peak stream flows from Spodosol areas.

Since Spodosols are quite acid and often have sandy surface horizons, they are generally low in plant nutrients. As a consequence, cultivated Spodosols are commonly quite unproductive unless they receive significant quantities of lime and manure or chemical fertilizers. The amount of lime needed is normally not high since the acidity of sandy soils can easily be neutralized. Other nutrients can be provided by animal manures on dairy farms, but on land where cash crops are grown fertilizers are essential. In fact, where potatoes and other vegetable crops are grown, applications of 1–2 Mg of fertilizers per hectare may be needed to obtain optimum crop yields. Excessive applications of chemical nutrients must be avoided, however, to prevent buildup of chemicals in drainage waters to undesirable levels.

3.16   HISTOSOLS (ORGANIC SOILS)

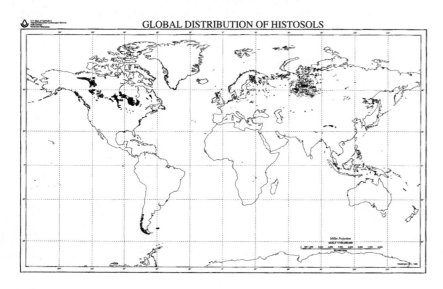

GLOBAL DISTRIBUTION OF HISTOSOLS

SUBORDERS

>   Fibrists (fibers of plants obvious)
>
>   Folists (leaf mat accumulations)
>
>   Hemists (fibers partly decomposed)
>
>   Saprists (fibers not recognizable)

Histosols are organic soils. These soils are generally brown to black in color and are characterized by high organic carbon contents (Plate 5). All of them must have at least 12% organic carbon (about 20% organic matter). This minimum carbon content for Histosols is even higher in soils with appreciable clay content and must be at least 18% (30% organic matter) in soils with 60% or more of clay.

While not all wetlands contain Histosols, virtually all Histosols occur in wetland environments. They can form in any climate, from the equator to permafrost areas, so long as water is available and plants can grow. Organic deposits accumulate in marshes, bogs, and swamps, which are habitats for water-loving plants such as pondweeds, cattails, sedges, reeds, mosses, shrubs, and even some trees. For generations, the residues of these plants have sunk into the water, which by reducing oxygen availability has inhibited their oxidation and, consequently, has acted as a partial preservative (see Figure 2.19).

Histosols are light in weight when dry. Their bulk density is only 0.2–0.3 Mg/m^3 compared to 1.25–1.45 Mg/m^3 for mineral surface soils. Thus, a hectare of dry peat to plow depth weighs only 15 to 20% as much as a dry mineral soil.

Histosols also have high water-holding capacities on a mass basis. While a mineral soil will absorb and hold from 20 to 40% of its weight of water, a cultivated Histosol will hold 200 to 400% of its dry weight of moisture. Because of the light weight of the organic soil, however, on a volume basis the Histosols will not hold any more available water for plants than a mineral soil.

Distribution and Use

Even though they cover only about 1% of the world's land area, peat deposits are found all over the world. They are most extensive in areas with cold climates, but peat beds are found even in the tropics. They comprise significant areas in cold, wet areas of Alaska, Canada, Finland, and Russia, as well as swampy areas such as Iceland.

Of approximately 200 million ha of Histosols worldwide, about 7.5 million ha are found in the United States. Three-quarters of this peat land is in glaciated areas such as Wisconsin, Minnesota, New York, and Michigan (see Figure 3.18). Other United States peat lands (about one-quarter) are found in near-coastal areas such as the Everglades in Florida, the tule-reed beds of California, and similar water-endowed areas in Louisiana and Texas.

In the United States, the most extensive use of cleared peat is as a field soil, especially for vegetable and flower production. Peat products are also used widely for potted plants, and for home flower and vegetable gardens, and as a mulch on home gardens and lawns as well as for commercial greenhouse production. Fiber pots made of compressed peats are used to start plants, which can later be transplanted to soil without being removed from the pot. Peat utilization is a profitable business in Europe and North America.

FIGURE 3.18 Profile of a drained Histosol on which onions are being produced in New York State. The organic soil rests on lake-laid (lacustrine) mineral material. (Photo courtesy of R. Weil)

Peat deposits are used extensively for fuel, especially in Russia where energy for some power stations is derived from peat. The use of peat for domestic fuel is declining but is still prevalent in remote areas of Russia, parts of Ireland, and elsewhere in Europe.

Management

With over 50% of the original wetland area in the United States now drained for agricultural use, much attention needs to be focused on managing the remaining natural wetlands, including areas of Histosols, for their ecological benefits under natural marsh vegetation. In some cases, it would be wise to permit existing agricultural areas to return to their natural undrained state.

For the large areas of Histosols that have already been drained and developed for agricultural use, the productivity of these organic soils depends very much on proper management.

DRAINAGE AND WATER TABLE. Because of the ease with which water flows through organic soils (i.e., high hydraulic conductivity), the water table in most Histosols can be easily controlled by adjustments in the drainage system. For most crops, the water level should be maintained between 45 and 75 centimeters from the soil surface. This level assures a ready water supply for vegetables and other shallow-rooted crops, and also keeps the surface moist enough to minimize wind erosion.

Keeping the water tables as high as possible also reduces the rate of oxidation of the organic matter and the associated subsidence of the soil. Subsidence can cause as much as 5 cm of soil a year to disappear in hot climates, but if the water table is kept within about 50 cm of the surface, the subsidence rate is commonly only around 1 cm a year. To slow the loss of valuable soil resources and avoid unnecessarily aggravating the global greenhouse effect (see Chapter 12), the water table in forested or agricultural Histosols should be kept no lower than is needed to assure adequate root aeration.

STRUCTURAL MANAGEMENT. Plowing is not ordinarily necessary because peat is porous and open unless it contains considerable silt and clay. In fact, a cultivated Histosol may need packing rather than loosening. Cultivation tends to destroy the original granular structure, leaving the soil in a powdery condition when dry. It is then susceptible to wind erosion, a serious problem in some locations.

USE OF LIME AND FERTILIZERS. Lime, which often must be used on mineral soils, is rarely needed on Histosols, unless the soils contain appreciable quantities of mineral matter that can supply toxic quantities of iron, aluminum, and manganese. Generally, the pH of a Histosol at a given percentage base saturation is lower than that of a representative mineral soil; therefore a lower pH can be tolerated than would be the case for mineral soils. Also, aluminum toxicity is typically less of a problem in organic soils. Overliming should be avoided because at high pH values the organic colloids bind micronutrients such as Zn, Cu, and Mn, leading to deficiencies of these elements.

Except on freshly cleared peats, the release of plant nutrients from the organic matter is generally too slow to preclude the need for chemical fertilizers in meeting crop demands for nitrogen. As many Histosols are very low in phosphorus and potassium, these elements must often also be added. In addition, crops grown on Histosols also commonly suffer from deficiencies of micronutrient elements such as copper, zinc, manganese, and boron, unless these elements are added as fertilizers.

3.17 KEY TO SOIL ORDERS

Having considered each soil order individually, it is helpful to have some simple means of quickly differentiating among the 11 soil orders of *Soil Taxonomy*.

A simple key to help with this differentiation is shown in Figure 3.19. This key, as well as a more detailed key in Appendix A helps illustrate the relationships among the soil orders. The keys should be studied carefully.

The approximate land area for each of the 11 soil orders in the United States and the world is shown in Table 3.6. Note that the percentages of Mollisols and Inceptisols are higher than the others in the United States, whereas the percent of land area on which Aridisols and Alfisols predominate is higher worldwide.

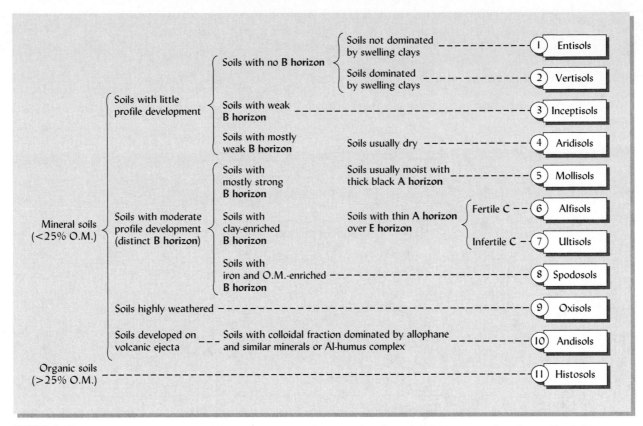

FIGURE 3.19 Simplified key to help differentiate among the 11 soil orders of *Soil Taxonomy* (see Appendix A for a more detailed key). (Reprinted by permission from *Soil Science Simplified,* second edition, by M. Harpstead, F. D. Hole, and W. F. Bennett, © 1988 by Iowa State University Press, Ames, IA 50010)

3.18 SOIL SUBORDERS, GREAT GROUPS, AND SUBGROUPS

Suborders

The 11 orders just described are subdivided into 55 suborders, as shown in Table 3.7. The characteristics used as a basis for subdividing into the suborders are those that give the class the greatest genetic homogeneity. Thus, soils formed under wet conditions generally are identified under separate suborders (e.g., Aquents, Aquerts, Aquepts), as are the less moist soils (e.g., Ustalfs, Ustults). This arrangement also provides a convenient mechanism for grouping soils outside the classification system (e.g., the **wet** and **dry** soils).

To determine the relationship between suborder names and soil characteristics, refer to Table 3.8. Here the formative elements for suborder names are identified and their connotations given. Thus, the **Borolls** (from the Greek *boreas,* northern) are cool Mollisols. Likewise, soils in the **Udults** suborder (from the Latin *udus,* humid) are moist Ultisols. Identification of the primary characteristics of each of the other suborders can be made by cross-reference to Tables 3.7 and 3.8.

Great Groups

The **great groups** are subdivisions of suborders. They are defined largely by the presence or absence of diagnostic horizons and the arrangements of those horizons. These horizon designations are included in the list of formative elements for the names of great groups shown in Table 3.9. Note that these formative elements refer to epipedons such as umbric and ochric (see Table 3.1 and Figure 3.2), to subsurface horizons such as argillic and natric, and to impervious layers (pans) such as duripan and fragipan (see Figure 3.3).

Remember that the great group names are made up of these formative elements attached as prefixes to the names of suborders in which the great groups occur. Thus, a *Ustoll* with a *natric* horizon (high in sodium) belongs to the *Natrustolls* great group. As might be expected, the number of great groups is high, some 238 having been identified.

TABLE 3.7 Soil Orders and Suborders in *Soil Taxonomy*

Note that the ending of the suborder names identifies the order in which the soils are found.

Order	Suborder	Order	Suborder	Order	Suborder
Entisols	Aquents	Alfisols	Aqualfs	Aridisols	Argids
	Arents		Boralfs		Calcids
	Fluvents		Udalfs		Cambids
	Orthents		Ustalfs		Cryids
	Psamments		Xeralfs		Durids
					Gypsids
Inceptisols	Aquepts	Ultisols	Aquults		Salids
	Ochrepts		Humults		
	Plaggepts		Udults	Spodosols	Aquods
	Tropepts		Ustults		Cryods
	Umbrepts		Xerults		Humods
					Orthods
Mollisols	Aquolls	Oxisols	Aquox		
	Albolls		Perox	Histosols	Fibrists
	Borolls		Torrox		Hemists
	Rendolls		Udox		Saprists
	Udolls		Ustox		Folists
	Ustolls				
	Xerolls	Vertisols	Aquerts	Andisols	Aquands
			Cryerts		Cryands
			Uderts		Torrands
			Usterts		Udands
			Xererts		Ustands
					Xerands
					Vitrands

TABLE 3.8 Formative Elements in Names of Suborders in *Soil Taxonomy*

Formative element	Derivation	Connotation of formative element
alb	L. *albus*, white	Presence of albic horizon (a bleached eluvial horizon)
and	Modified from Ando	Ando-like
aqu	L. *aqua*, water	Characteristics associated with wetness
ar	L. *arare*, to plow	Mixed horizons
arg	L. *argilla*, white clay	Presence of argillic horizon (a horizon with illuvial clay)
bor	Gk. *boreas*, northern	Cool
cry	Gk. *kruos*, icy cold	Cold
fibr	L. *fibra*, fiber	Least decomposed stage
fluv	L. *fluvius*, river	Floodplains
fol	L. *folia*, leaf	Mass of leaves
hem	Gk. *hemi*, half	Intermediate stage of decomposition
hum	L. *humus*, earth	Presence of organic matter
ochr	Gk. base of *ochros*, pale	Presence of ochric epipedon (a light surface)
orth	Gk. *orthos*, true	The common ones
perud	Continuously humid	Of year-round humid climates
plagg	Modified from Ger. *plaggen*, sod	Presence of plaggen epipedon
psamm	Gk. *psammos*, sand	Sand textures
rend	Modified from Rendzina	Rendzina-like
sapr	Gk. *sapros*, rotten	Most decomposed stage
torr	L *torridus*, hot and dry	Usually dry
trop	Gk. *tropikos*, turning	Continually warm
ud	L. *udus*, humid	Of humid climates
umbr	L. *umbra*, shade	Presence of umbric epipedon (a dark surface)
ust	L. *ustus*, burnt	Of dry climates, usually hot in summer
vitr	L. *vitreus*, glass	Resembling glass
xer	Gk. *xeros*, dry	Annual dry season

TABLE 3.9 Formative Elements for Names of Great Groups and Their Connotation

These formative elements combined with the appropriate suborder names give the great group names.

Formative element	Connotation	Formative element	Connotation	Formative element	Connotation
acr	Extreme weathering	fulvi	light-colored melanic horizon	plagg	Plaggen horizon
agr	Agric horizon	geli	Very cold	plinth	Plinthite
ala	Very low iron	gibbs	Gibbsite	psamm	Sand texture
alb	Albic horizon	gyps	Gypsic horizon	quartz	High quartz
and	Ando-like	gloss	Tongued	rhod	Dark red colors
arg	Argillic horizon	hal	Salty	sal	Salic horizon
bor	Cool	hapl	Minimum horizon	sider	Free iron oxides
calc	Calcic horizon	hum	Humus	sombr	Dark horizon
camb	Cambic horizon	hydr	Water	sphagn	Sphagnum moss
chrom	High chroma	kand	Low activity clay	sulf	Sulfur
cry	Cold	lithic	Near stone	torr	Usually dry
dur	Duripan	luv, lu	Illuvial	trop	Continually warm
dystr, dys	Low base saturation	med	Temperate climates	ud	Humid climates
endo	Fully water saturated	melan	Melanic epipedon	umbr	Umbric epipedon
epi	Perched water table	nadur	See *natr* and *dur*	ust	Dry climate, usually hot in summer
eutr, eu	High base saturation	natr	Natric horizon		
ferr	Iron	ochr	Ochric epipedon	verm	Wormy, or mixed by animals
fluv	Floodplain	pale	Old development	vitr	Glass
frag	Fragipan	pell	Low chroma	xer	Annual dry season
fragloss	See *frag* and *gloss*	plac	Thin pan		

The names of selected great groups from three orders are given in Table 3.10. This list illustrates again the usefulness of *Soil Taxonomy,* especially the nomenclature it employs. The names identify the suborder and order in which the great groups are found. Thus, Argiudolls are Mollisols of the Udolls suborder characterized by an argillic horizon. Cross-reference to Table 3.9 identifies the specific characteristics separating the great group classes from each other.

Note from Table 3.10 that not all possible combinations of great group prefixes and suborders are used. In some cases a particular combination does not exist. For example,

TABLE 3.10 Examples of Names of Great Groups of Selected Suborders of the Mollisol, Alfisol, and Ultisol Orders

The suborder name is identified as the italicized portion of the great group name.

	Dominant feature of great group		
	Argillic horizon	Minimum horizon development	Old land surfaces
Mollisols			
1. Aquolls (wet)	Argi*aquolls*	Hapl*aquolls*	—
2. Udolls (moist)	Argi*udolls*	Hapl*udolls*	Pale*udolls*
3. Ustolls (dry)	Argi*ustolls*	Hapl*ustolls*	Pale*ustolls*
4. Xerolls (Med.)[a]	Argi*xerolls*	Hapl*oxerolls*	Pale*xerolls*
Alfisols			
1. Aqualfs (wet)	—		—
2. Udalfs (moist)		Hapl*udalfs*	Pale*udalfs*
3. Ustalfs (dry)	—	Hapl*ustalfs*	Pale*ustalfs*
4. Xeralfs (Med.)[a]	—	Hapl*oxeralfs*	Pale*xeralfs*
Ultisols			
1. Aqults (wet)	—	—	Pale*aqults*
2. Udults (moist)	—	Hapl*udults*	Pale*udults*
3. Ustults (dry)	—	Hapl*ustults*	Pale*ustults*
4. Xerults (Med.)[a]	—	Hapl*oxerults*	Pale*xerults*

[a]Med. = Mediterranean climate; distinct dry period in summer.

Aquolls occur in lowland areas but not on old landscapes. Hence, there are no "Paleaquolls." Also, since all Alfisols and Ultisols contain an argillic horizon, the use of terms such as "Argiudults" would be redundant.

Subgroups

Subgroups are subdivisions of great groups. There are more than 1240 subgroups recognized. Subgroups permit characterization of the core concept of a given great group and of gradations from that central concept to other units of the classification system. The subgroup most nearly representing the central concept of a great group is termed *Typic.* Thus, the **Typic Hapludolls** subgroup typifies the Hapludolls great group. A Hapludoll with restricted drainage would be classified as an **Aquic Hapludoll.** One with evidence of intense animal (earthworm) activity would fall in the **Vermic Hapludolls** subgroup.

Some intergrades may have properties in common with other orders or with other great groups. Thus, the **Entic Hapludolls** subgroup intergrades toward the Entisols order. The subgroup concept illustrates very well the flexibility of this classification system.

3.19 SOIL FAMILIES AND SERIES

Families

The family category of classification is based on properties important to the growth of plant roots. The criteria used include broad classes of **particle size, mineralogy, temperature,** and **depth** of the soil penetrable by roots. Table 3.11 gives an example of the classes used (a more complete listing is given in Appendix B). Terms such as *loamy, sandy,* and *clayey* are used to identify the broad textural classes. Terms used to describe the mineralogical classes include *smectitic, kaolinitic, siliceous, carbonatic,* and *mixed.* For temperature classes, terms such as *frigid, mesic,* and *thermic* are used (see Table 3.3).

Thus, a Typic Argiudoll from Iowa, loamy in texture, having a mixture of clay minerals and with annual soil temperatures (at 50 cm depth) between 8 and 15°C, is classed in the **Typic Argiudolls loamy, mixed, mesic** family. In contrast, a sandy-textured Typic Cryorthod, high in quartz and located in a cold area in eastern Canada, is classed in the **Typic Cryorthods sandy, siliceous, frigid** family.

The terms **shallow** and **micro** are sometimes used at the family level to indicate unusual soil depths. About 7500 families have been identified.

Series

The families are subdivided into the lowest (most specific) category of the system called *series.* One series may be differentiated from another in a family through differences in one or more, but not necessarily all, of the soil properties. Series are also established on the basis of profile characteristics. This requires a careful study of the various horizons

TABLE 3.11 **Commonly Used Examples of Particle Size, Mineralogy, and Soil Temperature Used to Differentiate Families**

		Soil temperature class	
Particle-size class	Mineralogical class		Mean annual temperature (°C)
Fragmental	Carbonatic	Pergelic	<0
Sandy	Micaceous	Cryic	0–8
Loamy	Siliceous	Frigid	<8
Fine-loamy	Kaolinitic	Mesic	8–15
Loamy skeletal[a]	Smectitic	Isomesic[b]	8–15
Clayey	Oxidic	Thermic	15–22
	Mixed	Hypothermic	>22

[a]Skeletal refers to presence of up to 35% rock fragments.
[b]"Iso" prefix refers to soils where the difference between summer and winter temperatures is less than 5°C; in the Frigid to Hyperthermic soil temperature classes, this difference is greater than 5°C.

as to number, order, thickness, texture, structure, color, organic content, and reaction (acid, neutral, or alkaline). Features such as a hardpan at a certain distance below the surface, a distinct zone of calcium carbonate accumulation at a certain depth, or striking color characteristics greatly aid in series identification.

In the United States, each series is given a name, usually from some city, village, river, or county, such as Fargo, Muscatine, Cecil, Mohave, or Ontario. There are more than 18,800 soil series in the United States.

The complete classification of a Mollisol, the Brookston series, is given in Figure 3.20. This figure illustrates how *Soil Taxonomy* can be used to show the relationship between *the* soil, a comprehensive term covering all soils, and *a* specific soil series. The figure deserves study.

3.20 FIELD CONFIGURATIONS

In reviewing different units in the *Soil Taxonomy* classification system we have made little reference to the fact that soils exist alongside each other in the field. A given tract of land even as small as a single farm field will likely contain several different kinds of soils. On soil survey maps they will show up as different soil series. They may also be of different families, subgroups, etc. In some cases they may even be classified in different soil orders. An example of such a field situation is shown in Figure 3.21 where three soil series from the Mollisol order are found in association with two soil series of the Alfisol order. Such field associations will be given more attention in Chapter 19, where soil surveys and the use of geographic soils information will receive attention.

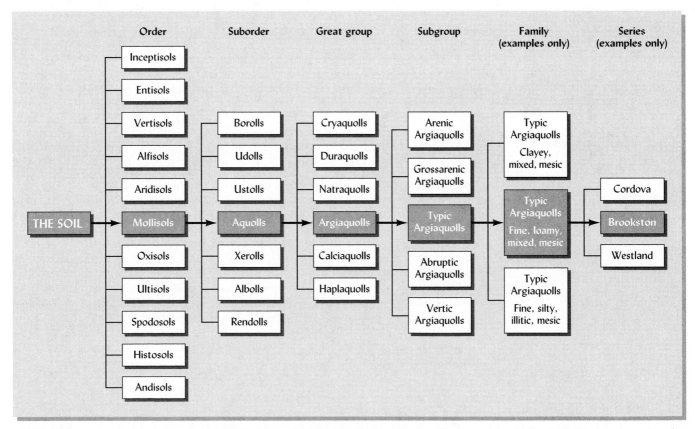

FIGURE 3.20 Diagram illustrating how one soil (Brookston) fits into the overall classification scheme. The shaded boxes show that this soil is of the Mollisols order, Aquolls suborder, Argiaquolls great group, and so on. In each category other classification units are shown.

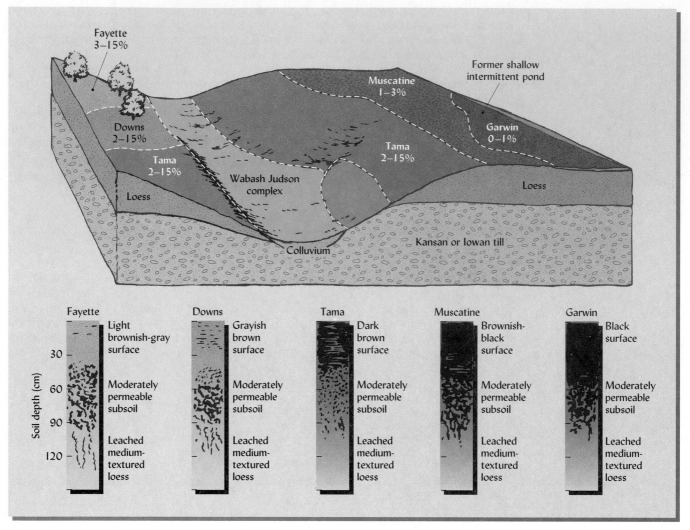

FIGURE 3.21 Association of soils in Iowa. Note the relationship of soil type to (1) parent material, (2) vegetation, (3) topography, and (4) drainage. Two Alfisols (Fayette and Downs) and three Mollisols (Tama, Muscatine, and Garwin) are shown. [From Riecken and Smith (1949)]

3.21 CONCLUSIONS

It is fortunate that scientists have recognized that the soil which covers the earth is in fact comprised of a large number of individual soils, each with distinctive properties. Among the most important of these properties are those associated with the horizontal layers, or horizons, found in a soil profile. These horizons reflect the physical, chemical, and biological processes that have stimulated the genesis of individual soils, and are among the most significant properties that influence how soils can and should be used.

Knowledge of the kinds and properties of soils around the world is critical to humanity's struggle for survival and well-being. A soil classification system based on these properties is equally critical if we expect to use knowledge gained at one location to solve problems at other locations where similarly classed soils are found. *Soil Taxonomy*, a classification system based on measurable soil properties, helps fill this need in some 45 countries. It is a system that is being constantly updated to better serve these needs as scientists learn more about the nature and properties of the world's soils and the relationships among them.

REFERENCES

Eswaran, H. 1993. "Assessment of global resources: Current status and future needs," *Pedologie*, XL111-1 19–39.

McCracken, R. J., et al. 1985. "An appraisal of soil resources in the USA," in R. F. Follet and B. A. Stewart (eds.), *Soil Erosion and Crop Productivity* (Madison, Wis.: Amer. Soc. Agron.).

Riecken, F. F., and G. D. Smith. 1949. "Principal upland soils of Iowa, their occurrence and important properties," *Agron.,* **49** revised, Iowa Agr. Exp. Sta.

SSSA. 1984. *Soil Taxonomy, Achievements and Challenges,* SSSA Spec. Publ. No. 14 (Madison, Wis.: Soil Sci. Soc. Amer.).

Soil Survey Staff. 1975. *Soil Taxonomy: A Basic System of Soil Classification for Making and Interpreting Soil Surveys* (Washington, D.C.: USDA Natural Resources Conservation Service).

Soil Survey Staff. 1994. *Keys to Soil Taxonomy* (Washington, D.C.: USDA Natural Resources Conservation Service).

4

SOIL ARCHITECTURE AND PHYSICAL PROPERTIES

And when that crop grew, and was harvested, no man had crumbled a hot clod in his fingers and let the earth sift past his fingertips.
—JOHN STEINBECK, GRAPES OF WRATH

Soil physical properties profoundly influence how soils function in an ecosystem and how they can best be managed. Success or failure of both agricultural and engineering projects often hinges on the physical properties of the soil used. The occurrence and growth of many plant species and the movement of water and solutes over and through the soil are closely related to soil physical properties. Soil color, texture, and other physical properties are used extensively in classifying soil profiles. Moreover, most of the physical properties discussed in this chapter can be readily determined in the field, often allowing the soil scientist to make on-the-spot judgments concerning the suitability of soils for proposed uses, as well as concerning problems likely to be encountered. Knowledge of the basic physical properties of soils is of great practical value and will help in understanding many other aspects of soils considered in later chapters.

The physical properties discussed in this chapter relate to the solid particles of the soil and the manner in which they are aggregated together. If we think of the soil as a house, the solid particles in soil are the building blocks from which the house is constructed. Soil **texture** describes the *size* of the soil particles. The larger mineral fragments usually are embedded in, and coated with, clay and other colloidal materials. Where the larger mineral particles predominate, the soil is gravelly, or sandy; where the mineral colloids are dominant, the soil is claylike. All gradations between these extremes are found in nature.

In building a house, the manner in which the building blocks are put together determines the nature of the walls, rooms, and passageways. Soil **structure** describes the manner in which soil particles are aggregated together. This property, therefore, defines the nature of the system of pores and channels in a soil. Organic matter acts as a binding agent between individual particles, encouraging the formation of clumps or aggregates of soil.

Soil is a complex system consisting of solid, liquid, and gaseous substances, in which the liquids and gases occupy the pore spaces between solid particles. The physical properties considered in this chapter directly describe the nature of soil solids, but also deal with impacts on the soil water and air. Together, soil texture and structure help determine the nutrient-supplying ability of soil solids, as well as the ability of the soil to hold and conduct the water and air necessary for plant root activity. These factors also determine how soils behave when used for highways, building construction and foundations, or when manipulated by tillage. In fact, through their influence on water movement through and off soils, soil physical properties also exert considerable control over the destruction of the soil itself by erosion.

4.1 SOIL COLOR[1]

When we examine a soil, the first thing we are likely to notice is its color. In and of themselves, soil colors have little effect on the behavior and use of soils. An important exception to this statement is the fact that dark-colored surface soils absorb more solar energy than lighter-colored soils, and therefore may warm up faster.

The main reason for studying soil colors is that color provides valuable clues to the nature of other soil properties and conditions. Because of the importance of accurate color description in soil classification and interpretation, a standard system for color description has been developed using Munsell color charts (Figure 4.1). In this system, a small piece of soil is compared to standard color chips in a soil color book. Each color chip is described by the three components of color: the **hue**, the **chroma** (intensity or brightness), and the **value** (lightness or darkness).

Causes of Soil Colors

Soils display a wide range of reds, browns, yellows, and even greens (see Plate 13). Some soils are nearly black, others nearly white. Some soil colors are very bright, oth-

---

[1]For an excellent collection of papers on the causes and measurement of soil colors, see Bigham and Ciolkosz (eds.), 1993.

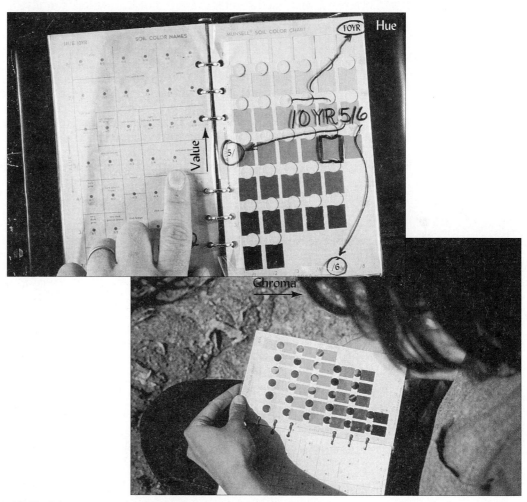

FIGURE 4.1 Determining soil color by comparison to color chips in a *Munsell Color Book*. Each page shows a different hue, ranging from 5R (red) to 5Y (yellow). On a single page, higher-value (lighter) colors are nearer the top and lower-value (darker) colors are near the bottom. Higher-chroma (brighter) colors are nearer the right-hand side and duller, grayish (low-chroma) colors are near the left-hand side. The complete Munsell color description of the soil pictured is 10YR 5/6 (yellowish brown), moist. (Photos by R. Weil)

ers are dull grays. Soil colors may vary from place to place in the landscape, as when adjacent soils have differing surface horizon colors (e.g., see Figure 3.11, bottom). Colors also typically change with depth through the various layers (horizons) within a soil profile. In many soils, the horizons in a given profile have colors that are similar in hue, but vary with respect to chroma and value. Even within a single horizon or clod of soil, colors may vary from spot to spot (Plate 12).

Most soil colors are derived from the colors of iron oxides and organic matter that coat the surfaces of soil particles. Organic coatings tend to darken and mask the colors derived from iron oxides (Plate 7). Subsoil horizons with little organic matter, therefore, often most clearly display the colors of the various iron oxides that coat the soil particles. Some examples of these oxides are the yellow of goethite, the red of hematite, and the brown of maghematite.

Other minerals that sometimes give soils distinctive colors are manganese oxide (black) and glauconite (green). Carbonates, such as calcite, that typically accumulate in soils of semiarid regions may impart a whitish color (Plate 19). Prolonged, very wet, anaerobic conditions (such as in a marsh) can cause iron oxide coatings to become chemically reduced, changing the red or brown color to gray or bluish colors, a condition referred to as *gley*.

Interpreting Soil Colors

As a result of the above influences, colors can tell us a great deal about a soil. We have already discussed several examples of how colors can give some indication of soil mineralogy. Extensive use is made of color in classifying soils. Colors often help us distinguish the different horizons of a soil profile. Typically an A horizon is darker and a B horizon is brighter (and sometimes redder) than adjacent horizons. In some cases, color is a diagnostic criterion for classification. For example, a mollic epipedon (characteristic of prairie soils; see Chapter 3 for definition) is defined as having a color so dark when moist that both its value and chroma are 3 or less (i.e., in the lower left or blackish corner of a Munsell color book page). Another example is that of the Rhodic subgroups of certain soil orders, which have B horizons characterized by very red colors having hues between 2.5YR (the most red of the yellowish reds) and 10R (the most red of the colors in the Munsell color book).

Because of the color changes that take place when various iron-containing minerals undergo oxidation and reduction, soil colors can provide extremely valuable insights into the hydrologic regime or drainage status of a soil. Bright (high-chroma) colors throughout the profile are typical of well-drained soils through which water easily passes and in which air is generally plentiful. The presence of gray, low-chroma colors, either alone or mixed in a mottled pattern with brighter colors (see Plates 12 and 13), is indicative of waterlogged conditions during at least a major part of the plant growing season. The depth in the profile at which gley colors (low chroma) are found helps to define the drainage class of the soil (see Figure 19.3). Color can also provide qualitative information about the current moisture status of a soil, dry soils generally having lighter (higher-value) colors than moist soils.

One common misconception about soil color is the idea that there is a consistent relationship between soil color and soil texture (see Section 4.2). For example, people often speak of "red clay," but reddish colors do *not* necessarily indicate clayey soils. Very sandy materials also commonly have reddish colors because iron minerals coat the sand particles (e.g., Plate 9).

Finally, it is worth mentioning that soils, with their distinctive colors, are important aesthetic components of the landscape. For example, warm, reddish colors are characteristic of many tropical and subtropical landscapes, while dark grays and browns typify cooler, more temperate regions. Thus, a native of Georgia would probably have a very different mental image of soil than a person from New York.

4.2 SOIL TEXTURE (SIZE DISTRIBUTION OF SOIL PARTICLES)

The size of mineral particles in soil may seem too mundane a subject to warrant much attention, yet knowledge of the proportions of different-size particles in a soil (i.e., the soil **texture**) is critical to understanding soil behavior and management. When investigating soils on a site, the texture of various soil horizons is often the first and most important property to determine, for a soil manager can draw many conclusions from this information. Furthermore, the texture of a soil in the field is not readily subject to change, so it is considered a basic property of a soil.

Nature of Soil Separates

Diameters of soil particles range over six orders of magnitude, from boulders (1 m) to submicroscopic clays ($<10^{-6}$ m). Scientists group these particles into *soil separates* according to several classification systems, as shown in Figure 4.2. The classification established by the U.S. Department of Agriculture is used in this text. The size ranges for these separates are not purely arbitrary, but reflect major changes in how the particles behave and in the physical properties they impart to soils.

Particles larger than 2 mm display very little cohesion or aggregation into the soil matrix. In sufficient quantity these *coarse fragments* (gravels, cobbles, boulders, etc.) may effect the behavior of a soil, but they are not considered to be part of the **fine earth fraction** to which the term *soil texture* properly applies. A major effect of coarse fragments is to reduce the volume of the soil available for the movement of air, water, and roots. The larger classes of coarse fragments, especially if consisting of hard minerals such as quartz, often pose a serious impediment to the use of tillage implements.

Sand is defined as particles smaller than 2 mm but larger than 0.05 mm. Sand particles may be rounded or angular (Figure 4.3), depending on the extent to which they have been worn down by abrasive processes during soil formation. Sand feels gritty between the fingers and the particles are generally visible to the naked eye (Figure 4.4). As sand particles are relatively large, so too the voids between them are relatively large and promote free drainage of water and the entry of air into the soil. Because of their large size, particles of sand have relatively little surface area for a given weight or volume (Figure 4.5). Therefore, sand particles can hold little water, and soils dominated by sand are prone to drought.

If smaller than 0.05 mm, individual particles are not visible to the unaided eye, nor do they feel gritty when rubbed between the fingers. Particles smaller than 0.05 mm but larger than 0.002 mm in diameter are classified as **silt**. These are essentially microsand particles, with quartz generally the dominant mineral. Although silt is composed of particles similar in shape to sand, it feels smooth, like flour. The pores between silt particles are much smaller than those in sand, so silt has a much higher water-holding capacity. However, even when wet, silt itself is not sticky or malleable (plastic). The silt separate, because it usually has an adhering film of clay, possesses some plasticity, cohesion (stickiness), and adsorptive capacity, but much less than the clay separate. Silt (like very fine sand) may cause the soil surface to be compact and crusty unless it is supplemented by adequate amounts of sand, clay, and organic matter.

Particles smaller than 0.002 mm are classified as **clay** and have a very large surface-area-to-volume ratio, giving clay a tremendous capacity to adsorb water and other sub-

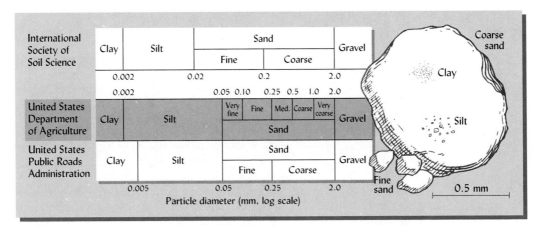

FIGURE 4.2 Classification of soil particles according to their size. The shaded scale in the center and the names on the drawings of particles follow the United States Department of Agriculture system, which is widely used throughout the world. The USDA system is also used in this book. The other two systems shown are also widely used by soil scientists and by highway construction engineers. The drawing illustrates the size of soil separates (note scale).

(a)

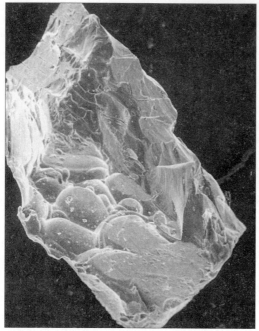

(b)

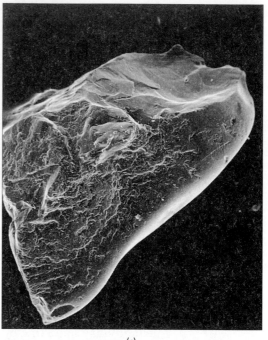

(c)

FIGURE 4.3 (a) Sand grains from soil. Note that the particles are irregular in size and shape. Quartz usually predominates, but other minerals may occur. Silt particles have about the same shape and composition, differing only in size. Scanning electron micrographs of sand grains show quartz sand (b) and a feldspar grain (c). The grains have been magnified about 40 times. (Photos courtesy J. Reed Glasmann, Union Oil Research)

stances on its surfaces. This large adsorptive surface area also causes clay particles to cohere together in a sticky, plastic mass that can be easily molded when wet. Fine colloidal clay has about 10,000 times as much surface area as the same weight of medium-sized sand. The specific surface (area per unit weight) of colloidal clay ranges from about 10 to 1000 square meters per gram (m^2/g), compared to 1 and 0.1 m^2/g for the smallest particles of silt and fine sand, respectively.

Since the adsorption of water, nutrients, and gas, as well as the attraction of particles to each other, are all surface phenomena, the very high specific surface for clay is significant in determining soil properties. This relationship is shown graphically in Figure 4.6.

Particles of clay size are so small that they behave as **colloids**—if suspended in water they do not readily settle out. Unlike most sand and silt particles, clay particles tend to be shaped like tiny flakes or flat platelets.

The presence of clay in a soil gives it a *fine texture* and slows water and air movement. While such a soil becomes sticky when wet, it can also be hard and cloddy when dry, unless properly handled. Some types of clay expand and contract greatly on wetting and drying, and the water-holding capacity of soils high in clay is generally high.

(a)

(b)

FIGURE 4.4 Separation of soil particles by size during sedimentation. The soil in the cylinder was agitated in suspension, then allowed to settle out. After about one minute, the sand particles have settled out and the amount of silt and clay remaining in suspension can be determined by floating a hydrometer (a). After about a day the clay remains in suspension, but the sand and silt have formed layers on the bottom (b). The boundary between silt and sand can be seen as a line between the zone in which the individual particles are visible and the zone in which they are not (see arrow). (Photos by R. Weil)

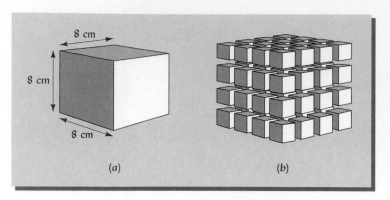

FIGURE 4.5 The relationship between the surface area of a given mass of material and the size of its particles. In the single large cube (a) each face has 64 cm² of surface area. The cube has six faces, so the cube has a total of 384 cm² surface area (6 faces × 64 cm² per face). If the same cube of material was cut into smaller cubes (b) so that each cube was only 2 cm on each side, then the same mass of material would now be present as 64 smaller cubes (4 × 4 × 4). Each face of each small cube would have 4 cm² (2 × 2 cm) of surface area, giving 24 cm² of surface area for each cube (6 faces × 4 cm² per face). The total mass would therefore have 1536 cm² (24 cm² per cube × 64 cubes) of surface area. This is four times as much surface area as the single large cube. Since clay particles are very very small and usually platelike in shape, their surface area is thousands of times greater than that of sand particles.

Influence of Texture on Other Soil Properties

SURFACE AREA.   As will be discussed in Section 4.3, clay-sized particles typically consist of different minerals than do the larger particles, but the principal reason the separates impart such differing properties to soils is the relationship between the size of particle and their *surface area.* Surface area has far-reaching implications for soil behavior because many fundamental soil processes are surface phenomena.

First, water is held in soils as thin films on the surfaces of soil particles; the greater the surface area, the greater the soil's capacity for holding water.

Second, both gases and dissolved chemical ions are attracted to and adsorbed by mineral particle surfaces; the greater the surface area, the greater the soil's capacity to hold plant nutrients.

Third, weathering takes place at the surface of mineral particles, releasing constituent elements into the soil solution. The greater the surface area, the greater the rate of release of plant nutrients from weatherable minerals.

Fourth, the surfaces of mineral particles often carry both negative and some positive electromagnetic charge so that particle surfaces and the water films between them tend to attract each other (see Section 4.10). The greater the surface area, the greater the propensity for soil particles to stick together in a coherent mass, or as discrete aggregates when shrinking water films draw particles together. This influence of surface area will be readily apparent if you compare the hardness and coherence of dry clods from a sandy soil to those from a clayey soil.

Fifth, microorganisms tend to grow on and colonize particle surfaces. For this and other reasons, microbial reactions in soils are greatly affected by surface area. Table 4.1 illustrates some of the ways in which surface phenomena allow soil texture to influence other soil properties.

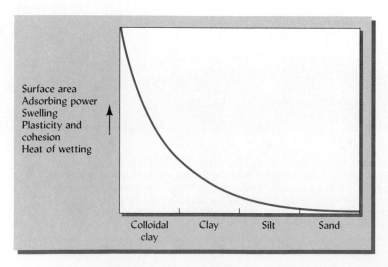

FIGURE 4.6 The finer the texture of a soil, the greater is the effective surface exposed by its particles. Note that adsorption, swelling, and the other physical properties cited follow the same general trend and that their intensities go up rapidly as the colloidal size is approached.

TABLE 4.1 Generalized Influence of Soil Separates on Some Properties and Behavior of Soils

Properties typical of very sandy soils	Properties typical of very silty soils	Properties typical of very clayey soils
Low water-holding capacity	Medium to high water-holding capacity	High water-holding capacity
Well aerated, rapid drainage	Moderate aeration, slow to medium drainage	Poorly aerated, very slow drainage (unless cracked)
Low in organic matter, rapid decomposition	Medium to high in organic matter, medium rate of decomposition	High to medium in organic matter, slow decomposition
Warm quickly in spring	Warm somewhat slowly in spring	Warm slowly in spring
Resists compaction (if coarse sands)	Easily compacted	Easily compacted
Easily blown by wind (if fine sands)	Very easily blown by wind	Resists wind erosion
Poor supply of plant nutrients, little capacity to hold them	Usually good supply of plant nutrients and medium capacity to hold them	Medium to excellent supply of plant nutrients, large capacity to hold them
Acidity easily raised or lowered	Moderately resists changes in acidity	Resists change in acidity
Allows leaching of most pollutants	Moderately retards leaching of pollutants	Retards leaching of most pollutants (unless cracking clay)
Easily tilled shortly after rain	Moderately difficult to till after rain	Very difficult to till after rain
Resists erosion by water (unless fine sand)	Very susceptible to erosion by water	Aggregated clay resists erosion by water, dispersable clay is easily eroded
Poor sealing properties for dams or ponds	Poor sealing properties, prone to rapid "piping," by which water washes out large channels in a soil mass	Good to excellent sealing properties for dams and ponds
Little or no shrinkage and swelling	Little shrinkage and swelling	Moderate to high shrinkage and swelling, depending on clay type

Exceptions to these generalizations do occur, especially as a result of the influences of soil structure and clay mineralogy.

4.3 MINERALOGICAL AND CHEMICAL COMPOSITIONS OF SOIL SEPARATES

Although the focus of this chapter is on the physical properties of soil particles, the mineralogical and chemical composition of these particles is important as well.

Mineralogical Characteristics

The coarsest sand particles often are fragments of rocks as well as individual minerals. Quartz (SiO_2) commonly dominates sand, especially the fine sands, as well as the silt separate (see Figure 4.7). In addition, differing quantities of other primary minerals, such as the various feldspars (aluminosilicates) and micas (iron and aluminum silicates), usually occur. Gibbsite (hydrous oxide of aluminum), hematite, and goethite (hydrous iron oxides) also are found, usually as coatings on the sand grains.

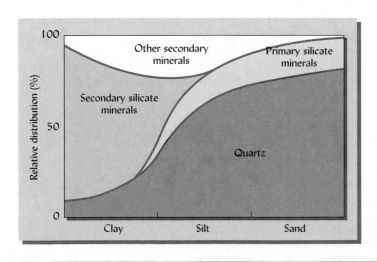

FIGURE 4.7 General relationship between particle size and kinds of minerals present. Quartz dominates the sand and coarse silt fractions. Primary silicates such as the feldspars, hornblende, and micas are present in the sands and, in decreasing amounts, in the silt fraction. Secondary silicates dominate the fine clay. Other secondary minerals, such as the oxides of iron and aluminum, are prominent in the fine silt and coarse clay fractions.

Some clay-size particles, especially those in the coarser clay fractions, are composed of minerals such as quartz and the hydrous oxides of iron and aluminum. The latter are particularly important in the tropics and other warmer climates. Of greater importance in temperate regions, however, are the silicate clays, which occur as platelike particles arranged in thin layers, much like the pages of a book. When some of these silicate clays are wetted, water can move in between layers within a particle as well as between particles, resulting in extensive swelling of the clay. Upon drying, the process is reversed and the clay shrinks. Cracks seen in clay soils when they dry after a rain are evidence of this swelling and shrinking phenomenon. Silicate clays vary considerably in their plasticity, cohesion, and swelling/shrinking capacity.

It is therefore important to know which clay type dominates or codominates any particular soil. Clay types will be considered in greater detail in Chapter 8.

Chemical Makeup

Since sand and silt are composed primarily of quartz (SiO_2) and other primary minerals known for their resistance to weathering, these two fractions have relatively low chemical activity. The primary minerals that may contain nutrient elements in their chemical makeup are generally so insoluble as to make their nutrient-supplying ability essentially nonexistent. However, some silt-sized micas and feldspars do release nutrients such as potassium and calcium rapidly enough to be of some use to plants.

Chemically, the silicate clays vary widely. Some are relatively simple aluminosilicates. Others contain in their crystal structures varying quantities of iron, magnesium, potassium, and other elements. As pointed out in Section 1.15, the surfaces of all the silicate clays hold small but significant quantities of cations such as Ca^{2+}, Mg^{2+}, K^+, H^+, Na^+, NH_4^+, and Al^{3+}. The cations are exchangeable and can be released for absorption by plants.

Weathering can have a profound effect on the chemical and mineralogical composition of soil separates. In highly weathered soils, such as those found in the hot, humid tropics, oxides of iron and aluminum are prominent, if not dominant, in the clay-size fraction.

Since various-sized soil separates differ so markedly in crystal form and chemical composition, it is not surprising that they also vary in their content of essential mineral nutrients. Logically, the sands, being mostly quartz, would be expected to be lowest and the clay separate to be highest. This inference is substantiated by the data in Table 4.2. The general relationships shown by these data hold true for most soils, although some exceptions may occur.

4.4 SOIL TEXTURAL CLASSES

Soil textural class names are used to convey an idea of the textural makeup of soils and to give an indication of their physical properties. Three broad groups of these classes are recognized: *sandy soils, loamy soils,* and *clayey soils.* Within each group specific textural class names have been devised (see Table 4.3).

Sandy Soils

These soils are dominated by the properties of sand, for the sand separate comprises at least 70% of the material by weight (less than 15% of the material is clay).

Two specific textural classes are recognized—*sand* and *loamy sand*—although in practice two subclasses are also used: *loamy fine sand* and *loamy very fine sand.*

TABLE 4.2 **Total Phosphorus, Potassium, and Calcium Contents of Sand, Silt, and Clay Separates Commonly Found in Temperate Humid Region Soils**

Separate	P (%)	K (%)	Ca (%)
Sand	0.05	1.4	2.5
Silt	0.10	2.0	3.4
Clay	0.30	2.5	3.4

TABLE 4.3   General Terms Used to Describe Soil Texture in Relation to the Basic Soil Textural Class Names

U.S. Department of Agriculture Classification System

Common names	General terms	Texture	Basic soil textural class names
Sandy soils	Coarse		Sands
			Loamy sands
Loamy soils		Moderately coarse	Sandy loam
			Fine sandy loam[a]
		Medium	Very fine sandy loam[a]
			Loam
			Silt loam
			Silt
		Moderately fine	Sandy clay loam
			Silty clay loam
			Clay loam
Clayey soils	Fine		Sandy clay
			Silty clay
			Clay

[a]Although not included as class names in Figure 4.8, these soils are usually treated separately because of their fine sand content.

Clayey Soils

Characteristics of the clay separate are distinctly dominant in these soils. Class names in this group and the amount of clay separate required to be classified in each of the classes are *clay* (at least 40% clay), *sandy clay* (at least 35% clay), and *silty clay* (at least 40% clay).

Loamy Soils

The loam group contains many subdivisions. An ideal loam may be defined as a mixture of sand, silt, and clay particles that exhibits the *properties* of those separates in about equal proportions. This definition does not mean that in a loam the three separates are present in equal *amounts* (as will be revealed by careful study of Figure 4.8). This anomaly exists because only a relatively small percentage of clay is required to engender clayey properties in a soil, whereas small amounts of sand and silt have a lesser influence on how a soil behaves. Thus, the *clay* modifier is used in the class name of soils with as little as 20% clay; but to qualify for the modifiers *sandy* or *silt*, a soil must have at least 40 or 45% of these separates, respectively.

Most soils are some type of loam. They may possess the ideal makeup of equal proportions described above and be classed simply as **loam.** Usually, however, the varying quantities of sand, silt, and clay in the soil require a modified textural class name. Thus, a loam in which sand is dominant is classified as a *sandy loam;* in the same way, there may occur *silt loams, silty clay loams, sandy clay loams,* and *clay loams.*

Variations in the Field

The textural class names—*sand, loamy sand, sandy loam, loam, silt loam, silt, sandy clay loam, silty clay loam, clay loam, sandy clay, silty clay,* and *clay*—form a graduated sequence from soils that are coarse in texture and easy to handle to the clays that are very fine and difficult to manage physically (Table 4.3).

While these textural classes are determined by particle-size distribution, they markedly affect other physical properties such as soil aeration, water percolation, and ease of tillage (review Table 4.1).

For some soils, qualifying factors such as stone, gravel, and the various grades of sand become part of the textural class name. Fragments that range from 2 to 75 millimeters (0.08 to 3 inches) along their greatest diameter are termed *gravel* or *pebbles;* those ranging from 75 to 250 mm (3–10 in.) are called *cobbles* (if round) or *flags* (if flat); and those more than 250 mm across are called *stones* or *boulders.* Thus, one might have a cobbly fine sandy loam, for example.

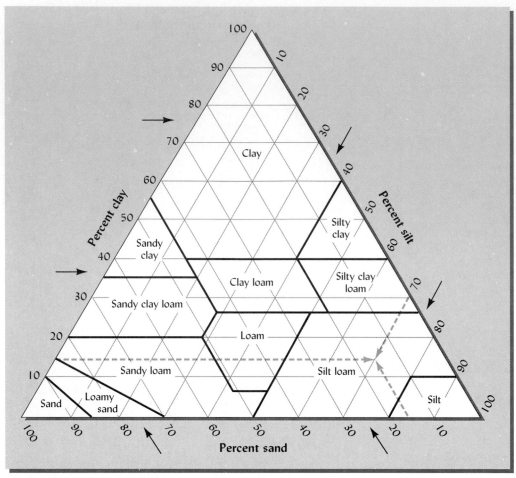

FIGURE 4.8 The major soil textural classes are defined by the percentages of sand, silt, and clay according to the heavy boundary lines shown on the textural triangle. If these percentages have been determined for a soil sample by particle size analysis, then the triangle can be used to determine the soil textural class name that applies to that soil sample. To use the graph, first find the appropriate clay percentage along the left side of the triangle, then draw a line from that location across the graph going parallel to the base of the triangle. Next find the sand percentage along the base of the triangle, then draw a line inward going parallel to the triangle side labeled "silt." The small arrows indicate the proper direction in which to draw the lines. The name of the compartment in which these two lines intersect indicates the textural class of the soil sample. Percentages for any two of the three soil separates is all that is required. Because the percentages for sand, silt, and clay add up to 100%, the third percentage can easily be calculated if the other two are known. If all three percentages are used, the three lines will all intersect at the same point. Consider, as an example, a soil which has been determined to contain 15% sand, 15% clay, and 70% silt. This example is indicated by the light dashed lines that intersect in the compartment labeled "silt loam." What is the textural class of another soil sample which has 33% clay, 33% silt, and 33% clay? The lines (not shown) for this second example would intersect in the center of the "clay loam" compartment.

4.5   DETERMINATION OF SOIL TEXTURAL CLASS

"Feel" Method

The common field method of determining the textural class of a soil is by its *feel*. This is ascertained by rubbing a sample of the soil, usually in a moist to wet condition, between the thumb and fingers (Figure 4.9). The way a wet soil "slicks out"—that is, develops a continuous ribbon when pressed between the thumb and fingers—indicates the amount of clay present. The longer and smoother the ribbon formed, the higher the clay content. Sand particles are gritty and can be heard grinding together if held close to one's ear; silt feels like flour or talcum powder when dry and is only slightly plastic and sticky when wet. Persistent cloddiness of dry soils generally is characteristic of silt and clay. (Table 4.4 gives some general criteria for field determination of soil texture.)

FIGURE 4.9 The "feel" method as used to distinguish between a sand (a), a silt loam (b), and a clay (c) soil. Moist samples are rubbed between the thumb and forefinger. Note the shiny appearance for the clay (c), the lack of cohesion in the sand (a), and the intermediate status of the silt loam. An estimate of percent clay is often a very useful first step in determining textural class by feel. Percent clay in a sample is best estimated by squeezing a ribbon of moist soil (d).

TABLE 4.4 Criteria Used with the Field Method of Determining Soil Texture Classes

Criterion	Sand	Sandy loam	Loam	Silt loam	Clay loam	Clay
1. Individual grains visible to eye	Yes	Yes	Some	Few	No	No
2. Stability of dry clods	Do not form	Do not form	Easily broken	Moderately easily broken	Hard and stable	Very hard and stable
3. Stability of wet clods	Unstable	Slightly stable	Moderately stable	Stable	Very stable	Very stable
4. Stability of "ribbon" when wet soil rubbed between thumb and fingers	Does not form	Does not form	Does not form	Broken appearance	Thin, will break	Very long, flexible

The feel method is used in soil survey and land classification. Accuracy in such a determination is of great practical value and depends largely on experience. Facility in texture class determination is one of the first skills a field soil scientist should develop.

Laboratory Particle-Size Analyses

A particle-size analysis is done by using sieves to mechanically separate out the very fine sand and larger separates from the finer particles. The weight of each separate is then measured. For particles smaller than fine sand, separation with sieves is not practical, so the amounts of silt and clay are determined by measuring the weight of particles still in suspension a certain number of minutes after an agitated suspension of soil in water is allowed to begin settling out.

The principle involved is simple. Because soil particles are more dense than water, they tend to sink. Because there is little variation in the density of most soil particles, the velocity (V) of settling is proportional to the square of the radius (r) of each particle. The equation that describes this relationship is referred to as Stokes' law:

$$V = kr^2$$

where k is a constant related to the density and viscosity of water and the acceleration due to gravity. With knowledge of the velocity of settling, Stokes' law can be rearranged to calculate the radius of the particles as they settle as well as the percentage of each size fraction in a sample. These percentages are used to identify the soil textural class, such as sandy loam, silty clay, or loam.

Although stone and gravel are considered in the practical examination and evaluation of a field soil, they do not enter into the analysis of the finer particles. Their amounts are rated separately. Organic matter is usually removed from a soil sample by oxidation before the mechanical separation. It is important to note that soils are assigned to textural classes solely on the basis of the mineral particles; therefore the percentages of sand, silt, and clay always add up to 100%.

The relationship between such analyses and textural class names is shown diagrammatically in Figure 4.8. This triangular graph enables us to use laboratory particle-size analysis data to check the accuracy of the field soil scientist's class designations as determined by feel. A working knowledge of how the textural triangle is used to name soils is essential. The caption of Figure 4.8 explains the use of this soil texture triangle.

The summation curves in Figure 4.10 illustrate the particle-size distribution in soils representative of three textural classes. Note the gradual change in percentage composition in relation to particle size. This figure emphasizes that there is no sharp line of demarcation in the distribution of sand, silt, and clay fractions and suggests a gradual change of properties with change in particle size.

Changes in Soil Texture

The textural classes of soils are not easily modified in the field. The texture of a given soil can be changed only by mixing it with another soil of a different textural class. For example, the incorporation of large quantities of sand to change the physical properties of a clayey soil for use in greenhouse pots or for turf would bring about such a change.[2]

It should be noted, however, that adding peat to a soil while mixing a potting medium does not constitute a change in texture, since this property refers only to the mineral particles. The term *soil texture* is not relevant to artificial media that contain mainly perlite, peat, styrofoam, or other nonsoil materials. For most large-scale agricultural, forestry, and wild land areas, soil texture is not changed by cultural management. Of course, over very long periods of time, pedologic processes (see Chapter 2) such as erosion, deposition, illuviation, and weathering can alter the textures of various soil horizons.

---

[2]Great care must be exercised in attempting to ameliorate physical properties of fine-textured soils by adding sand. If the sand is not of the proper size and not added in sufficient amounts, it may make matters worse rather than better. While adjacent coarse sand grains form large pores between them, sand grains embedded in a silty or clayey matrix do not. Therefore, moderate amounts of fine sand mixed into a fine-textured soil may yield a product more akin to concrete than to a sandy soil!

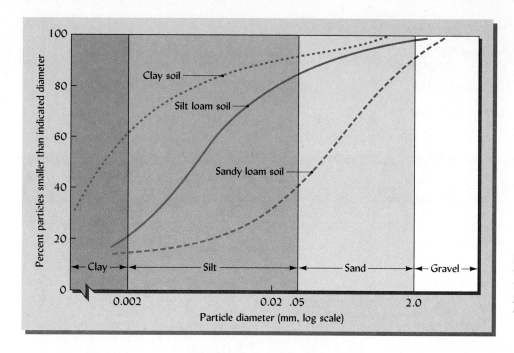

FIGURE 4.10 Particle-size distribution in three soils varying widely in their textures. Note that there is a gradual transition in the particle-size distribution in each of these soils.

4.6   STRUCTURE OF MINERAL SOILS

The term *structure* relates to the grouping or arrangement of soil particles. It describes the gross, overall combination or arrangement of the primary soil separates into secondary groupings called *aggregates* or *peds.*

Soil conditions and characteristics such as water movement, heat transfer, aeration, and porosity are much influenced by structure. To a large degree, it is through changes in soil structure that farming practices such as plowing, cultivating, draining, liming, and manuring influence soil properties.

Many types of structural peds occur in soils. A particular soil profile may be dominated by a single type of structure. More often, however, several types are encountered in the different horizons. The structure of the surface horizon is generally most subject to alteration by soil management practices.

Types of Soil Structure

The dominant shape of peds or aggregates in a horizon determines its structural *type.* The four principal types of soil structure are **spheroidal, platy, prismlike,** and **blocklike.** A brief description of each of these structural types (and appropriate subtypes) with schematic drawings is given in Figure 4.11. A more detailed description of each type follows.

1. *Spheroidal.*   **Granular** and **crumb** subtypes. Rounded peds or aggregates are placed in this category. They usually lie loosely and are separated from each other. Relatively nonporous aggregates are called *granules* and their shape *granular.* However, when the granules are especially porous, the term *crumb* is applied.

   Granular and crumb structures are characteristic of many surface soils, particularly those high in organic matter, and are especially prominent in grassland soils and soils that have been worked by earthworms. They are the only types of aggregation that are commonly influenced by practical methods of soil management.

2. *Platelike.*   **Platy.** In this structural type the aggregates (peds) are arranged in relatively thin horizontal plates, leaflets, or lenses. Platy structure is found in the surface layers of some virgin soils, but may characterize the lower horizons as well.

   Although most structural features are usually a product of soil-forming forces, the platy type is often inherited from soil parent materials, especially those laid down by water or ice. Platy structure can also be created by compaction by heavy machinery on clayey soils.

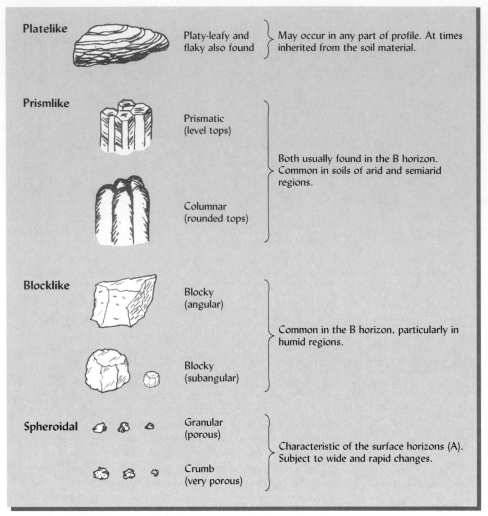

Platelike — Platy-leafy and flaky also found } May occur in any part of profile. At times inherited from the soil material.

Prismlike — Prismatic (level tops) / Columnar (rounded tops) } Both usually found in the B horizon. Common in soils of arid and semiarid regions.

Blocklike — Blocky (angular) / Blocky (subangular) } Common in the B horizon, particularly in humid regions.

Spheroidal — Granular (porous) / Crumb (very porous) } Characteristic of the surface horizons (A). Subject to wide and rapid changes.

FIGURE 4.11 Various structural types found in mineral soils. Their typical location in the profile is suggested. In arable topsoils, a stable granular structure is prized.

3. *Prismlike.* **Columnar** and **prismatic** subtypes. These subtypes are characterized by vertically oriented aggregates or pillars that vary in height among different soils and may reach a diameter of 15 cm or more. They commonly occur in subsurface horizons in arid and semiarid regions and, when well developed, are a very striking feature of the profile. They also occur in some poorly drained soils of humid regions, and they often are associated with the presence of swelling types of clay.

When the tops of the prisms are rounded, the term ***columnar*** is used. Columnar structure is especially common in subsoils high in sodium (e.g., natric horizons; see Section 3.2). When the tops of the prisms are angular and relatively flat horizontally, the structural pattern is designated ***prismatic.***

4. *Blocklike.* **Angular blocky** and **subangular blocky** subtypes. In this case the aggregates have been reduced to blocks, irregularly six-faced, with their three dimensions more or less equal. In size these peds range from about 1–10 cm in thickness. In general, the shape is so unique that identification is easy.

When the edges of the cubes are sharp and the rectangular faces distinct, the subtype is designated ***angular blocky.*** When some rounding has occurred, the aggregates are referred to as ***subangular blocky.*** These types usually are confined to the subsoil, and their stage of development and other characteristics have much to do with soil drainage, aeration, and root penetration.

When soil scientists describe a soil for classification, they note not only the type (shape) of the structural peds present, but also the relative size (fine, medium, coarse) and degree of development or distinctness of the peds (strong, moderate, or weak). Thus, a soil horizon might be described as having "moderate, fine, subangular blocky structure." An example is given in Table 19.1. Generally, the structure of a soil is easier to observe when the soil is relatively dry; when wet, structural peds may swell and press closer together, making the individual peds less well defined.

As already emphasized, two or more of the structural types listed usually occur in the same soil profile. In humid and subhumid regions, a granular aggregation in the surface horizon with a blocky, subangular blocky or platy type in the lower horizons is usual, although granular subsurface horizons and subangular blocky surface horizons sometimes occur. In low-rainfall areas, granular or crumb surface soils are often underlain by horizons with prismatic or columnar structure.

Genesis of Soil Structure

The mechanics of soil structure formation are not well known. As plant roots push their way through the soil, they tend to compress soil particles into small aggregates and to break apart larger aggregates already present in the soil. Similar compression and contraction results from the swelling and shrinking that occurs as soils are wetted and dried. Filamentous fungi are also thought to play an important role in binding together soil particles in many soils.

Plant roots (particularly root hairs) and fungal hyphae excrete sticky organic chemicals that help bind the soil particles together into aggregates. Microbial decomposition of plant residues produces other such organic materials, which interact with clays and cement the aggregates together. Thus, organic materials, stimulated by microbial processes, not only enhance the formation of soil aggregates but also help ensure their stability. These processes are most notable in surface soils where root and animal activities and organic matter accumulation are most pronounced.

The mechanism for the creation and stabilization of those structural types common in lower horizons is less certain. The downward migration of silicate clays, oxides of iron and aluminum, soluble salts, and calcium carbonate is thought to encourage aggregate formation under different climatic and soil conditions (see Figure 4.12). However, the exact mechanisms for specific aggregate formation in subsoils remains obscure.

FIGURE 4.12 Strong, medium angular blocky peds in the B horizon of a Alfisol (Ustalf) in a semiarid region. The knifepoint is prying loose an individual blocky ped. Note the lighter-colored surface coatings of illuvial clay that help define and bind together the ped. (Photo courtesy of R. Weil)

4.7   PARTICLE DENSITY OF MINERAL SOILS

One measure of soil mass (weight) in terms of the density of the solid particles is **particle density**. Particle density D_p is defined as the mass of a unit volume of soil solids. In the metric system, particle density can be expressed in terms of megagrams per cubic meter (Mg/m^3). Thus, if 1 m^3 of soil solids weighs 2.6 Mg, the particle density is 2.6 Mg/m^3 (which can also be expressed in grams per cubic centimeter^3).

Particle density depends on the chemical composition and crystal structure of the mineral particle and is not affected by pore space. Therefore particle density is *not* related to particle size or to the arrangement of particles (structure).

Although there is considerable range in the density of individual soil minerals, the figures for most mineral soils usually vary between the narrow limits of 2.60–2.75 Mg/m^3 because quartz, feldspar, micas, and the colloidal silicates that usually make up the major portion of mineral soils all have densities within this range. When unusual amounts of minerals with high particle density (such as magnetite, garnet, epidote, zircon, tourmaline, and hornblende) are present, the particle density may exceed 3.0 Mg/m^3.

Organic matter weighs much less than an equal volume of mineral solids, having a particle density of 0.9–1.3 Mg/m^3. Consequently, the amount of this constituent in a soil markedly affects the particle density. For example, mineral surface soils, which almost always have higher organic matter content than the subsoils, usually possess lower particle densities than do subsoils. Some mineral topsoils high in organic matter (15–20%) may have particle densities as low as 2.4 Mg/m^3 or even lower. Typically, the particle density of organic soils (Histosols) is between 1.3 and 2.0 Mg/m^3. Nevertheless, for general calculations, the average arable mineral surface soil (3–5% organic matter) may be considered to have a particle density of about 2.65 Mg/m^3.

4.8   BULK DENSITY OF MINERAL SOILS

A second important mass (weight) measurement of soils is **bulk density** D_b, which is defined as the mass (weight) of a unit volume of dry soil. This volume *includes both solids and pores*. The comparative calculations of bulk density and particle density are shown in Figure 4.13. A careful study of this figure should make clear the distinction between these two methods of expressing soil mass.

---

^{3}Since 1 Mg = 1 million grams and 1 m^3 = 1 million cubic centimeters, this particle density is equivalent to 2.6 g/cm^3.

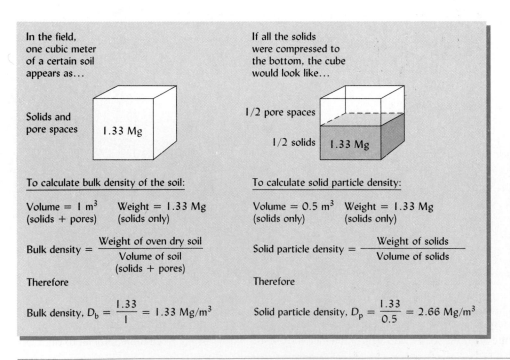

In the field, one cubic meter of a certain soil appears as...

Solids and pore spaces   1.33 Mg

If all the solids were compressed to the bottom, the cube would look like...

1/2 pore spaces
1/2 solids   1.33 Mg

To calculate bulk density of the soil:

Volume = 1 m^3 (solids + pores)   Weight = 1.33 Mg (solids only)

$$\text{Bulk density} = \frac{\text{Weight of oven dry soil}}{\text{Volume of soil (solids + pores)}}$$

Therefore

$$\text{Bulk density, } D_b = \frac{1.33}{1} = 1.33 \text{ Mg/m}^3$$

To calculate solid particle density:

Volume = 0.5 m^3 (solids only)   Weight = 1.33 Mg (solids only)

$$\text{Solid particle density} = \frac{\text{Weight of solids}}{\text{Volume of solids}}$$

Therefore

$$\text{Solid particle density, } D_p = \frac{1.33}{0.5} = 2.66 \text{ Mg/m}^3$$

FIGURE 4.13   Bulk density, D_b, and particle density, D_p, of soil. Bulk density is the weight of the solid particles in a standard volume of field soil (solids plus pore space occupied by air and water). Particle density is the weight of solid particles in a standard volume of those solid particles. Follow the calculations through carefully and the terminology should be clear. In this particular case the bulk density is one-half the particle density, and the percent pore space is 50.

There are several practical methods[4] of determining soil bulk density, all of which involve obtaining a known volume of soil, drying it to remove the water, and weighing the dry mass. The volume of soil may be determined using a special coring instrument (Figure 4.14) to obtain a sample of known volume without disturbing the natural soil structure. For surface soils, perhaps the simplest method is to dig a small hole, collecting all the excavated soil, and then line the hole with plastic film and fill it completely with a measured volume of water. It should be noted that both types of density calculations refer only to the mass of the *solids* in a soil, and therefore any water present is excluded from consideration.

Factors Affecting Bulk Density

Unlike particle density, which is a characteristic of solid particles only, bulk density relates to the combined volumes of the pore and solid spaces. Soils with a high proportion of pore space to solids have lower bulk densities than those that are more compact and have less pore space. Consequently, any factor that influences soil pore space will affect bulk density.

Fine-textured surface soils such as silt loams, clays, and clay loams generally have lower bulk densities than sandy soils.[5] The solid particles of the fine-textured soils tend to be organized in porous granules, especially if adequate organic matter is present. Thus, in these soils pores exist both between and within the granules. This condition assures high total pore space and a low bulk density. In sandy soil, however, organic matter contents generally are low, the solid particles less likely to be aggregated together, and the bulk densities are commonly higher than in the finer-textured soils. Figure 4.15 illustrates the concept that sandy soils have amounts of large pores similar to those in well-aggregated fine-textured soils but have few fine, within-ped pores, and so have less total porosity.

The bulk densities of clay, clay loam, and silt loam surface soils normally range from 1.00 Mg/m^3 to as high as 1.55 Mg/m^3, depending on their condition. A variation of from $1.20–1.75 \text{ Mg/m}^3$ may be found in sands and sandy loams. Very compact subsoils may have bulk densities of 2.0 Mg/m^3 or even greater. In such compact layers there are essentially no macropores. Root growth is greatly impaired, the constraint becoming most noticeable at bulk densities of 1.6 Mg/m^3 or above. Typical ranges of bulk density for various soil materials and conditions are illustrated in Figure 4.16.

Even in soils of the same surface textural class, significant differences in bulk density are found, as is shown by data on soils from several locations in Table 4.5. Moreover, the bulk density generally is higher in lower-profile layers. This increase apparently results

TABLE 4.5 Bulk Density (D_b) Data for Several Soil Profiles

Note that sandy soils generally have higher bulk densities than those high in silt and clay. The Erie soil has a very dense subsoil, whereas the Oxisol from Brazil is loose and open.

Horizon	Grandfield fine sandy loam (Oklahoma)	Miami silt loam (Wisconsin)	Erie channery silt loam (New York)	Houston clay (Texas)	Oxisol clay (Brazil)
Plow layer	1.72	1.28	1.33	1.24	0.95
Upper subsoil	1.74	1.41	1.46	1.36	—
Lower subsoil	1.80	1.43	2.02	1.51	1.00
Parent material	1.85	1.49	—	1.61	—

Data for Grandfield soil from Dawud and Gray (1979), Miami from Nelson and Muckenhirn (1941), Houston from Yule and Ritchie (1980). Erie calculated from Fritton and Olson (1972), Oxisol from Larson et al. (1980).

---

[4]An instrumental method that measures the soil's resistance to penetration by gamma rays is used in soil research, but is not discussed here.

[5]This fact may seem counterintuitive at first because sandy soils are commonly referred to as *light* soils, while clays and clay loams are referred to as *heavy* soils. The terms *heavy* and *light* in this context refer not to the weight per unit volume of the soils, but to the amount of effort that must be exerted to manipulate these soils with tillage implements—the sticky clays being much more difficult to till.

(a)

(b)

FIGURE 4.14 A special sampler designed to remove a cylindrical core of soil without causing disturbance or compaction (a). The sampler head contains an inner cylinder and is driven into the soil with blows from a drop hammer. The inner cylinder (b) containing an undisturbed soil core is then removed and trimmed on the end with a knife to yield a core whose volume can easily be calculated from its length and diameter. The weight of this soil core is then determined after drying in an oven. (Photo courtesy of R. Weil)

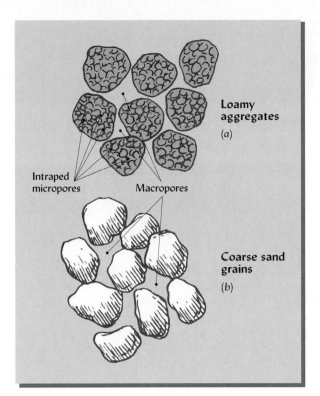

Loamy
aggregates
(a)

Intraped
micropores   Macropores

Coarse sand
grains
(b)

FIGURE 4.15 A schematic comparison of sandy and clayey soils showing the relative amounts of large (macro-) pores and small (micro-) pores in each. There is less total pore space in the sandy soils than in the clayey one because the clayey soil contains a large number of fine pores within each aggregate (a), but the sand particles (b), while similar in size to the clayey aggregates, are solid and contain no pore spaces within them. This is the reason why, among surface soils, those with coarse texture are usually more dense than those with finer textures.

from a lower content of organic matter, less aggregation and root penetration, and compaction caused by the weight of the overlying layers.

The system of land management employed on a given soil also influences its bulk density. While comparisons of bulk densities among different soils are difficult to interpret, changes in bulk density for a given soil are easily measured and can alert soil managers to changes in soil quality and ecosystem function. Increases in bulk density usually indicate a poorer environment for root growth, as well as generally undesirable changes in hydrologic function. Figure 4.17 illustrates the increased bulk density, especially in the surface layers, that results from the removal of forest trees by clear-cutting. Intensive recreational use of forests can also lead to increased bulk densities in areas with heavy foot traffic, such as around campsites and on trails (Figure 4.18).

Increased bulk densities are often indicative of reduced infiltration of rainwater and restricted root growth. For cropland, the addition of crop residues or farm manure in large amounts tends to lower bulk densities of surface soils, as does a grass sod. Intensive cultivation operates in the opposite direction, as shown in Table 4.6. These data are from long-term studies in different locations, the soils having been under cultivation for from 20 to 90 years. In all cases, cropping increased the bulk density of the topsoils.

TABLE 4.6 Bulk Density and Pore Space of Some Surface Soils from Cultivated and Nearby Uncultivated Areas (One Subsoil Included)

The bulk density was increased, and the pore space proportionately decreased, in every case.

Soil	Texture	Years cropped	Bulk density (Mg/m³)		Pore space (%)	
			Cultivated soil	Uncultivated soil	Cultivated soil	Uncultivated soil
Udalf (Pennsylvania)	Loam	58	1.25	1.07	50	57.2
Udoll (Iowa)	Silt loam	50+	1.13	0.93	56.2	62.7
Aqualf (Ohio)	Silt loam	40	1.31	1.05	50.5	60.3
Boroll (Canada)	Silt loam	90	1.30	1.04	50.9	60.8
Orthid (Canada)	Clay	70	1.28	0.98	51.7	63.0
Orthid, subsoil (Canada)	Clay	70	1.38	1.21	47.9	54.3
Mean of 3 Ustalfs (Zimbabwe)	Clay	20–50	1.44	1.20	54.1	62.6
Mean of 3 Ustalfs (Zimbabwe)	Sandy loam	20–50	1.54	1.43	42.9	47.2

Data for Canadian soils from Tiessen et al. (1982), for Zimbabwe soil from Weil (unpublished), and for other soils from Lyon et al. (1952).

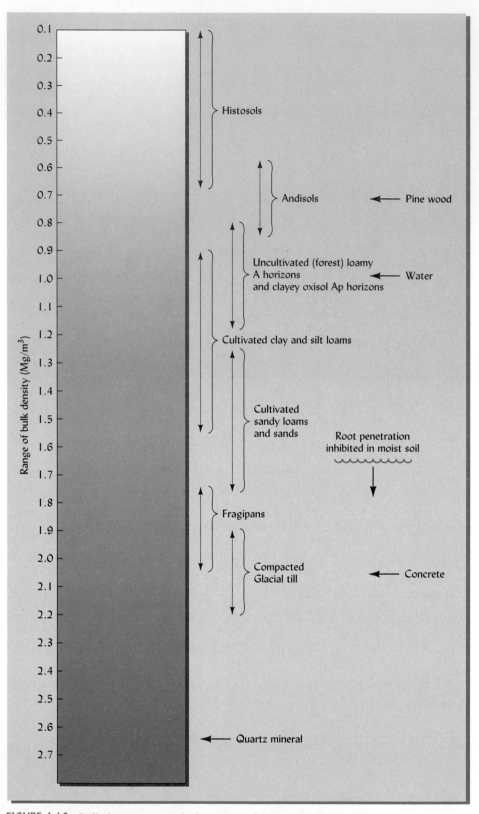

FIGURE 4.16 Bulk densities typical of a variety of soils and soil materials.

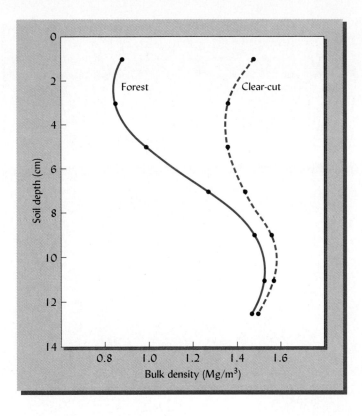

FIGURE 4.17 Effect of cutting trees (clear-cut) on the bulk density of the soil at different depths. [From McIntyre et al. (1987)]

Other Mass (Weight) Figures

For engineers involved with moving soil for construction purposes, or for landscapers bringing in topsoil by the truckload, a knowledge of the bulk density of various soils is useful in estimating the weight of soil to be moved. For this purpose, units of kg/m^3 in the metric system or lb/yd^3 and lb/ft^3 in the English system are most useful. A typical medium-textured mineral soil might have a bulk density of 1.25 Mg/m^3, which would convert to 1250 kg/m^3 or 2108 lb/yd^3. In practical terms, this information would tell a landscaper that his or her half-ton (1000-lb) load capacity pickup truck could carry less than ½ yd^3 of this soil (even though the truck could hold about six times this volume of a lightweight material).

The design of rooftop gardens is another practical application of these weight figures. In this instance, the weight of soil involved must be known to design a structure of sufficient strength to carry the load.[6]

Another figure of interest is the mass of soil in one hectare to a depth of normal plowing (15 cm). Such a **hectare–furrow slice** of a typical surface soil weighs about 2.2 million kilograms (kg). A comparable figure in the English system is 2 million pounds per **acre–furrow slice** 6.5 inches deep. These estimates of the weight of surface soil in an acre or hectare of land are very useful in calculating lime and fertilizer application rates and organic matter mineralization rates (see Chapters 9 and 12 for detailed examples).

4.9   PORE SPACE OF MINERAL SOILS

For the growth of plant roots and the movement of air, water, and solutes in soils, the volume and configuration of spaces between the solid particles are of critical importance. One of the main reasons for measuring soil bulk densities is that this value can be used to calculate pore space. The pore space is inversely related to the bulk density; for soils with the same particle density, the higher the bulk density, the lower the percent pore space (see Box 4.1 for derivation of the formula expressing this relationship).

---

[6]To convert values of bulk density given in units of Mg/m^3 into values in units of lb/yd^3, multiply by 1686. This conversion factor is derived as follows: $(1 \ Mg/m^3) \times (1000 \ kg/Mg) \times (2.205 \ lb/kg) \times (1 \ m^3 \div 1.308 \ yd^3) = 1686 \ lb/yd^3$.

FIGURE 4.18 Impact of campers on the bulk density of forest soils, and the consequent effects on rain infiltration rates and runoff losses (see white arrows). At most campsites the high-impact area extends for about 10 m from the fire circle or tent pad. Managers of recreational land must carefully consider how to protect sensitive soils from compaction that may lead to death of vegetation and increased erosion. [Data from Vimmerschmidt et al. (1982)].

The *pore space* of a soil is that portion of the soil volume occupied by air and water. The amount of pore space is determined largely by the arrangement of the solid particles. If the particles lie close together, as in compact subsoils, the total porosity is low. If they are arranged in porous aggregates, as is often the case in medium- to fine-textured soils high in organic matter, the pore space per unit volume will be high.

Factors Influencing Total Pore Space

The total pore space varies widely among soils. Sandy surface soils show a range of 35–50%, whereas medium- to fine-textured soils vary from 40–60%—even more in cases of high organic matter content and marked granulation (see Table 4.6 and Figure 4.19). In Chapter 1 (Figure 1.15) we noted that for a medium-textured, well-granulated surface soil in good condition for plant growth, approximately 50% of the soil volume consists of pore space, and that pore space is about half filled with air (mainly in the macropores) and half filled with water (mainly in the micropores). Pore space also varies with depth. Some compact subsoils have as little as 25–30%, accounting, in part, for the inadequate aeration and resistance to root penetration of such horizons. Figure 4.19 shows that the decrease in organic matter and increase in clay that occur with depth in many profiles are associated with a shift from macropores to micropores.

BOX 4.1 CALCULATION OF PERCENT PORE SPACE IN SOILS

The bulk density of a soil can be easily and inexpensively measured. Particle density is somewhat more difficult to measure, but can usually be assumed to be approximately the same as the particle density of the dominant soil minerals—2.65 Mg/m³ for most silicate-dominated mineral soils. On the other hand, direct measurement of the pore space in soil requires the use of much more tedious and expensive techniques. Therefore, when information on the percent pore space is needed, it is often desirable to calculate the pore space from data on bulk and particle densities.

The derivation of the formula used to calculate the percentage of total pore space in soil follows:

$$
\begin{aligned}
\text{Let} \quad D_b &= \text{bulk density} \\
D_p &= \text{particle density} \\
W_s &= \text{Weight of soil (solids)} \\
V_s &= \text{volume of solids} \\
V_p &= \text{volume of pores} \\
V_s + V_p &= \text{total soil volume } V_t
\end{aligned}
$$

By definition,

$$
\frac{W_s}{V_s} = D_p \qquad \text{and} \qquad \frac{W_s}{V_s + V_p} = D_b
$$

Solving for W_s gives

$$
W_s = D_p \times V_s \qquad \text{and} \qquad W_s = D_b(V_s + V_p)
$$

Therefore

$$
D_p \times V_s = D_b(V_s + V_p)
$$

and

$$
\frac{V_s}{V_s + V_p} = \frac{D_b}{D_p}
$$

Since

$$
\frac{V_s}{V_s + V_p} \times 100 = \% \text{ solid space}
$$

then

$$
\% \text{ solid space} = \frac{D_b}{D_p} \times 100
$$

Since % pore space + % solid space = 100, and % pore space = 100 − % solid space, then

$$
\% \text{ pore space} = 100 - \left(\frac{D_b}{D_p} \times 100 \right)
$$

EXAMPLE

As an example, let us consider the cultivated clay subsoil from Canada in Table 4.6 (the Orthid). The bulk density was determined to be 1.28 Mg/m³. Since we have no information on the particle density, we assume that the particle density is approximately that of the common silicate minerals (i.e., 2.65 Mg/m³). We calculate the percent pore space using the formula derived above:

$$
\% \text{ pore space} = 100 - \left(\frac{1.28}{2.65} \times 100 \right)
$$

$$
\% \text{ pore space} = 100 - (0.483 \times 100)
$$

$$
= 100 - 48.3
$$

$$
= 51.7
$$

(continued)

BOX 4.1 (*Cont.*)   CALCULATION OF PERCENT
PORE SPACE IN SOILS

This value of pore space, 51.7%, is quite close to the typical percentage of air and water space described in Figure 1.15 for a well-granulated, medium- to fine-textured soil in good condition for plant growth. This simple calculation tells us nothing about the relative amounts of large and small pores, however, and so must be interpreted with caution.

For certain soils it is inaccurate to assume that the soil particle density approximates that of common silicate minerals (i.e., 2.65 Mg/m³). For instance, organic matter has a particle density very much lower than that of quartz, and so a soil with a high organic matter content can be expected to have a particle density somewhat lower than 2.65. Similarly, a soil rich in iron oxide minerals will have a particle density greater than 2.65 because these minerals have particle densities as high as 3.5. As an example of the latter type of soil, let us consider the uncultivated clay soils from Zimbabwe (Ustalfs) described in Table 4.6. These are red-colored clays, high in iron oxides. The particle density for these soils was determined to be 3.21 Mg/m³ (not shown in Table 4.6). Using this value and the bulk density value from Table 4.6, we calculate the pore space as follows:

$$\% \text{ pore space} = 100 - \left(\frac{1.20}{3.21} \times 100 \right)$$

$$\% \text{ pore space} = 100 - (0.374 \times 100)$$

$$= 100 - 37.4$$

$$= 62.6$$

Such a high percentage pore space is an indication that this soil is in an uncompacted, very well granulated condition typical for soils found under undisturbed natural vegetation.

As is the case for bulk density, agricultural management can exert a decided influence on the pore space of soils (Table 4.6). Data from a wide range of soils show that cultivation tends to lower the total pore space compared to that of uncultivated soils. This reduction usually is associated with a decrease in organic matter content and a consequent lowering of granulation. Pore space in the subsoil also has been found to decrease with cropping, although to a lesser degree.

Size of Pores

Two types of individual pore spaces—*macro* and *micro*—occur in soils. Although there is no clear-cut demarcation, pores less than about 0.06 mm in diameter are considered as micropores and those larger as macropores. In many soils the micropores are primarily found within structural peds (intraped), while most of the macropores occur between the peds (interped). Thus, soil structure as well as texture influences the balance between macro- and micropores in a soil (Figures 4.15 and 4.19).

The macropores characteristically allow the ready movement of air and percolating water. In contrast, micropores are mostly filled with water in a moist field soil and do not permit much air movement into or out of the soil. Water movement also is slow. Thus, even though a sandy soil has relatively low total porosity, the movement of air and water through such a soil is surprisingly rapid because of the dominance of the macropores.

Fine-textured soils, especially those without a stable granular structure, allow relatively slow gas and water movement, despite the unusually large volume of total pore space. Here the dominating micropores often remain full of water. Aeration, especially in the subsoil, often is inadequate for satisfactory root development and desirable microbial activity. Clearly, the size of individual pore spaces rather than their combined volume is the important factor determining soil drainage and aeration. The loosening and granulating of fine-textured soils promotes aeration, not so much by increasing the total pore space as by raising the proportion of the macrospaces.

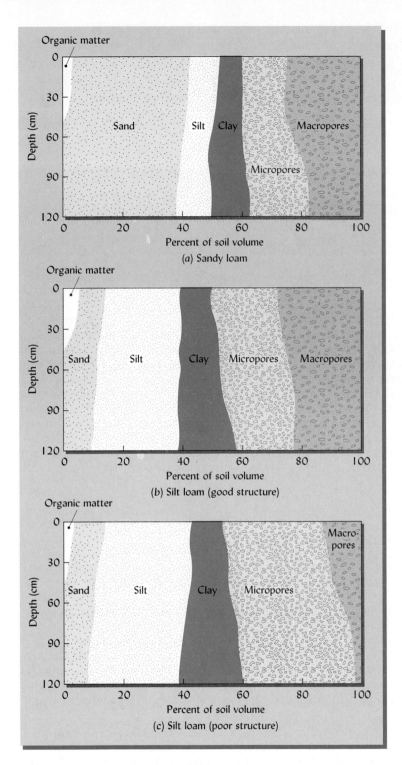

FIGURE 4.19 Volume distribution of organic matter, sand, silt, clay, and pores of macro- and microsizes in a representative sandy loam (a) and in two representative silt loams, one with good soil structure (b) and the other with poor soil structure (c). Both silt loam soils have more total pore space than the sandy loam, but the silt loam with poor structure has a smaller volume of larger (macro-) pores than either of the other two soils. Note that at the lowest depths in both silt loams, about one-third of the mineral matter is clay, giving the lower horizons enough clay to be classified as silty clay loams.

Cultivation and Pore Size

Continuous cropping, particularly of soils originally high in organic matter, often results in a reduction of macropore spaces. Data from a fine-textured soil in Texas (Table 4.7) clearly illustrate this effect. Cropping with plow tillage significantly reduced soil organic matter content and total pore space. But most striking is the effect of cropping on the size of the soil pores: The amount of macropore space necessary for ready air movement was reduced by about one-half. This reduction in pore size extended so deep into the soil profile that samples taken from as deep as 107 cm showed the same trend.

TABLE 4.7   Effect of About 50 Years of Continuous Cropping on the Macropore and Micropore Spaces of a Fine-Textured Ustert (Houston Black Clay) in Texas.

Compared to the undisturbed prairie (virgin soil), the cultivated soil has far less macropore space, but has gained some micropore space as aggregates were destroyed, changing large interped pores into much smaller micropores. The loss of macropore space probably resulted from the loss of organic matter that is evident.

Soil history	Organic matter (%)	Total pore space (%)	Macropore space (%)	Micropore space (%)	Bulk density (Mg/cm³)
		0–15 cm depth			
Virgin prairie	5.6	58.3	32.7	25.6	1.11
Tilled for 50 years	2.9	50.2	16.0	34.2	1.33
		15–30 cm depth			
Virgin prairie	4.2	56.1	27.0	29.1	1.16
Tilled 50 years	2.8	50.7	14.7	36.0	1.31

Data from Laws and Evans (1949).

In recent years, **conservation tillage** practices, which minimize plowing and associated soil manipulations, have been widely adopted in the United States (Sections 4.13 and 17.12). Such practices keep crop residues on the soil surface and are remarkably effective in reducing soil erosion losses. Because of increased accumulation of organic matter near the soil surface and the development of a long-lived network of macropores, some conservation tillage systems lead to greater macroporosity and lower bulk density in the surface soil. These benefits are particularly likely to accrue in soils with extensive production of earthworm burrows, which may remain undisturbed in the absence of tillage. Unfortunately, such improvements in porosity do not always occur. Under some circumstances, less pore space has been found with conservation tillage than with conventional tillage, a matter of some concern in soils with poor internal drainage.

4.10   AGGREGATION AND ITS PROMOTION IN ARABLE SOILS

In a practical sense, two sets of factors are of concern in dealing with soil aggregation: (1) those responsible for aggregate formation and (2) those that give the aggregates stability once they are formed. Since both sets of factors are operating simultaneously, it is sometimes difficult to distinguish their relative effects on the development of stable aggregates in soils.

GENESIS OF GRANULES AND CRUMBS.   Several specific factors influence the genesis of granules and crumbs, which are of great importance in surface soils. These factors include (1) physical processes that encourage particle contact, (2) the influence of decaying organic matter, (3) the modifying effects of adsorbed cations, and (4) soil tillage.

Physical Processes

Any action that will shift the particles back and forth and force contacts between particles will encourage aggregation. Alternate wetting and drying, freezing and thawing, the physical effects of root extension, and the mixing action of soil organisms and of tillage implements encourage such contacts, and therefore stimulate aggregate formation.

Farmers have long recognized that large, hard clods of soil will break down into a "mellow" seedbed if left to overwinter under conditions in which they will alternately freeze and thaw or become dry and wet (under gentle rains). In both cultivated and uncultivated soils, animals such as earthworms and various minute soil insects move soil particles around, often ingesting them and forming them into fecal pellets (see Chapter 11). Plant roots not only move soil particles about while pushing their way through the soil, they also have a distinct drying effect as they take up soil moisture in their immediate vicinity.

Influence of Organic Matter

Organic matter is the major agent stimulating the formation and stabilization of granular and crumb-type aggregates (see Figure 4.20). As organic residues decompose, gels and other viscous microbial products, along with associated bacteria and fungi, encourage crumb formation. Organic exudates from plant roots also participate in this aggregating action. Organic compounds such as polysaccharides then chemically interact with particles of silicate clays and iron and aluminum oxides. These compounds orient the clays in a common plane and then form bridges between individual soil particles, thereby binding them together in water-stable aggregates.

The overall aggregating influence of organic materials has long been known. However, only since scientists have been able to use ultrathin sectioning have they been able to produce photographs that show direct evidence of the organo-mineral bridges that bind soil particles (see Figure 4.21). The complexity of organic materials found in humus makes this binding process possible.

Effect of Adsorbed Cations

Most types of clay particles possess electronegative charges, which are balanced to some degree by positive-charged ions attracted to the clay surface from the soil solution (see Section 8.10). Aggregate formation is definitely influenced by the nature of the cations adsorbed by soil colloids. For instance, when Na^+ is a prominent adsorbed ion, as in some soils of arid and semiarid areas, the loosely held Na^+ ions do not effectively reduce the electronegativity of the soil colloids. The still negatively charged colloids therefore continue to repel each other, causing a dispersed condition. In this dispersed condition the soil particles do not come together to form structural aggregates. Such a relatively structureless, dispersed condition causes the soil to be almost impervious to water and air, and is very undesirable from the standpoint of plant growth. (See Section 10.3.)

By contrast, the adsorption of ions such as Ca^{2+}, Mg^{2+}, or Al^{3+} may encourage aggregate formation, starting with a process called *flocculation.* By forming electropositive links between electronegative particles, these ions encourage the individual colloidal particles to come together in small aggregates called *floccules* (see Figure 4.22). Flocculation itself, however, is only the first step in the process because it alone does not provide for the *stabilization* of the aggregates.

Influence of Tillage and Compaction

Tillage has both favorable and unfavorable effects on aggregation. If the soil is not too wet or too dry when the tillage is performed, the short-term effect of tillage is generally favorable. Tillage implements break up clods, incorporate organic matter into the soil, kill weeds, and create a more favorable **seed bed.** (See Box 4.2.) Modern advances in herbicides and planting equipment allow good weed control and seedling establishment

FIGURE 4.20 Puddled soil (left) and well-granulated soil (right). Plant roots and especially humus play the major role in soil granulation. Thus a sod tends to encourage development of a granular structure in the surface horizon of cultivated land. (Courtesy USDA Natural Resources Conservation Service)

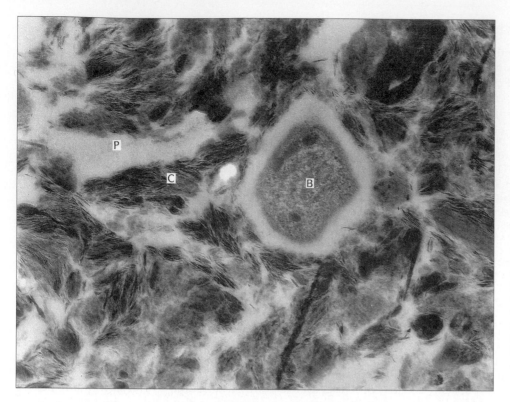

FIGURE 4.21 An ultrathin section illustrating the interaction among organic materials and silicate clays in a water-stable aggregate. The dark-colored materials (C) are groups of clay particles that are interacting with organic polysaccharides (P). Note the generally horizontal orientation of the clay particles, an orientation encouraged by the organic materials. [From Emerson et al. (1986); photograph provided by R. C. Foster, CSIRO, Glen Osmond, Australia])

without tillage, but many soil managers still consider tillage to be normal and necessary for the agricultural use of many soils.

Over longer periods, tillage operations have detrimental effects on surface soil structure. In the first place, by mixing and stirring the soil, tillage greatly hastens the oxidation of soil organic matter, thus reducing the positive effects of this soil component on soil structure. Second, tillage operations, especially those involving heavy equipment, tend to crush or break stable soil aggregates. Compaction occurs from repeatedly driving heavy farm equipment over fields. The greatest damage is done when the soil is relatively moist. An indication of the effect of such traffic on bulk density of soil is given

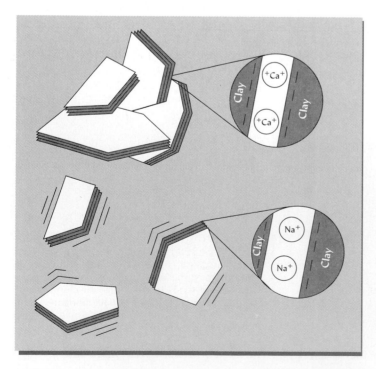

FIGURE 4.22 The role of cations in the flocculation of soil clays. The di- and trivalent cations, such as Ca^{2+} and Al^{3+}, are tightly adsorbed and can effectively neutralize the negative surface charge on clay particles. These cations can also form bridges that bring clay particles close together. Monovalent ions, especially Na^+, with relatively large hydrated radii, can cause clay particles to repel each other and create a dispersed condition. Two things contribute to the dispersion: (1) the large hydrated sodium ion does not get close enough to the clay to effectively neutralize the negative charges, and (2) the single charge on sodium is not effective in forming a bridge between clay particles.

BOX 4.2 PREPARING A GOOD SEEDBED

Early in the growing season, one of the main activities of a farmer or gardener is the preparation of a good seedbed. A good seedbed will help assure that the sowing operation goes smoothly, and that the plants come up quickly, evenly, and well spaced.

A good seedbed consists of soil loose enough to allow easy elongation of the young roots and emergence of the seedling (see photo a). At the same time, a seedbed should be packed firmly enough to ensure good contact between the seed and moist soil so that the seed can easily imbibe water to begin the germination process. The seedbed should also be relatively free of large clods. Seeds could fall between such clods and become lodged too deeply for proper emergence, and without sufficient contact with moist soil.

Tillage may be desirable to counteract the effects of previous compaction. Tillage may also help the soil dry out and warm up more rapidly, an important consideration in cool climates. Finally, tillage may help free the seedbed of growing weeds so as to give the crop a head start in its competition with these unwanted plants. On the other hand, the objectives of seedbed preparation may be achieved with little or no tillage if soil and climatic conditions are favorable and a mulch or herbicide is used to control early weeds.

(a) A bean seed emerges from a seedbed. (Photo courtesy R. Weil)

Mechanical, tractor-drawn planters can assist in maintaining a good seedbed. Most planters are equipped with **coulters** (sharp steel disks) designed to cut a path through plant residues on the soil surface so that these materials will not pile up and clog the machine as it is pulled through the soil. No-till planters usually follow the coulter with a pair of sharp cutting wheels called a *double disk opener* that opens a groove in the soil into which the seeds can be dropped (see photos b and c). Most planters also have a press wheel that follows behind the seed dropper

(continued)

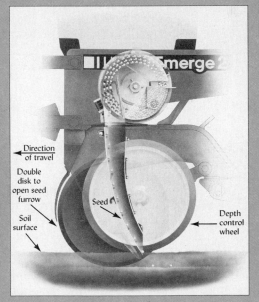

(b) A no-till planter in action and (c) a diagram showing how it works. (Photo b and diagram c courtesy of Deere & Company, Moline, Ill.)

BOX 4.2 (Cont.)   PREPARING A GOOD SEEDBED

and packs the loosened soil just enough to ensure that the groove is closed and the seed is pressed into contact with moist soil.

Ideally, only a narrow strip in the seed row is packed down to create a seed *germination* zone, while the soil between the crop rows is left as loose as possible to provide a good *rooting zone*. The surface of the interrow rooting zone may be left in a rough condition to encourage water infiltration and discourage erosion.

The above principles also apply to the home gardener who may be sowing seeds by hand. It is often not necessary to extensively till the whole garden, but rather, only a narrow strip for seed placement need be stirred. In order to assure proper seed placement, soil may need to be raked to break up clods where small seeded crops are to be sown. After ensuring that the seed is covered with the proper depth of soil (generally the cover soil should be three times as deep as the diameter of the seed), the loose soil should be pressed down gently, just as the press wheel on a mechanical planter would do.

In humid regions, excessively wet soil in early spring is problematic for both farmers and gardeners. Gardeners sowing seed manually can get their crops off to an early start by a simple technique that allows planting while the soil is still wet, and yet avoids causing undue soil compaction. Two boards (e.g., $15 \times 150 \times 2$ cm thick) are placed alongside the intended crop row and used as a walkway while opening the seed furrow and sowing the seed. The idea is simply to spread the weight of the gardener over such a large area that the soil is subjected to very little compactive force. The technique is also useful to avoid recompacting a freshly tilled garden (see photo d).

(d) Preparing a seedbed in a garden using a board to avoid compacting the soil. (Photo courtesy R. Weil)

in Figure 4.23. These data explain the increased interest in farming systems that drastically reduce the number of necessary tillage operations (Sections 4.13 and 17.2).

4.11   AGGREGATE STABILITY

The stability of aggregates is of great practical importance. Some aggregates readily succumb to the beating of rain and the rough and tumble of plowing and tilling the land. Others resist disintegration, thus making the maintenance of a suitable soil structure comparatively easy (Figures 4.24 and 4.25). Three major factors appear to influence aggregate stability:

1. The temporary mechanical binding action of microorganisms, especially the threadlike **hyphae** of fungi that enmesh soil aggregates. Cablelike bundles of hyphae are called ***mycelia*** and can often be seen with the unaided eye. These effects are pronounced when fresh organic matter is added to soils and are at a maximum a few weeks or months after this application. Filamentous fungi living in association with plant roots (called *mycorrhizae;* see Chapter 11) are especially

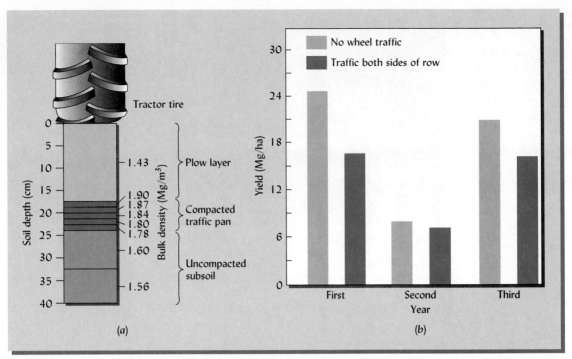

FIGURE 4.23 Heavy equipment used for tillage and other purposes compacts the soil, increases bulk density, and reduces crop yields. (a) Such equipment compacted the zone just below the plow layer on an Udult (Norfolk) soil and increased bulk density to more than 1.8 Mg/m³, the limit of penetration for cotton roots. (b) The yield of potatoes was reduced markedly for two out of three years in this test on a clay loam soil in Minnesota. No yield decrease in the second year, which was very dry. [Data from Camp and Lund (1964) and Voorhees (1984)]

important in forming the large (>2 mm) stable aggregates that characterize many uncultivated grassland soils (Figure 4.26).

2. The cementing action of the intermediate products of microbial synthesis and decay, such as microbially produced gums and certain polysaccharides. Bacterial polysaccharides are shown intermixed with clay in Figure 4.21. Root hairs and fungal hyphae also produce polysaccharides that help bind soil particles.

3. The cementing action of the more resistant stable humus components, aided by similar action of certain inorganic compounds such as iron oxides. These materials provide most of the long-term aggregate stability for smaller (<0.25 mm) aggregates.

It should be emphasized that aggregate stability is not entirely an organic phenomenon. Organic and inorganic components interact continually. Polyvalent inorganic cations that cause flocculation (e.g., Ca^{2+}, Mg^{2+}, Fe^{2+}, and Al^{3+}) are also thought to provide mutual attraction between the organic matter and soil clays, encouraging the development of clay–organic matter complexes. In addition, films of clay called *clay skins* often surround the soil peds and help provide stability. The largest granules or crumbs (often 2 to 5 mm in diameter) are usually associated with the short-lived organic

FIGURE 4.24 The aggregates of soils high in organic matter are much more stable than are those low in this constituent. The low organic matter soil aggregates fall apart when they are wetted; those high in organic matter maintain their stability

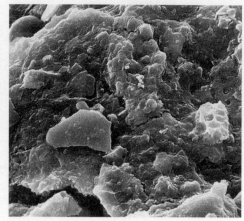

FIGURE 4.25 Scanning electron micrographs of aggregates from two topsoils, one (left) with low aggregate stability, the other (right) with high aggregate stability. Note that individual particles in the low-stability aggregate are not well associated, whereas those in the stable aggregate are bound together. (Photos from Craig Ross, New Zealand Soil Bureau; used with permission)

stabilizing compounds and are the most easily broken down when soil organic matter decreases. Many clayey red and yellow soils of humid tropical and semitropical areas have very stable small aggregates sometimes called *pseudo sand*. The stability of these aggregates is due largely to the hydrated oxides of iron they contain.

Soil Crusting

Aggregates exposed at the soil surface are most vulnerable to destruction, especially by heavy rains. Once these aggregates become dispersed, small particles tend to wash into and clog the soil pores. The result is a soil surface that is sealed over and prevents water infiltration. When the soil dries, a hard crust is formed. Seedlings, if they emerge at all, can do so only through cracks in the crust. A crust-forming soil is compared to one with stable aggregates in Figure 4.27. Formation of a crust soon after a crop is sown may allow so few seeds to emerge that the crop has to be replanted. In arid and semiarid regions soil crusting and sealing can have disastrous consequences because high runoff losses leave little water available to support plant growth.

Crust formation can be minimized by keeping some vegetative or mulch cover on the land to reduce the impact of raindrops. Also a light tillage (as with a rotary hoe) while the soil is still moist will break up the crust before it hardens.

Soil Conditioners

Certain synthetic chemicals (polymers) can stabilize soil structure in much the same way as do natural organic polymers such as polysaccharides and polyuronides. Synthetic polymers were commercially introduced with great enthusiasm in the early 1950s, since they had remarkable ability to stabilize aggregates. Unfortunately, their high cost in relation to their benefits made them uneconomical, and their use was essentially abandoned.

Recently, however, new synthetic organic compounds have been developed that are effective in helping to stabilize aggregates when applied at rates as low as 5 to 20 μg/L of irrigation water or sprayed on at rates as low as 10 to 20 kg/ha. These low rates substantially reduce materials costs and may well ensure continued interest in artificial soil conditioners. Figure 4.28 shows the stabilizing effect of one of these synthetic polyacrylamides.

Several species of algae that live near the soil surface are known to produce quite effective aggregate-stabilizing compounds. Application of small quantities of commercial preparations containing such algae may bring about a significant improvement in surface soil structure. The amount of amendment required is very small because the algae, once established in the soil, can multiply.

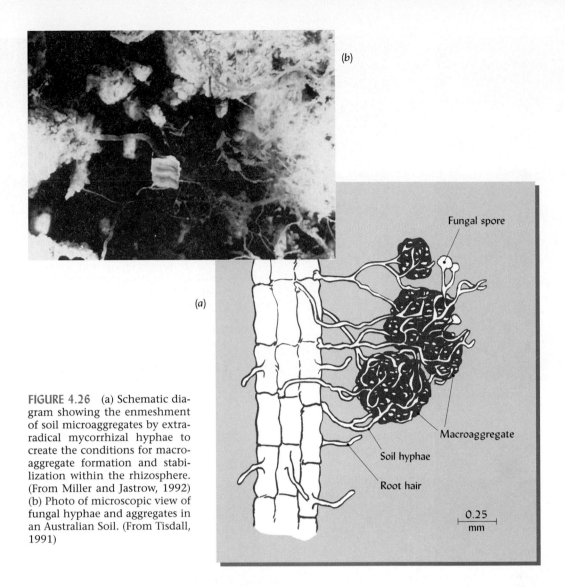

FIGURE 4.26 (a) Schematic diagram showing the enmeshment of soil microaggregates by extraradical mycorrhizal hyphae to create the conditions for macroaggregate formation and stabilization within the rhizosphere. (From Miller and Jastrow, 1992) (b) Photo of microscopic view of fungal hyphae and aggregates in an Australian Soil. (From Tisdall, 1991)

4.12 STRUCTURAL MANAGEMENT OF SOILS

When protected under dense vegetation and undisturbed by tillage, most soils (except perhaps some sparsely vegetated soil in arid regions) possess a surface structure sufficiently stable to allow rapid infiltration of water and to prevent crusting. However, for the manager of cultivated soils, the development and maintenance of stable surface soil structure is a major challenge.

While any tillage tends to hasten the loss of organic stabilizers, tillage when a soil is too wet mechanically breaks down aggregates so that the individual particles tend to act independently, filling in the macropores and creating a *puddled* condition. In this condition the soil is nearly impervious to air and water. Wet soil is also more susceptible to compaction and loss of structure caused by walking or driving on the soil.

Clayey soils are especially prone to puddling and compaction because of their high plasticity and cohesion. When these soils dry, they usually become dense and hard. On the other hand, if a clayey soil is tilled when too dry, large hard clods are turned up, which are difficult to work into a good seedbed. Tillage must be timed carefully with respect to soil moisture content. Proper timing is more difficult for clayey than for sandy soils, because the former take much longer to dry to a suitable moisture content and may also become too dry to work easily.

Some clay soils of humid tropical regions are much more easily managed than those just described. The clay fraction of these soils is dominated by hydrous oxides of iron and aluminum, which are not as sticky, plastic, and difficult to work. These soils may have very favorable physical properties, since they hold large amounts of water but have such stable aggregates that they respond to tillage after rainfall much like sandy soils.

(c)

FIGURE 4.27 Scanning electron micrographs of soil aggregated in the upper 1 mm of a soil with stable aggregation (a) compared to one with unstable aggregates (b). Note the particles in the immediate surface have been destroyed and a surface crust has formed. The bean seedling (c) must break the soil crust as it emerges from the seedbed. [Photos (a) and (b) from O'Nofiok and Singer (1984), used with permission of Soil Science Society of America; photo (c) courtesy of R. Weil.]

In tropical and subtropical regions with a long dry season, soil often must be tilled in a very dry state in order to prepare the land for planting with the onset of the first rains. Tillage under such dry conditions can be very difficult and can result in hard clods if the soils contain much silicate clay. Thus farmers in temperate regions typically find their soils too wet for tillage just prior to planting time (early spring), while farmers in tropical regions may face the opposite problem of soils too dry for easy tillage just prior to planting (end of dry season).

FIGURE 4.28 The remarkable stabilizing effect of a synthetic polyacrylamide is seen in the row on the right compared to the left, untreated, row. Irrigation water broke down much of the structure of the untreated soil but had no effect on the treated row. [From Mitchell (1986)]

Although each soil presents unique problems and opportunities, the following principles are generally relevant to managing soil structure:

1. Minimizing tillage, especially moldboard plowing and rototilling, reduces the loss of aggregate-stabilizing organic matter (see Section 4.13).
2. Tillage and traffic activities are least destructive to soil structure if restricted to periods of optimum soil moisture conditions.
3. Keeping the soil surface covered with crop residues or plant litter adds organic matter while protecting aggregates from the beating rain.
4. Incorporation of crop residues and animal manures into the soil is also effective in supplying the decomposition products that help stabilize soil aggregates.
5. Inclusion of sod crops in the rotation favors stable aggregation by helping to maintain soil organic matter and assuring a period without tillage.
6. Cover crops and green manure crops, where practical, are another good source of organic matter for structural management.

Figure 4.29 shows the enhanced aggregation resulting from the incorporation of poultry manure into a clay loam soil.

The data in Table 4.8 illustrates the importance of growing other than row crops if water-stable aggregates are to be maintained. The detrimental effect of continuously growing tilled corn is obvious.

4.13 TILTH AND TILLAGE

Although our discussion has focused on plowing, maintenance of stable soil structure throughout the crop-growing season also requires careful attention to seedbed preparation, cultivation to kill weeds, and trafficking for application of agrochemicals and crop harvest. The term *tilth* is central to such a discussion.

(a)

(b)

FIGURE 4.29 The incorporation of animal manures and other types of readily decomposable organic material can have dramatic effects on soil aggregation, especially the formation of large, stable aggregates. The soil pictured is a Davidson clay loam (Paleudults) from Virginia which had been fertilized with mineral fertilizer (a) or poultry manure (b) for five years. While both soils have a high proportion of water-stable aggregates, the aggregates and associated interped pores are much larger in the manure-treated soil. (From Weil and Kroontje, 1979)

TABLE 4.8 Water-Stable Aggregation of a Udoll (Marshall Silt Loam) near Clarinda, Iowa, under Different Cropping Systems

Crop	Water-stable aggregates (%)	
	Large (1 mm and above)	Small (less than 1 mm)
Corn continuously	8.8	91.2
Corn in rotation	23.3	76.7
Meadow in rotation	42.2	57.8
Bluegrass continuously	57.0	43.0

From Wilson et al. (1947).

Tilth

Simply defined, *tilth* refers to the physical condition of the soil in relation to plant growth, and hence includes all soil physical conditions that influence plant development. Tilth depends not only on granule formation and stability, but also on such factors as bulk density, soil moisture content, degree of aeration, rate of water infiltration, drainage, and capillary-water capacity. As might be expected, tilth often changes rapidly and markedly. For instance, the workability of fine-textured soils may be altered abruptly by a slight change in moisture.

Conventional Tillage and Crop Production

Tillage involves the mechanical manipulation of the soil. For centuries farmers have tilled the soil for three primary reasons: (1) to control weeds, (2) to present a suitable seedbed for crop plants, and (3) to incorporate organic residues into the soil.

Since the Middle Ages, the moldboard plow has been the **primary tillage** implement most used in the Western world.[7] Its purpose is to lift, twist, and turn the soil while incorporating crop residues and animal wastes into the plow layer (Figure 4.30).

In more recent times, the moldboard plow has been supplemented by the disk plow, which is used to cut up the residues and partially incorporate them into the soil. In conventional practice, such primary tillage has been followed by a number of **secondary tillage** operations, such as harrowing to kill weeds and to break up clods, thereby preparing a suitable seedbed.

[7]For an early, but still valuable, critique of the moldboard plow, see Faulkner (1943).

FIGURE 4.30 While the action of the moldboard plow lifts, turns, and loosens the upper 15–20 cm of soil (the furrow slice), the counterbalancing downward force compacts the next lower layer of soil. This compacted zone can develop into a "plowpan." Compactive action can be understood by imagining that you are lifting a heavy weight—as you lift the weight your feet press down on the floor below. (Photo courtesy of R. Weil)

After the crop is planted, the soil may receive further secondary tillage to control weeds and to break up crusting of the immediate soil surface. In mechanized agriculture, all conventional tillage operations are performed with tractors and other heavy equipment that may pass over the land several times before the crop is finally harvested. In many parts of the world, farmers use hand hoes or animal-drawn implements to stir the soil. Although humans and draft animals are not as heavy as tractors, their weight is applied to the soil in a relatively small area (foot- or hoofprint), and so can also cause considerable compaction.

Short-term Versus Long-term Effects on Soil Tilth

The immediate effects of most conventional tillage operations are beneficial. Crop residues are broken down more quickly if they are cut up and incorporated into the soil by tillage implements. Immediately after plowing, the soil is loosened and the total pore space is higher than before plowing. Tillage can provide excellent seedbeds and can be an effective means of weed control. However, conventional tillage also leaves the soil surface bare and subject to wind and water erosion.

The long-term effects of conventional tillage, especially plowing, are generally undesirable. Rapid breakdown of organic residues can hasten the reduction of soil organic matter content and, in turn, of aggregate stability (see Figure 4.24). By passing over the field frequently, tractors and other heavy equipment compact the soil. Certain implements, such as the moldboard plow and the disk harrow, compact the soil below their working depth as they lift and loosen the soil above. Use of these implements and heavy machinery encourages the development of a dense zone (**plow pan**) immediately below the plowed layer (Figure 4.31).

Other tillage implements, such as the chisel plow and spring-tooth harrow, do not press down upon the soil beneath them, and so are useful in breaking up plow pans and stirring the soil with a minimum of compaction. Large chisel-type plows can be used to break up dense subsoil layers, but in many soils the effects are quite temporary (Figure 4.32).

In recent years, land management systems have been developed that minimize the need for soil tillage. Since these systems also leave considerable plant residues on or near the soil surface, they protect the soil from erosion. For this reason the tillage practices followed in these systems are called *conservation tillage.* Figure 4.33 illustrates a *no-till* operation, where one crop is planted in the residue of another with no primary tillage. Other minimum-tillage systems permit some stirring of the soil, but still leave a high proportion of the crop residues on the surface. These organic residues protect the soil from the beating action of raindrops and the abrasive action of the wind, thereby reducing water and wind erosion and maintaining soil structure. Because of this association with erosion control, conservation tillage will be considered further in section 17.12.

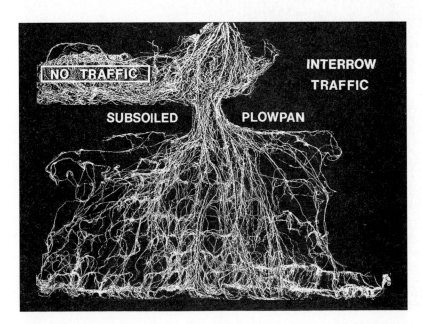

FIGURE 4.31 Root distribution of a cotton plant. On the right, interrow tractor traffic and plowing have caused a plowpan that restricts root growth. Roots are more prolific on the left where there had been no recent tractor traffic. The roots are seen to enter the subsoil through a loosened zone created by a subsoiling chisel-type implement. (Courtesy USDA National Tillage Machinery Laboratory)

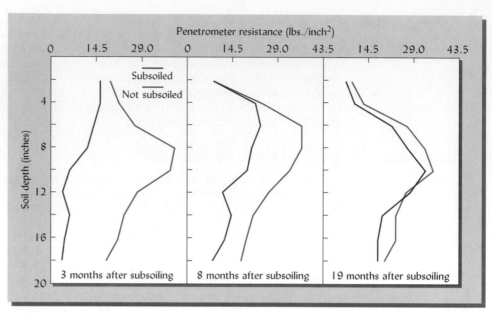

FIGURE 4.32 Deep tillage with a large chisel plow, sometimes called *subsoiling*, can break up a plowpan in the subsoil, reducing the soil strength and density as measured by resistance to penetration. Such subsoiling is most effective when a clayey subsoil is dry enough that pulling the implement through it shatters the mass of soil, creating a network of cracks. Even then, the effects are generally short-lived. In the Florida coastal plain Ultisol represented here, the subsoiling operation resulted in much-reduced penetration resistance for a few months. However, the difference between soil that had been subsoiled and soil that had not was less noticeable after 8 months and virtually unapparent after 27 months. Accordingly, in this experiment, subsoiling resulted in higher corn yield in the first year, but not in the second year after the deep tillage was performed. [From Wright et al. (1980)]

FIGURE 4.33 An illustration of one conservation tillage system. Wheat is being harvested (background) and soybeans are planted (foreground) with no intervening tillage operation. This no-tillage system permits double-cropping, saves fuel costs and time, and helps conserve the soil. (Courtesy Allis-Chalmers Corporation)

4.14 SOIL CONSISTENCE AND ENGINEERING IMPLICATIONS

Soil Consistence

Soil *consistence* is a term used to describe the resistance of a soil at various moisture contents to mechanical stresses or manipulations. This property is a composite expression of those forces of mutual attraction among soil particles that determine the ease with which a soil can be reshaped or ruptured. Consistence is commonly determined in the field by feeling and manipulating the soil by hand. As a clod of soil is squeezed between the thumb and forefinger, observations are made on the amount of force needed to deform the soils and on the manner in which the soil responds to the force.

The consistence of soils is generally described at three soil moisture levels: wet, moist, and dry. Terms used to describe soil consistence at these three moisture levels are shown in Table 4.9. Generally, the terms near the top of the table apply to very sandy or silty soils, while those near the bottom describe soils with high contents of clay. For example, a dry clod that could not be broken by hand would be called *very hard* or *extremely hard*. When wet, soils with a high content of silicate clays are likely to be termed *very sticky* and *very plastic*. Moisture content greatly influences how a soil responds to stress. A *hard* dry soil has greater resistance to deformation than a wet *plastic* soil, especially if the silicate clay content is high. Anyone who has gotten a vehicle stuck up to its axles in clayey mud has firsthand experience with these principles.

While most of the terms in Table 4.9 are self-explanatory, the term *friable* requires some comment. If a clod of moist soil crumbles into aggregates when crushed with only light pressure, it is said to be friable. *Friable* and *very friable* conditions are prized by farmers and gardeners, as such soils allow easy, effective tillage and root penetration.

The degree of cementation of soil materials, independent of soil moisture, also is considered in identifying soil consistence. Terms such as *weakly cemented, strongly cemented,* and *indurated* are used to define categories of increasing cementation.

Engineering Uses

The concept of soil consistence is as relevant to engineering uses of soils as it is to soil tillage. For engineering purposes, more precise measurements of how a soil responds to applied forces can be made using laboratory instruments.

Perhaps the most important property of a soil for engineering uses is its **soil strength.** This is a measure of the capacity of a soil to withstand stresses without giving way to those stresses by collapsing or becoming deformed. The failure of a soil to withstand stress can be seen where building foundations are not well supported or where earthen dams give way under pressure from the impounded reservoir waters.

Soil strength is determined by a number of related soil characteristics, such as soil **compressibility,** soil **compactability,** and **bearing resistance.** For the purpose of plant growth, soil compaction is to be avoided; however, compactability is desirable for most engineering purposes. Soils are purposely compacted prior to being used as a roadbed or for a building foundation, as compaction occurring later under the heavy load would result in uneven settling or a cracked pavement or foundation (Figure 4.34).

Some soil particles, such as those of certain colloidal silicate clays and of micas of all sizes, can be compressed when a load is placed upon them. When that load is removed,

TABLE 4.9 **Terms Used to Describe the Consistence of Soils**

Conditions of least coherence are represented by terms at the top of each column. Greater coherence characterizes terms as you move down the column.

Wet soils		Moist soils	Dry soils
Stickiness	*Plasticity*	*Moist soils*	*Dry soils*
Nonsticky	Nonplastic	Loose	Loose
Slightly sticky	Slightly plastic	Very friable	Soft
Sticky	Plastic	Friable	Slightly hard
Very sticky	Very plastic	Firm	Hard
		Very firm	Very hard
		Extremely firm	Extremely hard

FIGURE 4.34 Soils must be compacted before being used to bear heavy loads such as in highway subpavement or a building foundation. Often water is added to a soil to make it more easily compacted by a heavy "sheep's foot" roller as shown here. Note that this treatment is just the opposite of that applied to soils intended for plant growth. (Photo courtesy of R. Weil)

however, these particles tend to regain their original shape, in effect reversing their compression. As a result, these particles are not conducive to a more permanent compaction state—a property desirable in providing the soil strength needed for most engineering purposes.

Soil compressibility and compaction also influence the bearing resistance or the ability of a soil to resist the penetration of an object (such as a building foundation) into that soil. Bearing resistance is essential in preventing the penetration of the surface materials of a highway into the underlying roadbed. This is a property of great practical importance which evidently depends on some of the more fundamental physical properties of the individual soil particles.

Soil strength is affected by a number of factors, such as soil moisture content, particle-size distribution (soil texture), and the mineralogy of the different soil particles. In general, coarser-textured materials have higher soil strength than those with smaller particle size. For example, quartz sand grains are subject to little compressibility, while the silicate clays are more easily compressed. Consequently, most materials high in clay are not prized as being suitable for highway beds, dams (except the impermeable dam core), or building foundations. The effects of freezing and thawing on materials high in clay (and moisture) further complicate the use of these materials for engineering purposes. This subject will receive more attention in Chapter 8.

Influence of Mineralogy

The mineralogical makeup of soil particles also affects their use for engineering purposes. The equidimensional grains of minerals such as quartz and feldspars, of which the sand fraction is composed, are conducive to high soil strength. In contrast, the flakelike nature of micas and silicate clays provides an ease of compressibility, with subsequent expansion when the load is removed. This behavior in turn results in low soil strength.

Soil colloids differ widely in their physical properties, including plasticity, cohesion, swelling, and shrinkage. These properties greatly influence the usefulness of soils for both agricultural and engineering purposes.

Plasticity

Soils containing more than about 15% clay exhibit plasticity—that is, pliability and the capacity to be molded. This property most likely results from the platelike nature of the clay particles and the combined lubricating/binding influence of the adsorbed water. Thus, the particles easily slide over each other, much like panes of glass with films of water between them.

Plasticity is exhibited over a range of moisture contents referred to as *plasticity limits*. At the lower of these moisture levels, or **plastic limit**, the soil begins to be malleable, but molded pieces crumble easily when a little pressure is applied. The plastic limit is the lowest moisture content at which a soil can be deformed without cracking. Agricultural soils should not be tilled at moisture contents above the plastic limit because smearing and puddling will result.

The upper plasticity limit, or **liquid limit**, is the moisture content at which soil ceases to be plastic but becomes semifluid (like softened butter) and tends to flow like a liquid when jarred. These limits provide the main criteria in the classification of soils for engineering purposes. Obviously, a soil that tends to flow when wet will not make a suitable foundation for a building or for a highway (see Figure 1.8).

The plastic and liquid limits of several soils and three clay samples are shown in Table 4.10. These data illustrate the levels to be expected in a variety of soils. The difference between the liquid limit and plastic limit is the range of water content over which the soil is plastic, and is called its *plasticity index*. Soils with high plasticity indices are difficult to handle in the field and are unstable in bearing loads. The soils with highest plasticity indices (Susquehanna and Bashaw in Table 4.10) contain high levels of expanding clays. Some clays (such as smectites; see Section 8.5) generally have high liquid limits and plasticity indices, especially if saturated with sodium. In contrast, other clays (e.g., kaolinite), more typically found in highly weathered soils of warm, humid regions, have low liquid limit values. Engineers have devised several systems for classifying the stability of soil materials. In the Unified System of Classification, 14 soil classes are defined primarily on the basis of particle-size distribution, liquid limit, and plasticity index.

Cohesion

A second characteristic, **cohesion**, indicates the tendency of soil particles to stick together. This tendency results primarily from attraction of clay particles for water molecules held between them. Hydrogen bonding between clay surfaces and water, and also among water molecules, is the attractive force responsible for the cohesion. It accounts for the presence of hard clods in some clayey soils that resist being broken down, even with repeated tillage.

The various types of clay minerals found in soils are discussed in Chapter 8. Soils high in clays of the smectite and fine-grained mica type exhibit cohesion to a much more noticeable degree than do those dominated by the kaolinite or hydrous oxide type of clays. Humus, by contrast, tends to reduce the attraction of individual clay particles for each other.

Swelling and Shrinkage

The third and fourth major soil characteristics affected by mineralogy are *swelling* and *shrinkage*. Some clays, such as the smectites, swell when wet and shrink when dry (see Chapter 8). After a prolonged dry spell, soils high in smectites often are crisscrossed by

TABLE 4.10 Plastic and Liquid Limits of Several Soils and of Na- and Ca-Saturated Smectite

Clay soils with large amounts of smectites (Susquehanna and Bashaw) have high liquid limits as do Na-saturated clays.

Soil	Location	Plastic limit	Liquid limit	Plasticity Index
Davidson (Paleudult)	Georgia	19	27	8
Cecil (Hapludult)	Georgia	29	49	20
Putnam (Albaqualf)	Georgia	24	37	13
Susquehanna (Paleudalf)	Georgia	29	57	28
Sliprock (Haplumbrept)	Oregon	46	59	13
Jory (Haplohumult)	Oregon	30	45	15
Bashaw (Pelloxerert)	Oregon	18	71	53
Na-saturated smectite	—	—	950	—
Ca-saturated smectite	—	—	360	—
Na-saturated kaolinite	—	—	36	—

Data for Georgia soils from Hammel et al. (1983), Oregon soils from McNabb (1979), clays from Warkentin (1961).

wide, deep cracks that at first allow rain to penetrate rapidly (Figure 4.35). But later, because of swelling, such a soil is likely to close up and become much more impenetrable than one dominated by clays that swell less. This moisture-content-related swelling and shrinkage causes sufficient movement of the soil to crack building foundations, burst pipelines, and buckle pavements. This phenomenon also greatly influences the formation and stability of soil aggregates (review Sections 4.10 and 4.11). Therefore, both the content and mineralogy of clay must be taken into account when using soils for both agricultural and engineering purposes.

The adverse effects of some minerals on soil strength are not limited to the clay-size fractions. For example, because of their platelike nature, mica particles of all sizes are associated with low soil strength, poor compactability, and high compressibility. Soil materials high in mica do not make good roadbeds.

It should be emphasized that the various factors influencing soil strength do not act independently. For example, we have just discussed how mineralogy can interact with particle size. Another factor discussed earlier is the interaction between soil moisture and particle size, as reflected in soil consistence. Thus, a moist sand is fairly stable for vehicle traffic, while on a dry sand or a completely saturated sand, the tires of the vehicles may sink into the soil (see, for example, Figure 5.24). Apparently, a small amount of moisture results in sufficient cohesion between particles to give some soil strength. With either no moisture or with particles completely saturated with water, sand particles tend to act independently and do not resist traffic stress.

A detailed description of soil engineering properties is beyond the scope of this text.

4.15 CONCLUSION

Physical properties to a marked extent control the behavior of soils with regard to plant growth, hydrology, agricultural management, and engineering uses. The nature and properties of the individual particles, their size distribution, and their arrangement in

FIGURE 4.35 Certain types of clays have the capacity to change volume significantly as their moisture content changes. Here smectitic clays, or some other swelling type of clays, have shrunk during a dry period and have caused large cracks to open in the soil surface. (Courtesy of USDA Natural Resources Conservation Service)

soils have profound effects on root growth and on soil manipulations and use for all purposes. These effects in turn influence total volume of nonsolid pore space as well as pore size, thereby imparting water and air relationships.

The properties of individual particles and their proportionate distribution (soil texture) are subject to little human control in field soils. However, it is possible to exert some control on the arrangement of these particles into aggregates (soil structure) and on the stability of these aggregates. Proper plant species selection, crop rotation, and management help assure this control. In recent years, these management practices have been augmented by conservation tillage, which minimizes soil manipulations while it decreases soil erosion and water runoff. The stability of soil in response to loading forces from traffic, tillage, or foundations is greatly influenced by particle size, moisture content, and plasticity of the colloidal fraction. The physical properties presented in this chapter greatly influence nearly all other soil properties and uses, as discussed throughout this book.

REFERENCES

Bigham, J. M., and E. J. Ciolkosz (eds.). 1993. *Soil Color,* SSSA Special Publication No. 31. Soil Sci. Soc. of America, Madison, Wis. 172 pp.

Camp, C. R., and J. F. Lund. 1964. "Effects of soil compaction on cotton roots," *Crops and Soils,* **17**:13–14.

Dawud, A. Y., and F. Gray. 1979. "Establishment of the lower boundary of the sola of weakly developed soils that occur in Oklahoma," *Soil Sci. Soc. Amer. J.,* **43**:1201–1207.

Emerson, W. W., R. C. Foster, and J. M. Oades. 1986. "Organomineral complexes in relation to soil aggregation and structure," in P. M. Huang and M. Schnitzer (eds.), *Interaction of Soil Minerals with Natural Organics and Microbes,* SSSA Special Publication No. 17, Soil Sci. Soc. of America, Madison, Wis.

Faulkner, E. H. 1943. *Plowman's Folly* (Norman, Okla.: Univ. of Oklahoma Press).

Fritton, D. D., and G. W. Olson. 1972. "Bulk density of a fragipan soil in natural and disturbed profiles," *Soil Sci. Soc. Amer. Proc.,* **36**:686–689.

Hammel, J. E., M. E. Summer, and J. Burema. 1983. "Atterberg limits as indices of external areas of soils." *Soil Science Society of America Journal,* **47**:1054–1056.

Larson, W. E., S. C. Gupta, and R. A. Useche. 1980. "Compression of agricultural soils from eight soil orders," *Soil. Sci. Soc. Amer. J.,* **44**:450–457.

Laws, W. D., and D. D. Evans. 1949. "The effects of long-time cultivation on some physical and chemical properties of two rendzina soil," *Soil Sci. Soc. Amer. Proc.,* **14**:15–19.

Lyon, T. L., H. O. Buckman, and N. C. Brady. 1952. *The Nature and Properties of Soils,* 5th ed. (New York: Macmillan), p. 60.

McIntyre, S. C., J. C. Lance, B. L. Campbell, and R. L. Miller. 1987. "Using cesium-137 to estimate soil erosion on a clear-cut hillside," *J. Soil and Water Conservation,* **42**:117–120.

McNabb, D. H. 1979. "Correlation of soil plasticity with amorphous clay constituents," *Soil Science Society of America Journal,* **46**:450–456.

Miller, R. M., and J. D. Jastrow. 1992. "The role of mycorrhizal fungi in soil conservation," in G. J. Bethlenfalvay and R. J. Linderman (eds.), *Mychorrhizae in Sustainable Agriculture,* ASA Special Publication No. 54, ASA/SSSA/CSSA, Madison, Wis.

Mitchell, A. R. 1986. "Polyacrylamide application in irrigation water to increase infiltration," *Soil Sci.,* **141**:353–358.

Nelson, L. B., and R. J. Muckenhirn. 1941. "Field percolation rates of four Wisconsin soils having different drainage characteristics," *J. Amer. Soc. Agron.,* **33**:1028–36.

O'Nofiok, O., and M. J. Singer. 1984. "Scanning electron microscope studies of surface crusts formed by simulated rainfall," *Soil Sci. Soc. Amer. Jour.,* **48**:1137–1143.

Tiessen, H., J. W. B. Stewart, and J. R. Bettany. 1982. "Cultivation effects on the amounts and concentration of carbon, nitrogen, and phosphorus in grassland soils," *Agron. J.,* **74**:831–835.

Tisdall, J. M. 1991. "Fungal hyphae and structural stability of soil," *Aust. J. Soil Res.,* **29**:729–743.

Unger, P. W., and T. M. McCalla. 1980. "Conservation tillage systems," *Adv. in Agron.,* **33**:1–58.

Vimmerstadt, J., F. Scoles, J. Brown, and M. Schmittgen. 1982. "Effects of use pattern, cover, soil drainage class and overwinter changes on rain infiltration on campsites," *J. Environ. Qual.,* **11**:25–28.

Voorhees, W. B. 1984. "Soil compaction, a curse or a cure?" *Solutions,* **28**:42–47 (Peoria, Ill.: Solutions Magazine Inc.).

Warkentin, B. P. 1961. "Interpretation of the upper plastic limits of clays," *Nature,* **190**:287–288.

Weil, R. R., and W. Kroontje. 1979. "Physical condition of a Davidson clay loam after 5 years of heavy poultry manure applications," *J. Environ. Qual.,* **8**:389–392.

Wilson, H. A., R. Gish, and G. M. Browning. 1947. "Cropping systems and season as factors affecting aggregate stability," *Soil Sci. Soc. Amer. Proc.,* **12**:36–43.

Wright, D. L., F. M. Rhoads, and R. L. Stanley, Jr. 1980. "High level of management needed on irrigated corn," *Solutions,* May/June 1980:24–36 (Peoria, Ill.: Solutions Magazine Inc.).

Yule, D. F., and J. T. Ritchie. 1980. "Soil shrinkage relationships of Texas Vertisols: I. Small cores," *Soil Sci. Soc. Amer. J.,* **44**:1285–1291.

5

SOIL WATER: CHARACTERISTICS AND BEHAVIOR

When the earth will . . . drink up the rain as fast as it falls.
—*H. D. THOREAU, THE JOURNAL*

Water is a vital component of every living thing. Although it is one of nature's simplest chemicals, water has unique properties that promote a wide variety of physical, chemical, and biological processes. These processes greatly influence almost every aspect of soil development and behavior, from the weathering of minerals to the decomposition of organic matter, from the growth of plants to the pollution of groundwater. The characteristics and behavior of water in the soil is a common thread that interrelates nearly every chapter in this book. The principles contained in this chapter will help us understand why mud slides occur in water-saturated soils (Chapter 4), why earthworms improve soil quality (Chapter 11), why rice paddies contribute to global ozone depletion (Chapter 13), and why famine stalks humanity in certain regions of the world (Chapter 21).

We are all familiar with water. We drink it, wash with it, swim in it, and irrigate our crops with it. But water in the soil is something quite different from water in a drinking glass. In the soil, water is intimately associated with solid particles, particularly those that are colloidal in size. The interaction between water and soil solids changes the behavior of both.

Attraction to solid surfaces restricts some of the free movement of water molecules, making it less liquid and more solidlike in its behavior. In the soil, water can flow up as well as down. Plants may wilt and die in a soil whose profile contains a million kilograms of water in a hectare. A layer of sand and gravel deep in a soil profile may cause the upper horizons to become muddy and saturated with water during much of the year. These and other soil water phenomena seem to contradict our intuition about how water ought to behave.

Water causes soil particles to swell and shrink, to adhere to each other, and to form structural aggregates. Water participates in innumerable chemical reactions that release or tie up nutrients, create acidity, and wear down minerals so that their constituent elements eventually contribute to the saltiness of the oceans.

Soil–water interactions determine much about the ecological function of soils and practices of soil management. These interactions determine how much rainwater runs into and through the soil and how much runs off the surface. Control of these processes in turn determines the movement of chemicals to the groundwater, and of both chemicals and eroded soil particles to streams and lakes. The interactions affect the rate of water loss through leaching and evapotranspiration, the balance between air and water in soil pores, the rate of change in soil temperature, the rate and kind of metabolism of soil organisms, and the capacity of soil to store and provide water for plant growth.

143

5.1 STRUCTURE AND RELATED PROPERTIES OF WATER

Water participates directly in dozens of soil and plant reactions and indirectly affects many others. Its ability to do so is determined primarily by its structure. Water is a simple compound, its individual molecules containing one oxygen atom and two much smaller hydrogen atoms. The elements are bonded together covalently, each hydrogen atom sharing its single electron with the oxygen.

The resulting molecule is not symmetrical, however. Instead of the atoms being arranged linearly (H—O—H), the hydrogen atoms are attached to the oxygen in sort of a V-shaped arrangement at an angle of only 104.5°. As shown in Figure 5.1, this results in an asymmetrical molecule with the shared electrons being closer to the oxygen than to the hydrogen. Consequently, the side on which the hydrogen atoms are located tends to be electropositive and the opposite side electronegative. Molecules whose positive and negative charge centers do not coincide are termed *polar molecules*. The polarity of water accounts for many reactions of great importance to soil and environmental science.

Polarity

The property of polarity helps explain how water molecules interact with each other. Each water molecule does not act completely independently but rather is coupled with other neighboring molecules. The hydrogen (positive) end of one molecule attracts the oxygen (negative) end of another, resulting in a chainlike (polymer) grouping.

Polarity also accounts for a number of other important properties of water. For example, it explains why water molecules are attracted to electrostatically charged ions and to colloidal surfaces. Cations such as H^+, Na^+, K^+, and Ca^{2+} become hydrated through their attraction to the oxygen (negative) end of water molecules. Likewise, negatively charged clay surfaces attract water, this time through the hydrogen (positive) end of the molecule. Polarity of water molecules also encourages the dissolution of salts in water since the ionic components have greater attraction for water molecules than for each other.

When water molecules become attracted to electrostatically charged ions or clay surfaces, they are more closely packed than in pure water. In this packed state their freedom of movement is restricted and their energy status is lower than in pure water. Thus, when ions or clay particles become hydrated, energy must be released. That released energy is evidenced as **heat of solution** when ions hydrate or as **heat of wetting** in the case of hydrating clay particles. The phenomenon can be demonstrated by placing some dry, fine clay in the palm of the hand and then adding a few drops of water. A slight rise in temperature can be felt.

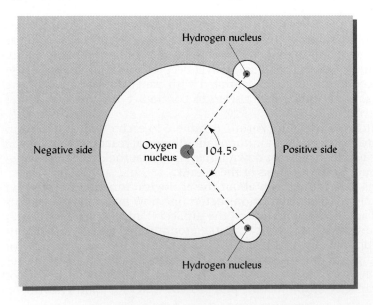

FIGURE 5.1 Two-dimensional representation of a water molecule showing a large oxygen atom and two much smaller hydrogen atoms. The HOH angle of 104.5° results in an asymmetrical arrangement. One side of the water molecule (that with the two hydrogens) is electropositive; the other is electronegative. This accounts for the polarity of water.

Hydrogen Bonding

The phenomenon by which hydrogen atoms act as links between water molecules is called *hydrogen bonding.* This is a relatively low energy coupling in which a hydrogen atom is shared between two electronegative atoms such as O and N. Because of its high electronegativity, an O atom in one water molecule exerts some attraction for the H atom in a neighboring water molecule. This type of bonding accounts for the polymerization of water. Hydrogen bonding also accounts for the relatively high boiling point, specific heat, and viscosity of water compared to the same properties of other hydrogen-containing compounds, such as H_2S, which has a higher molecular weight but no hydrogen bonding. It is also responsible for the structural rigidity of some clay crystals and for the structure of some organic compounds, such as proteins.

COHESION VERSUS ADHESION.   Hydrogen bonding accounts for two basic forces responsible for water retention and movement in soils: the attraction of water molecules for each other (**cohesion**) and the attraction of water molecules for solid surfaces (**adhesion**). By adhesion (also called *adsorption*), some water molecules are held rigidly at the soil solid surfaces. In turn, these tightly bound water molecules hold by cohesion other water molecules farther removed from the solid surfaces. Together, the forces of adhesion and cohesion make it possible for the soil solids to retain water and control its movement and use. Adhesion and cohesion also make possible the property of plasticity possessed by clays (see Section 4.14).

Surface Tension

One other important property of water that markedly influences its behavior in soils is that of **surface tension.** This property is commonly evidenced at liquid–air interfaces and results from the greater attraction of water molecules for each other (cohesion) than for the air above (Figure 5.2). The net effect is an inward force at the surface that causes water to behave as if its surface were covered with a stretched elastic membrane, an observation familiar to those who have seen insects walking on water in a pond (Figure 5.3). Because of the relatively high attraction of water molecules for each other, water has a high surface tension compared to that of most other liquids. As we shall see, surface tension is an important property, especially as a factor in the phenomenon of capillarity which determines how water moves and is retained in soil.

5.2 CAPILLARY FUNDAMENTALS AND SOIL WATER

The movement of water up a wick when the lower end is immersed in water exemplifies the phenomenon of capillarity. Two forces cause capillarity: (1) the attraction of

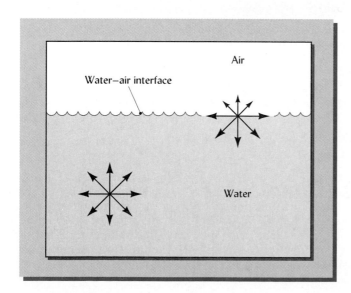

FIGURE 5.2 Comparative forces acting on water molecules at the surface and beneath the surface. Forces acting below the surface are equal in all directions since each water molecule is attracted equally by neighboring water molecules. At the surface, however, the attraction of the air for the water molecules is much less than that of water molecules for each other. Consequently, there is a net downward force on the surface molecules, and the result is something like a compressed film or membrane at the surface. This phenomenon is called *surface tension.*

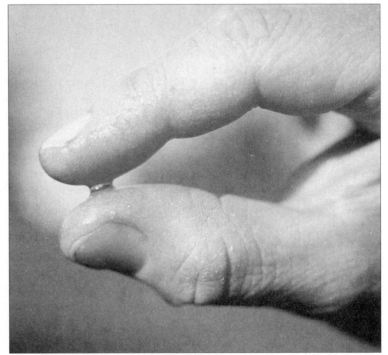

FIGURE 5.3 Everyday evidences of water's surface tension (above) as insects land on water and do not sink, and of forces of cohesion and adhesion (below) as a drop of water is held between the fingers. (Photos courtesy of R. Weil)

water for the solid walls of channels through which it moves (adhesion or adsorption) and (2) the surface tension of water, which is due largely to the attraction of water molecules for each other (cohesion).

Capillary Mechanism

Capillarity can be demonstrated by placing one end of a fine glass tube in water (Figure 5.4). The water rises in the tube, and the smaller the tube bore, the higher the water rises. The water molecules are attracted to the sides of the tube and start moving up the tube in response to this attraction. The cohesive force between individual water molecules ensures that water not directly in contact with the side walls is also pulled up the tube. This continues until the weight of water in the tube counterbalances the cohesive and adhesive forces.

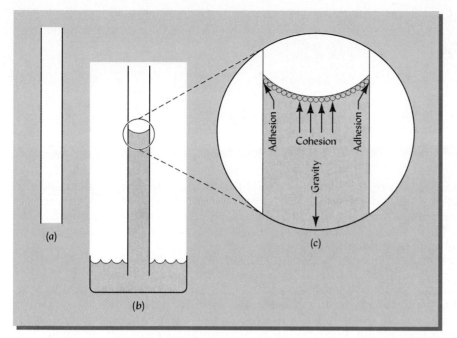

FIGURE 5.4 Diagrams illustrating the phenomenon of capillarity. (a) The situation just before lowering a fine glass tube to the water surface. (b) When the tube is inserted in the liquid, water moves up the tube owing to (c) the attractive forces between the water molecules and the wall of the tube (adhesion) and to the mutual attraction of the water molecules for each other (cohesion). Water will move up the tube until the downward pull of gravity equals the attractive forces of cohesion and adhesion.

The height of rise in a capillary tube is inversely proportional to the tube diameter and directly proportional to the surface tension, which, in turn, is determined largely by cohesion between water molecules. In other words, the smaller the tube bore, the higher the water rise in the tube (Figure 5.5). The capillary rise can be approximated as

$$h = \frac{2T}{rdg}$$

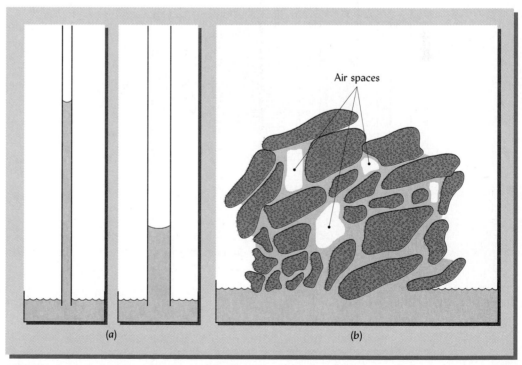

FIGURE 5.5 Upward movement by capillarity (a) in glass tubes of different sizes and (b) in soils. Although the mechanism is the same in the tubes and in the soil, adjustments are extremely irregular in soil because of the tortuous shape and variability in size of the soil pores and because of entrapped air.

where h is the height of capillary rise in the tube, T is the surface tension, r is the radius of the tube, d is the density of the liquid, and g is the force of gravity. For water, this equation reduces to the simple expression

$$h = \frac{0.15}{r}$$

This equation emphasizes the inverse relation between height of rise and size of the tube through which the water rises.

Height of Rise in Soils

Capillary forces are at work in all moist soils. However, the rate of movement and the rise in height are less than one would expect on the basis of soil pore size. One reason is that soil pores are not straight, uniform openings like glass tubes. Furthermore, some soil pores are filled with air, which may be entrapped, slowing down or preventing the movement of water by capillarity (Figure 5.5).

The upward movement due to capillarity in soils is illustrated in Figure 5.6. Usually the height of rise resulting from capillarity is greater with fine-textured soils if sufficient time is allowed and the pores are not too small. This is readily explained on the basis of the capillary size and the continuity of the pores. With sandy soils the adjustment is rapid, but so many of the pores are large in size that the height of rise cannot be great.

Capillarity is traditionally illustrated as an upward adjustment. But movement in any direction takes place since the attractions between soil pores and water are as effec-

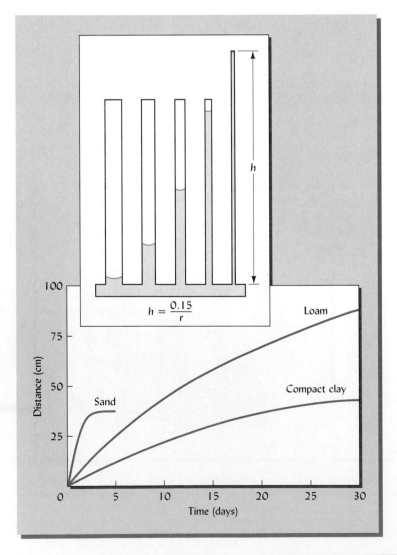

FIGURE 5.6 Upward movement of moisture from a water table through soils of different textures and structures. Note the very rapid rise in sand but the moderate height attained. Apparently, the pores of loam are more favorable for movement than those in compact clay.

tive with horizontal pores as with vertical ones (Figure 5.7). The significance of capillarity in controlling water movement in small pores will become evident as we turn to soil water energy concepts.

5.3 SOIL WATER ENERGY CONCEPTS

The retention and movement of water in soils, its uptake and translocation in plants, and its loss to the atmosphere are all energy-related phenomena. Different kinds of energy are involved, including potential, kinetic, and electrical. However, since potential energy is of greatest importance in determining the status and movement of soil water, it will be the energy referred to in this text. For the sake of simplicity, however, we will use the term *energy* to refer to potential energy.

As we consider energy, we should keep in mind that all substances, including water, have a tendency to move or change from a state of higher to one of lower energy. For example, all other conditions being equal, water movement in soils will generally be from a zone where the energy level of the water is high (wet soil) to one where its energy level is low (dry soil). Therefore, if we know the pertinent energy levels at various points in a soil, we can predict the direction of water movement. It is the *differences* in energy levels from one contiguous site to another that influence this water movement.

Forces Affecting Potential Energy

The discussion of the structure and properties of water in the previous section suggests three important forces affecting the energy level of soil water. First, adhesion, or the attraction of the soil solids (matrix) for water, provides a **matric** force (responsible for adsorption and capillarity) that markedly reduces the energy state of the adsorbed water molecules, and to a lesser degree of those held by cohesion. Second, the attraction of ions and other solutes for water, resulting in **osmotic** forces, tends to reduce the energy level of water in the soil solution. Osmotic movement of pure water across a semipermeable membrane into a solution is evidence of the lower energy state of the solution.

The third major force acting on soil water is **gravity**, which tends to pull the water downward. The energy level of soil water at a given elevation in the profile is thus higher than that of pure water at some lower elevation. This difference in energy level causes water to flow downward.

Soil Water Potential

The *difference* in energy level of water at one site or one condition (e.g., wet) from that at another site or condition (e.g., dry) will determine the direction and rate of water

FIGURE 5.7 As this field irrigation scene in Arizona shows (left), water has moved up by capillarity from the irrigation furrow toward the top of the ridge. The photo on the right illustrates some horizontal movement to both sides and away from the irrigation water.

movement in soils and in plants. Water always moves from where it has a high energy level to where it has a lower one. Since water in a wet soil is not held very tightly by the soil solids (matrix), the water molecules have considerable freedom of movement, so the energy level is not much lower than that of pure water not influenced by soil. In a drier soil, however, the water that remains is very tightly held by the soil solids, the water molecules have little freedom of movement, and the energy of the water is much lower than that of the water in the wet soil. If the wet and dry soil samples are brought in touch with each other, water will move from the wet soil (higher energy state) to the drier soil (lower energy).

For years scientists have sought viable means of comparing the energy levels of water in different soils. Recognizing the difficulties of ascertaining the absolute energy levels of soil water, they decided to compare the energy levels of soil water with that of a sample of pure water outside the soil in an environment that is maintained at standard pressure, temperature, and elevation levels. The *difference* in energy levels between this pure water and that of soil water is termed **soil water potential** (Figure 5.8).

Since all soil water values have a common reference point (the energy state of pure water), differences in the soil water potential of two soil samples in fact reflect differences in their energy levels. This means that water will move from a soil zone having a high soil water potential to one having a lower soil water potential. This fact should be kept in mind in considering the behavior of water in soils.

Research has shown that the soil water potential is due to several forces, each of which accounts for a component of the total soil water potential. These components are due to differences in energy levels resulting from gravitational, matric, and osmotic forces and are termed **gravitational potential** (ψ_g), **matric potential** (ψ_m), and **osmotic potential** (ψ_o), respectively. Each of these components results from a different force, and they are technically not strictly additive. They act simultaneously, however, to influence water behavior in soils. The general relationship of soil water potential to potential energy levels is shown in Figure 5.8.

Gravitational Potential

The force of gravity acts on soil water the same as it does on any other body (see Figure 5.9), the attraction being toward the earth's center. The gravitational potential (ψ_g) of soil water may be expressed mathematically as

$$\psi_g = gh$$

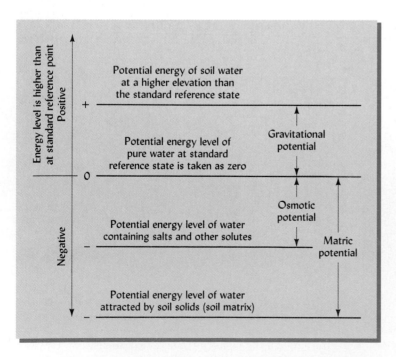

FIGURE 5.8 Relationship between the potential energy of pure water at a standard reference state (pressure, temperature, and elevation) and that of soil water. If the soil water contains salts and other solutes, the mutual attraction between water molecules and these chemicals reduces the potential energy of the water, the degree of the reduction being termed *osmotic potential*. Similarly, the mutual attraction between soil solids (soil matrix) and soil water molecules also reduces the water's potential energy. In this case the reduction is called *matric potential*. Since both of these interactions reduce the water's potential energy level compared to that of pure water, the changes in energy level (osmotic potential and matric potential) are both considered to be negative. In contrast, differences in energy due to gravity (gravitational potential) are always positive. This is because the reference elevation of the pure water is purposely designated at a site in the soil profile below that of the soil water. These three components of the total soil water potential are not strictly additive, since there is some overlap in their effects. Nevertheless, they encompass the major forces influencing water behavior in soils. A plant root attempting to remove water from a moist soil would have to overcome all three forces simultaneously.

where g is the acceleration due to gravity and h is the height of the soil water above a reference elevation. The reference elevation is usually chosen within the soil profile or at its lower boundary to ensure that the gravitational potential of soil water above the reference point will always be positive.

Gravity plays an important role in removing excess water from the upper rooting zones following heavy precipitation or irrigation. It will be given further attention when the movement of soil water is discussed (see Section 5.6).

Pressure Potential

This component accounts for the effects on soil water potential of all factors other than gravity and solute levels. It includes the positive pressure potential relating to (1) the weight of the overburden of soil, (2) hydrostatic pressures due to the weight of water in saturated soils and aquifers, and (3) the pressures stemming from air in the soil. The pressure potential also includes the effects of the attractive forces between the water and the soil solids or the soil matrix. These forces give rise to the **matrix potential** (ψ_m), which is always negative because the water attracted by the soil matrix has an energy state lower than that of pure water.

While each of these pressures is significant in specific field situations, the matrix potential is universally important because there are omnipresent interactions between soil solids and water. The movement of soil water, the availability of water to plants, and the solution of many civil engineering problems are determined to a considerable extent by matrix potential. Consequently, matric potential will receive primary attention in this text, along with gravitational and osmotic potential.

Matric potential (ψ_m), which results from the phenomena of adhesion (or adsorption) and of capillarity, influences soil moisture retention as well as soil water movement. Differences between the ψ_m of two adjoining zones of a soil encourage the movement of water. Water moves, for example, from a moist zone (high energy state) to a dry zone (low energy state). Although this movement may be slow, it is extremely important, especially in supplying water to plant roots.

FIGURE 5.9 Whether concerning matric potential, osmotic potential, or gravitational potential (as shown here), water always moves to where its energy will be lower. In this case the energy lost by the water is used to turn the historic Mabry Mills waterwheel and grind flour. (Photo courtesy of R. Weil)

Osmotic Potential

The osmotic potential (ψ_o) is attributable to the presence of solutes in the soil solution. The solutes may be inorganic salts or organic compounds. They reduce the potential energy of water, primarily because the solute ions or molecules attract the water molecules. The process of osmosis is illustrated in Figure 5.10. In this example, water moves toward the zone of lower energy because a membrane prevents the counter-movement of the dissolved substances.

Unlike the matric potential (ψ_m), the osmotic potential (ψ_o) has little effect on the mass movement of water in soils. Its major effect is on the uptake of water by plant root cells that are isolated from the soil solution by their semipermeable membranes. In soils high in soluble salts, ψ_o may be lower in the soil solution than in the plant root cells. This leads to constraints in the uptake of water by the plants. Since water vapor pressure is lowered by the presence of solutes, ψ_o also affects the movement of water vapor. The relationship between the matric and osmotic components of total soil water potential is shown in Figure 5.11.

Methods of Expressing Energy Levels

Several units have been used to express differences in energy levels of soil water. One is the height in centimeters of a unit water column whose weight just equals the potential under consideration. A second is the standard atmosphere pressure at sea level, which is 760 mm Hg or 1020 cm of water. The unit termed *bar* approximates that of a standard atmosphere. Energy may be expressed per unit of mass (joules/kg) or per unit of volume (newtons/m²). The latter (newtons/m²), expressed as the term *pascal* (Pa) in the International System of Units (SI), will be used in this text along with *kilopascal* kPa. The equivalency among common means of expressing soil water potential is seen in Table 5.1.

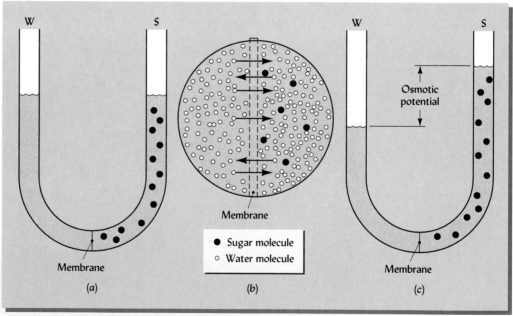

FIGURE 5.10 Illustration of the process of osmosis and of osmotic pressure. (a) A ∪ tube containing water (W) in the left arm and a solution (S) of sugar in water in the right arm. These liquids are separated by a membrane that is permeable to water molecules but not to the dissolved sugar. (b) Enlarged portion of the membrane with water molecules moving freely from the water side to the solution side and vice versa. The sugar molecules, in contrast, are unable to penetrate the membrane. Since the effect of sugar is to decrease the free energy of the water on the solution side, more water passes from left to right than from right to left. (c) At equilibrium sufficient water has passed through the membrane to bring about a significant difference in the heights of liquid in the two arms. The difference in the levels in the W and S arms represents the osmotic potential. [Modified from Keeton (1972)]

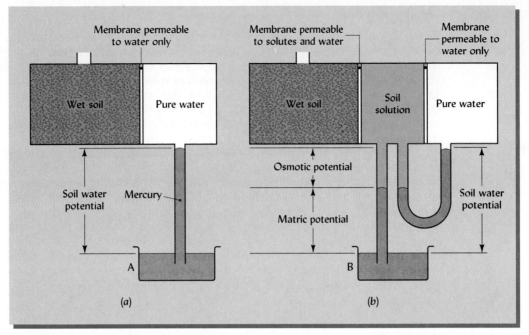

FIGURE 5.11 Relationship among osmotic, matric, and combined soil water potentials. Assume a container of soil separated from pure water by a membrane permeable only to the water; the pure water is connected with a vessel of mercury through a tube as shown (left). Water will move into the soil in response to the attractive forces associated with soil solids (matric) and with solutes (osmotic). At equilibrium the height of rise in the mercury in the tube above vessel A is a measure of this combined soil water potential (matric plus osmotic). If a second container were to be placed between the pure water and the soil and if it were separated from the soil by a membrane permeable to *both* water and solutes (right), ions would move from the soil into this container, eventually giving a concentration not too different from the soil solution. The difference between the potential energies of the pure water and of the soil solution gives a measure of the *osmotic potential*. The *matric potential* is the difference between the combined and osmotic potentials and is measured by the height of rise of the mercury in vessel B. The gravity potential is not shown in this diagram. [Modified from Richards (1965)]

5.4 SOIL MOISTURE CONTENT AND SOIL WATER POTENTIAL

The previous discussions suggest an inverse relation between the water content of soils and the tenacity with which the water is held in soils. Water is more likely to flow out of a wet soil than from one low in moisture. Many factors affect the relationship between soil water potential (ψ) and moisture content (θ). A few examples will illustrate this point.

TABLE 5.1 Approximate Equivalents Among Expressions of Common Differences in Energy Levels of Soil Water

Height of unit column of water (cm)	Soil water potential (bars)	Soil water potential (kPa)[a]
0	0	0
10.2	−0.01	−1
102	−0.1	−10
306	−0.3	−30
1,020	−1.0	−100
15,300	−15	−1,500
31,700	−31	−3,100
102,000	−100	−10,000

[a]The SI unit kilopascal (kPa) is equivalent to 0.01 bars.

Soil Moisture versus Energy Curves

The relationship between soil water potential (ψ) and moisture content (θ) of three soils of different textures is shown in Figure 5.12. The absence of sharp breaks in the curves indicates a gradual change in the water potential with increased soil water and vice versa. The clay soil holds much more water at a given potential level than does loam or sand. Likewise, at a given moisture content, the water is held much more tenaciously in the clay than in the other two soils. As we shall see, however, as much as half or even more of the water held by clay soils is held so tightly that it cannot be removed by growing plants. Soil texture clearly exerts a major influence on soil moisture retention.

Soil structure also influences soil water content–energy relationships. A well-granulated soil has more total pore space than one with poor granulation or one that has been compacted. The reduced pore space may result in a lower water-holding capacity. The compacted soil also may have a higher proportion of small- and medium-sized pores, which tend to hold water with greater tenacity than do larger pores.

The soil water versus potential curves in Figure 5.12 have marked practical significance. They illustrate water retention–energy relationships, which influence various field processes, including the movement of water in soils and the uptake and use of water by plants. The curves may be referred to frequently as the applied aspects of soil water behavior are considered in the following sections.

Hysteresis

If one measures the soil water content–soil water potential relationship as a soil dries, it will be found to differ slightly from the relationship measured as the soil is rewetted. This phenomena, known as *hysteresis,* is illustrated in Figure 5.13. Hysteresis is caused by a number of factors, including the nonuniformity of soil pores. As soils are wetted some of the smaller pores are bypassed, leaving entrapped air that prevents water penetration. Likewise, as a saturated soil dries, some of the macropores that may be surrounded only by micropores may not lose their water until the matric potential is low enough to remove the water from the smaller pores (see Figure 5.13). Also, the swelling and shrinking of clays as the soil is dried and rewetted brings about changes in soil

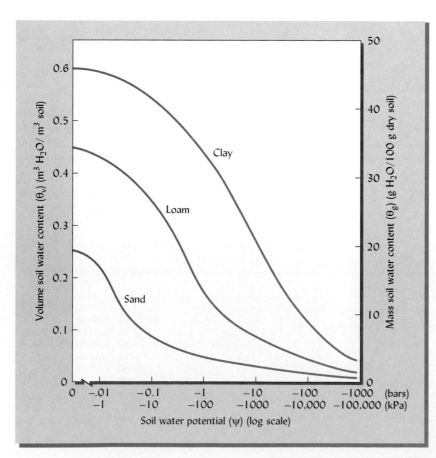

FIGURE 5.12 Soil water/potential curves for three representative mineral soils. The curves show the relationship obtained by slowly drying completely saturated soils. The soil water potential ψ (which is negative) is expressed in terms of bars (upper scale) and kilopascals (kPa) (lower scale). Note that the soil water potential is plotted on a log scale.

structure that affect the soil–water relationships. Hysteresis is important because one must know whether soils are being wetted or dried when properties of one soil are compared with those of another.

5.5 MEASURING SOIL WATER CONTENT AND WATER POTENTIAL

Two general types of measurements are applied to soil water. The water *content* may be measured, directly or indirectly, or the soil water *potential* energy status may be determined.

Water Content

The most common means of expressing soil water content is the mass or volume of water associated with a given mass or volume of soil. The **volumetric water content** (θ) is defined as the volume of water associated with a given volume (usually a cubic meter) of dry soil (see Figure 5.12). A comparable expression is the **mass water content,** (θ$_m$) or the mass of water associated with a given mass (usually 1 kg) of soil. Either of these two means of expression is acceptable; however, we shall use the volumetric water content (θ) in this text.[1]

---

[1]Another practical means of measuring water added to soil (especially by irrigation) is that of the **acre-foot.** This is the amount of water needed to cover an acre of land to a depth of one foot. Similarly, an acre-inch represents the water needed to cover an acre to a depth of one inch. Such terminology is used commonly in the United States in ascertaining the amount of irrigation water needed or used.

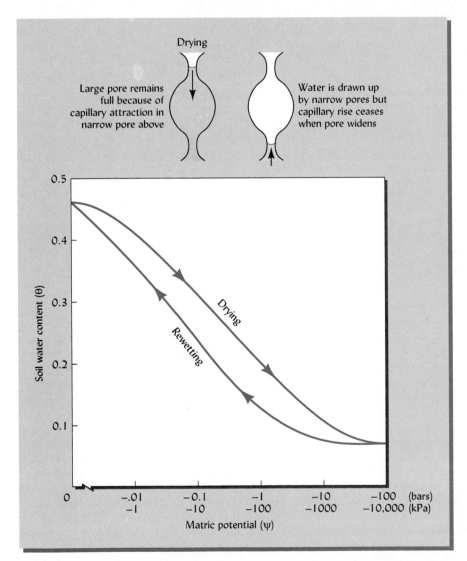

FIGURE 5.13 The relationship between soil water content and matric potential of a soil upon being dried and then rewetted. The phenomenon, known as *historesis,* is apparently due to factors such as the nonuniformity of individual soil pores, entrapped air, and the swelling and shrinking that might affect soil structure. The drawings show the effect of nonuniformity of pores.

A number of methods are used to measure the amount of water in a soil. *Gravimetric* procedures permit the direct measurement of the amount of water associated with a given mass (and, indirectly, a given volume) of dry soil solids. A sample of moist soil, usually taken in cores in the field, is weighed and then dried in an oven at a temperature of 100–110°C, and finally weighed again. The water lost by the soil represents the soil moisture in the moist sample. Box 5.1 provides examples of how θ and θ_m can be calculated.

BOX 5.1   GRAVIMETRIC DETERMINATION OF SOIL WATER CONTENT

The gravimetric procedures for determining soil water content are relatively simple. Assume that you want to determine the water content of a 100-g sample of moist soil. You dry the sample in an oven kept at 100–110°C and then weigh it again. Assume that the dried soil now weighs 70 g, which indicates that 30 g of water has been removed from the moist soil. Expressed in kilograms this is 30 kg water associated with 70 kg dry soil.

Since the mass soil water content (θ_m) is commonly expressed in terms of kg water associated with 1 kg dry soil (*not* 1 kg of wet soil), it can be calculated as follows:

$$\frac{30 \text{ kg water}}{70 \text{ kg dry soil}} = \frac{X \text{ kg water}}{1 \text{ kg dry soil}}$$

$$X = \frac{30}{70} = 0.428 \text{ kg water/kg dry soil} = \theta_m$$

To calculate the volume soil water content (θ), we need to know the bulk density of the dried soil, which in this case we shall assume to be 1.3 Mg/m³. In other words, a cubic meter of this soil has a mass of 1300 kg. From the above calculations we know that the mass of water associated with this 1300 kg is 0.428 × 1300 or 556 kg.

Since 1 m³ of water has a mass of 1000 kg, the 556 kg of water will occupy 556/1000 or 0.556 cubic meters. Thus, the volume water content is 0.556 m³/m³ of dry soil.

The relationship between the mass and volume water contents for a soil can be summarized as:

$$\theta = D_b \times \theta_m$$

There are several indirect methods of measuring soil water content, some of which are very useful in the field. The *electrical resistance block* method takes advantage of the fact that the electrical resistance of certain porous materials such as gypsum, nylon, and fiberglass is related to their moisture contents. When small blocks of these materials, suitably embedded with electrodes, are placed in moist soil, they absorb water in proportion to the soil moisture content. Measuring the electrical resistance in the block gives an estimate of the content of water in the surrounding soil (Figure 5.14). These blocks are limited in their accuracy and in the range of soil moisture contents they measure (–100 to –1500 kPa). However, they are inexpensive and can be used to measure approximate changes in soil moisture during one or more growing seasons.

Another indirect method of determining soil moisture in the field involves *neutron scattering*. The principle of the neutron moisture meter is based on the ability of hydrogen atoms to reduce drastically the speed of fast-moving neutrons and to scatter them (Figure 5.15). These meters are versatile and give accurate results in mineral soils, where water is the primary source of combined hydrogen. In organic soils, however, the method has some constraints, since many of the hydrogen atoms in these soils is combined in organic substances rather than in water.

A relatively new technique has been developed to measure both soil moisture and salinity. Known as *time-domain reflectometry (TDR)* it measures two elements: (1) the time it takes for the electromagnetic impulse provided by a buried parallel wire probe to move a few centimeters in the soil, and (2) the degree of dissipation of the impulse as it impacts with the soil. The transit time is related to the soil's dielectric constant. Since the dielectric constant of water is 15 to 20 times that of soil solids, the TDR values are proportional to the amount of water in the soil. Simultaneously, the dissipation of the

FIGURE 5.14 A cutaway view of commercial gypsum electrical resistance block placed about 45 cm below the soil surface. Thin wires lead from the block to the surface where they can be connected to a resistance meter. For most applications several blocks should be buried at different depths throughout the root zone. (Photo courtesy of R. Weil)

signal is related to the level of salts in the soil solution. Thus, both soil moisture content and salinity can be measured using TDR.

Field Tensiometers

The tenacity with which water is held in soils is an expression of soil water potential (ψ). *Field tensiometers,* such as the one shown in Figure 5.16, measure this attraction. The tensiometer is filled with water and then placed in the soil. Water in the tensiometer is drawn through a fine porous cup into the adjacent soil until equilibrium is reached, at which time the potential in the soil is the same as that in the tensiometer. Tensiometers are used successfully in determining the need for irrigation when the soil is to be kept well supplied with water. Their range of usefulness is between 0 and −80 kPa potential, a range comparable with that attained using laboratory tensiometers called *tension plates.*

A *pressure membrane* apparatus (Figure 5.17) is used to measure matric potential–moisture content relations at potential values as low as −10,000 kPa. This important laboratory tool makes possible simultaneous accurate measurement of energy–soil moisture relations of a number of soil samples over a wide energy range in a relatively short time. It is used to obtain the data to construct soil moisture/potential curves such as those shown in Figure 5.12.

We have described several methods for determining water *content* and the water *potential* in soils. Both types of information are needed to understand, predict, and manage changes in soil water. The behavior of soil water is most closely related to the energy status of the water, not the amount of water in a soil. Thus, a clay loam and a loamy sand will both feel moist and will easily supply water to plants when the ψ_m is, say, −10 kPa. However, the amount of water held by the clay loam, and thus the length of time it could supply water to plants, would be far greater at this potential than would be the case for the loamy sand.

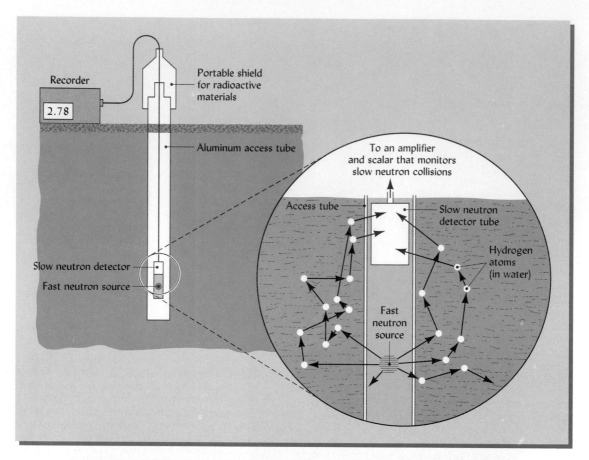

FIGURE 5.15 How a neutron moisture meter operates. The probe, containing a source of fast neutrons and a slow neutron detector, is lowered into the soil through an access tube. Neutrons are emitted by the source (e.g., radium or americium-beryllium) at a very high speed (fast neutrons). When these neutrons collide with a small atom such as the hydrogen contained in soil water, their direction of movement is changed and they lose part of their energy. These "slowed" neutrons are measured by a detector tube and a scalar. The reading is related to the soil moisture content. The photograph shows a neutron probe in the field. The heavy metal cylinder is a shield to protect the operator from irradiation. It will be placed over the aluminum lined hole (extreme lower right) and the neutron source will then be lowered down into the hole for measurement. (Photo courtesy of R. Weil)

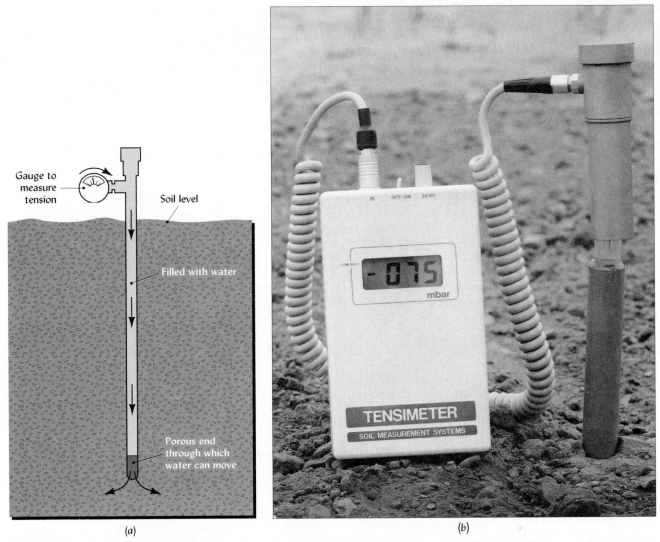

FIGURE 5.16 Tensiometer method of determining water potential in the field. (a) Cross section showing the essential components of a tensiometer. Water moves through the porous end of the instrument in response to the pull (matric potential) of the soil. (b) A tensiometer in place in the field showing a portable meter to measure the water potential in millibars (mbar). (Photo courtesy A. M. Wierenga, Soil Measurement Systems, Las Cruces, N.Mex.)

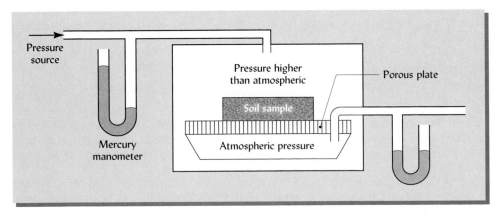

FIGURE 5.17 Pressure membrane apparatus used to determine water content–matric potential relations in soils. An outside source of gas creates a pressure inside the cell. Water is forced out of the soil through a porous plate into a cell at atmospheric pressure. The applied pressure when the downward flow ceases gives a measure of the water potential in the soil. This apparatus will measure much lower soil water potential values (drier soils) than will tensiometers or tension plates.

5.6 TYPES OF SOIL WATER MOVEMENT

Three types of movement within the soil are recognized—*saturated flow, unsaturated flow,* and *vapor movement.* Both saturated and unsaturated flow involve liquid water in contrast to vapor flow. We shall consider liquid water flow first.

The flow of liquid water is due to a *gradient* in potential from one soil zone to another. The direction of flow is from a zone of higher to one of lower water potential. Saturated flow takes place when the soil pores are completely filled (or saturated) with water. Unsaturated flow occurs when the larger pores in the soil are filled with air, leaving only the smaller pores to hold and transmit water. As we consider the three types of movement, it will be evident that in all cases water flows in response to energy gradients.

5.7 SATURATED FLOW THROUGH SOILS

Under some conditions, at least part of a soil profile may be completely saturated; that is, all pores, large and small, are filled with water. The lower horizons of poorly drained soils are often saturated, as are portions of well-drained soils above stratified layers of clay. During and immediately following a heavy rain or irrigation, pores in the upper soil zones are often filled entirely with water.

The quantity of water (Q) that flows through a column of saturated soil can be expressed by Darcy's law as follows:

$$Q = K_s A \Delta P / L$$

where K_s is the saturated hydraulic conductivity, A is area of the column through which the water flows, P the hydrostatic pressure difference from the top to the bottom of the column, and L is the length of the column. Since area A and length L of a given column are fixed, the rate of flow is seen to be determined by the **hydraulic force** ΔP driving the water through the soil (commonly gravity) and the **hydraulic conductivity** K_s or ease with which the soil pores permit water movement.

The saturated hydraulic conductivity K_s of a uniform saturated soil remains fairly constant over time, and is dependent on the size and configuration of the soil pores, all of which are filled with water. This is in contrast to the value of K in an unsaturated soil, which decreases as the water content decreases.

An illustration of vertical saturated flow is shown in Figure 5.18. The driving force, known as the *hydraulic gradient,* is the difference in height of water above and below the soil column. The volume of water moving down the column will depend on this force as well as on the saturated hydraulic conductivity of the soil.

It should not be inferred from Figure 5.18 that saturated flow occurs only down the profile. The hydraulic force can also cause horizontal and even upward flow as occurs when groundwater wells up under a stream. The rate of such flow is usually not quite as rapid, however, since the force of gravity does not assist horizontal flow and hinders upward flow. Downward and horizontal flow is illustrated in Figure 5.19, which records the flow of irrigation water into two soils, a sandy loam and a clay loam. Some of the water movement was likely by saturated flow. The water moved down much more rapidly in the sandy loam than in the clay loam. On the other hand, horizontal movement (which would have been largely by unsaturated flow) was much more evident in the clay loam.

Factors Influencing the Hydraulic Conductivity of Saturated Soils

Any factor affecting the size and configuration of soil pores will influence hydraulic conductivity. The total flow rate in soil pores is proportional to the fourth power of the radius. Thus, flow through a pore 1 mm in radius is equivalent to that in 10,000 pores with a radius of 0.1 mm even though it takes only 100 pores of radius 0.1 mm to give the same cross-sectional area as a 1-mm pore. As a result, the macropore spaces account for most of the saturated water movement.

The texture and structure of soils are the properties to which hydraulic conductivity is most directly related. Sandy soils generally have higher saturated conductivities than finer-textured soils. Likewise, soils with stable granular structure conduct water much more rapidly than do those with unstable structural units, which break down upon

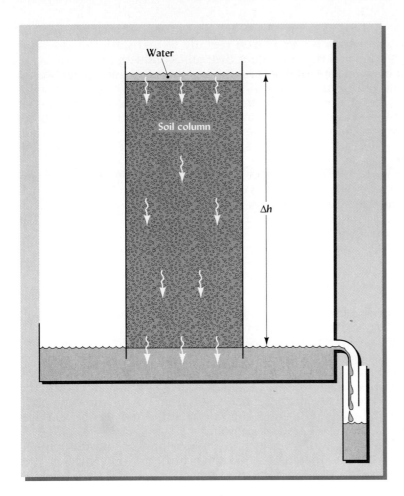

FIGURE 5.18 Saturated (percolation) flow in a column of soil. All soil pores are filled with water. The force drawing the water through the soil is Δh, the difference in the heights of water above and below the soil layer. This same force could be applied horizontally. The water is shown running off into a side container to illustrate that water is actually moving down the profile.

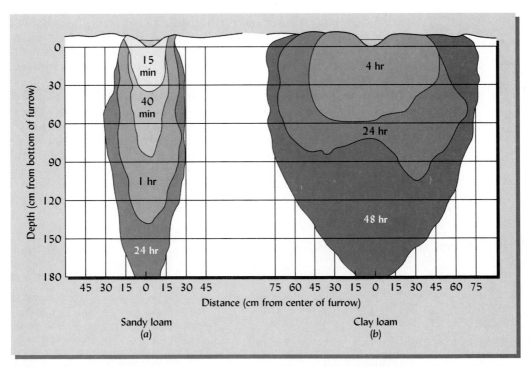

FIGURE 5.19 Comparative rates of irrigation water movement into a sandy loam and a clay loam. Note the much more rapid rate of movement in the sandy loam, especially in a downward direction. [Redrawn from Cooney and Peterson (1955)]

being wetted. Fine clay and silt can clog the small connecting channels of even the larger pores. Entrapped air, which is common in recently wetted soils, can also slow down the movement of water and thereby reduce hydraulic conductivity.

Concern over the movement of pesticides and other chemicals from organic wastes through the soil and into the groundwater has called attention to the heterogeneity of some soils that can markedly affect their hydraulic conductivity. Natural fissures or cracks in soils such as those created by burrowing animals and large root channels are pathways for rapid downward movement of the chemical-laden water which can then move with the groundwater to nearby streams and rivers. Likewise, large and deep cracks developed during dry spells in soils that are high in swelling-type clays permit ready downward movement of rainwater and associated chemicals before the surrounding soil can become wetted and the cracks closed. Figure 5.20 illustrates how such downward movement can take place and emphasizes the importance of hydraulic conductivity. The movement of water through macropores must not be underestimated in environmental-quality considerations.

5.8   UNSATURATED FLOW IN SOILS

Most of the time, water movement takes place when upland soils are unsaturated. Such movement occurs, however, in a more complicated environment than that which characterizes saturated water flow. In saturated soils essentially all the pores are filled with water, although the most rapid water movement is through the large and continuous pores. But in unsaturated soils, these macropores are filled with air leaving only the finer pores to accommodate water movement. Also, the water content and, in turn, the tightness with which water is held (water potential) varies from time to time and place to place in unsaturated soils. This influences the rate of water movement and also makes it more difficult to measure the flow of soil water.

As was the case for saturated water movement, the driving force for unsaturated water flow is differences in water potential. This time, however, differences in the matric potential, not gravity, is the primary driving force. This *matric potential gradient* is the difference in the matric potential of the moist soil areas and nearby drier areas into which the water is moving. Movement will be from a zone of thick moisture films (high matric potential, e.g., –1 kPa) to one of thin films (lower matric potential, e.g., –100 kPa).

Figure 5.21 shows the general relationship between matric potential (ψ_m) (and, in turn, water content level) and hydraulic conductivity of a sandy loam and clay soil. Note that at or near zero potential (which characterizes the saturated flow region), the hydraulic conductivity is thousands of times greater than at potentials that characterize typical unsaturated flow (–10 kPa and below).

At high potential levels (high moisture contents), hydraulic conductivity is higher in the sand than in the clay. The opposite is true at low potential values (low moisture contents). This relationship is to be expected since the dominance of large pores in the coarse-textured soil encourages saturated flow, whereas the prominence of finer (capillary) pores in the clay soil encourages more unsaturated flow than in the sand.

The influence of potential gradient on water movement from a soil at three different moisture levels to a drier soil is illustrated by the moisture curves shown in Figure 5.22. Researchers measured in the laboratory the rate of water movement from the three

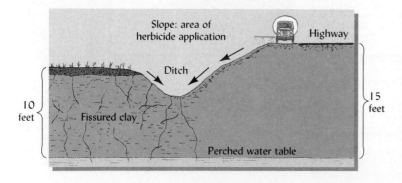

FIGURE 5.20 An illustration of the effects of natural cracks in soils on the conductivity of water and pesticides downward to the water table. An herbicide (weed killer) was applied alongside a highway (right) with the expectation that downward movement into the water table would not be a serious problem since the surrounding soils were fine-textured and would not be expected to readily permit infiltration of the chemical. Because of the wide cracks in this swelling-type clay, however, the first heavy rain carried the chemicals into the groundwater before the soil could swell and shut the cracks. Through the groundwater the herbicide could move into nearby streams. [From DeMartins and Cooper (1994) with permission of Lewis Publishers]

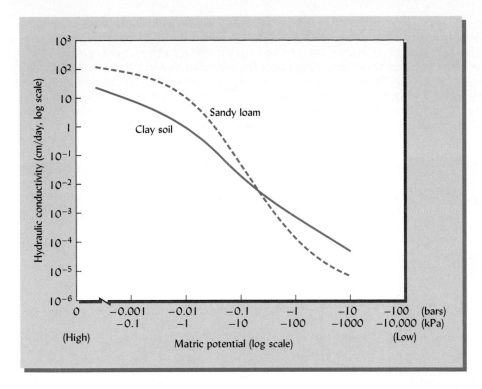

FIGURE 5.21 Generalized relationship between matric potential and hydraulic conductivity for a sandy soil and a clay soil (note log scales). Saturation flow takes place at or near zero potential, while much of the unsaturated flow occurs at a potential of −0.1 bar (−10 kPa) or below.

moist soil samples to an adjacent dry soil. Water moved more rapidly from the sample with highest moisture content. The higher the water content in the moist soil, the greater the matric potential gradient between the moist and dry soil and, in turn, the more rapid the flow.

5.9   WATER MOVEMENT IN STRATIFIED SOILS

The discussion up to now has assumed in most cases that soils are uniform in texture and structure. In the field, however, subsurface layers differing in physical makeup from the overlying horizons are common. These have a profound influence on water movement and deserve specific attention.

Various kinds of stratification are found in many soils. Impervious silt or claypans are common, as are sand and gravel lenses or other subsurface layers. In all cases, the effect on water movement is similar—that is, the downward movement is impeded. The

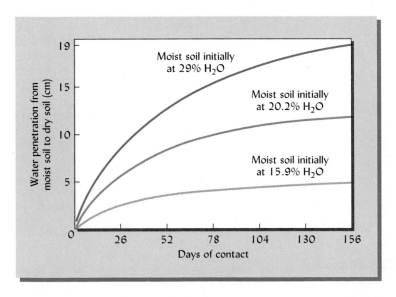

FIGURE 5.22   Rate of water movement from a moist soil at three moisture levels to a drier one. The higher the water content of the moist soil, the greater is the potential gradient and the more rapid the delivery. Water adjustment between two slightly moist soils at about the same water content will be exceedingly slow. [After Gardner and Widtsoe (1921)]

influence of layering is represented in Figure 5.23, where a layer of sand is seen to impede downward movement of water in an otherwise fine-textured soil. The macropores of the sand offer less attraction for the water than does the finer-textured material. Only when the downward-moving water builds up at the interlayer interface will the water be so loosely held by the fine-textured material that the sand below can attract the water and permit it to move downward. The fact that coarse-textured layers (e.g., gravel or sand) can *hinder* the downward flow of water must be considered when using such materials in planting containers or landscape drainage schemes.

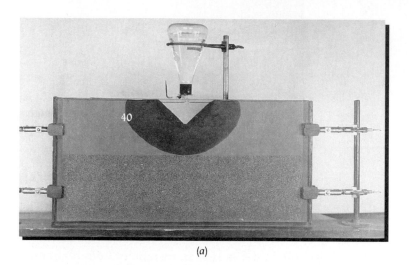

(a)

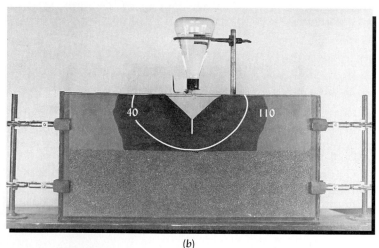

(b)

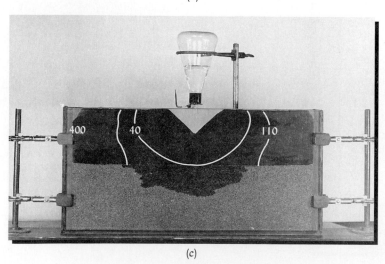

(c)

FIGURE 5.23 Downward water movement in soils with a stratified layer of coarse material. (a) Water is applied to the surface of a medium-textured topsoil. Note that after 40 min, downward movement is no greater than movement to the sides, indicating that in this case the gravitational force is insignificant compared to the matric potential gradient between dry and wet soil. (b) The *downward* movement stops when a coarse-textured layer is encountered. After 110 min, no movement into the sandy layer has occurred. The macropores of the sand provide less attraction for water than the finer-textured soil above. Only when the water content (and in turn the matric potential gradient) is raised sufficiently will the water move into the sand. (c) After 400 min, the water content of the overlying layer becomes sufficiently high to give a water potential of about −1 kPa or more, and downward movement into the coarse material takes place. Thus sandy layers, as well as compact silt and clay, influence downward moisture movement in soils. (Courtesy W. H. Gardner, Washington State University)

The effects of stratification have great practical consequences. For example, because it retards downward movement, stratification definitely influences the amount of water the upper part of the soil holds in the field. The stratified layer acts as a moisture barrier until a relatively high water content is built up. This gives a much higher field-moisture level than that normally encountered in freely drained soils (Figure 5.24).

There are other examples of the effects of stratification on water movement in soils. A layer of coarse sand or gravel underneath the soil in a flowerpot can impede the drainage, the water content building up immediately above the gravel layer. Also, the upward movement of capillary water is impeded by layers of sand or gravel. The large pores in these layers do not support capillary movement. Consequently, water moves up to the stratified layer but cannot cross it to supply moisture to overlying layers. Thus, plants growing on some stratified soils are subject to drought since the plants are able to exploit only the upper soil layers.

5.10 WATER VAPOR MOVEMENT IN SOILS

Two types of water vapor movement occur in soils, *internal* and *external*. Internal movement takes place within the soil, that is, in the soil pores. External movement occurs at the land surface, and water vapor is lost by *surface evaporation* (see Section 14.4).

Water vapor moves from one point to another within the soil in response to differences in vapor pressure. Thus, water vapor will move from a moist soil where the soil air is nearly 100% saturated with water vapor (high vapor pressure) to a drier soil where the vapor pressure is somewhat lower. Also, water will move from a zone of low salt content to one with a higher salt content (for example, around a fertilizer granule). The salt lowers the vapor pressure of the water and encourages water movement from the surrounding soil.

If the temperature of one part of a uniformly moist soil is lowered, the vapor pressure will decrease and water vapor will tend to move toward this cooler part. Heating will have the opposite effect in that heating will increase the vapor pressure and the water vapor will move away from the heated area. Figure 5.25 illustrates the relationships.

The actual amount of water vapor in a soil at optimum moisture for plant growth is surprisingly small, being at one time perhaps no more than 10 kg in the upper 15 cm of a hectare of a silt loam soil. This compares with some 600,000 kg of liquid water in the same soil volume.

Because the amount of water vapor is small, its movement in soils is of limited practical importance if the soil moisture is kept near optimum for plant growth. In dry soils, however, vapor water movement may be of considerable significance, especially in sup-

FIGURE 5.24 One result of soil layers with contrasting texture. This North Carolina soil has about 50 cm of loamy sand coastal plain material atop deeper layers of silty clay-loam-textured material derived from the piedmont. Rainwater rapidly infiltrates the sandy surface horizons, but its downward movement is arrested at the finer-textured layer, resulting in saturated conditions near the surface and a quicksandlike behavior. (Photo courtesy of R. Weil)

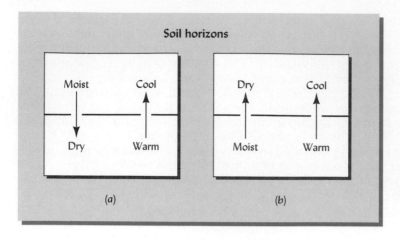

Soil horizons

(a)

(b)

FIGURE 5.25 Vapor movement tendencies that may be expected between soil horizons differing in temperature and moisture. In (a) the tendencies more or less negate each other, but in (b) they are coordinated and considerable vapor transfer might be possible if the liquid water in the soil capillaries does not interfere.

plying water to drought-resistant desert plants, many of which can exist at extremely low soil water content.

5.11   RETENTION OF SOIL WATER IN THE FIELD

Keeping in mind the energy–soil water relations covered in previous sections, we now turn to some more practical considerations. We shall start by following the moisture and energy relations of a soil during and after a heavy rain or the application of irrigation water.

Maximum Retentive Capacity

When all soil pores are filled with water from rainfall or irrigation, the soil is said to be *saturated* with respect to water (Figure 5.26) and at its *maximum retentive capacity*. The matric potential is high, being nearly the same as that of pure water. Maximum retentive capacities of soils in a watershed are useful in predicting how much rainwater can be stored in the soil temporarily, thus possibly avoiding downstream floods.

FIELD CAPACITY.   Following the rain or irrigation, some of the water will drain downward quite rapidly in response to the hydraulic gradient (mostly gravity) (Figure 5.27). After one to three days, this rapid downward movement will become negligible. The soil then is said to be at its *field capacity*. At this time, water has moved out of the **macropores**, and its place has been taken by air. The **micropores** or **capillary pores** are still filled with water and will supply the plants with needed water. The matric potential will vary slightly from soil to soil but is generally in the range of –10 to –30 kPa, assuming drainage into a less moist zone of similar porosity. Water movement will continue to take place, but the rate of movement (unsaturated flow) is slow since it now is due primarily to capillary forces, which are effective only in micropores (Figure 5.26).

Permanent Wilting Percentage or Wilting Coefficient

As plants absorb water from a soil, they lose most of it through evaporation at the leaf surfaces (transpiration). Some water also is lost by evaporation directly from the soil surface. These two losses occur simultaneously, and the combined loss is termed *evapotranspiration.*

As the soil dries, plants begin to wilt to conserve moisture during the daytime. At first the plants will regain their vigor at night, but ultimately they will remain wilted night and day. Although not dead, the plants are now in a permanently wilted condition and will die if water is not provided. Under this condition, a measure of soil water potential (ψ) shows a value of about –15 bars (–1500 kPa) for most crop plants. Some xerophytes (desert-type plants) can continue to remove water at even more negative potentials.

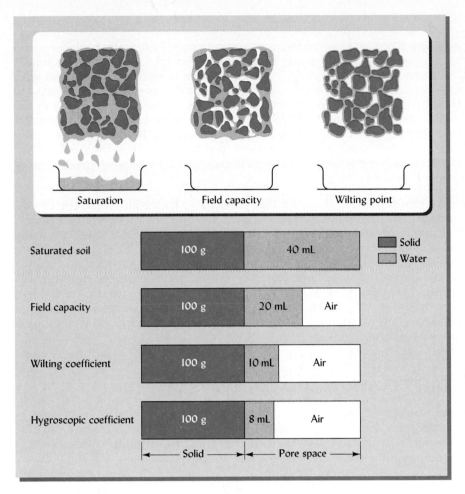

FIGURE 5.26 Volumes of water and air associated with 100 g of a well-granulated silt loam at different moisture levels. The top bar shows the situation when a representative soil is completely saturated with water. This situation will usually occur for short periods of time during a rain or when the soil is being irrigated. Water will soon drain out of the larger pores (*macropores*). The soil is then said to be at the *field capacity*. Plants will remove water from the soil quite rapidly until they begin to wilt. When permanent wilting of the plants occurs, the soil water content is said to be at the *wilting coefficient*. There is still considerable water in the soil, but it is held too tightly to permit its absorption by plant roots. A further reduction in water content to the *hygroscopic coefficient* is illustrated in the bottom bar. At this point the water is held very tightly, mostly by the soil colloids. (Top drawings modified from *Irrigation on Western Farms* published by the U.S. Departments of Agriculture and Interior)

The water content of the soil at this stage is called the ***wilting coefficient*** or ***permanent wilting percentage.*** The water remaining in the soil is found in the smallest of the micropores and around individual soil particles (Figure 5.26). Obviously, a considerable amount of the soil water is not available to higher plants, especially in fine-textured soils and those high in organic matter.

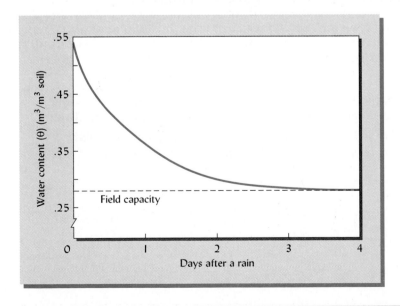

FIGURE 5.27 The water content of a soil drops quite rapidly by drainage following a heavy, steady rainstorm. After two or three days the rate of water movement out of the soil is quite slow and the soil is said to be at the *field capacity.*

Hygroscopic Coefficient

When soil moisture is further lowered below the wilting point, the water molecules that remain are very tightly held, mostly being adsorbed by colloidal soil surface. This state is approximated when the atmosphere above a soil sample is essentially saturated with water vapor (98% relative humidity) and equilibrium is established. The water is held so tightly (–3100 kPa) that much of it is considered nonliquid and can move only in the vapor phase. The moisture content of the soil at this point is termed the *hygroscopic coefficient.* Soils high in colloidal materials (clay and humus) will hold more water under these conditions than will sandy soils that are low in clay and humus (Table 5.2).

Water Potential and Content

As soil water content is reduced from the field capacity to the hygroscopic coefficient, the potential/moisture curves described in Section 5.4 are pertinent. This relationship is illustrated in Figure 5.28, which shows the water content–matric potential relationship for a loam soil and identifies the ranges in potential for each of the field soil conditions just described. The diagram at the right of this figure also suggests physical and biological classification schemes for soil water. However, this diagram is coupled intentionally with the moisture/potential curve to emphasize the fact that there are no clearly identifiable forms of water in soil. There is only a gradual change in potential with water content. This should be kept in mind in the next section as we discuss some commonly used soil water classification schemes.

5.12   CONVENTIONAL SOIL WATER CLASSIFICATION SCHEMES

On the basis of observations of soil–water–plant relations just described, two types of soil water classification schemes have been developed: *physical* and *biological*. These schemes are useful in a practical way even though they lack the scientific basis that characterizes the preceding moisture/energy discussions.

Physical Classification

From a physical point of view, the terms *gravitational, capillary,* and *hygroscopic* waters are identified in Figure 5.28. Water in excess of the field capacity (–10 to –30 kPa) is termed *gravitational*. Even though energy of retention is low, gravitational water is of limited use to plants because it is present in the soil for only a short time, and, while in the soil, it occupies the larger pores, thereby reducing soil aeration. The removal of gravitational water from the soil by drainage is generally a requisite for optimum plant growth. However, gravitational water is the form in which most of soil leaching occurs, including that which transports chemicals such as nutrient ions, pesticides, and organic wastes into the groundwater and, ultimately, streams and rivers.

As the name suggests, *capillary* water is held in the pores of capillary size and behaves according to laws governing capillarity. Such water includes most of the water taken up by growing plants and is held with potentials between –10 to –30 kPa and, at the lower extreme, –3100 kPa.

TABLE 5.2   Volumetric Water Content (θ) at Field Capacity and Hygroscopic Coefficient for Three Representative Soils and the Calculated Capillary Water

Note that the clay soil retains most water at the field capacity, but much of that water is held tightly in the soil at –31 bars potential by soil colloids (hygroscopic coefficient).

	Volume % (θ)		
Soil	Field capacity (10–30 kPa)	Hygroscopic coefficient (–3100 kPa)	Capillary water (col. 1 – col. 2)
Sandy loam	12	3	9
Silt loam	30	10	20
Clay	35	18	17

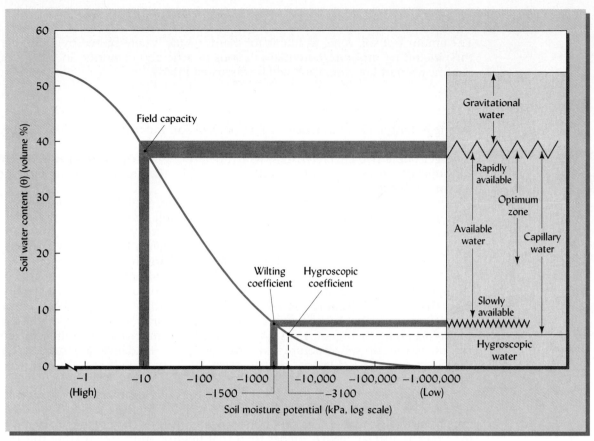

FIGURE 5.28 Water content–matric potential curve of a loam soil as related to different terms used to describe water in soils. The wavy lines in the diagram to the right suggest that measurements such as field capacity are not very quantitative. The gradual change in potential with soil moisture change discourages the concept of different "forms" of water in soils. At the same time, such terms as *gravitational* and *available* assist in the qualitative description of moisture utilization in soils.

Hygroscopic water is that bound tightly by the soil solids at water potential values lower than −3100 kPa. It is essentially nonliquid and moves primarily in the vapor form. Higher plants cannot absorb hygroscopic water, but some microbial activity has been observed in soils containing only hygroscopic water.

Biological Classification

There is a definite relationship between moisture retention and its use by plants. Gravitational water is of limited use to plants and may be harmful. In contrast, moisture retained in the soil between the field capacity (−10 to −30 kPa) and the permanent wilting percentage (−1500 kPa) is said to be usable by plants, and as such is **available water.** Water held at a potential lower than −1500 kPa bars is called *unavailable* to most plants (Figure 5.28).

In most soils, optimum growth of plants takes place when the soil moisture content is kept near the field capacity with a moisture potential near −0.1 to −0.3 bar (−10 to −30 kPa). Thus, the moisture zone for optimum plant growth does not extend over the complete range of moisture availability.

The various terms employed to describe soil water physically and biologically are useful in a practical way, but at best they are only semiquantitative. For example, measurement of field capacity tends to be somewhat arbitrary because the value obtained is affected by such factors as the initial soil moisture in the profile before wetting, the removal of water by plants, and surface evaporation during the period of downward flow. Also, determination of the point at which the downward movement of water due to gravity has essentially ceased is rather arbitrary. These facts stress once again that there is no clear line of demarcation between different forms of soil water.

5.13 FACTORS AFFECTING AMOUNT OF PLANT-AVAILABLE SOIL WATER

The amount of soil water available for plant uptake is determined by a number of factors, including moisture/potential relations (matric and osmotic), soil depth, and soil stratification or layering. Each will be discussed briefly.

Matric Potential

Matric potential (ψ_m) influences the amount of soil water plants can take up because it affects the amounts of water at the field capacity and at the permanent wilting percentage. These two characteristics, which determine the quantity of water a given soil can supply to growing plants, are influenced by the texture, structure, and organic matter content of the soil.

The general influence of texture is shown in Figure 5.29. Note that as fineness of texture increases, there is a general increase in available moisture storage from sands to loams and silt loams. However, clay soils frequently provide less available water than do well-granulated silt loams since the clays tend to have a high wilting coefficient. The comparative available water-holding capacities also are shown by this graph.

The influence of organic matter deserves special attention. The available moisture content of a well-drained mineral soil containing 5% organic matter is generally higher than that of a comparable soil with 3% organic matter. There has been considerable controversy over the degree to which this favorable effect was due directly to the water-supplying ability of organic matter and how much came from the indirect effects of soil organic matter on soil structure and total pore space. Evidence now suggests that both the direct and indirect factors contribute to the favorable effects of organic matter on soil water availability.

The direct effects are due to the very high water-holding capacity of organic matter which, when the soil is at the field capacity, is much higher than that of an equal volume of inorganic matter. Even though the water held by the organic matter at the wilting point is also somewhat higher than that held by inorganic matter, the amount of water available for plant uptake still appears to be greater from the organic fraction of the soil than from an equal volume of inorganic matter. Figure 5.30 provides data from a series of experiments to justify this conclusion.

Organic matter has a second and indirect effect on water available to plants. This is through its influence on soil structure and total pore space. We learned in Section 4.10 that organic matter helped stabilize soil structure and that it increased the total volume as well as the size of the pores. This results in an increase in the amount of water infil-

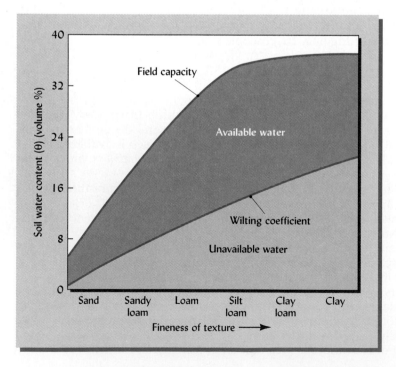

FIGURE 5.29 General relationship between soil water characteristics and soil texture. Note that the wilting coefficient increases as the texture becomes finer. The field capacity increases until we reach the silt loams, then levels off. Remember these are representative curves; individual soils would probably have values different from those shown.

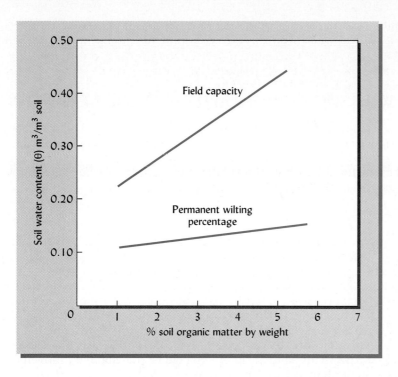

FIGURE 5.30 The effects of organic content on the field capacity and permanent wilting percentage of a number of soils with a silt loam texture. The differences between the two lines shown is the available soil moisture content, which was obviously greater in the soils with higher organic matter levels. [Redrawn from Hudson (1994); used with permission of the Soil & Water Conservation Society.]

trating into a soil as well as the amount of water a soil can hold without simultaneously increasing the quantity of water held at the wilting coefficient. The wisdom of maintaining the organic matter content of soils at reasonable levels is further demonstrated.

Osmotic Potential

The presence of salts in soils, either from applied fertilizers or as naturally occurring compounds, can influence soil water uptake. Osmotic potential (ψ_o) effects in the soil solution tend to reduce the range of available moisture in such soils by increasing the amount of water left in the soil at the time the plants wilt permanently (wilting coefficient). For soils high in salts, the total moisture stress will include the matric potential as well as the osmotic potential of the soil solution. In most humid region soils these osmotic potential effects are insignificant. In contrast, they become of considerable importance in some soils of arid and semiarid regions with significant salt contents.

Soil Depth and Layering

All other factors being equal, deep soils will have greater total available moisture-holding capacities than shallow ones. For deep-rooted plants, this is of practical significance, especially in those subhumid and semiarid regions where supplemental irrigation is not possible.

Soil stratification or layering will influence markedly the available water and its movement in the soil. Impervious layers slow down drastically the rate of water movement and also restrict the penetration of plant roots, thereby reducing the soil depth from which moisture is drawn. Sandy layers also act as barriers to soil moisture movement from the finer-textured layers above, as explained in Section 5.9 and Figure 5.23.

The presence of macropores from rodent tunnels, root channels, and earthworm burrows can greatly increase the downward movement of water. Likewise, development of wide cracks which extend downward into the profile of dry soils that are high in clay (especially Vertisols) markedly affects water infiltration into these soils. Rain or irrigation water literally pours downward into the holes and cracks. Such movement permits a considerable amount of water to enter the soil and to move down deep into the profile (Figure 5.20).

The capacity of soils to store available water determines to a great extent their usefulness for plant growth. The productivity of forest sites is often related to soil water-holding capacity. This capacity is often the buffer between an adverse climate and crop production. It becomes more significant as the use of water for all purposes—industrial and domestic, as well as irrigation—begins to tax the supply of this all-important natural resource.

5.14 HOW PLANTS ARE SUPPLIED WITH WATER: CAPILLARITY AND ROOT EXTENSION

At any one time, only a small proportion of the soil water is adjacent to the absorptive plant root surfaces. How then do the roots get access to the immense amount of water (see Section 6.7) needed to offset transpiration by vigorously growing plants? Two phenomena seem to account for this access: the capillary movement of the soil water to plant roots and the growth of the roots into moist soil.

Rate of Capillary Movement

When plant rootlets absorb water, they reduce the soil moisture content, thus reducing the water potential in the soil immediately surrounding them (see Figure 5.31). In response to this lower potential, water tends to move toward the plant roots. The rate of movement depends on the magnitude of the potential gradients developed and the conductivity of the soil pores. With some sandy soils, the adjustment may be comparatively rapid and the flow appreciable. In fine-textured and poorly granulated clays, the movement will be sluggish and only a meager amount of water will be delivered.

The total distance that water flows by capillarity on a day-to-day basis may be only a few centimeters (Figure 5.32). This would lead one to believe that capillary movement is not a significant means of enhancing moisture uptake by plants. However, if the roots have penetrated much of the soil volume, movement over greater distances may not be necessary. Even during periods of hot, dry weather when evaporative demand is high, capillary movement can be an important means of providing water to plants. It is of special significance during periods of low moisture content when plant root extension is minimized.

Rate of Root Extension

Capillary movement of water is complemented by rapid rates of root extension, which ensure that new root–soil contacts are constantly being established. Such root penetration may be rapid enough to take care of most of the water needs of a plant growing in a soil at optimum moisture. The mat of roots, rootlets, and root hairs in a meadow is an example of the viable root systems of plants. Table 5.3 provides data on the length of roots of soybeans in one experiment. To the figures given must be added the length of thousands of root hairs which penetrate the soil.

The primary limitation of root extension is the small proportion of the soil with which roots are in contact at any one time. Even though the root surface is considerable, as shown in Table 5.3, root–soil contacts commonly account for less than 1% of the total soil surface area. This suggests that most of the water must move from the soil to the root even though the distance of movement may be no more than a few mil-

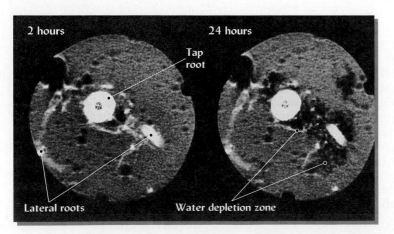

FIGURE 5.31 The intimate soil–root relationship and the rapid depletion of water near roots is illustrated in these two magnetic resonance images (MRIs) of a 2-mm slice of a 25-mm cross-sectional view of moist sand surrounding the roots of a loblolly pine seedling. (Left) Image of a taproot (large white circle) and two woody lateral roots (elongated bright area to the right and L-shaped somewhat bright area to the left) only two hours after water was supplied. Note some water depletion immediately around the taproot (dark area to the right of the taproot). (Right) After 24 hours, the zone of water depletion (dark area around the roots) has expanded around the taproot and the lateral root to the left. Ready movement of water to the roots is obvious. (Photos courtesy Dr. Janet S. MacFall, Duke University Medical Center)

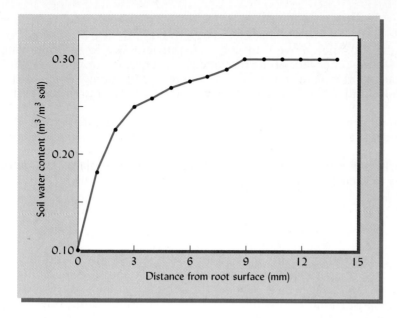

FIGURE 5.32 The drawdown of soil water levels surrounding a radish root after only two hours of transpiration. Water has moved by capillarity from a distance of at least 9 millimeters around the upper part of the radish root. [Modified from Hamza and Aylmore (1992); used with permission of Kluwer Academic Publishers, The Netherlands]

TABLE 5.3 Length of Soybean Roots at Different Soil Depths in a Captina Silt Loam (Typic Fragiudult) in Arkansas

Soil depth (cm)	Root length (km/m³)	
	Nonirrigated	Irrigated
0–16	76	89
16–32	30	37
32–48	21	27
48–64	14	16

Calculated from Brown et al. (1985).

limeters. It also suggests the complementarity of capillarity and root extension as means of providing soil water for plants.

Root Distribution

The distribution of roots in the soil profile determines to a considerable degree the plant's ability to absorb soil water. Most plants, both annuals and perrenials, have the bulk of their roots in the upper 25 to 30 cm of the profile (Table 5.4). Perennial plants such as alfalfa and trees have some roots that grow very deeply (>3 m) and are able to absorb a considerable proportion of their moisture from subsoil layers. Even in these cases, however, it is likely that much of the root absorption is from the upper layers of the soil, provided these layers are well supplied with water. On the other hand, if the upper soil layers are moisture-deficient, even annual plants such as corn and soybeans will absorb much of their water from the lower horizons.

TABLE 5.4 Percentage of Root Mass of Three Crops and Two Trees Found in the Upper 30 cm Compared with Deeper Depths (30–180 cm)

Crop	Percentage of roots	
	Upper 30 cm	30–180 cm
Soybeans	71	29
Corn	64	36
Sorghum	86	14
Pinus radiata	82	18
Eucalyptus marginata	86	14

Data for crops from Mayaki et al. (1976); for trees, estimated from Bowen (1985).

As roots grow into the soil, they move into pores of sufficient size to accommodate them. Contact between the outer cells of the root and the soil permits ready movement of water from the soil into the plant in response to differences in energy levels (Figure 5.33). When the plant is under moisture stress, however, the roots tend to shrink in size in response to this stress. Such conditions exist during a hot, dry spell and are most severe during the daytime, when transpiration from plant leaves is at a maximum. The diameter of roots under these conditions may shrink by 30 to 50%. This reduces considerably the direct root–soil contact as well as the movement of liquid water and nutrients into the plants. While water vapor can still be absorbed by the plant, its rate of absorption is too low to keep any but the most drought-tolerant plants alive.

5.15   CONCLUSION

Water impacts all life. The interaction and movement of this simple compound in soils helps determine whether these impacts are positive or negative. An understanding of the principles which govern the attraction of water for soil solids and for dissolved ions can help maximize the positive impacts while minimizing the less desirable ones.

The water molecule has a polar structure that stimulates the electrostatic attraction of water to both soluble cations and soil solids. These attractive forces tend to reduce the potential energy level of soil water below that of pure water. The extent of this reduction, called *soil water potential* (ψ), has a profound influence on a number of soil properties, but especially on the movement of soil water and its uptake by plants.

The water potential due to the attraction of soil solids (matrix) for water (matric potential, ψ_m) combines with the force of gravity (ψ_g) largely to control water movement. This movement is relatively rapid in soils high in moisture and with an abundance of macropores. In drier soils, however, the adsorption of water on the soil solids is so strong that its movement in the soil and its uptake by plants are greatly reduced. As a consequence, plants die for lack of water—even though there are still significant quantities of water in the soil—because that water is unavailable to plants.

Water is supplied to plants by capillary movement toward the root surfaces and by growth of the roots into moist soil areas. Both processes are important. Vapor movement takes place in soils but is of significance only as a supply of water for drought-resistant desert plants (xerophytes).

The osmotic potential (ψ_o) becomes significant in soils with high soluble salt levels that can impede plant uptake of water from the soil. Such conditions occur most often in soils with restricted drainage in areas of low rainfall.

The characteristics and behavior of soil water are very complex. As we have gained more knowledge, however, it has become apparent that soil water is governed by relatively simple, basic physical principles. Furthermore, researchers are discovering the similarity between these principles and those governing the movement of groundwater and the uptake and use of soil moisture by plants—the subject of the next chapter.

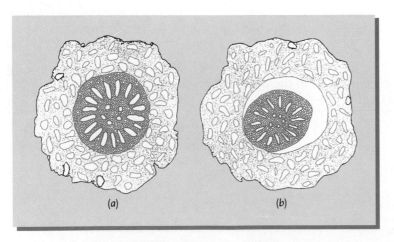

(a)   (b)

FIGURE 5.33 Cross section of a corn root surrounded by soil. (a) During periods of low soil moisture stress on the plant, the root completely fills the soil pore. (b) When the plant is under severe moisture stress, such as during a hot dry period, the root shrinks, significantly reducing root–soil contact. Such shrinkage of roots may occur on a hot summer day, even when the soil water content is high. [From Huck et al. (1970)]

REFERENCES

Bowen, G. D. 1985. "Roots as components of tree productivity," in M. G. R. Cannell and J. E. Jackson (eds.), *Attributes of Trees as Crop Plants* (Midlothian, Scotland: Institute of Terrestrial Ecology).

Brown, E. A., C. E. Caviness, and D. A. Brown. 1985. "Response of selected soybean cultivars to soil moisture deficit," *Agronomy J.,* **77**:274–278.

Cooney, J. J., and J. E. Peterson. 1955. *Avocado Irrigation,* Leaflet 50, California Agr. Ext. Serv.

DeMartinis, J. M., and S. C. Cooper. 1994. "Natural and man-made modes of entry in agronomic areas," in R. C. Honeycutt and D. J. Schabacker (eds.), *Mechanisms of Pesticide Movement in Ground Water* (Boca Raton, Fla.: Lewis Publishers), pp. 165–175.

Gardner, W., and J. A. Widtsoe. 1921. "The movement of soil moisture," *Soil Sci.,* **11**:230.

Hamza, M., and L. A. G. Aylmore. 1992. "Soil solute concentrations and water uptake by single lupin and radish plant roots I. Water extraction and solute accumulation," *Plant Soil,* **145**:187–196.

Huck, M. G., B. Klepper, and H. M. Taylor. 1970. "Diurnal variations in root diameter," *Plant Physiol.,* **45**:529.

Hudson, B. D. 1994. "Soil organic matter and available water capacity," *J. Soil and Water Cons.,* **49**:189–194.

Keeton, W. T. 1972. *Biological Science,* 2d ed. (New York: Norton).

MacFall, J. S., and G. A. Johnson. 1994. "Use of magnetic resonance imaging of plants and soils," pp. 99–113, in S. H. Anderson and J. W. Hopmans (eds.), *Tomography of Soil–Water–Root Processes,* SSSA Special Publication No. 36 (Madison, Wis.: Soil Sci. Soc. Amer.).

Mayaki, W. C., L. R. Stone, and I. D. Teare. 1976. "Irrigated and nonirrigated soybean, corn and grain sorghum root systems," *Agron. J.,* **68**:532–534.

Richards, L. A. 1965. "Physical condition of water in soil," in *Agron. 9: Methods of Soil Analysis,* Part I (Madison, Wis.: American Society of Agronomy).

6

THE SOIL AND THE HYDROLOGIC CYCLE

He meteth out the waters by measure . . .

—JOB 28:25

Worldwide there is ample water to satisfy most needs of humans and other forms of animal life and plants. If the 750 mm of water that, on an average, falls annually on the earth's land areas were evenly distributed and properly managed, there would be enough water to support a world human population several times the 5 billion living today. Unfortunately, however, water is not evenly distributed geographically or temporally (throughout the year). South America and the Caribbean receive nearly one-third of the annual global precipitation, Australia only 1%. Africa's total precipitation exceeds that of Europe, but most of it falls in the Congo River basin in west central Africa, leaving the north, east, and south with grossly inadequate water levels.

The timing of the precipitation also brings serious inequities from one region to another. Floods or droughts occur during years with abnormal rain or snowfall. Even in normal years, some areas have long dry spells alternated with periods of heavy rainfall. Droughts and floods accompany these precipitation extremes. These fluctuations in water supply have serious impacts on food production, water and sanitation for human health, and industrialization throughout the world.

The soil and the plants and animals it supports play key roles in helping to moderate the adverse effects of excesses and deficiencies of water. The soil serves as a massive reservoir to take in water during times of surplus and then release it in due time, either to satisfy the transpiration requirements of plants or to replenish groundwater that moves slowly into nearby streams or that replenishes even deeper reservoirs under the earth. The soil is also used as a recipient of animal, domestic, and industrial wastewaters that are later carried downward through the soil and through drainage channels to rivers and lakes downstream. The flow of water through the soil thereby connects the chemical pollution of soils to the groundwater.

In this chapter we shall first consider sources of the earth's water and how it is cycled from the atmosphere and back to the land and ocean areas. We will then examine the unique role of the soil in water resource management and will determine how well-managed soils can help make this cycle most useful to every living creature.

6.1 SOURCES OF WATER

There are nearly 1400 million cubic kilometers (km^3) of water on the earth—enough (if it were all aboveground at a uniform depth) to cover the earth's surface to a depth of some 3 kilometers. Most of this water, however, is relatively inactive or inaccessible and is not involved in the annual cycling of water that is so essential for all life forms in the upper layers of the earth. More than 97% is found in oceans (Figure 6.1) where the water has an average residence time of several thousand years. Only the near-surface layers take part in annual water cycling. Two percent of the earth's water is in glaciers and ice caps of mountains with similar long residence times (10,000 years ±). Some 0.7% is found in groundwater, most of which is more than 750 meters underground. Except where it is tapped for pumping by humans, it too has long average residence times of several hundred years or more.

The more active sources of water are the upper surfaces of the oceans, the shallow groundwater, lakes and rivers, the atmosphere, and soil moisture (Figure 6.1). Although their combined volume is insignificant compared to that of the inactive sources, they are accessible for movement in and out of the atmosphere and from one surface area to another. The average residence time for water in the atmosphere is about 10 days, that for the longest rivers is 20 days or less, and for soil moisture about one month. While water in large lakes such as the Great Lakes in the United States have longer residence times (90–200 years), the smaller lakes and reservoirs have much shorter periods of turnover. The accessibility of water in these latter sources makes them primary participants in the global water cycle, which will now receive our attention.

6.2 THE HYDROLOGIC CYCLE

The general principle of cycling water from the earth's surface to the atmosphere and back again to the earth is simple. The driving force behind what is termed the **hydrologic cycle** (Figure 6.2) is solar energy. About one-third of the solar energy that reaches the earth is absorbed by water on or near the earth's surface. The absorbed energy stimulates evaporation of the water, and the resultant water vapors move up into the atmosphere, eventually forming clouds that can move from one area to another. Within an average of about 10 days, pressure and temperature differences in the atmosphere cause the water vapor to condense into liquid droplets or solid particles, in which forms the water returns to the earth as rain or snow.

As Figure 6.2 illustrates, about 500,000 km^3 of water are evaporated from the earth surfaces and vegetation each year. More than 85% (430,000 km^3) of this water comes from the oceans, the remainder from the lakes, rivers, land, and vegetation on the continents. There is also a net migration of about 40,000 km^3 of water in the clouds above the oceans to the land areas.

The bulk of the precipitation also occurs in the ocean areas, some 390,000 km^3 or 78% of the total falling there each year. About 110,000 km^3 of water falls as rain or snowfall on the continents. This includes a return of the 70,000 km^3 evaporated from these areas plus the 40,000 km^3 coming from cloud movement from the ocean areas referred to above. Precipitation falling on the land stimulates water movement over the surface of the soil (runoff) and infiltration into and through the soil (drainage) into the groundwater. Both the runoff and groundwater seepage enter streams and rivers that, in

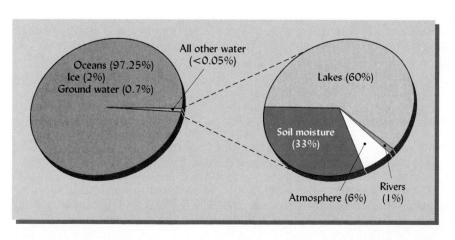

FIGURE 6.1 The sources of the earth's water. The preponderance of water associated with the earth is found in the oceans, glaciers and ice caps, and deep groundwater (left) but most of these waters are inaccessible for rapid exchange with the atmosphere and the land. The sources on the right, though much smaller in quantity, are actively involved in water movement through the hydrologic cycle. (Data from several sources)

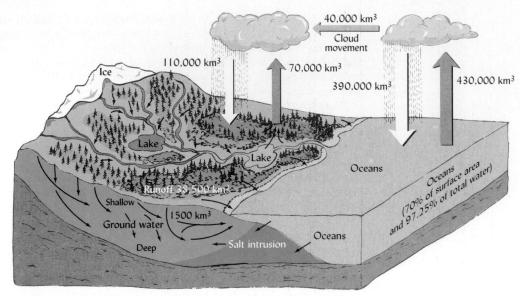

FIGURE 6.2 The hydrologic cycle upon which all life depends is very simple in principle. Water evaporates from the earth's surface, both the oceans and continents, and returns in the form of rain or snowfall. The net movement of clouds brings some 40,000 km³ of water to the continents and an equal amount of water is returned through runoff and groundwater seepage that is channeled through rivers to the ocean. About 86% of the evaporation and 78% of the precipitation occurs in the ocean areas. However, the processes occurring on land area where the soils are influential have impacts not only on humans but on all other forms of life, including those residing in the sea.

turn, flow downstream into the oceans. The volume of water returned in this way is about 40,000 km³, which balances the same quantity of water that moves annually by cloud flow from the ocean areas to the continents.

Role of the Soil

The soil plays a significant role in several of the processes involved in the hydrologic cycle. It provides direct and indirect support for evaporation in the continental areas. It is the direct source of water vapor coming from the evaporation of soil moisture. It also provides water for all growing plants, water that is then lost to the atmosphere by **transpiration** (evaporation from leaf surfaces).

Some of the precipitation that falls on soil is absorbed, some runs off, and some moves into the groundwater as just described. The downward movement of the gravitational water carries with it inorganic and organic chemicals coming from the decay of organic residues, from land-applied wastes, and from fertilizer and pesticide applications. These chemicals find their way into streams and downstream ponds, reservoirs, lakes, and the ocean. While some of these chemicals have beneficial effects, others are toxic to various forms of life, including humans. Soil properties as well as the management of soils can modify the kinds and quantities of chemicals that move through the hydrologic cycle.

We shall now give more detailed consideration to the role of soils in the hydrologic cycle and to the effects of this cycle on soils. The fate of precipitation water as it falls on land areas will be discussed first.

6.3   FATE OF PRECIPITATION AND IRRIGATION WATER

Rain and snowfall, as well as irrigation water supplied to soils, move in by a number of pathways. In areas covered with vegetation, 5 to 40% of the precipitation is *intercepted* by plant foliage (Table 6.1) and returns to the atmosphere by evaporation without ever reaching the soil. In some evergreen or deciduous forested areas, one-third to one-half of the precipitation is intercepted and does not reach the soil. Even in agricultural areas, the quantity of such plant interception is significant.

TABLE 6.1 Percentage Interception of Precipitation by Several Crop and Tree Species at Different Locations in the United States

Note the high interception for the forest species and close-growing crops such as alfalfa.

Species	Location	Percent of precipitation intercepted by plant
Alfalfa	Mo.	22
Corn	Mo.	7
Soybeans	N.J.	15
Ponderosa pine	Idaho	22
Douglas fir	Wash.	34
Maple, beech	N.Y.	43

Crop data from Haynes (1954); forest data from Kittridge (1948).

In level areas with friable soils, most of the water that reaches the soil penetrates downward. But in rolling to hilly areas, especially if the soils are not loose and open, considerable **runoff** and **erosion** take place, thereby reducing the proportion of the water that can move into the soil. In extreme cases, more than one-third of the precipitation will be lost in this manner. While some of this loss may be desirable to prevent excessive water saturation of the soil, most of it is undesirable, especially if the runoff carries with it considerable amounts of detached soil particles (soil erosion).

Once the water penetrates the soil, some of it is subject to downward percolation and eventual loss from the root zone by **drainage.** In humid areas, and in some irrigated areas of arid and semiarid regions, up to 50% of the precipitation may be lost as drainage water. However, during periods of low rainfall, some of this downward-percolating water may later move back up into the plant root zone by **capillary rise,** and thereby become available for plant absorption. Such movement is of practical importance in areas with deep soils, especially in semiarid to arid climates. Water also may reenter the plant root zone from the groundwater if the water table is sufficiently high.

The water remaining in the soil, sometimes referred to as *soil storage water,* is subject to two major vapor losses. Some moves upward by capillarity and is lost by **evaporation** from the soil surface. The remainder is absorbed by plants and then moves through the roots and stems to the leaves, where it is lost by **transpiration,** that is, evaporation through the stomata of the leaf surfaces. The moisture in the atmosphere is later returned to the soil's surface through precipitation or irrigation water, and the cycle starts again.

Effects of Precipitation

The proportion of water moving through the various channels just discussed is influenced greatly by the timing, rate, and form of the precipitation (Figure 6.3). Heavy rainfall arriving in only a few minutes or even in an hour or so can exceed the capacity of most soils to absorb the water. This accounts for the fact that in some arid regions with only 20 cm of annual rainfall, a cloudburst that brings 2 to 5 cm of water in a few minutes can result in serious runoff and gully erosion. A larger amount of precipitation spread over a period of several days could move gently into the soil, thereby increasing the stored water available for plant absorption as well as replenishing the underlying groundwater.

The timing of snowfall can also affect the proportion of water that infiltrates into the soil (Figure 6.4). In regions where winter temperatures are sufficiently low, the upper horizons of soil are commonly frozen in the wintertime. If significant snowfall occurs only after the soil is frozen, the snow insulates the underlying soil, causing it to remain frozen even when the air temperature rises to above the freezing level. In the spring when the snow melts the water, it cannot move into the still-frozen soil and is forced to run off the land into streams and rivers. Although this type of runoff does not commonly involve significant soil erosion, it does remove valuable water that would otherwise enter the soil.

In contrast to the above situation, when heavy snows occur before the ground is frozen, and if at least some of the snow remains on the soil during the winter, it insulates the underlying soil and thereby prevents its freezing. In this case, water from the

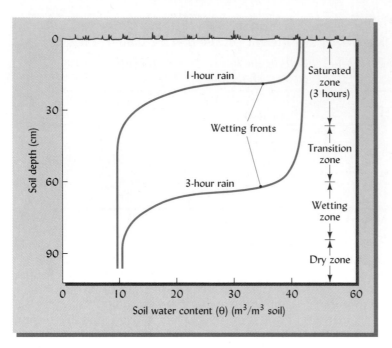

FIGURE 6.3   Water infiltration into a relatively dry soil after one and three hours of a steady rain. The wetting fronts indicate the depth of water penetration. After three hours, the upper 30–40 cm of soil is saturated with water. There is a transition zone of near saturation with water above the wetting zone which, in turn, is above the dry zone.

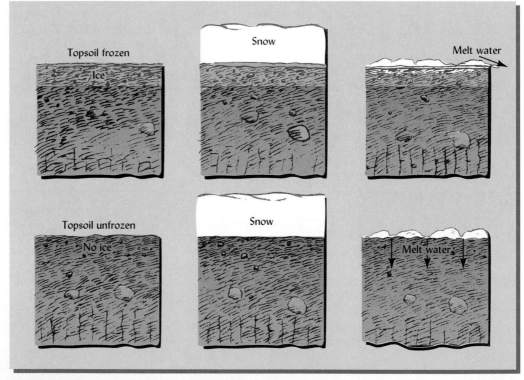

FIGURE 6.4   The timing of cold weather and snowfall in the winter months of some temperate regions with heavy snowfall drastically influences the runoff and infiltration of water in soils. The upper three diagrams show what takes place when cold weather precedes the heavy snowfall. (Upper left) The topsoil is frozen and ice fills the soil pores. (Upper center) The snow has come and is insulating the soil, which remains frozen. (Upper right) In the spring, the snow melts but the meltwater cannot enter the frozen soil and must run off. The lower three diagrams illustrate the situation when heavy snowfall covers the soil before the onset of very cold weather. (Lower left) The topsoil remains unfrozen early in the winter season, and when the snow comes (lower center) it insulates the soil and prevents its being frozen—even if the air temperatures are considerably below freezing. In the spring (lower right) the snow melts and most of the meltwaters can enter the soil.

melting snow in the spring slowly moves into the soil with little runoff or erosion, thereby increasing stored water for plant growth and replenishing the groundwater.

Effects of Vegetation and Soil Properties

The type of vegetation in a given area will definitely affect the rate of water penetration into soils. In addition to intercepting rain and snowfall before they reach the soil, plants help determine the proportion of water that runs off and that which penetrates the soil. Perennial grasses and dense forests protect the soil from the beating action of raindrops and the resultant damage to soil structure and porosity. Furthermore, surface residues from these plants will encourage water infiltration, and if runoff does occur, it is less likely to carry soil with it. Figure 6.5 illustrates differences in runoff from two grasses commonly used in golf fairways: bent grass and perennial ryegrass. The bent grass allows less runoff and spreads it out over a longer period of time.

The use of bulldozers and other heavy equipment to clear tropical forest lands for agricultural purposes can sharply reduce water infiltration rates. Figure 6.6 shows the results of one comparison of the effects of bulldozing and traditional hand methods of clearing the land. The traditional methods did little to disturb the soil, the infiltration rate remaining essentially the same as before clearing. While other research suggests that the ill effects of the compaction can be partially overcome by deep tillage, these results caution against indiscriminate use of heavy equipment to clear lands in the tropics.

Soil properties will also affect the fate of precipitation falling on land. If the soil is loose and open (e.g., sands and well-granulated silt loams), a high proportion of the incoming water will infiltrate the soil, and relatively less will runoff. In contrast, heavy clay soils with unstable soil structures resist infiltration and much of the water is lost through runoff. These differences, attributable to soil properties as well as vegetation, are illustrated in Figure 6.7.

Having considered the overall hydrologic cycle, we will now focus our attention on the component of the cycle for which the soil and plants play the most prominent roles.

6.4 THE SOIL–PLANT–ATMOSPHERE CONTINUUM

The flow of water through the soil–plant–atmosphere continuum (SPAC) is shown in Figure 6.8. This figure illustrates plant interception of precipitation, surface runoff, percolation and drainage, evaporation from the soil surface, and plant uptake of the water, its movement to the plants' leaves from which evaporation (transpiration) occurs. The SPAC is a major component of the overall hydrologic cycle.

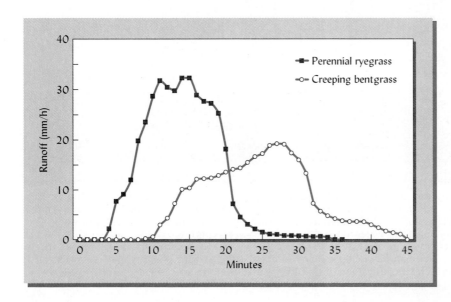

FIGURE 6.5 Runoff from two golf fairway grasses following irrigation at the rate of 150 mm per hour. Note that the runoff peak was much lower on the creeping bent grass that presented a dense thatch of plant stems near the soil surface and associated channels through which the water can move downward. [From Linde et al. (1995); used with permission of the American Society of Agronomy]

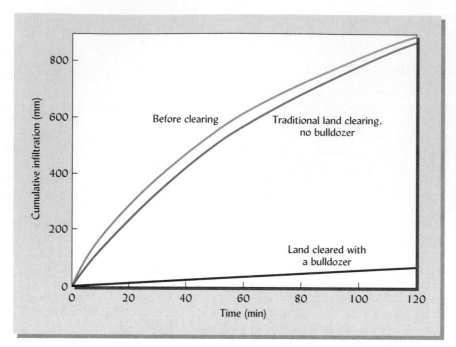

FIGURE 6.6 The effect of traditional and mechanical means of land clearing on the cumulative infiltration rate into an Ultisol located in the Amazon region of Peru. The measurements were made 14 weeks after land clearing. Apparently, the surface soil disturbance and compaction by the bulldozer reduced the quantity and size of pores, thereby drastically reducing the infiltration rate. The traditional land clearing was by hand, leaving the soil uncompacted. [Redrawn from Alegre et al. (1986); used with permission of the Soil Science Society of America]

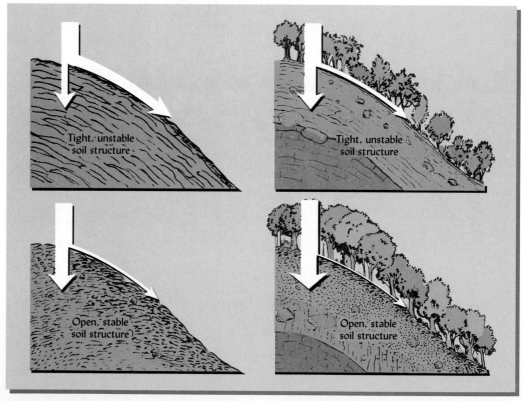

FIGURE 6.7 Illustration of the effect of soil properties and vegetation on runoff and infiltration. The upper two diagrams show the partitioning of precipitation between infiltration and runoff in a soil that has unstable structure and is tight and relatively impermeable to downward movement of water. Even with forest cover, much of this water runs off because the soil accepts it too slowly (upper right). The two lower diagrams show much greater infiltration into a soil with stable structure having significant macropore space (open). The more favorable soil environment coupled with good vegetation cover greatly reduces runoff and increases infiltration. It should be noted that the trees intercept some of the precipitation, which then evaporates and returns to the atmosphere without reaching the soil.

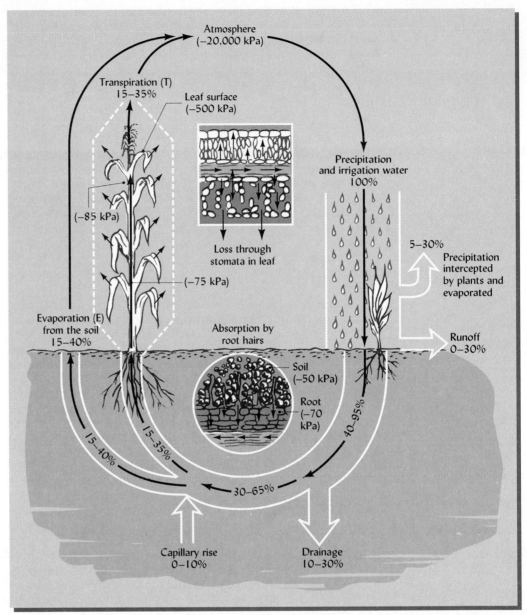

FIGURE 6.8 Soil–plant–atmosphere continuum (SPAC) showing water movement from soil to plants to the atmosphere and back to the soil in a humid to subhumid region. Water behavior through the continuum is subject to the same energy relations covering soil water that were discussed in Chapter 5. Note that the moisture potential in the soil is −50 kPa, dropping to −70 kPa in the root, declining still further as it moves upward in the stem and into the leaf, and is very low (−500 kPa) at the leaf–atmosphere interface, from whence it moves into the atmosphere where the moisture potential is −20,000 kPa. Moisture moves from a higher to lower moisture potential. Note the suggested ranges for partitioning of the precipitation and irrigation water as it moves through the continuum.

In studying the SPAC, scientists have discovered that the same basic principles apply for the retention and movement of water whether it is in soil, plants, or in the atmosphere. In Chapter 5, the **potential energy** level of water was seen to be a major controlling factor in determining soil water behavior. The same can be said for soil–plant and plant–atmosphere relations (see Figure 6.8). As water moves through the soil to plant roots, into the roots, across cells into stems, up the plant stems to the leaves, and is evaporated from the leaf surfaces, its tendency to move is determined by differences in potential energy levels of the water or by the soil water potential.

The soil water potential must be higher in the soil than in the plant roots if water is to be absorbed from the soil. Likewise, movement up the stem to the leaf cells is in response to differences in moisture potential, as is the movement from leaf surfaces to the atmosphere. Note from Figure 6.8 that the moisture potential drops from −50 kPa in the soil to −500 kPa at the leaf surfaces and finally to −20,000 kPa in the atmosphere.

Two Points of Resistance

Changes in water potential illustrated in Figure 6.9 suggest major resistance at two points as the water moves through the SPAC: the root–soil water interface and the leaf cell–atmosphere interface. This means that two primary factors determine whether plants are well supplied with water: (1) the rate at which water is supplied by the soil to the absorbing roots and (2) the rate at which water is evaporated from the plant leaves. Factors affecting the soil's ability to supply water were discussed in Section 5.13. Attention will now be given to the loss of water by evaporation and to factors affecting this loss.

6.5   EVAPOTRANSPIRATION

As shown in Figure 6.8, vapor losses of water from soils occur by **evaporation** (E) at the soil surface and by **transpiration** (T) from the leaf surfaces. The combined loss resulting from these two processes, termed *evapotranspiration* (ET), is responsible for most of the water removal from soils during the period of plant growth. On irrigated soils located in arid regions, for example, ET commonly accounts for the loss of 75 to 100 cm of water during the growing season of a plant such as alfalfa. Some ET is critical for growing plants as a means of cooling, turgor maintenance, and nutrient transport.

Evaporation from the soil surface (E) at a given temperature is determined to a large degree by soil surface wetness and by the ability of the soil to replenish this surface water as it evaporates. A high water table will help replenish the surface water through capillary action. In the absence of such a replenishment source, however, the upward movement of water is limited and evaporation is greatly reduced.

The proportion of total water vapor loss by evaporation from the soil (E) and that from the plant leaves (T) will vary from one climatic and vegetative situation to another. For example, in the corn plant, transpiration takes place only during a short 3½–4-month growing period, whereas evaporation from the soil surface continues to occur year-round (Figure 6.10). In a dense forest stand, by contrast, transpiration (T) could well exceed evaporation from the soil surface (E) throughout the year. In both cases, however, it is the total loss or evapotranspiration (ET) that is significant in the hydrologic cycle, and ET determines the overall rate of depletion of soil moisture. Evapotranspiration rates are influenced by a number of factors which we will now consider.

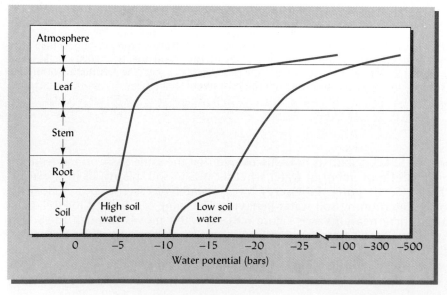

FIGURE 6.9 Change in water potential as water moves from the soil through the root, stem, and leaf to the atmosphere. Note that water potential decreases as the water moves through the system. [Adapted from Hillel (1980)]

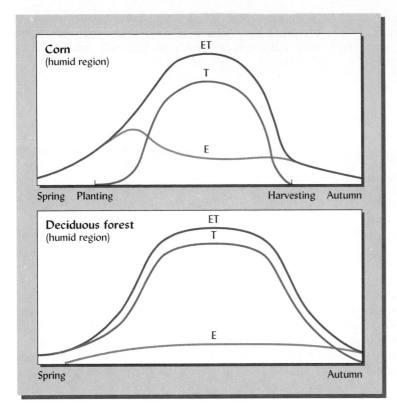

FIGURE 6.10 Relative rates of evaporation from the soil surface (E), of transpiration from the plant leaves (T), and the combined vapor loss (ET) for two field situations. (Upper) a field of corn in a humid region. Until the plants are well established, most of the vapor loss is from the soil surface (E), but as the plants grow transpiration (T) soon dominates. The soil surface is shaded and E may actually decrease somewhat since most of the moisture is moving through the plants. As the plants reach maturity, T declines as does the combined ET. For a nearby deciduous forested area (lower) the same general trend is illustrated, except there is relatively less evaporation from the soil surface and a higher proportion of the vapor loss is by transpiration. The soil is shaded by the leaf canopy most of the growing season in the forested area. It should be noted that these figures relate to a representative situation and that the actual field losses would be influenced by rainfall distribution, temperate fluctuations, and soil properties.

Radiant Energy

Solar radiant energy provides the 2260 joules (J) (540 calories) needed to evaporate each gram of water, whether from the soil (E) or from leaf surfaces (T). On a cloud-free day solar radiation is high, and evapotranspiration is stimulated. On cloudy days the solar radiation striking the soil and plant surfaces is reduced, and the evaporative potential is not as great.

Shading by plant leaves reduces evaporation (E) from a soil surface. The degree of this shading, known as *leaf area index*, significantly affects the radiant energy that reaches the soil surface and, in turn, the evaporation that takes place. Thus, newly emerging plants have a very low leaf area index, and they intercept very little solar energy. In contrast, perennial species such as alfalfa and forest stands have very high leaf area indices both early and late in the growing season; they shade the soil efficiently and thereby reduce markedly the soil surface evaporation. Where the forest floor (leaf litter) is undisturbed even less direct sunlight strikes the soil and the evaporation (E) is very low throughout the year.

The angle of the sun's rays as they strike the earth also influences evaporation. South-facing slopes in North America, for example, receive the sun's rays at nearly a 90° angle. The comparable angles for a north-facing slope may be 30°. As we shall see in Section 7.9, soil temperatures are highest when solar radiation strikes the earth at a right angle (90°). Consequently, ET is generally higher on the south-facing slopes. The opposite, of course, would be true for soils in areas south of the equator such as those in South America, southern Africa, or Australia.

Atmospheric Vapor Pressure

Evaporation occurs when the atmospheric vapor pressure is low compared to the vapor pressure at the plant and soil surfaces. Consequently, at comparable temperature levels evaporation is high from irrigated soils in arid climates (low atmospheric vapor pressure) and much lower in humid regions (higher atmospheric vapor pressures).

Temperature

A rise in temperature increases the vapor pressure at the leaf and soil surfaces but has much less effect on the vapor pressure of the atmosphere. As a result, on hot days there is a sharp difference in vapor pressure between leaf or soil surfaces and the atmosphere, and evaporation proceeds rapidly. Plants and especially soils may be warmer than the atmosphere on bright, clear days. This temperature difference definitely enhances the rate of evaporation.

Wind

A dry wind will continually sweep away moisture vapor from a wet surface. Hence, high winds intensify evaporation from both soils and plants. Farmers of the Great Plains of the United States dread the hot winds characteristic of that region. The eolian deposits so common in that area owe their presence to hot dry winds that dried out the soil and permitted it to be transported by wind.

Soil Moisture Supply

Evapotranspiration (ET) is higher where plants are grown on soils with modest moisture levels at or near the field capacity than where soil moisture is low (Table 6.2). This is to be expected because at low soil moisture levels water uptake by plants is restricted.

In most cases, the upper 15–25 cm of soil provides most of the water for surface evaporation (E). However, because of the much larger volume of the subsoil layers, they can provide a significant portion of the overall moisture needed for evapotranspiration (ET) (see Figure 6.11). This situation prevails in many regions having alternating moist and dry conditions during the year. Subsoil moisture stored during moist periods is available for evapotranspiration during the dry period (see Figure 1.5). Subsoil moisture is a major source of water for crops in subhumid and semiarid areas such as the Great Plains area of the United States.

Plant Characteristics

Some plant characteristics can affect ET over a growing season. Leaf characteristics and the length of the growth period of the plant will have an influence. Likewise, the depth of rooting will help determine the subsoil moisture available for absorption and eventually for transpiration. These interactions illustrate how the soil scientist and plant breeder must work together to determine the plant characteristics desired in improved varieties to enhance water conservation.

6.6 MAGNITUDE OF EVAPORATION LOSSES

Evapotranspiration (ET) is markedly influenced by climate. During a given growing season ET may vary from as little as 30 cm (12 in.) in cool mountain valleys where the growing seasons are short to as much as 220 cm (87 in.) or more in irrigated desert areas. The ranges more commonly encountered are 35–75 cm (14–30 in.) in unirrigated, humid to semiarid areas and 50–125 cm (20–50 in.) in hot, dry regions where irrigation is used.

TABLE 6.2 **Effect of Soil Moisture Level on Evapotranspiration Losses**

Where the surface moisture content was kept high, total evapotranspiration losses were greater than when medium level of moisture was maintained.

	Evapotranspiration (cm)	
Moisture condition of soil[a]	Corn	Alfalfa
High	45	62
Medium	32	52

Calculated from Kelly (1954).
[a]High moisture—irrigated when upper soil layers were 50% depleted of available water.
Medium moisture—irrigated when upper layers were 85% depleted of available water.

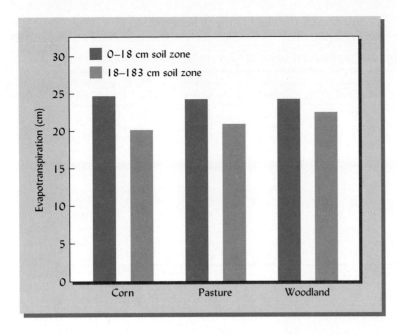

FIGURE 6.11 Evapotranspiration loss from the surface layer (0–18 cm) compared with loss from the subsoil (18–183 cm). Note that more than half the water loss came from the upper 18 cm and only half from the *next 165 cm* of depth. Periods of measurement: corn, May 23–September 25; pasture, April 15–August 23; woodland, May 25–September 28. [Calculated from Dreibelbis and Amerman (1965)]

Daily ET figures are also high during the hot dry periods in the summer. For example, daily ET rates for corn may be as high as 1.25 cm (0.5 in.). Even in deep soils having reasonably high capacities for storing available water, this rapid loss of moisture soon depletes the plant root zone of easily absorbed water. Sandy soils, of course, may lose most of their available moisture in a matter of a few days under these conditions. The importance of reducing vaporization losses is obvious.

Surface Evaporation versus Transpiration

A number of factors determine the relative water losses from the soil surface and from transpiration: (1) plant cover in relation to soil surface (leaf area index); (2) efficiency of water use by different plants; (3) proportion of time the plant is on the land, especially during the summer months; and (4) climatic conditions. Figure 6.10 illustrates E, T, and ET losses in cropped and forested areas of humid regions.

6.7 EFFICIENCY OF WATER USE

The dry matter produced by a given plant from a given amount of water is an important figure, especially in areas where moisture is scarce. This efficiency may be expressed in terms of dry matter yield per unit of water transpired (T efficiency) or the dry matter yield per unit of water lost by evapotranspiration (ET efficiency).

Another expression of water use efficiency is the **transpiration ratio**. The transpiration ratio for any given plant is markedly affected by climatic conditions. It is expressed in kilograms of water transpired to produce 1 kg of dry matter and generally ranges from 200 to 500 in humid regions and almost twice as much in arid regions. The data in Table 6.3 illustrate the water transpired by different plants at different locations around the world.

T Efficiency

At a given location there are differences in T efficiency of different plant species. The data in Figure 6.12 illustrate this point. These data suggest that plants such as corn, sorghum, and millet have relatively high T efficiencies, that is, they require relatively small quantities of water to produce 1 kg of dry matter. In contrast, some legume forages such as alfalfa have low T efficiencies, requiring considerably higher quantities of water to produce 1 kg of dry matter. The cereal crops, such as wheat, oats, and barley, and vegetables such as potatoes are intermediate in their T efficiencies.

It can be calculated from Table 6.3 that the amount of water necessary to bring a plant to maturity is very large. For example, a representative crop of wheat containing

Kilograms of water used in production of 1 kg of dry matter.

Crop	Harpenden, England	Dahme, Germany	Madison, Wisconsin	Pusa, India	Akron, Colorado
Barley	258	310	464	468	534
Beans	209	282	—	—	736
Buckwheat	—	363	—	—	578
Clover	269	310	576	—	797
Maize	—	—	271	337	368
Millet	—	—	—	—	310
Oats	—	376	503	469	597
Peas	259	273	477	563	788
Potatoes	—	—	385	—	636
Rape	—	—	—	—	441
Rye	—	353	—	—	685
Wheat	247	338	—	544	513

Data compiled by Lyon et al. (1952).

5000 kg/ha of dry matter (about 4500 lb/acre) and having a transpiration ratio of 500 will withdraw water from the soil during the growing season equivalent to about 25 cm (10 in.) of rain. The corresponding figure for corn, assuming the dry matter at 10,000 kg/ha and the transpiration ratio at 350, would be 35 cm (14 in.). These amounts of water *in addition* to that evaporated from the surface, must be supplied during the growing season. It is not surprising that moisture is often the most critical factor in plant growth. Likewise, these data well illustrate the importance of the soil as a reservoir of water to sustain plants during the periods of high use rate.

For years researchers have tried to increase the T efficiency of a plant by changing the management practices used to grow the plant. These efforts have in general not succeeded. The T efficiency of a given plant seems to be a characteristic that is not subject to easy change by management or even plant breeding. The amount of water required to produce a unit of dry matter of a plant species in a given climatic area is relatively constant if serious growth constraints, toxicities, or deficiencies are absent. This suggests that the major control on vapor loss from a soil/plant complex must be focused on reductions in evaporation from the soil surface rather than from the leaf surfaces.

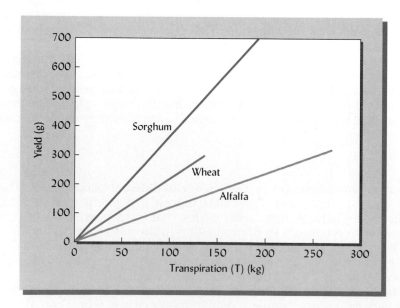

FIGURE 6.12   Relationship between the yield of three crops and the amount of water transpired. These data were obtained at several locations in the Great Plains of the United States. The plants were grown in containers, and evaporation from the soil surface was prevented by covering the soil. [Redrawn from graphs in Hanks (1983)]

ET Efficiency

Since evapotranspiration (ET) includes both transpiration (T) and evaporation (E) from the soil surface, ET efficiency is somewhat more variable than T efficiency. Furthermore, certain soil and crop management practices that influence evaporation from the soil can affect ET and, in turn, ET efficiency. As was the case with T efficiency, ET efficiency is affected by climate, being higher in moist areas than in dry areas.

ET efficiency also responds to factors that drastically affect plant growth. Highest efficiency is attained where moisture and nutrient status are generally satisfactory for plant growth, neither deficient nor in excess. Figure 6.13 illustrates the effect on ET efficiency of adding phosphorus and nitrogen fertilizers to a nutrient-deficient soil.

Research suggests that this increase in water use efficiency is likely not due to increased yield per unit of water transpired (T), but rather to an increase in the ratio of T to E. This simply means that a higher proportion of the water being lost from the soil through the evaporative process is used for plant growth than for mere evaporation from the soil surface. Apparently, the added fertility increases the leaf area per unit of land, called the *leaf area index* (LAI). This brings about greater shading of the soil and concomitant reduction in evaporation from the soil surface. In any case, the wisdom of maintaining optimum soil fertility to make most efficient use of available water is illustrated.

6.8   CONTROL OF EVAPOTRANSPIRATION (ET)

Preceding discussions remind us that transpiration (T) is a plant process subject to only minor control in a given climate if other plant processes are proceeding near normally. In contrast, evaporation (E) from the soil surface is not essential for plant growth and is subject to some degree of control by soil and crop management. Consequently, major focus should be on those practices that reduce E, so as to maximize the water remaining to accommodate the plant process, T. We will first consider how managing the plants to be grown can influence E.

Plant Characteristics and Management

The particular plant species being grown and the time of year it is grown will definitely influence both T and ET. For example, corn and sorghum have lower water requirements than alfalfa, and the requirement for wheat and other cereals is generally inter-

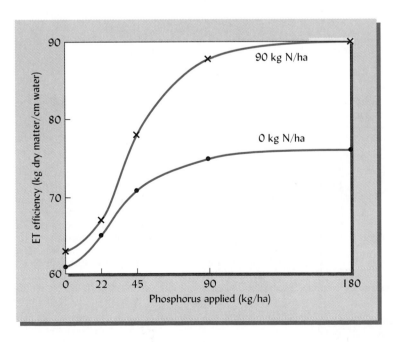

FIGURE 6.13 Water-use efficiency of wheat as affected by application of phosphorus fertilizer with and without nitrogen fertilizer (12-year average). ET efficiency increased with increased fertilizer rates. [Data from Black (1982)]

mediate. Also, both T and ET are lower for plants growing during cool seasons when the vapor pressure gradients are apt to be low.

A dense vegetative cover (high leaf area index) also can be effective in reducing E. Figure 6.14 illustrates this point. Woodlands and forested areas in humid regions generally have high leaf indices. Plant cover is also maintained by the use of mixtures of legumes and nonlegumes in pastures. In dryland areas, however, there are some limitations to maintaining high leaf area indices. While dense cover can reduce E, the high plant populations needed to produce the cover can deplete soil moisture prematurely and thereby drastically reduce crop yields.

Fallow Cropping

Farming systems that alternate summer fallow (noncropped) one year with traditional cropping the next are sometimes used to conserve soil moisture in semiarid and subhumid areas such as the Great Plains of the United States. The objective of fallow systems is to eliminate T every other year, thereby increasing the soil moisture storage for the subsequent crop. The previous year's stubble is commonly allowed to stand until the spring of the fallow year. The soil may then be disked, leaving most of the stubble near the surface. During the summer, weed growth is minimized by either light cultivation or the use of herbicides. The top part of Figure 3.11 shows typical alternating strips of summer fallow and wheat. During the fallow season, vapor loss is a result primarily of evaporation at the soil surface in the year when no crop is grown.

The effectiveness of summer fallow has varied greatly. However, moisture levels at planting time are commonly higher following a summer fallow than following a harvested crop, and yields have generally been increased through the use of the summer fallow. Data in Table 6.4 illustrate this point. Summer fallow certainly reduces the risk of crop failure from droughty conditions.

TABLE 6.4 **Wheat Yields at Seven Locations in the Great Plains Area of the United States Where Continuous Cropping and Fallow Cropping were Compared**

Location	Years of record	Yield (kh/ha)	
		After fallow	After wheat
Havre, Mont.	35	2100	540
Dickinson, N.Dak.	44	1400	780
Newell, S.Dak.	40	1420	910
Akron, Colo.	60	1420	500
North Platte, Nebr.	56	2140	830
Colby, Kans.	49	1320	620
Bushland, Tex.	29	1010	640

From USDA (1974).

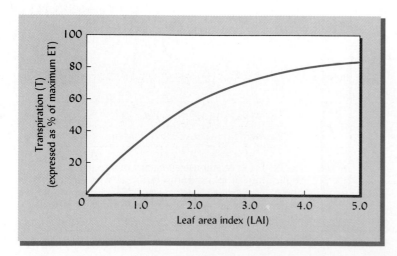

FIGURE 6.14 Relationship between the leaf area per unit of land area (leaf area index, LIA) of cotton and transpiration (T). Transpiration is expressed as a percentage of the maximum evapotranspiration (ET). Soil water was not limiting in these trials. Note that when the soil was effectively shaded (high LAI), most of the moisture vapor loss was by transpiration (T), evaporation from the soil surface (E) making up a small portion of the total, essentially none where the surface soil was dry. [Redrawn from Ritchie (1983)]

6.9 CONTROL OF SURFACE EVAPORATION (E)

More than half the precipitation in dryland areas is usually returned to the atmosphere by evaporation (E) directly from the soil surface. Evaporation losses are also high in arid region irrigated agriculture. Even in humid region rain-fed areas, E losses are of significance during hot rainless periods. Such moisture losses rob the plant of much of its growth potential.

The most effective practices aimed at controlling E are those that provide some cover to the soil. This cover can best be provided by mulches and by selected conservation tillage practices.

Mulches

Any material used at the surface of a soil primarily to reduce evaporation or to reduce weed growth may be designated a *mulch*. Examples are sawdust, manure, straw, leaves, crop residues, and plastic sheets. Mulches are highly effective in checking evaporation and are most practical for home garden and for high-valued plants, including strawberries, blackberries, fruit trees, and such other plants that require infrequent, if any, cultivation. Intensive gardening justifies the use of these moisture-saving mulches.

Mulches comprised of plant residues are effective in reducing E and, in turn, in conserving soil moisture. Combining summer fallow with residue mulches saves soil moisture in dry areas, as shown in Table 6.5. Unfortunately, plant growth on these soils is not sufficiently high to leave adequate residues to provide the organic mulch level that is needed to reduce E.

Paper and Plastic Mulches

Specially prepared paper and plastics are also used as mulches. This cover is spread and fastened down either between the rows or over the rows (Figure 6.15). The plants in the latter case grow through suitable slits or other openings. Paper and plastic mulches can be used only with crops planted in rows or in hills. As long as the ground is covered, evaporation and weeds are checked, and in some cases remarkable increases in plant growth have been reported. Unless the rainfall is very heavy, the paper or plastic does not seriously interfere with the infiltration of rainwater into the soil.

Paper and plastic mulches have been employed with considerable success in the culture of pineapples in Hawaii and with vegetable crops in desert areas of the Middle East, where water conservation is critical. This type of mulch is also being used to some extent in truck farming and vegetable gardening. The cost of the materials and the difficulty of keeping it in place limit the use of these materials to high-value crops. Furthermore, the temperature of the soil under the plastic is commonly 8 to 10°C higher than with a straw mulch. While this may be helpful for crops established early in the spring, it can reduce the yields of temperature-sensitive plants (see Section 7.14).

Crop Residue and Conservation Tillage

In recent years, conservation tillage practices, which leave a high percentage of the residues from the previous crop on or near the surface (Figure 6.16), have been widely

TABLE 6.5 **Gains in Soil Water from Different Rates of Straw Mulch During Fallow Periods at Four Great Plains Locations**

Location	Av. annual precipitation (cm)	Soil water gain (cm) at each mulch rate			
		0 Mg/ha	2.2 Mg/ha	4.4 Mg/ha	6.6 Mg/ha
Bushland, Tex.	50.8	7.1	9.9	9.9	10.7
Akron, Colo.	47.6	13.4	15.0	16.5	18.5
North Platte, Nebr.	46.2	16.5	19.3	21.6	23.4
Sidney, Mont.	37.9	5.3	6.9	9.4	10.2
Average		10.7	12.7	14.5	15.7
Average gain by mulching			2.0	3.8	5.0

From Greb (1983).

FIGURE 6.15   For crops with high cash value, plastic mulches are being used. The plastic is installed by machine (left) and at the same time the plants are transplanted (right). Plastic mulches help control weeds, conserve moisture, encourage rapid early growth, and eliminate need for cultivation. The high cost of plastic makes it practical only with the highest-value crops. (Courtesy K. Q. Stephenson, Pennsylvania State University)

adopted. These practices reduce markedly both vapor and liquid losses. In conservation tillage (see Section 17.12) the use of the traditional moldboard plow, which incorporates essentially all the residues into the soil, is replaced or minimized.

Among the most widely used conservation tillage practices is **stubble mulch** tillage. In this method, used mostly in subhumid and semiarid regions, the residues from the previous crop are uniformly spread on the soil surface. The land is then tilled with special tillage implements that permit much of the plant residue to remain on or near the surface. Wheat stubble, straw, and cornstalks are among the residues used.

Other conservation tillage systems that leave residues on the soil surface include **no-tillage** (Figure 6.16), where the new crop is planted directly in the sod or on residues of the previous crop with no plowing or disking. Another practice, called *till planting,* is a strip-tillage system where residues are swept away from the immediate area of planting, but most of the soil and surface area remain untilled. Like the other methods, till planting leaves most of the soil covered with crop residues to help reduce water vapor losses from soils. These systems will receive further attention in Section 17.12.

FIGURE 6.16   Planting soybeans in wheat stubble with no tillage (left). (Courtesy John Deere and Company, Moline, Illinois) In another field corn has been planted in wheat stubble (right). In both cases the wheat residue will help reduce evaporation losses from the soil surface and will reduce erosion. (Courtesy USDA Natural Resources Conservation Service)

Soil Mulch

One of the early contentions about the control of evaporation was that the formation of a **natural** or **soil mulch** was a desirable moisture-conserving practice. Years of experimentation showed, however, that a natural soil mulch does not necessarily conserve moisture, especially in humid regions. In fact, in some cases a soil mulch may encourage moisture loss. Only in regions with rather distinct wet and dry seasons, as in some tropical areas, can a soil mulch conserve moisture.

Cultivation and Weed Control

In addition to seedbed preparation, the most important reason for cultivation of the soil is to control weeds. Transpiration by the unwanted plants can extract soil moisture far in excess of that used by the crop itself, especially a row crop. In recent years chemical herbicides have been used to provide weed control without cultivation of the soil. Tillage may thereby be limited to situations where this practice is needed to maintain the proper physical conditions for plant growth. Unfortunately, some herbicides have accumulated to undesirable levels in downstream waterways or in underground water sources. Consequently, this means of weed management is controversial. It may be necessary to resume the use of cultivation to help control weeds, thereby minimizing the need for chemical herbicides.

6.10   CLIMATIC ZONE MANAGEMENT PRACTICES

There is variation in the need for practices that limit E in different climatic regions. Brief mention will be made of these regional differences.

Humid Regions

For high-valued specialty crops and for home gardens and nurseries, paper and plastic mulches are in common use. Likewise, the use of conservation tillage practices that maximize residue coverage is spreading throughout humid areas of the United States.

Weed control is a practical means of reducing losses of soil moisture by ET in humid regions. Tillage and relatively safe herbicides can help remove this source of moisture loss.

Semiarid and Subhumid Regions

A major objective of water management systems in semiarid and subhumid regions is to encourage water infiltration rather than runoff (Figure 16.17). In addition, these regions sorely need means of managing water vapor loss. The two primary methods used are the **summer fallow** and **stubble mulch** systems. While evaporation is still high even with these water-conserving systems, years of field experience show that they take at least part of the risk out of dryland farming.

Irrigated Fields in Arid Regions

Evapotranspiration in irrigated areas of arid regions is very high, being stimulated by the high solar radiation and frequent high wind velocities. Practical control over such evapotranspiration is not easy, but some irrigation and residue management practices have been found to be beneficial. For example, the use of *trickle* or *drip* irrigation has been effective. Through this system water is dripped or trickled out of pipes or tubes alongside each plant without wetting the soil between the plants. In some situations perforated tubes through which water can flow are buried several centimeters below the soil surface and immediately under the row of plants. These systems provide water for the plants and minimize evaporation from the soil surface. Figure 6.18 illustrates the use of drip irrigation to supply water for rose plants in a backyard garden.

The use of conservation tillage practices that involve plant residues has helped reduce evaporation from fields receiving sprinkle irrigation. The water acts very much like rainwater, moving through the residues and into the soil. The surface residues help reduce surface evaporation just as was the case in unirrigated soils of humid regions.

FIGURE 6.17 The effectiveness of small furrow dikes (right) in preventing water runoff and encouraging water infiltration. (Courtesy O. R. Jones, USDA Agricultural Research Service, Bushland, Tex.)

6.11 TYPES OF LIQUID LOSSES OF SOIL WATER

In our discussion of the hydrologic cycle we noted two types of liquid losses of water from soils: (1) the percolation or drainage water and (2) the runoff water (see Figure 6.2). The percolation water recharges the groundwater and moves chemicals out of the soil. Runoff losses generally include not only water but also appreciable amounts of soil (erosion).

FIGURE 6.18 Trickle or drip irrigation is commonly used in arid regions to provide water to plants while minimizing the evaporation of the water from the surrounding soil surface. The figure on the left shows water being trickled around a young rose bush in Arizona. Small amounts of water are being applied daily through small tubes alongside each plant. Note that only the soil immediately around the plant is being watered. (Right) A nearby olive tree is being watered through three somewhat larger tubes.

6.12 PERCOLATION AND LEACHING: METHODS OF STUDY

Two general methods are used to study percolation and leaching losses: (1) underground pipe or **tile drains** specially installed for the purpose and (2) specially constructed **lysimeters**[1] (from the Greek word *lysis,* meaning loosening, and *meter,* to measure). For the first method, an area should be chosen where the tile drain receives only the water from the land under study and where the drainage is efficient. The advantage of the tile method is that water and nutrient losses can be determined from relatively large areas of soil under normal field conditions.

The lysimeter method involves the measurement of percolation and nutrient losses under somewhat more controlled conditions. Soil is removed from the field and placed in concrete or metal tanks, or a small volume of field soil is isolated from surrounding areas by concrete or metal dividers (Figure 6.19). In either case, water percolating through the soil is collected and measured. The advantages of lysimeters over a tile-drain system are that the variations in a large field are avoided, the work of conducting the study is not so great, and the experiment is more easily controlled.

6.13 PERCOLATION LOSSES OF WATER

When the amount of rainfall entering a soil becomes greater than the water-holding capacity of the soil, losses by percolation will occur. Percolation losses are influenced by the amount of rainfall and its distribution, by runoff from the soil, by evaporation, by the character of the soil, and by the nature of the vegetation.

Percolation–Evaporation Balance

The relationship among precipitation, runoff, soil storage, and percolation for representative humid and semiarid regions and for an irrigated arid region is illustrated in Figure 6.20. In the humid temperate region, the rate of water infiltration into the soils (precipitation less runoff) is commonly greater, at least at some time of year, than the rate of evapotranspiration. As soon as the soil field capacity is reached, percolation into the substrata occurs.

---

[1] For review of lysimeter work, see Aboukhaled et al. (1982).

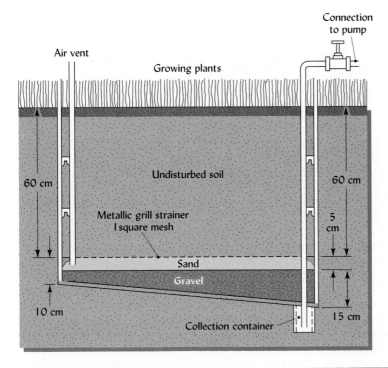

FIGURE 6.19 Field lysimeter used to collect percolation water from an undisturbed soil. The water moves down through the soil to the sand and gravel, then down the inclined slope to the container (lower right) from which it can be pumped and collected. Some more sophisticated systems are equipped with devices to weigh the entire lysimeter, thereby permitting the additional measurement of water intake and evapotranspiration. [Redrawn from UNDP/WMO (1974)]

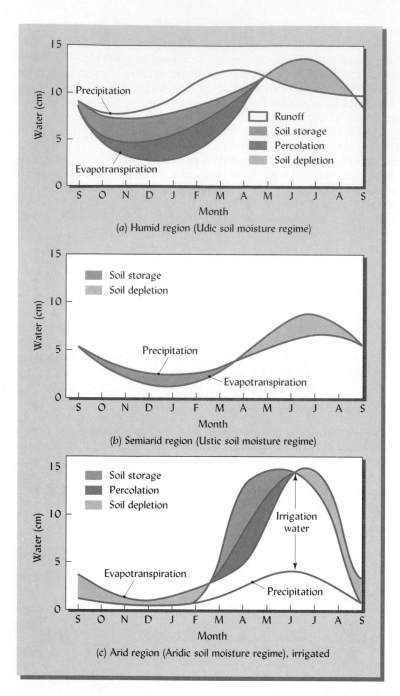

FIGURE 6.20 Generalized curves for precipitation and evapotranspiration for three temperate zone regions: (a) a humid region, (b) a semiarid region, and (c) an irrigated arid region. Note the absence of percolation through the soil in the semiarid region. In each case water is stored in the soil. This water is released later when evapotranspiration demands exceed the precipitation. In the semiarid region evapotranspiration would likely be much higher if ample soil moisture were available. In the irrigated arid region soil, the very high evapotranspiration needs are supplied by irrigation. Soil moisture stored in the spring is utilized by later summer growth and lost through evaporation during the late fall and winter.

In the example shown in Figure 6.20a, maximum percolation occurs during the winter and early spring, when evaporation is lowest. During the summer little percolation occurs, and evapotranspiration that exceeds the infiltration results in a depletion of soil water. Normal plant growth is possible only because of moisture stored in the soil the previous winter and early spring.

The general trends in the temperate zone semiarid region are the same as for the humid region. Soil moisture is stored during the winter months and used to meet the moisture deficit in the summer. But because of the low rainfall, essentially no percolation out of the profile occurs. Water may move to the lower horizons, but is absorbed by plant roots and ultimately lost by transpiration.

The irrigated soil in the arid region shows a unique pattern. Irrigation in the early spring, along with a little rainfall, provides more water than is being lost by evapotranspiration. The soil is charged with water and some percolation may occur. During the summer, fall, and winter months, this stored water is depleted because the amount being added is less than the very high evapotranspiration that takes place in response to large vapor pressure gradients.

Figure 6.20 depicts situations common in temperate zones. In the tropics one would expect evapotranspiration to be somewhat more uniform throughout the year, although it will vary from month to month depending on the distribution of precipitation. In very high rainfall areas of the tropics, the percolation and especially the runoff would be higher than shown in Figure 6.20. In irrigated areas of the arid tropics, the relationships would likely not be too different from those in arid temperate regions.

The comparative losses of water by evapotranspiration and percolation through soils of different climatic regions are shown in Figure 6.21. These differences should be kept in mind as attention is given to percolation and groundwaters.

6.14 PERCOLATION AND GROUNDWATERS

When drainage water moves downward and out of the soil it eventually encounters a zone of water saturation that lies above a layer of rocks or clays that is essentially impermeable to water movement. The upper surface of this zone of saturation is known as the **water table,** and the water so held is termed **groundwater.** The water (Figure 6.22) is commonly only 1 to 10 meters below the soil surface in humid regions but may be several hundred or even thousands of meters deep in arid regions. In swamps it is essentially at the land surface.

The unsaturated zone above the water table makes up what is termed the **vadose zone** (see Figures 6.22 and 6.23). This zone may be considerably deeper than the soil and may include underlying materials that are not saturated with water. In some cases, however, the water table may be sufficiently high to include at least the lower soil horizons. In those instances, the unsaturated vadose zone would include only the upper soil horizons.

Shallow groundwater serves as a recipient of downward-percolating drainage water, which in turn replenishes the deeper groundwater as it seeps horizontally through porous geological materials (aquifers) toward springs, streams, and rivers, or as it is removed by pumping for domestic and irrigation uses. The water table level will move up or down in response to the amount of drainage water coming from the soil. In humid temperate regions, the water table is commonly highest in the spring following winter rains and snow melt, and before evapotranspiration can pump water out of the soil above.

Groundwater is a significant source of water for domestic, industrial, and agricultural use. Shallow wells commonly provide water for farm and rural dwellings (Figure 6.22). Deeper wells are used as sources of water that is pumped to the surface for municipal, industrial, and agricultural purposes. Some 20% of all water used in the United States

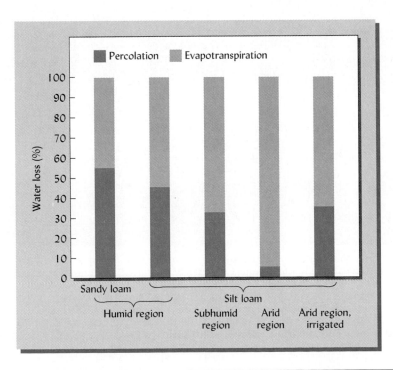

FIGURE 6.21   Percentage of the water entering the soil that is lost by downward percolation and by evapotranspiration. Representative figures are shown for different climatic regions.

comes from groundwater sources. While drainage water from soils is a major source of replenishment for these underground water resources, it is insufficient to compensate for the very extensive withdrawals that are being made in some of these deep reservoirs. They are being grossly overexploited, the rate of water removal far exceeding the rate of recharge. Water that filtered downward into the deep groundwater pools hundreds of years ago is being removed far more rapidly than it can be replaced. Consequently, these

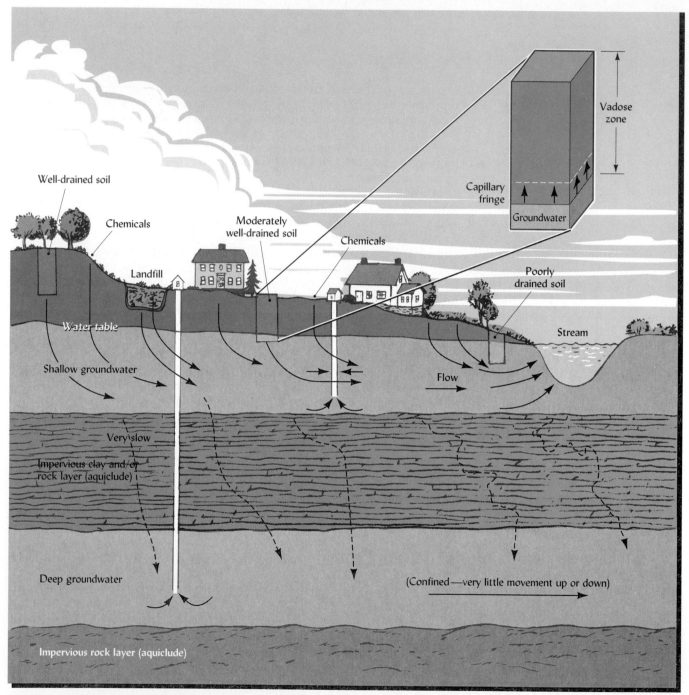

FIGURE 6.22 Relationship of water table and groundwater to water movement into and out of the soil. Precipitation and irrigation water move down the soil profile under the influence of gravity (gravitational water), ultimately reaching the water table and underlying shallow groundwater. The unsaturated zone above the water table is known as the *vadose zone*. As water is removed from the soil by evapotranspiration, groundwater moves up from the water table by capillarity in what is termed the *capillary fringe*. Groundwater also moves horizontally down the slope toward a nearby stream, carrying with it chemicals that have leached through the soil, including essential plant nutrients (N, P, Ca, etc.) as well as pesticides and other pollutants from domestic, industrial, and agricultural wastes. Groundwaters are major sources of water for wells, the shallow ones removing water from the groundwater near the surface, while deep wells exploit deep and usually large groundwater reserves.

Vadose zone

Capillary fringe

Water table

Ground-water

FIGURE 6.23 The water table, capillary fringe, zone of unsaturated material above the water table (vadose zone), and groundwater are illustrated in this photograph. The groundwater can provide significant quantities of water for plant uptake. (Photo courtesy of R. Weil)

deep water tables are becoming even deeper. Near ocean coastlines, as the water is removed by pumping, its place is being taken by nearby seawater. This possible intrusion of saltwater is illustrated in Figure 6.2.

Groundwater that is nearer the surface can also serve as a reciprocal water reservoir for the soil. As plants remove moisture from the soil, water can move upward from the water table by capillarity. The zone of wetting by capillary movement is known as the *capillary fringe* (see Figure 6.23). Such movement provides a steady and significant supply of water that enables plants to survive during periods of low rainfall.

The percolation/groundwater interchange are of significance in the hydrologic cycle, not only from the standpoint of water movement but in relation to the chemicals they transport from the soil to sources of water that are consumed by humans and other forms of life. Brief consideration will be given to the movement of these chemicals.

6.15 MOVEMENT OF CHEMICALS IN THE DRAINAGE WATER[2]

Drainage water carries with it a variety of inorganic and organic chemicals. Included are compounds containing elements such as nitrogen, phosphorus, potassium, calcium, magnesium, and others that are essential for plant growth. They may have originated in the soil through the dissolution of minerals or the breakdown of organic matter coming from plant residues. They may also come from chemical fertilizers and animal manures that have been added to the soil to stimulate plant growth. In any case, their downward movement through the soil and into underlying groundwaters has two serious implications. First, the shift of these chemicals represents a depletion of plant nutrients from the soil and from the root environment of plants. This has implications for plant health as well as for the replenishment of these nutrients in the soil solution (see Chapter 16).

A second problem is the accumulation of these chemical nutrients in ponds, lakes, reservoirs, and groundwater downstream. Such accumulation can stimulate the excess production of algae and other aquatic organisms (including weeds) in the water bodies. This process, called *eutrophication,* ultimately depletes the oxygen content of the water, with disastrous effects on fish and other aquatic life (see also Section 14.2). Also, in some areas underground sources of drinking water may become contaminated with excess nitrates to levels unsafe for human consumption (see Section 13.8).

---

[2]For a review of field studies of transport of chemicals through soil see Jury and Fluhler (1992).

While drainage water is only the means of transport of these chemicals, mechanisms for their movement into the underground water is under continuing surveillance and investigation.

Organic Chemicals

Some organic compounds formed in nature as organic residues decompose are soluble and can move downward with the drainage water. In recent years, however, other synthetic chemicals have been added to the soil/water systems. For example, pesticides that have become widely used in agriculture along with their breakdown products are found in most agricultural soils. Another source of organic chemicals is land application of organic wastes from domestic, industrial, and agricultural operations that have become more common in recent years. While these potential pollutants will be given more detailed consideration in Chapter 19, they are mentioned here since the drainage/groundwater continuum is their primary mode of transport to streams, rivers, and lakes. Figure 6.22 illustrates how the groundwater is charged with the chemicals, and how it transports them to downstream bodies of water.

Chemical Movement Through Macropores

Studies of the movement of chemicals in soils and from soils to the groundwater and to downstream bodies of water have called attention to the critical role played by large macropores in hydraulic conductivity. The pore configuration in most soils is not very uniform. Channels made in the soil through burrowing rodents and earthworms as well as root channels provide paths for water movement that are far more open than one would suspect from the average pore size of the soil. Also, differential swelling and shrinking of some types of clays leave cracks that at least temporarily are open for rapid downward transport of water and of the chemicals it carries.

The presence of large macropores just described makes possible more rapid downward movement of both organic and inorganic chemicals than had earlier been realized. This means that chemicals that are normally broken down in the soil by microorganisms within a few weeks, for example, may still have enough time to move down through large macropores to the groundwater before their degradation can occur.

In some cases, the infiltration of either inorganic or organic chemicals into soils is most serious if the chemicals are merely applied on the soil surface. As shown in Figure 6.24, heavy rainfall could easily wash these chemicals into the large pores through which they could quickly move downward. Research has shown that if the chemicals are worked into the upper few centimeters of soil, movement into the larger pores is reduced and the zone of penetration is greatly curtailed. This provides us with only one example of the importance of soil water management in helping to improve environmental quality.

6.16 HUMAN ENHANCEMENT OF SUPPLIES AND OUTFLOW OF SOIL WATER

Through the ages, people have taken steps to modify some aspects of the hydrologic cycle. We have already considered steps being taken to influence the infiltration of water into soils and to maximize biomass production from the water which evaporates from land areas and returns to the atmosphere. The picture would not be complete if we did not consider briefly two other anthropogenic modifications in the hydrologic cycle: (1) the use of artificial drainage to enhance the downward percolation of water from those soils having an excess of water for optimum plant growth, and (2) the application of irrigation water to supplement that present in soils, especially those of the semiarid and arid regions. We shall consider artificial land drainage first.

6.17 LAND DRAINAGE

Excess soil water is not conducive to the optimum growth of most upland plants. Except for water-tolerant plants such as rice, cattails, and wetland forest species, most plants flourish best in a soil environment containing both air and water, usually in about equal proportions. Soils with excessive moisture are poorly aerated, and the air they contain is low in the oxygen so essential for the respiration of most plant roots.

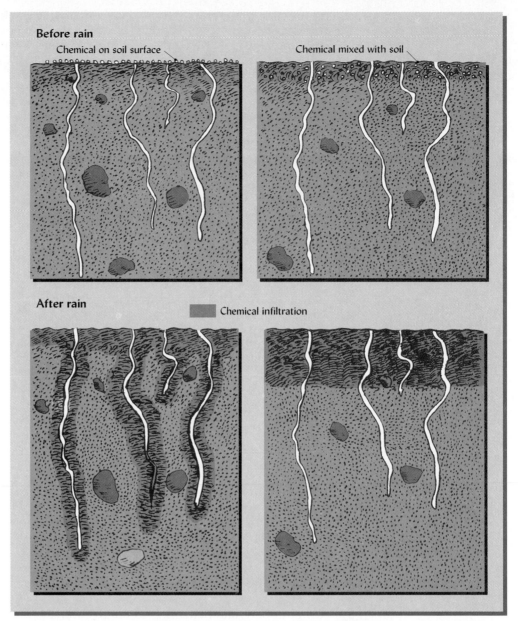

Before rain

Chemical on soil surface

Chemical mixed with soil

After rain

Chemical infiltration

FIGURE 6.24 Role of macropores in transport of a soluble chemical downward through a soil profile. Where the chemical is on the soil surface (left), and can dissolve in surface-ponded water when it rains, it may rapidly be transported down cracks, earthworm channels, and other macropores. Where the chemical is dispersed within the soil matrix in the upper horizon (right), most of the water moving down through the macropores will bypass the chemical, and thus little of the chemical will be carried downward.

Some soils with high water tables are excessively wet or waterlogged at least some time during the year. The water table and its accompanying groundwater may be of great benefit in providing water through capillarity for the plant roots just above the water table. But if the water is too high (at or near the surface) there is too little soil volume in which most roots can grow. Drainage systems can effectively lower the water table and accelerate the removal of excess water from such poorly drained soils.

Maintaining a low water table by artificial drainage is essential to prevent the flooding of basements in some urban areas and to discourage the breakup of roads and highways in some rural areas. Heavy trucks traveling over a paved road underlain by a high water table may lead to the creation of chuckholes and the destruction of asphalt pavement.

If a poorly drained soil is to provide a satisfactory habitat for most upland plants, the excess soil moisture must be removed by the installation of a *land-drainage system*. The objective is to lower the moisture content of the upper layers of the soil so that oxygen can be available to the crop roots, and carbon dioxide can diffuse from these roots into the atmosphere.

Land drainage is practiced in select areas in almost every climatic region, but it is most effective in enhancing plant growth in areas such as the flat coastal-plain sections of eastern United States. Likewise, the fine-textured soils of river deltas, such as those of the Mississippi and the Nile, and lake-laid soils throughout the world produce best if some form of artificial drainage is used. Even irrigated lands of arid regions often require extensive drainage systems to remove excess salts or to prevent their buildup.

Two general types of land-drainage systems are used: (1) surface field drains and (2) subsurface drains. Each will be discussed briefly.

6.18 SURFACE FIELD DRAINS

The most extensive means of removing excess water from soils is through the use of surface drainage systems. Their purpose is to remove the water from the land before it infiltrates the soil and moves into the groundwater. Surface drainage may involve deep and rather narrow field ditches such as those used to remove large quantities of water from peat soil areas (Figure 6.25). More often, however, shallow ditches with gentle side slopes are used to help remove the water. Shallow ditches are commonly constructed with simple equipment and are not too costly. If there is some slope on the land, the shallow ditches are usually constructed across the slope and across the direction of planting and cultivating, thereby permitting the interception of runoff water as it moves down the slope (Figure 6.26).

Land Forming

Surface drain ditches are combined with *land forming* or smoothing to rapidly remove water from soils. Depressions or ridges that prevent water movement to the drainage outlet are filled in with precision, using field-leveling equipment. The resulting land configuration permits excess water to move slowly over the soil surface to the outlet ditch and then on to a natural drainage channel (Figure 6.27).

Land smoothing is a common practice in irrigated areas. Surface irrigation is made possible, as is the removal of excess water by the outlet ditch. In humid areas, the same methods are being put to use to remove excess surface water. In combination with selectively placed tile drains, smoothing the land permits orderly water removal.

6.19 SUBSURFACE (UNDERGROUND) DRAINS

Systems of underground channels to remove water from the zone of maximum water saturation provide the most effective means of draining land. The underground network can be created in several ways, including the following.

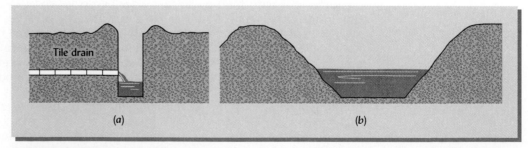

FIGURE 6.25 Two common types of drainage ditches. (a) A small ditch with vertical sides such as is used to drain organic soil areas. Note the tile outlet from one side. (b) A larger sloping-sided ditch of the type commonly used to transport drainage water to a nearby stream. The source of drainage water may be either tile or open drain systems.

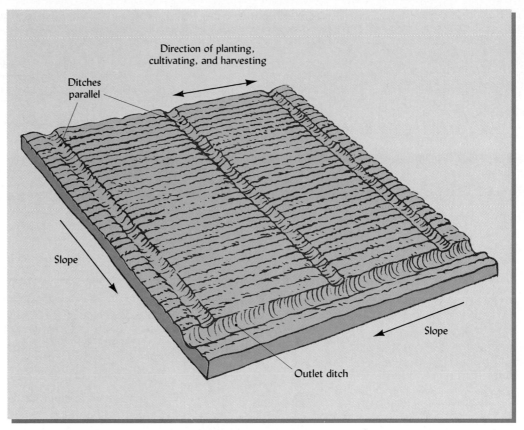

FIGURE 6.26 Example of an open ditch drainage system on a field with gentle slope. [From Hughes (1980); used with permission of Deere & Company, Moline, Ill.]

1. A **mole-drainage** system can be created by pulling through the soil at the desired depth a pointed cylindrical plug about 7 to 10 cm in diameter. The compressed wall channel thus formed provides a mechanism for the removal of excess water (Figure 6.28).

2. A **perforated plastic pipe** can be laid underground using special equipment (Figure 6.29). Water moves into the pipe through the perforations and can be channeled to an outlet ditch. About 90% of the underground drain systems being installed today are of this type.

3. A **clay-tile system** made up of individual clay-pipe units 30 to 40 cm long can be installed in an open ditch. The tiles are then covered with a thin layer of straw, manure, or gravel, and the ditch is refilled with soil. Tile drains were most popular 20 to 30 years ago, but high installation costs make them less competitive than the perforated plastic systems.

The mole-drain system is the least expensive to install but is also more easily clogged than the other two systems, which involve a more stable channel for water flow.

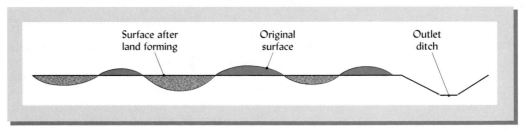

FIGURE 6.27 Land surface before and after land forming or smoothing. Note that soil from the ridges fills in the depressions. Land forming makes possible the controlled movement of surface water to the outlet ditch, which transports it to a nearby natural waterway.

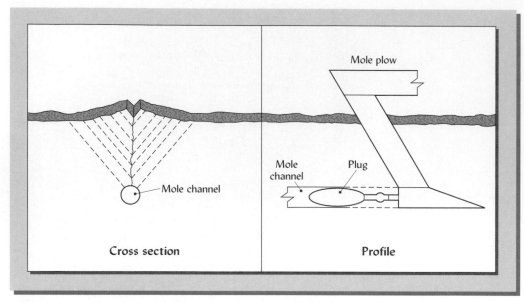

FIGURE 6.28 Diagrams showing how an underground mole drainage system is put in place. The plug is pulled through the soil, leaving a channel through which drainage water can move. [From Hughes (1980); used with permission of Deere & Company, Moline, Ill.]

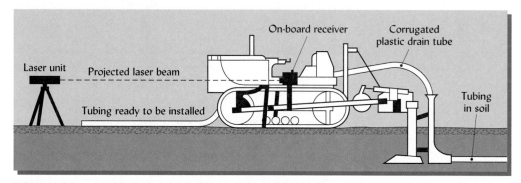

FIGURE 6.29 Drawing showing how corrugated plastic tubing may be placed in the soil. The flexible tubing is fed through a pipe as the tractor moves forward. Appropriate grade level is assured by a laser-beam-controlled mechanism. [From Fouss (1974); used with permission of the American Society of Agronomy]

Operation of Underground Systems

The same principles govern the function of all three types of underground drains. The tile, pipe, or mole channel is placed in the zone of maximum water accumulation. Water moves from the surrounding soil to the channel, enters through perforations in a pipe or joints between the tiles, and is then transported through the channel to an outlet ditch. An illustration of the water-flow pattern to the channel from the surrounding soil is shown in Figure 6.30. Different field layouts for underground systems are illustrated in Figure 6.31.

Care needs to be taken to place the tile or other underground channels in the zone of maximum water accumulation. The channels should be at least 75 cm below the soil surface because of the danger of damage from heavy machinery, and a depth of 1 m is usually recommended. The distance between the channel lines will vary with the soil conditions (Table 6.6). On heavy clay soils, the lines may be as close as 9–10 m, but a distance of 15–20 m is more common.

The grade or fall in the underground drain system is normally 12–25 cm per 50 m (about 3–6 in. per 100 ft), although a fall as great as 75 cm per 50 m is used for rapid water removal.

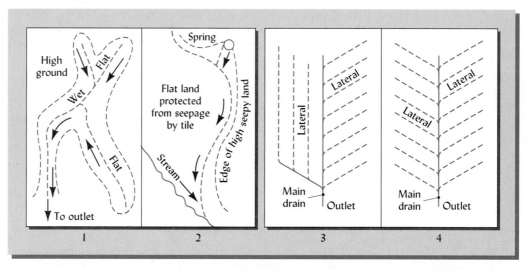

FIGURE 6.30 Demonstration of the saturated flow patterns of water toward a drainage tile. The water, containing a colored dye, was added to the surface of the saturated soil and drainage was allowed through the similation drainage tile shown on the extreme right. (Courtesy G. S. Taylor, The Ohio State University)

FIGURE 6.31 Tile drainage systems. (Top) Four typical systems for laying tile drain: (1) *natural*, which merely follows the natural drainage pattern; (2) *interception*, which cuts off water seeping into lower ground from higher lands above; (3) *gridiron* and (4) *fishbone* systems, which uniformly drain the entirety of an area. (Bottom) A freshly installed drainage system. Tile lines feed into main drains at the bottom, the direction of flow being shown by the arrows. (Photo courtesy U.S. Department of Agriculture)

TABLE 6.6 **Suggested Spacing Between Tile Laterals for Different Soil and Permeability Conditions**

Soil	Permeability	Spacing	
		Meters	Feet
Clay and clay loam	Very slow	9–18	30–70
Silt and silty clay loam	Slow to moderately slow	18–30	60–100
Sandy loam	Moderately slow to rapid	30–90	100–300
Muck and peat	Slow to rapid	15–61	50–200

Modified from Beauchamp (1955).

Special care must be taken to protect the outlet of an underground drainage channel. If the outlet becomes clogged with sediment, the whole system is endangered (see Figure 6.32). It is well to embed the drainage outlet in a masonry concrete wall or block. The last 2.5 to 3 m of drain may even be replaced by a galvanized iron pipe or sewer tile, thus ensuring against damage by frost. The outlet may be covered by a gate or by wire in such a way as to allow the water to flow out freely and to prevent the entrance of rodents in dry weather.

Drainage Around Building Foundations

Surplus water around foundations and underneath basement floors of houses and other buildings can cause serious and expensive damage. The removal of this excess water is commonly accomplished using underground drains similar in principle to those used for field drainage in agriculture.

There must be free movement of the excess water to a tile drain placed alongside and slightly below the foundation or underneath the floor. The water must move rapidly though the tile drains to an outlet ditch or sewer. Care must be taken to ensure that outlet openings are kept free from sediment or debris.

6.20 BENEFITS OF LAND DRAINAGE

Granulation, Heaving, and Root Zone

Draining the land promotes many conditions favorable to higher plants and soil organisms and provides greater stability of building foundations and roadbeds. Heaving—the bad effects of alternate expansion and contraction due to freezing and thawing of soil water—is alleviated (see Section 7.8). Such heaving can break foundations and seriously

FIGURE 6.32 This tile half-filled with sediment has lost much of its effectiveness. Such clogging can result from poorly protected outlets or inadequate slope of the tile line. In some areas of the western United States iron and manganese compounds may accumulate in the tile even when the system is properly designed. (Courtesy L. B. Grass, USDA Natural Resources Conservation Service)

disrupt roadbeds. Also, the heaving of small-grain crops and the disruption of such tap-rooted plants as alfalfa and sweet clover are especially feared (see Figure 7.12). By quickly lowering the water table at critical times, drainage maintains a sufficiently deep and effective root zone (see Figure 6.33). By this means, the volume of soil from which nutrients can be extracted is maintained at a higher level.

Soil Temperature

The removal of excess water also lowers the specific heat of soil, thus reducing the energy necessary to raise the temperature of the drained layers (see Section 7.9). At the same time, surface evaporation, which has a cooling effect, may be reduced. The two effects together tend to make the warming of the soil easier. The converse of the old saying that in the spring "a wet soil is a cold soil" applies. Good drainage is necessary if the land is to be satisfactorily cultivated, and is imperative in the spring if the soil is to warm up rapidly. In the spring a wet soil may be 3 to 8°C cooler than a moist one.

Aeration Effects

Perhaps the greatest benefits of drainage, direct and indirect, relate to aeration. Good drainage promotes the ready diffusion of oxygen to, and carbon dioxide from, plant roots. The activity of the aerobic organisms depends on soil aeration, which in turn influences the availability of nutrients such as nitrogen and sulfur. Likewise, the toxicity from excess iron and manganese in acid soils is decreased if sufficient oxygen is present, because the oxidized states of these elements are quite insoluble. Removal of excess water from soils is at times just as important for plant growth as is the provision of water when soil moisture is low.

Timeliness of Field Operations

A distinct advantage of tile-drain systems is the fact that drained land can be worked earlier in the spring. A well-functioning tile-drain system permits the farmer to start

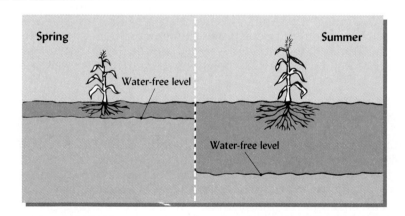

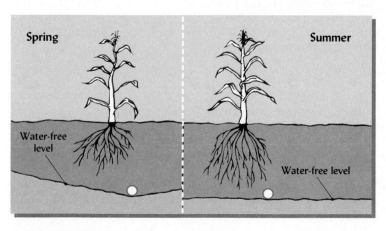

FIGURE 6.33 Illustration of water levels of undrained and tile-drained land in the spring and summer. Benefits of the drainage are obvious. [Redrawn from Hughes (1980); used with permission of Deere & Company, Moline, Ill.]

spring tillage operations one to two weeks earlier than where no drainage is installed. Efficient use of time in the spring of the year is critical in modern farming.

Benefits in Irrigated Areas

Drainage plays a critical role in maintaining the productivity of irrigated soils of arid regions, especially if the irrigation water contains significant salts. The provision of good drainage systems on such soils permits the removal of excess salts if they are already present in the soil, and prevents their buildup in the upper soil horizons if they are present in the irrigation water. Unfortunately, the essential need for adequate drainage was not recognized when some irrigation systems were first constructed. As a result, over time the level of salts in the soils has increased to the extent that crop productivity has been seriously curtailed. Some areas in Pakistan, the Middle East, and India have been so affected (see Figure 19.11).

This problem of salt buildup is accentuated by the practice of reusing water drained from one irrigated saline soil area to irrigate other fields farther down the irrigation system. This water is likely to be high in salts, and may contain sufficient sodium to encourage the formation of a sodic soil, which is physically unsuited for good crop production. A combination of good soil drainage and careful monitoring of the salt content of irrigation water is essential for sustained agricultural production in irrigated arid region soils.

6.21 IRRIGATION

For centuries, farmers and homeowners have used various means of supplementing nature's water supply to stimulate the production of their crops. Rice producers in Asia, wheat and barley producers in the Middle East, and corn producers in Central and South America were irrigating their crops long before the Christian era began. In some cases, rainfall was captured on a given field and then diverted to lower-lying areas where the crop of the farmer's choice was to be grown. In other cases, water was taken from a stream or pulled up from an open well, then hand-carried to the plants. Extensive and well-organized communal irrigation systems that depended upon water diverted into canals from streams and rivers were likewise under way thousands of years ago. The remarkable agricultural systems that developed in the Nile, Tigres, and Euphrates river valleys in the Middle East are examples of such systems that were initially quite successful.

Irrigation and Crop Productivity

Water use for irrigation increased tenfold during the past century. This trend is said to be justified by the markedly increased yields obtained with irrigation. In the western United States, irrigated land that comprises only about 3.5% of the farmed areas produces about 35% of the crop value each year. Such remarkable yield increases have stimulated the construction of reservoirs to store water that can be released later to supplement stream flow, thereby permitting downstream withdrawals for irrigation use. Such irrigation projects have been a major factor in the impressive increases in food production in Asia and Latin America in the past 30 years, and have helped the United States enjoy the world's most abundant food supplies.

Pump Irrigation

Irrigation water obtained from reservoirs and rivers is being complemented by dramatic increases in pump irrigation. In the United States, where some 40% of the total water use is for agriculture, the pumping of groundwater has increased greatly in the past three decades—and in several deep aquifers water is being removed at rates far in excess of the recharge. As a result, water tables are declining so rapidly that some farmers are forced to abandon their irrigation practices. Similar declines in groundwater levels are noted in China and India, where pump irrigation has expanded greatly in recent years.

6.22 METHODS OF IRRIGATION

Three primary methods are used to apply irrigation water: surface systems, sprinkler systems, and microirrigation systems. Each have their advantages and disadvantages, which we will discuss briefly.

Surface Systems

These systems involve supplying water to the surface of the soil, the water commonly being transported to the field in canals or ditches. Used for more than 5000 years, this system involves flooding the entire soil surface for crops such as alfalfa or wheat, or application in furrows for row crops (Figure 6.34). The irrigation water flows into the soil under the influence of gravity and capillarity. When the water is applied in the furrow, capillary attraction stimulates some horizontal movement, causing the water to move laterally into the ridges on which plants are growing.

Surface systems are inexpensive to operate and require little labor. However, the land must be leveled by machinery, so the initial leveling and land-preparation costs are sometimes quite high. Also, these systems are generally the least efficient in the use of the irrigation water. It is very difficult to control the water flow so that water infiltration will be reasonably constant from one side of the field to another. Often, one side receives too much water and the other side not enough.

Sprinkler Systems

Sprinkler systems are most widely used in areas where it is not economical to level the land, although these systems have become increasingly popular even for level fields.

FIGURE 6.34 Typical irrigation scene. The use of easily installed siphons or gated pipes reduces the labor of irrigation and makes it easier to control the rate of application of water. Note the upward capillary movement of water along the sides of the rows. (Courtesy USDA Natural Resources Conservation Service)

Used on more than one-third of the irrigated areas in the United States, these systems involve sprinkling water on the soil surface or on the plants if appropriate ground coverage has been attained. Center-pivot sprinklers that send water outward from a central point, wetting a large circle, are most common (Figure 6.35), although self-propelled traveling sprinklers that move across the field are also used.

The equipment costs for sprinkler systems are higher than those for surface flow systems. However, the water-use efficiency (plant productivity per unit of irrigation water applied) is generally higher for sprinkler systems, especially on sandy soils. Sprinklers also provide cooler water than surface irrigation since the sprinkler droplets tend to cool off as they fly through the air from the applicator to the soil and/or plants. The sprinkler system is generally not preferred for tree crops and vineyards.

Microirrigation

These systems, commonly referred to as *trickle* or *drip* irrigation (Section 6.10) apply very small amounts of water directly to or very near the plants without wetting the entire soil surface. They have already been referred to because of their ability to reduce evaporation from the soil surface (Figure 6.18). The irrigation water is limited to the root zone of the plant and is very efficient in stimulating plant growth.

The microsystems involve the installation of a system of tubing, often underground. Water is released either underground or on the surface alongside the plants. Initial costs of installing the small pipes are high, and the system needs constant monitoring because the small tubes can become plugged easily. Because of its high efficiency, however, microirrigation can be justified where water supplies are scarce and expensive, and where high-valued plants such as fruit trees are being grown.

6.23 WATER-USE EFFICIENCY

Most irrigation systems are very inefficient, with less than half the water that is taken from a stream ever reaching the plants. Much of the water loss occurs in the canal and ditch system used to deliver the water. If the conveyance is only an open ditch, water moves rapidly downward under the ditch and into the groundwater and horizontally into the soil alongside the ditches. While much of this water can be recycled as it moves farther downstream, the efficiency of using the water is initially very low. To reduce these drainage and evaporative losses from the canals and ditches, concrete and plastic are used to line the canals (see Figure 6.36).

FIGURE 6.35 A center-pivot irrigation system in eastern Colorado. The system is rotating toward the left (note the dry soil at the extreme left) and the center pivot of the system is located approximately 0.5 km in the distance. Each self-propelled tower moves in coordination with all the others. The shed at the right houses a diesel-powered water pump that taps a deep aquifer. The tank next to the shed holds liquid fertilizer that can be metered into the irrigation water. (Photo courtesy of R. Weil)

Water-use efficiency in the field is also low, especially if traditional surface systems are used. Only the microsystems can be said to be reasonably efficient in increasing plant production with only small quantities of water.

6.24 IRRIGATION WATER MANAGEMENT

The two most serious irrigation management problems relate to the quality of the water being applied and to the efficiency of the irrigation water in stimulating plant production. Most irrigation systems are located in semiarid and arid regions where the contents of soluble salts in the drainage water and, in turn, in the streams and rivers fed by these drainage waters are high. When this water is added to the soil and percolation takes place, still more salts are dissolved from the soil itself, making the drainage water even more saline than the originally added water. As this drainage water finds its way back into the stream, it adds to the stream's burden of salts. This scenario is repeated several times as one moves from the upper reaches of a river to the ocean. Such salt buildup in the water can be very damaging to both the physical and chemical properties of the soils to which the water is applied.

Practices to enhance water-use efficiency include the selection of the most efficient irrigation system (the microirrigation system being best) and the use of crop residues or mulches to reduce transpiration from the soil surface while simultaneously reducing the soil temperature. Any other practice that minimizes losses of surface-held water, thereby increasing the water available to the plants, should be encouraged.

6.25 CONCLUSION

The hydrologic cycle encompasses all movements of water on or near the earth's surface. It is driven by solar energy that evaporates water from the oceans and lands, which is then cycled into the atmosphere and finally returned to the soil and oceans in rain and snow.

The soil is an essential component of the hydrologic cycle. It receives precipitation from the atmosphere, rejecting some that is forced to run off into streams and rivers and absorbing the remainder that then moves downward, either to be transmitted to the groundwater, to be taken up and later transpired by plants, or to be evaporated directly from its surfaces and returned to the atmosphere.

The behavior and movement of water in soils and plants are governed by the same set of principles—water moves in response to differences in energy levels, moving from higher to lower moisture potential. These principles can be used to manage water more effectively and to increase its efficiency of use.

Management practices should encourage movement of water into well-drained soils while minimizing evaporative losses from the soil surface (E). These two objectives will provide as much water as possible for plant uptake, thereby satisfying the transpiration (T) requirements of healthy leaf surfaces. Practices which leave plant residues on the soil

FIGURE 6.36 Concrete-lined irrigation ditches (center) and standard-sized siphon pipes (right) can increase the efficiency of water delivery to the field. Unlined ditches (left) lose much of the water to adjacent soil areas or to the groundwater. Note the evidence of capillary movement above the water level in the unlined ditch.

surface and which maximize plant shading of this surface will help achieve high efficiency of water use.

Irrigation waters from streams or as pumped from groundwaters greatly enhance plant growth, especially in regions with scarce precipitation. It is important to so manage irrigation water resources that biomass production will be enhanced while soil surface water evaporation will be curtailed.

REFERENCES

Aboukhaled, A., A. Alfaro, and M. Smith. 1982. *Lysimeters*, FAO Irrigation and Drainage Paper 39 (Rome, Italy: UN Food and Agriculture Organization).

Allegre, J. C., D. K. Cassel, and D. E. Bandy. 1986. "Effects of land clearing and subsequent management on soil physical properties," *Soil Sci. Soc. Amer. J.*, **50**:1379–1384.

Beauchamp, K. H. 1955. "Tile drainage—its installation and upkeep," *The Yearbook of Agriculture (Water)* (Washington, D.C.: U.S. Department of Agriculture), p. 513.

Black, A. L. 1982, "Long-term N-P fertilizer and climate influences on morphology and yield components of spring wheat," *Agron J.*, **74**:651–57.

Dreibelbis, F. R., and C. R. Amerman. 1965. "How much topsoil moisture is available to your crops?," *Crops and Soils*, **17**:8–9.

Fouss, J. L. 1974. "Drain tube materials and installation," in J. Van Schilfgaarde (ed.), *Drainage for Agriculture*, Agronomy Series No. 17 (Madison, Wis.: Amer. Soc. Agron.), pp. 147–77.

Greb, B. W. 1983. "Water conservation: central Great Plains," in H. E. Dregue and W. O. Willis (eds.), *Dryland Agriculture*, Agronomy Series No. 23 (Madison, Wis.: Amer. Soc. of Agron.).

Hanks, R. J. 1983. "Yield and water-use relationships: an overview," in H. M. Taylor et al. (eds.), *Limitations to Efficient Water Use in Crop Production* (Madison, Wis.: Amer. Soc. Agron.).

Haynes, J. L. 1954. "Ground rainfall under vegetative canopy of crops," *J. Amer. Soc. Agron.*, **46**:67–94.

Hillel, D. 1980. *Applications of Soil Physics* (New York: Academic Press).

Hughes, H. S. 1980. *Conservation Farming* (Moline, Ill.: John Deere and Company).

Jury, W. A., and H. Fluhler. 1992. "Transport of chemicals through soil: mechanisms, models and field applications," *Adv. in Agron.*, **47**:141–201.

Kelly, O. J. 1954. "Requirement and availability of soil water," *Adv. Agron.*, **6**:67–94.

Kittridge, J. 1948. *Forest Influences. The Effects of Woody Vegetation on Climate, Water and Soil* (New York: McGraw-Hill).

Linde, D. T., T. L. Watschke, A. R. Jarrett, and J. A. Borger. 1995. "Surface runoff assessment from creeping bent grass and perennial ryegrass turf," *Agron. J.*, 176–182.

Lyon, T. L., H. O. Buckman, and N. C. Brady. 1952. *The Nature and Properties of Soils*, 5th ed. (New York: Macmillan).

Ritchie, J. T. 1983. "Efficient water use in crop production: discussion on the generality of relations between biomass production and evapotranspiration," in H. M. Taylor et al. (eds.), *Limitations to Efficient Water Use in Crop Production* (Madison, Wis.: Amer. Soc. Agron., Crop Sci. Soc. Amer., Soil Sci. Soc. Amer.).

UNDP/WMO. 1974. *Hydrometeorological Survey of the Catchments of Lakes Victoria, Kyoja and Albert*, Project RAF 66/025 (4 vols.), pp. 498–509.

USDA 1974. *Summer Fallow in the Western United States*, Cons. Res. Rept. No. 17 (Washington, D.C.: USDA Agricultural Research Service).

The naked earth is warm with Spring. . . .
—JULIAN GRENFELL, INTO BATTLE

We now turn to two other soil physical properties that influence not only plant life, but the quality and comfort of our homes and other buildings, and the stability of our highways. The growth of most plant species and of many microorganisms is affected by the amount and gaseous makeup of soil air that shares with soil water the nonsolid pore spaces. The color of soils and especially of subsoil horizons is influenced by soil aeration. Even the ion species of some nutrient elements that are available for plant uptake is determined by soil aeration.

Soil temperatures likewise affect plant and microorganism growth, and also help control the rate of evaporation of water from the soil surface. The formation of chuckholes in roads and highways is associated with a combination of excess water (low aeration) and changes in soil temperature. Even the heating of homes in the winter and cooling them in the summer can be enhanced by taking advantage of the resistance of soils to rapid changes in temperature. High soil temperatures resulting from forest fires can greatly decrease the rate of infiltration of water into soils. Soil temperature will be addressed after a consideration of soil aeration.

7.1 SOIL AERATION CHARACTERIZED

Soil aeration is a vital process because it largely controls the soil levels of two life-sustaining gases: oxygen and carbon dioxide. These gases, along with water, are primary participants in two vital biological reactions: (1) the *respiration* of all plant and animal cells and (2) *photosynthesis* that creates sugars, a fundamental building block for all food. Respiration involves the oxidation of organic compounds as follows (using sugar as an example of organic compounds):

$$C_6H_{12}O_6 + 6O_2 \longrightarrow 6CO_2 + 6H_2O$$
Sugar

Through photosynthesis this reaction is reversed. Carbon dioxide and water are combined by green plants to form sugars, and oxygen is released to be consumed by humans and animals.

Soil aeration is a critical component of this overall system. For respiration to continue in the soil, oxygen must be supplied and carbon dioxide removed. Through aeration, there is an exchange of these two gases between the soil and the atmosphere. In a well-aerated soil, this exchange is sufficiently rapid to prevent the deficiency of oxygen or the toxicity of excess carbon dioxide. For most upland plants, the supply of oxygen in the soil air must be kept above 10%. In turn, CO_2 concentration and that of other potentially toxic gases such as methane must not be allowed to build up excessively.

7.2 SOIL AERATION PROBLEMS IN THE FIELD

In the field, poor soil aeration occurs under two conditions: (1) when the moisture content is so high that there is little or no room for gases and/or (2) when the exchange of gases with the atmosphere is so slow that desirable levels of soil gases cannot be maintained. The latter may occur even when sufficient *total air space* is available.

Excess Moisture

The first case is characterized in the extreme by the waterlogged condition typical of wetlands. In agricultural fields, low spots or flat areas where water tends to stand for a short while provide examples of this condition. In well-drained soils saturated conditions may also occur temporarily during a heavy rainstorm, when excess irrigation water is applied, or if the soil has been compacted by plowing or by heavy machinery.

Such complete saturation of the soil with water is not a problem with some plant species that have unique means of obtaining oxygen even when their roots are surrounded with water. For example, rice plants transport oxygen down their stems to support the respiration of root cells growing in a water-saturated soil. A few other species have similar mechanisms (see Figure 7.1). For most plants, however, a saturated soil can be disastrous in only a short period of time—a matter of a few hours being critical in some cases. Most plants are dependent on a supply of oxygen from the soil, and so suffer dramatically if good soil aeration is not maintained by artificial drainage or other means.

Gaseous Interchange

The second case relates to gas exchange. The more rapid the usage of oxygen and the corresponding release of carbon dioxide, the greater is the need for the exchange of gases between the soil and the atmosphere. This exchange is facilitated by two mechanisms: **mass flow** and **diffusion.** Mass flow of air, which is due to pressure differences between the atmosphere and the soil air, is much less important than diffusion in determining the total exchange that occurs. It is enhanced, however, by fluctuations in soil moisture content. As water moves into the soil during a rain or from irrigation, air must be forced out. Likewise, when soil water is lost by evaporation from the soil surface, or is taken up by plants, air is drawn into the soil. Mass flow also is modified slightly by other factors such as temperature, barometric pressure, and wind.

The great bulk of the gaseous interchange in soils occurs by diffusion. Through this process, each gas moves in a direction determined by its own partial pressure (Figure 7.2). The partial pressure of a gas in a mixture is simply the pressure this gas would exert if it alone were present in the volume occupied by the mixture. Thus, if the pressure of air is 1 atmosphere (~1 bar), the partial pressure of oxygen, which makes up about 21% of the air by volume, is approximately 0.21 bar.

Diffusion allows extensive gas movement from one area to another even though there is no overall pressure gradient for the total mixture of gases. There is, however, a concentration gradient for each individual gas, which may be expressed as a *partial pressure gradient.* As a consequence, even though the total soil air pressure and that of the atmosphere may be the same, a higher concentration of oxygen in the atmosphere will result in a net movement of this particular gas into the soil. Carbon dioxide and water vapor normally move in the opposite direction, since the partial pressures of these two gases are generally higher in the soil air than in the atmosphere. A representation of the principles involved in diffusion is given in Figure 7.2.

FIGURE 7.1 Plants vary widely in their ability to withstand low oxygen conditions in their root zone. These Mangrove trees along the coast of the Indian Ocean are partially submerged during high tide, but have developed "knees" on their root system that enable them to access oxygen directly from the air. (Photos courtesy of R. Weil)

7.3 MEANS OF CHARACTERIZING SOIL AERATION

The aeration status of a soil can be characterized conveniently in several ways, including (1) the content of oxygen and other gases in the soil atmosphere, (2) the oxygen diffusion rate (ODR), and (3) the oxidation-reduction (redox) potential. Each of these three will receive brief attention.

Gaseous Oxygen of the Soil

The atmosphere above the soil contains nearly 21% O_2, 0.035% CO_2, and more than 78% N_2. In comparison, soil air has about the same level of N_2 but is consistently lower in O_2 and higher in CO_2. The O_2 content may be only slightly below 20% in the upper layers of a soil with a stable structure and an abundance of macropores. It may drop to less than 5% or even to near zero in the lower horizons of a poorly drained soil with few macropores.

Low O_2 contents also are found in low-lying areas where water has accumulated. Even in well-drained soils, marked reductions in the O_2 content of soil air may follow a heavy rain, especially if plants are growing rapidly or if large quantities of manure or other decomposing organic residues are present (Figure 7.3).

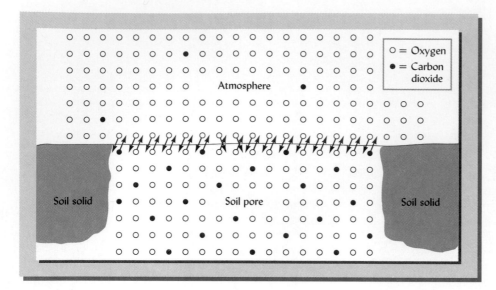

FIGURE 7.2 The process of diffusion between gases in a soil pore and in the atmosphere. The total gas pressure is the same on both sides of the boundary. The partial pressure of oxygen is greater, however, in the atmosphere. Therefore, oxygen tends to diffuse into the soil pore where fewer oxygen molecules per unit volume are found. The carbon dioxide molecules, on the other hand, move in the opposite direction owing to the higher partial pressure of this gas in the soil pore. This diffusion of O_2 into the soil pore and of CO_2 into the atmosphere will continue as long as the respiration of root cells and microorganisms consumes O_2 and releases CO_2.

It is fortunate that the water in many soils contains small but significant quantities of dissolved O_2. When all the soil pores are filled with water, soil microorganisms can extract most of the O_2 dissolved in the water for metabolic purposes. This small amount of dissolved O_2 soon is used up, however, and if the excess water is not removed, aerobic microbial activities as well as plant growth are jeopardized.

Since the N_2 content of soil air is relatively constant, there is a general inverse relationship between the contents of the other two major components of soil air—O_2 and CO_2—with O_2 decreasing as CO_2 increases. Although the actual differences in CO_2 amounts may not be impressive, they are significant, comparatively speaking. Thus, when the soil air contains only 0.35% CO_2, this gas is about 10 times as concentrated as it is in the atmosphere. In cases where the CO_2 content becomes as high as 10%, there is nearly 300 times as much present as is found in the air above. At such levels, carbon dioxide may be toxic to some plant processes.

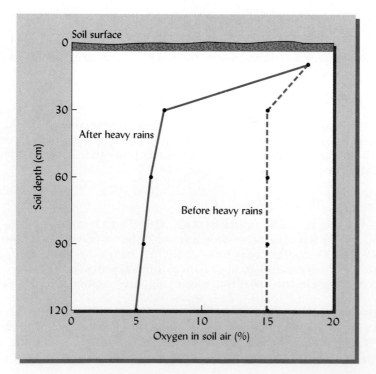

FIGURE 7.3 Oxygen content of soil air before and after heavy rains in a soil on which cotton was being grown. The rainwater replaced most of the soil air. The small amount of oxygen remaining was consumed in the respiration of roots and soil organisms. The carbon dioxide content (not reported) probably increased accordingly. [Redrawn from Patrick (1977); used with permission of the Soil Science Society of America]

Oxygen Diffusion Rates

A second means of measuring the aeration status of a soil is the *oxygen diffusion rate* (ODR). The ODR largely determines the rate at which oxygen can be replenished when it is used by respiring plant roots or by microorganisms, or when it is forced out of the pores by water. If the rate of replenishment is high, the soil's aeration status is satisfactory. Figure 7.4 is a graph showing how ODR decreases with soil depth. The ODR also decreases with increasing soil water content, largely because oxygen diffuses 10,000 times faster through a pore filled with air than through a similar pore filled with water.

Researchers have found that the ODR is important to growing plants. For example, the growth of roots of most plants ceases when the ODR drops to about 20×10^{-8} g/cm^2 per minute (see Figure 7.4). Top growth is normally satisfactory as long as the ODR remains above $30–40 \times 10^{-8}$ g/cm^2 per minute. In Table 7.1 are found some field measurements of ODR, along with comments about the condition of the plants. Note the general tendency for difficulty if the ODR gets below the critical level.

Unfortunately, measurements of ODR are not very accurate for dry soils and for sandy soils. Also, the conditions for making the measurements must be carefully standardized. Consequently, the measurements are most useful in heavier-textured soils and for comparing ODR levels at different depths in the same profile.

Oxidation-Reduction (Redox) Potential[1]

One important chemical characteristic of soils that is related to soil aeration is the reduction and oxidation states of the chemical elements in these soils. If a soil is well-aerated, oxidized states such as those of ferric iron (Fe^{3+}), manganic manganese (Mn^{4+}), nitrate (NO_3^-), and sulfate (SO_4^{2-}) are dominant. In poorly drained and poorly aerated soils, the reduced forms of such elements are found; for example, ferrous iron (Fe^{2+}), manganous manganese (Mn^{2+}), ammonium (NH_4^+), and sulfides (S^{2-}). The presence of these reduced forms is an indication of restricted drainage and poor aeration.

The reaction that takes place as the reduced state of an element is changed to the oxidized state may be illustrated by the oxidation of two-valent iron in FeO to the trivalent form in FeOOH.

---

[1]For a review of redox reactions in soils, see Bartlett and James (1993).

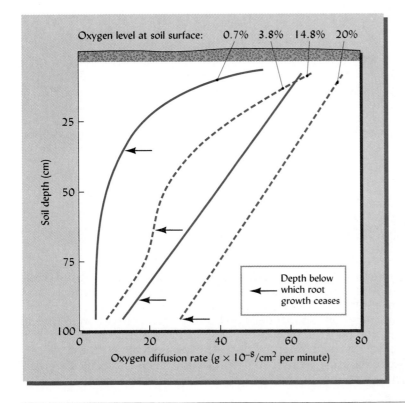

FIGURE 7.4 Effect of soil depth and oxygen concentration at the surface on the oxygen diffusion rate (ODR). Arrows indicate snapdragon root penetration depth. Even with a 20% O_2 level at the surface, the diffusion rate at ~95 cm is less than half that at the surface. Note that when the ODR drops to about 20×10^{-8} g/cm^2 per minute, root growth ceases. [Redrawn from Stolzy et al. (1961)]

TABLE 7.1   **Relationship Between Oxygen Diffusion Rate (ODR) and the Condition of Different Plants**

When the ODR drops below about 40×10^{-8} g/cm² per minute, the plants appear to suffer. Sugar beets require high ODR even at 30 cm depth. ODR values in 10^{-8} g/cm² per minute.

Plant	Soil texture	ODR at three soil depths			Remarks
		10 cm	20 cm	30 cm	
Broccoli	Loam	53	31	38	Very good growth
Lettuce	Silt loam	49	26	32	Good growth
Beans	Loam	27	27	25	Chlorotic plants
Sugar beets	Loam	58	60	16	Stunted taproot
Strawberries	Sandy loam	36	32	34	Chlorotic plants
Cotton	Clay loam	7	9	—	Chlorotic plants
Citrus	Sandy loam	64	45	39	Rapid root growth

From Stolzy and Letey (1964).

$$FeO + H_2O \rightleftharpoons FeOOH + H^+ + e \qquad (7.1)$$
$$(Fe^{2+}) \qquad\qquad (Fe^{3+})$$

Note that the divalent iron (Fe^{2+}) loses an electron (e) as it changes to the trivalent form (Fe^{3+}), and that the H^+ ions are formed in the process. The loss of an electron suggests that there are potentials for the transfer of electrons from one substance to another. This potential can be measured using a platinum electrode and may be referred to as platinum *electrode potential.* Since it deals with reduction/oxidation reactions, however, it is more commonly called **redox potential.**

The redox potential (E_h) is a measure of the tendency of a substance to accept or donate electrons. It is usually measured in volts or millivolts. As was the case for water potential, redox potential is related to a reference state, in this case the hydrogen couple $\frac{1}{2}H_2 \rightleftharpoons H^+ + e$, whose redox potential is arbitrarily taken as zero. If the E_h is high, the substance will accept electrons easily and is known as an *oxidizing agent.* All aerobic respiration requires molecular oxygen to serve as the electron acceptor as organic carbon is oxidized to release energy for life. Oxygen can oxidize both organic and inorganic substances. Thus, molecular oxygen rapidly accepts electrons and is a strong oxidizing agent.

Keep in mind, however, that as it oxidizes another substance, oxygen is in turn reduced. This reduction process can be seen in the following reaction.

$$O_2 + 4H^+ + 4e \rightleftharpoons 2H_2O \qquad (7.2)$$

Note that the oxygen has accepted four electrons which could have been donated by four molecules of FeO undergoing oxidation as shown in Equation (7.1) above. If we combined the two equations we can see the overall effect of oxidation and reduction.

$$4FeO + 4H_2O \rightleftharpoons 4FeOOH + 4H^+ + 4e$$
$$O_2 + 4H^+ + 4e \rightleftharpoons 2H_2O$$
$$\overline{4FeO + O_2 + 2H_2O \rightleftharpoons 4FeOOH}$$

The donation and acceptance of electrons (e) and the $4H^+$ ions on each side of the equation have balanced each other, but for the specific reduction and oxidation reactions they are both very important.

The redox potential (E_h) of a soil is dependent on both oxygen content and pH. The positive correlation in one soil between O_2 content of soil air and E_h (redox potential) is shown in Figure 7.5. In a well-drained soil, the E_h is in the 0.4–0.7 volt (V) range. As aeration is reduced, the E_h declines to a level of about 0.3–0.35 V when gaseous oxygen is depleted. Under flooded conditions E_h values as low as –0.3 V can be found.

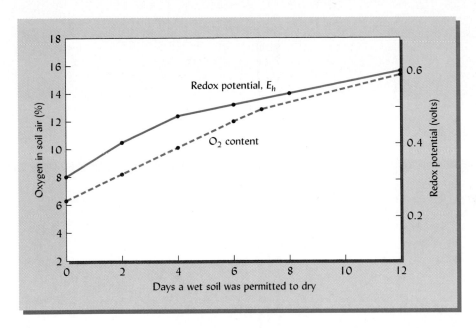

FIGURE 7.5 Relationship between oxygen content of soil air and the redox potential (E_h). Measurements were taken at a 28 cm-depth in a soil that had been irrigated continuously for 14 days prior to the drying. Note the general relationship between these two parameters. [Data selected from Meek and Grass (1975); used with permission of Soil Science Society of America]

The effect of pH on redox potentials relating to several important reactions taking place in soils is shown in Figure 7.6. Note that in all cases the E_h decreases as the pH rises from 2 to 8. Since both pH and E_h are easily measured, it is not too difficult to ascertain in a given soil whether a specified reaction would likely occur. For example, at pH 6 the E_h would need to be somewhat lower than +0.5 volts to encourage the reduction of nitrates to nitrites, and about +0.2 volts to stimulate the reduction of FeOOH to the Fe^{2+} form. For methane to form in a waterlogged soil at the same pH (6), an E_h of about –0.2 volts would be required.

The E_h value at which oxidation-reduction reactions occur varies with the specific chemical to be oxidized or reduced. In Table 7.2 are listed oxidized and reduced forms of several elements important in soils, along with the approximate redox potentials at which the oxidation-reduction reactions occur. Note that gaseous oxygen is reduced to water at E_h levels of 0.38 to 0.32 V. At lower E_h values, the only microorganisms which can function are those able to use elements other than oxygen as their metabolic electron acceptors. Thus at low E_h values, elements such as N (in NO_3^-), Mn, Fe, and S (in SO_4^{-2}) accept electrons and become reduced. Thus soil aeration helps determine the specific chemical species present, and in turn essential nutrient availability as well as chemical toxicities, in soils.

Other Gases

Soil air usually is much higher in water vapor than is the atmosphere, being essentially saturated except at or very near the surface of the soil. This fact already has been stressed in connection with the movement of water. Carbon dioxide is commonly 10 to 100 times higher in soil air than in the atmosphere and may be thousands of times higher in isolated pockets in wet soils. Also, under waterlogged conditions, the concentration of gases such as methane and hydrogen sulfide, which are formed as organic matter decomposes, is notably higher in soil air.

7.4 FACTORS AFFECTING SOIL AERATION

The drainage of excess water from a soil is a primary factor in determining its aeration status. Air can enter the soil pores only when at least some of the water is removed. The oxygen level in soil pores just above a water table is generally low.

The most important factors influencing the aeration of well-drained soils are those that determine the volume of the soil's macropores. Macropore content definitely affects the total air space as well as gaseous exchange and biochemical reactions. Soil texture, bulk density, aggregate stability, and organic matter content are among the soil properties that help determine macropore content and, in turn, soil aeration.

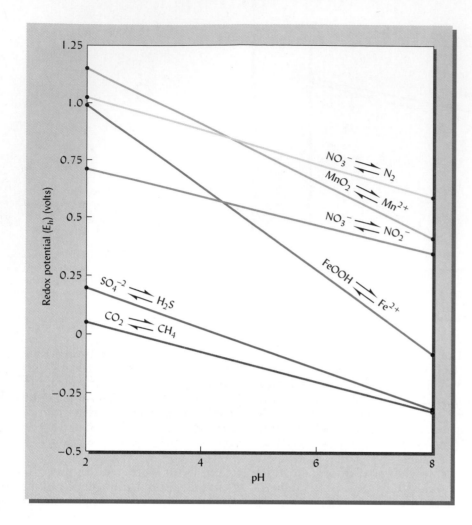

FIGURE 7.6 The effect of pH on the redox potential (E_h) at which several important reduction-oxidation reactions take place in soil. [From McBride (1994); used with permission of Oxford University Press]

The concentrations of both oxygen and carbon dioxide are definitely affected by microbial decomposition of organic residues. Incorporation of large quantities of manure, crop residues, or sewage sludge may alter the soil air composition appreciably. Respiration by higher plant roots and by soil organisms around the roots is also a significant process (Figure 7.7).

TABLE 7.2 Oxidized and Reduced Forms of Certain Elements in Soils and the Redox Potentials (E_h) at Which Changes in Form Occur in a Soil at pH 6.5

Note that gaseous oxygen is depleted at E_h levels of 0.38–0.32 V. At lower E_h levels microorganisms utilize elements other than oxygen as the electron acceptor in their metabolism. By donating electrons they transform these elements into their reduced valence state.

Oxidized form	Reduced form	E_h at which change of form occurs (V)
O_2	H_2O	0.38 to 0.32
NO_3^-	N_2	0.28 to 0.22
Mn^{4+}	Mn^{2+}	0.22 to 0.18
Fe^{3+}	Fe^{2+}	0.11 to 0.08
SO_4^{2-}	S^{2-}	−0.14 to −0.17
CO_2	CH_4	−0.2 to −0.28

From Patrick and Jugsujinda (1992).

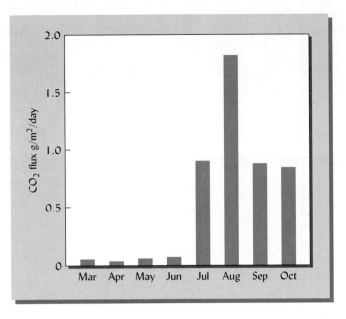

FIGURE 7.7   The rate of CO_2 movement (flux) from the surface of a soil within a northern hardwood forest ecosystem in New York State. Note the high rates from July to October when forest growth and microbical action were highest. [From Yavitt et al. (1995); used with permission of Soil Science Society of America]

Subsoil versus Topsoil

Subsoils are usually more deficient in oxygen than are top soils. The total pore space as well as the macropore space is generally much lower in the deeper horizons.

In some forested areas and in fruit orchards the deep rooted trees extend into layers that are very low in oxygen and high in carbon dioxide, especially if the subsoil is high in clay. Carbon dioxide levels of nearly 15% have been observed in some such subsoils.

Soil Heterogeneity

The aeration status varies greatly in different locations in a soil profile. Thus, poorly aerated zones or pockets may be found in an otherwise well-drained and well-aerated soil. The poorly aerated component may be a heavy-textured or compacted soil layer, or it may be merely the inside of a soil ped (structural unit) where tiny pores may limit ready air exchange (Figure 7.8). For these reasons, oxidation reactions may be occurring within a few centimeters from another location where reducing conditions exist. This heterogeneity of soil aeration should be kept in mind.

Seasonal Differences

There is marked seasonal variation in the composition of soil air. In the springtime in temperate humid regions the soils are often wet and opportunities for ready gas exchange are poor. But due to low soil temperatures, the respiration of plant roots and soil microorganisms is restricted, so the utilization of oxygen and release of CO_2 are also contained. In the summer months, the soils are commonly lower in moisture content, and the opportunity for gaseous exchange is increased. However, more favorable temperatures stimulate vigorous respiration of plant roots and breakdown of plant residues that release copious quantities of carbon dioxide. This is illustrated in Figure 7.9, which shows carbon dioxide levels in an Alfisol in Missouri during the corn-growing season and later into early winter. The effects of summer temperatures and vigorously growing plants are well illustrated.

7.5   EFFECTS OF SOIL AERATION

Effects on Soil Reactions and Properties

Soil aeration influences many soil reactions and, in turn, soil properties. The most obvious of these reactions are associated with the microbial breakdown of organic residues. Poor aeration slows down this rate of decay, as evidenced by the relatively high organic matter level of poorly drained soils, especially in swampy areas.

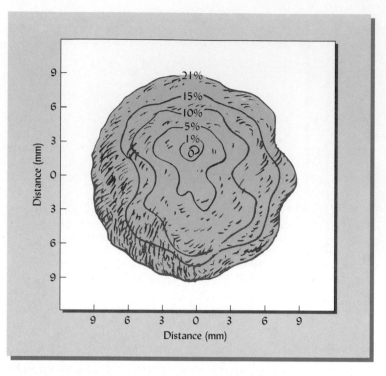

FIGURE 7.8 A "map" showing the oxygen content of soil air in a wet aggregate from an Aquic Hapludoll (Muscatine silty clay loam) from Iowa. The measurements were made with a unique microelectrode. Note that the oxygen content near the aggregate center was zero, while that near the edge of the aggregate was 21%. Thus pockets of oxygen deficiency can be found in a soil whose overall oxygen content may not be low. [From Sexstone et al. (1985)]

The nature as well as the rate of organic decay is determined by the oxygen gas content of the soil. Where gaseous oxygen is present, **aerobic** organisms are active, and oxidation reactions such as the following occur (sugar being used as an example of organic compounds):

$$C_6H_{12}O_6 + 6O_2 \longrightarrow 6CO_2 + 6H_2O$$
$$\text{(sugar)}$$

In the absence of gaseous oxygen, **anaerobic** organisms take over. Much slower breakdown occurs through reactions such as the following (see Section 12.6 for further details):

$$C_6H_{12}O_6 \longrightarrow 2CO_2 + 2C_2H_5OH$$
$$\text{(sugar)} \qquad\qquad\qquad \text{(ethanol)}$$

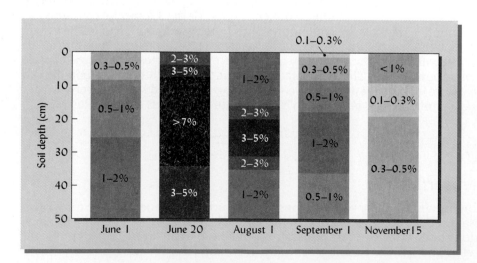

FIGURE 7.9 Seasonal changes in carbon dioxide content in the upper 50 centimeters of an Alfisol in Missouri on which corn was grown from early May until about September 1. By June 20, the corn was growing vigorously, the soil moisture level was still quite high, and the CO_2 level increased to more than 7% in the 10–30 cm zone. By August 1 gaseous exchange had probably increased and the CO_2 level had declined to 3–5% in the 20–30 cm layer, but even this is 100 times the level in the atmosphere. By November 15 activities of plants and microorganisms had declined because of lower temperatures and the CO_2 below 20 cm was only 10 times that in the atmosphere. [Data from Buyanovsky and Wagner (1983); used with permission of the Soil Science Society of America]

Poorly aerated soils therefore tend to contain a wide variety of only partially oxidized products such as alcohols, organic acids and ethylene gas (C_2H_4), which can be toxic to higher plants. Obviously, the presence or absence of oxygen gas completely modifies the nature of the decay process and its effect on plant growth.

Oxidation-Reduction of Elements

Through its effects on the redox potential, the level of soil oxygen largely determines the forms of several inorganic elements, as shown in Table 7.3. The oxidized states of the nitrogen and sulfur are readily utilizable by higher plants. In general, the oxidized conditions are desirable for iron and manganese nutrition of most plants in acid soils of humid regions because, in these soils, the reduced forms of these elements are so soluble that toxicities may occur.

In drier areas, the opposite is generally true, and reduced forms of elements such as iron and manganese are preferred. In the neutral to alkaline soils of these drier areas, oxidized forms of iron and manganese are tied up in highly insoluble compounds, resulting in deficiencies of these elements. Such differences illustrate the interaction of aeration and soil pH (soil acidity, or alkalinity) in supplying available nutrients to plants.

As was discussed in Section 4.1, **soil color** is also influenced markedly by the oxidation status of iron and manganese. Colors such as red, yellow, and reddish brown are encouraged by well-oxidized conditions. More subdued shades such as grays and blues predominate if insufficient oxygen is present. Soil color can be used in field methods for determining the status of soils drainage. Imperfectly drained soils are characterized by alternate streaks of oxidized and reduced materials. The **mottled** condition indicates a zone of alternate good and poor aeration, a condition not conducive to the optimum growth of most plants.

The production of the organic compound, methane, in submerged soils is of universal significance, since this gas is one of those contributing to the so-called greenhouse effect and global warming. Methane gas is produced by the reduction of CO_2. Its formation occurs when the E_h is reduced to about -0.2 V, a condition common in natural wetlands and in rice paddies. Because of the value of these areas as biodiversity-enhancing environments, soil scientists are seeking means of managing methane release without resorting to the drainage of the wetlands.

Effects on Activities of Higher Plants

Higher plants are adversely affected in at least three ways by conditions of poor aeration: (1) the growth of the plant, particularly the roots, is curtailed; (2) the absorption of nutrients and water is decreased; and (3), as discussed in the previous section, the formation of certain inorganic compounds toxic to plant growth is favored.

PLANT GROWTH. Different plant species vary in their ability to tolerate poor aeration (Table 7.4). Wheat and barley, for example, require high air porosities for best growth. In contrast, ladino clover can grow with very low air porosity, and rice can grow with its roots submerged in water. Furthermore, the tolerance of a given plant to low porosity may be different for seedlings than for rapidly growing plants. A case in point is the tolerance of red pine to restricted drainage during its early development and its poor growth or even death on the same site at later stages.

In spite of these wide variations in soil air porosity limitation, soil physicists generally consider that if the soil volume occupied by air is reduced below 10 to 12%, most upland plants are likely to suffer.

TABLE 7.3   Oxidized and Reduced Forms of Several Important Elements

Element	Normal form in well-oxidized soils	Reduced form found in waterlogged soils
Carbon	CO_2, $C_6H_{12}O_6$	CH_4, C_2H_4, CH_3CH_2OH
Nitrogen	NO_3^-	N_2, NH_4^+
Sulfur	SO_4^{2-}	H_2S, S^{2-}
Iron	Fe^{3+} (ferric oxides)	Fe^{2+} (ferrous oxides)
Manganese	Mn^{4+} (manganic oxides)	Mn^{2+} (manganous oxides)

TABLE 7.4 Comparative Tolerance by Different Plants of a High Water Table and Accompanying Restricted Aeration

Ladino clover grows well with a water table near the soil surface, but wheat and other cereals grow best when the water table is at least 100 cm below the surface.

	Plants tolerant to constant table at each depth		
15–30 cm	*40–60 cm*	*75–90 cm*	*100+ cm*
Ladino clover	Alfalfa	Corn	Wheat
Orchard grass	Potatoes	Peas	Barley
Fescue	Sorghum	Tomato	Oats
Loblolly pine	Mustard	Millet	Peas
Willow tree	Sycamore	Cabbage	Beans
	Willow oak	Snap beans	Horse beans
		Red maple	Sugar beets
		Silver maple	Colza
		Beech	Walnut
			White pine

Agricultural crops from several sources reviewed by Williamson and Kriz (1970).

NUTRIENTS AND WATER.   Oxygen deficiency is known to curtail the absorption by plants of both nutrients and water. This is because low oxygen levels constrain root respiration, a process that provides the energy needed for nutrient and water absorption. It is ironic that an oversupply of water in the soil can reduce the amount of water absorbed by plants. The effect of aeration on nutrient absorption by cotton is illustrated by the data in Table 7.5. It is no wonder that plants growing on poorly drained soils may show nutrient-deficiency symptoms even though the soils are fairly well supplied with available nutrient elements.

Soil Compaction and Aeration

The negative effects of soil compaction are not all owing to poor aeration. Soil layers can become so dense as to impede the growth of roots even if an adequate oxygen supply is available. For example, some compacted soil layers adversely affect cotton more by preventing root penetration than by lowering available oxygen content.

7.6 AERATION IN RELATION TO SOIL AND PLANT MANAGEMENT

Both surface and subsurface drainage are essential if an aerobic soil environment is to be maintained. The removal of excess quantities of water must take place if sufficient oxygen is to be supplied. The importance of surface runoff and tile drainage must be recognized.

TABLE 7.5 Effect of Water Table Level on Oxygen Content of Soil Air in a Typic Torrifluvent (Hoytville Silty Clay), Yield of Cotton, and Plant Uptake of N, P, and K

Oxygen content at a depth of 23 cm in this heavy soil was sharply reduced as the water table was raised from 90 to 30 cm depth, and cotton yields and nutrient uptake were decreased accordingly.

Depth of water table (cm)	Oxygen content at 23 cm[a] (%)	Yield of cotton (g)	Nutrient uptake by five plants (mg)		
			N	P	K
30	1.6	57	724	85	1091
60	8.3	108	1414	120	2069
90	13.2	157	2292	156	3174

From Meek et al. (1980).
[a]On June 23.

The maintenance of a stable soil structure is an important means of augmenting good aeration. Pores of macrosize, which are encouraged by large, stable aggregates, are soon drained of water following a rain, thereby allowing gases to move into the soil from the atmosphere. Maintenance of organic matter by addition of farm manure and crop residues and by growth of legumes is perhaps the most practical means of encouraging aggregate stability, which in turn encourages good drainage and better aeration.

In poorly drained, heavy-textured soils, however, it is often impossible to maintain optimum aeration without resorting to some cultivation of the soil. Thus, in addition to controlling weeds, cultivation of heavy-textured soils has a second very important function—that of aiding soil aeration. Consequently, no-tillage or minimum-tillage practices which are quite satisfactory on well-drained soils have limitations on poorly drained soils. Also, yields of row crops, especially those with large taproots such as sugar beets and rutabagas often are increased by frequent, light cultivations that do not injure the fibrous roots. Part of this increase, undoubtedly, is due to the effects of better aeration.

Crop–Soil Adaption

In addition to the direct methods of controlling soil aeration, the selection of plants tolerant of low oxygen is important. Alfalfa, fruit and forest trees, and other deep-rooted plants require deep, well-aerated soils, and are sensitive to a deficiency of oxygen, even in the lower soil horizons. In contrast, shallow-rooted plants such as some grasses and alsike and ladino clovers do well on soils that tend to be poorly aerated, especially in the subsoil. The rice plant flourishes even when the soil is submerged in water. These facts are significant in deciding which plants should be grown and how they are to be managed in areas where aeration problems are acute.

7.7 AREAS REQUIRING SPECIAL MANAGEMENT

Root zone aeration is just as critical for container-grown plants as it is for plants grown in the field. Potted plants frequently suffer from waterlogging because it is difficult to supply the exact amount of water the plant needs. To prevent such waterlogging most containers have holes at the bottom to drain excess water. As was the case with stratified soils in the field, water drains out of the holes at the bottom of the pot only when the soil at the bottom is saturated with water. If the potting medium is mostly mineral soil, the fine soil pores remain filled with water, leaving no room for air, and anaerobic conditions soon prevail. Use of taller pots will allow for better aeration in the upper part of the medium.

To manage these problems, potting mixes are engineered to meet the requirements of the containerized plants. Mineral soil generally makes up no more than one-third of the volume of most potting mixes, the remainder being composed of inert lightweight but coarse-grained materials such as perlite (expanded volcanic glass), vermiculite (expanded mica—not soil vermiculite), or pumice (porous volcanic rock). Most modern mixes also contain some stable organic material such as peat, shredded bark, wood chips, or sawdust.

Tree and Lawn Management

In transplanting a young tree seedling or any woody species, special caution must be taken to prevent waterlogging immediately around the young roots. Figure 7.10 illustrates the right and the wrong way to manage the transplanting of trees in compacted soils.

Aeration of well-established mature trees must also be safeguarded. If surplus soil from a nearby excavation is pushed immediately around the base of a tree, serious consequences are soon noticed. The tree's feeder roots near the original soil surface are soon deficient in oxygen even if the overburden is no more than 5 to 10 cm in depth. Protective brick walls around the base of the tree can prevent the excess soil's being pushed onto the tree roots and can thereby minimize adverse damage to the tree.

Management systems for heavy-trafficked lawns commonly have components relating to soil aeration. For example, one means of increasing the aeration in compacted

FIGURE 7.10 Providing a good supply of air to tree roots can be a problem, especially when trees are planted in fine-textured, compacted soils of urban areas. A machine-dug hole with smooth sides will act as a "tea cup" and fill with water, suffocating tree roots. Breather tubes, a larger rough-surfaced hole, and a layer of surface mulch in which some fine tree roots can grow are all measures that can improve the aeration status of the root zone. (Photo courtesy of R. Weil)

lawn areas is to use *core* cultivation that actually removes small cores of soil from the surface horizon, thereby permitting gas exchange to take place more easily (see Figure 7.11). Care must be taken not to merely punch holes in the soil, however, since such a practice would thereby increase soil compaction due to the pressure on the soil from the drilling instrument.

Natural Wetlands

Principles governing the soil air–soil water relations in natural wetlands are basically the same as for nonflooded agricultural systems. There is one important difference, however. In the case of wetlands, the primary management objective is to minimize aeration, to maintain a more or less permanently saturated soil condition, and to maintain very low E_h levels. While most domesticated plants grow poorly if at all on wetlands, other aquatic or water-tolerant species (mostly undomesticated) grow well under high water and low oxygen regimes. The satisfactory growth of these wild species is essential if the ecosystem so essential for a variety of plants and animals is to continue or to be reestablished. Bio-

FIGURE 7.11 One means of increasing the aeration of compacted soil is by core cultivation. The machine removes small cores of soil, leaving holes about 2 cm in diameter and 5–8 cm deep. This method is commonly used on high-traffic turf areas. Note that the machine *removes* the cores and does not simply punch holes in the soil, a process that would increase compaction around the hole and impede air diffusion into the soil. (Photos courtesy of R. Weil)

logical diversity of some species is dependent on the wetland environment. Many ecological functions of wetlands depend on low E_h levels which can be attained only if water-saturated conditions coincide with periods of warm temperatures that encourage the depletion of dissolved oxygen by root and microbial metabolism.

We have seen the influence of soil water on the aeration of the root zones of field, garden, and wild plants. We now turn to a third soil physical property, *soil temperature,* that is also greatly influenced by the level of water in a soil.

7.8 SOIL TEMPERATURE

The temperature of a soil greatly affects the physical, biological, and chemical processes occurring in that soil. In cold soils, rates of chemical and biological reactions are slow. Biological decomposition is at a near standstill, thereby limiting the rate at which nutrients such as nitrogen, phosphorus, sulfur, and calcium are made available. Also, absorption and transport of water and nutrient ions by higher plants are adversely affected by low temperatures.

Plant Processes

Plants vary widely in the soil temperature at which they grow best, and there is variability in the optimum temperature for different plant processes. For example, corn germination requires a soil temperature of at least 7–10°C (45–50°F), and optimum root growth occurs at about 25°C. Potato tubers develop best when the soil temperature is 16–21°C (60–70°F). Wheat grows best at about 24°C. Optimum vegetative growth of apples and peaches is obtained when the soil temperature is about 18°C. A comparable figure for citrus is 25°C.

In cool temperate regions, the yields of some vegetable and small fruit crops are increased by warming the soil (Figure 7.12). The life cycle of some flowers and ornamentals also is influenced greatly by soil temperature. For example, the tulip bulb requires chilling to develop flower buds in early winter, although flower development is suppressed until the soil warms up the following spring. In warm regions, root growth near the surface may be encouraged by shading the soil, either with vegetation or through the use of a mulch.

Microbial Processes

Some microbiological processes are influenced markedly by soil temperature changes. For example, the microbial oxidation of ammonium ions to nitrate ions which occurs most readily at temperatures of 27–32°C (80–90°F), is negligible when the soil temperature is lowered to about 10°C (50°F). Nitrogen fertilizer management takes advantage of this fact. Thus, anhydrous ammonia fertilizers can be injected into cold soils in the spring and the ammonium ions thus added will not be readily oxidized to nitrate ions until the soil temperature rises. This practice in effect can conserve nitrate ions, which (in contrast to ammonium ions) are subject to ready leaching from the soils. When the soil warms up later in the year, nitrates will begin to appear—but they may be more readily absorbed by rapidly growing plants (see Chapter 13).

Unfortunately there are some years with abnormally high soil temperatures, during which early applications of anhydrous ammonia can be troublesome. In these cases, oxidation of the ammonia to nitrates takes place before the plants are well established and during the period of maximum water infiltration and drainage. Much of the nitrate under these conditions is lost in the drainage water, to the detriment of both the farmer and the downstream users of the drainage water. Soil temperatures help control the impact of this and several other practices having environmental implications.

Advantage can be taken of the sensitivity of some microbes to high soil temperatures to control certain plant diseases. In environments with hot summers—maximum daily air temperatures >35°C (>95°F)—covering the ground with transparent plastic sheeting can raise the temperature of the upper few centimeters of soil to 50°C (122°F) and even higher. Such temperatures markedly reduce certain wilt-causing fungal diseases of vegetables and fruits, and adversely affect some weed seeds and insects. This heating process, called *soil solarization,* is used to control pests and diseases in some high-value crops.

Freezing and Thawing

In addition to the direct influence of temperature on plant and animal life, the effect of freezing and thawing must be considered. Alternate freezing and thawing subject the soil aggregates to pressures as ice crystals form and expand, and alter the physical structure of the soil. Freezing and thawing of the upper layers of soil can also result in **heaving** of perennial forage crops such as alfalfa (Figure 7.13). This action, which is most severe on bare, imperfectly drained soils, can drastically reduce the stand of alfalfa, some clovers, and trefoil. Changes in soil temperature have the same effect on shallow foundations for buildings, and on roads with fine material as a base. Such structures show the effects of heaving in the spring, especially if the soil is high in swelling-type clays. The combined effects of freezing and thawing and of wetting and drying make such soils very poor foundation bases.

Forest Fires

Forest fires can have significant effects on the temperature of the upper few centimeters of soil and, in turn, on the breakdown and movement of organic compounds. Figure

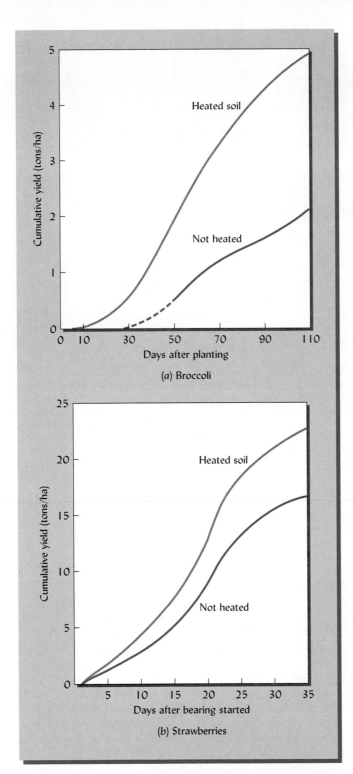

FIGURE 7.12 The influence of heating the soil in an experiment in Oregon on the yield of broccoli and strawberries. Heating cables were buried about 92 cm deep. They increased the average temperature of the 0–100 cm layer by about 10°C, although the upper 10 cm of soil was warmed by only about 3°C. [From Rykbost et al. (1975); used with permission of American Society of Agronomy]

7.14 shows the results of one such wildfire on a lodgepole pine stand in Oregon. The high temperatures of the burning organic material and of the underlying soil (more than 125°C being common) brought about the creation and/or volatilization of certain hydrocarbon compounds that move quickly through the soil pores to deeper, cooler areas. As these compounds interact with cooler soil particles in lower soil horizons, they condense (solidify) on the surface of the soil particles and fill some of the surrounding pore spaces. The condensed compounds are waxlike, water-repellent hydrocarbons that are unwettable. Consequently, when the rains come, water infiltration in even a sandy soil is greatly reduced over that of unburned areas. This effect of soil temperature on soil water illustrates once again the interdependency of various soil properties.

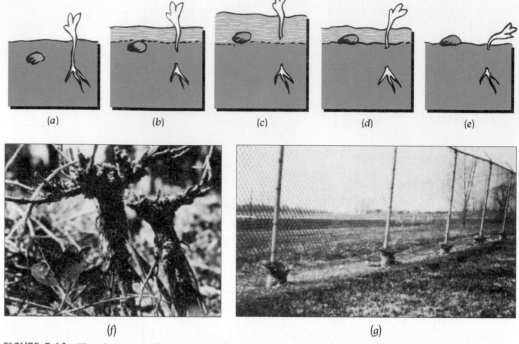

FIGURE 7.13 The diagrams illustrate how frost action in soils can move (heave) plants and stones upward. (a) Position of stone and plant before the soil freezes. (b) Frost action has started: thin lenses of ice have formed in the upper few centimeters of soil. Since water expands when it freezes, the lenses of ice expand and move the soil as well as the rock and plant upward, breaking the plant roots. (c) Maximum soil freezing with rock and upper plant parts moved farther up. (d) Partial thawing. (e) Final thawing, which leaves the stone on or near the soil surface and the plant upper roots exposed. (f) Alfalfa plants pushed out of the ground by frost action. (g) Fence posts encased in concrete "jacked" out of the ground by frost action. [Photos courtesy J. C. Henning, University of Missouri (f), and R. L. Berg, Corps of Engineers, Cold Regions Research and Engineering Laboratory, Hanover, N.H. (g)]

The temperature of soils in the field is dependent directly or indirectly on at least three factors: (1) the net amount of heat the soil absorbs; (2) the heat energy required to bring about a given change in the temperature of a soil; and (3) the energy required for processes such as evaporation, which are constantly occurring at or near the surface of soils. These factors will now be considered.

7.9 ABSORPTION AND LOSS OF SOLAR ENERGY

Solar radiation is the primary source of energy to heat soils. But clouds and dust particles intercept the sun's rays and absorb, scatter, or reflect most of the energy (Figure 7.15). Only about 35 to 40% of the solar radiation actually reaches the earth in cloudy humid regions, and 75% in cloud-free arid areas. The global average is about 50%.

Little of the solar energy reaching the earth actually results in soil warming. The energy is used primarily to evaporate water from the soil or leaf surfaces, or is radiated or reflected back to the sky. Only about 10% is absorbed by the soil and can be used to warm it. Even so, this energy is of critical importance to soil processes and to higher plants growing on the soils.

In addition to solar radiation, other factors influence the amount of energy absorbed by soils, including soil color, slope, and vegetative cover. Dark-colored soils absorb more energy than lighter-colored ones. This does not necessarily imply, however, that dark soils are always warmer. In fact, the opposite is often true. Dark soils usually are higher in organic matter and hold larger amounts of water. The water requires much energy to be warmed and it also cools the soil when it is evaporated.

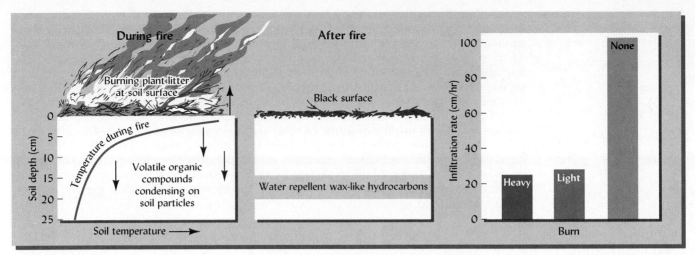

FIGURE 7.14 Wildfires of a lodgepole pine stand heat up the surface layers of this sandy soil (an Inceptisol) in Oregon. Note that the soil temperature was increased sufficiently near the surface to volatilize organic compounds, some of which then move down into the soil and condense (solidify) on the surface of cooler soil particles. These condensed compounds are waxlike hydrocarbons that are water repellent. As a consequence (right) the infiltration of water into the soil is drastically reduced and remains so for a period of at least six years. (From Dryness 1976)

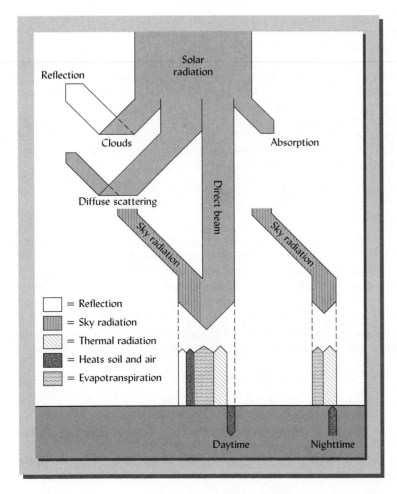

FIGURE 7.15 Schematic representation of the radiation balance in daytime and nighttime in the spring or early summer in a temperate region. About half the solar radiation reaches the earth, either directly or indirectly, from sky radiation. Most radiation that strikes the earth in the daytime is used as energy for evapotranspiration or is radiated back to the atmosphere. Only a small portion, perhaps 10%, actually heats the soil. At night the soil loses some heat, and some evaporation and thermal radiation occur.

The angle at which the sun's rays strike the soil also influences soil temperature. If the sun is directly overhead, the incoming path of the rays is perpendicular to the soil surface and energy absorption (and soil temperature increase) is greatest (Figure 7.16). As an example, three soils, one on a southerly slope of 20°, one on a nearby level field, and one on a northerly slope of 20°, will receive energy from the sun's rays on June 21 (at the 42d parallel north) in the proportion of 106:100:81. The effect of the direction of slope, or **aspect**, on forest species is illustrated in the photo in Figure 7.16.

Whether the soil is bare or is covered with vegetation or a mulch is another factor markedly influencing the amount of solar radiation reaching the soil. The effect of a dense forest is universally recognized. Even low-growing vegetation such as bluegrass has a very noticeable influence, especially on temperature fluctuations. Bare soils warm up more quickly and cool off more rapidly than those covered with vegetation or with plastic mulches. Frost penetration during the winter is considerably greater in bare, noninsulated land.

7.10   SPECIFIC HEAT OF SOILS

A dry soil is more easily heated than a wet one. This is because the amount of energy required to raise the temperature of water by 1°C (its *heat capacity*) is much higher than that required to warm soil solids by 1°C. When heat capacity is expressed per unit mass—for example, in calories per gram (cal/g)—it is called *specific heat.* The specific heat of pure water is about 1.00 cal/g or 1000 cal/kg [4.18 joules per gram (J/g)]; that of dry soil is about 0.2 cal/g (0.8 J/g).

The specific heat is an important soil property. It largely controls the degree to which soils warm up in the spring. For example, consider two soils with comparable characteristics, one a dry soil with only 10 kg water/100 kg soil solids, the other a wet soil with 30 kg water/100 kg soil solids. The drier soil has a specific heat of 273 cal/kg, whereas the wet soil has a specific heat of 385 cal/kg.[2] Obviously, the wet soil will warm up much more slowly than the drier one. Furthermore, if water cannot be drained freely from the wet soil, the water must be evaporated, a process that is very energy-consuming, as the next section will show.

Advantage is taken of the high specific heat of soils in temperature control systems (referred to as *heat pumps*) designed to both warm and cool buildings. A network of pipes are laid underground near the building to be heated/cooled to maximize contact with the soil. Advantage is taken of the fact that subsoils are generally warmer than the atmosphere in the winter and cooler in the summer. Water circulating through the network of pipes absorbs heat from the soil during the winter and releases it to the soil in the summer. The high specific heat of soils permits a large exchange of energy to take place without greatly modifying the soil temperatures (Figure 7.17).

7.11   HEAT OF VAPORIZATION

The evaporation of water from soil surfaces requires a large amount of energy, 540 kilocalories (kcal) or 2.257 joules for every kilogram of water vaporized. This energy must be provided by solar radiation or it must come from the surrounding soil. In either case, evaporation has the potential of cooling the soil. For example, if the amount of water

---

[2] These figures can be easily verified. For example, consider the above soil with 30 kg water/100 kg dry soil, or 0.3 kg water/kg dry soil. The number of calories required to raise the temperature of 0.3 kg of water by 1°C is

$$0.3 \text{ kg} \times 1000 \text{ cal/kg} = 300 \text{ cal}$$

The corresponding figure for the kilogram of soil solids is

$$1 \text{ kg} \times 200 \text{ cal/kg} = 200 \text{ cal}$$

Thus, a total of 500 cal is required to raise the temperature of 1.3 kg of the moist soil by 1°C. Since the **specific heat** is the number of calories required to raise the temperature of 1 kg of wet soil by 1°C, in this case it is

$$500/1.3 = 385 \text{ cal/kg}$$

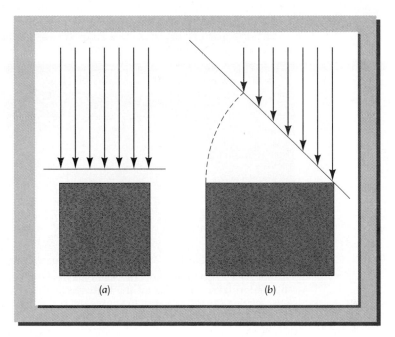

FIGURE 7.16 (Upper) Effect of the angle at which the sun's rays strike the soil on the area of soil that is warmed. (a) If a given amount of radiation from the sun strikes the soil at right angles, the radiation is concentrated in a relatively small area, and the soil warms quite rapidly. (b) If the same amount of radiation strikes the soil at a 45° angle, the area affected is larger by about 40%, the radiation is not so concentrated, and the soil warms up more slowly. This is one of the reasons why north slopes tend to have cooler soils than south slopes. It also accounts for the colder soils in winter than in summer. (Lower) A view looking eastward of a forested area in Virginia illustrates the temperature effect. The main ridge (left to right) is running north and south and the smaller side ridges east and west (up and down). The dark patches are pine trees in this predominantly hardwood deciduous forest. The pines dominate the southern (warmer and drier) slopes on each east-west ridge. (Photo courtesy of R. Weil)

FIGURE 7.17 This high school in Oklahoma has been built into the ground with only one side exposed. This design takes advantage of the high specific heat and low thermal conductivity of the overlying soils, keeping the school warm in winter and cool in summer with a minimum of energy used for heating or air-conditioning. (Photo courtesy of R. Weil)

associated with 100 g dry soil was reduced by evaporation from 25 g to 24 g (only about a 1% decrease) and if all the thermal energy needed to evaporate the water came from the moist soil, the soil would be cooled by about 12°C. Such a figure is hypothetical because only a part of the heat of vaporization comes from the soil itself. Nevertheless, it indicates the tremendous cooling influence of evaporation.

The low temperature of a wet soil is due partially to evaporation and partially to high specific heat. The temperature of the upper few centimeters of wet soil is commonly 3–6°C (6–12°F) lower than that of a moist or dry soil. This is a significant factor in the spring in a temperate zone when a few degrees will make the difference between the germination or lack of germination of crop seeds.

7.12   THERMAL CONDUCTIVITY IN SOILS

As shown in Section 7.9 some of the solar radiation that reaches the earth slowly penetrates the profile largely by **conduction**; this is the same process by which heat moves along an iron pipe when one end is placed in a fire. The movement of heat in soil is analogous to the movement of water (Section 5.7), the rate of flow being determined by a driving force and by the ease with which heat flows through the soil. This can be expressed as follows:

$$Q_h = K(T_1 - T_2)$$

where Q_h is the quantity of heat transferred across a unit cross-sectional area in unit time, K is the thermal conductivity of the soil, and $T_1 - T_2$ is the temperature gradient that serves as the driving force for the conductivity.

The thermal conductivity K of soil is influenced by a number of factors, the most important being the moisture content of the soil layers. Heat passes from soil to water about 150 times more easily than from soil to air. As the water content increases in a soil, the air content decreases, and the transfer resistance is lowered decidedly. When sufficient water is present to form a bridge between most of the soil particles, further additions will have little effect on heat conduction. Here again the major role of soil moisture is demonstrated. Relatively dry soil makes a good insulating material. Buildings built mostly underground can take advantage of both the low thermal conductivity and relatively high heat capacity of large volumes of soil (see Figure 7.17).

The significance of conduction with respect to field temperatures is not difficult to comprehend. It provides a means of temperature adjustments, but, because it is slow, changes in subsoil temperatures lag behind those of the surface layers. Moreover, temperature changes are always less in the subsoil. In temperate regions, surface soils in general are expected to be warmer in summer and cooler in winter than the subsoil, especially the lower horizons of the subsoil.

Mention also should be made of the effect of rain or irrigation water on soil temperature. For example, in temperate zones, spring rains definitely warm the surface soil as the water moves into it. Conversely, in the summer the rainfall cools the soil, since it is often cooler than the soil it penetrates. In practice, however, the spring rains, by increasing the amount of water to be removed by evaporation, often accentuate low-temperature problems.

7.13   SOIL TEMPERATURE DATA

The temperature of the soil at any time depends on the ratio of the energy absorbed to that being lost. The constant change in this relationship is reflected in the *seasonal, monthly,* and *daily* temperatures. The accompanying data (Figures 7.18 and 7.19) from College Station, Texas, and Lincoln, Nebraska, are representative of average seasonal and monthly temperatures in relation to soil depth in subhumid temperate regions.

It is apparent from these figures that considerable seasonal and monthly variations of soil temperature occur, even at the lower depths. The surface layer temperatures vary more or less according to the temperature of the air, although these layers are generally warmer than the air throughout the year.

In the subsoil, the seasonal temperature increases and decreases lag behind changes registered in the surface soil and in the air. Accordingly, the temperature data for March at College Station suggest that the surface soil temperatures have already begun to

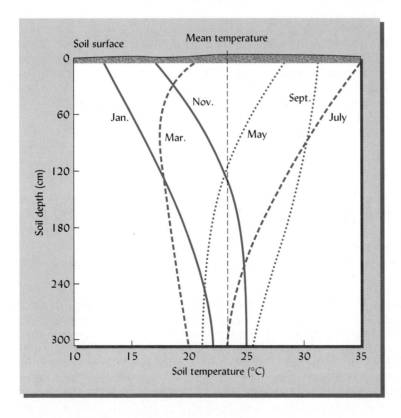

FIGURE 7.18   Average monthly soil temperatures for six of the twelve months of the year at different soil depths at College Station, Tex. (1951–1955). Note the lag in soil temperature change at the lower depths. [From Fluker (1958)]

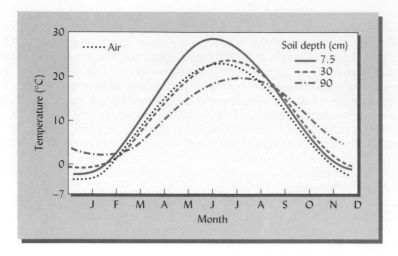

FIGURE 7.19 Average monthly air and soil temperatures at Lincoln, Nebr. (12 years). Note that the 7.5-cm soil layer is consistently warmer than the air above and that the 90-cm soil horizon is cooler in spring and summer, but warmer in the fall and winter, than surface soil.

respond to the warming of the spring, while temperatures of the deep subsoil seem to still be responding to the cold winter weather.

The subsoil temperatures were less variable than the air and surface soil temperatures, although there was some temperature variation even at the 300-cm depth. The subsoils were generally warmer in the late fall and winter and cooler in the spring and summer than the surface soil layers and the air. This is to be expected since the subsoils are not subject to the direct effects of solar radiation.

The low variation in subsoil temperatures is an advantage not only from the standpoint of plant growth but for other sectors as well. Thus, the deep subsoils are less apt to freeze in the winter months, thereby protecting the roots of perennial herbaceous plants and trees. Likewise, frost action on houses (Figure 7.13) or roads is not as apt to occur if the house or road foundation is based in the subsoil.

The temperature lag of subsoils is also of practical importance in temperature regulation in houses that use a soil heat pump, referred to in Section 7.10. The pumps circulate water in pipes embedded in a subsoil field surrounding a house, taking advantage of the fact that subsoils are cooler in summer and warmer in winter to help reduce temperature extremes in the house.

Daily Variations. With a clear sky, the air temperature in temperate regions rises from lowest in the morning to a maximum at about 2:00 P.M. The surface soil, however, does not reach its maximum until later in the afternoon because of the usual lag. This retardation is greater and the temperature change is less as the depth increases. The lower subsoil shows little daily or weekly fluctuation; the variation there, as already emphasized, is a slow monthly or seasonal change.

7.14 SOIL TEMPERATURE CONTROL

The temperature of field soils is not subject to radical human regulation. However, two kinds of management practice have significant effects on soil temperature: those that keep some type of cover or mulch on the soil, and those that reduce excess soil moisture. These effects have meaningful biological implications.

Organic Mulches and Tillage

Soil temperatures are influenced by soil cover and especially by organic residues or other types of mulch placed on the soil surface. Figure 7.20 shows that mulches are definitely soil temperature modifiers. In periods of hot weather, they keep the surface soil cooler than where no cover is used; in contrast, during cold spells in the winter, they moderate rapid temperature declines. They tend to buffer extremes in soil temperatures.

Until fairly recently, the significant use of mulches in modifying soil temperature extremes was limited mostly to home gardens and flower beds. Although these uses are still important, the effects of mulches have extended to field-crop culture in areas that

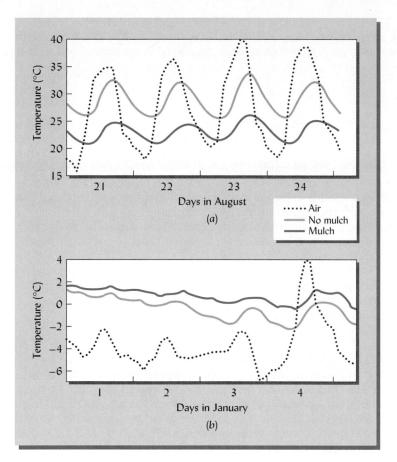

FIGURE 7.20 (a) Influence of straw mulch (8 tons/ha) on air temperature at a depth of 10 cm during an August hot spell in Bushland, Tex. Notice the soil temperatures in the mulched area are consistently lower than where no mulch was applied. (b) During a cold period in January the soil temperature was higher in the mulched than in the unmulched area. [Redrawn from Unger (1978); used with permission of American Society of Agronomy]

have adopted conservation tillage practices. Conservation tillage leaves much if not all the crop residues at or near the soil surface.

The influence of surface residues on soil temperature is illustrated by data shown in Figure 7.21. Soil temperatures were consistently lower during July and August when the no-tillage practice was followed. Crop residues left on the surface undoubtedly account for this effect.

The soil-temperature-depressing effect of some mulch practices has a serious negative impact on corn production in the northern Great Plains states. The lower temperatures in May and early June resulting from these practices inhibit seed germination, seedling performance, and often the yields of corn. Ridging the soil, permitting water to

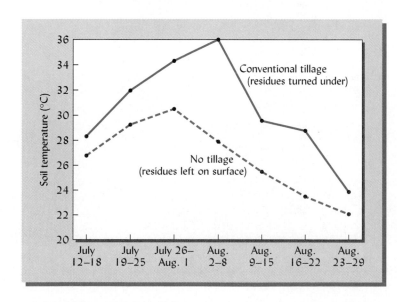

FIGURE 7.21 Effect of plowing under crop residues (conventional tillage) or leaving them on the surface (no tillage) on the temperature of an Ultic Hapludalf (Wheeling fine sandy loam) in West Virginia. [Data from Bennett et al. (1976)]

drain out of the ridge, and then planting on top of the ridge are innovative ways to alleviate this problem. In warm regions, mulching may provide the double benefit of increased rooting in the more fertile topsoil plus decreased evaporation of water from the soil surface.

Plastic Mulches

The use of plastic mulches for gardens and high-value specialty crops is becoming more common. One of the reasons for this increase in popularity is the effect of these plastic mulches on soil temperature. In contrast to the organic residue mulches, plastic mulches generally increase soil temperature. This has some advantages if they are used in the springtime in temperate regions. The soil warms up sufficiently to permit the early growth of specialty crops for early markets. Figure 7.22 shows the use of a plastic mulch for winter-grown strawberries in southern California.

Unfortunately, plastic mulches are not always so beneficial. Their use in warmer climates and during the summer months can lead to soil temperatures so high that plant root growth is adversely affected. In the tropics, midday soil temperatures under plastic mulches as high as 43°C are common (see Table 7.6). Although water management and weed control may be the primary reasons for selecting plastic mulch systems, their soil temperature effects can make them either beneficial or detrimental depending on whether increased soil temperature is desired.

Moisture Control

Another means of exercising some control over soil temperature is by controlling soil moisture. Poorly drained soils in temperate regions that are wet in the spring have temperatures 3 to 6°C lower than comparable well-drained soils. Only by removing this water can temperature depression be alleviated. Water can be removed by installing drainage systems using ditches and underground pipes (Sections 6.18 and 6.19). Where this is not feasible, the innovative ridging systems of tillage just referred to can be used. The excess water simply must be removed from the plant root zone.

As was the case with soil air, the controlling influence of soil water on soil temperature is apparent everywhere. Whether a problem concerns capturing solar energy, loss

FIGURE 7.22 These winter-grown southern California strawberries will come to market when prices are still high because of the effect of the clear plastic mulch on soil temperature. (Photo courtesy of R. Weil)

TABLE 7.6 Temperature of a Bare Tropical Soil at Different Times During the Day and the Increase or Decrease in Soil Temperature Resulting from the Use of a Straw Mulch and a Plastic Mulch

Note that the temperatures under the straw mulch were consistently lower and those under the plastic mulch considerably higher than those of the bare soil. Temperatures above 35°C often depress root function and plant growth.

	Soil Temperature (°C)		
		Change from using	
Hour of day	Bare soil check	straw mulch	plastic mulch
8	26	0	+2
10	28	1	+4
12	32	−1	+6
14	37	−4	+6
16	37	−4	+5
18	32	−1	+5
20	27	−1	+2

Data drawn from Maurya and Lal (1979).

of energy to the atmosphere, or the movement of heat within the soil, the amount of water present is always important. Water regulation seems to be the key to what little practical temperature control is possible for field soils.

7.15 CONCLUSION

Soil aeration and soil temperature critically affect the quality of soils as habitats for plants and other organisms. Most plants have definite requirements for soil oxygen along with limited tolerance for carbon dioxide, methane, and other such gases found in poorly aerated soils. Some microbes such as the nitrifiers and general-purpose decay organisms are also constrained by low levels of soil oxygen.

Plants as well as microbes are also quite sensitive to differences in soil temperature, particularly in temperate climates where low temperatures can limit essential biological processes. Soil temperature also impacts on the use of soils for engineering purposes, again primarily in the cooler climates. Frost action, which can move perennial plants such as alfalfa out of the ground, can do likewise for building foundations, fence posts, sidewalks, and highways.

Soil water can provide considerable control over both soil air and soil temperature. It competes with soil air for the occupancy of soil pores. Soil water also resists changes in soil temperature by virtue of its high specific heat and of its high energy requirement for evaporation. Moisture control in soils does more to influence soil aeration and soil temperature than any other soil management tool. Soil drainage, mulching, and innovative tillage practices are our best means of bringing about temperature control.

REFERENCES

Bartlett, R. J., and B. R. James. 1993. "Redox chemistry of soils," *Adv. in Agron.,* **50**: 151–208.

Bennett, O. L., E. L. Mathias, and C. B. Sperow. 1976. "Double cropping for hay and no-tillage corn production as affected by sod species with rates of atrazine and nitrogen," *Agron. J.,* **68**:250–254.

Buyanovsky, G. A., and G. H. Wagner. 1983. "Annual cycles of carbon dioxide level in soil air," *Soil Sci. Soc. Amer. J.,* **47**:1139–1145.

Dryness, C. T. 1976. "Effects of wildfire on soil wettability in the high cascades of Oregon," USDA Forest Serv. Res. Paper PNW-202.

Fluker, B. J. 1958. "Soil temperature," *Soil Sci.,* **86**:35–46.

Maurya, P. R., and R. Lal. 1979. "Effects of bulk density and soil moisture on radicle elongation of some tropical crops," in R. Lal and D. J. Greenland (eds.), *Soil Physical*

Properties and Crop Production in the Tropics (New York: John Wiley and Sons), pp. 339–347.

McBride, M. B. 1994. *Environmental Chemistry of Soils* (New York: Oxford Univ. Press), 406 pp.

Meek, B. D., and L. B. Grass. 1975. "Redox potential in irrigated desert soils as an indicator of aeration status," *Soil Sci. Soc. Amer. J.,* **39**:870–875.

Meek, B. D., E. C. Owen-Bartlett, L. H. Stolzy, and C. K. Labanauskas. 1980. "Cotton yield and nutrient uptake in relation to water table depth," *Soil Sci. Soc. Amer. J.,* **44**:301–305.

Patrick, W. H., Jr. 1977. "Oxygen content of soil air by a field method," *Soil Sci. Soc. Amer. J.,* **41**:651–652.

Patrick, W. H., Jr., and A. Jugsujinda. 1992. "Sequential reduction and oxidation of inorganic nitrogen, manganese, and iron in flooded soil," *Soil Sci. Soc. Amer. J.* **56**: 1071–1073.

Rykbost, K. A., L. Boersma, J. J. Mack, and W. E. Schmisseur. 1975. "Yield response to soil warming: vegetable crops," *Agron. J.,* **67**:738–743.

Sexstone, A. J., et al. 1985. "Direct measurement of oxygen profiles and denitrification rates in soil aggregates," *Soil Sci. Soc. Amer. J.,* **49**:645–651.

Stolzy, L. H., et al. 1961. "Root growth and diffusion rates as functions of oxygen concentration," *Soil Sci. Soc. Amer. Proc.,* **2**:463–67.

Stolzy, L. H., and J. Letey. 1964. "Characterizing soil oxygen conditions with a platinum microelectrode," *Adv. in Agron.,* **16**:249–79.

Unger, P. W. 1978. "Straw mulch effects on soil temperatures and sorghum germination and growth," *Agron. J.,* **70**:858–864.

Williamson, R. E., and G. J. Kriz. 1970. "Response of agricultural crops to flooding, depth of water table and soil gaseous composition," *Amer. Soc. Agr. Eng. Trans.,* **13**: 216–220.

Yavitt, J. B., T. J. Fahley, and J. A. Simmons. 1995. "Methane and carbon dioxide dynamics in a northern hardwood ecosystem," *Soil. Sci. Soc. Am. J.,* **59**:796–804.

8

SOIL COLLOIDS: THEIR NATURE AND PRACTICAL SIGNIFICANCE

The earth is covered thick with other clay,
Which her own clay shall cover, heaped and pent.
—*LORD BYRON, CHILDE HAROLD'S PILGRIMAGE*

Next to photosynthesis and respiration, probably no process in nature is as vital to plant and animal life as the exchange of ions between soil particles and growing plant roots. These **cation** and **anion exchanges** take place mostly on the surfaces of the finer or colloidal fractions of both the inorganic and organic matter—clays and humus.

Such colloidal particles serve very much as does a modern bank. They are the sites within the soil where ions of essential mineral elements such as calcium, potassium, and sulfur are held and protected from excessive loss by percolating rain or irrigation water. Subsequently, the essential ions can be withdrawn from the colloidal bank sites and taken up by plant roots. In turn, these elements can be deposited or returned to the colloids through the addition of commercial fertilizers, lime, manures, and plant residues.

But cation and anion exchange capabilities are not the only soil properties dependent upon soil colloids. Soil structure formation and stability, the retention and movement of water, with their implications for the control of soil air and soil temperature, are all influenced to a considerable extent by soil colloids. The ease of soil erosion is affected by the nature of soil colloids, which can move downstream to fill reservoirs that supply our domestic water needs or can clog up our water ports that must be dredged frequently. The use of soils for building sites and for highway construction is dependent to a considerable degree on the amount and nature of the soil colloids.

The colloidal particles (clays and humus) are the seat of most chemical, physical, and biological properties of soils. Consequently, no understanding of soils is complete without at least a general knowledge of the nature and constitution of these tiny particles. Their common properties will be covered first and then their individual characteristics will receive attention.

8.1 GENERAL PROPERTIES OF SOIL COLLOIDS

Size

The most important common property of inorganic and organic colloids is their extremely small size. They are too small to be seen with an ordinary light microscope.

Only with an electron microscope can they be photographed. Most are smaller than 2 micrometers (μm) in diameter.[1]

Surface Area

Because of their small size, all soil colloids expose a large **external** surface per unit mass. The external surface area of 1 g of colloidal clay is at least 1000 times that of 1 g of coarse sand. Some colloids, especially certain silicate clays, have extensive **internal** surfaces as well. These internal surfaces occur between platelike crystal units that make up each particle and commonly greatly exceed the external surface area. The total surface area of soil colloids ranges from 10 m^2/g for clays with only external surfaces to more than 800 m^2/g for clays with extensive internal surfaces. The colloidal surface area in the upper 15 cm of a hectare of a clay soil could be as high as 700,000 km^2 (270,000 mi^2)—greater than the area of France.

Surface Charges

Soil colloidal surfaces, both external and internal, characteristically carry negative and/or positive charges. For most soil colloids, electronegative charges predominate, although some mineral colloids in very acid soils have a net electropositive charge. The presence and intensity of the particle charges influence the attraction and repulsion of the particles toward each other, thereby influencing both physical and chemical properties. The source of these charges will be considered later.

Adsorption of Cations and Water

An important consequence of the charges on soil colloids is the attraction of ions of an opposite charge to the colloidal surfaces. Such attraction is of particular significance for negatively charged colloids. The colloidal particles, referred to as *micelles* (microcells), attract hundreds of thousands of positively charged ions, or **cations**,[2] such as H^+, Al^{3+}, Ca^{2+}, and Mg^{2+}. This gives rise to an **ionic double layer** (Figure 8.1). The colloidal particle constitutes the *inner* ionic layer, being essentially a huge anion, the external and internal layers of which are negative in charge. The *outer* layer is made up of a swarm of rather loosely held (adsorbed) cations that are attracted to the negatively charged surfaces. Thus, a colloidal particle is accompanied by a swarm of cations that are **adsorbed** or held on the particle surfaces.

In addition to the adsorbed cations, a large number of water molecules are associated with soil colloidal particles. Some are attracted to the adsorbed cations, each of which is hydrated. Others are held in the internal surfaces of the colloidal particles. These water molecules play a critical role in determining both the physical and chemical properties of soil.

8.2 TYPES OF SOIL COLLOIDS

There are four major types of colloids present in soils: layer silicate clays; iron and aluminum oxide clays; allophane and associated amorphous clays; and humus. Although each group possesses the general colloidal characteristics described in the preceding section, each has specific characteristics that make it distinctive and useful.

Layer Silicate Clays

The most important property of these clays is their crystalline, layerlike structures. Each particle is comprised of a series of layers much like the pages of a book (Figure 8.2). The layers comprise primarily horizontally oriented sheets of silicon, aluminum, magne-

---

[1]Since the upper limit in diameter for the colloidal state is generally considered to be about 1 μm, some clay particles (the upper limit for which is 2 μm) are technically not colloids. Even so, they have colloidlike properties. For a review of colloids in soils, see Dixon and Weed (1989).

[2]In water solutions the H^+ ion is always hydrated, giving rise to the **hydronium** ion H_3O^+ [sometimes written H^+ (aq)]. For simplicity, however, the H^+ ion designation will be used in this text.

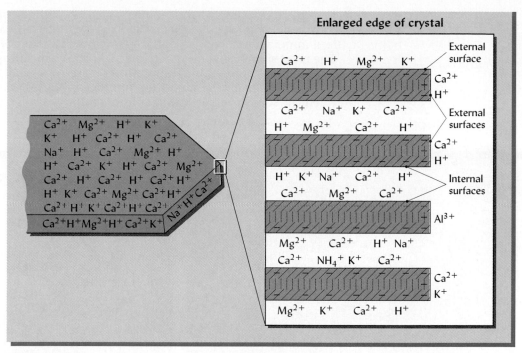

FIGURE 8.1 Diagrammatic representation of a silicate clay crystal (micelle) with its sheetlike structure, its innumerable negative charges, and its swarm of adsorbed cations. The enlarged schematic view of the edge of the crystal illustrates the negatively charged internal surfaces of this particular particle, to which cations and water are attracted. Note that each crystal unit has definite mineralogical structure. (See Figure 8.4 for a more detailed drawing of structure.)

sium, and/or iron atoms surrounded and held together by oxygen and hydroxy groups. The formula of one of these clays, kaolinite $(Si_2Al_2O_5(OH)_4)$, illustrates their general chemical makeup.

The exact chemical composition and the internal arrangement of the atoms in a given colloidal crystal account for its surface charge and for other associated properties. These include the capacity of each silicate colloid to hold and exchange ions as well as its physical properties. For example, some are highly sticky and plastic while others are only mildly so.

Silicate clay minerals are the dominant inorganic colloids in soils of temperate areas and are prominent in soils of the tropics as well. Each of the types of silicate clays will be given more detailed consideration later.

Hydrous Oxides of Iron and Aluminum

These clays are commonly dominant in the highly weathered soils of the tropics and semitropics and are also present in significant quantities in the soils of some temperate regions. Properties of the red and yellow soils (Ultisols) so common in the southeastern United States are controlled to a large degree by these clays.

Examples of iron and aluminum oxides common in soils are gibbsite $(Al_2O_3 \cdot 3H_2O)$ and goethite $(Fe_2O_3 \cdot H_2O)$. The formulas also may be written in the hydroxide form, that is, gibbsite as $Al(OH)_3$ and goethite as $FeOOH$. For simplicity, they will be referred to as the Fe, Al oxide clays.

Less is known about these clays than about the layer silicates. Some have definite crystalline structures, but others are amorphous. The Fe, Al clays are not as sticky and plastic as the layer silicate clays. At high pH values, the micelles carry a small negative charge and are surrounded by a corresponding equivalent of cations. In very acid soils, however, some Fe, Al oxides carry a net positive charge and attract negatively charged anions instead of cations (anion exchange is covered in Section 8.14).

Allophane and Other Amorphous Minerals

In many soils there are significant quantities of colloidal matter that is either amorphous or the crystalline structure is not sufficiently ordered to be detected by X rays.

Because they lack ordered three-dimensional crystalline structure they are sometimes referred to as *short-range-order minerals*. They are more difficult to study than well crystallized minerals and, consequently, less is known about them.

The most significant amorphous silicate colloids are **allophane** and its more weathered companion **imogolite.** These minerals are somewhat poorly defined aluminum silicates with a general composition approximating $Al_2O_3 \cdot 2SiO_2H_2O$. They are most prevalent in soils developed from volcanic ash (Andisols), such as those found in the

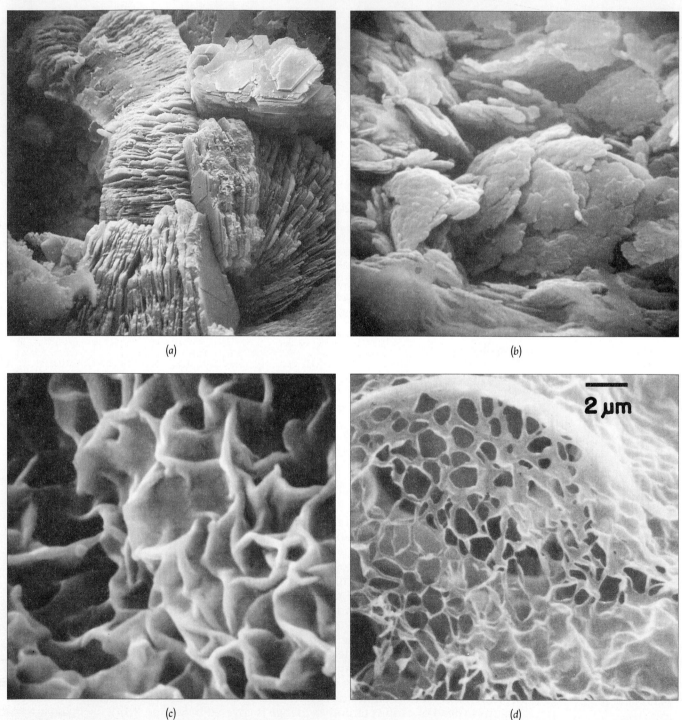

(a) (b)

(c) (d)

FIGURE 8.2 Crystals of three silicate clay minerals and a photomicrograph of humic acid found in soils. (a) Kaolinite from Illinois magnified about 1900 times (note hexagonal crystal at upper right). (b) A fine-grained mica from Wisconsin magnified about 17,600 times. (c) Montmorillonite (a smectite group mineral) from Wyoming magnified about 21,000 times. (d) Fulvic acid (a humic acid) from Georgia magnified about 23,000 times. [(a)–(c) Courtesy Dr. Bruce F. Bohor, Illinois State Geological Survey; (d) from Dr. Kim H. Tan, University of Georgia; used with permission of Soil Science Society of America]

northwestern part of the United States. Allophane has a high capacity to adsorb cations, but also adsorbs anions (see Section 8.14).

Organic Soil Colloids

The colloidal organization of humus has some similarities to that of clay. A highly charged micelle is surrounded by a swarm of cations. The humus colloids are not crystalline, however. They are composed basically of carbon, hydrogen, and oxygen rather than of silicon, aluminum, iron, oxygen, and hydroxy groups. The organic colloidal particles vary in size, but they may be at least as small as the silicate clay particles.

The negative charges of humus are associated with partially dissociated enolic (—OH), carboxyl (—COOH), and phenolic (◯-OH) groups; these groups in turn are associated with central units of varying size and complexity. The relationship is illustrated in Figure 8.3.

As is the case for the Fe, Al oxides, the negative charge associated with humus is dependent on the soil pH. Under very acid conditions, the negative charge is not very high—lower than that of some of the silicate clays. With a rise in pH, however, the hydrogen ions dissociate from first the carboxyl groups and then the enolic and phenolic groups. This leaves a greatly increased negative charge on the colloid. Under neutral to alkaline conditions, the electronegativity of humus per unit weight greatly exceeds that of the silicate clays. In these high pH soils, the adsorbed hydrogen is replaced by calcium, magnesium, and other cations (Figure 8.3).

8.3   ADSORBED CATIONS

In humid regions, the cations of calcium, aluminum, and hydrogen are by far the most numerous, whereas in an arid region soil, calcium, magnesium, potassium, and sodium predominate (Table 8.1). A colloidal complex may be represented in the following simple and convenient way for each region:

$$
\begin{array}{ll}
Ca^{2+} & \\
Al^{3+} & \boxed{\text{Micelle}} \\
H^{+} & \\
M & \\
\end{array}
\qquad
\begin{array}{ll}
e & Ca^{2+} \\
f & Mg^{2+} \\
g & K^{+} \\
h & M \\
\end{array}
\boxed{\text{Micelle}}
$$

(humid region)   (arid region)

The M stands for the small amounts of other base-forming cations (e.g., Na^+, NH_4^+) adsorbed by the colloids. The *a* through *h* indicate that the numbers of cations are variable.

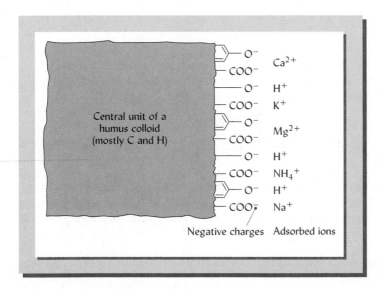

Central unit of a humus colloid (mostly C and H)

—O⁻ Ca^{2+}
—COO⁻
—O⁻ H^+
—COO⁻ K^+
—O⁻ Mg^{2+}
—COO⁻
—O⁻ H^+
—COO⁻ NH_4^+
—O⁻ H^+
—COO⁻ Na^+

Negative charges   Adsorbed ions

FIGURE 8.3   Adsorption of cations by humus colloids. The phenolic hydroxy groups are attached to aromatic rings,

◯— OH

Other —OH and the carboxyl (—COOH) groups are bonded to carbon atoms in the central unit. Note the general similarity to the adsorption situation in silicate clays.

TABLE 8.1 Typical Proportions of Major Adsorbed Cations in the Surface Layers of Different Soil Orders

The percentage figures in each case are based on the sum of the cation equivalents taken as 100.

Soil order[a]	Typical location	H^+ and Al^{3+} (%)[b]	Ca^{2+} (%)	Mg^{2+} (%)	K^+ (%)	Na^+ (%)
Oxisols	Hawaii	85	10	3	2	tr[c]
Spodosols	New England	80	15	3	2	tr
Ultisols	Southeast United States	65	25	6	3	1
Alfisols	Pennsylvania to Wisconsin	45	35	13	5	2
Vertisols	Alabama to Texas	40	38	15	5	2
Mollisols	Midwest United States	30	43	18	6	3
Aridisols	Southwest United States	—	65	20	10	5

[a]See Chapter 3 for soil descriptions.
[b]Al^{3+} adsorption includes that of complex aluminum hydroxy ions.
[c]tr = trace.

These examples illustrate that soil colloids and their associated exchangeable ions can be considered, although perhaps in an oversimplified way, as *complex salts* with the large anion (micelle) surrounded by numerous cations.

Cation Prominence

Two major factors will determine the relative proportion of the different cations adsorbed by clays. First, these ions are not all held with equal tightness by the soil colloids. The order of strength of adsorption[3] when they are present in equivalent quantities is $Al^{3+} > Ca^{2+} > Mg^{2+} > K^+ = NH_4^+ > Na^+$.

Second, the relative concentration of the cations in the soil solution will help determine the degree to which adsorption occurs. Thus, in the soil solution of very acid soils, the concentrations of both H^+ and Al^{3+} are high, and these ions dominate the adsorbed cations. At neutral pH and above, however, the concentrations in the soil solution of both H^+ and Al^{3+} are very low, and, consequently, the adsorption of these ions is minimal. In neutral to moderately alkaline soils, Ca^{2+} and Mg^{2+} dominate. In some poorly drained soils of arid regions, salts high in sodium accumulate, and the adsorption of Na^+ becomes much more prominent. The typical proportions of cations in different soils are shown in Table 8.1.

Cation Exchange

As pointed out in Chapter 1, the surface-held adsorbed cations are subject to exchange with other cations held in the soil solution. For example, a calcium ion held on the colloidal surface is subject to exchange with two H^+ ions in the soil solution.

$$\boxed{\text{Micelle}}\,Ca^{2+} \;+\; 2H^+ \;\rightleftharpoons\; \boxed{\text{Micelle}}\genfrac{}{}{0pt}{}{H^+}{H^+} \;+\; Ca^{2+}$$

    (colloid)   (soil solution)   (colloid)   (soil solution)

Thus, soil colloids are focal points for cation exchange reactions, which have profound effects on soil–plant relations. These will be discussed in greater detail (see Section 8.10) after consideration is given to the specific types and characteristics of soil colloids.

8.4 FUNDAMENTALS OF LAYER SILICATE CLAY STRUCTURE

Now that the general characteristics of soil colloids and their associated cations are understood, we turn to a more detailed consideration of silicate clays. The use of X-ray electron microscopy and other techniques has demonstrated that silicate clay particles are crystalline and that each particle is comprised of individual layers or sheets (Figure

---

[3]The strength of adsorption of the H^+ ion is difficult to determine, since hydrogen-dominated mineral colloids break down to form aluminum-saturated colloids.

8.1). The mineralogical organization of these layers varies from one type of clay to another and markedly affects the properties of the mineral. For this reason, some attention will be given to the fundamentals of silicate clay structure before consideration of specific silicate clay minerals.

Silica Tetrahedral and Alumina-Magnesia Octahedral Sheets

The most important silicate clays are known as *phyllosilicates* (Greek *phyllon*, leaf) because of their leaflike or platelet structure. As shown in Figure 8.4, they are comprised of two kinds of horizontal **sheets**, one dominated by silicon, the other by aluminum and/or magnesium.

The basic building block for the silica-dominated sheet is a unit composed of one silicon atom surrounded by four oxygen atoms. It is called the silica **tetrahedron** because of its four-sided configuration (Figure 8.4). An interlocking array of a series of these silica tetrahedra tied together horizontally by shared oxygen anions gives a **tetrahedral sheet.**

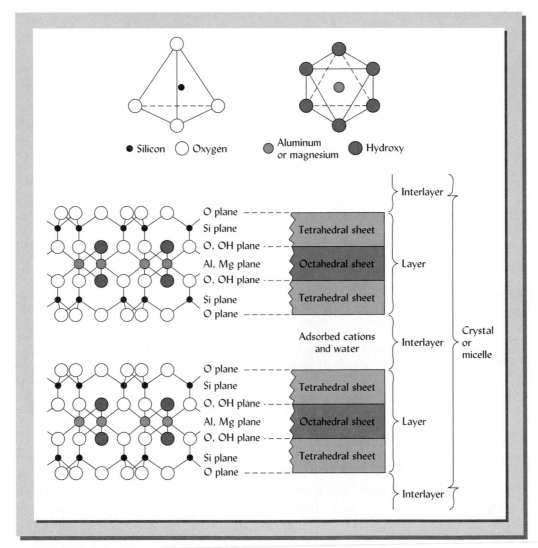

FIGURE 8.4 The basic molecular and structural components of silicate clays. (a) A single *tetrahedron*, a four-sided building block comprised of a silicon ion surrounded by four oxygen atoms, and a single eight-sided *octahedron*, in which an aluminum (or magnesium) ion is surrounded by six hydroxy groups or oxygen atoms. (b) In clay crystals thousands of these tetrahedral and octahedral building blocks are connected to give horizontal planes of silicon and aluminum (or magnesium) ions. These planes alternate with planes of oxygen atoms and hydroxy groups (heavy circles). The silicon plane and associated oxygen/hydroxy planes make up a *tetrahedral sheet*. Similarly, the aluminum/magnesium plane and associated oxygen/hydroxy planes comprise the *octahedral sheet*. Different combinations of tetrahedral and octahedral sheets are termed *layers*. In some silicate clays these layers are separated by *interlayers* in which water and adsorbed cations are found. Many layers are found in each *crystal* or *micelle* (microcell).

Aluminum and/or magnesium ions are the key cations in the second type of sheet. An aluminum (or magnesium) ion surrounded by six oxygen atoms or hydroxy groups gives an eight-sided building block termed *octahedron* (Figure 8.4). Numerous octahedra linked together horizontally comprise the *octahedral sheet.* An aluminum-dominated sheet is known as a *dioctahedral* sheet, whereas one dominated by magnesium is called a *trioctahedral* sheet. The distinction is due to the fact that *two* aluminum ions in a *di*octahedral sheet satisfy the same negative charge from surrounding oxygens and hydroxys as *three* magnesium ions in a *tri*octahedral sheet. As will be seen later, numerous intergrades occur where both cations are present.

The tetrahedral and octahedral sheets are the fundamental structural units of silicate clays. They, in turn, are bound together within the crystals by shared oxygen atoms into different **layers.** The specific nature and combination of sheets in these layers vary from one type of clay to another and largely control the physical and chemical properties of each clay. The relationship between sheets and layers shown in Figure 8.4 is important and should be well understood.

Isomorphous Substitution

The structural arrangements just described suggest a very simple relationship among the elements making up silicate clays. In nature, however, more complex formulas result. The weathering of a wide variety of rocks and minerals permits cations of comparable size to substitute for silicon, aluminum, and magnesium ions in the respective tetrahedral and octahedral sheets.

The ionic radii of a number of ions common in clays are listed in Table 8.2 to illustrate this point. Note that aluminum is only slightly larger than silicon. Consequently, aluminum can fit into the center of the tetrahedron in the place of the silicon without changing the basic structure of the crystal. The process, called *isomorphous substitution,* is common and accounts for the wide variability in the nature of silicate clays.

Isomorphous substitution also occurs in the octahedral sheet. Note from Table 8.2 that ions such as iron and zinc are not too different in size from aluminum and magnesium ions. As a result, these ions can fit into the position of either the aluminum or magnesium as the central ion in the octahedral sheet. It should be emphasized that some layer silicates are characterized by isomorphous substitution in either or both of the tetrahedral or octahedral sheets.

Source of Charges

Isomorphous substitution is of vital importance because it is the primary source of both negative and positive charges of silicate clays. For example, the substitution of one Al^{3+} for an Si^{4+} in the tetrahedral sheet leaves one unsatisfied negative charge. Likewise, the substitution of one Al^{3+} for one Mg^{2+} in a trioctahedral sheet results in one excess positive charge. The net charge associated with a clay micelle is the balance between the positive and negative charges. In most silicate clays, the negative charges predominate. This subject will receive more attention later (see Section 8.8).

8.5 MINERALOGICAL ORGANIZATION OF SILICATE CLAYS

On the basis of the number and arrangement of tetrahedral (silica) and octahedral (alumina-magnesia) sheets contained in the crystal units or layers, silicate clays are classified into three different groups: (1) 1:1-type minerals—one tetrahedral (Si) to one octahedral (Al) sheet; (2) 2:1-type minerals; and (3) 2:1:1-type minerals. For illustrative purposes, each of these is discussed briefly in terms of the main member of the group.

1:1-Type Minerals

The **layers** of the 1:1-type minerals are made up of one tetrahedral (silica) sheet combined with one octahedral (alumina) sheet—hence the terminology *1:1-type crystal* (Figure 8.5). In soils, **kaolinite** is the most prominent member of this group, which includes *halloysite, nacrite,* and *dickite.*

TABLE 8.2 Ionic Radii of Elements Common in Silicate Clays and an Indication of Which Are Found in the Tetrahedral and Octahedral Sheets

Note that Al, Fe, O, and OH can fit in either.

Ion	Radius (nm)[a]	Found in
Si^{4+}	0.042	
Al^{3+}	0.051	Silica tetrahedra
Fe^{3+}	0.064	
Mg^{2+}	0.066	Alumina octahedra
Zn^{2+}	0.074	
Fe^{2+}	0.070	
Na^+	0.097	Exchange sites
Ca^{2+}	0.099	
K^+	0.133	
O^{2-}	0.140	Both sheets
OH^-	0.155	

[a]1 nm = 10^{-9}m.

The tetrahedral and octahedral sheets in a layer of a kaolinite crystal are held together tightly by oxygen atoms, which are mutually shared by the silicon and aluminum cations in their respective sheets. These layers, in turn, are bound to other adjacent layers by **hydrogen bonding** (see Section 5.1). Consequently, the structure is *fixed,* and no expansion ordinarily occurs between layers when the clay is wetted. Cations and water do not enter between the structural layers of a 1:1-type mineral particle. The effective surface of kaolinite is thus restricted to its outer faces or to its external surface area. Also, there is little isomorphous substitution in this 1:1-type mineral. Along with the relatively low surface area of kaolinite, this accounts for its low capacity to adsorb cations.[4]

Kaolinite crystals usually are hexagonal in shape (Figure 8.2). In comparison with other clay particles, they are large in size, ranging from 0.10 to 5 μm across, with the majority falling within the 0.2- to 2-μm range. Because of the strong binding forces between their structural layers, kaolinite particles are not readily broken down into extremely thin plates.

In contrast with the other silicate groups, kaolinite exhibits very little plasticity (capacity of being molded), stickiness, cohesion, shrinkage, and swelling. Its restricted

---

[4]Since the adsorbed cations may be freely exchanged with other cations, the capacity to adsorb cations is usually referred to as ***cation exchange capacity*** (see Section 8.11).

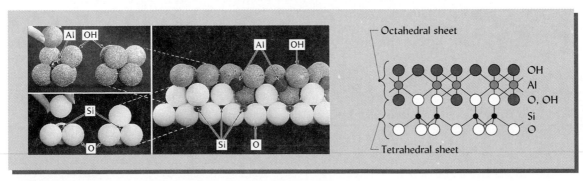

FIGURE 8.5 Models of ions that constitute a layer of the 1:1-type clay kaolinite. Note that each layer consists of alternating octahedral (alumina) and tetrahedral (silica) sheets—hence the designation 1:1. Aluminum ions surrounded by six hydroxy groups and oxygen atoms make up the octahedral sheet (upper left). Smaller silicon ions associated with four oxygen atoms constitute the tetrahedral sheet (lower left). The octahedral and tetrahedral sheets are bound together (center) by mutually shared oxygen atoms. The result is a layer with hydroxys on one surface and oxygens on the other. A schematic drawing of the ionic arrangement (right) shows a cross-sectional view of a crystal layer. The kaolinite mineral is comprised of a series of these flat layers tightly held together with no interlayer spaces.

surface and limited adsorptive capacity for cations and water molecules suggest that kaolinite does not exhibit colloidal properties of a high order of intensity (see Table 8.3). At the same time, kaolinite-containing soils make good bases for roadbeds and building foundations, and they are easy to manage for agriculture. With some nutrient supplementation from manure and fertilizer they can be very productive.

2:1-Type Minerals

The crystal units (layers) of these minerals are characterized by an octahedral sheet sandwiched between two tetrahedral sheets. Three general groups have this basic crystal structure. Two of them, **smectite** and **vermiculite**, include expanding-type minerals, while the third, **fine-grained micas (illite)**, is relatively nonexpanding.

EXPANDING MINERALS. The smectite group is noted for **interlayer expansion**, which occurs by swelling when the minerals are wetted, the water entering the interlayer space and forcing the layers apart. **Montmorillonite** is the most prominent member of this group in soils, although *beidellite, nontronite,* and *saponite* also are found.

The flakelike crystals of smectites (Figure 8.2) are composed of 2:1-type layers, as shown in Figure 8.6. In turn, these layers are loosely held together by very weak oxygen-to-oxygen and cation-to-oxygen linkages. Exchangeable cations and associated water molecules are attracted between layers (the interlayer space), causing **expansion** of the crystal lattice. The **internal surface** then exposed by far exceeds the **external surface** area of these minerals. For example, the **specific surface** or total surface area per unit mass (external and internal) of one smectite mineral (montmorillonite) is 650 to 800 m^2/g. A comparable figure for kaolinite is only 5 to 20 m^2/g (Table 8.3). Commonly, these smectite crystals range in size from 0.01 to 1 μm, much smaller than the average kaolinite particle.

Isomorphous substitution of Mg^{2+} for some of the Al^{3+} ions in the dioctahedral sheet accounts for most of the negative charge for smectites, although some substitution of Al^{3+} for Si^{4+} has occurred in the tetrahedral sheet. The smectites commonly show a high cation exchange capacity, perhaps 20–40 times that of kaolinite (Table 8.3).

Smectites also are noted for their high plasticity and cohesion and their marked swelling when wet and shrinkage on drying. Wide cracks commonly form as smectite-dominated soils (e.g., Vertisols) are dried (Figure 4.35). The dry aggregates or clods are very hard, making such soils difficult to till. Also soils high in smectite make very poor bases for building foundations or roads.

Vermiculites are also 2:1-type minerals, an octahedral sheet being found between two tetrahedral sheets. In most soil vermiculites, the octahedral sheet is aluminum-dominated (**dioctahedral**), although magnesium-dominated (**trioctahedral**) vermiculites are also common. In the tetrahedral sheet of most vermiculites, considerable substitution of aluminum for silicon has taken place. This accounts for most of the very high net negative charge associated with these minerals.

Water molecules, along with magnesium and other ions, are strongly adsorbed in the interlayer space of vermiculites (Figure 8.7). However, they act primarily as bridges holding the units together rather than as wedges driving them apart. The degree of swelling is, therefore, considerably less for vermiculites than for smectites. For this reason, vermiculites are considered **limited-expansion** clay minerals, expanding more than kaolinite but much less than the smectites.

TABLE 8.3 Major Properties of Selected Silicate Clay Minerals and of Humus

Property	Smectite	Vermiculite[a] (dioctahedral)	Fine mica	Chlorite	Kaolinite	Humus
Size (μm)	0.01–1.0	0.1–5.0	0.2–2.0	0.1–2.0	0.5–5.0	0.1–1.0
Shape	Flakes	Plates; flakes	Flakes	Variable	Hexagonal crystals	Variable
External surface (m²/g)	70–120	50–100	70–100	70–100	10–30	} 500–800
Internal surface (m²/g)	550–650	500–600	—	—	—	
Intralayer spacing[b] (nm)	1.0–2.0	1.0–1.5	1.0	1.4	0.7	—
Net negative charge (cmol_c/kg)[c]	80–120	100–180	15–40	15–40	2–5	100–550

[a]Dioctahedral vermiculite is more common in soils than trioctahedral vermiculite.
[b]From the top of one layer to the top of the next similar layer; 1 nm = 10^{-9} m.
[c]Centimoles charge per kilogram (1 cmol_c = 0.01 mol_c), a measure of the cation exchange capacity (see Section 8.11).

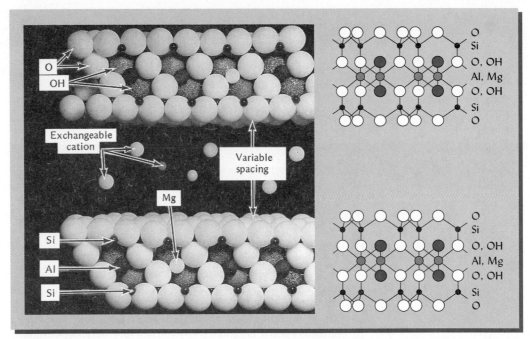

FIGURE 8.6 Model of two crystal layers and an interlayer characteristic of montmorillonite, a smectite expanding-lattice 2:1-type clay mineral. Each layer is made up of an octahedral (alumina) sheet sandwiched between two tetrahedral (silica) sheets. There is little attraction between oxygen atoms in the bottom tetrahedral sheet of one unit and those in the top tetrahedral sheet of another. This permits a ready and variable space between layers, which is occupied by water and exchangeable cations. This internal surface far exceeds the surface around the outside of the crystal. Note that magnesium has replaced aluminum in some sites of the octahedral sheet. Likewise, some silicon atoms in the tetrahedral sheet may be replaced by aluminum (not shown). These substitutions give rise to a negative charge, which accounts for the high cation exchange capacity of this clay mineral.

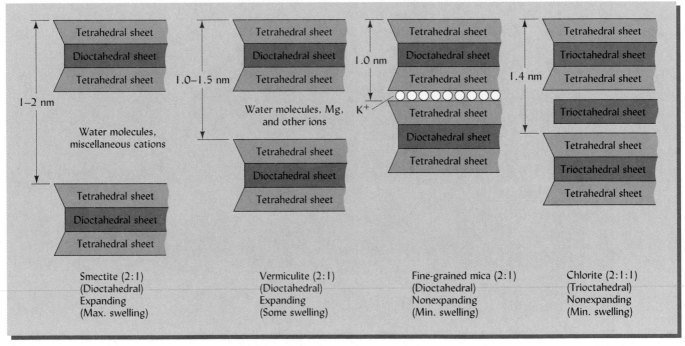

FIGURE 8.7 Schematic drawing illustrating the organization of tetrahedral and octahedral sheets in three 2:1-type minerals (smectite, vermiculite, and fine-grained mica) and one 2:1:1-type, chlorite. The aluminum-dominated dioctahedral minerals are illustrated for the 2:1 types because such minerals are most commonly found in soils. Trioctahedral chlorite in which magnesium ions dominate is the most common form of this mineral in soils. Note maximum interlayer expansion in smectite, with reduced expansion in vermiculite due to moderate binding power of numerous Mg^{2+} ions. Fine-grained mica and chlorite do not expand because K^+ ions (fine-grained mica) and an extra trioctahedral sheet (chlorite) tightly bind the 2:1 layers together. The interlayer spacings are shown in nanometers (1 nm = 10^{-9} m).

The cation exchange capacity of vermiculites usually exceeds that of all other silicate clays, including montmorillonite and other smectites (Table 8.3), because of the very high negative charge in the tetrahedral sheet. Vermiculite crystals are larger than those of the smectites but much smaller than those of kaolinite.

NONEXPANDING MINERALS. **Micas** are the type of minerals in this group. Muscovite and biotite are examples of unweathered micas often found in the sand and silt separates. Weathered minerals similar in structure to these micas are found in the clay fraction of soils. They are called *fine-grained micas.*[5]

Like smectites, fine-grained micas have a 2:1-type crystal. However, the particles are much larger than those of the smectites. Also, the major source of charge is in the tetrahedral sheet where about 20% of the silicon sites are occupied by aluminum atoms. This results in a high net negative charge in the tetrahedral sheet, even higher than that found in vermiculites. To satisfy this charge, potassium ions are strongly attracted in the interlayer space and are just the right size to fit snugly into certain spaces in the adjoining tetrahedral sheets (Figures 8.7 and 8.8). The potassium thereby acts as a binding agent, preventing expansion of the crystal. Hence, fine-grained micas are quite **nonexpansive.**

Such properties as hydration, cation adsorption, swelling, shrinkage, and plasticity are much less intense in fine-grained micas than in smectites (Table 8.3). The fine-grained micas exceed kaolinite with respect to these characteristics, but this may be due in part to the presence of interstratified layers of smectite or vermiculite. In size, too, fine-grained mica crystals are intermediate between the smectites and kaolinites (Table 8.3). Their specific surface area varies from 70 to 100 m^2/g, about one-eighth that for the smectites.

---

[5]In the past, the term *illite* has been used to identify these clay minerals.

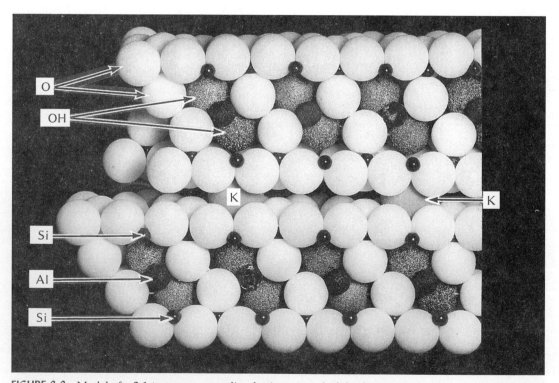

FIGURE 8.8 Model of a 2:1-type nonexpanding lattice mineral of the fine-grained mica group. The general constitution of the layers is similar to that in the smectites, one octahedral (alumina) sheet between two tetrahedral (silica) sheets. However, potassium ions are tightly held between layers, giving the mineral a more or less rigid type of structure that prevents the movement of water and cations into the space between layers. The internal surface and exchange capacity of fine-grained micas are thus far below those of the smectites.

2:1:1-Type Minerals

This silicate group is represented by **chlorites**, which are common in a variety of soils. Chlorites are basically iron-magnesium silicates with some aluminum present. In a typical chlorite clay crystal, 2:1 layers, such as are found in vermiculites, alternate with a magnesium-dominated trioctahedral sheet, giving a 2:1:1 ratio (Figure 8.7). Magnesium also dominates the trioctahedral sheet in the 2:1 layer of chlorites. Thus, the crystal unit contains two silica tetrahedral sheets and two magnesium-dominated trioctahedral sheets, giving rise to the term *2:1:1-* or *2:2-type* structure.

The negative charge of chlorites is about the same as that of fine-grained micas and considerably less than that of the smectites or vermiculites. Like fine micas, chlorites may be interstratified with vermiculites or smectites in a single crystal. Particle size and surface area for chlorites are also about the same as for fine-grained micas. There is no water adsorption between the chlorite crystal units, which is in keeping with the non-expansive nature of this mineral.

Mixed and Interstratified Layers

Specific groups of clay minerals do not occur independently of one another. In a given soil, it is common to find several clay minerals in an intimate mixture. Furthermore, some mineral colloids have properties and composition intermediate between those of any two of the well-defined minerals just described. Such minerals are termed **mixed layer** or **interstratified** because the individual layers within a given crystal may be of more than one type. Terms such as *chlorite-vermiculite* and *fine-grained mica-smectite* are used to describe mixed-layer minerals (Figure 8.9). In some soils, they are more common than single-structured minerals such as montmorillonite.

8.6 GENESIS OF SOIL COLLOIDS

Silicate Clays

The silicate clays are developed from the weathering of a wide variety of minerals by at least two distinct processes: (1) a slight physical and chemical **alteration** of certain primary minerals and (2) a **decomposition** of primary minerals with the subsequent **recrystallization** of certain of their products into the silicate clays. These processes will each be given brief consideration.

ALTERATION. The changes that occur as muscovite mica is altered to fine-grained mica represent a good example of alteration. Muscovite is a 2:1-type primary mineral with a nonexpanding crystal structure and a formula of $KAl_2(Si_3Al)O_{10}(OH)_2$. As weathering occurs, the mineral is broken down in size to the colloidal range, part of the potassium is lost, and some silicon is added from weathering solutions. The net result is a less rigid crystal structure and the assumption of an electronegative charge. The fine mica colloid that emerges still has a 2:1-type structure, only having been *altered* in the process. Some of these changes, perhaps oversimplified, can be shown as follows:

$$KAl_2(AlSi_3)O_{10}(OH)_2 + 0.2Si^{4+} + 0.1M^+ \xrightarrow{H_2O}$$

Muscovite (rigid crystal)   (soil solution)

$$M^+_{0.1}(K_{0.7})Al_2(Al_{0.8}Si_{3.2})O_{10}(OH)_2 + 0.3K^+ + 0.2Al^{3+}$$

Fine mica (semirigid crystal)   (soil solution)

RECRYSTALLIZATION. This process involves the complete breakdown of the crystal structure and recrystallization of clay minerals from products of this breakdown. It is the result of much more intense weathering than that required for the alteration process just described.

An example of recrystallization is the formation of kaolinite (a 1:1-type clay mineral) from solutions containing soluble aluminum and silicon that came from the breakdown of primary minerals having a 2:1-type structure. Such recrystallization makes possible the formation of more than one kind of clay from a given primary min-

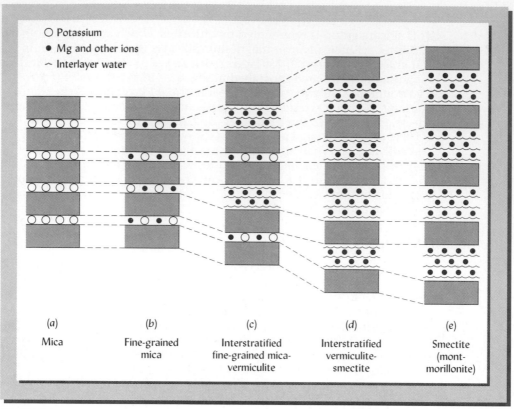

○ Potassium
● Mg and other ions
∼ Interlayer water

(a)
Mica

(b)
Fine-grained
mica

(c)
Interstratified
fine-grained mica–
vermiculite

(d)
Interstratified
vermiculite–
smectite

(e)
Smectite
(mont-
morillonite)

FIGURE 8.9 Structural differences among silicate minerals and their mixtures. Potassium-containing micas (a) with rigid crystals lose part of their potassium and weather to fine-grained mica, which is less rigid and attracts exchangeable cations to the interlayer space (b). At a more advanced stage of weathering (c), the potassium is leached from between some of the 2:1 layers; water and magnesium ions bind the layers together and an interstratified fine-grained mica–vermiculite is present. Further weathering removes more potassium (d), water and exchangeable ions push in between the 2:1 layers, and an interstratified vermiculite—smectite is formed. More weathering produces smectite (e), the highly expanded mineral. Smectite in turn is subject to weathering to kaolinite and iron and aluminum oxides. Although most smectites actually are formed by other processes, the sequence here illustrates the structural relationships of smectites to the other minerals.

eral. The specific clay mineral that forms depends on weathering conditions and the specific ions present in the weathering solution as crystallization occurs.

RELATIVE STAGES OF WEATHERING. The more specific conditions conducive to the formation of important clay types are shown in Figure 8.10. Note that fine-grained micas and magnesium-rich chlorites represent earlier weathering stages of the silicates, and kaolinite and (ultimately) iron and aluminum oxides the most advanced stages. The smectites (e.g., montmorillonite) represent intermediate stages.

GENESIS OF INDIVIDUAL SILICATE CLAYS. There are a variety of processes by which individual clays are formed. For example, fine-grained micas and chlorite are commonly formed from the alteration of muscovite and biotite micas, respectively. Vermiculites also can be formed through this process, although they as well as the smectites can be products of weathering of the fine-grained micas and chlorites. Smectites may also result from the process of recrystallization under neutral to alkaline weathering conditions. Kaolinite is formed by recrystallization under reasonably intense acid weathering conditions, which remove most of the metallic cations. Under hot, humid conditions of the tropics, the intense weathering commonly produces the oxides of iron and aluminum (Figure 8.10).

Hydrous Oxides

Gibbsite $(AlOH)_3$, the most common hydrous oxide of aluminum found in soils, is a product of weathering of a variety of primary aluminosilicates. Hydrogen ions replace

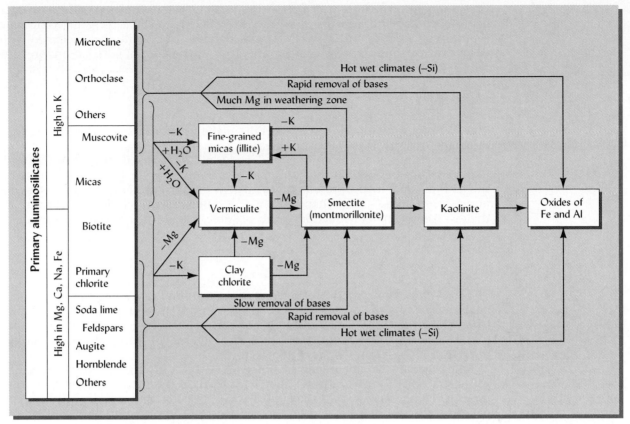

FIGURE 8.10 General conditions for the formation of the various layer silicate clays and oxides of iron and aluminum. Fine-grained micas, chlorite, and vermiculite are formed through rather mild weathering of primary aluminosilicate minerals, whereas kaolinite and oxides of iron and aluminum are products of much more intense weathering. Conditions of intermediate weathering intensity encourage the formation of smectite. In each case silicate clay genesis is accompanied by the removal in solution of such elements as K, Na, Ca, and Mg.

the base-forming cations and the mineral framework breaks down, releasing the aluminum and silicon. The breakdown of some basic rocks such as gabbro and basalt commonly yields gibbsite directly. The weathering of acid rocks such as granite and gneiss may first produce kaolinite or halloysite that, upon further weathering, yields gibbsite. Gibbsite represents the most advanced stage of weathering in soils.

Goethite (FeOOH) and hematite (Fe_2O_3) are the two most common iron oxides in soils. They are both produced by the weathering of iron-containing primary silicate minerals. The iron is commonly in the low-valent state that upon the breakdown of the mineral framework as above is released but quickly oxidized to the very insoluble three-valent form. Goethite tends to be dominant in the temperate or more moist zones, while hematite with its deep red color is more prominent (even dominant) in tropical or drier conditions.

It is interesting to note that isomorphic substitution of Al for Fe and vice versa takes place to some degree among hydrous oxides of these elements. However, since Fe^{3+} and Al^{3+} have the same valence, this substitution has no effect on the charge on the micelle.

Allophane, Imogolite, and Humus

Relatively little is known of factors influencing the formation of allophane and imogolite except that they are commonly associated with materials of volcanic origin. Apparently, volcanic ashes release significant quantities of Si $(OH)_x$ and Al $(OH)_x$ materials that precipitate as amorphous gels (allophane) in a relatively short period of time. These minerals are generally noncrystalline in nature, although imogolite has some unidimensional organization and is considered to be the product of a more advanced state of weathering than that which produces allophane.

Humus is formed from the breakdown and alteration of plant residues by microorganisms and by the subsequent synthesis of new, more stable, organic compounds in the process (see Section 12.9). The various organic structural units associated with the decay and synthesis provide charged sites for the attraction of both cations and anions.

8.7 GEOGRAPHIC DISTRIBUTION OF CLAYS

The clay of any particular soil is generally made up of a mixture of different colloidal minerals. In a given soil, the mixture may vary from horizon to horizon. This occurs because the kind of clay that develops depends not only on climatic influences and profile conditions but also on the nature of the parent material. The situation may be further complicated by the presence in the parent material itself of clays that were formed under a preceding and perhaps an entirely different type of climatic regime. Nevertheless, some general deductions seem possible, based on the relationships shown in Figure 8.10.

Zonal Soil Differences

The well-drained and well-weathered Oxisols of humid and subhumid tropics tend to be dominated by the oxides of iron and aluminum. These clays are also prominent in Ultisols such as those in the southeastern part of the United States. Kaolinite is commonly the dominant silicate mineral in Ultisols (Table 8.4) and is also found along with the hydrous oxide clays in more tropical areas.

The smectite, vermiculite, and fine-grained mica groups are more prominent in Alfisols, Mollisols, and Vertisols where weathering is less intense. These clays are common in the northern part of the United States, Canada, and regions with similar temperatures throughout the world. Where either the parent material or the soil solution surrounding the weathering minerals is high in potassium, fine-grained micas are apt to be formed. Parent materials high in metallic cations (particularly magnesium) or subject to restricted drainage, which discourages the leaching of these cations, encourage smectite formation. Thus, fine-grained micas and smectites are more likely to be prominent in Aridisols than in the more humid areas.

The strong influence of parent material on the geographic distribution of clays can be seen in the black-belt Vertisols of Alabama, Mississippi, and Texas. These soils, which are dark in color, have developed from base-rich marine parent materials and are dominated by smectite clays. The surrounding soils, which have developed from different parent materials, are high in kaolinite and hydrous oxides, clays that are more representative of this warm, humid region. Similar situations exist in central India and Sudan.

Table 8.4 shows the dominant clay minerals in different soil orders, the descriptions of which were given in Chapter 3. These data tend to substantiate the generalizations

TABLE 8.4 Prominent Occurrence of Clay Minerals in Different Soil Orders in the United States and Typical Locations for These Soils

Soil order[a]	General weathering intensity	Typical location in U.S.	Hydrous oxides	Kaolinite	Smectite	Fine-grained mica	Vermiculite	Chlorite	Intergrades
Aridisols	Low	Dry areas			XX	XX		X	X
Vertisols[b]	↑	Alabama, Texas			XXX				X
Mollisols		Central		X	XX	X	X	X	X
Alfisols		Ohio, Pennsylvania, New York		X	X	X	X	X	X
Spodosols		New England	X	X					
Ultisols	↓	Southeast	XX	XXX			X	X	X
Oxisols	High	Hawaii, Puerto Rico	XXX	XX					

[a]See Chapter 3 for soil descriptions.
[b]By definition these soils have swelling-type clays, which account for the dominance of smectites.

just discussed. For example, Oxisols and Ultisols are characteristic of areas of intense weathering, and Aridisols are found in desert areas. The dominant clay minerals for these areas are as expected on the basis of what is seen in Figure 8.10.

Although a few broad generalizations relating to the geographic distribution of clays are possible, these examples suggest that local parent materials and weathering conditions tend to dictate the kinds of clay minerals found in soils.

8.8 SOURCE OF CONSTANT CHARGES ON SILICATE CLAYS

In Section 8.4, it was noted that isomorphous substitution of one cation for another within the crystal structures leads to a charge imbalance in silicate clays—an imbalance that accounts for the ability of clays to attract ions to particle surfaces. Examples of specific substitutions will now be considered.

Negative Charges

A net negative charge is found in minerals where there has been an isomorphous substitution of a lower-charged ion (e.g., Mg^{2+}) for a higher-charged ion (e.g., Al^{3+}). Such substitution commonly occurs in aluminum-dominated dioctahedral sheets. As shown in Figure 8.11, this leaves an unsatisfied negative charge. The substitution of Mg^{2+} for Al^{3+} is an important source of the negative charge on the smectite, vermiculite, and chlorite clay micelles.

A second example is the substitution of an Al^{3+} for an Si^{4+} in the tetrahedral sheet, which also leaves one negative charge unsatisfied. This can be illustrated as follows, assuming that two silicon atoms are associated with four oxygens in a charge-neutral unit:

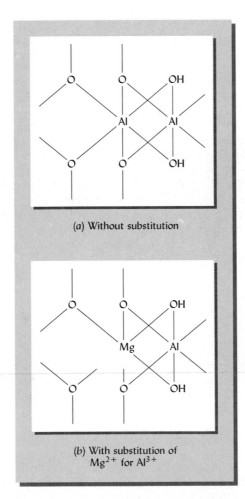

(a) Without substitution

(b) With substitution of Mg^{2+} for Al^{3+}

FIGURE 8.11 Atomic configuration in the octahedral sheet of silicate clays (a) without substitution and (b) with a magnesium ion substituted for one aluminum. Where no substitution has occurred, the three positive charges on aluminum are balanced by an equivalent of three negative charges from six oxygens or hydroxy groups. With magnesium in the place of aluminum, only two of those negative charges are balanced, leaving one negative charge unsatisfied. This negative charge can be satisfied by an adsorbed cation.

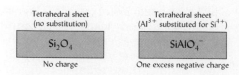

Such a substitution is common in several of the important soil silicate clay minerals such as the fine-grained micas, vermiculites, and even some smectites.

Positive Charges

Isomorphous substitution can also be a source of positive charges if the substituting cation has a higher charge than the ion for which it substitutes. In a trioctahedral sheet, there are three magnesium ions surrounded by oxygen and hydroxy groups, and the sheet has no charge. However, if an Al^{3+} ion substitutes for one of the Mg^{2+} ions, a net positive charge results.

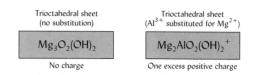

Such positive charges are characteristic of the trioctahedral sheet in some vermiculites and chlorites.

In some silicate clays (e.g., smectites), substitutions in both the tetrahedral and octahedral sheets can occur. The net charge in these clays is the balance between the negative and positive charges. In essentially all silicate clays, however, the net charge is negative since those substitutions leading to negative charges far outweigh those producing positive charges.

The isomorphous substitutions just described have taken place slowly through past weathering cycles and are not subject to easy modification. Consequently, these charges are termed **constant** or **permanent charges** because they are related to the chemical composition of the silicate clays.

Chemical Composition and Charge

Because of the numerous ionic substitutions just discussed, simple chemical formulas cannot be used to identify specifically the clay in a given soil. However, type formulas for the major silicate clays shown in Table 8.5 can be used to illustrate the sources of both positive and negative charges in the tetrahedral and octahedral sheets. These formulas are commonly referred to as **unit layer formulas.**

Note that Table 8.5 shows no ionic substitution in kaolinite. Although some such substitution does occur, much of the negative charge on this colloid is provided by another mechanism. All other clay minerals shown in Table 8.5 owe their charge (negative and/or positive) to the isomorphous substitutions. This table should be studied to ascertain the source of these charges as well as the net charge on the micelles of each of the minerals listed.

8.9 pH-DEPENDENT CHARGES

There is a second source of charges on some silicate clays (e.g., kaolinite), humus, allophane, and Fe, Al hydroxides. Because these charges are dependent on the soil pH, they are termed **variable** or **pH-dependent,** in contrast to the constant or more permanent charges resulting from isomorphous substitution.

Negative Charges

Some of the pH-dependent charges are associated primarily with hydroxy (OH) groups on the edges and surfaces of the inorganic and organic colloids (Figure 8.12). The OH groups are attached to iron and/or aluminum in the inorganic colloids (e.g., >Al—OH)

TABLE 8.5 **Typical Unit Layer Formulas of Several Clay and Other Silicate Minerals Showing Octahedral and Tetrahedral Cations as Well as Coordinating Anions, Charge per Unit Formula, and Fixed and Exchangeable Interlayer Components**

Note that the negative charges in the crystal units are counterbalanced by equivalent positive charges in interlayer areas.

Mineral	Octahedral sheet	Tetrahedral sheet	Coordinating anions	Charge per unit formula	Interlayer components	
					Fixed	Exchangeable[a]
Kaolinite	Al_2	Si_2	$O_5(OH)_4$	0	None	None
Serpentine	Mg_3	Si_2	$O_5(OH)_4$	0	None	None
2:1-Type Dioctahedral Minerals						
Pyrophyllite	Al_2	Si_4	$O_{10}(OH)_2$	0	None	None
Montmorillonite	$Al_{1.7}Mg_{0.3}$ $\underset{-0.3}{}$	$Si_{3.9}Al_{0.1}$ $\underset{-0.1}{}$	$O_{10}(OH)_2$	-0.4	None	$M_{0.4}^+$
Beidellite	Al_2	$Si_{3.6}Al_{0.4}$ $\underset{-0.4}{}$	$O_{10}(OH)_2$	-0.4	None	$M_{0.4}^+$
Nontronite	Fe_2	$Si_{3.6}Al_{0.4}$ $\underset{-0.4}{}$	$O_{10}(OH)_2$	-0.4	None	$M_{0.4}^+$
Vermiculite	$Al_{1.7}Mg_{0.3}$ $\underset{-0.3}{}$	$Si_{3.6}Al_{0.4}$ $\underset{-0.4}{}$	$O_{10}(OH)_2$	-0.7	xH_2O	$M_{0.7}^+$
Fine mica (illite)	Al_2	$Si_{3.2}Al_{0.8}$ $\underset{-0.8}{}$	$O_{10}(OH)_2$	-0.8	$K_{0.7}^+$	$M_{0.1}^+$
Muscovite	Al_2	Si_3Al $\underset{-1.0}{}$	$O_{10}(OH)_2$	-1.0	K^+	None
2:1-Type Trioctahedral Minerals						
Talc	Mg_3	Si_4	$O_{10}(OH)_2$	0	None	None
Vermiculite	$Mg_{2.7}Fe_{0.3}^{3+}$ $\underset{+0.3}{}$	Si_3Al $\underset{-1.0}{}$	$O_{10}(OH)_2$	-0.7	xH_2O	$M_{0.7}^+$
2:1:1-Type Trioctahedral Minerals						
Chlorite	$Mg_{2.6}Fe_{0.4}^{3+}$ $\underset{+0.4}{}$	$Si_{2.5}(Al,Fe)_{1.5}$ $\underset{-1.5}{}$	$O_{10}(OH)_2$	-1.1	$Mg_2Al(OH)_6^+$	$M_{0.1}^+$

[a]Exchangeable cations such as Ca^{2+}, Mg^{2+}, and H^+ are indicated by the singly charged cation M^+.

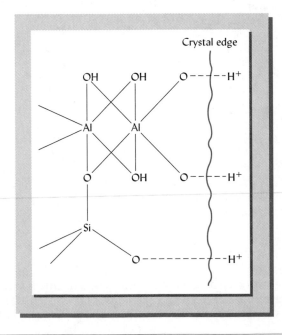

FIGURE 8.12 Diagram of a broken edge of a kaolinite crystal showing oxygens as the source of negative charge. At high pH values the hydrogen ions tend to be held loosely and can be exchanged for other cations.

and to CO groups in humus (e.g., —CO—OH). Under moderately acid conditions, there is little or no charge on these particles, but as the pH increases, the hydrogen dissociates from the colloid OH group, and negative charges result.

$$
\diagdown \!\!\!Al - OH + OH^- \rightleftharpoons \diagdown \!\!\!Al - O^- + H_2O
$$

$$
-CO - OH + OH^- \rightleftharpoons -CO - O^- + H_2O
$$

| No charge (soil solids) | (soil solution) | Negative charge (soil solids) | (soil solution) |

As indicated by the $\rightleftharpoons$ arrows, the reactions are reversible. If the pH increases, more OH^- ions are available to force the reactions to the right, and the negative charge on the particle surfaces increases. If the pH is lowered, OH^- ion concentrations are reduced, the reaction goes back to the left, and the negativity is reduced. It is easy to see the pH dependency of the charge on these colloids.

Another source of increased negative charges as the pH is increased is the removal of positively charged complex aluminum hydroxy ions (e.g., $Al(OH)_2^+$). At low pH levels, these ions block negative sites on the silicate clays (e.g., vermiculite) and make them unavailable for cation exchange. As the pH is raised, the $Al(OH)_2^+$ ions react with the OH^- ion in the soil solution to form insoluble $Al(OH)_3$, thereby freeing the negatively charged sites.

$$
\diagdown \!\!\!Al - (OH)_2^- \cdot Al(OH)_2^+ + OH^- \longrightarrow \diagdown \!\!\!Al - (OH)_2^- + Al(OH)_3
$$

(negative charged site is blocked)   (negative charge site is freed)   (no charge)

This mechanism of increasing negative charges is of special importance in soils high in hydrous oxides of iron and aluminum such as are found in the southeastern United States and in the tropics.

Positive Charges

Under moderate to extreme acid soil conditions, some silicate clays, iron, and aluminum hydrous oxides may exhibit positive charges. Once again, exposed OH groups are involved. In this case, however, as the soils become more acid, protonation—the attachment of H^+ ions to the surface OH groups—takes place. The reaction for silicate clays may be shown simply as

$$
\diagdown \!\!\!Al - OH \quad + \quad H^+ \rightleftharpoons \diagdown \!\!\!Al - OH_2^+
$$

| No charge (soil solids) | (soil solution) | Positive charge (soil solids) |

Thus, in some cases, the same site on the inorganic soil colloid may be responsible for negative charge (high pH), no charge (intermediate pH), or positive charge (very low pH). The reaction as the pH of a highly alkaline soil is reduced may be illustrated as follows:

$$
\diagdown \!\!\!Al - O^- + H^+ \rightleftharpoons \diagdown \!\!\!AlOH + H^+ \rightleftharpoons \diagdown \!\!\!AlOH_2^+
$$

| Negative charge (high pH) | No charge (intermediate pH) | Positive charge (very low pH) |

Since a soil mixture of humus and several inorganic colloids is usually found in soil, it is not surprising that positive and negative charges may be exhibited at the same time. In most soils of temperate regions, the negative charges far exceed the positive ones (Table 8.6). However, in some acid soils of the tropics high in iron and aluminum

TABLE 8.6 **Charge Characteristics of Representative Colloids Showing Comparative Levels of Permanent (Constant) and pH-Dependent Negative Charges as Well as pH-Dependent Positive Charges**

Colloid type	Negative charge			Positive charge $(cmol_c/kg)$
	Total at pH 7 $(cmol_c/kg)$	Constant (%)	Variable (%)	
Organic	200	10	90	0
Smectite	100	95	5	0
Vermiculite	150	95	5	0
Fine-grained micas	30	80	20	0
Chlorite	30	80	20	0
Kaolinite	8	5	95	2
Gibbsite (Al)	4	0	100	5
Goethite (Fe)	4	0	100	5

hydroxides, the overall net charge may be positive. The effect of soil pH on positive and negative charges on such soils is illustrated in Figure 8.13.

The charge characteristics of selected soil colloids are shown in Table 8.6. Note the high percentage of constant negative charges in some 2:1-type clays (e.g., smectites and vermiculites). Humus, kaolinite, allophane, and Fe, Al oxides have mostly variable (pH-dependent) negative charges and exhibit modest positive charges at very low pH values.

Cation and Anion Adsorption

The charges associated with soil particles attract simple and complex ions of opposite charge. Thus, a given colloidal mixture may exhibit not only a maze of positive and negative surface charges, but an equally complex group of simple cations and anions such as Ca^{2+} and SO_4^{2-} that are attracted by the particle charges.

Figure 8.14 illustrates how soil colloids attract mineral elements so important for plant growth.[6] The adsorbed anions are commonly present in smaller quantities than the

---

[6]In addition to the simple inorganic cations and anions, more complex organic compounds as well as charged organomineral complexes are adsorbed by the charged soil colloids.

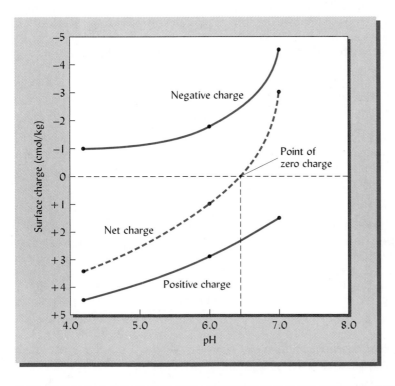

FIGURE 8.13 Surface charge on the colloids from the subsoil of an Oxisol. Note the presence of both negative and positive charges. The negative charge increases and the positive charge decreases with an increase in soil pH. The point of zero charge, which in this case is about pH 6.5, is characteristic of the colloids in this soil. [From Van Raij and Peech (1972)]

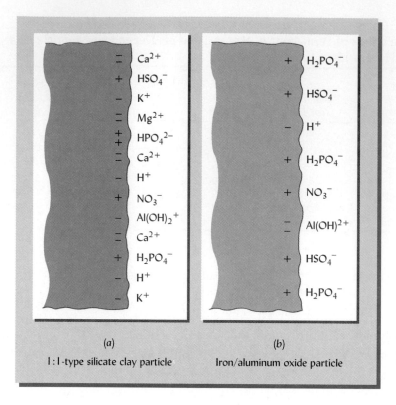

FIGURE 8.14 Illustration of adsorbed cations and anions in an acid soil with (a) 1:1-type silicate clay and (b) iron and aluminum oxide particles. Note the dominance of the negative charges on the silicate clay and of the positive charges on the aluminum oxide particle.

(a)
1:1-type silicate clay particle

(b)
Iron/aluminum oxide particle

cations because the negative charges generally predominate on the soil colloid, especially in soils of temperate regions. Consideration now will be given to the exchange of ions between micelles and the soil solution, starting first with cation exchange.

8.10 CATION EXCHANGE

In Section 8.3, the various cations adsorbed by the exchange complex[7] were shown to be subject to replacement by other cations through a process called *cation exchange.* For example, hydrogen ions generated as organic matter decomposes (see Section 9.6) can displace calcium and other metallic cations from the colloidal complex. This can be shown simply as follows, where only one adsorbed calcium ion is being replaced:

$$\text{Ca}^{2+}\boxed{\text{Micelle}} + 2\text{H}^+ \rightleftharpoons \frac{\text{H}^+}{\text{H}^+}\boxed{\text{Micelle}} + \text{Ca}^{2+}$$

The reaction takes place rapidly, and the interchange of calcium and hydrogen is chemically equivalent. As shown by the double yield arrows, the reaction is reversible and will go to the left if calcium is added to the system.

Cation Exchange Under Natural Conditions

As it usually occurs in temperate region surface soils the cation exchange reaction is somewhat more complex, but the principles illustrated in the simple equation apply. Assume, for the sake of simplicity, that the numbers of Ca^{2+}, Al^{3+}, H^+, and other metallic cations such as Mg^{2+} and K^+ (represented as M^+) are in the ratio 40, 20, 20, and 20 per micelle, respectively. (The metallic cations represented as M^+ will be considered monovalent in this case.) Hydrogen from carbonic acid (H_2CO_3) reacts with the micelle as follows:

---

[7]This term includes all soil colloids, inorganic and organic, capable of holding and exchanging cations.

$$\begin{array}{c} Ca_{40} \\ Al_{20} \\ H_{20} \\ M_{20} \end{array} \boxed{\text{Micelle}} + 5H_2CO_3 \rightleftharpoons \begin{array}{c} Ca_{38} \\ Al_{20} \\ H_{25} \\ M_{19} \end{array} \boxed{\text{Micelle}} + 2Ca(HCO_3)_2 + M(HCO_3)$$

(soil solids)   (in soil solution)   (soil solids)   (in soil solution)

Where sufficient rainfall is available to leach calcium and other metallic cations, the reaction tends to go toward the right and the soils tend to become more acid. In regions of low rainfall, however, calcium and other salts are more plentiful since they are not easily leached from the soil. The metallic ions drive the reaction to the left, keeping the soil at pH 7 or above. In addition, some of the calcium will be precipitated as $CaCO_3$ especially in the subsoil at the lower limit of rainfall penetration. The interaction of climate, biological processes, and cation exchange thus helps determine the chemical properties of soils.

Influence of Lime and Fertilizer

Cation exchange reactions are reversible; therefore, if some basic calcium compound such as limestone is applied to an acid soil, the preceding reaction will be driven to the left. Calcium ions will replace the hydrogen and other cations; the H^+ ions will be neutralized by OH^- or CO_3^{2-} ions; and the soil pH is raised. If, on the other hand, acid-forming chemicals, such as sulfur, are added to an alkaline soil in a dryland area, the H^+ ions would replace the metal cation on the soil colloids, and the soil pH would decrease.

One more illustration of cation exchange is the reaction that may occur when a fertilizer containing potassium chloride is added to soil.

$$\begin{array}{c} Ca_{40} \\ Al_{20} \\ H_{40} \\ M_{20} \end{array} \boxed{\text{Micelle}} + 7KCl \rightleftharpoons \begin{array}{c} K_7 \\ Ca_{38} \\ Al_{20} \\ H_{39} \\ M_{18} \end{array} \boxed{\text{Micelle}} + 2CaCl_2 + HCl + 2MCl$$

(soil solids)   (in soil solution)   (soil solids)   (in soil solution)

The added potassium is adsorbed on the colloid and replaces an equivalent quantity of calcium, hydrogen, and other cations that appear in the soil solution. The adsorbed potassium remains largely in an available condition, but is less subject to leaching than are most fertilizer salts. Hence, cation exchange is an important consideration, not only for nutrients already present in soils but also for those applied in manures, crop residues, or commercial fertilizers.

8.11 CATION EXCHANGE CAPACITY

Previous sections have dealt qualitatively with exchange. We now turn to a consideration of the quantitative cation exchange or the **cation exchange capacity** (CEC). This property, which is defined simply as "the sum total of the exchangeable cations that a soil can adsorb," is easily determined. By standard methods all the adsorbed cations in a soil are replaced by a common ion, such as barium, potassium, or ammonium; then the amount of adsorbed barium, potassium, or ammonium is determined (Figure 8.15).

Means of Expression[8]

The cation exchange capacity (CEC) is expressed in terms of moles of positive charge adsorbed per unit mass. For the convenience of expressing CEC in whole numbers, we

---

[8]In the past, cation exchange capacity generally has been expressed as milliequivalents per 100 g soil. In this text, the International System of Units (SI) is being used. Fortunately, since one milliequivalent per 100 g soil is equal to 1 centimole ($cmol_c$) of positive or negative charge per kilogram of soil, it is easy to compare soil data using either of these methods of expression.

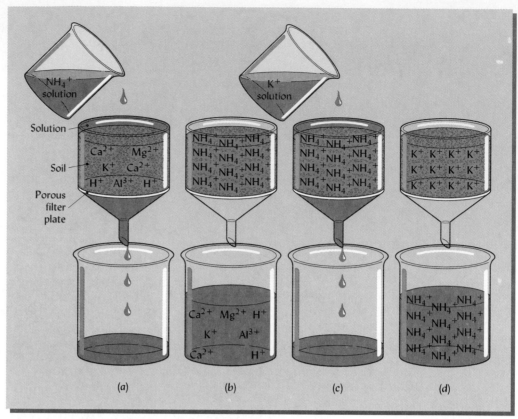

FIGURE 8.15 Illustration of a method for determining the cation exchange capacity of soils. (a) A given mass of soil containing a variety of exchangeable cations is leached with an ammonium (NH_4^+) salt solution. (b) The NH_4^+ ions replace the other adsorbed cations, which are leached into the container below. (c) After the excess NH_4^+ salt solution is removed with an organic solvent, such as alcohol, a K^+ salt solution is used to replace and leach the adsorbed NH_4^+ ions. (d) The amount of NH_4^+ released and washed into the lower container can be determined, thereby measuring the chemical equivalent of the cation exchange capacity (i.e., the negative charge on the soil colloids).

shall use *centimoles of positive charge per kilogram of soil* ($cmol_c/kg$). Thus, if a soil has a cation exchange capacity of 10 $cmol_c/kg$, 1 kg of this soil can adsorb 10 cmol of H^+ ion, for example, and can exchange it with 10 cmol of another monovalent cation, such as K^+ or Na^+, or with 5 cmol of divalent cation, such as Ca^{2+} or Mg^{2+}. In each case, the 10 cmol of negative charge associated with 1 kg of soil is attracting 10 cmol of positive charges, whether they come from H^+, K^+, Na^+, NH_4^+, Ca^{2+}, Mg^{2+}, Al^{3+}, or any other cation.

Note that the cations are adsorbed and exchanged on a *chemically equivalent* basis. One mole of charge is provided by 1 mole of H^+, K^+, or any other monovalent cation; by ½ mole of Ca^{2+}, Mg^{2+}, or other divalent cation; and by ⅓ mole of Al^{3+} or other trivalent cation.

With this chemical equivalency in mind, it is easy to express cation exchange in practical field terms. For example, when an acid soil is limed and calcium ions replace part of the adsorbed hydrogen ions, the reaction that occurs between the monovalent H^+ ion and the divalent Ca^{2+} ion is

$$\boxed{Micelle}\begin{matrix}H^+\\H^+\end{matrix} + Ca^{2+} \rightleftharpoons \boxed{Micelle}\, Ca^{2+} + 2H^+$$

Note that the 2 moles of charge associated with two H^+ ions are replaced by the equivalent charge associated with one Ca^{2+}. In other words, 1 mole of H^+ ion (1 g) would exchange with ½ mole of Ca^{2+} ion (40/2 = 20 g). Accordingly, to replace 1 *centimole* H^+/kg

would require 20/100 = 0.2 g calcium/kg soil.[9] The practical implications of these relationships for soil acidity management will be discussed in Chapter 9.

Cation Exchange Capacities of Soils

The cation exchange capacity (CEC) of a given soil is determined by the relative amounts of different colloids in that soil and by the CEC of each of these colloids. Thus, sandy soils have lower CECs than clay soils because the coarse-textured soils are commonly lower in both clay and humus content. Likewise, a clay soil dominated by 1:1-type silicate clays and Fe, Al oxides will have a much lower CEC than will one with similar humus content dominated by smectite clays. See Table 8.6 for the charge characteristics that determine the exchange capacities of the major soil colloids.

These generalizations of the preceding paragraph can be verified by examination of Table 8.3. Scientists at the National Soil Survey Laboratory of the U.S. Soil Conservation Service determined the cation exchange capabilities and pH values of soil samples taken from 3045 sites around the country. Data in Table 8.7 show the average CEC values for the nine different soil orders included in the survey. Note the very high CEC for the Histosols, verifying the high CEC of the organic colloids. The Vertisols, which are very high in swelling-type clays (mostly smectite), had the highest average CEC of the mineral soils. Next came the Aridisols and Mollisols that are also commonly high in 2:1-type clays. The Ultisols, whose clays are dominantly kaolinite and hydrous oxides of iron and aluminum, had relatively low CEC values. The Entisols and Inceptisols likely include soils developed on recent alluvium and lacustrine materials that are quite high in clay. These data are what might be expected considering the probable quantity and kinds of soil colloids found in the soils. They confirm some of the generalizations of the previous paragraph.

pH and Cation Exchange Capacity

In previous sections it was pointed out that the cation exchange capacity of most soils increases with pH. At very low pH values, the cation exchange capacity is generally low (Figure 8.16). Under these conditions, only the permanent charges of the 2:1-type clays

---

[9]The amount of Ca^{2+} required for a hectare–furrow slice (2.2 million kg) is $0.2 \times 2.2 \times 10^6 = 440,000$ g or 440 kg. This can be expressed in terms of the quantity of limestone ($CaCO_3$) to supply the Ca^{2+} by multiplying by the ratio $CaCO_3/Ca = 100/40 = 2.5$. Thus, $440 \times 2.5 = 1100$ kg limestone per hectare–furrow slice would exchange with 1 mole H^+/kg soil (if complete exchange took place). A comparable figure in the English system is 1000 pounds per acre–furrow slice.

TABLE 8.7   The Average Cation Exchange Capacities (CEC) and pH Values of More Than Three Thousand Soil Samples Representing Nine Different Soil Orders

Note the very high CEC of the Histosols and among the mineral soils of the Vertisols. Ultisols have the lowest CEC. The low average pH values of the Spodosols and high values for the Aridisols and Entisols (many of which were from low-rainfall areas) are noteworthy.

Soil order	CEC (cmol$_c$/kg)	pH
Ultisols	3.5	5.60
Alfisols	9.0	6.00
Spodosols	9.3	4.93
Mollisols	18.7	6.51
Vertisols	35.6	6.72
Aridisols	15.2	7.26
Inceptisols	14.6	6.08
Entisols	11.6	7.32
Histosols	128.0	5.50

From Holmgren et al. (1993).

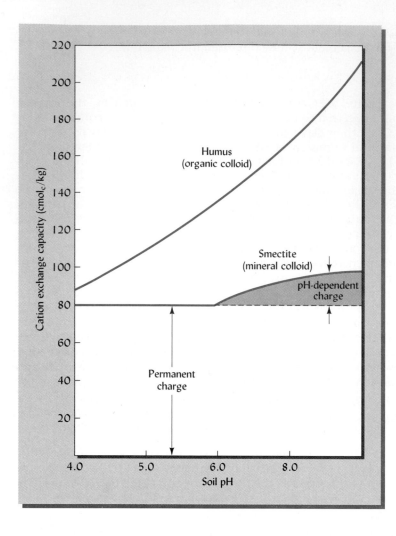

FIGURE 8.16 Influence of pH on the cation exchange capacity of smectite and humus. Below pH 6.0 the charge for the clay mineral is relatively constant. This charge is considered permanent and is due to ionic substitution in the crystal unit. Above pH 6.0 the charge on the mineral colloid increases slightly because of ionization of hydrogen from exposed hydroxy groups at crystal edges. In contrast to the clay, essentially all of the charges on the organic colloid are considered pH dependent. [Smectite data from Coleman and Mehlich (1957); organic colloid data from Helling et al. (1964)]

(see Section 8.8) and a small portion of the pH-dependent charges of organic colloids, allophane, and some 1:1-type clays hold exchangeable ions. As the pH is raised, the negative charges on some 1:1-type silicate clays, allophane, humus, and even Fe, Al oxides increases, thereby increasing the cation exchange capacity. To obtain a measure of this maximum retentive capacity, the CEC is commonly determined at a pH of 7.0 or above. At neutral or slightly alkaline pH, the CEC reflects most of those pH-dependent charges as well as the permanent ones.

Representative Figures

From the preceding discussions, it is possible to approximate the CEC of a soil if the type and contents of the dominant clay colloids and the humus are known. For example, the CEC of the dominant clays in surface soils of the humid temperate region in the United States averages about 0.5 $cmol_c/kg$ for each 1% clay in the soil. A comparable figure for humus is 2.0 $cmol_c/kg$ for each 1% of the humus content. By using these figures, it is possible to calculate roughly the cation exchange capacity of a representative humid temperate region surface soil from the percentages of clay and organic matter present. For soils dominated by kaolinite and Fe, Al oxides, comparable figures might be 0.1 for each 1% of clay and 2.0 for each 1% of organic matter. Such approximations can be made in any area where the types and contents of the dominant clays and of humus are known.

8.12 EXCHANGEABLE CATIONS IN FIELD SOILS

The specific exchangeable cations associated with soil colloids differ from one climatic region to another—Ca^{2+}, H^+, Al^{3+}, and complex aluminum hydroxy ions being most prominent in humid regions, and Ca^{2+}, Mg^{2+}, and Na^+ dominating in low-rainfall areas

TABLE 8.8   Cation Exchange Data for Representative Mineral Surface Soils in Different Areas

Characteristics	Warm humid region soil (Ultisol)	Cool humid region soil (Alfisol)	Semiarid region soil (Aridisol)	Arid region soil (Natrargids)[a]
Exchangeable calcium ($cmol_c$/kg)	2–5	6–9	14–17	12–14
Other exchangeable bases ($cmol_c$/kg)	1–2	2–3	5–7	8–12
Exchangeable hydrogen and/or aluminum ($cmol_c$/kg)	3–7	4–8	0–2	0
Cation exchange capacity ($cmol_c$/kg)	3–12	12–18	20–26	20–26
Base saturation (%)	25	50–75	90–100	100
Probable pH	5.0–5.4	5.6–6.0	7	8–10

[a]Significant sodium saturation.

(Table 8.8). The cations that dominate the exchange complex have a marked influence on soil properties.

The proportion in any soil of the cation exchange capacity satisfied by a given cation is termed the *percentage saturation* for that cation. Thus, if 50% of the CEC were satisfied by Ca^{2+} ions, the exchange complex is said to have a *percentage calcium saturation* of 50.

This terminology is especially useful in identifying the relative proportions of sources of acidity and alkalinity in the soil solution. Thus, the percentage saturation with H^+ and Al^{3+} ions gives an indication of the acid conditions, while increases in the nonacid cation percentage (commonly referred to as the **percentage base saturation**[10]) indicate the tendency toward neutrality and alkalinity. The percentage base saturation is an important soil property, especially since it is inversely related to soil acidity. These relationships will be discussed further in Chapter 9.

The percentage cation saturation of essential elements such as calcium and potassium also greatly influences the uptake of these elements by growing plants. This now will receive attention along with other cation exchange–plant nutrition interactions.

8.13   CATION EXCHANGE AND THE AVAILABILITY OF NUTRIENTS

Exchangeable cations generally are available to both higher plants and microorganisms. By cation exchange, hydrogen ions from the root hairs and microorganisms replace nutrient cations from the exchange complex. The nutrient cations are forced into the soil solution, where they can be assimilated by the adsorptive surfaces of roots and soil organisms, or they may be removed by drainage water. Cation exchange reactions affecting the mobility of organic and inorganic pollutants in soils will be discussed in Chapter 18. Here we focus on the plant nutrition aspects.

Cation Saturation and Nutrient Availability

Several factors operate to expedite or retard the release of nutrients to plants. First, there is the percentage saturation of the exchange complex by the nutrient cation in question. For example, if the percentage calcium saturation of a soil is high, the displacement of this cation is comparatively easy and rapid. Thus, 6 cmol/kg of exchangeable calcium in a soil whose exchange capacity is 8 cmol/kg (75% calcium saturation) probably would mean ready availability, but 6 cmol/kg when the total exchange capacity of a soil is 30 cmol/kg (20% calcium saturation) produces quite the opposite condition. This is one reason that, for calcium-loving plants such as alfalfa, the calcium saturation of at least part of the soil should approach or even exceed 80%.

Influence of Complementary Adsorbed Cations

A second factor influencing the plant uptake of a given cation is the complementary ions held on the colloids. As was discussed in Section 8.3, the strength of adsorption of different cations on most colloids is in the following order.

---

[10]Technically speaking, nonacid cations such as Ca^{2+}, Mg^{2+}, K^+, and Na^+ are not bases. When adsorbed by soil colloids in the place of H^+ ions, however, they reduce acidity and increase the soil pH. For that reason, they are referred to as *bases* and the portion of the CEC that they satisfy is usually termed *percentage base saturation*.

$$Al^{3+} > Ca^{2+} > Mg^{2+} > K^+ = NH_4^+ > Na^+$$

Consequently, a nutrient cation such as K^+ is less tightly held by the colloids if the complementary ions are Al^{3+} and H^+ (acid soils) than if they are Mg^{2+} and Na^+ (neutral to alkaline soils). The loosely held K^+ ions are more readily available for absorption by plants or for leaching in acid soils (see also Section 14.13).

There are also some nutrient antagonisms, which in certain soils cause inhibition of uptake of some cations by plants. Thus, potassium uptake by plants is limited by high levels of calcium in some soils. Likewise, high potassium levels are known to limit the uptake of magnesium even when significant quantities of magnesium are present in the soil.

Effect of Type of Colloid

Third, differences exist in the tenacity with which several types of colloidal micelles hold specific cations and in the ease with which they exchange cations. At a given percentage base (nonacid) saturation, smectites—which have a high charge density per unit of colloid surface—hold calcium much more strongly than does kaolinite (low charge density). As a result, smectite clays must be raised to about 70% base saturation before calcium will exchange easily and rapidly enough to satisfy most plants. In contrast, a kaolinite clay exchanges calcium much more readily, serving as a satisfactory source of this constituent at a much lower percentage base saturation. The need to add limestone to the two soils will be somewhat different, partly because of this factor.

8.14 ANION EXCHANGE

Anion exchange was discussed briefly in Section 8.9. The positive charges associated with hydrous oxides of iron and aluminum, some 1:1-type clays, and amorphous materials such as allophane make possible the adsorption of anions (Figure 8.14). In turn, these anions are subject to replacement by other anions, just as cations replace each other. The **anion exchange** that takes place, although it may not approach cation exchange quantitatively, is important as a means of providing to higher plants readily available nutrient anions.

The basic principles of cation exchange apply also to anion exchange, except that the charges on the colloids are positive and the exchange is among negatively charged anions. A simple example of an anion exchange reaction is

$$\boxed{Micelle}\ NO_3^- \ + \ Cl^- \ \rightleftharpoons \ \boxed{Micelle}\ Cl^- \ + \ NO_3^-$$

(soil solid) (in soil solution) (soil solid) (in soil solution)

Just as in cation exchange, *equivalent* quantities of NO_3^- and Cl^- are exchanged, the reaction can be reversed, and plant nutrients can be released for plant adsorption.

Although simple reactions such as this are common, note that the adsorption and exchange of some anions (including phosphates, molybdates, and sulfates) are somewhat more complex. The complexity is owing to specific reactions between the anions and soil constituents. For example, the $H_2PO_4^-$ ion may react with the protonated hydroxy group rather than remain as an easily exchanged anion.

$$\diagdown\!\!Al - OH_2^+ \ + \ H_2PO_4^- \ \longrightarrow \ \diagdown\!\!Al - H_2PO_4 \ + \ H_2O$$

(soil solid) (in soil solution) (soil solid) (soil solution)

This reaction actually reduces the net positive charge on the soil colloid. Also, the H_2PO_4 is held very tightly by the soil solids and is not readily available for plant uptake.

Despite these complexities, anion exchange is an important mechanism for interactions in the soil and between the soil and plants. Together with cation exchange it largely determines the ability of soils to hold nutrients in a form that is accessible to plants.

8.15 PHYSICAL PROPERTIES OF COLLOIDS

Soil colloids differ widely in their physical properties, including plasticity, cohesion, swelling, shrinkage, dispersion, and flocculation. These properties greatly influence the usefulness of soils for both agricultural and nonagricultural purposes.

The effects of colloids on soil physical properties were discussed in Chapter 4. But having become better acquainted with the structural framework of each of these col-

FIGURE 8.17   The different swelling tendencies of two types of clay are illustrated in the lower left. All four cylinders initially contained dry, sieved clay soil, the two on the left from the B horizon of a soil high in kaolinite, the two on the right from one of a soil high in montmorillonite. An equal amount of water was added to the two center cylinders. The kaolinitic soil settled a bit and was not able to absorb all the water. The montmorillonitic soil swelled about 25% in volume and absorbed nearly all the added water. The scenes to the right and above show a practical application of knowledge about these clay properties. (Upper) Soils containing large quantities of smectite undergo pronounced volume changes as the clay swells and shrinks with wetting and drying. Such soils (e.g., the California Vertisol shown here) make very poor building sites. The normal-appearing homes (upper) are actually built on deep, reinforced-concrete pilings (lower right) that rest on nonexpansive substrata. Construction of the 15 to 25 such pilings needed for each home more than doubles the cost of construction. (Photo courtesy of R. Weil)

loids, it is easier to appreciate how these materials affect our lives. Figure 8.17 gives an example of steps that must be taken in the building construction business to overcome the adverse effects of the shrinking and swelling properties of the soil colloid, smectite. Fortunately, at alternate sites, other colloids are more user friendly and sound housing footings are possible. If preventative measures are not taken before the houses are constructed on smectite clays, homeowners will likely pay dearly in the future. This is just one example of the critical role soil colloids play in determining the usefulness of our soils.

8.16   CONCLUSION

Colloidal materials control most of the chemical and physical properties of soils. These materials include five major classes of crystalline silicate clays, as well as the hydrous oxides of iron and aluminum, amorphous aluminosilicates such as allophane, and the ever-present humus. Due to their extremely small particle sizes these colloids possess surface areas far in excess of those of sand and silt. The platelike structure of some silicate clays contains internal surfaces that are much like the pages of a book. The total internal surface areas for some of these clays (e.g., smectite) far exceeds the area of their external surfaces.

The mineralogical and chemical constitutions of these colloids result in electrostatic charges on or near the surfaces of the colloidal particles. Negative charges predominate on most silicate clays, but positive charges characterize some colloids such as the hydrous oxides of Fe and Al, especially if the pH is low. The charges on colloids permit them to attract ions and other substances with opposite charges: cations, for example, being attracted to negatively charged sites and anions to the positive ones.

At the charge sites on soil colloids, exchanges between one ion or substance and others can take place. Cation and anion exchanges are important processes that enable plants to exchange H^+ ions for nutrient ions such as Ca^{2+} and K^+, and OH^- ions for SO_4^-. The capacity of a colloid to hold cations and to exchange them, known as the *cation exchange capacity,* is a measure of a soil's ability to hold nutrients in reserve for ultimate withdrawal (through exchange) by plant roots. Cation exchange joins photosynthesis as a fundamental life-supporting process. Without this property of soils terrestrial ecosystems would not be able to retain sufficient nutrients to support natural or introduced vegetation, especially in the event of such disturbances as timber harvest, fire, or cultivation.

REFERENCES

Buseck, P. R. 1983. "Electron microscopy of minerals," *Amer. Scientist,* **71**:175–185.

Coleman, N. T., and A. Mehlich. 1957. "The chemistry of soil pH," *The Yearbook of Agriculture (Soil)* (Washington, D.C.: U.S. Department of Agriculture).

Dixon, J. B., and S. B. Weed. 1989. *Minerals in Soil Environments* (2d ed.) Madison, Wis.: Soil Sci. Soc. Amer.

Helling, C. S., et al. 1964. "Contribution of organic matter and clay to soil cation exchange capacity as affected by the pH of the saturated solution," *Soil Sci. Soc. Amer. Proc.,* **28**:517–20.

Holmgren, G. G. S., M. W. Meyer, R. L. Chaney, and R. B. Daniels. 1993. "Cadmium, lead, zinc, copper and nickel in agricultural soils in the United States of America," *J. Environ. Qual.,* **22**:335–348.

Parker, J. C., D. F. Amos, and L. W. Zelanzny. 1982. "Water adsorption and swelling of clay minerals in soil systems," *Soil Sci. Soc. Amer. J.,* **46**:450–456.

Van Raij, B., and M. Peech. 1972. "Electrochemical properties of some Oxisols and Alfisols of the tropics," *Soil Sci. Soc. Amer. Proc.,* **36**:587–593.

9

SOIL REACTION: ACIDITY AND ALKALINITY

What have they done to the rain?
—PETER, PAUL, AND MARY

Just like adding salt to soup, it is easier to add a little more (lime to your soil) later, than to try and take out any excess.
—ENCYCLOPEDIA OF ORGANIC GARDENING

The degree of acidity or alkalinity (i.e., the *soil reaction*) is considered a *master variable* that affects nearly all soil properties—chemical, physical, and biological. Expressed as the soil pH, this variable is a major factor in determining which trees, shrubs, or grasses will dominate the landscape under natural conditions. It certainly affects which cultivated crop will grow well or even grow at all in a given field site.

Soil reaction influences aggregate stability as well as air and water movement. It determines the fate of many soil pollutants, affecting their breakdown and possible movement through the soil and into the groundwater. Finally, soil reaction is a major controller of plant nutrient availability and of microbial reaction in soils.

Many human activities can influence soil reaction. For example, certain chemical fertilizers and organic wastes react in the soil to form strong inorganic acids such as HNO_3 and H_2SO_4. These lead to increased soil acidity. These same two acids are found in **acid rain**, which originates from gases emitted into the atmosphere primarily by the combustion of fossil fuel (power plants, automobiles, etc.) and the burning of trees and other biomass. Concerns are increasing about the damage the acid rain may cause to soils, crops, and lakes. As soils become excessively acidic, cationic plant nutrients are depleted and toxic quantities of aluminum and other metals are mobilized. Certain forest trees are weakened and eventually killed. The productivity of some lakes receiving water from the affected areas has been decimated.

Nature also tends to stimulate either acidity or alkalinity in soils. In humid regions, soils tend to be quite acid—too acid in some cases for healthy growth of most crops and many ornamental plants. The acidity develops because there is sufficient rainfall to leach out much of the base-forming cations (Ca^{2+}, Mg^{2+}, K^+, and Na^+) leaving the colloidal complex dominated by H^+ and Al^{3+} ions.

In low-rainfall areas the opposite is true. Leaching is not very intense, and the base-forming cations are left to dominate the exchange complex in the place of Al^{3+} and H^+. As we noted in Section 8.12 this leads to neutrality or even alkalinity.

Luckily, both soil acidity and alkalinity are subject to relatively simple human controls. Materials such as limestone (to reduce acidity) and sulfur (to reduce alkalinity) can be used as soil amendments where they are economically available. But in many areas of the world even these remedies are unavailable or they are too expensive. This makes it imperative that we learn as much as we can about soil reaction and how it can be controlled or accommodated in sustainable agriculture systems. We shall give greater emphasis to soil acidity in this chapter, and will cover alkaline and salt-affected soils and their management in Chapter 10.

271

9.1 SOURCES OF HYDROGEN AND HYDROXIDE IONS[1]

Before we consider sources of H⁺ and OH⁻ ions in nature, reference to the chart in Figure 9.1 will remind us of the pH values of some substances we use daily. These are contrasted with the pH of different acid and alkaline soils to which we will be referring in this and the following chapter.

Two adsorbed cations—hydrogen and aluminum—are largely responsible for soil acidity. The mechanisms by which these two cations exert their influence depends on the degree of soil acidity and on the source and nature of the soil colloids.

Strongly Acid Soils

Under very acid soil conditions (pH less than 5.0), much aluminum[2] becomes soluble (Figure 9.2) and is either tightly bound by organic matter or is present in the form of aluminum or aluminum hydroxy cations. These exchangeable ions are adsorbed in preference to other cations by the negative charges of soil colloids, which at low pH values are dominantly permanent charges associated with 2:1-type silicate clays.

The adsorbed aluminum is in equilibrium with aluminum ions in the soil solution, and aluminum ions contribute to soil acidity through their tendency to hydrolyze. Two simplified reactions illustrate how adsorbed aluminum can increase acidity in the soil solution.

---

[1]In water hydrogen ions are always hydrated, existing as hydronium ions (H_3O^+) rather than as simple H⁺ ions. However, for simplicity this text will use the unhydrated H⁺ to represent this ion.

[2]Iron (Fe^{3+}) also is solubilized under acid conditions and forms hydroxy cations just as does aluminum. However, since iron is solubilized only in extremely acid soils, only the aluminum involvement will be considered.

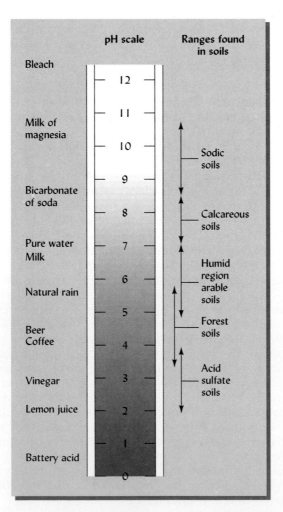

FIGURE 9.1 Chart showing a range of pH from 1 to 12 and the approximate pH of products commonly used in our society every day (left). Comparable pH ranges are shown (right) for soils we will be studying in this text.

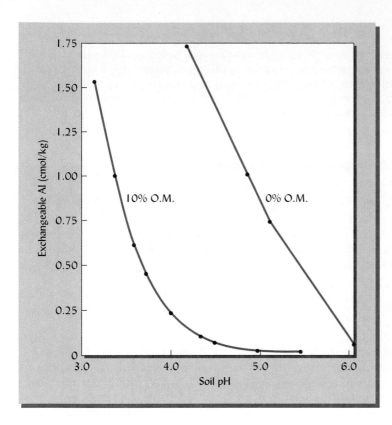

FIGURE 9.2 The effect of soil pH on exchangeable aluminum in a soil–sand mixture (0% O.M.) and a soil–peat mixture (10% O.M.). Apparently, the organic matter binds the aluminum in an unexchangeable form. This organic matter–aluminum interaction helps account for better plant growth at low pH values on soils high in organic matter. [Modified from Hargrove and Thomas (1981); used with permission of American Society of Agronomy]

$$\boxed{\text{Micelle}}\;Al^{3+} \;\rightleftharpoons\; Al^{3+}$$

Adsorbed   Aluminum ion
aluminum   (in soil solution)

The aluminum[3] ions in the soil solution are then hydrolyzed as follows:

$$Al^{3+} + H_2O \;\rightleftharpoons\; Al(OH)^{2+} + H^+$$

The H^+ ions thus released lower the pH of the soil solution and are the major source of hydrogen in most very acid soils.

Adsorbed hydrogen ions are a second but much more limited source of H^+ ions in very acid soils. Much of the hydrogen in very acid soils, along with some iron and aluminum, is bound so tightly by covalent bonds in the organic matter and on clay crystal edges that it contributes only modestly to the soil solution acidity (see Section 8.9). Only on the strong acid groups of humus and some of the permanent charge exchange sites of the clays are the H^+ ions held in an exchangeable form. These H^+ ions are in equilibrium with the soil solution. A simple equation to show the release of adsorbed hydrogen to the soil solution is

$$\boxed{\text{Micelle}}\;H^+ \;\rightleftharpoons\; H^+$$

Adsorbed   Hydrogen ion
hydrogen   (in soil solution)

Thus, it can be seen that the effect of both adsorbed hydrogen and aluminum ions is to increase the H^+ ion concentration in the soil solution.

---

[3]The Al^{3+} ion is actually highly hydrated, being present in forms such as $Al(H_2O)_6^{3+}$. For simplicity, however, it will be shown as the simple Al^{3+} ion.

Moderately Acid Soils

Aluminum and hydrogen compounds also account for soil solution H^+ ions in moderately acid soils, with pH values between 5.0 and 6.5, but by different mechanisms. Moderately acid soils have somewhat higher percentage base saturations than the strongly acid soils. The aluminum can no longer exist as Al^{3+} ions but is converted to aluminum hydroxy ions[4] by reactions such as

$$Al^{3+} + OH^- \longrightarrow Al(OH)^{2+}$$
$$Al(OH)^{2+} + OH^- \longrightarrow Al(OH)_2^+$$

Aluminum hydroxy ions

Some of the aluminum hydroxy ions are adsorbed and act as exchangeable cations. As such, they may be in equilibrium with the soil solution just as is the Al^{3+} ion in very acid soils. In the soil solution they produce hydrogen ions by the following hydrolysis reactions, using again as examples the simplest of the aluminum hydroxy ions:

$$Al(OH)^{2+} + H_2O \longrightarrow Al(OH)_2^+ + H^+$$
$$Al(OH)_2^+ + H_2O \longrightarrow Al(OH)_3 + H^+$$

In some 2:1-type clays, particularly vermiculite, the aluminum hydroxy ions (as well as iron hydroxy ions) play another role. They move into the interlayer space of the crystal units and become very tightly adsorbed, preventing intracrystal expansion and blocking some of the exchange sites. Raising the soil pH results in the removal of these ions and the freeing up of some of the exchange sites. Thus aluminum (and iron) hydroxy ions are partly responsible for the **pH-dependent charge** of soil colloids.

In moderately acid soils, the small amount of readily exchangeable hydrogen contributes to soil acidity in the same manner as shown for very acid soils. In addition, with the rise in pH, some hydrogen atoms that at low pH were bound tenaciously through covalent bonding by the organic matter, Fe, Al oxides, and 1:1-type clays are now subject to release as H^+ ions. These hydrogen ions are associated with the pH-dependent sites previously mentioned (see Section 8.9). Their contribution to the soil solution might be illustrated as follows:

$$\boxed{Micelle}\!\begin{array}{c}H\\H\end{array} + Ca^{2+} \longrightarrow \boxed{Micelle}\; Ca^{2+} + 2H^+$$

| Bound hydrogen (not dissociated) | Calcium ion (in soil solution) | Calcium ion (exchangeable) | Hydrogen ion (in soil solution) |

Again, the colloidal control of soil solution pH has been demonstrated, as has the dominant role of the aluminum and base-forming cations.

Neutral to Alkaline Soils (pH 7 and Above)

Soils that are neutral to alkaline are not dominated by either hydrogen or aluminum ions. The permanent charge exchange sites are now occupied mostly by exchangeable Ca^{2+}, Mg^{2+}, and other base-forming cations. Both the hydrogen and aluminum hydroxy ions have been largely replaced. Most of the aluminum hydroxy ions have been converted to gibbsite by reactions such as

$$Al(OH)_2^+ + OH^- \longrightarrow Al(OH)_3$$

Gibbsite (insoluble)

More of the pH-dependent charges have become available for cation exchange, and the H^+ ions released therefrom move into the soil solution and react with OH^- ions to

---

[4]The actual aluminum hydroxy ions are much more complex than those shown. Formulas such as $[Al_6(OH)_{12}]^{6+}$ and $[Al_{10}(OH)_{22}]^{8+}$ are examples of the more complex ions.

form water. The place of hydrogen on the exchange complex is taken by Ca^{2+}, Mg^{2+}, and other base-forming cations.

Soil pH and Cation Associations

Figure 9.3 summarizes the distribution of ions in a hypothetical soil as affected by pH. It illustrates changes in soil pH associated with changes in cation dominance around soil colloids. Study it carefully, keeping in mind that for any particular soil the distribution of ions might be quite different.

The effect of pH on the distribution of Ca^{2+}, Mg^{2+}, and other base-forming cations and of H^+ and Al^{3+} ions in a muck (organic soil) and in a soil dominated by 2:1 clays is shown in Figure 9.4. Note that permanent charges dominate the exchange complex of the mineral soil, whereas the pH-dependent charges account for most of the adsorption in the muck soil. Consequently, the exchange capacity of the muck declines rapidly as the pH is lowered, whereas there is little such decline in the case of the 2:1-type clay. The effect of pH on the cation exchange capacity of kaolinite and related clays is similar to that shown by the organic soil.

Note that in Figures 9.3 and 9.4 two forms of hydrogen and aluminum are shown: that tightly held by the pH-dependent sites (*bound*) and that associated with permanent negative charges on the colloids (*exchangeable*). Only the exchangeable ions have an immediate effect on soil pH, but, as we shall see later, both forms are very much involved in determining how much lime or sulfur is needed to change soil pH (see Section 9.2).

While the factors responsible for soil acidity are far from simple, two dominant groups of elements are in control. The H^+ ions and different aluminum ions (Al^{3+}, $Al(H_2O)^{3+}$, and so on) generate acidity, and most of the other cations combat it. This simple statement is worth remembering.

Source of Hydroxide Ions

In arid and semiarid areas, base-forming cations dominate the colloidal complex of soils. These adsorbed cations enhance the OH^- ion concentration of the soil solution in two ways. First, they take the place of exchangeable hydrogen and aluminum ions that are the primary sources of H^+ ions as already described. This results in an increase in OH^- ions since there is an inverse relationship between H^+ and OH^- ions in water-based solutions.

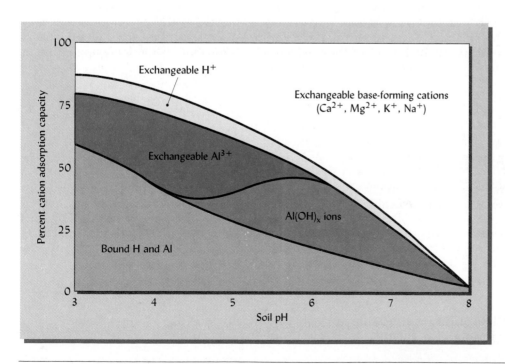

FIGURE 9.3 General relationship between soil pH and the cations held by soil colloids. Under very acid conditions, exchangeable aluminum and hydrogen ions and bound hydrogen and aluminum dominate. At higher pH values, the exchangeable bases predominate, while at intermediate values, aluminum hydroxy ions such as $Al(OH)^{2+}$ and $Al(OH)_2^+$ are prominent. This diagram is for average conditions; any particular soil would likely give a modified distribution.

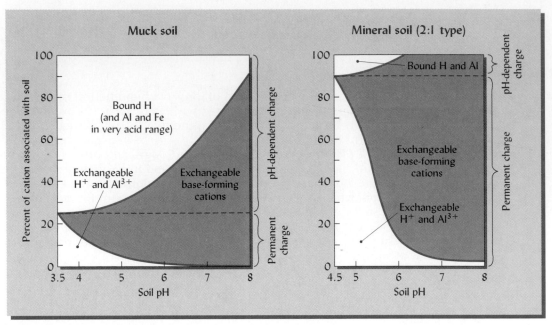

FIGURE 9.4 Relationship between pH and the association of hydrogen, aluminum, and base-forming cations with an organic (muck) soil and a mineral soil dominated by 2:1-type silicate clays. Note the dominance of the permanent charge in the mineral soil and of the pH-dependent charge in the organic soil. The CEC of the muck soil (the sum of all exchangeable ions) declined rapidly as the pH was reduced. [Redrawn from Mehlich (1964)]

The Ca^{2+}, Mg^{2+}, K^+, and Na^+ ions also have a second and more direct effect on the OH^- ion concentration in the soil solution. The hydrolysis of colloids saturated with these cations releases OH^- ions as follows:

$$Ca^{2+} \boxed{Micelle} + 2H_2O \rightleftharpoons \begin{matrix} H^+ \\ H^+ \end{matrix} \boxed{Micelle} + Ca^{2+} + 2OH^-$$

(soil solid)   (in soil solution)   (soil solid)   (in soil solution)

In a soil that is highly saturated with base-forming cations, OH^- ion levels are high. The ultimate soil pH will be determined by the relative proportions of the different base-forming cations and of H^+ and associated aluminum ions. As will be discussed in Section 9.3, these relative proportions are best expressed as the *percentage base saturation*. The presence of certain salts in the soil solution also contributes to high pH values. This is due to the hydrolysis of these salts to yield OH^- ions, as the following example illustrates:

$$Na^+ + HCO_3^- + H_2O \rightleftharpoons Na^+ + OH^- + H_2CO_3$$

9.2 CLASSIFICATION OF SOIL ACIDITY

Research suggests three pools of acidity: (1) **active acidity** due to the H^+ ion in the soil solution; (2) **salt-replaceable acidity**, involving the hydrogen and aluminum that are *easily exchangeable* by other cations in a simple unbuffered salt solution such as KCl; and (3) **residual acidity**, which can be neutralized by limestone or other alkaline materials but cannot be detected by the salt-replaceable technique. These types of acidity all add up to the **total acidity** of a soil.

Active Acidity

The active acidity is a measure of the H^+ ion activity in the soil solution at any given time. However, the quantity of H^+ ions owing to active acidity is very small compared to that in the exchangeable and residual acidity forms. For example, only about 2 kg of calcium carbonate would be required to neutralize the active acidity in the upper 15 cm of 1 hectare of an average mineral soil at pH 4 and 20% moisture. Even so, the active acidity is extremely important, since the soil solution is the environment to which plant roots and microbes are exposed.

Salt-Replaceable (Exchangeable) Acidity

This type of acidity is primarily associated with the exchangeable aluminum and hydrogen ions that are present in largest quantities in very acid soils (Figure 9.3). These ions can be released into the soil solution by an unbuffered salt such as KCl.

$$\boxed{\text{Micelle}\begin{array}{c}Al^{3+}\\H^+\end{array}} + 4KCl \rightleftharpoons \boxed{\text{Micelle}\begin{array}{c}K^+\\K^+\\K^+\\K^+\end{array}} + AlCl_3 + HCl$$

(soil solid)   (soil solution)   (soil solid)   (soil solution)

The chemical equivalent of salt-replaceable acidity in strongly acid soils is commonly thousands of times that of active acidity in the soil solution. Even in moderately acid soils, where the quantity of easily exchangeable aluminum and hydrogen is quite limited, the limestone needed to neutralize this type of acidity is commonly more than 100 times that needed for the soil solution (active acidity).

At a given pH value, exchangeable acidity is generally highest for smectites, intermediate for vermiculites, and lowest for kaolinite. In any case, however, it accounts for only a small portion of the total soil acidity, as the next section will verify.

Residual Acidity

Residual acidity is that which remains in the soil after the active and exchange acidity have been neutralized. Residual acidity is generally associated with the aluminum hydroxy ions and with hydrogen and aluminum ions that are bound in nonexchangeable forms by organic matter and silicate clays (see Figure 9.3). If lime is added to an acid soil, the pH increases and the aluminum hydroxy ions are changed to uncharged gibbsite as follows:

$$Al(OH)^{2+} \xrightarrow{OH^-} Al(OH)_2{}^+ \xrightarrow{OH^-} Al(OH)_3$$

Likewise, as the pH increases, bound hydrogen and aluminum can be released by calcium and magnesium in the lime materials [$Ca(OH)_2$ is used as an example of a reactive calcium liming material].

$$\boxed{\text{Micelle}\begin{array}{c}Al\\|\\H\end{array}} + 2Ca(OH)_2 \longrightarrow \boxed{\text{Micelle}\begin{array}{c}Ca^{2+}\\Ca^{2+}\end{array}} + Al(OH)_3 + H_2O$$

Bound H and Al   Exchangeable Ca^{2+}
(not exchangeable)

The residual acidity is commonly far greater than either the active or salt-replaceable acidity. Conservative estimates suggest the residual acidity may be 1000 times greater than the soil solution or active acidity in a sandy soil and 50,000 or even 100,000 times greater in a clayey soil high in organic matter. The amount of ground limestone recommended to at least partly neutralize residual acidity in the upper 15 cm of soil is commonly 5–10 metric tons (Mg) per hectare (2.25–4.5 tons/acre). It is obvious that the pH of the soil solution is only the tip of the iceberg in determining how much lime is needed.

9.3   COLLOIDAL CONTROL OF SOIL REACTION

The previous sections clearly illustrate the colloidal control of soil pH. This control is exerted through a number of chemical mechanisms. Below pH 5.5, aluminum ions and compounds, and their association with silicate clays, and humus provide primary control. In the pH range for the most productive agricultural soils (5.5–7.5) the control is through (1) the percentage base saturation, (2) the nature of the micelles, (3) the kind of adsorbed bases, and (4) the level of soluble salts in the soil solution. Each of these factors will be considered briefly.

Percentage Base Saturation

The percentage of the CEC that is satisfied by the base-forming cations is termed *percentage base saturation.*

$$\% \text{ base saturation} = \frac{\text{exchangeable base-forming cations (cmol}_c/\text{kg)}}{\text{CEC(cmol}_c/\text{kg)}}$$

Figure 9.5 should be helpful in illustrating this relationship. A low percentage base saturation is associated with acidity, whereas a percentage base saturation of 50–80 will result in neutrality or alkalinity. In general, soils dominated with appreciable amounts of 2:1-type **silicate clays** and **humus** have pH values below 7 if their percentage base saturation is much below 80. This figure may be 50% or even lower for soils high in 1:1-type silicate clays and hydrous oxides.

Figure 9.5 also emphasizes that pH is related more to percentage base saturation than to the absolute amounts of exchangeable base-forming cations. The clay loam soil with a relatively high CEC has more exchangeable base-forming cations at pH 5.5 than the sandy loam soil (lower CEC) at pH 6.5.

Nature of the Micelle

There is considerable variation in the pH of different types of colloids at the same percentage base saturation. This is due to differences in the ability of the colloids to furnish hydrogen ions to the soil solution. For example, the dissociation of adsorbed H^+ ions from the smectites is much higher than that from the hydrous Fe and Al oxide clays. Consequently, the pH of soils dominated by smectites is appreciably lower than that of the oxides at the same percentage base saturation. The dissociation of adsorbed hydrogen from 1:1-type silicate clays and organic matter is intermediate between that from smectites and from the hydrous oxides.

In acid soils, the interaction among the soil colloids, especially those involving iron and aluminum compounds, markedly affects the pH-base saturation relations. This relationship will receive attention when buffering is considered (see Section 9.4).

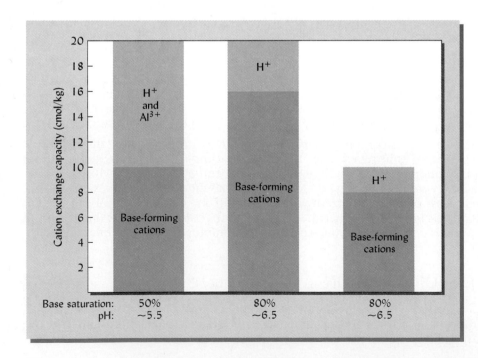

FIGURE 9.5 Three soils with percentage base saturations of 50, 80, and 80, respectively. The first is a clay loam (left); the second, the same soil satisfactorily limed; and the third (right), a sandy loam with a CEC of only 10 cmol$_c$/kg. Note especially that soil pH is correlated more or less closely with percentage base saturation. Also note that the sandy loam (right) has a higher pH than the acid clay loam (left), even though clay contains more exchangeable bases.

Kind of Adsorbed Base-Forming Cations

The comparative quantities of each of the base-forming cations present in the colloidal complex is another factor influencing soil pH. Soils with high sodium saturation have much higher pH values than those dominated by calcium and magnesium. Some highly alkaline soils of arid and semiarid regions have significant sodium saturation. These will be considered in Chapter 10.

Neutral Salts in Solution

The presence in the soil solution of neutral salts, such as the sulfates and chlorides of sodium, potassium, calcium, and magnesium, has a tendency to increase the activity of the H^+ ion in solution and, consequently, to decrease the soil pH. For example, if $CaCl_2$ were added to a slightly acid soil, the quantity of H^+ ion in the soil solution would increase.

$$\begin{array}{c} H^+ \\ H^+ \end{array} \boxed{Micelle} + Ca^{2+} + 2Cl^- \rightleftharpoons Ca^{2+} \boxed{Micelle} + 2H^+ + 2Cl^-$$

(soil solid)   (soil solution)   (soil solid)   (soil solution)

The presence of neutral salts in an alkaline soil also will result in lower pH but by a different mechanism (see Section 10.4). In alkaline soil the salts depress the hydrolysis of colloids saturated with base-forming cations. Thus, the presence of sodium chloride (NaCl) in a soil with high sodium saturation will tend to force the following reaction to move to the left.

$$Na^+ \boxed{Micelle} + H_2O \rightleftharpoons H^+ \boxed{Micelle} + Na^+ + OH^-$$

(soil solid)   (soil solution)   (soil solid)   (soil solution)

The Na^+ ion in the salt will drive the reaction to the *left*, thereby reducing the concentration of OH^- ion in the soil solution and lowering the pH. This reaction is of considerable practical importance in certain alkaline soils of arid regions because it keeps the pH from rising to levels toxic to plants.

Since the reaction of the soil solution is influenced by the four relatively independent factors just discussed, a close correlation would not always be expected between percentage base saturation and pH when comparing widely diverse soils. Yet with soils of similar origin, texture, and organic content, a reasonably good correlation does exist.

9.4 BUFFERING OF SOILS

As previously indicated, soils tend to resist changes in the pH of the soil solution. This resistance is called **buffering**. For soils with intermediate pH levels, buffering can be explained in terms of the equilibrium that exists among the active, salt-replaceable, and residual acidities (see Figure 9.6). This general relationship can be illustrated as follows:

If just enough lime (a base) is applied to neutralize the hydrogen ions in the soil solution, they are replenished as the reactions move to the right. This buffering action means that the rise in soil pH with the addition of base is negligibly small until enough lime is added to deplete appreciably hydrogen and aluminum of the exchangeable and residual acidities (see Figure 9.7).

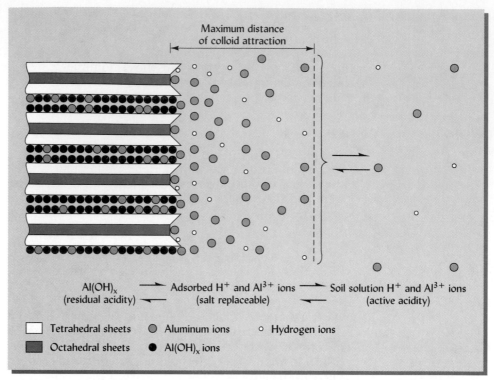

FIGURE 9.6 Equilibrium relationship among residual, salt-replaceable (exchangeable), and soil solution (active) acidity on a 2:1 colloid. Note that the adsorbed and residual ions are much more numerous than those in the soil solution even when only a small portion of the clay crystal is shown. The $Al(OH)_x$ ions are held tightly in internal spaces and are not exchangeable. Remember that the aluminum ions, by hydrolysis, also supply hydrogen ions to the soil solution. It is obvious that neutralizing only the hydrogen and aluminum ions in the soil solution will be of little consequence. They will be quickly replaced by ions associated with the colloid. The soil, therefore, demonstrates high buffering capacity.

Buffering is equally important in preventing a rapid lowering of the pH of soils. Hydrogen ions are generated as organic and inorganic acids are formed during organic decay, and a temporary increase in the hydrogen ion concentration in the soil solution occurs. In this case, the equilibrium reaction shown above would immediately shift to the left so that more hydrogen ions would become adsorbed or bound on the micelle. Again, because of buffering, the resultant drop in the pH of the soil solution would be

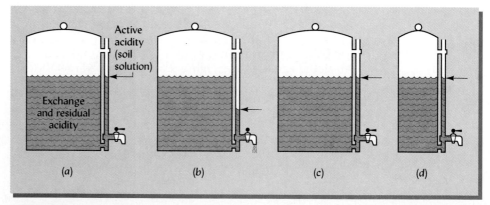

FIGURE 9.7 The buffering action of a soil can be likened to that of a coffee dispenser. (a) The active acidity, which is represented by the coffee in the indicator tube on the outside of the urn, is small in quantity. (b) When hydrogen ions are removed, this active acidity falls rapidly. (c) The active acidity is quickly restored to near the original level by movement from the exchange and residual acidity. By this process, there is considerable resistance to the change of active acidity. A second soil with the same active acidity (pH) level but a much smaller exchange and residual acidity (d) would have a lower buffering capacity.

very small. Fortunately such buffering resists reduction in soil pH from acid rain, especially in soils with medium to high cation exchange capacities.

In acid soils, equilibrium reactions involving aluminum and iron hydroxy ions are also major factors in resisting changes in soil pH. Reactions such as the following occur.

$$Al(OH)^{+2} + H_2O \rightleftharpoons Al(OH)_2^+ + H^+$$
$$Al(OH)_2^+ + H_2O \rightleftharpoons Al(OH)_3 + H^+$$

The addition of H^+ ions would drive these reactions to the left so relatively few H^+ ions would accumulate in the soil solution and the pH would be reduced only modestly. In other words, buffering would occur.

Similar buffer reactions involving carbonates, bicarbonates, carbonic acid, and water take place in alkaline soils.

$$CaCO_3 + H_2CO_3 \rightleftharpoons Ca(HCO_3)_2$$
$$Ca(HCO_3)_2 + H_2O \rightleftharpoons Ca^{2+} + 2OH^- + H_2CO_3$$

The addition of H^+ ions would shift these reactions to the right while increased OH^- ions would shift them to the left. In any case resistance to pH change (buffering) would occur. Thus, throughout the entire pH range there are mechanisms that buffer the soil solution and prevent rapid changes in soil pH.

9.5   BUFFER CAPACITY OF SOILS

The higher the cation exchange capacity of a soil, the greater is its buffer capacity, other factors being equal. This relationship exists because more reserve acidity must be neutralized to effect a given increase in the percentage base saturation. Thus, the higher the clay and organic matter contents, the more lime is required for a given change in soil pH.

Buffer Curves

Buffer capacity of soils is normally not the same throughout the percentage base saturation range. Titration curves, such as the one shown in Figure 9.8, which is based on a large number of Florida soils, illustrate this point. Note that for these soils the degree of buffering is highest between pH 4.5 and 6.5 and is relatively uniform in this range. (The buffer capacity drops off below pH 4.5 and above pH 6.5.) This means that the same amount of lime would be required to raise the pH of these soils from 5.0 to 5.5 as from 5.5 to 6.5.

The curve in Figure 9.8 is a composite for many soils and does not represent the buffer capacity for any particular soil. However, it does illustrate the practical importance of the buffering curves in estimating the amount of lime needed to bring about change in the pH of a soil, and it illustrates the primary buffering agents for low pH (aluminum compounds), for intermediate pH (cation exchange), and higher pH (carbonates).

Variation in Titration Curves

The relationship of pH to percent base saturation varies from one colloid to another (see Figure 9.9). At a given percent base saturation, the pH is normally highest in hydrous oxide clay materials, intermediate in kaolinite and humus, and lowest in 2:1-type minerals. Obviously, the type of clay definitely affects the pH base saturation percentage relationship.

As previously mentioned, aluminum and iron compounds can have major effects on the buffering of soils. For example, at low pH values, Al^{3+} and hydroxy aluminum ions react with and bind or block exchange sites in silicate clays and in humus, thereby reducing the CEC of the colloids. If the soil is then limed, these ions are removed, the CEC increases, and more calcium, magnesium, and other base-forming cations can be adsorbed. The net result is an increase in the quantity of lime needed to increase the soil pH.

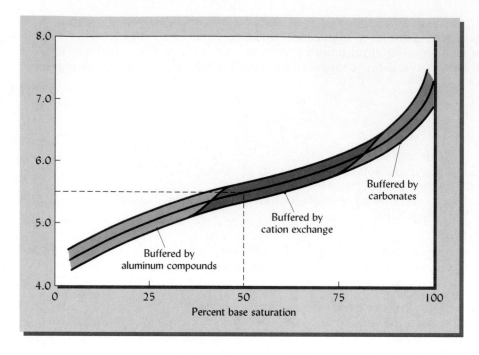

FIGURE 9.8 Theoretical titration curve based on a large number of Florida soils and the approximate pH range where three major buffering mechanisms are effective: aluminum compounds, cation exchange, and carbonate–bicarbonate interaction. The maximum buffering occurs on these soils at about 50% base saturation. [Titration curve from Peech (1941)]

Importance of Soil Buffering

Soil buffering is important for two primary reasons. First, it tends to ensure reasonable stability in the soil pH, preventing drastic fluctuations that might be detrimental to higher plants and to soil microorganisms. Resistance to the acidifying effect of acid rain on both the soil and drainage water is a good example of such a benefit. Second, it influences the amount of amendments, such as lime or sulfur, required to effect a desired change in soil pH. Buffering is indeed a significant soil property.

9.6 CHANGES IN SOIL pH

Natural Changes

The weathering and organic decay processes in nature result in the formation of both acid- and base-forming chemicals. The base-forming cations (Ca^{2+}, Mg^{2+}, etc.) are

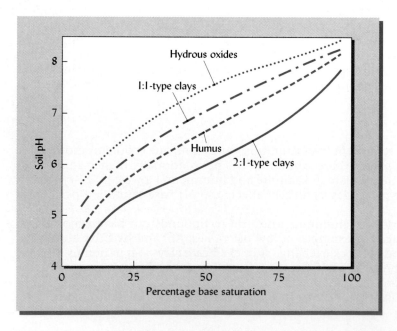

FIGURE 9.9 Theoretical titration curves of representative soil colloids. The hydrous oxides exhibit the highest pH values at a given percent base saturation; the 2:1-type clays, the lowest; and the 1:1 type and organic matter in minerals soils, in between. The reaction of Al^{3+} and Fe^{3+} ions and of Fe, Al oxides with silicate clays and organic matter markedly influence the titration curves of these colloids.

released from rocks and minerals as they break down. Hydrogen ions are generated as a net result of a complex series of reactions as organic matter decomposes:

$$[C_2H_4ONS] \; + \; 5O_2 \; + \; H_2O \; \longrightarrow \; H_2CO_3 \; + \; RCOOH \; + \; H_2SO_4 \; + \; HNO_3$$

Organic matter (generalized)   Carbonic acid   Strong organic acids   Strong inorganic acids

As these acids disassociate, they become direct sources of hydrogen ions. They also help solubilize aluminum from mineral surfaces, a key factor in the more acid soils.

In low-rainfall areas there is little leaching of the base-forming cations to lower soil horizons. Consequently, these cations are able to compete with the hydrogen and aluminum ions for the exchange sites on the colloidal complex. This results in a relatively high degree of base saturation and pH values of 7 and above.

In more humid areas, leaching depletes the upper horizons of calcium and other base-forming cations. Their places on the exchange complex are taken by hydrogen and aluminum ions generated through organic matter decomposition and other weathering processes. As a result, the percentage of base saturation declines and the pH is lowered. Figure 9.10 illustrates this effect of increasing rainfall and other forms of precipitation on soil acidity levels in both natural forested and grassland areas. Such natural changes have been going on for centuries and will likely continue to do so.

Human-Induced Changes

In recent times, these long-term natural changes have been increasingly augmented by changes stemming from human activities. Rising demands of the world's rapidly increasing human populations along with widespread industrialization have stimulated some activities that have significant impacts on soil pH. A few of these will be considered.

CHEMICAL FERTILIZER USAGE. Chemical fertilizers, while stimulating remarkable increases in global food supplies, have also brought about significant changes in soil pH. The widely used ammonium-based fertilizers such as $(NH_4)_2SO_4$ and $(NH_4)_2HPO_4$ are oxidized in the soil by microbes to produce strong inorganic acids by reactions such as the following:

$$(NH_4)_2SO_4 \; + \; 4O_2 \; \longrightarrow \; 2HNO_3 \; + \; H_2SO_4 \; + \; 2H_2O$$

These strong acids provide H^+ ions that result in lower pH values. Since worldwide fertilizer use increased tenfold between 1950 and 1990, the effects of this type of reaction are not minor. In the United States alone it would require at least 20 million metric tons of limestone to neutralize the acidity produced by nitrogen fertilizers every year.

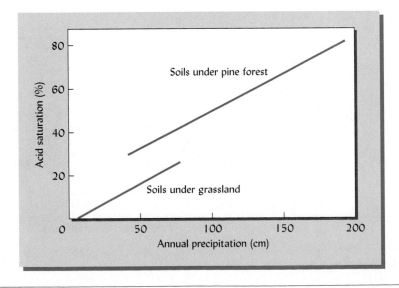

FIGURE 9.10 Effect of annual precipitation on the percent acid saturation of untilled California soils under grassland and pine forests. Note that the degree of acidity goes up as the precipitation increases. Also note that the forest produced a higher degree of soil acidity than did the grassland. [From Jenny et al. (1968)]

TILLAGE PRACTICES. These influence the pH of the upper soil horizons. In no-tillage systems, the breakdown of organic residues that are concentrated on the soil surface produces organic and inorganic acids that acidify the soil. Likewise, nitrogen fertilizer applications are made primarily to the surface layers in the no-tillage systems, thereby localizing the acidifying effect of the fertilizer. The combined effects on pH of tillage practices and fertilizer applications are shown in Figure 9.11. Keep in mind that the soil layers affected are those near the soil surface where plant root density is the highest. Fortunately, those are also the horizons most affected by surface-applied liming materials.

ACID DEPOSITION FROM THE ATMOSPHERE. Acid rain is a second significant worldwide source of nitric and sulfuric acids. Nitrogen and sulfur-containing gases are emitted into the atmosphere from the combustion of coal, gasoline, and other fossil fuels, as well as from the burning of forests and crop residues (see Figure 9.12). These gases react with water and other substances in the atmosphere to form HNO_3 and H_2SO_4 that are then returned to the earth in rainwater. This precipitation is called *acid rain* since its pH is commonly between 4.0 and 4.5 and may be as low as 2.0. Normal rainfall that is in equilibrium with atmospheric carbon dioxide has a pH of 5.0 to 6.0. A map showing the pH of acid rain in North America is shown in Figure 9.13.

The quantity of H_2SO_4 and HNO_3 brought to the earth in acid rain globally is enormous, but the amount falling on a given hectare in a year is not enough to significantly change soil pH immediately. In time, however, the cumulative effects of annual deposition can influence both the soils and the plants growing in them.

Acid rain is particularly worrisome in humid forested areas near urbanized and industrialized sites that provide the S- and N-containing pollutants.[5] Forests in the eastern part of the United States and much of Europe are affected (see Figure 9.14). Studies have shown that the leaching of base-forming cations and the mobilization of aluminum is enhanced by the presence in the soil solution of the strong acid anions (SO_4^{2+} and NO_3^-) along with the H^+ ion provided by the acid rain. Figure 9.15 shows the reduction in Ca^{2+}, Mg^{2+}, and K^+ and in pH levels that occurred over a period of 20 years in two forested watersheds that were subject to acid rain deposits. While natural processes undoubtedly account for part of such changes, it is thought that at least half were due to acid rain caused by human activities.

The leaching of calcium and the mobilization of aluminum result in Ca/Al ratios in the soil solution of less than 1.0, the ratio considered a threshold for restrained plant growth and associated nutrient uptake. Higher aluminum concentrations can lead to toxicities and to reduced uptake of calcium and other elements by trees and other plants.

---

[5]See Joslin et al. (1992) and Robarge and Johnson (1992) for discussions of chemical changes associated with forest soil acidification.

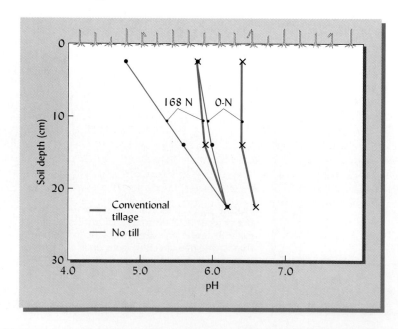

FIGURE 9.11 The effects of conventional (plow) tillage and no-tillage (no-till), and two rates of nitrogen application (0 and 168 kg/ha/yr) on the pH of the upper horizons of an Alfisol in Kentucky on which corn was grown every year for 10 years. Note the decided increase in soil acidity on the high-nitrogen plots and especially on those where no-till was used. [Drawn from data from Blevins et al. (1983); used with permission of Elsevier Science Publishing Co.]

$$SO_2 + H_2O \longrightarrow H_2SO_3 \longrightarrow H^+ + HSO_3^-$$

$$SO_2 \xrightarrow[O_2]{Sunlight} SO_3 \xrightarrow{H_2O} H_2SO_4 \rightleftharpoons 2H^+ + SO_4^{2-}$$

$$2N_2O + O_2 \longrightarrow 4NO \xrightarrow{2O_2} 4NO_2 \xrightarrow{2H_2O} 2HNO_3 + 2HNO_2$$

$$HNO_3 \longrightarrow H^+ + NO_3^-$$

FIGURE 9.12 Illustration of the formation of nitrogen and sulfur oxides from the combustion of fuel in sulfide ore processing and from motor vehicles. The further oxidation of these gases and their reaction with water to form sulfuric acid and nitric acid are shown. These help acidify rainwater, which falls on the soil as acid rain. NO_x indicates a mixture of nitrogen oxides, primarily N_2O and NO. [Modified from NRC (1983)]

Base depletion and acidification of humid forest soils is of vital concern. These soils are already quite acid. Further intensification of the acidity could result in aluminum toxicity and micronutrient imbalances. Acid rain is suspected as being one of the causes of the decline of some forest areas in Europe and the United States. Unfortunately it is impractical to try to replenish the base-forming cations in forest soils because of the expense and difficulty of adding liming materials to areas with standing trees.

DISPOSAL OF ACID-FORMING ORGANIC WASTES. Deposition of wastes such as sewage sludge on agricultural and forested lands is another human activity that affects soil reaction. A common consequence of this practice is soil acidification. Organic and inorganic acids are formed as the large quantities of sludge are decomposed and soil pH is lowered. Figure 9.16 provides an example of this effect and reminds us that waste disposal can have some troublesome effects. Fortunately government regulations are in place in some countries to ensure that these wastes are treated with lime before they are applied. In fact, some *lime-stabilized* sludges may have a final pH of 7.5–8.5, making them useful in combating soil acidity. Control of soil pH after amendment with sewage sludge is also regulated with the objective of minimizing the mobility of toxic metals found in some sewage sludges (see Section 18.8).

IRRIGATION PRACTICES. In arid and semiarid regions, irrigation can result in the buildup of *excess salts* that stimulate undesirable increases in soil pH as well as soil salinity. Irrigation waters bring salts from upstream watersheds and irrigation sites into the downstream irrigated areas. When the water evaporates, the salts are left behind and will accumulate in the surface horizon if the internal drainage is not rapid enough to permit their being leached to lower horizons in the soil profile. Included are salts containing the Na^+ ion that not only can increase the soil pH to undesirable levels, but can destabilize the soil structure as well. This subject will receive more detailed attention in Chapter 10.

DRAINAGE OF SOME COASTAL WETLANDS. This practice has resulted in the formation of so-called **acid sulfate** soils. Among the areas so affected are some coastal areas of Southeast United States and comparable areas of some countries in Southeast Asia and West

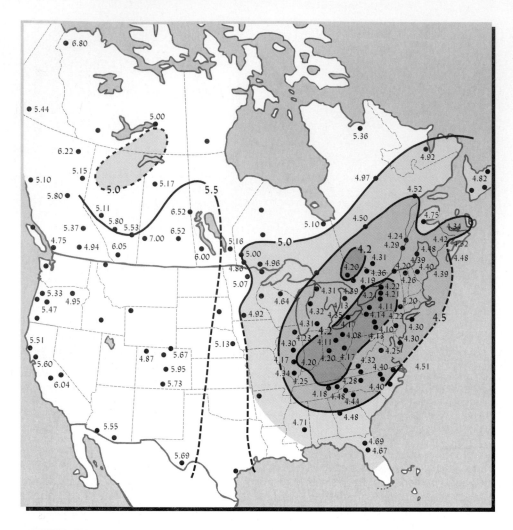

FIGURE 9.13 Annual mean value of pH in precipitation weighted by the amount of precipitation in the United States and Canada for 1980. [From U.S./Canada Work Group No. 2 (1982)]

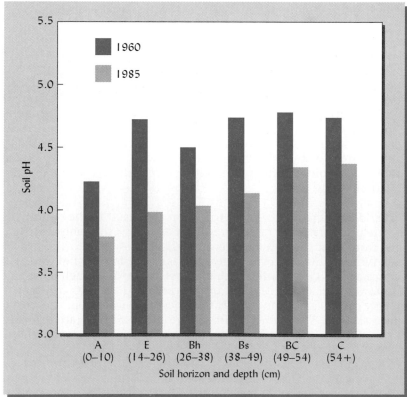

FIGURE 9.14 Reduction in soil pH of 67 forested Spodosol sites in northern Belgium thought to be due to 25 years of atmospheric deposition as well as natural acidification. [From Ronse et al. (1988); used with permission of Williams and Wilkins, Baltimore, Md.]

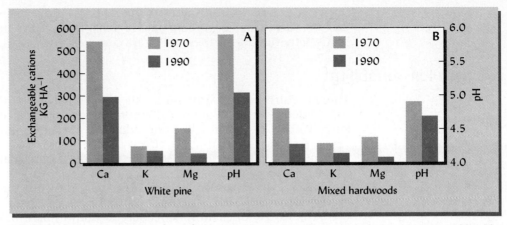

FIGURE 9.15 Exchangeable Ca^{2+}, Mg^{2+}, and K^+ soil pH declined appreciably from 1970 to 1990 in the A horizon of soils of two forested watersheds in the mountains of North Carolina, one (a) producing white pine, the other mixed hardwood trees (b). Such reductions have been ascribed to cation uptake by the tree biomass and to leaching of the Ca^{2+}, Mg^{2+}, and K^+ that is enhanced by acid rain. [Modified from Knoepp and Swank (1994); used with permission of the Soil Science Society of America]

Africa. The soils in these areas contain large quantities of iron sulfide (FeS) and elemental sulfur (S) that have been formed by the microbial reduction of sulfates. When these areas are drained, the FeS and S are oxidized, ultimately resulting in the formation of sulfuric acid, which accounts for their being called *acid sulfate* soils.

$$4FeS + 9O_2 + 4H_2O \longrightarrow 2Fe_2O_3 + 4H_2SO_4$$

$$2S + 3O_2 + 2H_2O \longrightarrow 2H_2SO_4$$

Under these conditions, the soil acidity is extreme, with pH values as low as 1.5 having been noted. The amount of $CaCO_3$ required to neutralize this acidity is enormous and generally uneconomical. The soils would best be returned to their undrained conditions, not only for agricultural and economic reasons, but to maintain these areas as natural habitats for wildlife and natural flora.

Minor Fluctuations in Soil pH

Minor fluctuations in soil pH result from the movement of salts into and out of different soil zones as soil moisture moves up and down through the profile. Similarly, the pH of mineral soils declines during the crop-growing season as a result of the acids pro-

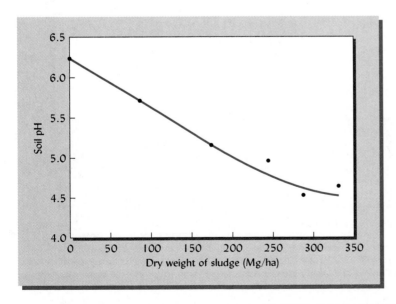

FIGURE 9.16 Effect of large applications of sewage sludge over a period of six years on the pH of a Paleudult (Orangeburg fine sandy loam). The reduction was due to the organic and inorganic acids, such as HNO_3, formed during decomposition and oxidation of the organic matter. [Data from Robertson et al. (1982)]

duced by microorganisms and by the roots of higher plants. When soil temperatures decline, as in the fall in temperate regions, an increase in pH often is noted because biotic activities during these times are considerably slower.

Hydrogen Ion Variability

There is considerable variation in the pH of the soil solution in adjacent parts of the soil. Differences in pH are noted among sites in the soil only a few centimeters apart. Such variation may result from microbial decay of unevenly distributed organic residues in the soil or from the effects of plant roots (see Section 11.7).

The variability of the soil solution is important in many respects. For example, it affords microorganisms and plant roots a great variety of solution environments. Organisms unfavorably influenced by a given hydrogen ion concentration may find, at an infinitesimally short distance away, a different environment that is more satisfactory. The variety of environments may account in part for the great diversity in microbial species present in normal soils.

9.7   SOIL REACTION: CORRELATIONS

Information on soil pH is very useful to the soil manager because pH affects such a wide variety of chemical and biological phenomenon in soils.

Soil pH significantly affects other soil chemical properties, as well as biological activity. Figure 9.17 shows one example: the effect of soil acidity (and aluminum toxicity) on the growth of cotton roots in a mineral soil. Only a few other examples will be considered here to illustrate this point further.

Nutrient Availability and Microbial Activity

Figure 9.18 shows in general terms the relationship between soil pH and the availability of plant nutrients, as well as the activities of soil microorganisms. Note that in strongly acid soils the availability of the macronutrients (Ca, Mg, K, P, N, and S) as well as molybdenum and boron is curtailed. In contrast, the availability of micronutrient cations (Fe, Mn, Zn, Cu, and Co) is increased by low soil pH, even to the extent of toxicity to higher plants and/or microorganisms.

In slightly to moderately alkaline soils, molybdenum and all the macronutrients (except phosphorus) are amply available, but levels of available Fe, Mn, Zu, Cu, and Co

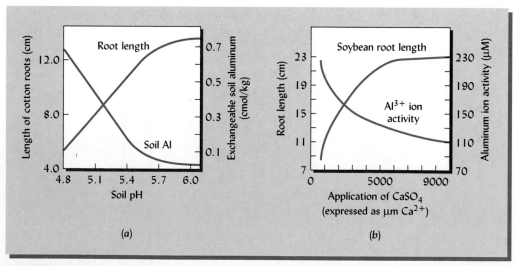

FIGURE 9.17   High concentrations of exchangeable or soil solution aluminum in acid soils are toxic to plant roots, a toxicity that can be reduced somewhat by adding $CaSO_4$. (a) Cotton root length is increased as soil pH is increased and exchangeable aluminum is decreased. (b) Adding $CaSO_4$ to a nutrient solution (pH = 2) reduces the Al^{3+} activity and permits an increase in soybean root length while having minimal effect on the solution pH. [(a) From Adams and Lund (1966); (b) drawn from data in Nobel et al. (1988)]

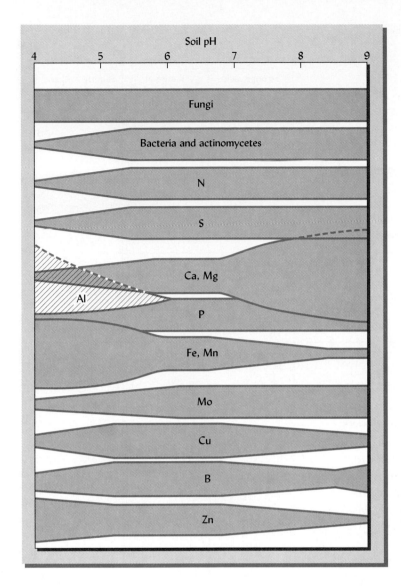

FIGURE 9.18 Relationships existing in mineral soils between pH on the one hand and the activity of microorganisms and the availability of plant nutrients on the other. The wide portions of the bands indicate the zones of greatest microbial activity and the most ready availability of nutrients. When the correlations as a whole are taken into consideration, a pH range of about 5.5–7.0 seems to promote best the availability of plant nutrients. In short, if soil pH is suitably adjusted for phosphorus, the other plant nutrients, if present in adequate amounts, will be satisfactorily available in most cases.

are so low that plant growth is constrained. Phosphorus and boron availability is likewise reduced in alkaline soil—commonly to a deficiency level.

It is difficult to generalize about nutrient/pH relationships. However, it appears from Figure 9.18 that the pH range of 5.5 to 6.5 or perhaps 7.0 may provide the most overall satisfactory plant nutrient levels. However, this generalization may not be valid for all soil and plant combinations. For example, certain micronutrient deficiencies are common on some sandy soils when the soils are limed to pH values of only 5.5 to 6.0.

Figure 9.18 suggests that most general-purpose bacteria and actinomycetes function well at intermediate and high pH values. Fungi seem to be particularly versatile, flourishing satisfactorily over a wide pH range. Therefore, fungal activity tends to predominate in most acid soils, whereas at intermediate and higher pH they meet stiff competition from actinomycetes and bacteria.

Higher Plants and pH

Plants vary considerably in their tolerance to acid and/or alkaline conditions (see Figure 9.19). For example, forage legume crops such as alfalfa and sweet clover grow best in near-neutral or alkaline soils, and most humid region mineral soils must be limed to grow these crops satisfactorily. Generally, plants that originally evolved in relatively dry regions where soils are near neutral or alkaline have less tolerance for soil acidity than plants originating in humid regions where soils are acidic.

Rhododendrons and azaleas are at the other end of the scale. They apparently require a considerable amount of iron, which is abundantly available only at low pH

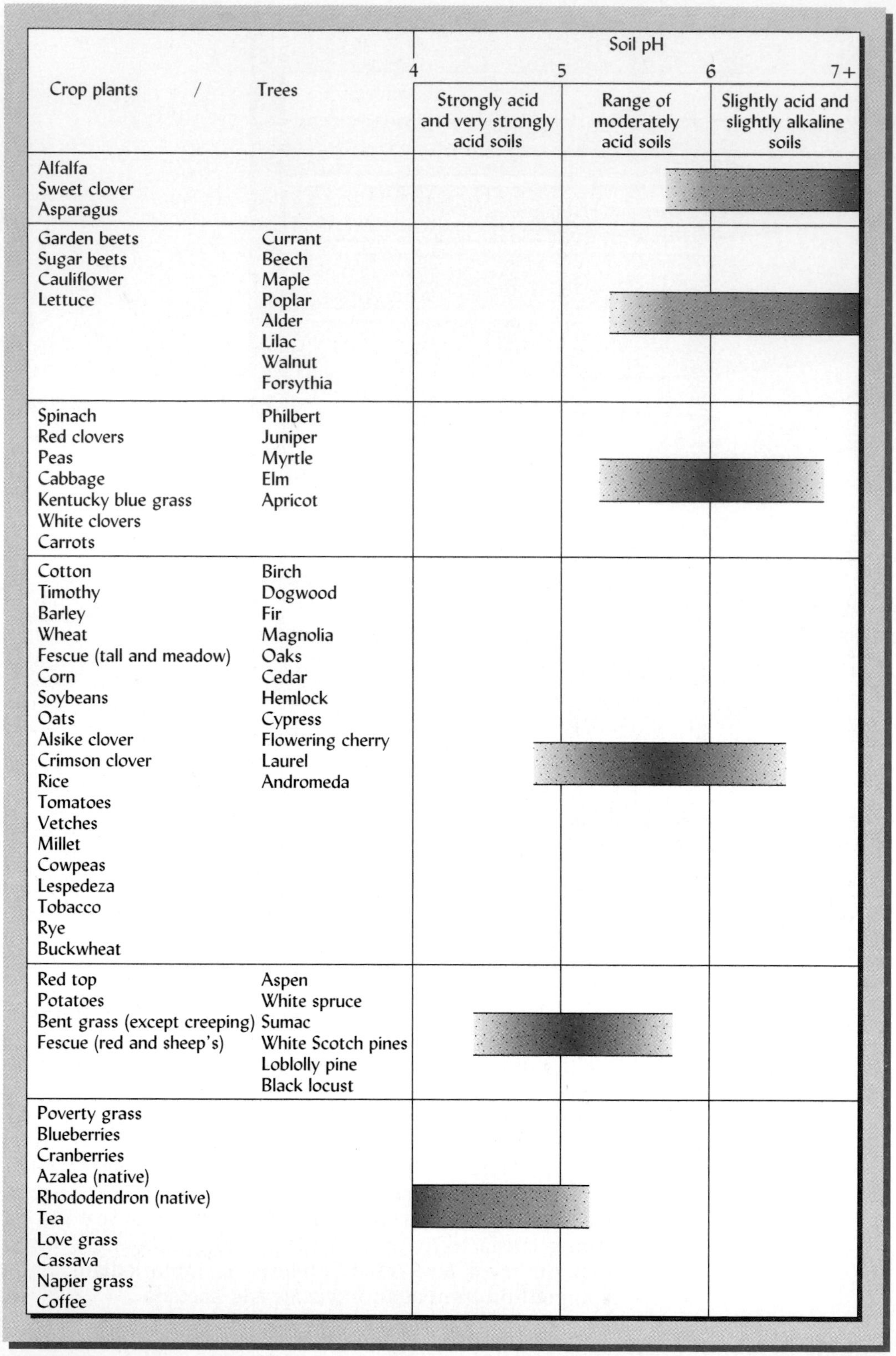

FIGURE 9.19 Relation of higher plants to the physiological conditions presented by mineral soils of different reactions. Note that the correlations are very broad and are based on pH ranges. The fertility level will have much to do with the actual relationship in any specific case. Such a chart is of great value in deciding whether or not to apply chemicals such as lime or sulfur to change the soil pH.

The figure shows a chart with the following structure:

Crop plants	Trees	Soil pH: 4 — 5 — 6 — 7+
		Strongly acid and very strongly acid soils / Range of moderately acid soils / Slightly acid and slightly alkaline soils
Alfalfa, Sweet clover, Asparagus		(band in slightly acid and slightly alkaline soils, ~6–7+)
Garden beets, Sugar beets, Cauliflower, Lettuce	Currant, Beech, Maple, Poplar, Alder, Lilac, Walnut, Forsythia	(band ~5.5–7+)
Spinach, Red clovers, Peas, Cabbage, Kentucky blue grass, White clovers, Carrots	Philbert, Juniper, Myrtle, Elm, Apricot	(band ~5.5–6.5)
Cotton, Timothy, Barley, Wheat, Fescue (tall and meadow), Corn, Soybeans, Oats, Alsike clover, Crimson clover, Rice, Tomatoes, Vetches, Millet, Cowpeas, Lespedeza, Tobacco, Rye, Buckwheat	Birch, Dogwood, Fir, Magnolia, Oaks, Cedar, Hemlock, Cypress, Flowering cherry, Laurel, Andromeda	(band ~5–6)
Red top, Potatoes, Bent grass (except creeping), Fescue (red and sheep's)	Aspen, White spruce, Sumac, White Scotch pines, Loblolly pine, Black locust	(band ~4.5–5.5)
Poverty grass, Blueberries, Cranberries, Azalea (native), Rhododendron (native), Tea, Love grass, Cassava, Napier grass, Coffee		(band ~4–5)

values. If the pH and the percentage base saturation are not low enough, these plants will show chlorosis (yellowing of the leaves) and other symptoms indicative of iron deficiency (see Figure 15.8).

Most forest trees seem to grow well over a wide range of soil pH levels, which indicates they have some tolerance to soil acidity. This is to be expected since forest trees (especially conifers) generally tend to enhance soil acidity, and forests exist in humid regions where acid soils are common. There are some differences, however, among tree species in their tolerance of high soil acidity and aluminum. For example, the pines are generally very tolerant to highly acid soils. The spruces are somewhat less tolerant, but more so than most hardwoods (e.g., maples, oaks, and beeches). The red spruce and honey locust are known to be less tolerant to acid soils than are most forest species. Plants that grow poorly in very acid soils (pH < 5) are generally affected by *aluminum toxicity*, which causes plant roots to become short, thick, and stubby.

The most productive arable soils in use today have intermediate pH values, being not too acidic and not too alkaline. Most cultivated crop plants grow well on these soils. Since pasture grasses, many legumes, small grains, intertilled field crops, and a large number of vegetables are included in this broadly tolerant group, mild soil acidity or alkalinity is not a deterrent to their growth. In terms of pH, a range of perhaps 5.5 to 7.0 is most suitable for most crop plants.

Soil pH and Environmental Quality

There are many examples of the influence of soil pH on environmental quality, but only one will be cited—the influence of pH on groundwater contamination by herbicides. Chemical groupings such as $-NH_2^+$ and $-COOH$ on some herbicide molecules encourage reactions with H^+ and/or OH^- ions associated with soil colloids. These reactions may bind the herbicide molecule to the organic matter or clay in the soil, or they may influence its surface charge and consequently its ability to exchange with anions or cations on the soil colloid. An example of such adsorption is that of Atrazine, an herbicide that is widely used on corn. If the soil is quite acid (pH less than about 5.7) protonation or the attraction of a proton (the H^+ ion) takes place, and the herbicide molecule becomes positively charged. The molecule is then attracted to the negatively charged soil colloids. In this condition Atrazine is held by the soil until it can be decomposed by soil organisms and is less likely to move downward and into the groundwater. The strong adsorption of the herbicide molecule also renders the chemical ineffective in killing weeds. At pH values above 5.7, the adsorption is greatly reduced and the tendency for the herbicide to move downward in the soil and into the drainage water is increased. Once again we are reminded of the universal importance of soil pH.

9.8   DETERMINATION OF SOIL pH

Soil pH is tested routinely, and the determination is easy and rapid to make. Soil samples are collected in the field and the pH is measured directly, or the samples are brought to the laboratory for more accurate pH determination.

Electrometric Method

The most accurate method of determining soil pH is with a pH meter (Figure 9.20). In this method a sensing glass electrode (referenced with a standard calomel electrode) is inserted into a soil–water mixture that simulates the soil solution. The difference between the H^+ ion activities in the wet soil and in the glass electrode gives rise to an electrometric potential difference that is related to the soil solution pH. The instrument gives very consistent results and its operation is simple.

Most soil testing laboratories in the United States measure the pH of a suspension of soil in water (usually a ratio of 1:1 or 1:2). The measurement also can be made in a similar suspension of soil in a dilute, unbuffered 0.01 molar $CaCl_2$ solution. The latter method eliminates the effects of small differences in electrolyte concentrations that may be caused by upward salt movement in soils. The pH of $CaCl_2$ suspensions are usually about 0.2 to 0.4 pH unit lower than the measurements with water.

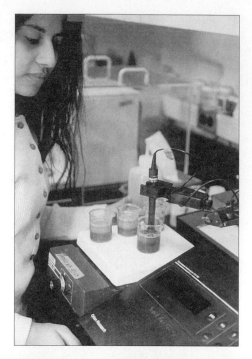

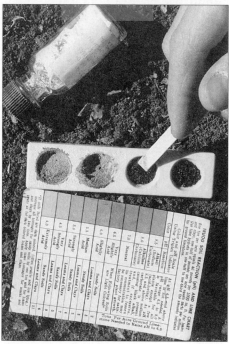

FIGURE 9.20 (Upper) Soil pH is measured quickly and inexpensively in the laboratory using an electrometric pH meter. Hundreds of analyses can be made in a day. (Lower) In the field, advantage is taken of the fact that pH-sensitive dyes give shades of color that can be calibrated against standard color charts, providing an equally simple and inexpensive means of measuring pH. Although the dye method is not quite as accurate as the electrometric method, it is quite satisfactory for most field management purposes. (Photos courtesy of R. Weil)

Dye Methods

Dye methods take advantage of the fact that certain organic compounds change color as the pH is increased or decreased. Mixtures of such dyes provide significant color changes over a wide pH range (3–8). A few drops of the dye solutions are placed in contact with the soil, usually on a white spot plate (see Figure 9.20). After standing a few minutes, the color of the dye is compared to a color chart that indicates the approximate pH.

In other dye methods, porous strips of paper are impregnated with the dye or dyes. When brought in contact with a mixture of water and soil, the paper absorbs the water and by color change indicates the pH. Such dye methods are accurate within about 0.2 to 0.5 pH unit.

Limitations of pH Values

Laboratory pH measurements can be made rather precisely. However, the interpretation of this measurement must be applied in the field with some caution. First, there is considerable variation in the pH from one spot to another in a given field. Localized effects of fertilizers may give sizable pH variations within the space of a few inches. Also, interpretations with respect to the management of a whole field, involving several million kilograms of soil, are made on the basis of the pH of a mere 10-g sample of soil. Obviously, care must be taken to minimize sampling errors (see Section 16.17).

Despite these limitations, soil pH is a very important characteristic and it is easily measured. It is correlated with many chemical and biological factors that influence both soils and plants. Thus, more may be inferred regarding the chemical and biological conditions in a soil from the pH value than from any other single measurement.

9.9 METHODS OF INTENSIFYING SOIL ACIDITY

There are some acid-loving plants that cannot tolerate moderate to high pH values. Among them are a few highly valued ornamental species that are the favorites of gardeners around the world. Fortunately, there are soil amendments and practices that can help reduce the pH of soils to levels even lower than those stimulated by nature. We will focus on only a few species that are found in acid humid region soils, but the principles relating to the reduction in soil pH are valid at all pH ranges.

Among the plants to receive our attention are rhododendrons and azaleas that grow best on soils at pH of 5.0 and below. To accommodate these plants and to control some soilborne diseases, it is sometimes necessary to increase the acidity of even acid soils. This is done by adding acid-forming organic and inorganic materials.

Acid Organic Matter

As organic matter decomposes, organic and inorganic acids are formed that can reduce the soil pH, especially if the organic matter is low in base-forming cations. Leaf mold, pine needles, tanbark, sawdust, and acid moss peat are quite satisfactory organic materials to add around ornamental plants. Farm manures, particularly poultry manures, may be alkaline, however, and should be used with caution.

Inorganic Chemicals

When the addition of acid organic matter is not feasible, inorganic chemicals may be used. For rhododendrons, azaleas, and other plants that require considerable iron (such as blueberries and cranberries), ferrous sulfate sometimes is recommended. This chemical provides available iron (Fe^{2+}) for the plant and, upon hydrolysis, enhances acidity by reactions such as the following:

$$Fe^{2+} + H_2O \rightleftharpoons Fe(OH)_2 + 2H^+$$

$$4Fe^{2+} + 6H_2O + O_2 \rightleftharpoons 4Fe(OH)_2^+ + 4H^+$$

The H^+ ions released lower the pH locally and liberate some of the iron already present in the soil. Ferrous sulfate thus serves a double purpose by supplying available iron directly and by reducing the soil pH—a process that may cause a release of fixed iron present in the soil.

Other materials that are often used to increase soil acidity are elemental sulfur and, in some irrigation systems, sulfuric acid. Sulfur usually undergoes rapid microbial oxidation in the soil (see Section 11.25), and sulfuric acid is produced.

$$2S + 3O_2 + 2H_2O \longrightarrow 2H_2SO_4$$

Under favorable conditions, sulfur is four to five times more effective, kilogram for kilogram, in developing acidity than is ferrous sulfate. Although ferrous sulfate brings about more rapid plant response, sulfur is less expensive, is easy to obtain, and is often used for other purposes. The quantities of ferrous sulfate or sulfur that should be applied will depend upon the buffering capacity of the soil and its original pH level.

CONTROL OF PLANT DISEASES.   Soilborne pathogens are commonly sensitive to soil acidity that sulfur can stimulate. For example, potato scab is caused by an actinomycete that attacks the potato tuber surfaces, leaving a rough discolored appearance that makes the products quite unmarketable. When sulfur is added to reduce the pH to about 5.3 or below, the virulence of actinomycetes is much reduced. However, using sulfur for this purpose to increase soil acidity affects the management of the land, especially the crop rotation. Consideration must be given to choosing succeeding crops that will not be adversely affected by a low soil pH regime.

9.10   DECREASING SOIL ACIDITY: LIMING MATERIALS

Soil acidity is commonly decreased by adding carbonates, oxides, or hydroxides of calcium and magnesium, compounds that are referred to as **agricultural limes**.[6]

Carbonate Forms

Sources of carbonate of lime include marl, oyster shells, basic slag, and precipitated carbonates, but ground limestone is the most common and is by far the most widely used of all liming materials. The two important minerals carried by limestones are **calcite**, which is mostly calcium carbonate ($CaCO_3$), and **dolomite**, which is primarily calcium-magnesium carbonate [$CaMg(CO_3)_2$]. These minerals occur in varying proportions in limestones. When little or no dolomite is present, the limestone is referred to as *calcitic*. As the magnesium increases, this grades into a *dolomitic* limestone. Finally, if the stone is almost entirely composed of calcium-magnesium carbonate, the term *dolomite* is used. Most of the crushed limestone on the market today is either calcitic and/or dolomitic.

Dolomitic limestones and the oxides or hydroxides made from them have one advantage over the calcitic stones. They provide both calcium and magnesium to the soil/plant system. In situations where available magnesium is low, dolomite or dolomitic limestones would be the products of choice. The nutrient-supplying ability of all liming materials should not be overlooked.

Ground limestone is effective in increasing crop yields (see Figure 9.21). It varies in the amount of calcium and magnesium carbonates from approximately 75 to 99%. The total carbonate level of the representative crushed limestone is about 94%.

---

[6]For a discussion of lime and its use, see Adams (1984).

FIGURE 9.21 Alfalfa will not grow on very acid soils (foreground) but responds well to adequate liming (background).

Oxide and Hydroxide Forms

Two other forms of lime are important, especially for small-scale use such as on home gardens where a very rapid change in pH may be required. One is calcium oxide (CaO), sometimes called *burned lime* or *quicklime;* the other is calcium hydroxide ($Ca(OH)_2$), commonly referred to as *hydrated lime.* These are more irritating to handle than limestone and are more expensive, but they continue to find a market where rapid adjustment of soil pH is needed. Box 9.1 shows how they are prepared and indicates some of their properties.

9.11 REACTIONS OF LIME IN THE SOIL

The calcium and magnesium compounds, when applied to an acid soil, react with carbon dioxide and with the acid colloidal complex. These reactions will be considered in order.

Reaction with Carbon Dioxide

All liming materials—whether the oxide, hydroxide, or the carbonate—when applied to an acid soil react with carbon dioxide and water to yield the bicarbonate form. The carbon dioxide partial pressure in the soil, usually several hundred times greater than that in atmospheric air, is generally high enough to drive such reactions. For the purely calcium limes, the reactions are as follows.

$$CaO + H_2O + 2CO_2 \longrightarrow Ca(HCO_3)_2$$
$$Ca(OH)_2 + 2CO_2 \longrightarrow Ca(HCO_3)_2$$
$$CaCO_3 + H_2O + CO_2 \longrightarrow Ca(HCO_3)_2$$

Reaction with Soil Colloids

These liming materials react directly with acid soils, the calcium and magnesium replacing hydrogen and aluminum on the colloidal complex. The adsorption with respect to calcium may be indicated as follows, assuming hydrogen ions are replaced.

$$\begin{array}{l} H^+ \\ H^+ \end{array} \boxed{Micelle} + Ca(OH)_2 \rightleftharpoons Ca^{2+} \boxed{Micelle} + 2H_2O$$

$$\begin{array}{l} H^+ \\ H^+ \end{array} \boxed{Micelle} + \underset{\text{(in solution)}}{Ca(HCO_3)_2} \rightleftharpoons Ca^{2+} \boxed{Micelle} + 2H_2O + 2CO_2 \uparrow$$

$$\begin{array}{l} H^+ \\ H^+ \end{array} \boxed{Micelle} + \underset{\text{(solid phase)}}{CaCO_3} \rightleftharpoons Ca^{2+} \boxed{Micelle} + H_2O + CO_2 \uparrow$$

As these reactions proceed, carbon dioxide is freely evolved, pulling the reaction to the right. In addition, the adsorption of the calcium and magnesium ions raises the percentage base saturation of the colloidal complex, and the pH of the soil solution increases correspondingly.

Depletion of Calcium and Magnesium

As the soluble calcium and magnesium compounds are removed from the soil by the growing plants or by leaching, the percentage base saturation and pH are gradually reduced; eventually another application of lime is necessary. This type of cyclic activity is typical of much of the calcium and magnesium added to arable soils in humid regions (see Figure 9.22).

The need for repeated applications of limestone in humid regions suggests significant losses of calcium and magnesium from the soil. Table 9.1 illustrates losses of these elements by leaching compared with those from crop removal and soil erosion. Note that the total loss from all three causes (leaching, erosion, and crop removal), expressed

BOX 9.1   MANUFACTURE AND USE OF LIMING MATERIALS

LIMESTONE

Eons ago, calcium and magnesium carbonates were deposited in the bottom of ancient lakes or seas. Under pressure of overlying materials these carbonates were compressed into layers of limestone rocks. Today, these are broken up, ground very finely, and made available for application to soils. Calcite ($CaCO_3$) and dolomite ($CaMg(CO_3)_2$) are the minerals contained in limestones. They provide the least expensive means of combating soil acidity and are usually stored in bulk and applied by trucks.

OXIDE OF LIME: BURNED LIME

Oxide of lime is produced by heating limestone to about 850°C and driving off carbon dioxide through reactions such as the following:

$$CaCO_3 + heat \longrightarrow CaO + CO_2 \uparrow$$

$$CaMg(CO_3)_2 + heat \longrightarrow CaO + MgO + CO_2 \uparrow$$

Oxide of lime is caustic to handle and is normally stored in waterproof bags. It is much more expensive than limestone, but reacts more quickly with the soil and gives a high pH after it is applied.

HYDROXIDE OF LIME: HYDRATED LIME

By adding hot water to oxide of lime, hydroxide of lime [$Ca(OH)_2 + Mg(OH)_2$] is formed:

$$CaO + MgO + 2H_2O \longrightarrow Ca(OH)_2 + Mg(OH)_2$$

It is a white powder that is even more caustic than burned lime. It, too, must be kept in bags, not only for user safety, but to keep it from reverting to $CaCO_3 + MgCO_3$ by reaction with CO_2 in the atmosphere. Also more expensive than limestone, it, too, reacts quickly with the soil. The use of both hydroxide and oxide of lime is confined largely to home gardens and specialty crops.

Ruins of an 19th century lime-kiln in Maryland used to convert limestone into burned lime and, in turn, hydrated lime. (Photo courtesy of R. Weil)

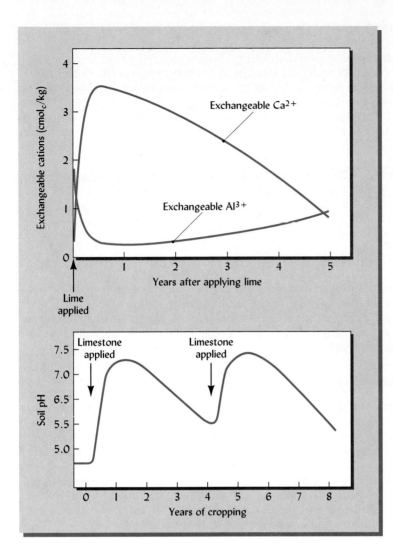

FIGURE 9.22 Diagram to illustrate why repeated applications of limestone are needed to maintain the appropriate chemical balances in the soil (upper). When a soil is limed, the exchangeable calcium increases and the exchangeable aluminum declines, as this diagram of data from an Oxisol in Brazil shows. Within a few years, however, the situation is reversed, the calcium having declined and the aluminum having increased. (Lower) Soil pH likewise increases after 4–8 Mg/ha (3.5–4.5 tons/acre) of limestone is applied to a temperate region soil. The pH reaches a peak after about one year. Leaching and crop removal deplete the calcium and magnesium and, in time, the pH decreases until a renewed application of limestone is necessary. [Upper modified from Smyth and Cravo (1992); used with permission of the American Society of Agronomy]

in the form of carbonates, approaches 1 Mg/ha per year. While there is considerable variation in these losses among different farming systems and in different locations, it is well to recognize that the liming of soils is not a one-time venture but must be repeated with some degree of regularity. The data in Table 9.1 suggest that magnesium as well as calcium must be provided by liming or by other means if appropriate nutrient balance for crop production is to be maintained. The data also suggest that timber harvest methods that cause severe disturbance lead to rapid losses of these cations from forest ecosystems.

TABLE 9.1 Calcium and Magnesium Losses from a Soil by Erosion, Crop Removal, and Leaching in a Humid Temperate Region

Values are in kilograms per hectare per year.

	Calcium expressed as		Magnesium expressed as	
Manner of removal	Ca	CaCO$_3$	Mg	MgCO$_3$
By erosion, Missouri experiments, 4% slope	95	238	33	115
By the average crop of a standard rotation	50	125	25	88
By leaching from a representative silt loam	115	288	25	88
Total (Agricultural areas)		651		291
Loss in streams draining Douglas fir watersheds				
Clear-cut and slash/burned	81	203	26	91
Undisturbed forest	26	65	8	28

9.12 LIME REQUIREMENTS: QUANTITIES NEEDED

The amount of liming material required to bring about a desired pH change is determined by several factors, including (1) the change in pH required, (2) the buffer capacity of the soil, (3) the chemical composition of the liming materials to be used, and (4) the fineness of the liming materials.

The range of soil pH optimum for various plants discussed in Section 9.7 suggests the increase in soil pH that may be desired. Clearly, a higher pH is needed for alfalfa than for corn or soybeans. In Section 9.4 differences in buffer capacity of soils were covered. The relationship of buffering to limestone requirements of soils with different textures is shown in Figure 9.23. Because of greater buffering capacity, the lime requirement for an acid fine-textured clay is much higher than that for a sand or a loam with the same pH value.

9.13 INFLUENCE OF CHEMICAL COMPOSITION AND FINENESS OF LIMING MATERIALS

The chemical composition of liming materials affects the rate of reaction of these compounds with soils. For example, burned and hydrated limes react quickly with soils, bringing about significant changes in soil pH in only a few weeks. In contrast, limestones (particularly those high in dolomite) react much more slowly, taking a year and even longer to fully react with soil colloids.

Chemical Guarantees and Chemical Equivalency

The chemical composition of liming materials also determines their long-term effects on soil pH and how much of each material is needed to achieve these effects. Likewise, it informs the user about the chemical equivalency and nutrient contents of different materials, thereby preventing misinterpretations regarding their comparative chemical values.

To be certain that users are properly informed of the contents of liming materials, laws governing their sale require *guarantees* as to their chemical composition. Box 9.2 illustrates the different means of expressing chemical composition and how they can be used to compare the neutralizing ability of different liming materials.

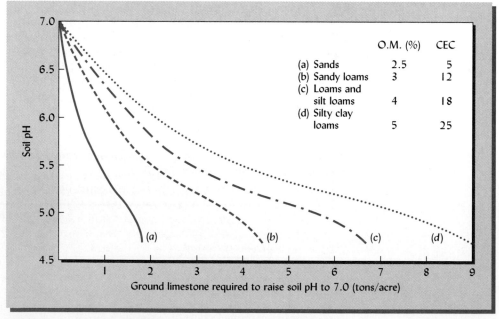

FIGURE 9.23 Relationship between soil texture and the amount of limestone required to raise the pH of New York soils to 7.0. Representative organic matter (O.M.) and cation exchange capacity (CEC) levels are shown. [From Peech (1961)]

BOX 9.2   CHEMICAL GUARANTEES AND EQUIVALENCES

Laws require that the chemical composition of liming materials be guaranteed. The means of expressing this composition varies from one governing body to another, but will usually include guarantees of the contents as expressed by one or more of the following:

1. As elemental Ca and Mg
2. As oxides of these cations (CaO, MgO)
3. As CaO equivalent (neutralizing ability of all compounds expressed as CaO)
4. As total carbonates ($CaCO_3$ plus $MgCO_3$)
5. As $CaCO_3$ equivalent (neutralizing ability of all compounds expressed in terms of $CaCO_3$)

There are advantages and disadvantages to each of these means of expressing the liming effectiveness. However, it is relatively easy to convert one means of expression to another using the concept of *chemical equivalency*. In other words, one atom (or molecule) of Ca, Mg, CaO, MgO, $CaCO_3$, or $MgCO_3$ will neutralize the same amount of acidity as another. Consequently, to make comparisons among different liming materials we need merely multiply by the ratios of their molecular masses. For example, to calculate the $CaCO_3$ equivalent of a pure burned lime (CaO) we need simply multiply by the molecular ratio of $CaCO_3$ to CaO:

$$\frac{CaCO_3}{CaO} = \frac{100}{56} = 1.786$$

Thus, 1 Mg (1000 kg) of this pure burned lime will neutralize as much acidity as 1786 kg of a pure limestone. Some adjustment will need to be made, of course, for the percentage purity of the respective liming materials.

Using chemical equivalencies it is possible to compare a different means of expressing the liming abilities of four materials: a burned lime (oxide), a hydrated lime (hydroxide), and two limestones (one dolomitic, the other calcitic). Each is considered to be 95% pure.

Material	Actual chemicals (%)	Element (%) Ca	Mg	Conventional oxide (%) CaO	MgO	CaO equivalent (%)	Total carbonates (%)	CaCO₃[a] equiv. (%)
Burned lime	77 CaO 18 MgO	55.0 —	— 10.9	77.0 —	— 18.0	102.2	—	182.5
Hydrated lime	75 Ca(OH)₂ 20 Mg(OH)₂	40.5 —	— 8.3	58.6 —	— 13.8	76.0	—	135.8
Calcitic limestone	95 CaCO₃	38.0	—	53.2	—	53.2	95	95
Dolomitic limestone	35 CaCO₃ 60 CaMg(CO₃)₂	(14.0) (13.0)	— (7.9)	(19.6) (18.2)	— (13.1)	(19.6) (36.4)	95	(35) (65)
		27.0	7.9	37.8	13.1	56.0		100

Means of expressing composition

[a] Sometimes referred to as total neutralizing power.

Limestone Fineness and Reactivity

The finer a liming material, the more rapidly it will react with the soil. The oxide and hydroxide of lime usually appear on the market as powders, so their fineness is always satisfactory, but different limestones may vary considerably in particle size as well as hardness. The inefficiency of coarser and harder limestones has necessitated legal requirements for a *fineness guarantee*. Box 9.3 explains how fineness is measured and indicates its effect on reaction of the limestone.

Crop response to liming is influenced markedly by limestone fineness. Figure 9.24 presents a summary of data from many locations to illustrate this point. To obtain at least 80% of the highest yields, 8 Mg/ha of limestone would be required if only 20 to 30% of the particles passed through a 60 mesh sieve. Less than 4 Mg/ha would be needed if 81 to 100% of the limestone particles could pass through a 60 mesh sieve. These data and others obtained from field experiments throughout the humid regions of the United States suggest that *limestones with at least 50% of the particles passing*

The fineness of limestones is measured by passing a sample through a series of standard sieves (screens) with openings of designated size. A number 10 mesh sieve has 10 wires per inch and opening sizes of 2 mm. Similarly, a 60 mesh screen opening is 0.25 mm, and a 200 mesh screen opening 0.075 mm. Remember that the size opening of a given screen represents the maximum diameter of particles that can pass through; thus, all smaller particles will also pass through the cscreen.

The effect of the limestone particle size on the rates of reaction of calcite and dolomite with soils is shown in the accompanying graph. Note that at the end of 3 months nearly 90% of the calcite in the finest fraction (100 mesh) has reacted with the soil, whereas less that 20% of the 20-mesh-size particles had reacted. Also note that at all of the size fractions, dolomite was less reactive than calcite. The data emphasize the interaction between chemical compositions and fineness in ascertaining liming rates. [From Schollenberger and Salter (1943).]

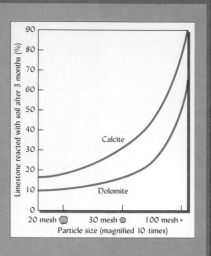

through a 60 mesh screen are quite satisfactory for most agricultural purposes. Although finer materials may give slightly higher yields, the additional cost of grinding to produce the finer limestone may negate the benefits achieved.

9.14   PRACTICAL CONSIDERATIONS

The soil and lime characteristics along with cost factors largely determine the quantity and nature of lime to be applied to agricultural fields. However, in ordinary practice it is seldom wise to apply more than about 7–9 Mg/ha (3–4 tons/acre) of finely ground limestone to a mineral soil at any one time—and even lower rate limits may be appropriate on sandy soils. Data given in Table 9.2 suggest reasonable rates that may be

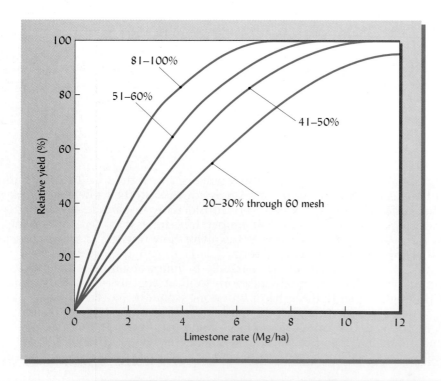

FIGURE 9.24   The effect of limestone fineness on the response of crops to increased rates of limestone application. These are average data from a number of field experiments. [Redrawn from Barber (1984)]

applied on coarse- and fine-textured soils. Box 9.4 provides some step-by-step calculations to help you understand how liming recommendations can be developed.

In the choice of liming materials, attention should be given to the need for magnesium. Everything else being equal, a magnesium-containing limestone should be favored to help maintain appropriate nutritional balance for the crops to be grown.

The lime should be spread with or ahead of the crop that gives the most satisfactory response. Thus, in a rotation with corn, wheat, and two years of alfalfa, the lime may be applied after the wheat harvest to favorably influence the growth of the alfalfa crop that follows. However, since most lime is bulk-spread using heavy trucks (Figure 9.25), applications commonly take place on the sod or hay crop. This prevents adverse soil compaction by the truck tires, which might occur if the lime were applied on recently tilled land.

TABLE 9.2 Approximate Lime Requirement to Raise the pH of Soils of Different Textural Classes from Different Starting Levels to pH 6.5 (Typical of Temperate Humid Region)

Starting pH	Limestone requirement (kg/ha)					
	Sands	Sandy loams	Loams	Silt loams	Clay loams	Clays
4.5	4,000	6,000	9,500	12,500	15,000	17,000
5.0	3,000	5,000	7,500	10,000	11,500	13,500
5.5	2,000	3,000	5,000	6,500	7,500	9,000
6.0	750	1,200	1,800	2,500	3,000	3,500

BOX 9.4 CALCULATING LIME NEEDS

Assume that you must provide a recommendation as to how much limestone should be applied for the production of a high-pH-requiring crop—asparagus—to be grown in a loam soil that is currently quite acid (pH = 5.0). The cation exchange capacity (CEC) is about 10 $cmol_c/kg$, and you want to raise the pH to about 6.8. Assuming the pH–base saturation relationship is about the same as that shown in Figure 9.8, the soil is now at 50% base saturation and will need to be brought to about 90% base saturation to reach the goal of pH = 6.8. How much of a dolomitic limestone with a $CaCO_3$ equivalent of 90 will you need to apply to bring about the desired pH changes?

First, we need to know the $cmol_c$ of Ca/kg soil needed to bring about the change. Since the CEC is 10 $cmol_c/kg$ and we need a 40% change in base saturation (from 50 to 90%) the Ca^{2+} required is

$$10 \text{ cmol}_c/kg \times 40/100 = 4 \text{ cmol}_c \text{ Ca/kg soil}$$

Second, since each Ca^{2+} has two charges, 4 $cmol_c$ can be expressed in grams of Ca^{2+} by multiplying by the molecular mass of Ca (40) and then dividing by 2, the Ca^{2+} charge, and by 100 since we are dealing with a *centimol_c*.

$$4 \text{ cmol}_c \text{ Ca}^{2+} \times \left(\frac{40}{2} \times \frac{1}{100} \right) g\text{Ca}^{2+}/\text{cmol}_c = 0.8 g\text{Ca}$$

Third, we calculate the amount of $CaCO_3$ needed to provide the $0.8 gCa^{2+}$ by multiplying by the ratio of the molecular masses of $CaCO_3$ and Ca.

$$0.8 g/kg \text{ soil} \times \frac{100 \text{ (CaCo}_3)}{40 \text{ (Ca)}} = 2 g\text{CaCO}_3/kg \text{ soil}$$

Fourth, to express the 2 grams $CaCO_3/kg$ soil in terms of the mass needed to change the pH of a hectare–furrow slice (HFS) we must multiply by $2.2 \times 10^6 kg/HFS$

$$2 g\text{CaCO}_3/kg \times 2.2 \times 10^6 kg/HFS = 4.4 \times 10^6 g \text{ or } 4.4 Mg\text{CaCO}_3/HFS$$

Fifth, since the $CaCO_3$ equivalency of our limestone is only 90, some 100 kg would be required to match 90 kg of pure $CaCO_3$. Consequently, the amount of limestone needed would be adjusted upward by a factor of 100/90.

$$4.4 \times \frac{100}{90} = 4.9 \text{ Mg/HFS or about 2.2 tons per acre–furrow slice.}$$

FIGURE 9.25 Bulk application of limestone by specially equipped trucks is the most widespread method of applying liming materials. Because of the weight of such machinery, much limestone is applied to sod land and plowed under. In many cases this same method is used to spread commercial fertilizers.

Special Situations

The spreading of limestone on forested watersheds requires ingenuity. These areas are not accessible to ground-based spreaders, and manual application is too tedious and expensive. In some cases land owners have turned to the use of helicopters for aerial spreading of small quantities of these liming materials. For very acid sandy soils such small applications can ameliorate the ill effects of soil acidity.

Some soil–crop systems make it difficult to incorporate and mix the limestone with the soil. For example, in no-till farming systems the soil is not plowed or cultivated. Furthermore, these systems tend to increase the acidity of the upper soil layers where the crop residues are deposited or remain on the land. Even though earthworms are known to be effective in incorporating surface-applied limestone into the soil (see Figure 9.26), growers may find it necessary to break with the no-till routine every four to six years to permit the plowing and/or discing in of the limestone. Such a practice permits mixing of the limestone with the upper 10 to 15 cm of soil without seriously undermining the objectives of conservation tillage.

Lawns, golf greens, and other turf grass areas face liming difficulties even more constraining than those of the no-till farmer. When such areas are first seeded, care must be taken to ensure that the soil pH is at a satisfactory level. By proper timing of future liming applications with annual aeration tillage operations that leave openings down into the soil (Section 7.7), some downward movement of the lime can be affected. This means more frequent but smaller applications of the lime.

Choice of Adapted Plants

It is often more judicious to solve soil acidity and alkalinity problems by changing the plant to be grown rather than by trying to change the soil pH. As was pointed out in Section 9.7, there are marked differences in the acidity and alkalinity tolerances of different plant species. Furthermore, there is considerable natural variability within a species in the tolerance of different strains to low or high pH. Plant breeders have taken advantage of this variability to develop cultivars of widely grown food crops that are quite tolerant of very acid or alkaline conditions. For example, high-yielding wheat varieties that are quite tolerant of soil acidity and high aluminum levels have been developed and are in use in some areas of the tropics where even modest liming applications are economically impractical. Plant and soil scientists must collaborate to enhance the ability of low-income families to meet their food needs.

Another example of the role of soil–plant interaction in overcoming the adverse effects of soil acidity is the use of green manure crops (crops grown specifically to provide organic residues) to overcome aluminum toxicity. Al^{3+} ions in the soil solution are strongly attracted to the surfaces of decaying organic matter where, through ligand

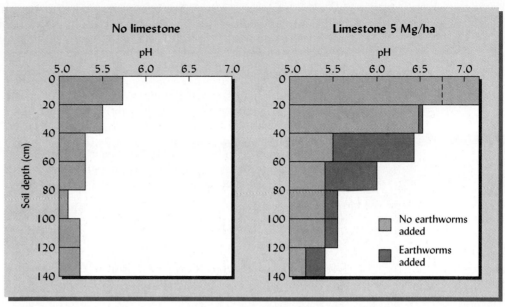

FIGURE 9.26 The effect of surface-applied limestone with and without the introduction of an active earthworm, *Allolobophora longa*, on the pH at different depths of an Alfisol. On the left, no limestone was applied. The limestone increased the pH (right), but without the introduced earthworm the change in pH was mostly in the upper 40 cm. The earthworm apparently moved the limestone into lower layers, thereby making them more suitable for plant root growth. [Redrawn from Springett (1983); used with permission of Blackwell Scientific Publications Ltd.]

bonding, they are tightly held (see Figure 9.2). Green manure crops can provide the organic residues needed to stimulate such interactions and thereby reduce the level of Al^{3+} in the soil solution. Al-sensitive crops can then be grown following the green manure crop. Once again, this type of interaction can be most helpful to farmers in low-income countries where limestone availability and cost factors constrain attempts to adjust the soil pH.

Another practical consideration is the danger of *overliming*—the application of so much lime that the resultant pH values are too high for optimal plant growth. Overliming is not very common on fine-textured soils with high buffer capacities, but it can occur easily on coarse-textured soils that are low in organic matter. The detrimental results of excess lime include deficiencies of iron, manganese, copper, and zinc; reduced availability of phosphate; and restraints on the plant absorption of boron from the soil solution. It is an easy matter to add a little more lime later, but quite difficult to counteract the results of applying too much. Therefore, liming materials should be added conservatively to poorly buffered soils.

9.15 AMELIORATING ACIDITY IN SUBSOILS[7]

As discussed in previous sections, surface-applied limestone can easily ameliorate soil acidity in the upper soil layers. The subsoils of limed areas remain acid, however, since the limestone does not move readily down the profile. Root growth and, in turn, crop yields are often constrained by excessively acid subsoils. Aluminum toxicity and/or calcium deficiency have been found in some of these acid subsoils.

Amelioration by Gypsum

Researchers in the southeastern part of the United States and in some countries in the tropics have found that gypsum ($CaSO_4 \cdot 2H_2O$) can ameliorate aluminum toxicity in subsoils and thereby increase crop yields. Applied on the soil surface, the gypsum slowly dissolves and is leached into the subsoil. About one year after applying the gypsum,

---

[7]For a review of this subject, see Sumner (1993).

favorable yield responses have been obtained with common crops such as corn, soybeans, alfalfa, beans, wheat, apples, and cotton. Penetration of crop roots into the subsoil is increased when gypsum is applied (Figure 9.27), making the crop less sensitive to subsequent droughts. Gypsum is fairly widely available in natural deposits, but is also an industrial by-product of the manufacture of high-analysis fertilizers and of flue gas desulfurization.

The mechanisms for an amelioration of subsoil acidity by gypsum may vary from soil to soil. However, the applied gypsum is known to increase the level of calcium and reduce the level of aluminum both in the soil solution and on the exchange complex. The calcium increases come directly from the added gypsum. The reduced aluminum levels are thought to be stimulated by reaction involving the sulfate ion. For example, the SO_4^{2-} can replace certain OH^- ions associated with Fe, Al sesquioxides through reactions such as the following:

$$\boxed{\begin{array}{c}\text{Fe, Al}\\\text{complexes}\end{array}}\begin{array}{l}-OH\\-OH\end{array} + CaSO_4 \rightleftharpoons \boxed{\begin{array}{c}\text{Fe, Al}\\\text{complexes}\end{array}}\!\!>\!SO_4 + Ca(OH)_2$$

The $Ca(OH)_2$ can then react with Al^{3+} ions in the soil solution to form insoluble $Al(OH)_3$, thereby reducing the concentration of the Al^{3+} while increasing that of Ca^{2+}:

$$3Ca(OH)_2 + 2Al^{3+} \longrightarrow 2Al(OH)_3 + 3Ca^{2+}$$

Gypsum and Physical Properties

Gypsum can influence the physical properties of some highly acid soils of the tropics and subtropics. For example, by-product gypsum has been found to prevent the formation of surface crusts that can increase runoff and soil erosion and decrease water penetration. Apparently the crusts result from the dispersion of the clay in response to the beating action of raindrops. The easily dissolved by-product gypsum can provide enough electrolytes (cations and anions) to keep the clay from dispersing and thereby prevent surface crusting. Field trials have demonstrated that gypsum-treated soils permit greater water infiltration and are less subject to erosion than untreated soils. Similarly, gypsum can reduce the resistance of hard subsurface layers to root penetration, thereby increasing plant uptake of water from the subsoil.

9.16   LIME AND SOIL FERTILITY MANAGEMENT

The maintenance of satisfactory soil fertility levels in humid regions depends considerably on the judicious use of lime and, in some cases, gypsum (see Figure 9.28). Liming

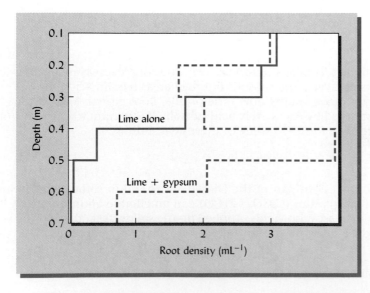

FIGURE 9.27 Surface-applied gypsum moved down the profile to help ameliorate the extreme acidity of the subsoil of this Ultisol in South Africa, and to permit an increase in the density of corn roots. This extension of the roots into the subsoil can increase the availability of both water and nutrients. [Redrawn from Farina and Channon (1988); used with permission of the Soil Science Society of America]

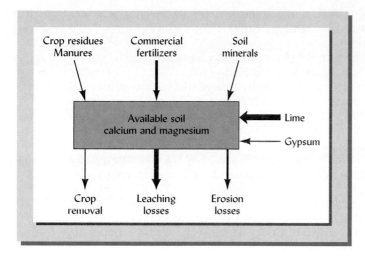

FIGURE 9.28 Important ways by which *available* calcium and magnesium are supplied to and removed from soils. The major losses are through leaching and erosion. These losses are largely replaced by lime and fertilizer applications. Fertilizer additions are much higher than is generally realized, for some phosphate fertilizers contain large quantities of calcium.

not only maintains the levels of exchangeable calcium and magnesium but in so doing also provides a chemical and physical environment that encourages the growth of most common plants. Furthermore, lime counteracts the acid-forming tendency of nitrogen-containing fertilizers, the use of which has increased markedly in the past quarter century. Lime is truly a foundation for much of modern humid region agriculture.

9.17   CONCLUSION

No other single chemical soil characteristic is more important in determining the chemical environment of higher plants and soil microbes than the pH. There are few reactions involving any component of the soil or of its biological inhabitants that are not sensitive to soil pH. This sensitivity must be recognized in any soil management system.

Soil pH is largely controlled by soil colloids and their associated exchangeable cations. Aluminum and hydrogen enhance soil acidity, whereas calcium and other base-forming cations (especially sodium) encourage soil alkalinity. The colloids are also the primary mechanism for soil buffering, which resists rapid changes in soil reaction, giving stability to most plant–soil systems. Knowing how pH is controlled in nature, how it influences the supply and availability of essential plant nutrients as well as toxic elements, how it affects higher plants and human beings, and how it can be ameliorated is essential for the conservation and sustainable management of soils throughout the world.

REFERENCES

Adams, Fred (ed.). 1984. *Soil Acidity and Liming,* 2d ed. (Madison, Wis.: Amer. Soc. Agron.).

Adams, F., and Z. F. Lund. 1966. "Effect of chemical activity of soil solution aluminum on cotton root penetration of acid subsoils," *Soil Sci.,* **101**:193–198.

Barber, S. A. 1984. "Liming material and practices," in F. Adams (ed.), *Soil Acidity and Liming,* 2d ed. (Madison, Wis.: Amer. Soc. Agron.)

Blevins, R. L., G. W. Thomas, M. S. Smith, W. W. Frye, and P. L. Cornelius. 1983. "Changes in soil properties after 10 years continuous non-tilled and conventionally tilled corn," *Soil Tillage Res.,* **3**:135–146.

Farina, M. P. W., and P. Channon. 1988. "Acid subsoil amelioration II. Gypsum effects on growth and subsoil chemical properties," *Soil Sci. Soc. Amer. J.* **52**:175–180.

Hargrove, W. L., and G. W. Thomas. 1981. "Effect of organic matter on exchangeable aluminum and plant growth in acid soils," pp. 151–166 in *Chemistry in the Soil Environment,* ASA Special Publication No. 40 (Madison, Wis.: Amer. Soc. Agron. and Soil Sci. Soc. Amer.).

Jenny, H., et al. 1968. "Interplay of soil organic matter and soil fertility with state factors and soil properties," in *Organic Matter and Soil Fertility,* Pontificiae Academia Scientiarum Scripta Varia 32 (New York: Wiley).

Joslin, J. D., J. M. Kelly, and H. van Miegroet. 1992. "Soil chemistry and nutrition of North American spruce-fir stands: Evidence for recent change." *J. Environ. Qual.,* **21**:12–30.

Knoepp, J. D., and W. T. Swank. 1994. "Long-term soil chemistry changes for aggrading forest systems," *Soil Sci. Soc. Amer. J.,* **58**:325–331.

Mehlich, A. 1964. "Influence of adsorbed hydroxyl and sulfate on neutralization of soil acidity," *Soil Sci. Soc. Amer. Proc.,* **28**:492–96.

National Res. Council. 1983. *Acid deposition: atmospheric processes in Eastern North America* (Washington: National Research Council).

Noble, A. D., M. E. Sumner, and A. K. Alva. 1988. "The pH dependency of aluminum phyto-toxicity alleviation by calcium sulfate," *Soil Sci. Soc. Amer. J.,* **52**:1398–1402.

Peech, M. 1941. "Availability of ions in light sandy soils as affected by soil reaction." *Soil Sci.,* **51**:479–480.

Peech, M. 1961. "Lime Requirements vs. soil pH curves for soils of New York State," mimeographed (Ithaca, N.Y.: Agronomy, Cornell University).

Robarge, W. P., and D. W. Johnson. 1992. "The effects of acidic deposition on forested soils," *Adv. in Agron.,* **47**:1–84.

Robertson, W. K., M. C. Lutrick, and T. L. Yuan. 1982. "Heavy applications of liquid-digested sludge on three Ultisols: I. effects on soil chemistry," *J. Environ. Qual.,* **11**:278–282.

Ronse, A., L. De Temmerman, M. Guns, and R. De Borger. 1988. "Evaluation of acidity, organic matter content and CEC in uncultivated soils of Northern Belgium during the past 25 years," *Soil Sci.,* **146**:453–460.

Schollenberger, C. J., and R. M. Salter. 1943. "A chart for evaluating agricultural limestone," *J. Amer. Soc. Agron.,* **35**:995–66.

Smyth, T. J., and M. S. Cravo. 1992. "Aluminum and calcium constraints to continuous crop production in a Brazilian Amazon Oxisol," *Agron. J.,* **84**:843–850.

Springett, J. A. 1983. "Effect of five species of earthworms on some soil properties," *J. Applied Ecol.,* **20**:865–872.

Sumner, M. E. 1993. "Gypsum and acid soils: The world scene," *Adv. in Agron.,* **51**:1–32.

US/Canada Work Group No. 2, 1982. *Atmospheric Science and Analysis*, Final report. H. L. Ferguson and L. Machita (Co-chairmen) (Washington, D.C. US Environ. Protection Agency).

*. . . alkali has accumulated on them to such an
extent that they are mere bogs and swamps and
alkali flats, and the once fertile lands are thrown out
as ruined and abandoned tracts.*
—MILTON WHITNEY, FIRST CHIEF OF THE DIVISION
OF SOILS, USDA, BULLETIN NO. 14. 1898

Alkaline and salt-affected soils are found mostly in arid and semiarid regions, where more than half of the earth's arable lands are located. These dry areas receive less than 500 mm annual precipitation. They include some relatively barren deserts, but also extensive areas of rangeland and dryland farming. They are home to a wide variety of wild plants and animals that contribute to biological diversity. These soils have also played a unique role in world history as the rise and fall of several ancient civilizations were tied to the irrigation and subsequent mismanagement of alkaline and salt-affected soils. By discussing some of the mistakes that led to the downfall of these civilizations, it is hoped that we can learn to avoid these errors in the future.

The soils of arid and semiarid areas are mostly alkaline, and some are also salt-affected. The rain or snowfall is insufficient to leach out the base-forming cations (Ca^{2+}, Mg^{2+}, K^+, Na^+, etc.) that are slowly released as the rocks and minerals weather. As a result, the percentage base saturation is high and pH values above 7.0 dominate. In some areas with poor internal soil drainage, the leaching of even soluble salts such as $NaCl$, $CaCl_2$, $MgCl_2$, and KCl is constrained, leading to *saline,* as well as alkaline conditions.

The agricultural potential of many soils of arid and semiarid regions is markedly increased if they are irrigated. The availability of irrigation water from either reservoir impoundments or groundwater pumping systems can transform these dry areas into some of the world's most productive farmlands. At the same time, some irrigation systems have brought about significant increases in the salt level of already saline soils. Irrigation water transports salts from upstream watershed areas to the cultivated fields. If the irrigation systems do not provide good internal drainage, soil salinity can increase to intolerable levels, as can the exchangeable sodium level. The latter engenders chemical and physical problems that, if not corrected, will render a soil virtually useless as a habitat for plants. Saline soils will receive our attention after a summary of the properties and management of normal alkaline soils.

10.1 NORMAL ALKALINE SOILS OF DRY AREAS

Consideration of soil reaction in Chapter 9 identified a number of important characteristics of alkaline soils in addition to their relatively low levels of H^+ ions and high levels of OH^- ions. We will reemphasize only a few of these characteristics.

Nutrient Deficiencies

We learned in Chapter 9 that nutrient elements such as iron, manganese, and zinc that are freely available in acid soils (sometimes at toxic levels) are sparingly available or even deficient in many alkaline soils. The low solubility of these micronutrients may result in plant deficiencies unless steps are taken to improve their availability or to increase their supply in the soil using soil amendments. Unfortunately, applications of chemical fertilizers containing these micronutrients are sometimes ineffective because the elements are quickly tied up in insoluble forms in high pH soils. Special protective organic complexes called *chelates* can help meet the nutrient needs of plants growing on these soils (see Section 15.6). Also, for the production of high-valued plants, micronutrients are often sprayed directly on the foliage to avoid interaction between the micronutrients and the high pH soils.

The availability of other nutrients such as phosphorus and boron is reduced, and that of molybdenum is increased by the high pH levels of alkaline soils. In these soils phosphorus is held tightly in the form of highly insoluble calcium and magnesium phosphates. Even the phosphorus added in commercial fertilizers is soon converted to these insoluble forms, rendering the phosphorus only sparingly available for plant use. Likewise, the availability of boron is decreased at high pH levels because above pH 7 the boron tends to be tightly adsorbed by the soil colloids. In contrast, molybdenum availability is high in the alkaline range and molybdenum toxicity is a problem in localized areas. These nutrient-by-nutrient interactions in alkaline soils must be carefully considered in managing these soils.

Micronutrient deficiencies in alkaline soils are especially troublesome for home landscaping and gardening enthusiasts. Unfortunately, some of the most popular ornamental plants are not well adapted to alkaline soils, especially if the soils are calcareous (contain $CaCO_3$ in the upper horizons). Figure 10.1 illustrates the difficulty of trying to grow ornamental oak in such a soil in Denver, Colorado. Gardeners must either select locally adapted varieties of plants that are tolerant of alkalinity or add acid-forming chemicals such as sulfur to change the soil pH to suit the plant.

Cation Exchange Capacity

The cation exchange capacities (CECs) of alkaline soils are commonly higher than those of acid soils with comparable soil textures. The high CEC is due to types of clays present and to increases in pH-dependent negative charges on the colloidal complex at high pH levels. The 2:1-type silicate clays that have relatively high CECs are prominent in arid region soils, although 1:1-type clays, hydrous oxides, and allophane are of importance locally. The negative-charge density on humus and 1:1-type clays is much higher at high pH values than with acid soils (see Figure 9.4). These factors contribute to the higher cation exchange capacities of soils in low-rainfall areas.

Calcareous Layers

Another common characteristic of many well-developed soils of low-rainfall areas is the presence at some level in the soil profile (usually the C horizon) of calcium carbonate accumulation (Figure 10.2). These horizons are referred to as being *calcareous*. The high carbonate concentrations in these horizons are not conducive to optimum root growth for many crops. Where carbonate concentrations are found at or near the soil surface, which is often the case in areas of very low rainfall, serious micronutrient deficiencies can result for plants that are not adapted to these conditions (Plate 19). The soil orders and suborders of low-rainfall-region soils are described in Section 3.14.

Water Management

Water management must receive high priority in the cultivation of arid and semiarid-region soils. Tillage practices that increase water infiltration during periods of rain or snowfall should be encouraged (see Figure 6.17). Likewise, it is often wise to use tillage practices such as stubble mulch (see Section 6.9) that help reduce evaporative losses and soil erosion by keeping some cover at or near the soil surface. Irrigated soils of dry areas have moisture relations during periods of plant growth that are not too different from those in humid regions. However, moisture losses by evaporation in the dry areas are

FIGURE 10.1 Ornamental oak (upper, with dying branches in foreground) that thrives well in mildly acid soils of humid regions grows poorly on this alkaline soil in Colorado. The nutrient deficiency (probably of iron) is illustrated by the light green color between the veins of the leaves (lower). A soil test verifies the fact that the soil is alkaline. Gardeners must either select species known to tolerate alkaline conditions, acidify the soil, or supply the micronutrients that the alkaline-sensitive plants need. (Photos courtesy of R. Weil)

very high and salt buildup in irrigated soils must be monitored to ensure successful plant production. The characteristics and management of salt-affected soils will be covered in the sections that follow.

10.2 DEVELOPMENT OF SALINE AND SODIC SOILS[1]

Natural Salt Accumulation

Salts accumulate naturally in some surface soils of arid and semiarid regions because there is insufficient rainfall to flush them from the upper soil layers. In the United States, about one-third of the soils in these regions suffer from some degree of salinity. The salts are primarily chlorides and sulfates of calcium, magnesium, sodium, and potassium. They may be formed during the weathering of rocks and minerals or brought to the soils through rainfall and irrigation. Other localized but important sources are fossil deposits of salts laid down during geological time in the bottom of now extinct lakes or oceans or in underground saline water pools. These fossil salts can

---

[1]See Bressler et al. (1982) for a discussion of these soils.

FIGURE 10.2 Calcium carbonate accumulation in the lower part of the B horizon (B_k) and the upper part of the C horizon (C_k) (the knife is inserted at the upper boundary) characterizes this Ustoll and many other soils of arid and semiarid regions. This calcareous layer helps maintain high pH levels and constrains the availability of micronutrients such as iron, manganese, and zinc. (Photo courtesy of R. Weil)

be dissolved in underground waters that move horizontally over underlying impervious geological layers and ultimately rise to the surface of the soil in the low-lying parts of the landscape, often forming **saline seeps.** The water then evaporates, leaving salts in place at or near the soil's surface. Figure 10.3 illustrates how salts can accumulate in lower elevations from salts that originate up the slopes from either the natural soils or fossil deposits. Unfortunately, high levels of these salts cannot be tolerated by most plants, a fact that severely limits the use of some salt-affected soils.

Salt-affected soils are widely distributed throughout the world (Table 10.1), the largest areas being found in Australia, Africa, Latin America, and the Near and Middle East. They are most often found in areas with precipitation-to-evaporation ratios of 0.75 or less, and

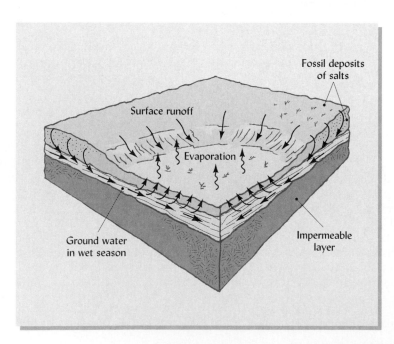

FIGURE 10.3 Field cut showing conditions that encourage salt buildup in soils. Mineral containing waters from fossil deposits of salt or from soil weathering in upland areas move down the slope toward lower basins, either on the surface during torrential storms or as groundwater on top of impervious geological layers. When the water in the low-lying areas evaporates, it carries the salts up to or near the surface where the salts accumulate.

TABLE 10.1 **Area of Salt-Affected Soils in Different Regions**

Region	Area (million ha)
Africa	69.5
Near and Middle East	53.1
Asia and Far East	19.5
Latin America	59.4
Australia	84.7
North America	16.0
Europe	20.7
World Total	322.9

From Beek et al. (1980).

in low, flat areas with high water tables that may be subject to seepage from higher elevations. While in most of these areas salts have accumulated naturally, irrigation-induced salinity of the type that brought disaster to ancient cultures centuries ago has become increasingly important in recent years.

Irrigation-Induced Salinity and Alkalinity

Irrigation waters, whether from upstream watersheds or underground pumping, carry significant quantities of soluble salts. In areas where the land is well drained, sufficient irrigation water can percolate through the soil to leach the soluble salts from the upper soil layers, thereby preventing excessive salt accumulation. If the soil is poorly drained, however, downward movement of the salts to the groundwater is impaired, and the salts are left in the soil and are later brought up to the surface as the irrigation water evaporates. A **saline** soil is thus created.

The degradation of irrigated soils in arid regions is even more severe if the colloidal complex becomes significantly saturated with the Na^+ ion. This results in extremely high pH values and the dispersal of soil colloids that then plug the soil's drainage pores, preventing the downward percolation of the irrigation water. The soils are rendered virtually useless for commercial agriculture. The term *sodic* (derived from sodium) is used to describe such soils (see Section 10.4).

During the past three decades, low-income countries in the dry regions of the world have greatly expanded their areas of irrigated lands in order to produce the food needed by their rapidly growing human populations. As a result, the proportion of arable land that is irrigated has increased dramatically, reaching about 45% in China, 25% in India, 72% in Pakistan, and 28% in Indonesia. Initially, phenomenal increases in food crop production were stimulated by the expanded irrigation. However, in many project areas the need for good soil drainage was overlooked and the process of salinization has been accelerated. As a consequence, salts have accumulated to levels that are already adversely affecting crop production. In some areas, *sodic* soils that are high in exchangeable Na^+ ions have been created. They present farmers with serious soil problems, both chemical and physical in nature.

These events remind the world of the flat wastelands of southeastern Iraq that have been barren since the 12th century. In biblical times, the soils of these areas were fertile and productive. The Euphrates and Tigris rivers provided irrigation water that stimulated such high crop productivity that the overall region was called the Fertile Crescent. Unfortunately, in broad areas the soils were not well drained and the process of salinization proceeded. The salts accumulated to such a high level that crop production declined and the area had to be abandoned. Even today, this same process is being repeated at different locations around the world, and some irrigated farming areas are having to be abandoned (see, for example, Figure 19.11). Truly the world is justified in giving serious attention to salt-affected soils.

10.3 MEASURING SALINITY AND ALKALINITY

Plants are detrimentally affected by excess salts in some soils and by high levels of exchangeable sodium in others, the latter being detrimental to the soil both physically and chemically. Techniques have been developed to measure three primary properties that, along with pH, can be used to characterize salt-affected soils: (1) soil **salinity**, (2) **exchangeable sodium percentage** (ESP), and (3) the **sodium adsorption ratio** (SAR). Each will be discussed briefly.

Salinity[2]

First, the salt concentration is estimated by methods based on the ability of the salt in the soil solution to conduct electricity. Pure water is a poor conductor of electricity, but conductivity increases as more and more salt is dissolved in the water. The **electrical conductivity** (EC) is measured by both laboratory and field methods. The **saturation paste extract** method is the most commonly used laboratory procedure. The soil sample is saturated with distilled water to a paste consistency, allowed to stand overnight to dissolve the salts, and the electrical conductivity of the water extracted from the paste is measured (EC_e). A variant of this method involves the EC of the solution extracted from a 1:2 soil-to-water mixture after 0.5 hours of shaking (EC_w). The latter method takes less time but is often less accurate than the saturation paste extract method. The salinity is expressed in terms of the EC, which is measured in decisiemens per meter.[3]

Field methods in use today involve the measurement of bulk soil conductivity that in turn is directly related to soil salinity. One such method involves the insertion into the soil of four electrode sensors (Figure 10.4) and the direct measurement of apparent EC in the field (EC_a). The sensors can be mounted behind a tractor and moved quickly from one site in the field to another. This technique is rapid, simple, and practical, and gives values that can be correlated with EC_e.

A second field method is even more rapid. It employs **electromagnetic induction** of electrical current in the body of the soil, the level of which is related to electrical conductivity and, in turn, to soil salinity. A magnetic field is generated within the soil by a small transmitter coil that is placed on the soil surface and is energized by an alternat-

---

[2]For a recent discussion of these methods, see Rhoades (1993).

[3]Formerly expressed as millimhos per centimeter (mmho/cm). Since 1 S = 1 mho, 1 dS/m = 1 mmho/cm.

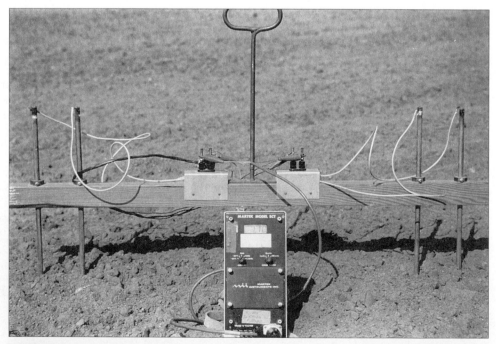

FIGURE 10.4   A four-electrode sensor used to measure electrical conductivity of bulk soil in the field. The electrodes are a fixed distance apart so they can be quickly inserted into the soil and the measurements made. A combination of a portable generator and resistance meter (foreground) makes this equipment very useful for field survey work. [From Rhoades (1993); used with permission of Academic Press]

ing current. This magnetic field, in turn, induces within the soil small electric currents whose values are related to the soil's conductivity. These small currents generate their own secondary magnetic fields that can be measured by a small receiving cell nearby. The instrument that provides these emitting and receiving coils thus can measure ground EC (designated E^*_a) without the use of probes, and is proving to be useful in surveying saline soils (see Section 19.3).

Since electrical conductivity is influenced by both salinity and soil water content, the field measurements should be made when the soil is near the field capacity. With irrigated soils, measurements are best made soon after the soil is irrigated to assure reasonably high soil moisture levels.

It should be noted that each of the methods discussed estimates soil salinity indirectly by measuring electrical conductivity. But the values of EC_e, EC_w, E_a, or E^*_a obtained by these procedures will not be identical. For example, for highly saline soils EC_a measured by the four-electrode-sensor method are about one-fifth of those measured using the standard saturated extract procedure (EC_e). In any case, however, these values are sufficiently well correlated with each other so that calculations can be made of soil salinity expressed conventionally in terms of the standard EC_e (saturation paste extract method).

Sodium Status

Two means are used to characterize the sodium status of highly alkaline soils. The **exchangeable sodium percentage** (ESP) identifies the degree to which the exchange complex is saturated with sodium.

$$ESP = \frac{\text{exchangeable sodium (cmol}_c\text{/kg)}}{\text{cation exchange capacity (cmol}_c\text{/kg)}}$$

ESP levels of 15 yield pH values of 8.5 and above. Higher levels may bring the pH to above 10.

The ESP is complemented by a second more easily measured characteristic, the **sodium adsorption ratio** (SAR), which gives information on the comparative concentrations of Na^+, Ca^{2+} and Mg^{2+} in soil solutions. It is calculated as follows:

$$SAR = \frac{[Na^+]}{\sqrt{[Ca^{2+}] + [Mg^{2+}]}}$$

where $[Na^+]$, $[Ca^{2+}]$, and $[Mg^{2+}]$ are the concentrations (in mmol$_c$/L) of the sodium, calcium, and magnesium ions in the soil solution. The SAR of a soil extract takes into consideration that the adverse effect of sodium is moderated by the presence of calcium and magnesium ions. The SAR also is used to characterize the irrigation water added to these soils. This property is becoming more widely used in characterizing salt-affected soils.

10.4 CLASSES OF SALT-AFFECTED SOILS

Using EC, ESP, SAR characteristics, and soil pH, salt-affected soils are classified as **saline, saline-sodic,** and **sodic** (Figure 10.5). These will be considered in order.

Saline Soils

The natural processes that result in the accumulation of neutral soluble salts are referred to as *salinization.* The salts are mainly chlorides and sulfates of sodium, calcium, magnesium, and potassium. Saline soils contain a concentration of these salts sufficient to interfere with the growth of most plants. The electrical conductivity (EC) of a saturation extract of the soil solution is more than 4 decisiemens per meter (dS/m).[4] Salts are com-

---

[4]Some fruit and vegetable crops are adversely affected when the EC is lower than 4 dS/m, leading some scientists to recommend that 2 dS/m be the level above which a soil could be classed as saline.

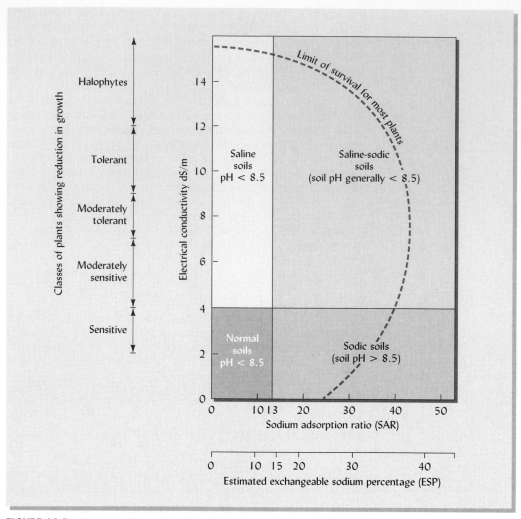

FIGURE 10.5 Diagram illustrating the classification of normal, saline, saline-sodic, and sodic soils in relation to soil pH, electrical conductivity, sodium adsorption ratio (SAR), and exchangeable sodium percentage (ESP). Also shown are the ranges for different degrees of sensitivity of plants to salinity.

monly brought to the soil surface by evaporating waters, creating a white crust which accounts for the name *white alkali* that is sometimes used to designate these soils (see Figure 10.6).

The exchange complex of saline soils is dominated by calcium and magnesium. As a consequence, the exchangeable sodium percentage (ESP) is less than about 15, and the pH usually is less than 8.5.

The Na^+ ion concentration in the soil solution of saline soils may be somewhat higher than that of Ca^{2+} or Mg^{2+} due to the presence of soluble salts that are commonly high in sodium. However, because of the greater affinity of the soil colloids for divalent cations such as Ca^{2+} or Mg^{2+}, the sodium adsorption ratio (SAR) of saline soils is less than 13.

Plant growth on saline soils is not generally constrained by poor soil physical conditions. The soluble salts help prevent dispersion of the soil colloids that otherwise would be encouraged by high sodium levels. Aggregate stability and good aeration result.

Saline-Sodic Soils

Saline-sodic soils have characteristics intermediate between those of saline and sodic soils (Figure 10.5). Like saline soils, they contain appreciable levels of neutral soluble salts, as shown by EC_e levels of more than 4 dS/m. But they have a higher exchangeable sodium percentage (ESP), greater than 15, and a higher sodium adsorption ratio (SAR), at least 13. Crop growth can be adversely affected by both excess salts and sodium levels.

The physical and chemical conditions of saline-sodic soils are relatively similar to those of saline soils. This is due to the moderating effects of the neutral salts. These salts

FIGURE 10.6 (Lower) White alkali spot in a field of alfalfa under irrigation. Because of upward capillarity and evaporation, salts have been brought to the surface where they have accumulated. (Upper) White crust on a saline soil from Colorado. The white salts are in contrast with the darker-colored soil underneath. (Scale is shown in inches and centimeters) [Lower photo Courtesy USDA Agricultural Research Service]

provide excess cations that move closely to the negatively charged colloidal particles, thereby reducing their tendency to repel each other, or to disperse. The salts help keep the colloidal particles associated with each other in aggregates. Unfortunately, however, this situation is subject to rather rapid change if the soluble salts are leached from the soil, especially if the leaching waters are high in Na^+ ions—that is, if the SAR of the water is high. The exchangeable sodium level will increase, as will the pH (to above 8.5). Since Na^+ ions are not attracted very closely to the soil colloids, the effective electronegativity of the colloidal particles causes them to repel each other or to disperse, thereby breaking up the soil aggregates. The dispersed colloids clog the soil pores as they move down the profile. Water infiltration is thus greatly reduced and the puddled condition, so characteristic of sodic soils, pertains.

Sodic Soils

The preceding brief description of how sodic soils can be formed suggests that they are the most troublesome of the salt-affected soils. Both their chemical and physical characteristics discourage the growth of plants. While their levels of neutral soluble salts are low (EC_e less than 4.0 dS/m), their ESP values are above 15 and their SARs above 13, suggesting a comparatively high level of sodium on the exchange complex. The pH values of sodic soils exceed 8.5, rising to 10 or higher in some soils. Plant growth on these soils is constrained by high levels of Na^+, OH^-, and HCO_3^- ions as well as by their very poor soil physical conditions and slow permeability to water.

The high pH is largely due to the hydrolysis of sodium carbonate.

$$2Na^+ + CO_3^{2+} + H_2O \rightleftharpoons 2Na^+ + HCO_3^- + OH^-$$

The sodium complex also undergoes hydrolysis.

$$Na^+ \boxed{Micelle} + H_2O \rightleftharpoons H^+ \boxed{Micelle} + Na^+ + OH^-$$

Owing to the deflocculating (dispersing) influence of the sodium, sodic soils usually are in an unsatisfactory physical condition. The aggregate structure is broken down, the pores are filled with fine materials, and hydraulic conductivity and water infiltration is reduced to a minimum. Few plants can tolerate these conditions. As already stated, the leaching of a saline-sodic soil can readily produce a similar physical condition unless the leaching waters are high in calcium salts.

Because of the extreme alkalinity resulting from the high sodium content, the surface of sodic soils usually is discolored by the dispersed humus that can be carried upward by the capillary water and deposited when it evaporates. Hence, the name **black alkali** has been used to describe these soils. Sometimes located in small areas called **slick spots,** sodic soils may be surrounded by soils that are relatively productive.

10.5   GROWTH OF PLANTS ON SALINE AND SODIC SOILS

Plants respond to the different salt-affected soils in different ways. High soluble salt concentrations through their effects on osmotic potentials (see Section 5.3) reduce plant growth on saline and saline-sodic soils. Root cells, as they come in contact with the soil solution high in salts, will lose water by osmosis to the more concentrated soil solution. The cell then collapses. The kind of salt, the plant species, and the rate of salinization are factors that determine the concentration at which the cell succumbs. The adverse physical condition of some saline-sodic soils also may be a factor in determining which plants can grow on these soils.

Sodic soils, dominated by active sodium, exert a detrimental effect on plants in five ways: (1) caustic influence of the high pH induced by the sodium carbonate and bicarbonate; (2) toxicity of the bicarbonate and other anions; (3) the adverse effects of the active sodium ions on plant metabolism and nutrition; (4) the low micronutrient availability due to high pH; and (5) oxygen deficiency due to breakdown of soil structure.

10.6   TOLERANCE OF HIGHER PLANTS TO SALINE AND SODIC SOILS

Satisfactory plant growth on salty soils depends on a number of interrelated factors, including the physiological constitution of the plant, its stage of growth, and its rooting habits. It is interesting to note that old alfalfa plants are more tolerant to salt-affected soils than young ones, and that deep-rooted legumes show a greater resistance to such soils than those that are shallow-rooted.

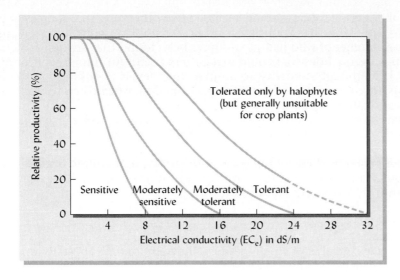

FIGURE 10.7 Relative productivity of four groups of plants classified by their sensitivity to salinity as measured by electrical conductivity. The plant groupings are shown in Table 10.2. [Generalized from data of Carter (1981)]

Soil properties—including the nature of the various salts, their proportionate amounts, their total concentration, and their distribution in the solum—must be considered. The structure of the soil and its drainage and aeration are also important.

Plant Sensitivity

While it is difficult to forecast accurately the tolerance of crops to salty soils, numerous tests have made it possible to classify many domestic plants into four general groups based on their salinity tolerance. Figure 10.7 shows the relative productivity of these general groups influenced by soil salinity that is measured by electrical conductivity (EC_e). Table 10.2 provides a listing of many plants according to this classification. Note that trees, shrubs, fruits, vegetables, and field crops are included in the different categories.

TABLE 10.2 Relative Tolerance of Certain Plants to Salty Soils

Tolerant	Moderately tolerant	Moderately sensitive	Sensitive
Barley, grain	Barley, forage	Alfalfa	Almond
Bermuda grass	Beet, garden	Arborvitae	Apple
Bougainvillea	Broccoli	Boxwood	Apricot
Cotton	Brome grass	Broad bean	Azalea
Date	Cow pea	Cauliflower	Bean
Elm	Fescue, tall	Cabbage	Blackberry
Locust	Fig	Celery	Boysenberry
Natal plum	Harding grass	Clover, alsike,	Burford holly
Nutall alkali grass	Honeysuckle	ladino, red,	Carrot
Oak	Hydrangea	strawberry,	Celery
Olive	Juniper	berseem	Dogwood
Rescue grass	Kale	Corn	Grapefruit
Rosemary	Mandarin	Cucumber	Hibiscus
Rugosa	Orchard grass	Dallis grass	Larch
Salt grass	Oats	Grape	Lemon
Sugar beet	Pomegranate	Juniper	Linden
Tamarix	Privet	Lettuce	Onion
Wheat grass, crested	Rye, hay	Pea	Orange
Wheat grass, fairway	Ryegrass, perennial	Peanut	Peach
Wheat grass, tall	Safflower	Radish	Pear
Wild rye, altai	Sorghum	Rice, paddy	Pineapple
Wild rye, Russian	Soybean	Squash	Plum, prune
Willow	Squash, zucchini	Sugarcane	Potato
	Sudan grass	Sweet clover	Raspberry
	Trefoil, birdsfoot	Sweet potato	Rose
	Wheat	Timothy	Star jasmine
	Wheat grass, western	Turnip	Strawberry
		Vetch	Sugar Maple
		Viburnum	Tomato

Selected from Carter (1981) and Maas (1984), see figure 10.7 for tolerance ranges.

To this list of domestic species should be added two other potential sources of plants to grow on salty soils: (1) wild halophytes and (2) salt-tolerant cultivars developed by plant breeders. A number of wild halophytes have been found that are quite tolerant to salts and that possess qualities that could make them useful for human and/or animal consumption.[5] Even though they may accumulate high levels of salt in their stems and leaves, the seeds are commonly not too different from domesticated crops in their content of protein and fat, for example.

Plant breeders have been able to develop new strains with salt tolerance greater than that possessed by conventional varieties. An example is shown by the data in Figure 10.8 that illustrates superior performance by improved strains of barley. Plant selection and improvement should be one of the goals of the future to maintain or even increase food production on salt-affected soils. However improved plant tolerance must not be viewed as a substitute for proper salinity control.

Special Cases

Urbanites may have concern for two types of salinity problems. Those living in cities where deicing salts are used in abundance during winter months may have concern for the salt tolerance of their trees, shrubs, or grass. Repeated application of the deicing salts can result in salinity levels sufficiently high to adversely affect plants growing alongside a treated highway or sidewalk. Figure 10.9 illustrates tree damage from such treatments. To avoid the specific chemical and physical problems associated with sodium salts, many municipalities have switched from NaCl to KCl for deicing purposes.

Salinity can also be a problem in the growing of potted plants, particularly perennials that remain in the same pot for long periods of time. Salts in the water as well as those applied in fertilizers can build up if care is not taken to flush them out occasionally.

10.7   MANAGEMENT OF SALINE AND SODIC SOILS

The first requisite for wise management of salt-affected soils is to know something about the amount and nature of soluble salts being added to and removed from the soil. In irrigated areas, this means knowing the quality of the irrigation water and the status of the soil drainage.

Water Quality Considerations

The chemical quality of water added to salt-affected soils is a prime management tool. For example if the water is too low in salts, such as in rainwater, it can hasten the

---

[5]For a review of such halophytes, see NAS (1990).

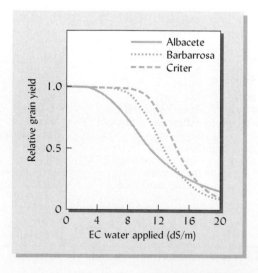

FIGURE 10.8   Three barley varieties differ significantly in their tolerance to the salinity of water applied to the soil in which they were grown. Such genetic differences can be useful in breeding varieties with greater tolerance of saline conditions. [Redrawn from Royo and Aragues (1993); used with permission of the American Society of Agronomy]

FIGURE 10.9 During the winter months in some temperate region areas, sodium chloride and other salts are used to melt the snow and ice from the streets and sidewalks. Heavy applications of these chemicals can result in salinity damage to plants similar to that experienced on saline soils of arid regions. (Left) Localized area of salt concentration resulting from the piling of salt-treated snow alongside a street. (Right) Leaf damage resulting from the excess salt uptake by the plants. (Photos courtesy of R. Weil)

change from a saline-sodic soil to a sodic soil. The rainwater not only leaches soluble salts from the upper few centimeters of soil but the impact of the raindrops encourages dispersion of the soil colloids, an initial step in the process of sodic soil development.

If the salt level of irrigation water is too high, and especially if the SAR of the water is also high, the process of forming sodic soils will accelerate. Also, the presence of bicarbonates in the irrigation water can reduce concentrations of Ca^{2+} and Mg^{2+} in the soil solution by precipitating these ions as insoluble carbonates. Such chemical characteristics of the irrigation water can increase the SAR of the soil solution and move the soil toward the sodic class.

The quality and disposition of *waste irrigation waters* must also be carefully monitored because of the potential harm to downstream users and habitats. Drainage water from one irrigation system that may be high in salt concentration or in the concentration of one or more specific elements may reenter a stream and be used again at lower elevations. At some locations in the western United States, for example, toxic levels of trace elements such as selenium, molybdenum, and boron have accumulated in downstream wetlands or evaporative ponds. Table 10.3 suggests that plants growing on these affected areas can accumulate levels of these elements that are unsafe for livestock and/or wildlife. Indeed, selenium toxicity is known to be damaging to waterfowl in parts of California. Such chemical accumulations in plants may be useful in helping to remove the toxic elements from the soil, but the process must be so managed as to minimize risks to grazing animals.

TABLE 10.3 Levels of Molybdenum and Selenium in Three Salinity-Tolerant Forage Species Grown on a Highly Salinized Soil That Had Received Large Quantities of Waste Irrigation Water in the San Joaquin Valley of California

The upper "safe" limit for animal consumption (potential toxicity level) is also shown.

Trace elements	Potential toxicity level (mg/kg)	Level in the forages (mg/kg)		
		Tall wheatgrass	Alkali sacation	Astragalus racemosus
Molybdenum	5	26	10	18
Selenium	5	12	9	670

Data selected from Retana et al. (1993).

The soluble salt level of different water sources varies greatly. Data for a number of water sources varying from high-quality irrigation water to that in the Pacific Ocean is shown in Table 10.4. Note even the Colorado River has sufficiently high salt content to alert users to the need for wise irrigation management if soil quality is to be maintained. In fact, by the time the Colorado River reaches the U.S./Mexico border it is so loaded with salts from upstream irrigation systems and from domestic and industrial uses that a huge desalinization plant is required to enable the United States to meet its treaty obligations with Mexico. Without this desalinization plant, the river water entering Mexico would be too salty to use for irrigation. The environmental effects of irrigation have similar implications in many parts of the world.

Management Objectives

The primary objective in managing salt-affected soils is to lessen the major constraints to plant production—excess soluble salts and exchangeable sodium. The removal of these constraints causes several chemical and physical changes that bring the soils back to a more productive state. Consequently, the overall processes are referred to as the **reclamation** of these soils. We will deal briefly with the reclamation of saline soils first, then sodic and saline-sodic soils, and will assume that ample supplies of good-quality irrigation water are available to bring about the reclamation.

10.8   RECLAMATION OF SALINE SOILS

The removal of excess salts from saline soils requires access to ample low-salt irrigation water and an effective internal soil drainage system that will quickly remove the salt-laden water once it leaches or flushes down through the soil. If the natural soil drainage is inadequate to accommodate the leaching water, an artificial drainage network must be installed. Intermittent applications of excess irrigation water may be required to effectively reduce the salt content to a desired level. The process can be monitored by measuring the soil's EC, using either the saturation extract procedure or one of the field instruments.

The amount of water needed to remove the excess salts from saline soils, called the *leaching requirement* (*LR*), is determined by characteristics of the crop to be grown, the irrigation water, and the soil. An approximation of the *LR* is given for relatively uniform salinity conditions by the ratio of the salinity of the irrigation water (expressed as its EC_w) to the maximum permissible salinity of the soil solution for the crop to be grown (expressed as EC_{dw}, the EC of the drainage water).

$$LR = EC_w/EC_{dw}$$

The *LR* is water added in excess of the moisture needed to thoroughly wet the soil.

Note that if EC_w is high and the high salt sensitivity of the crop to be grown dictates a low EC_{dw}, a very large leaching requirement *LR* would result. An example of how the *LR* values are calculated is given in Box 10.1.

Attention must be paid to the disposal of the drainage water once the soil leaching has taken place. If this water cannot be returned to a river or stream without overburdening the stream with salts, then other means must be sought to dispose of the water. As mentioned earlier, care must be taken to minimize burdens on downstream users.

Tillage and planting practices can influence crop production on saline soils. Any tillage or surface residue management practice (such as conservation tillage) that reduces evaporation from the soil surface should reduce the upward transport of soluble salts. Likewise, specific irrigation schemes such as those utilizing sprinkling and

TABLE 10.4   **The Comparative Salinity of Six Different Water Sources as Measured by Electrical Conductivity and Dissolved Solids Content**

Salinity measurement	Irrigation water quality (good)	(marginal)	Colorado River	Alamo River	Negev (Middle East) groundwater	Pacific Ocean
Electrical conductivity (dS/m)	0–1	1–3	1.3	4.0	4.0–7.0	46
Dissolved solids, ppm	0–500	500–1,500	850	3,000	3,000–4,500	35,000

Adapted from a number of sources by the U.S. National Academy of Sciences (NAS 1990).

trickle systems (see Section 6.22) as well as planting techniques for row crops can reduce the salt concentration immediately around the young plant roots (see Figure 10.10).

10.9   RECLAMATION OF SALINE-SODIC AND SODIC SOILS

The saline-sodic soils have some of the adverse properties of both the saline and sodic soils. If attempts are made to leach out the soluble salts in saline-sodic soils as was discussed for saline soils, the exchangeable Na^+ level as well as the pH would likely increase and the soil would take on adverse characteristics of the sodic soils. Consequently, for both saline-sodic and sodic soils, attention must first be given to reducing the level of exchangeable Na^+ ions and then to the problem of excess soluble salts.

Gypsum

Removing Na^+ ions from the exchange complex is most effectively accomplished by replacing them with either the Ca^{2+} or the H^+ ion. Providing Ca^{2+} in the form of gypsum ($CaSO_4 \cdot 2H_2O$) is the most practical way to bring about this exchange. When gypsum is added reactions such as the following take place:

$$2NaHCO_3 + CaSO_4 \longrightarrow CaCO_3 + Na_2SO_4 + CO_2$$

$$Na_2CO_3 + CaSO_4 \rightleftharpoons CaCO_3 + \underset{\text{(leachable)}}{Na_2SO_4}$$

$$\begin{matrix} Na^+ \\ Na^+ \end{matrix} \boxed{\text{Micelle}} + CaSO_4 \rightleftharpoons Ca^{2+} \boxed{\text{Micelle}} + \underset{\text{(leachable)}}{Na_2SO_4}$$

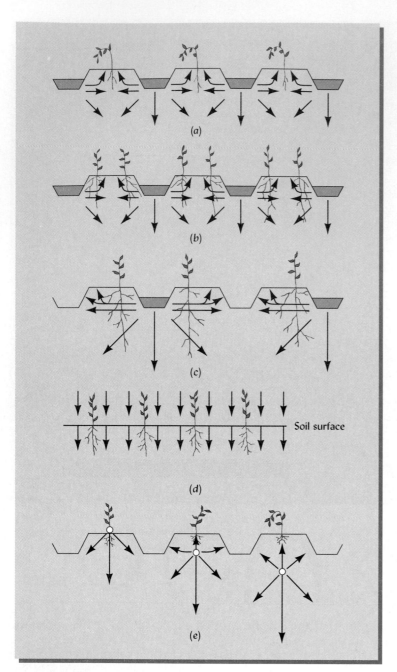

FIGURE 10.10 Effect of irrigation on salt movement and plant growth in saline soil (a) With irrigation water applied to furrows on both sides of the row, salts move to the center and damage plant root systems. (b) Placing plants on edges of the bed rather than the center helps avoid the damage seen in (a). (c) Application of water to every other furrow and placement of plants on the side of the bed nearest the water also helps plants avoid the highest salt concentrations. Sprinkle irrigation (d) or uniform flood irrigation moves salts downward and the problem is alleviated. (e) Drip or trickle irrigation results in salt removal or accumulation, depending on the placement of the trickle tube. (Courtesy of Wesley M. Jarrell)

Note that in each case the soluble salt Na_2SO_4 is formed that can be easily leached from the soil as was done in the case of the saline soils.

Several tons of gypsum per hectare are usually necessary. In Box 10.2, calculations are made to approximate the amount of gypsum that is theoretically needed to remove an acceptable portion of the Na^+ ion from the exchange complex. The soil must be kept moist to hasten the reaction, and the gypsum should be thoroughly mixed into the surface by cultivation—not simply plowed under. The treatment must be supplemented later by a thorough leaching of the soil with irrigation water to leach out most of the sodium sulfate.

Sulfur and Sulfuric Acid

Elemental sulfur and sulfuric acid can be used to advantage on salty lands, especially where sodium bicarbonate abounds. The sulfur, upon biological oxidation (see Sections 9.6 and 13.21), yields sulfuric acid, which not only changes the sodium bicarbonate to

BOX 10.2 CALCULATING THE THEORETICAL
GYPSUM REQUIREMENT

PROBLEM

How much gypsum is needed to reclaim a sodic soil with an exchangeable sodium percentage (ESP) of 25% and a cation exchange capacity of 18 $cmol_c/kg$? Assume that you want to reduce the ESP of the upper 30 cm of soil to about 5% so that a crop like alfalfa could be grown.

SOLUTION

First, determine the amount of Na^+ ions to be replaced by multiplying the CEC (18) by the change in Na^+ saturation desired (25 − 5 = 20%).

$$18 \; cmol_c/kg \times 0.20 = 3.6 \; cmol_c/kg$$

From the reaction that occurs when the gypsum is applied,

$$\boxed{Micelle} \; \begin{matrix} Na^+ \\ Na^+ \end{matrix} + CaSO_4 \cdot 2H_2O \; \rightleftharpoons \; \boxed{Micelle} \; Ca^{2+} + Na_2SO_4 + 2H_2O$$

We know that the Na^+ is replaced by a *chemically equivalent* amount of Ca^{2+} in the gypsum ($CaSO_4 \cdot 2H_2O$). In other words, 3.6 $cmol_c$ of $CaSO_4 \cdot 2H_2O$ will be needed to replace 3.6 $cmol_c$ of Na^+.

Second, calculate the weight in grams of gypsum needed to provide the 3.6 $cmol_c/kg$ soil. This can be done by first dividing the molecular weight of $CaSO_4 \cdot 2H_2O$ (172) by 2 (since Ca^{2+} has two charges and Na^+ only one) and then by 100 since we are dealing with $centimole_c$ rather than $mole_c$.

$$\frac{172}{2} = 86 \; g \; CaSO_4 \cdot 2H_2O/mol_c$$

$$and \; \frac{86}{100} = 0.86 \; g \; CaSO_4 \cdot 2H_2O/cmol_c \; required \; to \; replace \; 1 \; cmol_c \; Na/100 \; g \; soil$$

The 3.6 $cmol_c Na^+/kg$ would require

$$3.6 \; cmol_c/kg \times 0.86 \; g/cmol_c = 3.1 \; g \; CaSO_4 \cdot 2H_2O/kg \; of \; soil$$

Last, to express this in terms of the amount of gypsum needed to treat one hectare of soil to a depth of 30 cm, multiply by 4.4×10^6, which is twice the weight in kg of a 15-cm-deep hectare-furrow slice (see Section 4.8).

$$3.1 \; g/kg \times 4.4 \times 10^6 \; k/ha = 13,640,000 \; g \; gypsum/ha$$

This is 13,640 kg/ha, 13.6 Mg/ha or about 6.0 tons per acre.

Because of impurities in the gypsum and the inefficiency of the overall process, these amounts would likely be adjusted upward by 20–30% in actual field practice.

the less harmful sodium sulfate but also decreases the alkalinity. The reactions of sulfuric acid with the compounds containing sodium may be shown as follows:

$$2NaHCO_3 + H_2SO_4 \rightleftharpoons 2CO_2 \uparrow + 2H_2O + Na_2SO_4$$

$$Na_2CO_3 + H_2SO_4 \rightleftharpoons CO_2 \uparrow + H_2O + Na_2SO_4$$
(leachable)

$$\begin{matrix} Na^+ \\ Na^+ \end{matrix} \boxed{Micelle} + H_2SO_4 \rightleftharpoons \begin{matrix} H^+ \\ H^+ \end{matrix} \boxed{Micelle} + Na_2SO_4$$
(leachable)

Not only are the sodium carbonate and bicarbonate changed to sodium sulfate, a mild neutral salt, but the carbonate radical is removed from the system. When gypsum is used, however, a portion of the carbonate may remain as a calcium compound ($CaCO_3$).

In research trials sulfur and even sulfuric acid have proven to be very effective in the reclamation of sodic soils, especially if large amounts of $CaCO_3$ are present. In practice, however, gypsum is much more widely used than the acid-forming materials. Gypsum is less expensive, is widely available in both natural and in industrial by-product forms, and is more easily handled. Care must be taken, however, to be certain that the gypsum is finely ground and that it is well mixed with the upper soil horizons so that its solubility and rate of reaction are maximized.

Physical Condition

The effects of gypsum and sulfur on the physical condition of sodic soils is perhaps more spectacular than are the chemical effects. Sodic soils are almost impermeable to rainwater and irrigation water since the soil colloids are largely dispersed and the soil is essentially void of stable aggregates. When the exchangeable Na^+ ions are replaced by Ca^{2+} or H^+, soil aggregation and improved water infiltration results (see Figure 10.11). The neutral sodium salts (e.g., Na_2SO_4) formed when the exchange takes place can then be leached from the soil, thereby reducing both salinity and sodicity.

The reclamation effects of gypsum or sulfur are greatly accelerated by plants growing on the soil. Crops such as sugar beets, cotton, barley, sorghum, berseem clover, or rye that have some degree of tolerance to saline and sodic soils can be grown initially. Their roots help provide channels through which gypsum can move downward into the soil. Deep-rooted crops such as alfalfa are especially effective in improving the water conductivity of gypsum-treated sodic soils. Figure 10.11 illustrates the ameliorating effects of the combination of gypsum and deep-rooted crops.

Some research suggests that aggregate-stabilizing synthetic polymers may be helpful in at least temporarily increasing the water infiltration capacity of gypsum-treated sodic soils. Data in Table 10.5 from one experiment shows the possible potential of these soil conditioners.

Irrigation Timing

The timing of irrigation is extremely important on salty soils, particularly during the spring planting season. Because young seedlings are especially sensitive to salts, irrigation often precedes or follows planting to move the salts downward and away from the seedling roots (see Figure 10.10). After the plants are well established, their salt tolerance is somewhat greater.

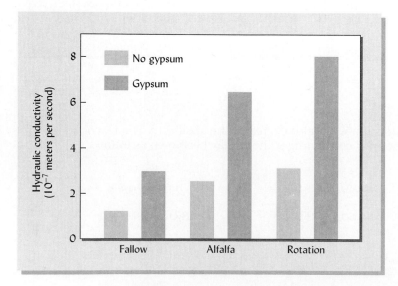

FIGURE 10.11 The influence of gypsum and growing crops on the hydraulic conductivity of the upper 20 centimeters of a saline-sodic soil (Natrustaff) in Pakistan. The use of gypsum to increase water conductivity in all plots was more effective when deep-rooted alfalfa and rotation of sesbenia-wheat were grown. [Drawn from selected data from Ilyas et al. (1993); used with permission of the Soil Science Society of America]

TABLE 10.5 Use of Gypsum and Synthetic Polymers on Reclamation of a Sodic Soil

Adding gypsum to samples of a fine-textured (clay) saline-sodic soil, a Mollisol from California, increased both the hydraulic conductivity and the salts leached in the experiment while decreasing the exchangeable sodium percentage (ESP). Adding two experimental synthetic polymers (T4141 and 21J) with the gypsum gave even greater increases in hydraulic conductivity and leached salts.

	Characteristic measured					
	Hydraulic conductivity (mm/hr)		Total salts leached (mg/kg)		ESP (%)	
Gypsum added→	No	Yes	No	Yes	No	Yes
Polymer treatment						
No polymer	0.0	0.06	0.0	4.7	22.9	9.6
T4141	0.0	0.28	0.0	10.1	25.4	9.6
21J	0.0	0.28	0.0	9.7	25.5	9.6

Data from Zahow and Amrhein (1992).

Ammonia Effects

The practice in recent years of adding nitrogen as anhydrous ammonia to irrigation water as it is applied to a field has created some soil problems. The high pH brought about by the ammonia causes some of the calcium in the water to precipitate, and thus raises the sodium adsorption ratio (SAR) and the hazard of increased exchangeable sodium percentage (ESP). To counteract these difficulties, sulfuric acid is sometimes added to the irrigation water to reduce its pH as well as that of the soil. This practice may well spread where there are economical sources of sulfuric acid and where the personnel applying it have been trained and alerted to serious hazards of using this strong acid.

10.10 MANAGEMENT OF RECLAIMED SOILS

Once salt-affected soils have been reclaimed, prudent management steps must be taken to be certain that the soils remain productive. For example, surveillance of the EC and SAR and other pertinent chemical characteristics of the irrigation water is essential. Management adjustment would be needed to accommodate any change in water quality that could affect the soil. The number and timing of irrigation episodes could help determine the balance of salts entering and leaving the soil. Likewise, the maintenance of good internal drainage is essential for the removal of excess salts.

Steps should also be taken to monitor appropriate chemical characteristics of the soils such as pH, EC, and SAR, as well as specific levels of elements such as boron and selenium that could lead to chemical toxicities. These measurements will help determine the need for subsequent remedial practices and/or chemicals.

Crop and soil fertility management to maintain satisfactory yield levels is essential to maintain the overall quality of salt-affected soils. The crop residues (roots and aboveground stalks) will help maintain organic matter levels and good physical condition of the soil. To maintain high yields, micronutrient and phosphorus deficiencies characteristic of other high pH soils will need to be overcome by appropriate organic and inorganic sources.

10.11 CONCLUSION

Alkaline and salt-affected soils are vast in extent. They dominate arid and semiarid regions where more than half the world's arable lands are found. They are widely used for rangeland and dryland farming. When irrigated, these soils can be among the most productive in the world. Since their pH levels are high, management practices used on these soils are quite different from those applied to acid soil regions. Deficiencies in alkaline soils of some essential micronutrients (such as Fe and Mn) and macronutrients (such as phosphorus) must be overcome.

There are three classes of salt-affected soils. *Saline soils* are dominated by neutral salts (electrical conductivity (EC) of 4 dS/m or higher) and by pH values less than 8.5. *Saline-sodic soils* have similar salt and pH levels, but exchangeable sodium percentages are 15 or above. *Sodic soils* also have high exchangeable sodium percentages (above 15), but their soluble salt concentrations are relatively low (EC < 4 dS/m) and their pH values are higher than 8.5. The physical conditions of saline and saline-sodic soils are satisfactory for plant growth, but the colloids in sodic soils are largely dispersed, the soil is puddled, and the water infiltration rate is extremely slow.

The reclamation of saline, saline-sodic, and sodic soils requires two major actions: (1) the removal by leaching of excess soluble salts from the upper levels of the soil profile and (2) the removal of exchangeable sodium ions, first from the exchange complex by replacing the Na^+ ion with either Ca^{2+} or H^+ ions and then from the soil solution by leaching the replaced Na^+ ion from the soil. Gypsum ($CaSO_4 \cdot 2H_2O$) and elemental sulfur (S) are chemicals that can supply the Ca^{2+} and H^+ ions needed. Monitoring the chemical content of both the irrigation water and the soil is essential to achieve this goal of removing the Na^+ ions from the exchange complex and, ultimately, the soil.

REFERENCES

Beek, K. L., W. A. Blokhuis, P. M. Driessen, N. Van Breeman, N. Brinkman, and L. J. Pons. 1980. "Problem soils: their reclamation and management," ILRI Publ. No. 27 (Wageningen, Netherlands: ILRI), pp. 47–72.

Bresler, E., B. L. McNeal, and D. L. Carter. 1982. *Saline and Sodic Soils, Principles—Dynamics—Modeling* (Berlin: Springer-Verlag).

Carter, D. L. 1981. "Salinity and plant productivity," in *CRC Handbook Series in Nutrition and Food* (Boca Raton, Fla.: CRC Press).

Ilyas, M., R. W. Miller, and R. H. Qureshi. 1993. "Hydraulic conductivity of saline-sodic soil after gypsum application," *Soil Sci. Soc. Amer. J.*, **57**:1580–1585.

Maas, E. U. 1984. "Crop tolerance," *Calif. Agric.* **38**:20–21.

National Academy of Sciences. 1990. *Saline Agriculture: Salt Tolerant Plants for Developing Countries* (Washington, D.C.: National Academy Press).

Retana, J., D. R. Parker, C. Amrhein, and A. L. Page. 1993. "Growth and trace element concentrations of 5 plant species grown on a highly saline soil," *J. Environ. Qual.*, **22**: 805–811.

Rhoades, J. D. 1993. "Electrical conductivity methods for measuring and mapping soil salinity," *Adv. in Agron.*, **49**:201–251.

Richards, L. A. (ed.). 1947. *Diagnosis and Improvement of Saline and Alkali Soils* (Riverside, Calif.: U.S. Regional Salinity Lab.).

Royo, A., and R. Aragues. 1993. "Validation of salinity crop production functions obtained with the triple line source sprinkler system," *Agron J.*, **85**:795–800.

Zahow, M. F., and C. Amrhein. 1992. "Reclamation of a saline sodic soil using synthetic polymers and gypsum," *Soil Sci. Soc. Amer. J.*, **56**:1257–1260.

11

ORGANISMS AND ECOLOGY OF THE SOIL[1]

> *Under the silent, relentless chemical jaws of the fungi, the debris of the forest floor quickly disappears. . . .*
> —A. FORSYTH AND K. MIYATA, TROPICAL NATURE

The terms *ecosystem* and *ecology* usually call to mind scenes of lions stalking vast herds of wildebeest on the grassy savannas of East Africa, or the interplay of phytoplankton, fish, and fishermen in some great estuary. Like a savanna or an estuary, a soil is an ecosystem in which thousands of different creatures interact and contribute to the global cycles that make all life possible. This chapter will introduce some of the organisms in the living drama that goes on largely unseen in the soil beneath our feet. If our bodies were small enough to enter the tiny passages in the soil, we would discover a world populated by great numbers of organisms fiercely competing for every leaf that falls to the forest floor, every root that sloughs off the growing grass, every fecal pellet, every dead body that reaches the soil. We would also find predators of every description lurking in the dark, some with fearsome jaws to snatch unwary victims, others whose jellylike bodies simply engulf and digest their microscopic prey.

We will learn how these organisms, both flora and fauna, interact with one another, what they eat, how they affect the soil, and how soil conditions affect them. The central theme will be how this community of organisms assimilates plant and animal residues and waste products, creating soil humus, recycling carbon and mineral nutrients, and supporting plant growth. We will emphasize the activities rather than the scientific classification of these organisms. Consequently, we will consider only very broad, simple categories.

The animals (**fauna**) of the soil range in size from **macrofauna** (such as moles, prairie dogs, earthworms, and millipedes) through **mesofauna** (such as tiny springtails and mites) to **microfauna** (such as nematodes and single-celled protozoans). Plant organisms (**flora**) include the roots of higher plants as well as microscopic algae and diatoms. Other microorganisms (too small to be seen without the aid of a microscope) include fungi, bacteria, and actinomycetes. A fertile soil is literally teeming with life. Three groups of organisms tend to predominate in terms of numbers, mass, and metabolic activity: fungi, actinomycetes, and bacteria. All the organisms of the soil, both large and small, cause important effects both on soil properties and on the functioning of larger ecosystems.

---

[1]For a review of soil biology see Paul and Clark (1989) and Killham (1994).

11.1 TYPES OF ORGANISMS IN THE SOIL

Soil organisms are creatures that spend all or part of their lives in the soil environment. Every handful of soil is likely to contain billions of organisms with representatives of nearly every phylum of living things. A simplified, general classification of these organisms is shown in Figure 11.1.[2]

Notice that we have classified the organisms according to their size (*macro* being larger than 2 mm in width, *meso* being between 0.2 and 2 mm, and *micro* being less than 0.2 mm) as well as by their ecological functions (what they eat). A typical, healthy soil might contain several species of vertebrate animals (mice, gophers, snakes, etc.), a half dozen species of earthworms, 20 to 30 species of mites, 50 to 100 species of insects (collembola, beetles, ants, etc.), dozens of species of nematodes, hundreds of species of fungi, and perhaps thousands of species of bacteria and actinomycetes.

This diversity is possible because of the nearly limitless variety of foods and the wide range of habitat conditions found in soils. Within a handful of soil there may be areas of good and poor aeration, high and low acidity, cool and warm temperatures, moist and dry conditions, and localized concentrations of dissolved nutrients, organic substrates, and competing organisms. A healthy soil is indeed one of nature's most diverse and complex ecosystems.

11.2 ORGANISMS IN ACTION

The activities of soil flora and fauna are intimately related in what ecologists call a *food chain* or, more accurately, a *food web*. Some of these relationships are shown in Figure 11.2, which illustrates how various soil organisms are involved in the degradation of residues from higher plants.

Primary Producers

As in all ecosystems, plants play the role of **primary producers** by using energy from the sun, water, and carbon from atmospheric carbon dioxide to make organic molecules and living tissues. These organic materials, and the energy they contain, are then utilized by other organisms, either directly or after having been passed on through intermediaries.

HERBIVORES AND DETRITIVORES. Certain soil organisms eat live plants and are called *herbivores.* Examples are parasitic nematodes and insect larvae that attack plant roots, as well as termites, ants, beetle larvae, woodchucks, and mice that devour aboveground plant parts. Because they attack living plants that may be of value to humans, many of these soil herbivores are considered pests. For the vast majority of soil organisms, however, the principal source of food is the debris of dead tissues left by plants and animals. This debris is called *detritus,* and the animals that directly feed on it are called *detritivores.*

Plant and animal detritus is continually being deposited in and on the soil. This deposition of organic materials is particularly dramatic in deciduous forests, as trees drop their leaves at the end of the growing season. Each fall the soil in temperate regions is covered with these leaves to a depth of 15 cm or more. Yet, by the following summer, most of these residues seem to have disappeared. The soil surface may show a few leaf petiole fragments, but little else as evidence of the mass of material which was deposited the previous fall. The assimilation of this mass of plant and animal debris is a critical function of both large and small soil organisms.

Primary Consumers

As soon as a leaf, a stalk, or a piece of bark drops to the ground, it is subject to coordinated attack by microflora and by detritivores, animals that live on dead and decaying plant tissues (Figure 11.2). These animals, which include mites, woodlice, and earthworms, chew or tear holes in the tissue, opening it up to more rapid attack by the microflora. The animals and microflora that use the energy stored in the plant residues are termed *primary consumers.*

---

[2]Some biologists prefer to classify living organisms into five kingdoms rather than two, since it has proven difficult to fit many microorganisms into either the plant or animal kingdoms.

Animals (fauna), all heterotrophs

Macrofauna: Largely herbivores and detritivores

Vertebrates	Gophers, mice, squirrels
Arthropods	Ants, beetles and their larvae, maggots, termites, grubs, spiders, millipedes, woodlice
Annelida	Earthworms
Mollusca	Snails, slugs

Largely predators

Vertebrates	Moles, snakes
Arthropods	Beetles, ants, centipedes

Mesofauna: Largely detritivores

Arthropods	Mites, collembola (springtails)
Annelida	Enchytraeid (pot) worms

Largely predators

Arthropods	Mites, protura

Microfauna: Detritivores, predators, fungivores, bacterivores

Nematoda	Nematodes
Roterifera*	Rotifers, water bears
Protozoa*	Amoebae, ciliata

Plants (flora)

Macroflora: Largely autotrophs

Vascular plants	Feeder roots
Bryophytes	Mosses

Microflora: Largely autotrophs

Vascular plants	Root hairs
Algae	Greens, yellow-greens, diatoms

Largely heterotrophs, aerobic

Fungi	Yeasts, mildews, molds, rusts, mushrooms
Actinomycetes†	Many kinds of actinomycetes

Autotrophs and heterotrophs

Bacteria†	Aerobes, anaerobes
Cyanobacteria†	Blue-green algae

*In some classification schemes these are placed in the kingdom Protista, rather than Animalia.
† Often classified together in the kingdom Monera.

FIGURE 11.1 General classification of some important groups of soil organisms. Some organisms subsist on living plants (herbivores), others on dead plant debris (detritivores). Some consume animals (predators); some devour fungi (fungivore) or bacteria (bacterivores); and some live off of, but do not consume, other organisms (parasites). Heterotrophs rely on organic compounds for their carbon and energy needs while autotrophs obtain their carbon mainly from carbon dioxide and their energy from photosynthesis or oxidation of various elements.

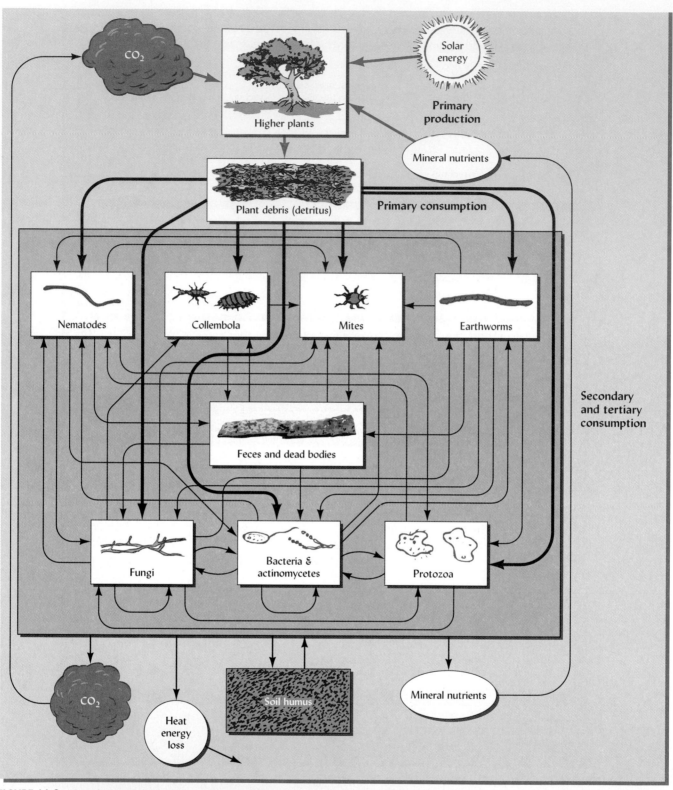

FIGURE 11.2 Greatly simplified diagram of the food web involved in the breakdown of higher-plant tissue. The boxes represent broad groups of organisms and pools of organic material, while the arrows represent transfers of carbon, energy, and nutrients between these pools. Because they capture carbon dioxide and energy, the higher plants are known as *primary producers*. Heavy arrows from the plant debris (detritus) to various organism groups represent *primary consumption*. The arrows within the large box represent *secondary* and *tertiary consumption*. Although all the groups shown play important roles in the process, the microorganisms represented by the lower three boxes account for 80 to 90% of the total metabolic activity. As a result of this metabolism, soil humus is synthesized, and carbon dioxide, heat energy, and mineral nutrients are released into the soil environment.

ACTION OF THE FAUNA. While the actions of the microflora are mostly chemical, those of the fauna are both physical and chemical. The mesofauna and macrofauna chew the plant residues and move them from one place to another on the soil surface and even into the soil. Earthworms literally eat their way through the soil as they incorporate plant residues into the mineral soil by passing them through their bodies along with mineral soil particles. Larger animals, such as gophers, moles, prairie dogs, and rats, also burrow into the soil and bring about considerable soil mixing and granulation.

The actions of these animals enhance the activity of the microflora in several ways. First, the chewing action fragments the litter, cutting through the resistant waxy coatings on many leaves to expose the more easily decomposed cell contents for microbial digestion. Second, the chewed plant tissues are thoroughly mixed with microorganisms in the animal gut where conditions are ideal for microbial action. Third, the mobile animals carry microorganisms with them and help the latter to disperse and find new food sources to decompose. The ability of the soil fauna to enhance plant residue decomposition is illustrated in Figure 11.3.

Secondary Consumers

The bodies of primary consumers are food sources for predators and parasites in the soil. These *secondary consumers* include small forms of plant life such as bacteria, fungi, algae, and lichens, as well as carnivores such as centipedes (which consume other animals such as small insects), spiders, nematodes, snails, and certain moles that feed primarily on earthworms. Examples of *microphytic feeders,* which use the microflora as sources of food, are certain collembola (springtail insects), mites, termites, nematodes, and protozoa (see Figure 11.2). Meso- and microfauna (especially those nematodes, mites, and collembola that feed on fungi and bacteria) exert considerable influence over the activity of fungi and bacteria. The grazing of these meso- and microfauna on microbial colonies may stimulate faster growth and activity among the microbes, in much the same way that grazing animals can stimulate the growth of pasture grasses. In other cases, the attack of the microphytic feeders may kill off so much of a microbial colony as to inhibit the work of the microorganisms.

Tertiary Consumers

Moving farther up the food chain, the secondary consumers are prey for still other carnivores, called *tertiary consumers*. For example, ants consume centipedes, spiders, mites, and scorpions—all of which can themselves prey on primary or secondary consumers.

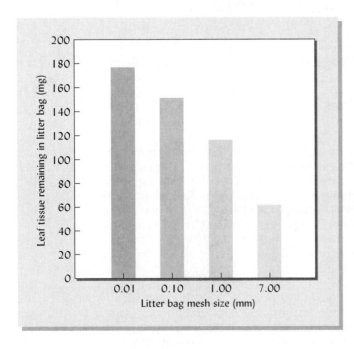

FIGURE 11.3 Influence of various sizes of soil organisms on the decomposition of corn leaf tissue buried in soil. Small bags made of nylon material with four different-size openings (mesh size) were filled with 558 mg (dry weight) of corn leaf tissue and buried in the soil for 10 weeks. The amount of corn leaf tissue remaining in the bags was considerably greater (less decomposition had taken place) where the meso- and macrofauna were excluded by the smaller mesh sizes. [Data from Weil and Kroontje (1979)]

The microflora are intimately involved in every level of the process. In addition to their direct attack on plant tissue (as primary consumers), microflora are active within the digestive tracts of many soil animals, helping these animals digest more resistant organic materials. The microflora also attack the finely shredded organic material in animal feces, and they decompose the bodies of dead animals. For this reason they are referred to as the *ultimate decomposers*.

As indicated in Figure 11.2, broad groups of organisms, such as the fungi or mites, include some species that are primary consumers, some that are secondary consumers, and others that are tertiary consumers. For example, some mites attack detritus directly, while others eat mainly fungi or bacteria that are growing on the detritus. Still others attack and devour the mites that eat fungi. Also note that there are two-way interactions between most groups. For example, some nematodes eat fungi, and some fungi attack nematodes.

Ecologists consider a community to be highly diverse when the individual organisms are fairly evenly distributed among a large number of species. In general, such complexity and diversity leads to a stable ecosystem because there are several organisms to carry out each task, and no single organism can become completely dominant. Aquatic ecologists have long considered a highly diverse community of aquatic organisms to indicate good water quality. In the same way, soil scientists are now using the concept of biological diversity as an indicator of *soil quality*.

11.3   ORGANISM ABUNDANCE, BIOMASS, AND METABOLIC ACTIVITY

Soil organism numbers are influenced by many factors, including climate, vegetation, and the physical and chemical characteristics of soils. The species composition in an arid desert will certainly be different from that in a humid forest, which, in turn, will be quite different from that in a cultivated field. Acid soils are populated by different species from those in alkaline soils. Likewise, species diversification and numbers are different in a tropical rain forest from those in a cool temperate area.

Despite these variations, there are a few generalizations that can be made. For example, forested areas support a more diverse soil fauna than do grasslands, although the total faunal weight per hectare and faunal activity is higher in the grasslands (see Table 11.1). Likewise, cultivated fields are generally lower than undisturbed native lands in numbers and biomass of the soil organisms, especially the fauna. The total soil biomass is generally related to the amount of organic matter present; on a dry weight basis the living portion is usually between 1 and 8% of the total soil organic matter. Also, scientists commonly observe that the ratio of soil organic matter to detritus to microbial biomass to faunal biomass is approximately 1000:100:10:1.

Comparative Organism Activity

The activities of specific groups of soil organisms are commonly identified by (1) their numbers in the soil, (2) their weight (biomass) per unit volume or area of soil, and (3) their metabolic activity (often measured as the amount of carbon dioxide given off in respiration). The numbers and biomass of groups of organisms that commonly occur in soils are shown in Table 11.2. Although the relative metabolic activities are not shown, they are generally related to the biomass of the organisms.

As might be expected, the microorganisms are the most numerous and have the highest biomass. Together with earthworms (or termites in the case of some tropical soils) the microflora also dominate the biological activity in most soils. It is estimated that 60 to 80% of the total soil metabolism is due to the microflora, although, as previously mentioned, their activity is enhanced by the actions of soil fauna. For these reasons, major attention will be given to the microflora, along with earthworms, termites, and certain other fauna.

Even those animals that account for only very small fractions of the total metabolism in the soil can play important roles in soil formation (see Chapter 2) and management. Rodents pulverize, mix, and granulate soil and incorporate surface organic residues into lower horizons. They provide large channels through which water and air can move freely. Ants are important in localized areas, especially in warm regions, for their exceptional ability to break down woody materials and turn over soil materials as they build their nests. Mesofauna detritivores (mostly mites and collem-

TABLE 11.1 Biomass of Groups of Soil Animals Under Grassland and Forest Cover

The mass and in turn the metabolism are greatest under grasslands. The spruce, with low-base-containing leaves, encourages acid conditions and slow organic matter decomposition.

| | Biomass (g/m²)[a] | | |
| | | Forest | |
Group of organisms	Grassland meadow	Oak	Spruce
Herbivores	17.4	11.2	11.3
Detritivores			
Large	137.5	66.0	1.0
Small	25.0	1.8	1.6
Predators	9.6	0.9	1.2
Total	189.5	79.9	15.1

Data from Macfadyen (1963).
[a]Depth about 15 cm.

bola) (see Figure 11.4) translocate and partially digest organic residues and leave their excrement for microfloral degradation. By living in the soil, many animals favorably affect its physical condition. Others use the soil as a habitat for destructive action against higher plants. In any case, all these organisms significantly impact both the soil and growing plants.

Source of Energy and Carbon

Soil organisms may be classified on the basis of their source of energy and carbon as either **autotrophic** or **heterotrophic** (see Figure 11.1). The heterotrophic soil organisms obtain their energy and carbon from the breakdown of organic soil materials. These organisms are far more numerous than the autotrophs and are responsible for the largest volume of decay. They include the soil fauna, most bacteria, and the fungi and actinomycetes.

The autotrophs obtain their energy from sources other than the breakdown of organic materials. Some use solar energy (*photoautotrophs*) while others use inorganic elements such as nitrogen, sulfur, and iron (*chemautotrophs*). Most obtain their carbon from carbon dioxide or carbonate minerals.

TABLE 11.2 Relative Numbers and Biomass of Fauna and Flora Commonly Found in Surface[a] Soils

Since metabolic activity is generally related to biomass, microflora and earthworms dominate the life of most soils.

| | Number | | Biomass[b] | |
Organisms	per m²	per gram	(kg/ha)	(g/m²)
Microflora				
Bacteria	10^{13}–10^{14}	10^8–10^9	400–5000	40–500
Actinomycetes	10^{12}–10^{13}	10^7–10^8	400–5000	40–500
Fungi	10^{10}–10^{11}	10^5–10^6	1,000–20,000	100–2,000
Algae	10^9–10^{10}	10^4–10^5	10–500	1–50
Fauna				
Protozoa	10^9–10^{10}	10^4–10^5	20–200	2–20
Nematodes	10^6–10^7	10–10^2	10–150	1–15
Mites	10^3–10^6	1–10	5–150	.5–1.5
Collembola	10^3–10^6	1–10	5–150	.5–1.5
Earthworms	10–10^3		100–1700	10–170
Other fauna	10^2–10^4		10–100	1–10

[a]Generally considered 15 cm (6 in.) deep, but in some cases (e.g., earthworms) a greater depth is used.
[b]Biomass values are on a liveweight basis. Dry weights are about 20–25% of these values.

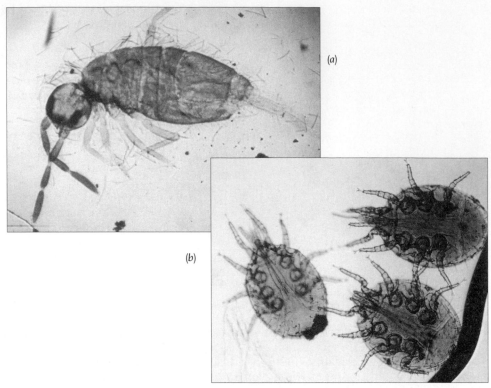

FIGURE 11.4 Springtails (collembala) (a) and mites (b) are mesofauna detritivores commonly found in soils. (Photos courtesy of R. Weil)

11.4 EARTHWORMS[3]

Earthworms are probably the most important macroanimal in soils (Figure 11.5). They are egg-laying hermaphrodites (no separate male and female organisms) that eat detritus, soil organic matter, and microorganisms found on these materials. They do not eat living plants or their roots, and so do not act as pests to crops.[4] Of the more than 1800 species known worldwide, *Lumbricus terrestris,* a deep-boring, reddish organism, and *Allolobophora caliginosa,* a shallow-boring, pale pink organism, are the two most common in Europe and in the eastern and central United States. *Lumbricus terrestris,* commonly called *nightcrawlers,* live in the soil head up, coming to the surface to pull detritus into their burrows, often creating middens of gathered detritus near the burrow entrance (see Figure 11.5). *Allolobophora caliginosa,* commonly called *red worms,* live head down in the soil, and expel their casts onto the soil surface. In the tropics and semitropics, other types of earthworms are prevalent, and dominate the soil fauna in regions with at least 800 mm of annual rainfall. Some tropical species are small and others surprisingly large. In Australia, aboriginal peoples have been known to hunt earthworms that grow up to 1 meter long.

Interestingly, *Lumbricus terrestris* is not native to America, but was brought over from Europe probably in fruit tree stock carried by the settlers. As forests and prairies were put under cultivation, this European-introduced worm rapidly replaced the native types, which could not withstand the change in soil environment brought about by tilling the soil. Virgin lands, however, still retain at least part of their native populations.

Influence on Soil Fertility and Productivity

Earthworms literally eat their way through the soil. They may ingest a weight of soil equal to 2 to 30 times their own weight in a single day. In a year, the earthworm popu-

---

[3]For an excellent review of earthworms, see Lee (1985).

[4]Earthworms should not be confused, in this regard, with wireworms, rootworms, or other soil-inhabiting insect larvae that are commonly referred to as *worms,* and may be serious plant pests.

lation of a hectare of land may ingest between 50 and 1000 Mg of soil (22 to 450 tons/acre), the higher figure occurring in moist tropical climates. In so doing, they create extensive systems of burrows and eject enormous quantities of partially digested soil and organic materials. The earthworm's wastes are called *casts,* and usually take the form of globular soil aggregates of high stability (see Table 11.3). The casts are deposited within the soil profile or on the soil surface, depending on the species of earthworm. Both the piles of casts on the soil surface and the circular holes of burrow entrances are important signs of earthworm activity. In many cases, pieces of plant litter may have to be moved aside to observe the burrow entrances. Another way to observe earthworms is to dig into the surface soil to a depth of about 30 cm. Under favorable conditions there should be several earthworms in every shovelful of upturned soil.

The activities of earthworms greatly enhance soil fertility and productivity by altering both physical and chemical conditions in the soil, especially in the upper 15 to 35 cm of soil. Earthworms increase the availability of mineral nutrients to plants in two ways. First, as soil and organic materials pass through the earthworm's body, they are ground up physically as well as attacked chemically by the digestive enzymes of the earthworm and its gut microflora. Second, as earthworms ingest detritus and soil organic matter of relatively low nitrogen, phosphorus, and sulphur concentrations they assimilate part of this material into their own body tissues. Consequently, their bodies contain high concentrations of these nutrients, which are then readily released to plants when the earthworm dies and decays.

Compared to the bulk soil, the casts are definitely higher in bacteria and organic matter and available plant nutrients (Table 11.3). The casts are mostly deposited in burrows that provide paths of low resistance to root growth, further enhancing the plant availability of these nutrients. The lush growth of grass around earthworm casts seen in lawns and pastures is evidence of the favorable effect of earthworms on soil productivity. Furthermore, physical incorporation of surface residues into the soil by earthworms and other soil animals reduces the loss of nutrients, especially nitrogen, by erosion and volatilization. This physical action is particularly important in grasslands and other uncultivated soils (including those where reduced tillage is practiced) and has earned the earthworm the title of "nature's tiller."

BENEFICIAL PHYSICAL EFFECTS.   Earthworms are important in other ways. The holes left in the soil serve to increase aeration and drainage, an important consideration in crop production and soil development. In turf grass, the mixing activity of earthworms can alleviate or reduce compaction problems and nearly eliminate the formation of an undesirable thatch layer. Under conditions of heavy rainfall, earthworm burrows may greatly increase the infiltration of water into the soils (see Figure 11.6), thus playing an important role in water conservation and prevention of soil erosion. Moreover, earthworms mix and granulate the soil, their casts forming a substantial portion of the stable aggregates in many soils. Earthworms increase both the size and stability of the soil

FIGURE 11.5   Earthworms are perhaps the most significant macroorganism in soils of humid temperate regions, particularly in relation to their effects on the physical conditions of soils. Surface feeding species such as *Lumbricus terrestris* incorporate large amounts of plant litter into the soil and often gather plant debris near their burrow entrances to form *middens.* (Photo courtesy of R. Weil)

TABLE 11.3   Comparative Characteristics of Earthworm Casts and Soils

Average of six Nigerian soils.

Characteristic	Earthworm casts	Soils
Silt and clay (%)	38.8	22.2
Bulk density (Mg/m³)	1.11	1.28
Structural stability[a]	849	65
Cation exchange capacity (cmol/kg)	13.8	3.5
Exchangeable Ca^{2+} (cmol/kg)	8.9	2.0
Exchangeable K^+ (cmol/kg)	0.6	0.2
Soluble P (ppm)	17.8	6.1
Total N (%)	0.33	0.12

From de Vleeschauwer and Lal (1981).
[a]Numbers of raindrops required to destroy structural aggregates.

aggregates. As much as 50 to 60% of the surface soil may consist of stable earthworm casts. Also, fungal hyphae that proliferate in earthworm burrows help bind soil particles into stable aggregates.

Influence on Chemical Leaching[5]

Not all effects of earthworms are beneficial. For example, *Lumbricus terrestris* has been observed to remove the crop residue cover from much of the soil surface in the process of building middens (see Figure 11.5). It is not yet known to what extent this action may leave the exposed soil susceptible to the impact of raindrops, crust formation, and increased erosion. Similarly, on closely cropped golf greens, earthworm casts on the soil surface may be considered a nuisance.

Another aspect of concern is that water rapidly percolating down vertical earthworm burrows may carry potential pollutants toward the groundwater. However, the material with which earthworms line their burrows is greatly enriched in organic matter. Compared to the surrounding soil, this lining material may have up to twice the capacity to adsorb certain herbicides, substantially retarding their transport through the soil profile. Clearly, on balance, earthworms are very beneficial and their activity is generally to be encouraged by soil managers.

---

[5]For recent advances in our understanding of this influence see Stehouwer et al. (1994).

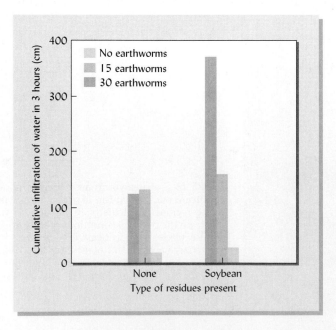

FIGURE 11.6   Earthworms may dramatically increase the infiltration of water into a soil, especially where they can feed on plant residues left on the soil surface. These data are for a Raub soil (Mollisol) in Illinois, in which a known number of earthworms were present. (From Kladivko et al., 1986)

Factors Affecting Earthworm Activity

Earthworms prefer a well-aerated but moist habitat. For this reason, they are found most abundantly in medium-textured upland soils where the moisture capacity is high rather than in droughty sands or poorly drained lowlands. Earthworms must have organic matter as a source of food. Generally they grow best on fresh, undecomposed organic matter. Consequently, they thrive where farm manure or plant residues have been added to the soil. A few species are reasonably tolerant to low pH, but most earthworms thrive best where the soil is not too acid (pH 5.5 to 8.5) and has an abundant supply of calcium (which is an important component of their slime excretions). However, most earthworms are quite sensitive to excess salinity.

Soil temperature affects earthworm numbers and their distribution in the soil profile. For example, a temperature of about 10°C (50°F) appears optimum for *Lumbricus terrestris*. This temperature sensitivity, together with preference for moist soil, probably accounts for the maximum earthworm activity noted in spring and autumn in temperate regions.

Some earthworms burrow deeply into the profile, thereby avoiding unfavorable moisture and temperature conditions. Penetration as deep as 1 to 2 meters is not uncommon. Unfortunately, in soils not insulated by residue mulches, a sudden heavy frost in the fall may kill the organisms before they can move lower in the profile. Under such circumstances soil cover is important in maintaining a high earthworm population.

Other factors that depress earthworm populations are predators (moles, mice, and certain mites and millipedes); very sandy soils (partly because of the abrasive effect of sharp sand grains); direct contact with ammonia fertilizer; application of certain insecticides (especially carbamates); and tillage. The last factor is often the overriding deterrent to earthworm populations in agricultural soils. Minimum tillage with plenty of crop residues left as a mulch on the soil surface is ideal for encouraging earthworms.

Because of their sensitivity to soil and other environmental factors, the numbers of earthworms vary widely in different soils. In very acid forest soils (Spodosols), an average of fewer than one organism per square meter is common. In contrast, more than 500 per square meter have been found on rich grassland soils (Mollisols) and high organic matter cropland managed with minimum tillage. The numbers commonly found in arable soils range from 30 to 300 per square meter, equivalent to from 300,000 to 3 million per hectare. The biomass or liveweight for this number would range from perhaps 110 to 1100 kg/ha (100–1000 lb/acre).

11.5   TERMITES[6]

Termites, or *white ants,* are major contributors to the breakdown of organic material in or at the surface of soils. There are about 2000 species of termites. They are found in about two-thirds of the land areas of the world but are most prominent in the grasslands (savannas) and forests of tropical and subtropical areas (both humid and semiarid). Their activity is on a scale comparable to that of earthworms. Up to 16 million individuals have been recorded in a hectare of tropical deciduous forest. In the drier tropics (less than 800 mm annual rainfall) termites surpass earthworms in dominating the soil fauna.

Termite Mounds

Termites are social animals that live in very complex labyrinths of nests, passages, and chambers that they build both below and above the soil surface. Termite mounds built of cemented soil particles are characteristic features of many landscapes in Africa, Latin America, Australia, and Asia (Figure 11.7). These mounds, and the network of underground passages and aboveground covered runways that typically spreads 20 to 30 meters beyond the mounds, are essentially termite "cities." Although a few species eat living woody plants and sound deadwood, most termites eat rotting woody materials and plant residues. Several species, such as *Macrotermes* spp. in Africa, use plant residues to grow fungi in their mounds as a source of food.

---

[6]For an excellent discussion of termites and ants in the tropics, see Lal (1987).

FIGURE 11.7 Termite mounds in cultivated fields in Africa. A termite mound in a cornfield (a). Dissected profile of another mound showing the size and depth of the underground chambers (b). Individual termites dragging cut pieces of leaves into their underground nest (c). [Photos (a) and (c) courtesy of R. Weil, photo (b) courtesy of R. Lal, Ohio State University]

In building their mounds, termites transport soil from lower layers to the surface, thereby extensively mixing the soil and incorporating into it the plant residues they use as food. Scavenging a large area around each mound, these insects remove up to 4000 kg/ha of leaf and woody material annually, a substantial portion of the plant litter produced in many tropical ecosystems. They also annually move 300 to 1200 kg/ha of soil in their mound-building activities. These activities have significant impacts on soil formation, as well as on current soil fertility and productivity.

The quantity of soil materials incorporated into termite mounds can be enormous; up to 2.4 million kg per hectare has been recorded (recall, from Chapter 4, that a 15-cm-deep layer of soil over 1 hectare weighs approximately 2.2 million kg). Depending on the species involved and on environmental conditions, termites may build mounds 6 meters or more in height, and may extend them to an even greater depth into the soil in search of water or clay layers. Each mound provides the home for 1 million or more termites. The mounds are abandoned after 10 to 20 years and can then be broken down to level the land for crop production. Attempts to level an occupied mound are usually frustrating as the termites rebuild very rapidly unless the queen termite is destroyed.

Effect of Termites on Soil Productivity

Unlike earthworms, termites do not generally have a beneficial effect on soil productivity. This is because the digestive processes of termites, aided by microorganisms in their gut, are generally more efficient than those of earthworms. Also, earthworms incorporate organic matter into the soil in a relatively uniform manner over a hectare of land. In contrast, termites mix plant residues into the soil only in the localized areas of their nests, while denuding the remaining land area of surface residues. This termite behavior can make it extremely difficult to provide cropland with the benefits of a protective crop residue mulch (see Figure 11.8).

PLANT GROWTH ON TERMITE MOUND MATERIAL. Termite mound material often has a lower organic matter and nutrient content than the surrounding undisturbed topsoil. This is because termites build their mounds mainly with subsoil, which is typically lower in organic matter content than topsoil. Crop growth in soil in areas where these mounds have existed is often poor, not only because of low nutrient content in the surface soil, but also because of the greater density of some of the mound material, the particles of which were cemented together by the termites as the mound was constructed. However, if the subsoil is richer in nutrients than the topsoil, or rich in clay compared to a very infertile, droughty, sandy surface soil, the material from abandoned mounds may provide islands of relatively high plant production, due to greater availability of phosphorus, potassium, calcium, and moisture. The same is true on very poorly drained soils with a high water table; the termite mounds provide islands of better drainage and aeration

FIGURE 11.8 Termites are quickly making these sorghum residues disappear, depriving the sandy, erosion-prone soil of the benefits of a protective mulch. However, the termites are also creating numerous stable macrochannels which will aid in capturing water when the rains finally come to this field in the Sahelian region of West Africa. Note the soil casings that the termites have built around parts of the plant residues. Under these casings the termites are protected from the drying sun and wind. The knife blade in the picture is 7 cm long. (Photo courtesy of R. Weil)

which produce much better plant growth. In certain semiarid and savanna regions, the stable macrochannels constructed by termites greatly increase water infiltration into soils that otherwise tend to form impermeable surface crusts (see Figure 11.8).

Termite activity is a significant factor in the formation of soils of tropical and subtropical areas (see Chapter 2). These insects also have both positive and negative effects on current land use in these areas. They accelerate the decay of dead trees and grasses, but also disrupt crop production, and even road construction, by the rapid development of their nests or mounds. Before leaving the subject of termites, it should be mentioned that the bacterial metabolism in the guts of these widespread soil animals accounts for a substantial fraction of the *global* production of methane (CH_4), an important greenhouse gas (see Chapter 12).

11.6 SOIL MICROANIMALS

From the viewpoint of microscopic animals, soils present many habitats that are essentially aquatic, at least intermittently so. For this reason the soil microfauna are closely related to the microfauna found in bodies of water. The two groups exerting the greatest influence on soil processes are the nematodes and protozoa.

Nematodes

Nematodes—commonly called *threadworms* or *eelworms*—are found in almost all soils, often in surprisingly large numbers (Table 11.2). They are tiny, unsegmented roundworms, seldom being large enough to be seen with the naked eye (Figure 11.9).

The nematodes comprise a highly diverse group. The most numerous and varied of the nematodes are those that live on decaying organic matter (saprophytes) or are predatory on other nematodes, fungi, bacteria, algae, protozoa, and insect larvae. Certain predatory nematodes are being studied for potential use as biological control agents for soilborne insect pests such as the corn rootworm.

Some nematodes, especially those of the genus *Heterodera*, can infest the roots of practically all plant species by piercing the roots with their sharp mouth parts. These wounds often allow infection by secondary pathogens and cause the formation of knot-like growths on the roots. Minor nematode infestations are nearly ubiquitous and often go unnoticed, but heavy infestations result in serious stunting and yield depression, especially to soybeans, solonaceous crops, and fruit trees (see Figure 11.10). Until recently the principal methods of controlling plant parasitic nematodes were long rotations with nonhost crops (often five years is required for the parasitic nematode populations to sufficiently dwindle), use of genetically resistant crop varieties, and soil fumigation with toxic chemicals. The use of soil nematicides, such as methyl bromide, has been sharply restricted because of undesirable environmental effects. New, less dangerous approaches to nematode control include the use of hardwood bark for containerized plants and interplanting susceptible crops with marigolds or canola (rapeseed), both of which produce root exudates with nematicidal properties.

Protozoa

Protozoa are mobile, single-celled creatures that capture and engulf their food. They are the most varied and numerous of the soil microfauna (though often they are classified as Protista rather than as animals). (See Figure 11.1 and Table 11.2.) They are considerably larger than bacteria, having a diameter range of 5–100 μm. Their cells also do not have true cell walls, and have a distinctly more complex organization than bacterial

FIGURE 11.9 A nematode commonly found in soil (magnified about 120 times). More than 1000 species of soil nematodes are known. Most nematodes feed on bacteria and fungi, but plant parasitic species can be very detrimental to plant growth. (Courtesy William F. Mai, Cornell University)

FIGURE 11.10 The stunted (foreground) and large (background) tobacco plants pictured above both received identical management. Examination of the root systems of the stunted plant showed that it was heavily infested with root-knot nematodes, which stunted the roots and produced knotlike deformities. (Photos courtesy of R. Weil)

cells. Soil protozoa include amoeba (which move by extending and contracting pseudopodia), ciliates (which move by waving hairlike structures, Figure 11.11), and flagellates (which move by waving a whiplike appendage called a *flagellum*).

More than 250 species of protozoa have been isolated in soils; sometimes as many as 40 or 50 of such groups occur in a single sample of soil. The liveweight of protozoa in surface soil ranges from 20 to 200 kg per hectare (Table 11.2). A considerable number of serious animal and human diseases are attributed to protozoan infections.

These organisms generally thrive best in moist, well-drained soils and are most numerous in surface horizons. Some protozoa are predators on soil bacteria and other microflora, especially in the area immediately around the plant roots (the rhizosphere). Overall, however, the protozoa are generally not sufficiently abundant in soils to be a major factor in organic matter decay and nutrient release.

11.7 ROOTS OF HIGHER PLANTS

Higher plants store the sun's energy and are the primary producers of organic matter (Figure 11.2). Their roots grow and die in the soil and are, for a number of reasons, clas-

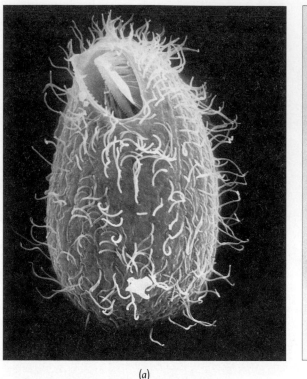

(a)   (b)

FIGURE 11.11 Two microanimals typical of those found in soils. (a) Scanning electron micrograph of a ciliated protozoan, *Glaucoma scintillaus*. (b) Photomicrograph of two species of rotifer, *Rotaria rotatoria* (stubby one at left) and *Philodina acuticornus* (slender one at right). [(a) Courtesy of J. O. Corliss, University of Maryland; (b) courtesy of F. A. Lewis and M. A. Stirewalt, Biomedical Research Institute, Rockville, Md.]

sified as soil organisms in this text. Roots supply the soil community of fauna and microflora with carbon and energy. The activities of plant roots greatly influence soil chemical and physical properties. As we shall see, plant roots interact with other soil organisms in varied and complex ways.

Morphology of Roots

Depending on their size, roots may be considered to be either meso- or macroorganisms. Fine feeder roots have diameters in the range of 100 to 400 μm, while root hairs are only 10 to 50 μm in diameter—similar in size to the strands of microscopic fungi. Root hairs are elongated single cells of the outer (epidermal) layer (see Figure 11.12). One function of root hairs is to anchor the root as it attempts to push its way through the soil. Another function is to increase the amount of root surface area available to absorb water and nutrients from the soil solution. Roots grow by forming and expanding new cells at the growing point (meristem), which is located just behind the root tip. The root tip itself is shielded by a protective cap of expendable cells that slough off as the root pushes through the soil. Root morphology is affected by both type of plant and soil conditions. For example, fine roots proliferate in localized areas of high nutrient concentrations. Root hair formation is stimulated by contact with soil particles and by low nutrient supply. Many roots become thick and stubby in response to high soil bulk density or high aluminum concentrations in soil solution (see Chapters 4 and 9).

Living roots physically modify the soil as they push through existing cracks and make new openings of their own (Figure 11.12). Tiny initial channels are increased in size as the roots swell and grow. By removing moisture from the soil, plant roots encourage soil shrinkage and cracking, which in turn stimulates soil aggregation. Root exudates also support a myriad of microorganisms, which help to further stabilize soil aggregates. In addition, when roots die and decompose, they provide building materials for humus, not only in the top few centimeters, but also to greater soil depths.

Amount of Organic Tissue Added

The importance of root residues in helping to maintain soil organic matter is often overlooked. In grasslands, fires may remove most of the aboveground biomass, so that the deep, dense root systems are the main source of organic matter added to these soils. In plantation and natural forests more than half of the total biomass production may be in the form of tree roots. In arable soils the mass of roots remaining in the soil after crop harvest is commonly 15 to 40% that of the aboveground crop. If an average figure of 25% is used, good crops of oats, corn, and sugarcane would be expected to leave about 2500, 4500, and 8500 kg/ha of root residues, respectively. The mass of organic compounds contributed by roots is seen to be even greater when the rhizosphere effects discussed below are considered.

Rhizosphere

Living roots of higher plants have a profound effect on the soil in their immediate vicinity. The **rhizosphere** consists of the soil zone within 1 or 2 mm of the root surface. The chemical and biological characteristics of this zone can be very different from those of the bulk soil. Soil acidity may be 10 times higher (or lower) in the rhizosphere than in the bulk soil. Roots greatly affect the nutrient supply in this zone by withdrawing soluble nutrients on one hand and by solubilizing nutrients from soil minerals on the other. By these and other means, roots affect the mineral nutrition of soil microbes, just as the microbes affect the nutrients available to the plant roots.

RHIZODEPOSITION.   Significant quantities of at least three broad types of organic compounds are exuded, secreted, or otherwise released at the surface of young roots (Figure 11.12):

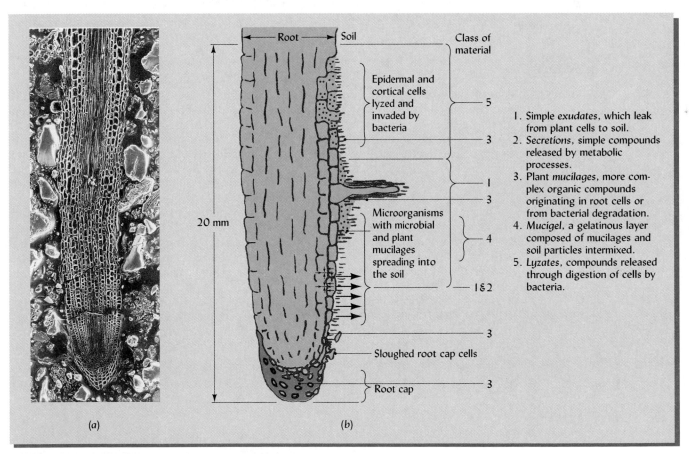

FIGURE 11.12   (a) Photograph of a root tip illustrating how roots penetrate soil and emphasizing the root cells through which nutrients and water move into and up the plant. (b) Diagram of a root showing the origins of organic materials in the rhizosphere. [(a) From Chino (1976), used with permission of Japanese Society of Soil Science and Plant Nutrition, Tokyo; (b) redrawn from Rovira et al. (1979), used with permission of Academic Press, London]

1. Low molecular weight organic compounds exuded by root cells include organic acids, sugars, amino acids, and phenolic compounds. Some of these root exudates, especially the phenolics, exert growth-regulating influences on other plants and soil microorganisms in a phenomenon called *allelopathy* (see Section 12.10).

2. High molecular weight mucilages secreted by root-cap cells and epidermal cells near apical zones form a substance called *mucigel* when mixed with microbial cells and clay particles. This mucigel appears to have several beneficial functions: it lubricates the root's movement through the soil; it improves root–soil contact, especially in dry soils when roots may shrink in size and lose direct contact with the soil (Figure 5.33); it may protect the root from certain toxic chemicals in the soils; and it provides an ideal environment for the growth of the rhizosphere microorganisms.

3. Cells from the root cap and epidermis continually slough off as the root grows, spill their cell contents into the rhizosphere, and enrich it with various cell tissues.

Taken together, these types of **rhizodeposition** may account for 1 to 30% of total dry matter production of young plants. Wheat plants three to eight weeks of age have been observed to rhizodeposit 20 to 40% of the organic substances translocated to them from the plant shoot. Sometimes when a plant is carefully uprooted from a loose soil, these root exudates cause a thin layer of soil, approximating the rhizosphere, to form a sheath around the actively growing root tips (see Figure 11.13). By the time annual plants mature, the amount of organic material lost to the rhizosphere during the growing season may be more than twice the amount remaining in the root system at the end of the growing season. Rhizodeposition decreases with plant age but increases with soil stresses such as compaction and low nutrient supply.

Because of the processes just described, plant roots are among the most important organisms in the soil ecosystem.

11.8 SOIL ALGAE

Like higher plants, algae consist of eukaryotic cells, those with nuclei organized within a nuclear membrane. (Organisms formerly called blue-green algae are prokaryotes and therefore will be considered with the bacteria.) Also like higher plants, algae are equipped with chlorophyll, enabling them to carry out photosynthesis. As photoautotrophs, algae need light, and are therefore found mostly very near the surface of the soil. Some species can also function as heterotrophs in the dark. Most grow best under moist to wet conditions. Under ideal conditions, the growth of algae may be so great that the soil surface becomes green or orange in color.

Several hundred species of algae have been isolated from soils, but a small number of species are the most prominent in soils throughout the world. Soil algae are divided into three general groups: green, yellow-green, and diatoms. The green algae are most evident in moist, but nonflooded acidic soils; diatoms are often numerous in well-drained, older gardens that are rich in organic matter. The green algae generally outnumber the diatoms.

Algal populations commonly range from 1 to 10 billion per square meter, 15 cm deep (10,000–100,000 cells per gram). The mass of live algae in soils may range from 10 to 500 kg per hectare–furrow slice (Table 11.2). In addition to producing a substantial amount of organic matter in some fertile soils, certain algae excrete polysaccharides which have very favorable effects on soil aggregation (see Section 4.10).

11.9 SOIL FUNGI

Soil fungi comprise an extremely diverse group of microorganisms. About 700 species have been identified in soils, representing some 170 genera. Their biomass in soils commonly ranges from 1000 to 20,000 kg per hectare–furrow slice (Table 11.2). Fungi have plantlike eukaryotic cells with a nuclear membrane and cell walls, but, unlike higher plants, have no capacity for photosynthesis. They are heterotrophs, meaning that they depend on living or dead plant or animal tissue for their carbon and energy. Fungi are

FIGURE 11.13 The zone of soil within 1 to 2 mm of living plant roots is termed the *rhizosphere*. This zone is greatly enriched in organic compounds excreted by the roots. These exudates and the microorganisms they support have caused the rhizosphere soil to adhere to some of these wheat roots as a sheath (roots on the right, see arrows). The roots on the left have been washed in water and the soil of the rhizosphere has been removed. (Photo courtesy of R. Weil)

aerobic organisms, although some can tolerate the rather low oxygen concentrations and high levels of carbon dioxide found in wet or compacted soils. Strictly speaking, fungi are not entirely microscopic, since some of these organisms, such as mushrooms, form macroscopic structures that can be easily seen without magnification.

For convenience of discussion, fungi may be divided into three groups: yeasts, molds, and mushroom fungi. *Yeasts,* which are single celled-organisms, live principally in waterlogged, anaerobic soils. *Molds* and *mushrooms* are both considered to be filamentous fungi because they are characterized by long, threadlike, branching chains of cells. Individual fungal filaments, called **hyphae** (see Figure 11.14a), are often twisted together in a mass that appears somewhat like a woven rope. Such a mass of hyphae is called a **mycelium.** Fungal mycelia are often visible as thin, white or colored strands

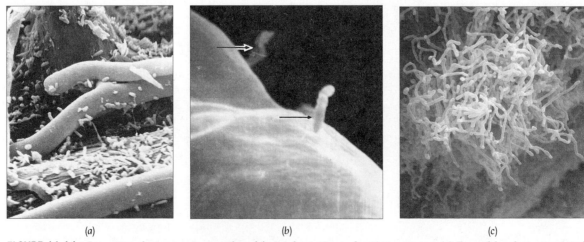

| (a) | (b) | (c) |

FIGURE 11.14 Scanning electron micrographs of fungi, bacteria, and actinomycetes. (a) Fungal hyphae associated with much smaller rod-shaped bacteria. (b) Rod-shaped bacteria attached to a plant root hair. (c) Actinomycete threads. [(a) Courtesy of R. Campbell, University of Bristol, used with permission of American Phytopathological Society; (b, c) courtesy of Maureen Petersen, University of Florida]

running through decaying plant litter (see Figure 11.15). The filamentous fungi reproduce by means of spores, often formed on fruiting bodies, which may be microscopic (i.e., molds) or macroscopic (i.e., mushrooms).

Molds

The molds are distinctly filamentous, microscopic, or semimacroscopic fungi that play a much more important role in soil organic matter breakdown than the mushroom fungi, approaching or even exceeding at times the influence of bacteria. Molds develop vigorously in acid, neutral, or alkaline soils. Some are favored, rather than harmed, by lowered pH (see Figure 9.18). Consequently, they are noticeably abundant in acid surface soils, where bacteria and actinomycetes offer only mild competition. The ability of molds to tolerate low pH is especially important in decomposing the organic residues in acid forest soils.

Many genera of molds are found in soils. Four of the most common are *Penicillium, Mucor, Fusarium,* and *Aspergillus*. Species from these genera occur in most soils. Soil conditions determine which species are dominant. The complexity of the organic compounds being attacked seems to determine which particular mold (or molds) prevail. Their numbers fluctuate greatly with soil conditions; perhaps 100,000 to 1 million individuals per gram of dry soil (10–100 billion per square meter) represent a more or less normal range in population.

Mushroom Fungi

These fungi are associated with forest and grass vegetation where moisture and organic residues are ample. Although the mushrooms of many species are extremely poisonous to humans, some are edible—and a few have been "domesticated." Edible mushrooms are grown in caves and specially designed houses, where composted organic materials (particularly horse manures) provide their source of food.

The aboveground fruiting body of most mushrooms is only a small part of the total organism. An extensive network of hyphae permeate the underlying soil or organic residue. While mushrooms are not as widely distributed as the molds, these fungi are very important, especially in the breakdown of woody tissue, and because some species form a symbiotic relationship with plant root (see mycorrhizae, below).

Activities of Fungi

In their ability to decompose organic residues, fungi are the most versatile and perhaps the most persistent of any group. Cellulose, starch, gums and lignin, as well as the more easily metabolized proteins and sugars, succumb to their attack. In affecting the pro-

FIGURE 11.15 The thin white strands (near the fingers) in this photo are fungal mycelium in decaying plant residues. The other strands (upper right of photo) are plant roots. (Photo courtesy of R. Weil)

cesses of humus formation (Section 12.9) and aggregate stabilization (Section 4.11), molds are even more important than bacteria. Fungi function more efficiently than bacteria in that they assimilate into their tissues a larger proportion of the organic materials they metabolize. Up to 50% of the substances decomposed by molds may become organism tissue, compared to about 20% for bacteria. Soil fertility depends in no small degree on nutrient cycling by molds, since they continue to decompose complex organic materials after bacteria and actinomycetes have essentially ceased to function. They are especially active in acid forest soils, but they play a significant role in all soils.

In addition to the breakdown of organic residues and the formation of humus, numerous other fungal activities have significant impact on soil ecology. Some fungi are predators of soil animals. For example, certain species even trap nematodes (see Figure 11.16). Soil fungi can synthesize a wide range of complex organic compounds in addition to those associated with soil humus. It was from a soil fungus, a *Penicillium* species, that the first modern antibiotic drug, penicillin, was obtained. By killing bacteria, such compounds probably help a fungus outcompete its rival microorganisms in the soil.

Unfortunately, not all the compounds produced by soil fungi can be seen as beneficial to humans or higher plants. A few fungi produce chemicals (mycotoxins) that are highly toxic to plants or animals (including humans). An important example of the latter is the production of highly carcinogenic aflatoxin by the fungus *Aspergillus flavus* growing on grains such as corn or peanuts, especially when grain is exposed to soil and moisture. Other fungi produce compounds that allow them to invade the tissues of higher plants, causing such serious plant diseases as wilts (e.g., *Verticillium*) and root rots (e.g., *Rhizoctonia*).

On the other hand, efforts are now under way to develop the potential of certain fungi as biological control agents against some insects and mites that damage higher plants. These examples merely hint at the impact of the complex array of fungal activities in the soil.

Mycorrhizae[7]

One of the most ecologically and economically important activities of soil fungi is the mutually beneficial (symbiotic) association between certain fungi and the roots of higher plants. This association is called **mycorrhizae,** a term meaning "fungus root." A mycorrhiza forms when the appropriate fungus invades a plant root in a process superficially similar to infection by pathogenic fungi. However, in the mychorrhizal association, the fungi and the plants have apparently coevolved so that both parties benefit from the relationship. In fact, in natural ecosystems many plants are quite dependent

---

[7]For excellent overviews of mycorrhizae and their effects on the soil-plant system, see Menge (1981) and Bethlenfalvay and Linderman (1992).

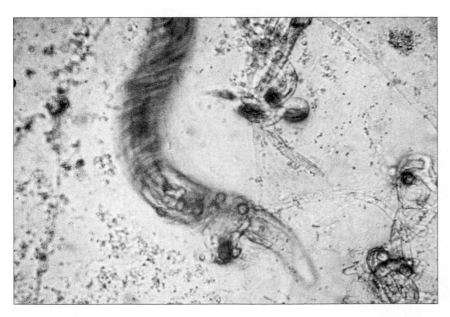

FIGURE 11.16 Several species of fungi prey on soil nematodes, often on those nematodes that parasitize higher plants. Some species of nematode-killing fungi attach themselves to and slowly digest the nematodes. Others make a loop with their hyphae and wait for a nematode to swim through this lasso-like trap. The loop is then constricted and the nematode is trapped. The nematode shown here is thrashing about in a vain effort to escape from such a trap. (Photo courtesy of the American Society of Agronomy)

on the mycorrhizal relationship and cannot survive without it. Mycorrhizae are the rule, not the exception, for most plant species, including most crops plants.

Mycorrhizal fungi derive an enormous survival advantage from teaming up with plants to take direct advantage of the latter's photosynthetic capabilities. Instead of having to compete with all the other soil heterotrophs for decaying organic matter, the mycorrhizal fungi obtain sugars directly from the plant's root cells. This represents an energy cost to the plant, which may lose as much as 5 to 10% of its total photosynthate production to its mycorrhizal fungal symbiont.

In return, plants receive some extremely valuable benefits from the fungi. The fungal hyphae grow out into the soil some 5 to 15 cm from the infected root, reaching farther and into smaller pores than the plant's own root hairs could. The network of mycorrhizal hyphae extend the plant's root system and provide perhaps 10 times as much absorptive surface as the root system of an uninfected plant. This extension also increases the efficiency of the root system. Mycorrhizae greatly enhance the ability of plants to take up nutrients that are relatively immobile and present in low concentrations in the soil solution. Phosphorus uptake is most commonly and dramatically increased, but mycorrhizae have also been shown to enhance uptake of many other nutrients, including zinc, copper, magnesium, calcium, iron, and manganese (see Table 11.4). In contrast, mycorrhizae also protect plants from excessive uptake of salts and toxic metals in saline, acid, or contaminated soils (see Chapter 18). Water uptake may also be improved by mycorrhizae, making plants more resistant to drought. There is evidence that mycorrhizae protect plants from certain soilborne diseases by producing antibiotics, altering the root epidermis, and competing with pathogens for infection sites.

ECTOMYCORRHIZA. Two types of mycorrhizal associations are of considerable practical importance: **ectomycorrhiza** and **endomycorrhiza**. The ectomycorrhiza group includes hundreds of different fungal species associated primarily with temperate or semiarid region trees, such as pine, birch, hemlock, beech, oak, spruce, and fir. These fungi, stimulated by root exudates, cover the surface of feeder roots with a fungal mantle. Their hyphae penetrate the roots and develop in the free space around the cells of the cortex but do not penetrate the cortex cell walls. Ectomycorrhizae cause the infected root system to consist primarily of short, stubby, white rootlets with a characteristic Y shape (see Figure 11.17). These Y-shaped rootlets provide visible evidence of mycorrhizal infection. For this reason, the first mycorrhizae discovered and studied were the ectomycorrhizae of forest trees.

Many ectomycorrhizal fungi are facultative symbionts (they can also live independently in the soil) and therefore can be cultured in large quantities on artificial media. Such commercial ectomycorrhizal fungal inoculants are widely used in tree nurseries. For some tree species, satisfactory growth and survival are dependent on proper inoculation with mycorrhizal fungi before planting if the proper fungi are not already in soils at the site.

TABLE 11.4 **Effect of Inoculation with Mycorrhiza and of Added Phosphorus on the Content of Different Elements in the Shoots of Corn**

Expressed in micrograms per plant.

Element in plant	No phosphorus		25 mg/kg phosphorus added	
	No mycorrhiza	Mycorrhiza	No mycorrhiza	Mycorrhiza
P	750	1,340	2,970	5,910
K	6,000	9,700	17,500	19,900
Ca	1,200	1,600	2,700	3,500
Mg	430	630	990	1,750
Zn	28	95	48	169
Cu	7	14	12	30
Mn	72	101	159	238
Fe	80	147	161	277

From Lambert et al. (1979).

ENDOMYCORRHIZA. The most important members of the endomycorrhiza group are called ***vesicular arbuscular (VA) mycorrhizae.***[8] When forming VA mycorrhizae, fungal hyphae actually penetrate the cortical root cell walls and, once inside the plant cell, form small, highly branched structures known as ***arbuscules.*** These structures serve to transfer mineral nutrients from the fungi to the host plants, and sugars from the plant to the fungus. Other structures, called ***vesicles,*** serve as storage organs for the mycorrhizae (Figure 11.17).

The VA mycorrhizae are the most common and widespread group. Nearly 100 identified species of fungi form these endomycorrhizal associations in soils from the tropics to the arctic. The roots of most agronomic crops, including corn, cotton, wheat, potatoes, beans, alfalfa, sugarcane, cassava, and dryland rice, form VA mycorrhizae, as do most vegetables and fruits, such as apples, grapes, and citrus. Many forest trees, including maple, yellow poplar, and redwood, as well as important tree crops such as apple, cacao, coffee, and rubber, also form VA mycorrhizae. Two important groups of crop plants that do *not* form mycorrhizae are the Cruciferae (cabbage, rape, canola, broccoli) and the Chenopodiaceae (sugar beet, beet, spinach).

---

[8]Sometimes the group is referred to simply as *arbuscular mycorrhizal* (AM) *fungi* because vesicles are not universally present.

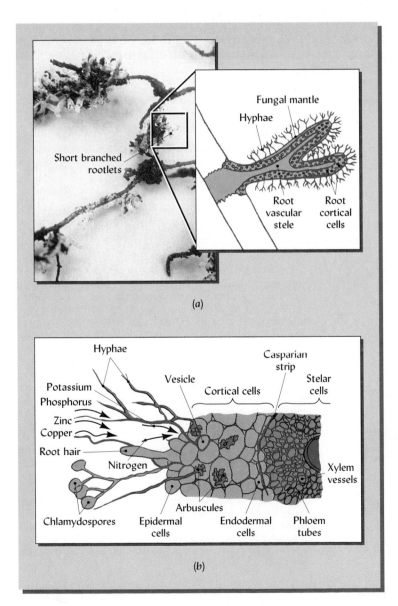

(a)

(b)

FIGURE 11.17 Diagram of ectomycorrhiza and vesicular arbuscular (VA) mycorrhiza association with plant roots. (a) The ectomycorrhiza association produces short branched rootlets that are covered with a fungal mantle, the hyphae of which extend out into the soil and between the plant cells but do not penetrate the cells. (b) In contrast, the VA mycorrhizae penetrate not only between cells but into certain cells as well. Within these cells, the fungi form structures known as *arbuscules* and *vesicles.* The former transfer nutrients to the plant, and the latter store these nutrients. In both types of association, the host plant provides sugars and other food for the fungi and receives in return essential mineral nutrients that the fungi absorb from the soil. [Redrawn from Menge (1981); Photo courtesy of R. Weil]

Research on VA mycorrhizae continues to reveal the ecological and practical significance of this symbiosis. The importance of mycorrhizal hyphae in stabilizing soil aggregate structure is becoming increasingly clear (see Figure 4.26). The presence of mycorrhizae is known to enhance nodulation and N fixation by legumes. Also, VA mycorrhizae have been observed to transfer nutrients from one plant to another via hyphal connections, sometimes resulting in a complex four-way symbiotic relationship (see Figure 11.18). Interplant transfer of nutrients has been observed in forest, grassland, and pasture ecosystems, but researchers are still trying to determine the extent and ecological significance of this VA-microrrhizal-mediated plant interaction.

The role VA mycorrhizae play in helping plants absorb nutrients is most agriculturally important where soils are relatively infertile and fertilizer inputs are limited (Figure 11.19). Evidence indicates that soil tillage disrupts hyphal networks, and therefore minimum tillage may increase the effectiveness of VA mycorrhizae. Also, certain crop rotations, as opposed to monocropping and periods of bare fallow, may favor the buildup of effective mycorrhizae in soils. Surface soil stockpiled during mining and construction activities (see Figure 1.14) may lose its VA mycorrhizal inoculum potential because of the long period without any host plants. Where such stockpiled soil is not inoculated with mycorrhizae-containing soil before use, revegetation may be very poor.

Although definite differences exist in mycorrhizal efficiency among various species and strains of VA mycorrhizal fungi, the ubiquitous distribution of native strains ensures natural infection in plants under most circumstances. Thus, gains from inoculation with efficient mycorrhizal fungi are usually quite small. Exceptions occur where native populations of fungi are low due to fumigation or other soil disturbance such as surface mining. Practical application of VA mycorrhizae technology has been limited by difficulties in producing large quantities of disease-free inoculum by culturing these obligate symbionts on artificial media.

11.10 SOIL ACTINOMYCETES

Actinomycetes resemble molds in that they are filamentous and are often profusely branched. Their mycelial threads are smaller, however, than those of fungi (Figure

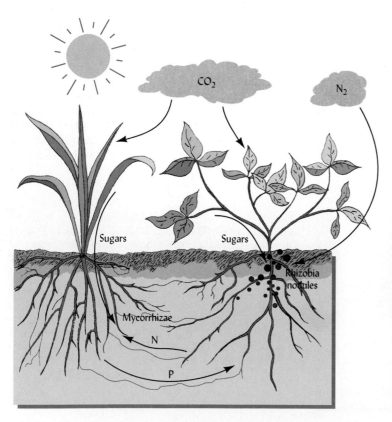

FIGURE 11.18 Mycorrhizal fungi, rhizobia bacteria, legume and nonlegume plants can all interact in a four-way, mutually beneficial relationship. Both the fungi and the bacteria obtain their energy from sugars supplied through photosynthesis by the plants. The rhizobia form nodules on the legume roots and enzymatically capture atmospheric nitrogen, providing the legume with nitrogen to make amino acids and proteins. The mycorrhizal fungi infect both types of plants and form hyphal interconnections between them. The mycorrhizae then not only assist in the uptake of phosphorus from the soil, but can also directly transfer nutrients from one plant to the other. Isotope tracer studies have shown that, by this mechanism, nitrogen is transferred from the nitrogen-fixing legume to the nonlegume (e.g., grass) plant, and phosphorus is mostly transferred to the legume from the nonlegume. The nonlegume grass plant has a fibrous root system and an extensive mycorrhizal network which is relatively more efficient in extracting P from soils than the root system of the legume. Research indicates that direct transfer of nutrients via mycorrhizal connections occurs in many mixed plant communities such as in forest understories, grass-legume pastures, and mixed cropping systems. The extent to which this phenomenon effects plant productivity is not yet well understood.

FIGURE 11.19 The effect of mycorrhizae on availability of phosphorus to a pasture legume, *Pueraria phaseoloides:* (a) no treatment, (b) rock phosphate, (c) rock phosphate plus mycorrhizae. (Courtesy Dr. Fritz Kramer, CIAT, Cali, Colombia)

11.14c). Actinomycetes are similar to bacteria in that they are unicellular and of about the same diameter. They break up into spores that closely resemble bacteria. Although historically they have often been classified with the fungi, actinomycetes, like bacteria, have no nuclear membrane (i.e., both have prokaryotic cells) and are now commonly classified with bacteria in the kingdom Monera. Actinomycetes are generally aerobic heterotrophs that live on decaying organic matter in the soil, or on compounds supplied by plants with which certain species form parasitic or symbiotic relationships. Many actinomycete species produce antibiotic compounds that kill other microorganisms. Actinomycin, neomycin, and streptomycin are examples of familiar antibiotic drugs based on actinomycete products. In forest ecosystems, much of the nitrogen supply depends on certain actinomycetes that are capable of fixing atmospheric nitrogen gas into ammonium nitrogen that is then available to higher plants (see Section 13.12).

Actinomycetes develop best in moist, warm, well-aerated soil. But in times of drought, they remain active to a degree not usually exhibited by either bacteria or molds. They are generally rather sensitive to acid soil conditions, with optimum development occurring at pH values between 6.0 and 7.5. Farmers take advantage of this sensitivity to soil reaction in controlling potato scab, an actinomycete-caused disease of wide distribution. By lowering the soil pH with sulfur, the disease can be kept under control (see Section 11.16). Some actinomycete species tolerate very high temperatures and play an important role in active compost piles where the temperature can reach 50°C (see Section 12.17).

NUMBERS AND ACTIVITIES OF ACTINOMYCETES. Actinomycete numbers in soil exceed those of all other organisms except bacteria; their numbers sometimes reach hundreds of millions per gram of soil, one-tenth those of bacteria. In actual liveweight, they often exceed bacteria, a biomass of up to 5000 kg per hectare being present in some surface soils. These organisms are especially numerous in soils high in humus, such as old meadows or pastures, where the acidity is not too great. The aroma of freshly plowed land that is so noticeable at certain times of the year is probably due to volatile products of actinomycetes and certain molds.

Actinomycetes undoubtedly are of great importance in the decomposition of soil organic matter and the liberation of its nutrients. Apparently, they reduce even the more resistant compounds, such as cellulose, chitin, and phospholipids to simpler forms. They often become dominant in the later stages of decay when the easily metabolized substrates have been used up. The abundance of actinomycetes in soils that have been under sod for many years is an indication of their capacity to attack complex compounds.

11.11   SOIL BACTERIA

CHARACTERISTICS.   Bacteria are single-celled organisms, one of the simplest and smallest forms of life known. Their extremely rapid reproductive potential enables them to increase their populations quickly in response to favorable changes in soil environment and food availability. They are perhaps the most diverse group of soil organisms; a gram of soil typically contains 20,000 different bacteria species.

Bacteria are very small, single-celled prokaryotic organisms. They range in size from 0.5 to 5 μm, the smaller ones approaching the size of the average clay particle. Bacteria are found in various shapes: nearly round, rodlike, or spiral. In the soil, the rod-shaped organisms seem to predominate (Figure 11.14b). Many bacteria are motile, swimming about in the soil water by means of hairlike cilia or flagella.

BACTERIAL POPULATIONS IN SOILS.   As with other soil organisms, the numbers of bacteria are extremely variable, but high, ranging from a few billion to more than a trillion in each gram of soil. A biomass of a 400 to 5000 kg/ha (0.2 to 2 tons/acre) liveweight is commonly found in the plow layer of fertile soils (see Table 11.2).

In the soil, bacteria exist as mats, filaments, or clumps, called *colonies.* Many soil bacteria are able to produce spores or similar resistant bodies. This ability to achieve a resting stage as well as a vegetative stage allows the organisms to survive unfavorable conditions. In terms of types of metabolism, range of substrates attacked, and environmental conditions tolerated, bacteria are the most diverse group of soil organisms.

SOURCE OF ENERGY.   Soil bacteria are either **autotrophic** or **heterotrophic** (see Section 11.3). The autotrophs obtain their energy from sunlight (photoautotrophs) or from the oxidation of inorganic constituents such as ammonium, sulfur, and iron (chemautotrophs), and most of their carbon from carbon dioxide or dissolved carbonates. Autotrophic bacteria are not very numerous but they play vital roles in controlling nutrient availability to higher plants. Most soil bacteria are heterotrophic—both their energy and their carbon coming directly from organic matter. Heterotrophic bacteria, along with fungi and actinomycetes, account for the general breakdown of organic matter in soils. Where oxygen supplies are depleted, as in wetlands, nearly all decomposition is bacterially mediated. Certain gaseous products of such anaerobic bacterial decomposition, such as methane and nitrous oxide, have major effects on the global environment (see Sections 12.2 and 13.9).

IMPORTANCE OF BACTERIA.   Bacteria as a group participate vigorously in virtually all of the organic transactions that characterize a healthy soil system. They rival or exceed both fungi and actinomycetes in the breakdown of organic compounds, formation of humus, stabilization of soil structure, and cycling of nutrients.

Soil bacteria possess such a broad range of enzymatic capabilities that they can digest most naturally occurring materials, plus a diverse array of such troublesome substances as crude oil, insecticides, and various other organic toxins. Scientists are now finding ways of harnessing, even improving, the metabolic activities of bacteria to help with the remediation of polluted soils (see Section 18.5).

In addition, bacteria hold near monopolies in several basic enzymatic transformations. One such process is the oxidation or reduction of selected chemical elements in soils (see Sections 7.3 and 7.5). Certain autotrophic bacteria obtain their energy from such inorganic oxidations, while anaerobic and facultative bacteria reduce a number of substances other than oxygen gas. Many of these biochemical oxidation and reduction reactions have significant implications for environmental quality as well as for plant nutrition. For example, through *nitrogen oxidation* (**nitrification**), selected bacteria oxidize relatively stable ammonium nitrogen to the much more mobile nitrate form of nitrogen. Likewise, other bacteria are responsible for *sulfur oxidation,* which yields plant-available sulfate ions, but also potentially damaging sulfuric acid (see Section 13.21). Also, bacterial oxidation and reduction of inorganic ions such as iron and manganese not only influence the availability of these elements (see Section 15.5), but also help determine the soil colors (see Section 4.1).

A second critical process in which bacteria are prominent is **nitrogen fixation**—the biochemical combining of atmospheric nitrogen with hydrogen to form organic nitrogen compounds usable by higher plants (see Section 13.10). The process can take place in soils independent of plants, but the amount of nitrogen fixed is much greater if the bacteria are intimately associated with plant roots, much as are the mycorrhizal fungi.

Cyanobacteria contain chlorophyll, allowing them to photosynthesize like plants. Previously classified as blue-green algae, cyanobacteria are especially numerous in rice paddies and other wetland soils. When such lands are flooded, appreciable amounts of atmospheric nitrogen are fixed by these organisms. Certain cyanobacteria, growing within the leaves of the aquatic fern **azolla** are able to fix nitrogen in quantities great enough to be of practical significance for rice production (see Section 13.12).

11.12 CONDITIONS AFFECTING THE GROWTH OF SOIL BACTERIA

Many soil conditions affect the growth of bacteria. Among the most important are the supplies of oxygen and moisture, the temperature, the amount and nature of the soil organic matter, the pH, and the amount of exchangeable calcium present. These effects are briefly outlined below.

1. Oxygen requirements
   a. Some bacteria use mostly oxygen gas (*aerobic*).
   b. Some use substances other than oxygen (e.g., NO_3^-, SO_4^{2-} or other electron acceptor) (*anaerobic*).
   c. Some use either of the above forms of metabolism (*facultative*).
   d. All three of the above types usually function simultaneously in a soil.

2. Moisture relationships
   a. Optimum moisture level for higher plants (moisture potential of -10 to -100 kPa) is usually best for most aerobic bacteria.
   b. The moisture content affects oxygen supply (see item 1 above).

3. Suitable temperature range
   a. Bacterial activity is generally greatest at 20–40°C (about 70–100°F).
   b. Ordinary soil temperature extremes seldom kill bacteria, and commonly only temporarily suppress activity.

4. Organic matter requirements
   a. Organic matter is used as an energy source for the majority of bacteria (*heterotrophic*).
   b. Organic matter is not required as an energy source for other bacteria (*autotrophic*).
   c. Many bacteria are stimulated by the energy-rich compounds excreted by plant roots, and certain bacteria are dependent on amino acids and other growth factors found in the rhizosphere.

5. Exchangeable calcium and pH relationships
   a. High calcium concentration and pH values from 6 to 8 generally are best for most bacteria.
   b. Calcium and pH values determine the specific bacteria present.
   c. Certain bacteria function at very low pH (<3.0), and a few at high pH values.
   d. Exchangeable calcium seems to be more important than pH.

11.13 COMPETITION AMONG SOIL MICROORGANISMS

The population of soil microbes is generally limited by the food supply, increasing rapidly whenever more organic materials become available. As a result, the soil presents an intensely competitive environment for its microscopic inhabitants. When fresh organic matter is added, the vigorous heterotrophic soil organisms (bacteria, fungi, and actinomycetes) compete with each other for this source of food. If sugars, starch, and amino acids are available in the organic material, the bacteria dominate initially because they reproduce most rapidly and prefer such simple compounds. As these simple compounds are broken down, the fungi, and particularly the actinomycetes, become more competitive.

The soil microbes use many different tactics in their deadly competition. Certain organisms alter the acidity level in their vicinity to the disadvantage of competitors. Others produce substances (**siderophores**) that so strongly bind to iron that other organisms in their immediate vicinity cannot grow for lack of this essential element.

Certain soil bacteria, fungi, and actinomycetes produce antibiotics, compounds that kill specific types of other organisms on contact (Figure 11.20). The discovery of this antibiotic production marked an epochal advance in medical science. Many antibiotics now on the market for the treatment of human and animal diseases, such as penicillin, streptomycin, and aureomycin, can be produced by organisms found in soils. When these and other antagonistic activities of soil microorganisms are turned against potentially plant-pathogenic organisms, significant suppression of plant diseases can occur. Microbial ecologists are attempting to harness some of these competitive mechanisms to favor desirable microorganisms in soils (Box 11.1).

11.14 EFFECTS OF AGRICULTURAL PRACTICE ON SOIL ORGANISMS

Changes in environment affect both the number and kinds of soil organisms. Clearing forested or grassland areas for cultivation drastically changes the soil environment. First, the quantity and quality of plant residues (food for the organisms) is markedly reduced. Also, the number of species of higher plants is reduced. Monoculture or even common crop rotations provide a much narrower range of original plant materials and rhizosphere environments than nature provides in forests or grasslands.

While agricultural practices have different effects on different organisms, a few generalizations can be made. Figure 11.22 illustrates some principles relating to total soil organism diversity and populations. For example, some agricultural practices generally reduce the species diversity as well as the total organism population (e.g., extensive tillage and monoculture). Adding lime, fertilizers, and manures to an infertile soil generally will increase the activities of the microflora. Reduced tillage tends to increase the role of fungi at the expense of the bacteria. Pesticides (especially the fumigants) can sharply reduce organism numbers, especially fauna, at least on a temporary basis. On the other hand, application of a particular pesticide often stimulates the population of specific microorganisms which can use the pesticide as food. Likewise, monoculture cropping systems reduce species numbers and diversity but may actually increase the organism count of some species.

Most changes in agricultural technology have ecological effects on soil organisms that can affect higher plants and animals, including humans. The effects of pesticides, both positive and negative, provide evidence of this. For this reason, the ecological side effects of modern technology must be carefully scrutinized.

11.15 BENEFICIAL EFFECTS OF SOIL ORGANISMS

In their influence on plant productivity and the ecological functioning of soils, the soil fauna and flora are indispensable. Of their many beneficial effects, only the most important can be emphasized here.

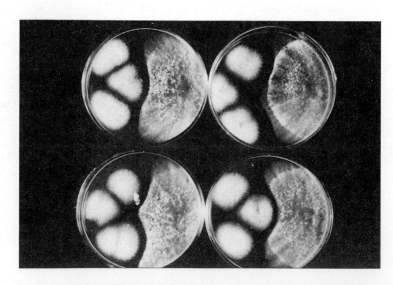

FIGURE 11.20 Fungal antagonism. The three smaller colonies in each petri dish are exuding an antibiotic compound that prevents the larger fungal colony from spreading over their half of the plate. Note the clear zone around each of the antibiotic-producing colonies. (Photo courtesy of American Society of Agronomy)

BOX 11.1 CHOOSING SIDES IN THE WAR AMONG THE MICROBES

Recent advances in our understanding of microbial genetics, combined with new techniques for transferring genetic material among different species of microorganisms, have opened many interesting and promising opportunities for scientists to join in the microbial wars in the soil.

Scientists are developing innovative approaches to harness the antagonisms long observed among microorganisms. Two examples are (1) the development of a bacterial seed treatment that helps ward off pathogenic fungi, and (2) a chemically armed *Rhizobium* bacteria that may enable farmers to efficiently use improved strains of nitrogen-fixing bacteria on leguminous crops.

The first biological-control seed treatments are now commercially available. These seed treatments use antagonistic soil bacteria instead of chemical fungicide to protect seed from rot-causing soil pathogens. The protective bacteria are specially selected and enhanced strains of *Pseudomonas flourescens*, a species known to inhabit rhizosphere soil and help protect plant roots from soilborne diseases. When a preparation of these bacteria was applied in the furrow as wheat seeds were sown, the seeds germinated safely in soil known to be infested with seed-rotting fungi. Seeds sown without the treatment decayed before they could germinate (Figure 11.21). Widespread use of such biologically based seed protection in the place of conventional antifungal chemicals should reduce the loading of toxic compounds into the environment.

The second example involves the use of genetic engineering to improve rhizobial inoculum effectiveness. Scientists have been quite successful in finding strains of rhizobia and bradyrhizobia bacteria that have superior nitrogen-fixing ability. Unfortunately, there has been little practical success in using these improved strains to enhance nitrogen fixation in the field. The problem is that most soils contain large populations of indigenous rhizobia strains that, while not as efficient in fixing nitrogen, are very competitive in forming nodules at infection sites on

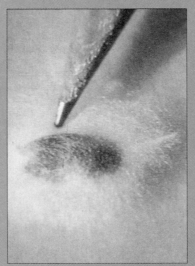 

FIGURE 11.21   *(Right) The wheat seed protected with antagonistic bacteria remain healthy, while (left) unprotected seed becomes heavily infected with pathogenic fungi. (Photo courtesy of Ecogen, Inc.)*

legume roots. Because indigenous rhizobia are successfully competing for survival in the soil and for nodule-formation sites on legume roots, improved strains of rhizobia bacteria generally have little success. Even when high levels of bacterial inoculum are applied to the legume seed or to the soils in which the seed is sown, the vast majority (generally 80 to 95%) of nodules found on the legume roots are formed by indigenous strains, not by the improved, introduced strains. In order to take advantage of the improved strains' greater nitrogen-fixing ability, it is necessary to give the improved strains a competitive edge over their indigenous cousins. Dr. Eric Triplett and coworkers in Wisconsin discovered that the strain, *Rhizobia leguminosarium* bv. *trifolii*, produced a substance they called *trifolitoxin* that inhibits other rhizobial strains but did not inhibit the strain that produced it (Fitzmaurice, 1995). After working out the sequence of the cluster of genes controlling the production and reaction to trifolitoxin, the scientists were able to transfer the relevant genes to other strains of rhizobia. Their hope is that, armed with these genes, improved rhizobia strains that are highly efficient in nitrogen fixation will be able to outcompete the less efficient indigenous strains for nodule formation sites on legume roots. If successful, this use of microbial antagonism should lead to enhanced nitrogen fixation by legumes, with consequent increased yields of the legumes and increased potential to substitute biologically fixed nitrogen for fertilizer nitrogen in the production of nonlegume crops grown together or in rotation with the legumes.

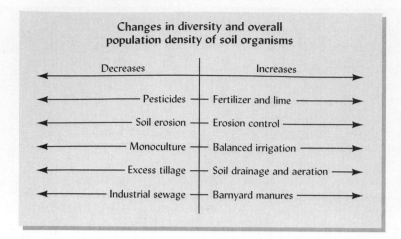

Changes in diversity and overall population density of soil organisms

Decreases	Increases
Pesticides	Fertilizer and lime
Soil erosion	Erosion control
Monoculture	Balanced irrigation
Excess tillage	Soil drainage and aeration
Industrial sewage	Barnyard manures

FIGURE 11.22 A simple diagrammatic illustration of the general effects of major agricultural practices on the diversity and population densities of soil organisms. Note that practices promoting sustainable crop production and the protection and improvement of soils generally favor both diversity and overall organism population numbers.

ORGANIC MATTER DECOMPOSITION.   Perhaps the most significant contribution of the soil fauna and flora to higher plants is that of organic matter decomposition. By this process, plant residues are broken down, releasing organically held nutrients for use by plants. Nitrogen is a prime example. At the same time, the stability of soil aggregates is enhanced not only by the gluelike intermediate products of decay, but by the more resistant portion, humus.

BREAKDOWN OF TOXIC COMPOUNDS.   Many organic compounds toxic to plants or animals find their way into the soil. Some of these toxins are produced by soil organisms as metabolic by-products; some are applied purposefully by humans as agrochemicals to kill pests; and some are deposited in the soil as a result of unintentional environmental contamination. If these compounds accumulated unchanged, they would do enormous ecological damage. Fortunately, there are soil organisms which not only are unharmed by these compounds, but can use them as food. Soil bacteria and fungi are especially important in helping maintain a nontoxic soil environment by breaking down toxic compounds (see Section 18.5 on **bioremediation**). The detoxifying activity of these microorganisms is by far the greatest in the surface layers of soil, where microbial numbers are concentrated in response to the greater availability of organic matter and oxygen (see Figure 11.23).

INORGANIC TRANSFORMATIONS.   The transformation of inorganic compounds is of great significance to higher plants. The presence in soils of nitrates, sulfates, and, to a lesser degree, phosphate ions is due primarily to the action of microorganisms. Organically bound forms of N, S, and P are converted by the microbes to these plant-available forms.

Likewise, the availability of the other essential elements such as iron and manganese is determined largely by microbial action. In well-drained soils, these elements are oxidized by autotrophic organisms to their higher valence states, in which forms their solubilities are quite low. This keeps the greater portion of iron and manganese in insoluble and nontoxic forms, even under fairly acid conditions. If such oxidation did not occur, plant growth would be jeopardized because of toxic quantities of these elements in solution. Microbial oxidation also controls the potential toxicity in soil contaminated with selenium or chromium.

NITROGEN FIXATION.   The fixation of elemental nitrogen, which cannot be used directly by higher plants, into compounds usable by plants is one of the most important microbial processes in soils. While actinomycetes in the genus *Frankia* fix major amounts of nitrogen in forest ecosystems, and cyanobacteria are important in flooded rice paddies, rhizobia bacteria are the most important group for the capture of gaseous nitrogen in agricultural soils. By far the greatest amount of nitrogen fixation by these organisms occurs in root nodules or in other association with plants. Worldwide, enormous quantities of nitrogen are fixed annually into forms usable by higher plants (see Section 13.10).

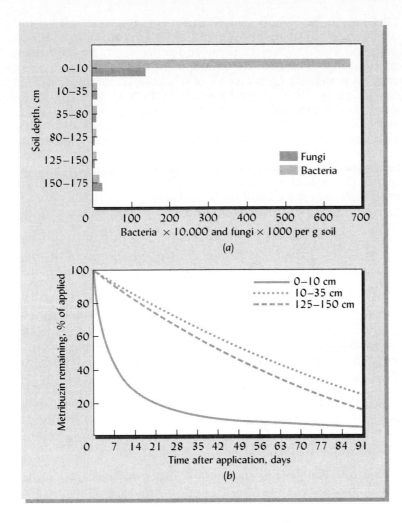

FIGURE 11.23 (a) Populations of aerobic bacteria and fungi at various depths in a Dundee soil (Aqualfs). (b) Breakdown of the herbicide metribuzin in the same soil at various depths. Soil was sampled from each depth and incubated with the herbicide. Note that the fungi and bacteria were concentrated in soil from the upper 10 cm, and that the breakdown of the herbicide was far more rapid in soil from this same layer than in soil from in deeper layers. [From Moorman and Harper (1989)]

11.16   INJURIOUS EFFECTS OF SOIL ORGANISMS ON HIGHER PLANTS

Although most of the activities of soil organisms are vital to a healthy soil ecosystem and economic plant production, some soil organisms affect plants in detrimental ways that cannot be overlooked.

SOIL FAUNA.   It already has been suggested that certain of the soil fauna are injurious to higher plants. For instance, some rodents and moles may severely damage crops. Snails and slugs in some climates are dreaded pests, especially of vegetables. Ants, which may transfer aphids onto certain plants, must be held in check by gardeners. Also, most plant roots are infested with nematodes, sometimes so seriously that the successful growth of certain crops is both difficult and expensive.

MICROFLORA AND PLANT DISEASES.   Among microorganisms it is the microflora that generally cause the most damage to higher plants. Although bacteria and actinomycetes contribute their quota of plant diseases, fungi are responsible for most of the common soil-borne diseases of crop plants. Included are wilts, damping-off, root rots, clubroot of cabbage and similar crops, and blight of potatoes. In short, disease infestations occur in great variety and are induced by many different organisms.

Soils are easily infested with disease organisms, which are transferred to the soil through farm implements and plants, and even through manure from animals that were fed infected plants. Soil erosion also can carry diseases from one field to another. Once a soil is infested, it is apt to remain so for a long time.

DISEASE CONTROL BY SOIL MANAGEMENT.   Prevention is the best defense against soilborne diseases. Strict quarantine systems will restrict the transfer of soilborne pathogens from one farm to another. Elimination of the host crop from the infested field will help in cases where no alternate hosts are available. In some cases, crop rotations and tillage practices can be used to help control a disease. These and other organic matter management practices seem also to be related to the development of **disease-suppressive soils.** Research on plant diseases ranging from take-all of wheat in Washington State to fusarium wilt of bananas in Central America has documented the existence of suppressive soils in which a particular disease fails to develop even though the pathogen is introduced in the presence of a susceptible host. The reason that certain soils become disease suppressive is not well understood, but much evidence points to a role for certain antagonistic bacteria and fungi that inhibit the pathogenic organisms.

Regulation of the pH is effective in controlling some diseases. Keeping the pH low can control actinomycete-caused potato scab (see Figure 11.24), while raising the pH to 7.0 and above helps control fungus-caused clubfoot in cabbage.

Wet, cold soils favor some seed rots and seedling diseases known as *damping-off*. Good drainage and ridging help control these diseases. Steam or chemical sterilization is a practical method of treating greenhouse soils for a number of pathogens. It should be remembered, however, that sterilization kills beneficial microorganisms, such as mycorrhizal fungi, as well as pathogens. The addition of certain types of organic matter to the soil has been shown to suppress damping-off fungi and other diseases.

RHIZOSPHERE CAMOUFLAGE.[9]   The breeding of plants that resist particular soilborne diseases has been successful in the case of many crops. According to recent research, many plants that are resistant to soilborne diseases have *camouflaged* rhizospheres. That is, although the rhizosphere of these resistant plants is enriched in microbial numbers, the types of microbes are more similar to those in the bulk soil than is the case for the rhizosphere of disease-susceptible plants. Having a rhizosphere microbial community similar to the community in the bulk soil may make the rhizosphere of these plants less attractive to pathogens. Cultural practices that add large quantities of fresh organic matter may cause the bulk soil community to more resemble the rhizosphere community, thus assisting in rhizosphere camouflage and suppressing root disease.

COMPETITION FOR NUTRIENTS.   Soil organisms may harm higher plants, at least temporarily, by competition for available nutrients. Soil organisms can quickly absorb essential nutrients into their own cells, leaving the more slowly growing higher plants to use only what is left. Nitrogen is the element for which competition is usually greatest, although similar competition occurs for phosphorus, potassium, calcium, and even the micronutrients. This subject is considered in greater detail in Chapters 12 and 13.

---

[9]The concept of rhizosphere camouflage was introduced by Gilbert, et al. (1994).

FIGURE 11.24 Soilborne pathogens damage roots as well as other belowground organs. This potato has been attacked by the potato scab actinomycete, which may be present in soils with pH above about 5.0. (Courtesy of F. E. Manzer, University of Maine)

OTHER DETRIMENTAL EFFECTS. Under conditions of poor drainage, active soil microflora may deplete the already limited soil oxygen supply. This oxygen depletion may affect plants adversely in two ways. First, the plant roots require a certain minimum amount of O_2 for normal growth and nutrient uptake. Second, oxidized forms of several elements, including nitrogen, sulfur, iron, and manganese, will be chemically reduced by further microbial action. In the cases of nitrogen and sulfur, some of the reduced forms are gaseous and may be lost to the atmosphere, thus reducing the nutrient supply for plants. In soils that are quite acid, bacterial reduction of iron and manganese may increase the solubility of these elements to such a degree that they become toxic. Thus, nutrient deficiencies and toxicities, both microbiologically induced, can result from the same basic set of conditions.

11.17 CONCLUSION

The soil is a complex ecosystem with a diverse community of organisms. Soil organisms are vital to the cycle of life on earth. They incorporate plant and animal residues into the soil and digest them, returning carbon dioxide to the atmosphere where it can be recycled through higher plants. Simultaneously, they create humus, the organic constituent so vital to good soil physical and chemical conditions. During digestion of organic substrates, they release essential plant nutrients in inorganic forms which can be absorbed by plant roots or leached from the soil. They also influence soil color through oxidation-reduction reactions that they stimulate.

Animals, particularly earthworms, are responsible for mechanically incorporating residues into the soil, and leave open channels through which water and air can flow. Microorganisms such as fungi, actinomycetes, and bacteria are the major general-purpose organisms responsible for most organic decay. Certain of these microorganisms form symbiotic associations with higher plants, playing special roles in plant nutrition and nutrient cycling. Competition among soil microbes, and between these organisms and higher plants for mineral nutrients can result in plant nutrient deficiencies. Microbial requirements are factors in determining the success of most soil management systems.

The organisms of the soil must have energy and nutrients if they are to function efficiently. To obtain these, soil organisms break down organic matter, aid in the production of humus, and leave behind compounds that are useful to higher plants. These biotic features and their practical significance are considered in the next chapter, which deals with soil organic matter.

REFERENCES

Bethlenthalvay, G. J., and R. G. Linderman (eds.). 1992. *Mycorrhizae in Sustainable Agriculture.* ASA Special Publication No. 54 (Madison, Wis.: American Society of Agronomy).

Chino, M. 1976. "Electron microprobe analysis of zinc and other elements within and around rice root growth in flooded soils," *Soil Sci. and Plant Nut. J.,* **22**:449.

de Vleeschauwer, D., and R. Lal. 1981. "Properties of worm casts under secondary tropical forest regrowth," *Soil Sci.,* **132**:175–181.

Fitzmaurice, L. 1995. "Giving Rhizobium the competitive edge," *Wisconsin Bioissues,* **6**:17.

Gilbert, G. S., J. Handelsman, and J. L. Parke. 1994. "Root camouflage and disease control," *Phytopathology,* **84**:222–225.

Hendrix, P. F., D. A. Crossley, Jr., J. M. Blair, and D. C. Coleman. 1990. "Soil biota as components of sustainable agricultural agroecosystems," in C. Edwards and R. Lal, (eds.), *Sustainable Agricultural Systems* (Akeny, Iowa: Soil Water Conserv. Soc. Am.), pp. 637–654.

Killham, Ken. 1994. *Soil Ecology.* (New York: Cambridge University Press), 242 pp.

Kladivko, E. J., A. D. Mackay, and J. M. Bradford. 1986. "Earthworms as a factor in the reduction of soil crusting," *Soil Sci. Soc. Amer. J.,* **50**:191–196.

Lal, R. 1987. *Tropical Ecology and Physical Edaphology* (New York: Wiley).

Lambert, D. H., D. E. Baker, and H. Cole, Jr. 1979. "The role of Mycorrhizae in the interactions of phosphorus with zinc, copper, and other elements," *Soil Sci. Soc. Amer. J.,* **43**:976–980.

Lee, K. E. 1985. *Earthworms, Their Ecology and Relationships with Soils and Land Use* (New York: Academic Press).

Macfadyen, A. 1963. In J. Doeksen and J. van der Drift (eds.), *Soil Organisms* (Amsterdam: North-Holland Publ. Co.).

Menge, J. A. 1981. "Mycorrhizae agriculture technologies," in *Background Papers for Innovative Biological Technologies for Lesser Developed Countries,* Paper No. 9., Office of Technology Assessment Workshop, Nov. 24–25, 1980 (Washington, D.C.: U.S. Government Printing Office), pp. 383–424.

Mooreman, T. B., and S. S. Harper. 1989. "Transformation and mineralization of Metribuzin in surface and subsurface horizons of a Mississippi Delta soil," *J. Environ. Qual.,* **18**:302–306.

Paul, E. A., and F. E. Clark. 1989. *Soil Microbiology and Biochemistry* (New York: Academic Press).

Rovira, A. D., R. C. Foster, and J. K. Martin. 1979. "Origin, nature and nomenclature of the organic materials in the rhizosphere," in J. L. Harley and R. S. Russell (eds.), *The Soil-Root Interface* (New York: Academic Press).

Stehouwer, R. C., W. A. Dick, and S. J. Traina. 1994. "Sorption and retention of herbicides in vertically oriented earthworm and artificial burrows," *J. Environ. Qual.,* **23**:286–292.

Weil, R. R., and W. Kroontje. 1979. "Organic matter decomposition in a soil heavily amended with poultry manure." *J. Environ. Qual.,* **8**:584–588.

12

SOIL ORGANIC MATTER

The earth lay rich and dark and fell apart lightly under the points of their hoes.
—P. S. BUCK, THE GOOD EARTH

All organic substances, by definition, contain carbon. Organic matter in the world's soils contains about three times as much carbon as is found in all the world's vegetation. Soil organic matter,[1] therefore, plays a critical role in the global carbon balance that is thought to be the major factor affecting global warming, or the ***greenhouse effect.*** Although organic matter comprises only a small fraction of the total mass of most soils, this dynamic soil component influences soil physical, chemical, and biological properties to an extent far out of proportion to the small quantities present.

Soil organic matter is a complex and varied mixture of organic substances. It provides much of the cation exchange and water-holding capacities of surface soils. Certain components of soil organic matter are largely responsible for the formation and stabilization of soil aggregates. Soil organic matter also contains large quantities of plant nutrients and acts as a slow-release nutrient storehouse, especially for nitrogen. Furthermore, organic matter supplies energy and body-building constituents for most of the microorganisms whose general activities were discussed in Chapter 11. In addition to enhancing plant growth through the above-mentioned effects, certain organic compounds found in soils have direct growth-stimulating effects on plants. For all these reasons, enhancing the quantity and quality of soil organic matter is a central challenge of long-term soil management.

We will first examine the role of soil organic matter in the *global carbon cycle*. Next we will focus on inputs and losses with regard to soil carbon in specific ecosystems. Finally, we will study the processes and consequences involved in soil organic matter management.

12.1 THE GLOBAL CARBON CYCLE

The element *carbon* is the foundation of all life. From cellulose to chlorophyll, the compounds which comprise living tissues are made of carbon atoms arranged in chains or rings and associated with many other elements. The cycle of carbon on earth is the story of life on this planet. The carbon cycle is all-inclusive because it involves not only the soil,

---

[1]For a detailed review of soil organic matter see Tate, 1987. For a less technical, more practical review for agricultural use see Magdoff, 1992.

its fauna and flora, and higher plants of every description, but also all animal life, including humans. Disruption of the carbon cycle would mean disaster for all living organisms.

The basic pathways involved in the global carbon cycle are shown in Figure 12.1. Plants take in carbon dioxide from the atmosphere. Then, through the process of photosynthesis, the energy of sunlight is trapped in the carbon-to-carbon bonds of organic molecules (such as those described in Section 12.4). Some of these organic molecules are used as a source of energy (via respiration) by the plants themselves (especially by the plant roots), and the carbon is returned to the atmosphere as carbon dioxide. The remaining organic materials are stored temporarily as constituents of the standing vegetation, most of which is eventually added to the soil as plant litter (including crop residues) or root deposition (see Section 11.7). Some plant material may be eaten by animals (including humans), in which case about half of the carbon eaten is exhaled into the atmosphere as carbon dioxide. The carbon not returned to the atmosphere is returned to the soil as bodily wastes or body tissues. Once deposited on or in the soil, these plant tissues are metabolized (digested) by soil organisms, which eventually return this carbon to the atmosphere as carbon dioxide.

Much smaller amounts of carbon dioxide react in the soil to produce carbonic acid (H_2CO_3) and the carbonates and bicarbonates of calcium, potassium, magnesium, and other base-forming cations. The bicarbonates are readily soluble and can be removed in drainage or used by higher plants. Eventually much of the carbon in the carbonates and bicarbonates is also returned to the atmosphere as carbon dioxide.

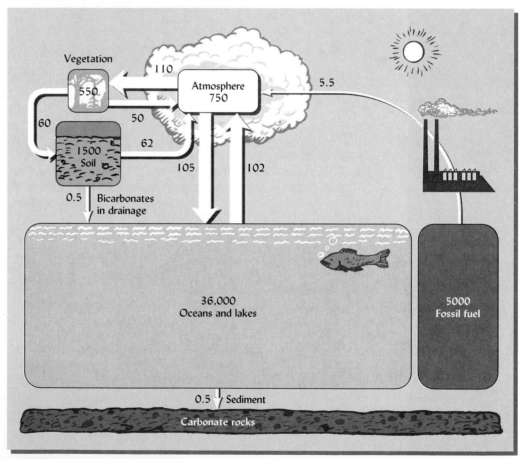

FIGURE 12.1   A simplified representation of the global carbon cycle emphasizing those pools of carbon which interact with the atmosphere. The numbers in boxes indicate the Pg (petagrams = 10^{15}g) of carbon stored in the major pools. The numbers by the arrows show the amount of carbon annually flowing (Pg/yr) by various pathways between the pools. Note that the soil contains more carbon than the vegetation and the atmosphere combined. Imbalances caused by human activities can be seen in the flow of carbon to the atmosphere from fossil fuel burning (5.5) and in the fact that more carbon is leaving (62 + 0.5) than entering (60) the soil. These imbalances are only partially offset by increased absorption of carbon by the oceans. The end result is that a total of 219.5 Pg/yr enters the atmosphere while only 215 Pg/yr of carbon is removed. It is easy to see why carbon dioxide levels in the atmosphere are rising. (Data from several sources)

Microbial metabolism in the soil produces some organic compounds of such stability that years or even centuries may pass before the carbon in them is returned to the atmosphere as carbon dioxide. Such resistance to decay allows organic matter to accumulate in soils.

The original source of the soil organic matter is plant tissue. Under natural conditions, the tops and roots of trees, shrubs, grasses, and other native plants annually supply large quantities of organic residues. Even with harvested crops, one-tenth to two-thirds of the aboveground part of the plant commonly falls to the soil surface and remains there or is incorporated into the soil. Except in the case of root crops such as carrots and sugar beets, all of the roots remain in the soil.

Animals are secondary sources of organic matter. As they attack the original plant tissues, they contribute waste products, and they leave their own bodies when they die (see Figure 11.2). Certain forms of animal life, especially earthworms, termites, and certain beetles, also play an important role in the incorporation and translocation of plant residues.

Globally, at any one time, approximately 1500 Pg (petagrams, or 10^{15} grams) of carbon are stored in soils as surface litter or soil organic matter. In fact, more carbon is stored in the soil than in the world's vegetation and atmosphere combined (see Figure 12.1). Of course, this carbon is not equally distributed among all types of soils (see Table 12.1). More than 40% of the total is contained in soils of just two orders, Histosols (mainly because of their extremely high concentration of carbon) and Inceptisols (mainly because of their vast areal extent combined with moderate concentrations of carbon). The reasons for the varying amounts of organic carbon in different soils will be detailed in Section 12.13.

In a mature natural ecosystem, the release of carbon as carbon dioxide by oxidation of soil organic matter (mostly by microbial respiration) is balanced by the input of carbon into the soil as plant residues (and, to a far smaller degree, animal residues). However, as discussed in Section 12.14, certain human land use practices, such as deforestation, soil drainage, and tillage, result in a net loss of carbon from the soil system. Figure 12.1 shows that, globally, the release of carbon from soils into the atmosphere is about 62 Pg/yr, while only about 60 Pg/yr enter the soils from the atmosphere via plant residues. This imbalance of about 2 Pg/yr, along with about 5 Pg/yr of carbon released by the burning of fossil fuels (carbon sequestered from the atmosphere millions of years ago) is only partially offset by increased absorption of atmospheric carbon dioxide by the ocean

TABLE 12.1 Mass of Organic C in the World's Soils

Values for the upper 1 meter represent most of the C in the soil profile. The upper 15 cm generally represents the surface soil which is most readily influenced by land use and soil management.

| Soil order | Global area 10^3 km² | Organic carbon[a] in upper 100 cm | | | Organic carbon[a] in upper 15 cm | |
		Mg/ha	Global Pg[b]	% of global	Range (%)[c]	Typical (%)[c]
Entisols	14,921	99	148	9	0.06–6.0	—[d]
Inceptisols	21,580	163	352	22	0.06–6.0	—
Histosols	1,745	2,045	357	23	12–57	47
Andisols	2,552	306	78	5	1.2–10	6
Vertisols	3,287	58	19	1	0.5–1.8	0.9
Aridisols	31,743	35	110	7	0.1–1.0	0.6
Mollisols	5,480	131	73	5	0.9–4.0	2.4
Spodosols	4,878	146	71	5	1.5–5.0	2.0
Alfisols	18,283	69	127	8	0.5–3.8	1.4
Ultisols	11,330	93	105	7	0.9–3.3	1.4
Oxisols	11,772	101	119	8	0.9–3.0	2.0
Misc. land	7,644	24	18	1	—	—
Total	135,215		1576	100		

Data calculated from Eswaran et al. (1993) and Brady (1990).

[a]Organic matter may be roughly estimated as 1.72 times this value. Organic nitrogen may also be estimated from organic carbon values by dividing by 12 for most soils. For aridisols and arid region soils this factor is somewhat lower, about 10. For Histosols and other humid, wetland soils this factor is somewhat higher, about 20.

[b]Petagram = 10^{15} grams.

[c]Percent on mass basis (i.e., g/100 g).

[d]These soils are too variable to suggest a typical value.

waters. As a result, the concentration of carbon dioxide in the atmosphere has been going up at an accelerating rate, increasing from 290 to 350 ppm during the past century.

12.2 SOILS AND THE GREENHOUSE EFFECT

Carbon dioxide is one of the greenhouse gases in the earth's atmosphere that make the atmosphere function like the glass in a greenhouse, causing the earth to be much warmer than it would otherwise be. Like the greenhouse glass, these gases allow short-wave solar radiation in, but trap much of the outgoing long wave radiation. This heat-trapping action of the atmosphere is a major determinant of global temperature and hence global climates. Relatively small increases in the concentration of *greenhouse gases* can result in a significantly higher average global temperature, which in turn can greatly alter climates around the globe. Many scientists believe that the average global temperature has increased by 0.5 to 1.0°C during the past century. If this increase has in fact taken place, and if the trend continues, major changes in rainfall distribution, growing season length, and sea level may occur during the next century. However, predicting changes in global climate is complicated by numerous factors, such as cloud cover and volcanic dust, which can counteract the heat-trapping effects of the greenhouse gases. While it is known that the concentrations of most greenhouse gases are increasing, it has not yet been scientifically demonstrated that global temperatures are actually rising.

Much effort and expense are currently being directed at reducing the anthropogenic (human-caused) contributions to the greenhouse effect and its potential for altering the earth's climate. Soil science has a major role to play in dealing with global warming. Gases produced by biological processes, such as those occurring in the soil, account for approximately half of the problem (see Figure 12.2). Of the five primary greenhouse gases, only CFCs (chlorofluorocarbons) are exclusively of industrial origin. As already mentioned, much of the increase in atmospheric carbon dioxide levels comes from a net loss of organic matter from the soil system.

When soils are poorly aerated, as is the case in rice paddies and wetlands, methane (CH_4), rather than carbon dioxide, is produced by organisms in decomposing soil organic matter (see Section 12.6). The rate of methane production in wet soils is very dependent on the availability of oxidizable carbon, either in the soil organic matter or in plant residues returned to the soil (Figure 12.3). The methane contribution to the atmosphere from such soils may also be regulated by the nature and management of the plants growing on them. For example, about 70 to 80% of the methane released from flooded soils escapes to the atmosphere through the stems of rice plants or natural marsh grasses. In well-aerated soils, soil-dwelling termites may produce large quantities of methane (see Section 11.5). These biological processes account for much of the methane emitted into the atmosphere. Methane makes a substantial contribution to the greenhouse effect, for although methane is found in the atmosphere in far smaller amounts than carbon dioxide, each molecule of methane is about 30 times as effective

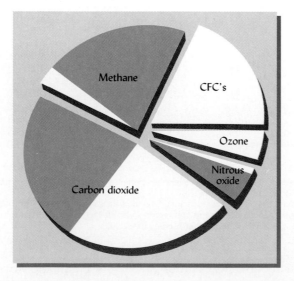

FIGURE 12.2 Relative contribution of different gases to the global greenhouse effect. The shaded portions indicate the emissions related to biological systems (in which soils play a role), the white portions being industrial contributions. [Modified from Dale et al. (1993)]

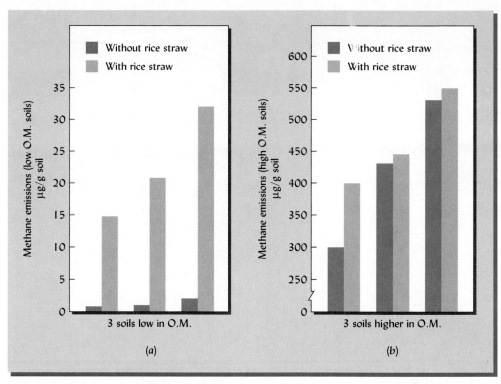

FIGURE 12.3 Global warming is enhanced by the release of methane from poorly drained soils. Methane is a greenhouse gas with atmospheric warming potential about 30 times that of CO_2. Oxygen-deficient, flooded rice soils supply food for over 2 billion people, but also account for up to one-fifth of the global emissions of CH_4. Methane release is increased by organic matter, either in the soil (compare left to right) or in crop residues (compare dark bars to light bars). Note the different scales for the low (left) and high (right) organic matter soils. [Data selected from Neue et al. (1994)]

as carbon dioxide in trapping outgoing radiation. Also, the amount of methane in the atmosphere has been increasing much more rapidly than other greenhouse gases.

Nitrous oxide (N_2O) is another greenhouse gas produced by microorganisms in poorly aerated soils, but since it is not directly involved in the carbon cycle it will be discussed in the next chapter (see Section 13.9).

Because the soil can act as a major source or sink for carbon dioxide, methane, and nitrous oxide, it is clear that, together with steps to modify industrial outputs, soil management can play a major role in controlling the emissions of greenhouse gases.

12.3 CARBON CYCLING IN THE SOIL–PLANT–ATMOSPHERE SYSTEM

Whether the goal is to reduce greenhouse gas emissions or to enhance soil quality and plant production, proper management of soil organic matter requires an understanding of the factors and processes influencing the cycling and balance of carbon in an ecosystem. Although each type of ecosystem, whether a deciduous forest, a prairie, or a wheat field, will emphasize particular compartments and pathways in the carbon cycle, consideration of a specific example, such as that described by Figure 12.4, can help us develop a general model that can be applied to many different situations (see Box 12.1).

Soil Organic Carbon Balance: An Agroecosystem Example

The rate at which soil organic matter either increases or decreases is determined by the balance between the process of **humification**, by which plant and animal residues are transformed into soil humus, and the process of **oxidation**, by which organic carbon is transformed into carbon dioxide gas.

Figure 12.4 describes the principal carbon pools and annual flows in a typical corn field in a warm temperate region. During the growing season the corn plants produce (by photosynthesis) 16,500 kg/ha of dry biomass containing 7500 kg/ha of carbon. This carbon is equally distributed among the roots, grain, and stover (unharvested stalks). In

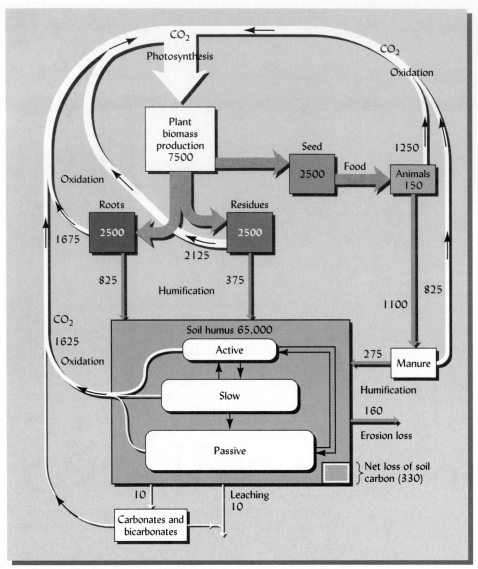

FIGURE 12.4 Carbon cycling in an agroecosystem. In the case shown a grain crop is grown, with the grain fed to cattle and the crop residues and cattle manure applied to the soil. The boxes represent the major compartments or pools of carbon in the soil-plant system (numbers in boxes represent kg C/ha). The arrows represent flows of carbon from one pool to another, with the white arrows representing carbon dioxide flows and the darker arrows representing organic carbon flows (numbers by arrows represent kg C/ha/year). The size of the boxes is roughly proportional to the relative amounts of carbon in the various pools. The thickness of the arrows is roughly proportional to the annual flow of carbon represented. Note that most of the carbon in residues is lost to the atmosphere as carbon dioxide, and that the portion so lost is greater for aboveground residues than for root residues. Transformations among active, slow, and passive organic matter within the soil humus compartment are discussed in Section 12.14.

this example, the harvested grain is used to fatten cattle. The corn stover and roots are left in the field and, along with the manure from the cattle, incorporated into the soil by tillage or by earthworms.

Of the carbon in the grain fed to the cattle, about 50% is respired by the cattle. Of the carbon not respired, a portion is assimilated by the cattle as weight gain; but a larger portion is excreted as cattle manure and applied to the soil. As the soil microbes decompose the crop residues and manure, their process of respiration converts much of the organic carbon in these residues to carbon dioxide. For aboveground plant residues and manure, about 85% of the carbon contained is lost as carbon dioxide. The corresponding figures for root residues and manure are about 67 and 75%. The remaining 15% (for aboveground plant residues) or 33% (for root residues) of the organic carbon is assimilated into the soil as humus. Thus, during the course of one year, the total organic car-

The Biosphere 2 structure, a huge, sealed, ecological laboratory. (Photo by C. Allen Morgan, ©1995 by Decisions Investments Corp. Reprinted with permission.)

An interesting experience with the global impact of soil organic matter dynamics was had by the eight biospherians who lived in Biosphere 2, a 1.3-hectare sealed glass building in the Arizona desert. This structure was designed to hold a self-contained, self-supporting ecosystem in which scientists could live and study as part of the ecosystem. Costing over $150 million to build, Biosphere 2 contains a miniature ocean, coral reef, marsh, forests, and farms designed to maintain the ecological balance. However, not only did the biospherians have a very hard time growing enough food for themselves, their atmosphere soon ran so low on

oxygen and became so enriched in carbon dioxide that engineers had to pump in oxygen and install a scrubber to cleanse carbon dioxide from the air. It turned out that the ecosystem was thrown out of kilter by the organic-matter-rich soil installed in the Biosphere. The designers had underestimated the rate at which soil organic matter would decompose when placed in the warm environment and aerated by garden tillage (see Section 12.2). The decomposition of the soil organic matter was carried out by aerobic soil microbes which use up oxygen and give off carbon dioxide as they respire.

Biospherians at work growing their own food supply in the intensive agriculture biome with compost-amended soils rich in organic matter. (Photo by Pascale Maslin, ©1995 by Decisions Investments Corp. Reprinted with permission.)

bon input into the pool of soil humus in this example is 1475 kg/ha (825 kg from roots, 375 from stover, and 275 from manure). These figures will vary widely among different soil conditions and ecosystems.

At the beginning of the year, the soil in our example contained 65,000 kg/ha organic carbon in humus. This is an amount corresponding to about 1.7% organic carbon (or about 2.8% organic matter) in the upper 30 cm of soil, a typical value for a medium-textured soil in a humid warm temperate region.[2] Such a soil used for row crops would

---

[2]The calculations are as follows for a *30-cm depth of soil.*

Mass of soil in one ha: (10,000 m²/ha) × (0.3 m deep) × (bulk density of 1.3 Mg/m³) = 3,900 Mg/ha

% organic C: (65,000 kg/ha) × (1 Mg/1000 kg) × (1 ha/soil mass of 3,900 Mg) × 100
= 1.67% organic carbon

% organic matter: 1.67% organic C × (1 g organic matter / 0.58 g organic C) = 2.88% organic matter

Note that 1 gram organic matter/0.58 g organic carbon reflects the assumption that soil organic matter is 58% C. Also, 1/0.58 = 1.72, the factor referred to in Table 12.1.

typically lose about 2.5% of its organic matter by soil respiration each year. In our example this loss amounts to some 1625 kg/ha of carbon. Smaller losses of soil organic carbon occur by soil erosion (160 kg/ha), leaching (10 kg/ha), and formation of carbonates and bicarbonates (10 kg/ha). Comparing total losses (1805 kg/ha) with the total additions (1475 kg/ha) for the pool of soil humus, we can see that the soil in our example suffered a *net loss* of 330 kg/ha of organic carbon (568 kg/ha organic matter) in one year. This amount represents an annual net loss of about 0.5% of the soil humus ($100 \times (330 \div 65,000) \simeq 0.5\%$), a rate of loss that suggests that under the management practices in our example, the soil is suffering degradation. Most likely, in the long term, soil productivity will decline, and the unit cost of crop production will rise as a result.

In order to halt or reverse the net carbon loss shown in Figure 12.4, management practices would have to be implemented that would increase the additions of carbon to the soil or decrease the loss of carbon from the soil. Since all crop residues and animal manures in our example are already being returned to the soil, additional carbon inputs could most practically be achieved by growing more plant material (i.e., increasing crop production or growing winter cover crops).

Specific practices to reduce carbon losses would include better control of soil erosion and reduced tillage, which would leave crop residues as a mulch on the soil surface. Residues left on the surface have less contact with the soil, dry out more often, and result in lower soil temperatures. All of these factors act to slow the residue decomposition and allow greater accumulation of soil carbon.

Natural Ecosystems

Those interested in natural ecosystems may want to compare the carbon cycle of a natural forest with that of the cornfield in Figure 12.4. If the forest soil fertility were not too low, the total annual biomass production would probably be similar to that of the cornfield. The standing biomass, on the other hand, would be much greater in the forest since the tree crop is not removed each year. While some litter would fall to the soil surface, most of the biomass produced would remain in the trees.

The rate of humus oxidation in the undisturbed forest would be considerably lower than in the tilled field because the litter would not be incorporated into the soil through tillage. Furthermore, losses of organic matter through soil erosion are much smaller on the forested site. Taken together, these factors allow annual net gains in soil organic matter in a young forest, and maintenance of high soil organic matter levels in mature forests. Similar trends occur in natural grasslands, but have a greater contribution from the roots.

12.4 COMPOSITION OF PLANT RESIDUES

Because the primary source of soil organic matter is plant residues, we will now consider the makeup of plant tissues and the products of their decomposition.

Green plant tissues consist mostly of water, varying in moisture content from 60 to 90%, with 75% a representative figure (Figure 12.5). If plant tissues are dried to remove all water, analyses of the *dry matter* remaining shows that, on a weight basis, the dry matter consists mostly (over 90 to 95%) of carbon, oxygen, and hydrogen. During plant photosynthesis these elements were obtained from carbon dioxide and water. If the dry matter is burned (oxidized), these elements become carbon dioxide and water once more. Of course, if dry plant residues are burned, some ash (and smoke) will also be formed, accounting for the remaining 5 to 10% of the dry matter. In the ash (and smoke) can be found the many nutrient elements originally taken up by the plants from the soil. Even though these elements are present in relatively small quantities, they play a vital role in plant nutrition and in meeting the requirements of microorganisms. The essential elements in the ash, such as nitrogen, sulfur, phosphorus, potassium, and micronutrients will be discussed in more detail in later chapters.

ORGANIC COMPOUNDS IN PLANT RESIDUES. The organic compounds in plant tissue can be grouped into broad classes, as shown in Figure 12.5. Representative percentages common in plant materials are shown. The carbohydrates, which range in complexity from simple sugars and starches to cellulose, are usually the most prominent of plant organic

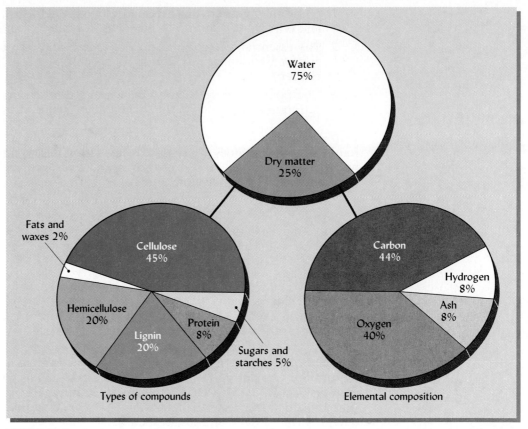

FIGURE 12.5 Composition of representative green-plant materials. The pie charts show typical composition. The major types of organic compounds are indicated at left and the elemental composition at right. The *ash* is considered to include all the constituent elements other than carbon, oxygen, and hydrogen (nitrogen, sulfur, calcium, etc.).

compounds. Lignins, which are complex compounds with ring-type structures, are found in mature, and especially woody, plant tissues. They are very resistant to decomposition. Fats and oils, which are somewhat more complex than carbohydrates, but less so than lignins, are found primarily in seeds.

Proteins contain about 6% nitrogen and smaller amounts of other essential elements such as sulfur, manganese, copper, and iron. The simple proteins are decomposed easily, while the complex crude proteins are more resistant to breakdown.

12.5   DECOMPOSITION OF ORGANIC COMPOUNDS

RATE OF DECOMPOSITION.   Organic compounds may be listed in terms of ease of decomposition as follows:

1. Sugars, starches, and simple proteins   Rapid decomposition
2. Crude proteins
3. Hemicelluloses
4. Cellulose
5. Fats, waxes, etc.
6. Lignins   Very slow decomposition

When organic tissue is added to soil, three general reactions take place:

1. Carbon compounds are enzymatically oxidized to release carbon dioxide, water, and energy.

2. The essential elements, such as nitrogen, phosphorus, and sulfur, are released and/or immobilized by a series of specific reactions relatively unique for each element.

3. Compounds very resistant to microbial action are formed either through modification of compounds in the original plant tissue or by microbial synthesis.

DECOMPOSITION: AN OXIDATION PROCESS. In a well-aerated soil, all of the organic compounds found in plant residues are subject to oxidation. Since the organic fraction of plant materials is composed largely of carbon and hydrogen, the oxidation of the organic compounds in soil can be represented as:

$$\underset{\substack{\text{Carbon- and hydrogen-}\\\text{containing compounds}}}{R\!-\!(C,4H)} + 2O_2 \xrightarrow[\text{oxidation}]{\text{enzymatic}} CO_2 + 2H_2O + \text{energy}$$

Many intermediate steps are involved in this overall reaction, and it is accompanied by important side reactions that involve elements other than carbon and hydrogen. Even so, this basic reaction accounts for most of the organic matter decomposition in the soil as well as for the oxygen consumption and CO_2 release.

BREAKDOWN OF PROTEINS. The plant proteins also succumb to microbial decay, yielding not only carbon dioxide and water, but amino acids such as glycine (CH_2NH_2COOH) and cysteine ($CH_2HSCHNH_2COOH$). In turn, these nitrogen and sulfur compounds are further broken down, eventually yielding simple inorganic ions such as ammonium (NH_4^+), nitrate (NO_3^-), and sulfate (SO_4^{2-}).

EXAMPLE OF ORGANIC DECAY. The process of organic decay in time sequence is illustrated in Figure 12.6. Assume the soil has not been disturbed or amended with plant residues for a long time. Initially, no readily decomposable materials are present. The principal microorganisms actively metabolizing are small populations of **autochthonous organisms**, those which can survive by slowly and steadily digesting the very resistant, humified soil organic matter. Competition for food is severe and microbial activity (as reflected in soil respiration) is relatively low. The supply of soil carbon is slowly but steadily being depleted.

Then, suddenly, an abundance of fresh, decomposable tissue is added to the soil. Maybe the trees are losing their leaves in fall, or a farmer has plowed in the residue of a harvested crop. In any case, the appearance of easily decomposable compounds such as sugars, starches, and cellulose stimulates increased metabolic activity among the microbes. Soon the slower-acting autochthonous populations are overtaken by rapidly multiplying populations of **zymogenous** (opportunist) organisms that have been dormant. Microbial numbers and carbon dioxide evolution from microbial respiration both increase exponentially in response to the new food resource.

Soon microbial activity is at its peak, at which point energy is being liberated rapidly, and carbon dioxide is being formed in large quantities, as are new organic compounds synthesized by the microbes. The microbial biomass at this point may account for as much as one-sixth of the organic matter in a soil. The peak of microbial activity often stimulates the breakdown of even the original, resistant, soil organic matter, a phenomenon known as the ***priming effect***.

With all this frenetic microbial activity, the easily decomposed food is soon exhausted, and microorganisms begin to die of starvation. As microbial populations plummet, living cells devour the dead cells and evolve carbon dioxide and water. As food supplies are further reduced, microbial activity continues to decline, and the general-purpose, opportunist zymogenous soil organisms again sink back into comparative quiescence. The decomposition of the dead microbial cells is associated with a release of simple products such as nitrates and sulfates. The organic matter now remaining has become soil **humus**, a dark-colored, heterogeneous, mostly colloidal mixture of newly synthesized organic compounds that are resistant to decay.

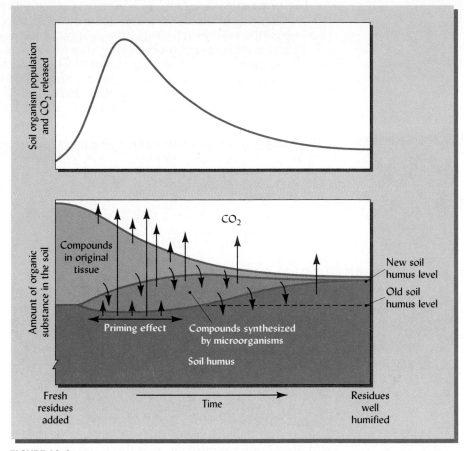

FIGURE 12.6 Diagrammatic illustration of the general changes that occur when fresh plant residues are added to soil. The time required for the process will depend on the nature of both the residues and the soil. Most of the carbon released during the initial rapid breakdown of the residues is converted to carbon dioxide, but the smaller amounts of carbon converted into microbially synthesized compounds and, eventually, soil humus should not be overlooked. Note that the humus level is increased by the end of the process, but that during the peak of microbial activity some of the original soil humus is lost, a phenomenon known as the *priming effect*. Arrows indicate transfer of carbon among compartments.

SIMPLE DECOMPOSITION PRODUCTS.   As proteins are attacked by microbes, the long chains of amino acids are broken, and individual amino acids appear in the soil solution along with dissolved CO_2. The amide $(-R-NH_2)$ and sulfide $(-R-S)$ groups of the amino acids, in turn, are broken off to produce first ammonium (NH_4^+) and sulfide (S^{2-}) compounds, and finally nitrates (NO_3^-) and sulfates (SO_4^{2-}). Similar decomposition of other organic compounds releases these and other inorganic nutrient ions. The term, ***mineralization,*** applies to the overall process that releases elements from organic compounds to produce these inorganic (mineral) forms. Most of the inorganic ions released by mineralization are readily available to higher plants and to microorganisms. The decay of organic tissues is an important source of these essential elements for plants.

12.6 DECOMPOSITION IN ANAEROBIC SOILS

Microbial decomposition proceeds most rapidly in the presence of plentiful supplies of oxygen gas, which aerobic organisms use to accept the electrons produced as they oxidize organic compounds. Oxygen supplies may become depleted when soil pores filled with water prevent the diffusion of this gas into the soil from the atmosphere. Without sufficient oxygen present, aerobic organisms cannot function, so anaerobic or facultative organisms become dominant. Under low-oxygen or anaerobic conditions, decomposition takes place much more slowly than when oxygen is plentiful. Hence wet, anaerobic soils tend to accumulate large amounts of organic matter in a partially decomposed condition.

The products of anaerobic decomposition include a wide variety of partially oxidized organic compounds such as organic acids, alcohols, and methane gas. Anaerobic decomposition releases relatively little energy for the organisms involved; therefore the end products still contain much energy. (For this reason alcohol and methane can serve as fuel.) Some of the products of anaerobic decomposition are of concern because they produce foul odors or inhibit plant growth, and the methane gas produced in wet soils is a major contributor to the greenhouse effect (Section 12.2). The following reactions are typical of those carried out in wet soils by various **methanogenic bacteria:**

$$4C_2H_5COOH + 2H_2O \xrightarrow{\text{Bacteria}} 4CH_3COOH + CO_2 + CH_4$$

Propionate   Acetate   Carbon   Methane
                                                dioxide

$$CH_3COOH \xrightarrow{\text{Bacteria}} CO_2 + CH_4$$

$$CO_2 + 4H_2 \xrightarrow{\text{Bacteria}} 2H_2O + CH_4$$

12.7 CARBON/NITROGEN RATIO OF ORGANIC MATERIALS

As indicated in Section 12.4, the carbon content of most plant dry matter is about 40%; that of soil organic matter is generally estimated to range from 45 to 58% (the traditional value used in calculations is 58%). In contrast, the nitrogen content of plant residues is much lower and varies widely (from <1 to >6%). The ratio of carbon to nitrogen (C/N) in soil organic matter and in organic residues applied to soils is important for two major reasons: (1) Intense competition among microorganisms for available soil nitrogen occurs when residues having a high C/N ratio are added to soils. (2) Because this ratio is relatively constant in soils, the maintenance of carbon—and hence soil organic matter—is constrained by the soil nitrogen level. The constancy of the C/N ratio in soils helps determine the rate of organic decay, the level of total organic matter, and the amount of nitrogen available to plants. The C/N ratio is therefore an important consideration in developing sound soil management schemes.

C/N Ratio in Soils.   The C/N ratio in the organic matter of arable (cultivated) surface (Ap) horizons commonly ranges from 8/1 to 15/1, the median being between 10/1 and 12/1. The ratio is generally lower for subsoils than for surface layers in a soil profile. Other variations largely correlate with climatic conditions, especially those affecting the leaching of calcium from soil profiles. In a given climatic region, little variation occurs in the C/N ratio for similarly managed soils. For instance, in calcium-rich soils of semi-arid grasslands (e.g., Mollisols and tropical Alfisols), the C/N ratio is relatively narrow. In more severely leached soils in humid regions, the C/N is relatively wide. For some types of highly leached, acidic soils, C/N ratios as high as 30/1 are not uncommon. When such soils are brought under cultivation and limed to increase their pH and calcium content, the enhanced decomposition tends to lower the C/N ratio to near 12/1 (see Table 12.2).

Ratio in Plants and Microbes.   The C/N ratio in plant residues ranges from between 10/1 to 30/1 in legumes and young green leaves to as high as 600/1 in some kinds of sawdust (see Table 12.3). Generally, as plants mature, the proportion of protein in their tissues declines, while the proportion of lignin and cellulose, and the C/N ratio, increase. As can be seen from the decay curves in Figure 12.7, these differences in composition have pronounced effects on the rate of decay when plant residues are added to the soil.

In the bodies of microorganisms, the C/N ratio is not only less variable than in plant tissues, but also much lower, ordinarily falling between 5/1 and 10/1. Among microorganisms, bacteria are generally somewhat richer in protein than fungi, and consequently have a lower C/N ratio.

12.8 INFLUENCE OF CARBON/NITROGEN RATIO ON DECOMPOSITION

Soil microbes, like other organisms, require a balance of nutrients from which to build their bodies and extract energy. The majority of soil organisms obtain carbon both for

TABLE 12.2 Organic Carbon, Nitrogen (in Percent), and Carbon to Nitrogen Ratio in Surface Soil (A Horizons) and Subsoil (B Horizons) in Ultisols and Mollisols

Except as noted, the Ultisols were under forest vegetation and the Mollisols were under grassland. In all cases the subsoil contains less organic carbon than the surface soils, and generally also has a lower ratio of carbon to nitrogen. In Ultisol surface soils the carbon-to-nitrogen ratio is generally near 20:1, except where the soil has been cultivated and limed. A carbon-to-nitrogen ratio of 12:1 is representative of most nonforested surface soils.

Location	Soil great group	Organic C A horizon	Organic C B horizon	Nitrogen A horizon	Nitrogen B horizon	C/N ratio A horizon	C/N ratio B horizon
Ultisols							
Ga.	Albaqult	2.2	0.54	0.095	0.060	23	9
P.R.	Plinthaqult	3.25	0.74	0.25	0.07	13[a]	10
Calif.	Haplohumult	6.69	2.16	0.26	0.103	26	21
Va.	Hapludult	3.40	0.71	0.127	0.056	27	13
Tenn.	Rhodudult	3.36	0.44	0.207	0.057	16	8
P.R.	Tropudult	2.40	0.38	0.205	0.044	12[a]	9
P.R.	Haplustult	1.64	0.49	0.149	0.053	11[a]	9
N.C.	Umbraqult	5.3	0.42	0.199	0.051	27	8
Calif.	Haploxerult	4.22	0.75	0.140	0.041	30	18
Md.	Fragiudult	3.58	0.35	0.124	0.041	29	9
Mollisols							
Kans.	Argiustoll	1.17	0.54	0.118	0.066	10[a]	8
Mont.	Argiboroll	2.06	0.74	0.192	0.074	11	10
Nebr.	Argiustoll	2.54	0.75	0.202	0.081	13	9
Kans.	Argiustoll	4.35	2.3	0.318	0.178	14	13
N. Dak.	Natriboroll	2.71	1.09	0.199	0.086	14[a]	13
Iowa	Haplaquoll	4.2	0.41	0.325	0.046	13[a]	9
Ill.	Haplaquoll	1.51	0.94	0.135	0.088	11[a]	11
Iowa	Hapludoll	2.86	0.49	0.247	0.057	12	8
Utah	Calcixeroll	3.69	2.07	0.291	0.177	13	12
Iowa	Argiaquoll	2.78	0.92	0.188	0.078	15[a]	12

Data compiled from Soil Survey Staff (1975).
[a]These data are for Ap horizons in cultivated soils. P.R. = Puerto Rico.

building essential organic compounds and for obtaining energy for life processes by metabolizing carbonaceous materials. However, they cannot multiply and grow on carbon alone. Organisms must also obtain sufficient nitrogen to synthesize nitrogen-containing cellular components, such as amino acids, enzymes, and DNA. On the average, soil microbes must incorporate into their cells about eight parts of carbon for every one part of nitrogen (i.e., the microbes have an average C/N ratio of 8/1). Because only about one-third of the carbon metabolized by microbes is incorporated into their cells (the remainder is respired and lost as CO_2) the microbes need to find about 24 parts carbon for every part nitrogen assimilated into their bodies (i.e., a C/N ratio of 24/1 in their "food"). This requirement results in two extremely important practical consequences: (1) If the C/N ratio of organic material added to soil exceeds about 25/1, the soil microbes will scavenge the soil solution to obtain enough nitrogen. Thus the incorporation of high C/N residues will deplete the soil's supply of soluble nitrogen, causing higher plants to suffer from nitrogen deficiency. (2) The decay of organic materials can be slowed if sufficient nitrogen to support microbial growth is neither present in the material undergoing decomposition nor available in the soil solution. These concepts are illustrated by the example in Figure 12.8.

EXAMPLES OF INORGANIC NITROGEN RELEASE DURING DECAY. The practical significance of the C/N ratio becomes apparent if we compare the changes that take place in the soil when residues of either high or low C/N ratio are added (see Figure 12.9). Consider a soil with a moderate level of soluble nitrogen (mostly nitrates). General-purpose decay organisms are at a low level of activity in this soil, as evidenced by low carbon dioxide production. If no plants are taking up nitrogen, the level of nitrates will very slowly increase as the native soil organic matter decays.

TABLE 12.3 Typical Carbon and Nitrogen Contents and C/N Ratios of Some Organic Materials Commonly Associated with Soils

Organic material	% C	% N	C/N
Spruce sawdust	50	0.05	600/1
Hardwood sawdust	46	0.1	400/1
Wheat straw	38	0.5	80/1
Corn stover	40	0.7	57/1
Sugarcane trash	40	0.8	50/1
Rye cover crop, anthesis	40	1.1	37/1
Bluegrass from fertilized lawn	40	1.3	31/1
Rye cover crop, vegetative stage	40	1.5	26/1
Mature alfalfa hay	40	1.8	25/1
Rotted barnyard manure	41	2.1	20/1
Finished household compost	40	2.5	16/1
Young alfalfa hay	40	3.0	13/1
Hairy vetch cover crop	40	3.5	11/1
Digested municipal sewage sludge	31	4.5	7/1
Soil microorganisms			
Bacteria	50	10.0	5/1
Actinomycetes	50	8.5	6/1
Fungi	50	5.0	10/1
Soil organic matter			
Mollisol Ap horizon	56	4.9	11/1
Ultisol A1 horizon	52	2.3	23/1
Average B horizon	46	5.1	9/1

Data calculated from many sources.

Now consider what happens when a large quantity of organic material is added to this soil. If these residues have a C/N ratio greater than 25, changes will occur according to the pattern shown in Figure 12.9a. In the example shown, the initial C/N ratio of the residues is about 55/1, typical for cornstalks or many kinds of leaf litter. As soon as the residues contact the soil, the microbial community responds to the new food supply (see Section 12.3). Heterotrophic zymogenous microorganisms become active, multiply rapidly, and yield carbon dioxide in large quantities. Because of the microbial demand for nitrogen, little or no mineral nitrogen (NH_4^+ or NO_3^-) is available to higher plants during this period.

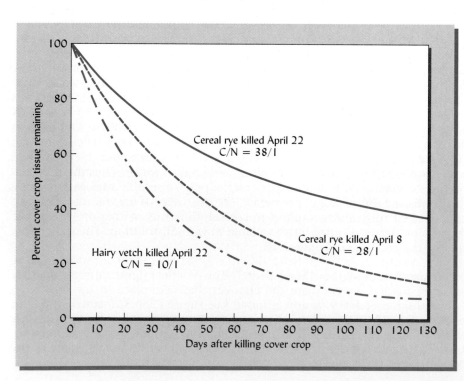

FIGURE 12.7 Rates of decomposition of various cover-crop residues. The cover crops were grown over the winter and early spring, then killed with a herbicide and the residues left as a mulch on the soil surface. Corn was planted into this mulch without tillage. The lower the initial C/N ratio of the residues, the more rapid was the decomposition process. Note that the legume (hairy vetch) had a much lower C/N ratio than the grass (cereal rye). Also note that a two-week delay in killing the rye cover crop resulted in more mature plants with a considerably higher C/N ratio and slower rate of decomposition. (Redrawn from S. Davis and R. Weil, unpublished)

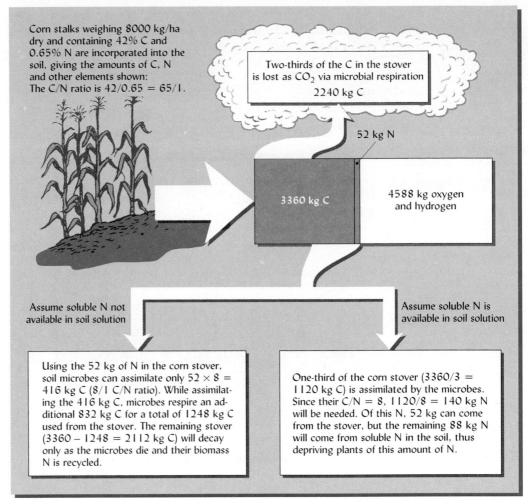

Corn stalks weighing 8000 kg/ha dry and containing 42% C and 0.65% N are incorporated into the soil, giving the amounts of C, N and other elements shown: The C/N ratio is 42/0.65 = 65/1.

Two-thirds of the C in the stover is lost as CO_2 via microbial respiration
2240 kg C

52 kg N

3360 kg C

4588 kg oxygen and hydrogen

Assume soluble N not available in soil solution

Assume soluble N is available in soil solution

Using the 52 kg of N in the corn stover, soil microbes can assimilate only 52 × 8 = 416 kg C (8/1 C/N ratio). While assimilating the 416 kg C, microbes respire an additional 832 kg C for a total of 1248 kg C used from the stover. The remaining stover (3360 − 1248 = 2112 kg C) will decay only as the microbes die and their biomass N is recycled.

One-third of the corn stover (3360/3 = 1120 kg C) is assimilated by the microbes. Since their C/N = 8, 1120/8 = 140 kg N will be needed. Of this N, 52 kg can come from the stover, but the remaining 88 kg N will come from soluble N in the soil, thus depriving plants of this amount of N.

FIGURE 12.8 A simplified, quantitative example of residue decay illustrating the fates of carbon and nitrogen and the consequences for decomposition and soil nitrogen availability.

This condition, often called the ***nitrate depression period,*** persists until the activities of the decay organisms gradually subside due to lack of easily oxidizable carbon. Their numbers decrease, carbon dioxide formation drops off, nitrogen demand by microbes becomes less acute, and the release of nitrates (nitrification) can proceed. As decay occurs, the C/N ratio of the remaining plant material decreases because carbon is being lost (by respiration) and nitrogen conserved (by incorporation into microbial cells). Generally, one can expect mineral nitrogen to begin to be released when the C/N ratio of the remaining material drops below about 20/1. Then, nitrates appear again in quantity, and the original conditions prevail, except that the soil is somewhat richer in both nitrogen and humus.

The nitrate depression period may last for a few days, a few weeks, or even several months. A longer, more severe period of nitrate depression is typical when added residues have a higher C/N ratio and greater ease of decomposition, and when a larger quantity of residues is added. To avoid producing seedlings that are stunted, chlorotic, and nitrogen-starved, planting should be delayed until after the nitrate depression period, or additional sources of nitrogen (e.g., fertilizer) should be applied to satisfy the nutritional requirements of both the microbes and the plants. These principles can be applied in many practical ways and should be considered whenever organic residues are added to a soil.

The effects on soil nitrate level will be quite different if the residues added have a C/N ratio smaller than 20/1, as in the case represented by Figure 12.9b. With organic materials of low C/N ratio, more than enough nitrogen is present to meet the needs of the decomposing organisms. Therefore, soon after decomposition begins, some of the nitrogen from organic compounds is released into the soil solution, augmenting the

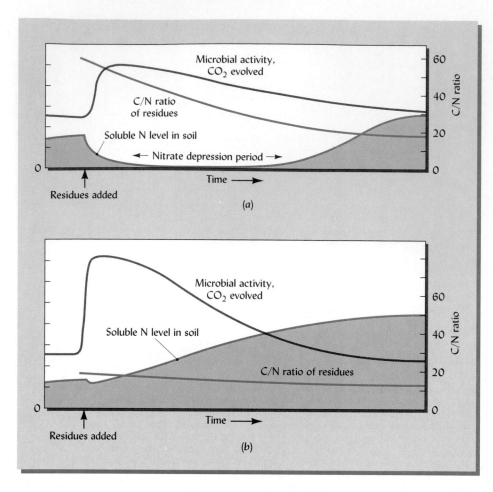

FIGURE 12.9 Changes in microbial activity, in soluble nitrogen level, and in residual C/N ratio following the addition of either high (a) or low (b) C/N ratio organic materials. Where the C/N ratio of added residues is above 25, microbes digesting the residues supplement the nitrogen contained in the residues with nitrogen from the soil. During the resulting nitrate depression period, competition between higher plants and microbes would be severe enough to cause nitrogen deficiency in the plants. Note that in both cases soluble N in the soil has increased from its original level once the decomposition process has run its course. The trends shown are for soils without growing plants, which, if present, would remove the soluble nitrogen as soon as it was released.

level of soluble nitrogen available for plant uptake. Generally, nitrogen-rich materials decompose quite rapidly, resulting in a period of intense microbial growth and activity and no nitrate depression period.

Because the nitrogen content of soil organic matter generally remains constant at about 5% (N/OM = 1:20), the amount of organic matter that can be maintained in any soil is largely dependent on the amount of nitrogen present. A soil's organic matter cannot, therefore, be increased without simultaneously increasing its organic nitrogen content and vice versa. Maintenance or accumulation of organic matter in a soil is enhanced by amending the soil with nitrogen-rich residues. A similar effect can be achieved by applying inorganic nitrogen fertilizer along with low-nitrogen residues. These principles are well illustrated by the field experiment results summarized in Table 12.4.

12.9 HUMUS: GENESIS AND NATURE[3]

Soil organic matter is a general term that encompasses all the organic components of the soil: (1) living biomass (intact plant and animal tissues and microorganisms), (2) dead roots and other recognizable plant residues, as well as (3) a largely amorphous and colloidal mixture of complex organic substances no longer identifiable as plant tissues. Only the third category of organic material is properly referred to as soil *humus* (see Figure 12.10). Most of the complex compounds that make up humus are not merely degraded plant materials. The bulk of them have resulted from two general types of biochemical reactions: *decomposition* and *synthesis*.

---

[3]For discussions of humus formation, see Haynes (1986) and Stevenson (1982) and Frimmel and Christman (1988).

TABLE 12.4 Soil Organic Carbon Content and C/N Ratio of the Soil Organic Matter After 23 Years of Various Organic Amendment and Nitrogen Fertilization Practices

Regardless of the type and amount of organic residues added to the soil, the addition of N from fertilizer increased the accumulation of soil organic matter. A high content of lignin and a low C/N ratio in the added residues favored organic matter accumulation. The C/N ratio of the soil organic matter was little affected by these factors, however.

Soil management practices followed for 32 years[a]	C/N ratio of added residues	Lignin in added residues (%)	Soil organic matter after 32 years	
			Total C (Mg/ha)	C/N ratio
No additions (just crop roots)	—	—	28.6	8.3/1
No additions (just crop roots) + 80 kg fertilizer N/ha	—	—	33.9	9.1/1
Green manure (grass), 3760 kg C/ha	17	6	40.4	9.3/1
Straw, 3850 kg C/ha	70	15	37.2	9.1/1
Straw, 3850 kg C/ha + 80 kg fertilizer N/ha	70	15	43.1	9.6/1
Sawdust, 3970 kg C/ha	450	30	42.0	9.9/1
Sawdust, 3970 kg C/ha + 80 kg fertilizer N/ha	450	30	48.4	10.2/1

Data from Paustian et al. (1992).

[a]Small-grain crops grown in 27 of the years; root crops grown in the other 5 years. Amounts of organic amendments indicated were applied in the autumn every second year and tilled into the soil. Initially the soil organic carbon content was 38 Mg/ha (1.5%). The soil was a sandy clay loam (35% clay) located near Uppsala, Sweden, where the mean annual temperature is 5.4°C.

As decomposition of plant residues occurs, compounds in the residues are broken down or drastically modified by soil organisms. Complex components of the plant residues, including lignin, are broken down into simpler compounds. The resulting simpler compounds are then metabolized by the soil microbes. Some of the carbon not lost as carbon dioxide in respiration, along with most of the nitrogen, sulfur, and oxygen from these compounds, is used by the microorganisms to synthesize new cellular components and biomolecules. The microbes also *polymerize* (link together) some of the simpler new compounds and break down products (especially those with phenolic ring structures)

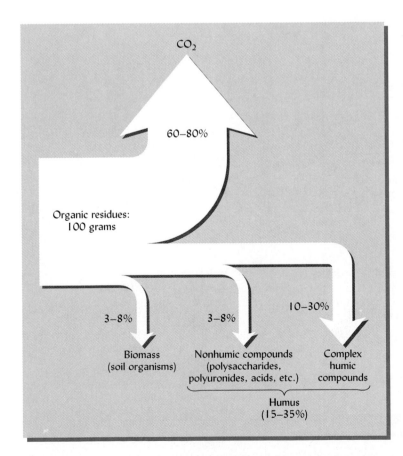

FIGURE 12.10 Disposition of 100 g of organic residues one year after they were incorporated into the soil. More than two-thirds of the carbon has been oxidized to CO_2, and less than one-third remains in the soil—some in the cells of soil organisms, but a larger component as soil humus. The amount converted to CO_2 is generally greater for aboveground residues than for belowground (root) residues. (Estimates from many sources)

into long, complex chains that resist further decomposition. These high molecular weight compounds interact with nitrogen-containing amino compounds and give rise to a significant component of resistant humus. The formation of these polymers is encouraged by the presence of colloidal clays. These ill-defined, complex, resistant, polymeric compounds are called **humic substances.** The less resistant, identifiable biomolecules produced by microbial action are grouped together as **nonhumic substances.**

Figure 12.10 shows the approximate proportions of organic carbon from plant residues that are found in humic and nonhumic compounds one year after the residues were added to the soil. While most of the carbon has returned to the atmosphere as CO_2, one-fifth to one-third remains in the soil either as live biomass ($\approx$5%) or as the humic ($\approx$20%) and nonhumic ($\approx$5%) fractions of soil humus. The proportion remaining from root residues tends to be somewhat higher than that remaining from aboveground plant litter. This figure should be compared with Figure 12.4.

Humic Substances

Humic substances comprise about 60–80% of the soil organic matter. They include the most complex materials, which are also the most resistant to microbial attack. Humic substances are characterized by aromatic, ring-type structures that include *polyphenols* (numerous phenolic compounds linked together) and comparable *polyquinones,* which are even more complex. They are formed by the decomposition and synthesis processes just described, including the polymerization into these more complex forms. Humic substances generally are dark-colored, amorphous substances with molecular weights varying from 2000 to 300,000 g/mol.

SOLUBILITY GROUPINGS.   Humic substances have been classified into three chemical groupings based on solubility (Figure 12.11): (1) **fulvic acid**, lowest in molecular weight and lightest in color, soluble in both acid and alkali and most susceptible to microbial attack; (2) **humic acid**, medium in molecular weight and color, soluble in alkali but

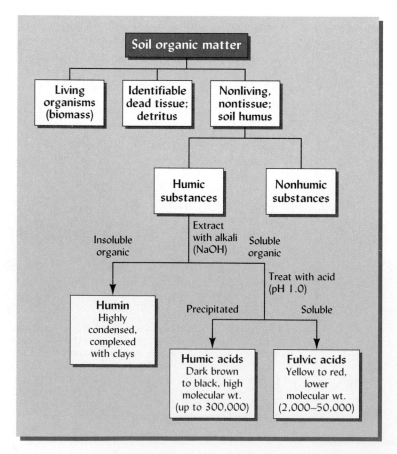

FIGURE 12.11   Classification of soil organic matter components separable by chemical and physical methods. Solubility in alkali and acid is a widely used criterion for grouping different fractions of soil humus. The classical scheme dividing soil humic substances into humin, fulvic acids, and humic acids fraction (shown in the lower part of the flowchart) is based on their insolubility in NaOH (humin), their subsequent solubility (fulvic acids), and insolubility (humic acids) in acid solutions (pH = 1).

insoluble in acid and intermediate in resistance to degradation; and (3) **humin**, highest in molecular weight, darkest in color, insoluble in both acid and alkali and most resistant to microbial attack.

All three groups of humic substances are relatively stable in soils. Even fulvic acid, the most easily degraded, is more resistant to microbial attack than most freshly applied crop residues. Depending on the environment, the *half-life* (the time required to destroy half the amount of a substance) of fulvic acid may be 10 to 50 years, while the half-life of humic acid is generally measured in centuries.

Nonhumic Substances

About 20–30% of the humus in soils consists of nonhumic substances. These substances are less complex and less resistant to microbial attack than those of the humic group. They are comprised of specific biomolecules with definite physical and chemical properties. Some of these nonhumic substances are microbially modified plant compounds, while others are compounds synthesized by soil microbes as by-products of decomposition.

Included among the nonhumic substances are *polysaccharides*, polymers which have sugarlike structures and a general formula of $C_n(H_2O)_m$, where n and m are variable. Polysaccharides are especially important in enhancing soil aggregate stability (see Section 4.11). Also included are *polyuronides*, which are not found in plants, but are synthesized by soil microbes.

Some yet simpler compounds are part of the nonhumic group. For example, low molecular weight organic acids and some proteinlike materials are included. Although none of these simpler materials are present in large quantities, their presence is critical, especially in their effect on the availability of some plant nutrients, such as nitrogen and iron, and in their direct effects on plant growth.

Stability of Humus

Studies using radioactive isotopes have shown that some organic carbon converted to humus thousands of years ago is still present in soils, evidence that some humic materials are extremely resistant to microbial attack. This resistance of humic substances to oxidation is important in maintaining organic matter levels and in protecting associated nitrogen and other essential nutrients against mineralization and loss from the soil. For example, during humus formation complex combinations that protect the protein nitrogen from microbial attack are formed. Despite its relative resistance to decay, humus is subject to continual microbial attack. As illustrated in Figure 12.4, without the annual addition of sufficient plant residues, microbial oxidation results in a reduction in soil organic matter levels.

CLAY–HUMUS COMBINATIONS.[4] Interaction with clay minerals provides another means of stabilizing soil nitrogen and organic matter. Certain clays are known to attract and hold substances such as amino acids, peptides, and proteins, forming complexes that protect the nitrogen-containing compounds from microbial degradation. Organic matter that is entrapped in the very small (<1 µm) pores formed by clay particles is physically inaccessible to decomposing organisms. It is also possible that layer silicate clays, such as vermiculite, may bind organic matter in their interlayers in forms that strongly resist decomposition.

Thus, clay, along with humic and polysaccharide polymers, can protect relatively simple nitrogen compounds from microbial attack. In many soils more than half the organic matter is associated with clay and other inorganic constituents. Although the extent and mechanisms are not yet fully understood, clay–humus interactions partially account for the high organic matter content of clay soils.

Colloid Characteristics of Humus

In Chapter 8, the colloidal nature of humus was emphasized. The surface area of humus colloids per unit mass is very high, generally exceeding that of silicate clays. The col-

---

[4]For detailed discussions of this subject, see Huang and Schnitzer (1986) and Loll and Bollag (1983).

loidal surfaces of humus are negatively charged, the sources of the charge being carboxylic (—COOH) or phenolic (—OH) groups. The extent of the negative charge is pH-dependent. At high pH values, the cation-exchange capacity of humus on a mass basis (150 to 300 cmol/kg) far exceeds that of most silicate clays. However, because the humus is much less dense than clay, the cation exchange capacity of humus on a volume basis (40 to 80 cmol/L) is comparable to that of silicate clays. Cation exchange reactions with humus are qualitatively similar to those occurring with silicate clays. The humus micelles, like the particles of clay, carry a swarm of adsorbed cations (Ca^{2+}, H^+, Mg^{2+}, K^+, Na^+, etc.), which are exchanged with cations from the soil solution (see Chapter 8).

The water-holding capacity of humus on a mass basis (but not on a volume basis) is four to five times that of the silicate clays. Humus plays a very important role in aggregate formation and stability. The black color of humus tends to distinguish it from most of the other colloidal constituents in soils.

Effect of Humus on Nutrient Availability

Humus enhances mineral breakdown and, in turn, nutrient availability in two ways. First, humic acids can attack soil minerals and bring about their decomposition, thereby releasing essential base-forming cations as follows.

$$KAlSi_3O_8 + H\boxed{Micelle} \longrightarrow HAlSi_3O_8 + K\boxed{Micelle}$$

Microcline   Humic acid   Acid silicate   Adsorbed K

The potassium is changed from a structural component to an adsorbed ion, which is more readily available to higher plants.

Second, humus forms stable organomineral complexes with certain nutrient ions. For example, nonhumic organic acids, polysaccharides, and fulvic acids all can form complexes with metallic ions such as Fe^{3+}, Cu^{2+}, Zn^{2+}, and Mn^{2+}. The cations are attracted from the minerals in which they are found and held in complex form by the organic molecule. Later they may be taken up by plants or may take part in the synthesis of clay and other inorganic constituents.

12.10 DIRECT INFLUENCES OF ORGANIC MATTER ON PLANT GROWTH

Long ago, the observation that plants generally grow better on organic-matter-rich soils led people to think that plants derive much of their nutrition by absorbing humus from the soil. We now know that higher plants derive their carbon from carbon dioxide and that most of their nutrients come from inorganic ions dissolved in the soil solution. In fact, plants can complete their life cycles quite well, growing totally without humus in quartz sand or aerated water, provided that sufficient mineral nutrients are applied. This is not to say that humus is less important to plants than once supposed, but rather that most of the many beneficial influences of humus on plant growth are indirect in nature (see Figure 12.12).

Direct Influence of Humus on Plant Growth

Although the concept of humus as plant food has long been discredited, it is well established that certain organic compounds are absorbed by higher plants. Plants can absorb a very small portion of their nitrogen and phosphorus needs as soluble organic compounds. Various growth-promoting compounds such as vitamins, amino acids, auxins, and gibberellins are formed as organic matter decays. These substances may at times stimulate growth in both higher plants and microorganisms.

Small quantities of both fulvic and humic acids in the soil solution are known to enhance certain aspects of plant growth (see Table 12.5). Components of these humic substances probably act as regulators of specific plant growth functions, such as cell elongation and lateral root initiation. The concentrations of these humic substances effective in stimulating plant growth are generally in the range of 50 to 100 mg/L (ppm), levels commonly present in the soil solution in humid regions. Commercial humate products have been marketed with claims that small amounts enhance plant growth, but

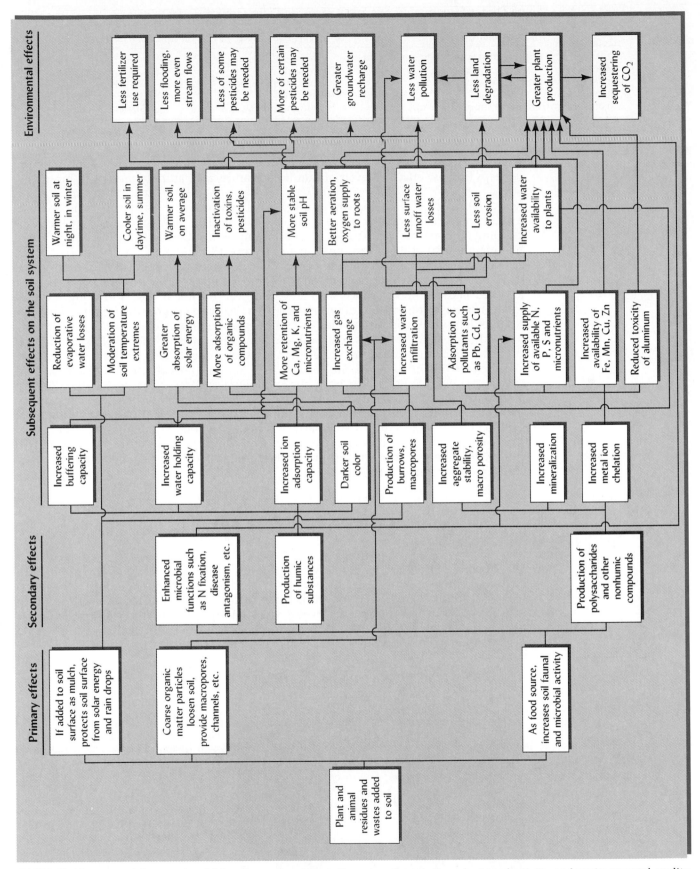

FIGURE 12.12 Some of the ways in which soil organic matter influences soil properties, plant productivity, and environmental quality. Many of the effects are indirect, the arrows indicating the cause-and-effect relationships. It can readily be seen that the influences of soil organic matter are far out of proportion to the relatively small amounts present in most soils. Many of these influences are discussed in this and other chapters in this book.

TABLE 12.5   Some Direct Effects of Humic Substances on Plant Growth

Effect on plant growth	Humic substance	Concentration range (mg/L)
Accelerated water uptake and enhanced germination of seeds	Humic acid	0–100
Stimulated root initiation and elongation	Humic and fulvic acids	50–300
Enhanced root cell elongation	Humic acid	5–25
Enhanced growth of plant shoots and roots	Humic and fulvic acids	50–300

From Chen and Aviad (1990).

scientific tests of many of these products have failed to show any benefit from their use, perhaps because effective levels of humic substances are naturally present in most soils.

ALLELOPATHIC EFFECTS.[5]   **Allelopathy** is the process by which one plant infuses the soil with a chemical that affects the growth of other plants. The plant may do this by directly exuding allelopathic compounds, or these compounds may be formed through microbial metabolism of the plant's dead tissues (Figure 12.13).

Allelopathic chemicals present in the soil are apparently responsible for many of the positive and negative effects observed when various plants grow in association with one another. Because they produce such chemicals, certain weeds (e.g., Johnsongrass, giant foxtail) damage crops far out of proportion to the size and number of weeds present. Crop residues left on the soil surface may inhibit the germination and growth of the next crop planted (e.g., wheat residues often inhibit sorghum plants). Other allelopathic interactions influence the succession of species as forest and grassland ecosystems mature.

Allelopathic interactions are usually very specific, involving only certain plant species, or even varieties, on both the producing and receiving ends. The effects of allelopathic chemicals are many and varied, and can be either positive (as in certain garden companion plantings) or negative. Although many different chemistries are represented, most allelochemicals are relatively simple phenolic or organic acid compounds that can be rapidly destroyed by soil microorganisms. Because of the susceptibility of the allelochemicals to degradation and leaching, effects are usually relatively short-lived once the source is removed.

12.11 INFLUENCE OF ORGANIC MATTER ON SOIL PROPERTIES AND THE ENVIRONMENT

Soil organic matter affects so many soil properties and processes that a complete discussion of the topic is beyond the scope of this chapter. Indeed, in almost every chapter in this book there is mention of the roles of soil organic matter. As background for discussion of the distribution and management of organic matter in soils, it will be useful to study Figure 12.12, which summarizes some of the more important effects of organic matter on soil properties and soil–environment interactions.

---

[5]For a comprehensive review of allelopathy, see Inderjit et al. (1995).

FIGURE 12.13 Positive and negative allelopathic effects of winged beans on grain amaranth plants. In the pot on the left (T4) amaranth is growing in fresh soil (no association with winged beans). In the center pot (T20) amaranth is growing in soil previously used to grow winged beans (positive effect). In the pot on the right (T28) the amaranth is growing in fresh soil, but the plant was watered three times with a water extract of winged bean tissue (negative effect). All pots were watered with a complete nutrient solution. [From Weil and Belmont (1987)]

As we have seen in the previous section, some soil organic compounds have deleterious effects on plant growth. However, the vast majority of effects stemming from soil organic matter are beneficial. Maintaining a sufficiently high level of organic matter is one of the most critical objectives of soil management.

INFLUENCE ON SOIL PHYSICAL PROPERTIES. Humus tends to give surface horizons dark brown to black colors. Granulation and aggregate stability are encouraged, especially by the non-humic fractions of soil organic matter. Plasticity, cohesion, and stickiness of clayey soils are reduced, making these soils easier to manipulate. Soil water retention is also improved, since organic matter increases both infiltration rate and water-holding capacity.

INFLUENCE ON SOIL CHEMICAL PROPERTIES. Because humus has a CEC two to thirty times as great (per kg) as that of the various types of clay minerals, it generally accounts for 20–90% of the cation-adsorbing power of mineral soils. Like clays, humus colloids hold nutrient cations (potassium, calcium, magnesium, etc.) in easily exchangeable form. Organic acids associated with humus also accelerate the release of nutrient elements from mineral structures. In addition, nitrogen, phosphorus, sulfur, and micronutrients are stored as constituents of soil organic matter until released by mineralization.

Certain components of soil humus **chelate** or otherwise form complexes with metal ions. Some of these metal ions are micronutrients (iron, zinc, etc.) and are made more available to plants because they are kept in soluble, chelated form (see Chapter 15). In the case of aluminum ions, which are toxic to plants in very acid soils, organic matter alleviates the toxicity by binding the aluminum ions in nonexchangeable forms.

Finally, soil organic matter greatly affects the biology of the soil because it provides the main food for the community of heterotrophic soil organisms described in Chapter 11.

12.12 AMOUNT OF ORGANIC MATTER[6] IN SOILS

The amount of organic matter in soils varies widely; mineral surface soils contain from a mere trace (sandy, desert soils) to as high as 20 or 30% (some forested A horizons). Some soils contain even more organic matter, but those that do are considered to be organic—not mineral—soils and will be discussed in Section 12.15. In practice, soil organic matter content is usually estimated from analysis of soil organic carbon content, because the latter can be determined more precisely. Therefore, for quantitative discussions we will usually refer to soil organic carbon. The ranges and representative organic carbon contents of different soil orders were given in Table 12.1. Even though within a single soil order, organic carbon contents vary as much as tenfold, it is possible to make a few generalizations about the organic carbon contents of different types of soils. Aridisols (dry soils) are generally the lowest in organic matter, and Histosols (organic soils) are definitely the highest (see Plates 3 and 5). Contrary to popular myth, forested soils in humid tropical regions (e.g., Oxisols and some Ultisols) contain similar amounts of organic carbon to those in humid temperate regions (e.g., Alfisols and Spodosols). Andisols (volcanic ash soils) generally have some of the highest organic carbon contents of any mineral soils, probably because association of organic matter with the allophane clay in these soils protects the organic carbon from oxidation (see Plate 2). Among cultivated soils in humid and subhumid regions, Mollisols (prairie soils) are known for their dark, organic, carbon-rich surface layers (see Plate 7).

The organic carbon contents of subsurface horizons are generally much lower than those of the surface soil (Figure 12.14). Since most of the organic residues in both cultivated and virgin soils are incorporated in, or deposited on, the surface, organic matter

---

[6]No precise figures can be given for the humus content of mineral soils because there is no satisfactory method for its determination. Likewise, there is no satisfactory method of measuring the exact quantity of soil organic matter. The usual procedure is to first measure the organic carbon content, which can be done accurately. The resulting figure is then multiplied by the factor 1.72 to give the approximate amount of organic matter present. This factor is based on the conventional (but not always accurate—see Table 12.4) assumption that organic matter in surface soils generally contains about 58% C (a ratio of OM/C = 1/0.58 = 1.72). The amount of organic nitrogen can be similarly estimated by dividing organic carbon content by a factor of 12, assuming the typical C/N ratio of 12/1 applies (see Section 12.7).

tends to accumulate in the upper layers. Also, note that the organic carbon content decreases less abruptly with depth in grassland soils than in forested ones, because much of the residue added in grasslands consists of fibrous roots throughout the profile.

12.13   FACTORS AFFECTING SOIL ORGANIC MATTER

As was indicated in Section 12.3, the level to which organic matter accumulates in soils is determined by the balance of gains and losses of organic carbon. The gains are principally governed by the amounts and types of plant and animal residues added to the soil each year, while the losses result from oxidation of existing soil organic matter, as well as from erosion. We will now consider the numerous factors which influence the rates of gain and loss.

Influence of Climate

Climatic conditions, especially temperature and rainfall, exert a dominant influence on the amounts of organic carbon (and nitrogen) found in soils (see Jenny, 1941, for a classical discussion of these influences). The effect of temperature results from the different manner in which the processes of organic matter production (plant growth) and organic matter destruction (microbial decomposition) respond to increases in this climatic variable. Figure 12.15a shows that at low temperatures plant growth outstrips decomposition, but that the opposite is true above approximately 25°C. In warm soils, nutrient release is accelerated but residual organic matter accumulation is lower than in cooler soils. Therefore, as one moves from a warmer to a cooler climate, the organic matter and associated nitrogen content of comparable soils tend to increase. Some of the most rapid rates of organic matter decomposition occur in irrigated soils of hot desert regions.

Within belts of uniform moisture conditions and comparable vegetation, the average total amounts of organic matter and nitrogen in soils increase from two to three times for each 10°C decline in mean annual temperature. This temperature effect can be readily observed by noting the darkening color of well-drained surface soils as one travels from south (Louisiana) to north (Minnesota) in the humid grasslands of the North American Great Plains region. Similar changes in soil organic matter are evident as one climbs from warm lowlands to cooler highlands in mountainous regions.

Soil moisture also exerts a major influence on the accumulation of organic matter and nitrogen in soils. Under comparable conditions, the nitrogen and organic matter

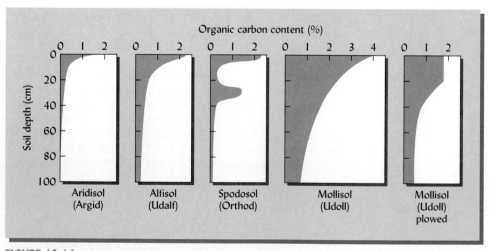

FIGURE 12.14   Vertical distribution of organic carbon in well-drained soils of four soil orders. Note the higher content and deeper distribution of organic carbon in the soils formed under grassland (Mollisols) compared to the Alfisol and Spodosol which formed under forests. Also note the bulge of organic carbon in the Spodosol subsoil due to illuvial humus in the spodic horizon (see Chapter 3). The Aridisol has very little organic carbon in the profile, as is typical of dry region soils.

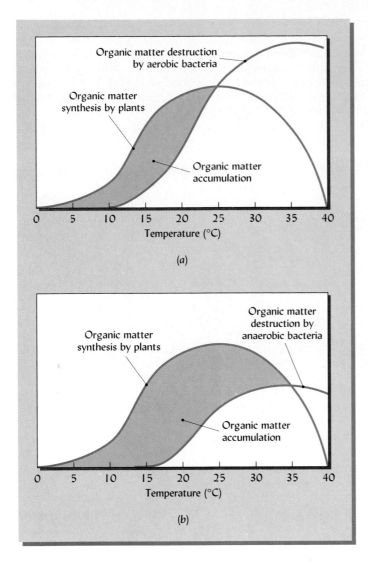

FIGURE 12.15 The balance between plant production and microbial oxidation of organic matter determines the effect that temperature has upon organic matter accumulation in soils. The shaded areas indicate organic matter accumulation under aerobic (a) and anaerobic (b) conditions. Soil organic matter will accumulate to higher levels in cool climates, especially in waterlogged, anaerobic soils. Notice that anaerobic accumulation is greater at most temperatures, and continues at higher temperatures than under aerobic conditions. This explains why subtropical areas in Florida can contain both organic soils (e.g., the Everglades) and soils containing very little organic matter (e.g., in better drained parts of the state). [Adapted from Mohr and van Baren (1954)]

content of soils increase as the effective moisture becomes greater. At the same time, the C/N ratio tends to be higher in the more thoroughly leached soils of the higher rainfall areas. These relationships are illustrated by the darker and thicker A horizons encountered as one travels across the North American Great Plains region within a belt of similar mean annual temperature, from the drier zones in the west (Colorado) to the higher rainfall east (Missouri and Illinois). The explanation lies mostly in the sparser vegetation of the drier regions. In determining this rainfall correlation, however, it must be remembered that the level of organic matter in any one soil is influenced by both temperature and precipitation, as well as by other factors.

The lowest natural levels of soil organic matter and the greatest difficulty in maintaining those levels are found where annual mean temperature is high and rainfall is low (see Figure 12.16). These relationships are extremely important to the productivity and conservation of soils and to the relative difficulty of sustainable natural resource management.

Influence of Natural Vegetation

Climate and vegetation usually act together to influence the soil contents of organic carbon and nitrogen. Grasslands generally dominate the subhumid and semiarid areas, while trees are dominant in humid regions. In climatic zones where the natural vegetation includes both forests and grasslands, the total organic matter is higher in soils developed under grasslands than under forests (Figure 12.14). With grassland vegetation, a relatively high proportion of the plant residues consist of root matter, which decomposes more slowly and contributes more effectively to soil humus than does forest leaf litter.

Effects of Texture and Drainage

While climate and natural vegetation affect soil organic matter over broad geographic areas, soil texture and drainage often are responsible for marked differences in soil organic matter within a local landscape. All else being equal, soils high in clay and silt are generally higher in organic matter than are sandy soils (Figure 12.17). The amount of organic residues returned to the soil is generally higher in finer-textured soils because the greater nutrient and water-holding capacities of these soils promote greater plant

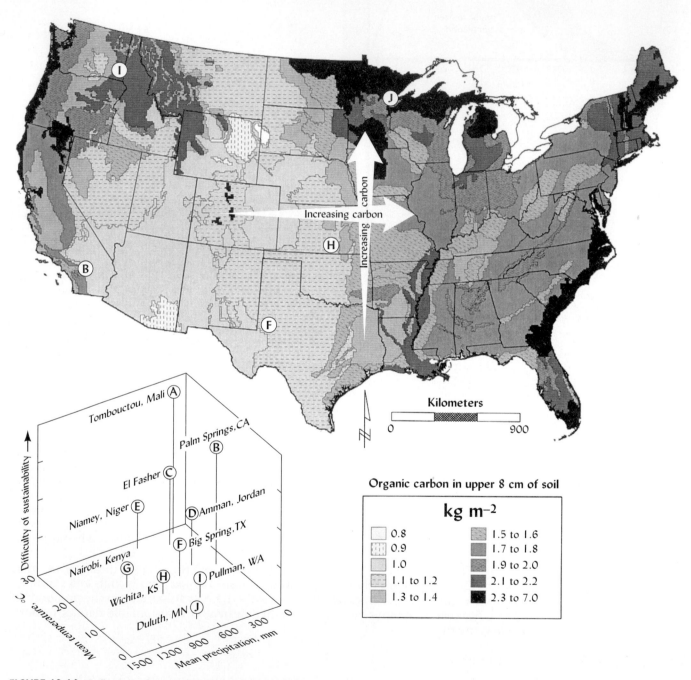

FIGURE 12.16 Influence of mean annual temperature and precipitation on organic matter levels in soils and on the difficulty of sustaining the soil resource base. The large white arrows on the map indicate that in the North American Great Plains region, soil organic matter increases with cooler temperatures going north, and with higher rainfall going east, provided that the soils compared are similar in texture, type of vegetation, drainage, and all other aspects except temperature and rainfall. These trends can be further generalized for global environments. Because of their influence on rates of plant growth and organic matter oxidation, temperature and precipitation largely determine the difficulty of preventing soil degradation and developing sustainable agricultural systems in various climatic regions of the world. Note that several of the locations represented on the graph are also on the U.S. map. [Kern (1994) and Stewart et al. (1991); Map courtesy of J. Kern, U.S. Environmental Protection Agency.]

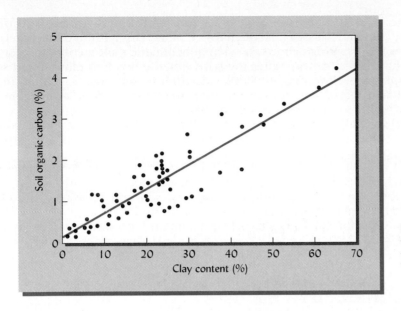

FIGURE 12.17 The effect of clay content on the soil organic carbon levels in the southern Great Plains area of the United States. [From Nichols (1984)]

production. At the same time, the generally wetter conditions of the fine-textured soils may restrict aeration and reduce the rate of organic matter oxidation. Another factor favoring the greater accumulation of organic matter in fine-textured soils is the formation of clay–humus complexes that protect the organic matter from degradation (refer to Section 12.9).

In poorly drained soils high moisture supply promotes plant dry matter production and relatively poor aeration inhibits organic matter decomposition (see Figure 12.15). Poorly drained soils therefore generally accumulate much higher levels of organic matter and nitrogen than similar, but better-aerated, soils (Figure 12.18). For instance, soils lying along streams are often quite high in organic matter, owing in part to their poor drainage. In very poorly drained environments, sufficient organic matter may accumulate to form Histosols. If the naturally waterlogged conditions of Histosols are altered by installation of an artificial drainage system, the resulting increased oxygen supply causes much of the accumulated organic matter to disappear from these soils (see Section 12.15).

Influence of Cropping and Tillage

Aside from artificial drainage of wetland soils, the management practices that most influence soil organic matter contents are cropping and tillage. A very rapid decline in soil organic matter content occurs when a virgin soil is brought under cultivation.

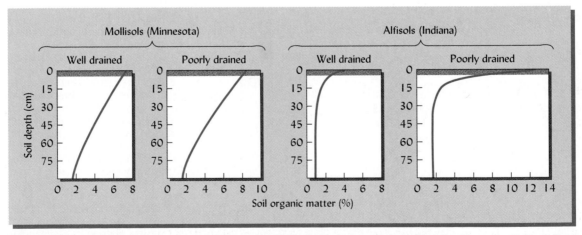

FIGURE 12.18 Distribution of organic matter in four soil profiles, two well drained and two poorly drained. Poor drainage results in higher organic matter content, particularly in the surface horizon.

Eventually, the gains and losses of organic carbon reach a new balance and the soil organic matter content levels off at a much lower value. This pattern of decline is illustrated in Figure 12.19a, which shows the changing organic matter contents of a grassland soil during the first century after the native prairie was first plowed and several different cropping systems imposed. Similar declines in soil organic matter are seen when tropical rain forests are cleared; however, the losses may be even more rapid because of the higher soil temperatures involved.

It is safe to generalize that cropped land contains much lower levels of both soil nitrogen and organic matter than do comparable virgin areas. This is not surprising; under natural conditions all the organic matter produced by the vegetation is returned to the soil, and the soil is not disturbed by tillage. In contrast, in cultivated areas much of the plant material is removed for human or animal food and relatively less finds it way back to the land. Also, soil tillage aerates the soil and breaks up the organic residues, making them more accessible to microbial decomposition.

Modern conservation tillage practices help maintain high surface soil organic matter levels (see Figure 12.20). Compared to conventional tillage, practices such as stubble

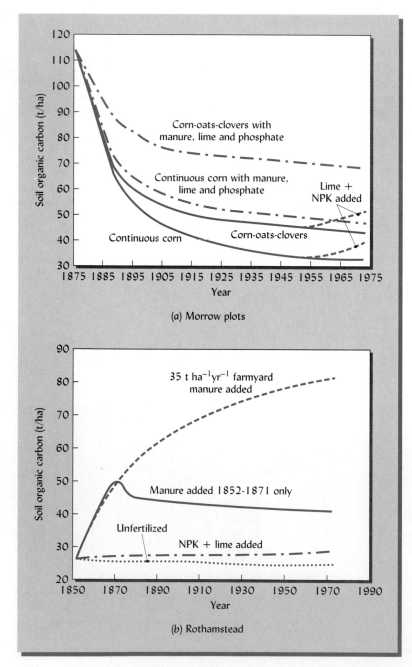

(a) Morrow plots

(b) Rothamstead

FIGURE 12.19 Soil organic carbon contents of selected treatments of (a) the Morrow plots at the University of Illinois and (b) of the classical experiments at Rothamstead Experiment Station in England. The Morrow plots were begun on virgin grassland soil in 1876 and so suffered rapid loss of organic carbon in the early years of the experiment. The Rothamstead plots were established on soils with a long history of previous cultivation. As a result, the soil at Rothamstead had reached an equilibrium level of organic carbon characteristic of the unfertilized small-grains (barley and wheat) cropping system traditionally practiced in the area. Therefore, little change in soil organic carbon occurred in the plots that simply continued this cropping system. Note that at both sites, when fertilization or manuring practices were altered, significant changes in soil organic carbon were observed within a few decades. For instance, lime and fertilizer were applied to half of the untreated fields in the Morrow plots after 1955. Note that this addition of lime and fertilizer began to reverse the trend of declining soil organic C (see dashed line). [Data recalculated from Odell et al. (1984) and Jenkinson and Johnson (1977), used with permission of the Rothamsted Experiment Station, Harpenden, England]

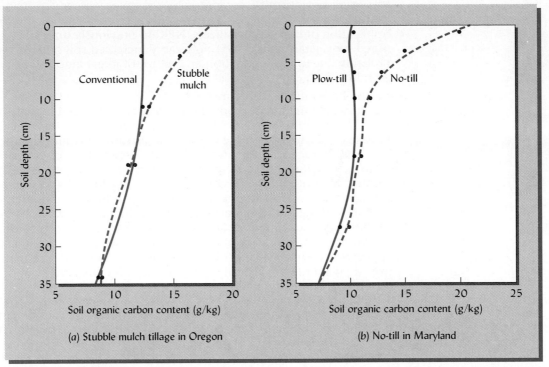

FIGURE 12.20 Effect of two conservation tillage systems on soil organic carbon content. The stubble mulch system was compared to conventional clean tillage on a Mollisol (Typic Haploxeroll) after 44 years of these practices for dryland wheat grown in a semiarid region of Oregon (data from Rasmussen and Rhode, 1988). The no-till system was compared to conventional plow tillage after eight years of growing corn on an Ultisol (Typic Hapludult) in the humid coastal plain of Maryland (data from Weil et al., 1988). Both conservation tillage systems leave much crop residue on or near the soil surface. The no-till system also leaves the soil almost completely unstirred, thus further slowing decomposition. In both cases, most of the increase in soil organic carbon occurred very near the soil surface.

mulching and no-till leave a higher proportion of the residues on or near the soil surface. These techniques protect the soil from erosion and also discourage the rapid destruction of crop residues.

Influence of Rotations, Residues, and Plant Nutrients

The nature and sequence of crops, as well as the use of lime, fertilizers, and animal manures, influence soil organic matter levels. The relationships are illustrated by the graphs in Figure 12.19 showing changes in the soil organic matter contents in two sets of famous long-term soil fertility experimental plots, the Morrow plots at the University of Illinois and the classical field experiments at Rothamstead Experiment Station, England. Different crop sequences and manure, lime, and fertilizer treatments were initiated on these two set of plots in 1876 and 1852, respectively. The Morrow plots were established on a virgin Mollisol, so initially soil organic matter declined according to the pattern typical for virgin soils brought under cultivation.

The Rothamstead plots, in contrast, were initiated on soils that had been under cereal cultivation for many years, with few nutrient inputs applied, and thus these soils had already reached a steady state at the low organic matter level characteristic of that cropping system.

From the data in Figure 12.19 we can draw the following conclusions:

1. From the Morrow plots
   a. A rotation of corn, oats, and clovers resulted in a higher soil organic matter level than did continuous corn, probably because nitrogen fixed by the clover decreased the C/N ratio and the corn-oat-clover rotation left greater crop residues.
   b. Application of manure, lime, and phosphorus helped maintain much higher organic matter levels, especially where a rotation was followed. Again, there

was a greater return of organic matter through the added manure and the increased residues from the higher-yielding crops.

    c. Application of lime and fertilizers (NPK) to previously unfertilized and unmanured plots (starting in 1955) noticeably increased soil organic matter levels, probably due to the production and return of larger amounts of crop residues with low C/N.

2. From the Rothamstead experiments

    a. Continuous small-grain production and harvest resulted in little change, or a slow decline, in the already low level of soil organic matter. This result could have been predicted because the soil was already in equilibrium with the carbon and nitrogen inputs and removals characteristic of unfertilized small-grain production.

    b. Annual applications of animal manures at rates sufficient to supply all needed nitrogen resulted in a dramatic initial rise in soil organic matter, until a new equilibrium state was approached at a much higher level of soil organic matter.

    c. In the plots where manure was applied for only the first 20 years of the experiment, the soil organic matter level began to decline as soon as the manure applications were suspended, but the positive effect of the manure on soil organic matter level was still evident some 100 years later.

Results from these world-famous experimental plots demonstrate that soils kept highly productive by supplemental applications of fertilizers, lime, and manure and by the choice of high-yielding crop varieties are likely to have more organic matter than comparable, less productive soils. High productivity means not only greater yields of economic crops, but also larger amounts of root and shoot residues returned to the soil. Nevertheless, the most productive cultivated soil will likely be considerably lower in organic matter than a similar soil carrying undisturbed natural vegetation, except in the case of irrigated soil in the desert where natural vegetation is sparse.

12.14 MANAGEMENT OF AMOUNT AND QUALITY OF SOIL ORGANIC MATTER

Consideration of the *quality* as well as the *quantity* of plant residues and soil organic matter is important in understanding the dynamics of carbon cycling. Perhaps the most useful approach to defining *soil organic matter quality* is to recognize different pools of organic carbon that vary in their susceptibility to microbial metabolism. A model identifying five such pools of carbon in plant residues and soil organic matter is illustrated in Figure 12.21. As discussed in Section 12.5, plant residues contain some components (**metabolic carbon**), such as sugars, proteins, and starches, that are quite readily metabolized by soil microbes. Other components (**structural carbon**) exist mostly in the structure of the plant cell walls; include lignin, cellulose, and waxes; and are resistant to decomposition. The model in Figure 12.21 denotes these groups as metabolic and structural pools within the plant residues.

Our previous discussions about the accumulation and loss of soil organic matter focused on the total amount of organic matter in the soil. However, it is useful to recognize that this total is the sum of several different pools of soil organic matter, namely, the *active, slow,* and *passive* fractions.

The **active fraction** of soil organic matter consists of materials with low C/N ratios and short half-lives (half of these materials can be metabolized in a matter of a few weeks to a few years). Components probably include the living biomass, some of the fine particulate detritus, most of the polysaccharides, and other nonhumic substances described in Section 12.9, as well as some of the more **labile** (easily decomposed) fulvic acids. This active fraction provides most of the readily accessible food for the soil organisms, most of the readily mineralizable nitrogen, and most of the beneficial effects on structural stability that lead to enhanced infiltration of water, resistance to erosion, and ease of tillage. The active fraction can be readily increased by the addition of fresh plant and animal residues, but is also very readily lost when such additions are reduced or tillage is intensified. This fraction rarely comprises more than 10 to 20% of the total soil organic matter.

The **passive fraction** of soil organic matter consists of very stable materials remaining in the soil for hundreds or even thousands of years. This fraction includes most of the humus physically protected in clay–humus complexes, most of the humin, and

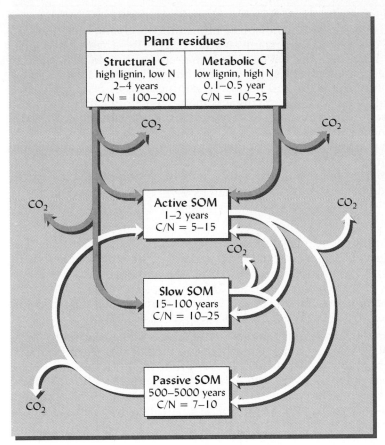

FIGURE 12.21 A conceptual model that recognizes various pools of soil organic matter differing by their susceptibility to microbial metabolism. Although effective methods are not yet available to actually isolate and measure the *active, slow,* and *passive* fractions of soil organic matter, models which incorporate these pools have proven very useful in explaining and predicting real changes in soil organic matter levels and in attendant soil properties. In particular, the existence of a pool of easily metabolized, active organic matter helps explain why even a relatively minor loss of soil organic matter often results in a major deterioration in soil structural stability, biological diversity, and nutrient cycling—all attributes associated primarily with the active fraction. By contrast, the passive fraction, which is much less susceptible to change, is associated with such soil properties as cation exchange capacity, water-holding capacity, and dark color. Note that microbial action can transfer organic carbon from one pool to another. For example, when the nonhumic substances and other components of the active fraction are rapidly broken down, some resistant, complex by-products may be formed, adding to the slow and passive fractions. [Adapted from Paustian et al. (1992)]

much of the humic acids described in Section 12.9. The passive fraction accounts for 60 to 90% of the organic matter in most soils, and its quantity is increased or diminished only slowly. The passive fraction is most closely associated with the colloidal properties of soil humus, and it is responsible for most of the CEC and water-holding capacity contributed to the soil by organic matter.

Intermediate in properties between the active and passive fractions is the **slow fraction** of soil organic matter. This fraction probably includes very finely divided plant tissues, high in lignin, and other slowly decomposable and chemically resistant components. The half-lives of these materials are typically measured in decades. The slow fraction is an important source of mineralizable nitrogen and other plant nutrients, and provides the underlying food source for the steady metabolism of the autochthonous soil microbes (see Section 12.8). The slow fraction also probably makes some contribution to the effects associated primarily with the active and passive fractions.

Effective methods are not yet available to accurately isolate and measure the active, slow, and passive fractions of soil organic matter. Despite these difficulties, models which assume the existence of these three pools have proven very useful in explaining and predicting real changes in soil organic matter levels and in attendant soil properties. Studies on the dynamics of soil organic matter have established that the different fractions of soil organic matter play quite different roles in soil management and in the carbon cycle. The presence of a resistant (structural) pool of carbon in plant residues, as well as an easily decomposed (metabolic) pool, explains the initially rapid but decelerating rate of decay that occurs when plant tissues are added to a soil (these decay patterns are illustrated in Figures 12.6 and 12.7). Similarly, the existence of a pool of complex, resistant, physically protected soil organic matter (passive fraction), as well as a pool of easily metabolized soil organic matter (active fraction), explains why cultivation of a virgin soil results in a very rapid decline in organic matter during the first few years, followed by a much slower decline thereafter (see, for example, Figure 12.19a).

In many cases, changes in soil management result in relatively small effects on soil organic matter, but pronounced effects on certain soil properties and aspects of soil productivity attributed to soil organic matter. Figure 12.22 explains this anomaly by showing how the different fractions of soil organic matter are affected by changing management (in this case, cultivating a previously undisturbed soil) and how they contribute to the overall change in soil organic matter level. The accumulated plant residues and the active fraction of the soil organic matter are the first to be affected by changes in land management, accounting for most of the early losses in soil organic matter when cultivation of the virgin soil begins. In contrast, losses from the passive fraction are very gradual. As a result, the soil organic matter remaining after some years is far less effective in promoting structural stability and nutrient cycling than the original organic matter in the virgin soil.

From the discussion of active and passive organic matter fractions, we can conclude that achieving a particular level of total soil organic matter is far less important[7] than maintaining a substantial proportion in the active fraction, so that biological metabolism will constantly enhance soil tilth and nutrient cycling.

The following general principles are useful in managing soil organic matter in many situations:

1. A continuous supply of organic materials must be added to the soil to maintain an appropriate level of soil organic matter, especially in the active fraction. Plant residues (roots and tops), animal manures, composts, and organic wastes are the primary sources of these organic materials. Cover crops can provide protective cover and add organic material to the soil as **green manure.** It is almost always preferable to keep the soil vegetated than to keep it in bare fallow.

2. No attempt should be made to try to maintain higher soil organic matter levels than the soil-plant-climate control mechanisms dictate. For example, 1.5% organic matter might be an excellent level for a sandy soil in a warm climate, but would be indicative of a very poor condition for a finer-textured soil in a cool climate. It would be foolhardy to try to maintain as high a level of organic matter in a well-drained Texas silt loam soil as might be desirable for a similar soil found in Minnesota.

---

[7]Except for greenhouse-effect considerations involving the sequestration of carbon in soils.

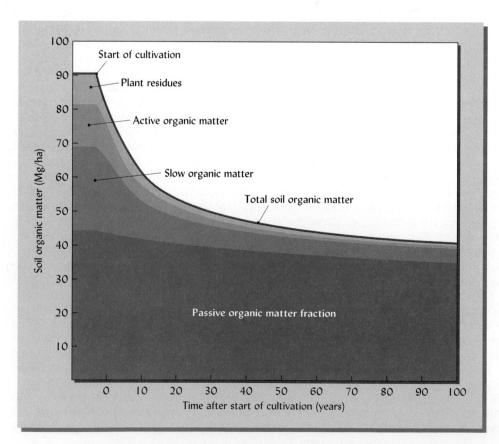

FIGURE 12.22 Changes in various fractions of organic matter in the upper 25 cm of a representative soil after bringing virgin land under cultivation. Initially, under natural vegetation, this soil contained about 91 Mg/ha of total organic matter. The resistant *passive fraction* accounted for about 42 Mg/ha, or nearly half of the total soil organic matter. The rapidly decomposing *active fraction* accounted for about 14.5 Mg, or about 16% of the total soil organic matter. After about 40 years of cultivation, the passive fraction had declined by about 6% to about 39 Mg/ha, while the active fraction had lost 90% of its mass, declining to only 1.4 Mg/ha. Note that much of the organic loss due to the change in land management came at the expense of the active fraction. The susceptibility of the active fraction to rapid change explains why even relatively small changes in total soil organic matter can produce dramatic changes in important soil properties, such as aggregate stability and nitrogen mineralization, associated with this soil organic matter fraction.

3. Because of the linkage between soil nitrogen and organic matter, adequate nitrogen inputs are requisite for adequate organic matter levels. Accordingly, the inclusion of legumes in the crop rotation and the judicious use of nitrogen-containing fertilizers to enhance high soil productivity are two desirable practices. At the same time, steps must be taken to minimize the loss of nitrogen by leaching, erosion, or volatilization (see Chapter 13).

4. Maximum plant growth will increase the amount of organic matter added to soil from crop residues. Even if some plant parts are removed in harvest, vigorously growing plants leave root and top residues as major sources of organic matter for the soil. Moderate applications of lime and fertilizers should help remove chemical toxicities and deficiencies that might constrain plant growth.

5. Because tillage accelerates the loss of organic matter both by increased oxidation of soil organic matter and by erosion, it should be limited to that needed to control weeds and to maintain adequate soil aeration. Conservation tillage practices that minimize tillage leave much of the plant residues on or near the soil surface, and thereby slow down the rate of residue decay and reduce erosion losses (see Sections 4.13 and 17.12). In time, conservation tillage can lead to higher organic matter levels (Figure 12.20).

6. Perennial vegetation, especially natural ecosystems, should be encouraged and maintained wherever feasible. Improved agricultural production on existing farmlands should be pursued to allow land currently supporting natural ecosystems to be left relatively undisturbed. In addition, there should be no hesitation about taking land out of cultivation and letting it return to its natural vegetation state where such a move is appropriate. Large areas of land under cultivation today never should have been cleared.

12.15 ORGANIC SOILS (HISTOSOLS)

Discussion up to this point has focused on organic matter in mineral soils. Under waterlogged conditions in bogs and marshes, the oxidation of plant residues is so retarded that organic matter may accumulate to the point that soil properties are dominated by the properties of the organic matter. If these soils contain greater than 20–30% organic matter, depending on clay content, they are termed *organic soils* and are classified in the **Histosols** order in *Soil Taxonomy* (see Chapter 3). The organic matter in Histosols may be either peat or muck. **Peat** is brownish, only partially decomposed, fibrous remains of plant tissues. **Muck**, on the other hand, is a black, powdery material in which decomposition is much more complete and the organic matter is highly humified.

Whether cultivated or in their natural state (often in flooded wetlands), Histosols have unique properties resulting from their high organic matter content (Figure 12.23). They are dark brown to black in color, have very low bulk density ($0.2–0.4$ Mg/m^3), and have water-holding capacities two to three times their dry weight. The cation exchange capacities of Histosols commonly exceed those of mineral clays on a weight basis, though not on a volume basis (Table 12.6). Despite relatively high C/N ratios (average 20/1 compared to 12/1 for a mineral soil), Histosols often show high levels of nitrification and accompanying nitrate release when drained. Apparently most of the carbon in peats is of the passive type that is very resistant to microbial attack. Consequently, the growth of general-purpose decay organisms is not stimulated by carbon in the peat, and nitrates released as organic material as a drained peat oxidizes are not tied up.

Histosols are important in the global carbon cycle because, though they cover only approximately 1% of the world's land area, they hold 20% of the global soil carbon. While drainage of Histosols makes these soils extremely valuable for vegetable and floriculture crops (see Chapter 3), it also speeds up oxidation of organic matter, which over time leads to Histosol destruction and increases release of CO_2 to the atmosphere. Histosols may in fact help regulate long-term global climate change. As increased atmospheric CO_2 causes the global temperature to rise, the polar ice caps melt more rapidly and cause the sea level to rise. The rise in sea level has two effects that help to counteract the impact of increased CO_2 content in the atmosphere. First, the greater volume of water in the world's oceans can absorb more CO_2. Second, the rising ocean waters flood more coastal land area, creating conditions conducive to the formation of Histosols in tidal marshes. The organic matter accumulation in these new Histosols represents a sig-

TABLE 12.6 Representative Maximum Cation Exchange Capacities of an Organic Colloid and Several Inorganic Colloids

On a weight basis, the CEC of humus greatly exceeds that of the minerals, but on a volume basis both vermiculite and smectites have higher capacities

	Cation exchange capacity	
Colloid	Weight basis (cmol$_c$/kg)	Volume basis (cmol$_c$/L)
Humus (organic)	300	75
Vermiculite	120	150
Smectite	90	113
Fine-grained micas	25	31
Kaolinite	5	6
Hydrous oxides	3	4

nificant sequestering of CO_2. This removal of CO_2 from the atmosphere mitigates the global warming trend.

12.16 ORGANIC MATERIALS FOR POTTING MEDIA

The economic importance of plants grown in containers (containerized production) has greatly increased in recent years. Container-grown plants are used in forest nurseries, for bedding-plant production, for indoor ornamental plantings, as architectural landscape elements, and in scientific greenhouse experiments. As noted in Chapters 4, 5, 7, and 10, growing plants in small pots or other containers presents special problems not encountered with field-grown plants. Most of the problems of containerized rooting media are physical in nature, principally involving water, air, and temperature. Peat and similar organic materials are lightweight, have great water-holding capacity, and provide large

FIGURE 12.23 A productive area of Histosols (Saprists) in central Michigan. Prior to being cleared and drained for agricultural production, this area was a partially forested bog. The rows of trees were planted to serve as windbreaks to protect the valuable organic soils from wind erosion. The areas of darker color have a moist soil surface due to maintenance of a high water table (field in background) or sprinkler irrigation (field in left foreground). A moist soil surface is effective in preventing these fluffy Histosols from blowing away. (Courtesy USDA Natural Resources Conservation Service)

air-filled pores. The organic materials also provide considerable cation exchange capacity for holding available plant nutrients. Some organic materials, such as composts, also release nitrogen and other nutrients as they decompose.

The peats used in potting mixes are mined from deposits in which Histosols have formed. Three general types of peats are recognized: *fibrous, woody,* and *sedimentary.* Fibrous peats are most desirable for potting mixes. These peats have unusually high water-holding capacities; they may hold 10 to 20 times their dry weight (2 to 4 times their volume) in water. Some fibrous peat (e.g., moss peats) may be quite acid because of their low ash content.

Peat utilization is a profitable business in Europe and North America. In addition to their use in potting mixes, peats are also used as mulch on home gardens and lawns. Fiber pots made of compressed peats are used to start plants, which are later transplanted to soil without being removed from the pot. Peat deposits are also used extensively for fuel (after drying), especially in the former Soviet Union and the Republic of Ireland, where energy for some electrical-power-generation stations is derived from burning peat. Unfortunately, these activities are not a sustainable use of the Histosol resource.

12.17 COMPOSTS AND COMPOSTING

Composting is the practice of creating humuslike organic materials outside of the soil by mixing, piling, or otherwise storing organic materials under conditions conducive to aerobic decomposition and nutrient conservation. The decomposition processes and organisms involved are very similar to those already described for the formation of humus in soils. The main difference is that with composting, decay occurs outside of the soil, and in such a concentrated fashion as to generate considerable heat. The finished product, **compost**, is popular as a mulch, as an ingredient for potting mixes, and as an organic slow-release fertilizer.

Composting provides a means of effectively and safely storing organic materials until it is convenient to apply them to soils. As a result of CO_2 losses and settling, the volume of composted organic materials decreases by about 30 to 50% during the composting process. The smaller volume of material may greatly ease the handling and eventual use of the organic matter as a soil amendment or potting medium. As raw organic materials are humified in a compost pile, the CEC of the organic matter increases to about 50 to 70 cmol$_c$/kg of compost.

During the composting process, the C/N ratio of organic materials in the pile decreases until a fairly stable ratio, in the range of 15:1 to 20:1, is achieved. For residues with a high initial C/N ratio, this reduction in C/N ratio ensures that any nitrate depression period will occur in the compost pile, not in the soil. Organic wastes with very low C/N ratios (such as poultry manure and sewage sludge) may be a pollution hazard if they are applied directly to the soil. High nitrogen levels in these materials may lead to environmentally damaging nitrate leaching (see Chapter 13). If low-C/N-ratio materials are composted with high-C/N-ratio materials (such as sawdust or wood chips) prior to soil application, microbes in the compost will have sufficient carbon to immobilize the excess nitrogen and minimize leaching hazard.

Because composting concentrates the decomposition process in space and time, the heat energy generated by oxidation of organic matter is usually sufficient to significantly raise the temperature of the compost pile. At the peak of microbial activity, thermophilic bacteria (adapted to high temperatures) dominate the decomposition process, and the center of the compost pile may reach temperatures as high as 65 to 72°C (149 to 161°F). This is hot enough to kill most weed seeds and pathogenic organisms in a matter of a few days. Less than ideal conditions will result in considerably lower maximum temperatures, often only 40 to 50°C (104 to 122°F), requiring weeks or months to achieve the same partial sterilization effects. Proper management of composted materials is essential if the finished product is to be desirable for use as a potting media or soil amendment (see Box 12.2).

Although much of the carbon in the composted material is lost by microbial respiration, mineral nutrients are mostly conserved. Finished compost is therefore generally more concentrated in nutrients than the combination of raw materials composted. Composting can be expected to safely break down many allelopathic compounds and pesticides prior to their application to the soil (e.g., certain herbicide residues in lawn clippings).

BOX 12.2   MANAGEMENT OF A COMPOST PILE

MATERIALS TO USE.   For homeowners, good materials for composting include leaves (preferably shredded), grass clippings, weeds (preferably before they have gone to seed), kitchen scraps, wood shavings, gutter cleanings, pine needles, spoiled hay (especially legume hay), straw (old construction bales are often available), and even vacuum cleaner dust. It is best to avoid using meat and wastes that would attract rodents or dogs. Large-scale commercial compost is often made from materials such as municipal refuse (primarily garbage and paper), sewage sludge, wood chips, animal manures, municipal leaves, and food processing wastes. Many industries and municipalities have found that composting is a viable means of managing their solid wastes.

It may be necessary to add some fertilizer or other concentrated sources of nutrients, especially nitrogen. Materials with a high C/N ratio, such as straw or paper, should be mixed or layered with enough nitrogen-rich material to lower the overall C/N ratio to below 30:1. Possible sources of nitrogen include chemical fertilizers, animal excreta, cottonseed meal, and blood meal. Other materials often added to compost to improve nutrient balance and content are mixed fertilizers (high in N, P, and K); wood ashes (high in K, Ca, and Mg); bone meal or phosphate rock powder (high in P and Ca); and seaweed (high in K, Mg, Ca, and micronutrients). Some of these materials contain enough soluble salts to necessitate leaching of the finished compost before using it on salt-sensitive plants.

COMPOSTING METHODS.   The design of the compost pile must allow for good aeration throughout the pile while still providing sufficient mass to generate heat and prevent excessive drying. Backyard compost piles usually measure 1 to 1.5 m square and 1 to 1.5 m high. Various commercially made compost bins are available to make turning the compost easier or save the trouble of building a bin. Large-scale composting is usually carried out in *windrows*, about 2 to 3 m wide, 2 m tall, and many meters long. In dry climates, composting is often done in pits dug about 1 m deep to protect the material from drying. The various materials to be composted can be mixed together or applied in thin layers. Often, a small amount of garden soil or finished compost is added to ensure that plenty of decomposer organisms will be immediately available. Microbial compost activators are commercially available, but while some may speed the initial heating of the pile, scientific tests rarely show any other advantage to using these preparations.

Control of moisture and aeration are critical to successful composting. The material should feel like a wrung-out sponge (50–70% moisture)—moist but not dripping wet when squeezed. Good aeration can be promoted by mixing in a bulking agent, such as wood chips, and by avoiding excessive packing. Decay can be speeded by turning the pile when the temperature begins to drop. Turning is most easily achieved for backyard composting by using the three-pile method (see Figure 12.24a). The efficiency of large-scale composting can be greatly increased by frequent turning or aeration (Figure 12.24b). In managing the compost pile, it is useful to monitor the internal temperature and oxygen content of the pile.

12.18   CONCLUSION

Organic matter is a complex and dynamic soil component that exerts a major influence on soil behavior, properties, and function in the ecosystem. Because of the enormous amount of carbon stored in soil organic matter and the dynamic nature of this soil component, soil management may be an important tool for moderating the global greenhouse effect.

Although present as only a few percent of the soil mass, organic matter greatly enhances the usefulness of soils for plant production. A large proportion of soil organic matter is in the passive fraction. This fraction provides cation exchange and water-holding capacities, but is relatively inert biologically. A smaller proportion of soil organic matter is in the biologically labile active fraction. The active fraction plays a major role in nutrient cycling, micronutrient chelation, maintenance of structural stability and soil tilth, and as a food source underpinning biological diversity and activity on soils. While soil organic matter has a direct influence on some soil physical properties, its indirect effects on soil structure, soil water, soil aeration, and soil temperature likely exceed in importance its function as a direct supplier of water and nutrients.

The maintenance of soil organic matter, especially the active fractions, in mineral soils is one of the great challenges in natural resource management around the world. By encouraging vigorous growth of crops or other vegetation, abundant residues (which contain nitrogen) can be returned to the soil directly or through feed-consuming animals. Also, the rate of destruction of soil organic matter can be minimized by restricting

(a)

(b)

FIGURE 12.24 An efficient and easily managed method of composting suitable for homeowners is the three-bin method (a), in which materials are moved from one bin to the next when they are turned. The bin on the right contains relatively fresh materials, while the one at the far left contains finished compost. These bins at the Rodale Research Institute in Pennsylvania are being used to compost mainly lawn cuttings and plant residues from a vegetable garden. Most large-scale composting operations use long windrows that can be turned by special machines or aerated by pulling air through the pile and into perforated pipes (b). In the case shown here, sewage sludge mixed with wood chips is being composted by the USDA Beltsville aerated pile method. Air is being drawn through the large composting windrow and expelled through the smaller piles of finished compost which absorb odors from the airstream. (Photos courtesy of R. Weil)

soil tillage, controlling erosion, and keeping most of the plant residues at or near the soil surface.

Organic soils (Histosols) contain a large share of the world's soil organic carbon. Histosols are far more permeable to air and water, and generally can hold somewhat more water and cations per hectare, than most mineral soils. While drainage of these soils allows them to be used for production of high-value vegetable and floriculture crops, it also leads to accelerated oxidation of the organic matter they contain and increased CO_2 release. To preserve these soils, the water table should be kept as near the surface as practical. Some Histosols, especially peats, are exploited to provide valuable organic materials for potting soil mixes, horticultural mulches, and even for fuel.

For some purposes it is advantageous to manage the decomposition of organic matter outside of the soil in a process known as composting. Composting transforms various organic waste materials into a humuslike product that can be used as a soil amendment or a component of potting mixes. The aerobic decomposition in a compost pile can conserve nutrients while avoiding certain problems, such as the presence of either excessive or deficient quantities of soluble nitrogen, which can occur if fresh organic wastes are applied directly to soils.

The decay and mineralization of soil organic matter is one of the main processes governing the economy of nitrogen and sulfur in soils—the subject of the next chapter.

REFERENCES

Bouwman, A. F. (ed.). 1990. *Soils and the Greenhouse Effect* (Chichester, U.K.: John Wiley and Sons).

Brady, N. C. 1990. *The Nature and Properties of Soils* (MacMillan: New York).

Chen, Y., and T. Aviad. 1990. "Effects of humic substances in plant growth," in P. MacCarthy, C. E. Clapp, R. L. Malcolm, and P. R. Bloom (eds.), *Humic Substances in Soil and Crop Sciences: Selected Readings* (Madison, Wis.: ASA Special Publications), pp. 161–186.

Dale, V. H., R. A. Houghton, A. Grainger, A. E. Lugo, and S. Brown. 1993. "Emissions of greenhouse gases from tropical deforestation and subsequent uses of the land," in National Research Council, *Sustainable Agriculture and the Environment in the Humid Tropics* (Washington, D.C.: National Academy Press), pp. 215–260.

Eswaran, H., E. Van Den Berg, and P. Reich. 1993. "Organic carbon in soils of the world," *Soil Sci. Soc. Am. J.,* **57**:192–194.

Frimmel, F. H., and R. F. Christman. 1988. *Humic Substances and Their Role in the Environment* (New York: John Wiley and Sons).

Haynes, R. J. 1986. "The decomposition process: mineralization, immobilization, humus formation and degradation," in Haynes, R. J. (ed.), *Mineral Nitrogen in the Plant-Soil System* (New York: Academic Press, Inc.).

Huang, P. M., and M. Schnitzer (eds.). 1986. *Interactions of Soil Minerals with Natural Organics and Microbes,* SSSA Special Publication No. 17 (Madison, Wis.: Soil Sci. Soc. Amer.).

Inderjit, K. M., M. Dakshini, and F. A. Einhellig. 1995. *Allelopathy—Organisms, Processes, and Applications,* ACS Symposium series 582, American Chemical Society, Washington, D.C.

Jenkinson, D. S., and A. E. Johnson. 1977. "Soil organic matter in the Hoosfield barley experiment," *Rep. Rothamstead Exp. Stn. for 1976,* Pt. **2**:87–102.

Jenny, H. 1941. *Factors of Soil Formation* (New York: McGraw-Hill).

Kern, J. S. 1994. "Spatial patterns of soil organic carbon in the contiguous United States," *Soil Science Society of America Journal,* **58**:439–455.

Loll, M. J., and J. M. Bollag. 1983. "Protein transformation in soil," *Adv. in Agron.,* **36**:352–382.

Magdoff, F. 1992. *Building Soils for Better Crops* (Lincoln: University of Nebraska Press).

Mohr, E. C. J., and van Baren, P. A. 1954. *Tropical Soils* (The Hague: N. V. Uitgeverij, W. Van Hoeve).

Neue, H. U., R. Wasserman, R. S. Lantin, M. C. Alberto, and J. B. Aduna. 1994. "Methane production potential of soils," *International Rice Research Notes,* **19**:37–38, Rice Research Institute, Manila, Philippines.

Nichols, J. D. 1984. "Relation of organic carbon to soil properties and climate in the southern Great Plains," *Soil Science Society of America Journal,* **48**:1382–1384.

Odell, R. T., S. W. Melsted, and V. M. Walker. 1984. "Changes in organic carbon and nitrogen of Morrow plots under different treatments 1904–1973." *Soil Sci. Soc. Amer. J.,* **137**:160–171.

Paustian, K., W. J. Parton, and J. Persson. 1992. "Modeling soil organic matter-amended and nitrogen-fertilized long-term plots," *Soil Sci. Soc. Am. J.,* **56**:476–488.

Rasmussen, P. E., and C. R. Rhode. 1988. "Long-term tillage and nitrogen fertilization effects on organic nitrogen and carbon in a semi-arid soil," *Soil Sci. Soc. Amer. J.,* **52:**1114–1117.

Soil Survey Staff. 1975. *Soil Taxonomy: A Basic System of Soil Classification for Making and Interpreting Soil Surveys,* Agric. Handbook No. 436 (Washington, D.C.: USDA Natural Resources Conservation Service).

Stevenson, F. J. 1986. *Cycles of Soil* (New York: John Wiley and Sons, Inc.).

Stevenson, F. J. 1982. *Humus Chemistry-Genesis, Composition Reactions* (New York: John Wiley and Sons, Inc.).

Stewart, B. A., R. Lal, and S. A. El-Swaify. 1991. "Sustaining the resource base of an expanding world agriculture," Chp. 11 in R. Lal and F. J. Pierce (eds.), *Soil Management for Sustainability* (Ankeny, Iowa: Soil and Water Conservation Soc.).

Tate, R. L. 1987. *Soil Organic Matter: Biological and Ecological Effects* (New York: Wiley).

Waksman, S. A. 1948. *Humus* (Baltimore: Williams & Wilkins), p. 95.

Weil, R. R., P. W. Benedetto, L. J. Sikora, and V. A. Bandel. 1988. "Influence of tillage practices on phosphorus distribution and forms in three Ultisols," *Agron. J.,* **80:** 503–509.

Weil, R. R., and G. S. Belmont. 1987. "Interactions between winged bean and grain amaranth," *Amaranth Newsletter,* **3(1):**3–6.

13

NITROGEN AND SULFUR ECONOMY OF SOILS

. . . the pulse and body of the soil . . .
—*D. H. LAWRENCE, THE RAINBOW*

As essential plant nutrients, the elements nitrogen and sulfur have many features in common. Nitrogen[1] will receive our attention first. More money and effort have been, and are being, spent on the study and management of nitrogen than on any other mineral nutrient. And for good reason. The pale yellowish-green foliage of nitrogen-starved crops forebodes crop failure, financial ruin, and hunger for people in all corners of the world. At the same time, excess nitrogen in the soil–water–atmosphere system can lead to dying forests, declining fisheries, increased risks of toxicity to humans, and other widespread environmental problems. As discussed in Chapter 12, the supply of nitrogen also influences the rate of decomposition, and thus soil organic matter accumulation and carbon cycling.

The productivity of most ecosystems, including agroecosystems, is limited by the availability of nitrogen. Very few soils can sustain satisfactory crop production without the addition of nitrogen from some source. Plants take up relatively large amounts of nitrogen, while soils generally contain relatively small amounts of this element. Hence nitrogen deficiency is more widespread among crop plants than the deficiency of any other nutrient. Fortunately, both agricultural and natural ecosystems take advantage of the biological nitrogen-fixation activities of certain microorganisms to supplement soil supplies of this crucial nutrient.

Nitrogen is closely associated with protein. Because of its nutritional importance and relative scarcity, protein is highly sought after by most animals, humans included. Supplying sufficient nitrogen often represents a major expense in agricultural production. Also, the manufacture of nitrogen fertilizer accounts for a large part of the fossil fuel energy used by the agricultural sector.

Management of nitrogen in various compounds presents one of the most critical environmental challenges related to soils. As in terrestrial systems, the growth of algae and other plants in aquatic systems (especially those with salty or brackish water) is commonly nitrogen-limited. Thus the addition of nitrogen to aquatic systems, often from soil sources, may disrupt the balance in these systems and result in eutrophication, and eventually in decline of fish and other desirable aquatic populations. Nitrogen in the form of nitrates can lead to toxicity in both ruminant animals and human infants. For this reason nitrate levels are monitored in wells, reservoirs, and other drinking water supplies.

---

[1]Nitrogen in soils and in crop production is covered extensively in the reviews edited by Stevenson (1982) and Hauck (1984). A comprehensive treatment of nitrogen in agriculture with an emphasis on environmental quality is given by Addiscott, et al. (1991).

Yet another way in which nitrogen links the soil to the wider environment is the ozone-destroying action of soil-generated nitrous oxide gas. Clearly, for the management of the soil nitrogen cycle, the ecological, financial, and environmental stakes are very high.

Many of the principles that govern nitrogen also apply to sulfur. Both nutrients are taken up by plants primarily in anionic form; both are found largely in the organic fraction of soils; and both link the soil to the wider environment by complex biogeological cycles involving solid, liquid, and gaseous phases. This chapter will reveal that the behaviors of both nitrogen and sulfur in soils are complex and difficult to predict with precision.

13.1 INFLUENCE OF NITROGEN ON PLANT GROWTH AND DEVELOPMENT

ROLES IN THE PLANT. Nitrogen is an integral component of many essential plant compounds. It is a major part of all amino acids, the building blocks of proteins, including the enzymes which control virtually all biological processes. Other critical nitrogenous plant components include the nucleic acids, in which hereditary control is vested, and chlorophyll, which is at the heart of photosynthesis. Nitrogen is also essential for carbohydrate use within plants. A good supply of nitrogen stimulates root growth and development, as well as the uptake of other nutrients.

Plants respond quickly to increased availability of nitrogen. This element particularly encourages aboveground vegetative growth and gives leaves a deep green color. It increases the plumpness of cereal grains, the protein content of both seeds and foliage, and the succulence of crops such as lettuce and radishes. Nitrogen can dramatically stimulate plant productivity, whether measured in tons of grain, volume of lumber, carrying capacity of pasture, or thickness of lawn. Healthy plant foliage generally contains 2.5 to 4.0% nitrogen, depending on the age of the leaves and whether the plant is a legume.

DEFICIENCY. Plants deficient in nitrogen tend to have a pale yellowish green color (**chlorosis**), have a stunted appearance, and develop thin, spindly stems. In a nitrogen-deficient plant, the protein content is low and the sugar content is high because there is insufficient nitrogen to combine with all the carbon chains that would normally be used to make proteins. Nitrogen is quite mobile (easily translocated) within the plant, and when plant uptake is inadequate supplies are transferred to the newest foliage, causing the older leaves to show the most pronounced chlorosis. The older leaves of nitrogen-starved plants are therefore the first to turn yellowish, possibly becoming prematurely senescent and dropping off (see Figures 13.1a and b). Nitrogen-deficient plants often have a low shoot-to-root ratio, and they mature more quickly than healthy plants. The negative effects of nitrogen deficiency on plant size and vigor are often dramatic (Figure 13.1c).

OVERSUPPLY. When too much nitrogen is applied, excessive vegetative growth occurs; the cells of the plant stems become enlarged but relatively weak, and the top-heavy plants are prone to falling over (lodging) with heavy rain or wind (Figure 13.1d). High nitrogen applications may delay plant maturity and cause the plants to be more susceptible to disease (especially fungal disease) and to insect pests. These problems are especially noticeable if other nutrients, such as potassium, are in relatively low supply.

When oversupply of nitrogen does not lead to the problems just mentioned, crop yields may be large, but crop quality often suffers. The color and flavor of fruits may be poor, as may the sugar and vitamin contents of certain vegetables and root crops. Flower production on ornamentals is reduced in favor of overabundant foliage. Under certain conditions, an oversupply may cause nitrogen to accumulate in plant tissues as nitrate instead of becoming assimilated into proteins. Such high-nitrate foliage has been known to be harmful to livestock in the case of forages, or to human babies in the case of leafy vegetables.

The amount of nitrogen that can be usefully assimilated by plants varies greatly among species. Many grasses and leafy vegetables require a great deal of nitrogen for optimum growth. Whether or not the plants themselves are harmed by high nitrogen levels, potential adverse effects on the environment must always be considered (see Sections 13.8, 13.9, and 16.13).

FIGURE 13.1 Some plant effects of too little and too much nitrogen. (a) The older, outer leaves have become chlorotic on these spinach plants growing on a low-nitrogen sandy Ultisol. (b) These older leaves from nitrogen-deficient corn plants (upper leaf is normal) have yellowed, beginning at the tip and continuing down the midrib—a pattern typical of nitrogen deficiency on corn. (c) The nitrogen-starved bean plant (right) shows the typical chlorosis of lower leaves and markedly stunted growth compared to the normal plant on the left. (d) The traditional tall Asian rice cultivar in the left field has lodged because of a large application of nitrogen fertilizer. In the right-hand field, the farmer has planted a modern rice cultivar, which yields well with high nitrogen inputs because of its short stature and stiff straw. (Photos courtesy of R. Weil)

FORMS OF NITROGEN TAKEN UP BY PLANTS.   Plant roots take up nitrogen from the soil solution principally as nitrate (NO_3^-) and ammonium (NH_4^+) ions.[2] Although certain plants grow best when provided mainly one or the other of these forms, a relatively equal mixture of the two ions gives the best results with most plants. Nitrate **anions** (negatively charged ions) move easily to the root with the flow of soil water and are thus readily available for plant uptake. In contrast, ammonium **cations** (positively charged ions) move less readily to the root because they are held back by predominantly negatively charged soil colloids.

13.2 ORIGIN AND DISTRIBUTION OF NITROGEN

By far the greatest amount of nitrogen in terrestrial ecosystems is contained in the soil. Whether forested or used for agriculture, the soil contains 10 to 20 times as much nitrogen as does the standing vegetation (including roots). The nitrogen content of surface mineral soils normally ranges from 0.02 to 0.5%, a value of about 0.15% being representative for cultivated soils. A hectare of such a soil would likely contain about 3.5 Mg of nitrogen in the A horizon and perhaps an additional 3.5 Mg in the deeper layers. In forested soils the litter layer (O horizons) might contain another 1 to 2 Mg of nitrogen. In contrast, the air above that hectare of soil would contain nearly 300,000 Mg of nitrogen. In this context, the atmosphere, which is composed of 78% N, appears to be a virtually limitless reservoir of this element.[3] In air, nitrogen occurs mainly in the form of dinitrogen gas, N_2. The very stable triple bond between the two nitrogen atoms makes this gas quite inert and not directly usable by plants or animals. However, as we shall see (Section 13.10), certain microorganisms can break this triple bond and form biological compounds from the dinitrogen gas found in the atmosphere.

Most soil nitrogen occurs as part of organic molecules. Soil organic matter typically contains about 5% nitrogen; therefore, the distribution of soil nitrogen closely parallels that of soil organic matter (see Section 12.11). Because association with certain silicate clays or resistant humic acids helps protect the nitrogenous organic compounds from rapid microbial breakdown, typically only about 2 to 3% of the nitrogen in soil organic matter is released annually as inorganic nitrogen.

Except where large amounts of chemical fertilizers have been applied, inorganic (i.e., mineral) nitrogen seldom accounts for more than 1 to 2% of the total nitrogen in the soil. Unlike most of the organic nitrogen, the mineral forms of nitrogen are mostly quite soluble in water, and may be easily lost from soils through leaching and volatilization.

13.3 THE NITROGEN CYCLE

As it moves through the **nitrogen cycle**, an atom of nitrogen may appear in many different chemical forms, each with its own properties, behaviors, and consequences for the ecosystem. The cyclic nature of these processes explains why vegetation (and animals, indirectly) can continue to remove nitrogen from a soil for centuries without depleting the soil of this essential nutrient—the biosphere does not run out of nitrogen because it uses the same nitrogen over and over again. The nitrogen cycle has long been the subject of intense scientific investigation, for understanding the translocations and transformations of this element is fundamental to solving many environmental, agricultural, and natural resource problems. The principal pools and forms of nitrogen, and the processes by which they interact, are illustrated in Figure 13.2. This figure deserves careful study; we will refer to it frequently as we discuss each of the major divisions of the nitrogen cycle.

---

[2]Nitrite (NO_2^-) can also be taken up, but this ion is toxic to plants and rarely occurs in greater than trace quantities in soils.

[3]About 98% of the earth's nitrogen is contained in the igneous rocks deep under the earth's crust, where it is effectively out of contact with the soil--plant--air--water environment in which we live. We will therefore focus our attention on the remaining 2% that cycles in the biosphere.

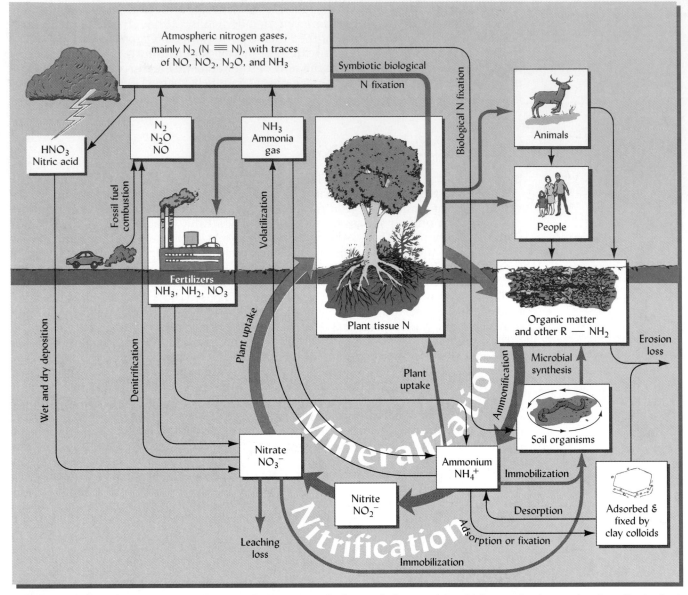

FIGURE 13.2 The nitrogen cycle, emphasizing the primary cycle (heavy, dark arrows) in which organic nitrogen is mineralized, plants take up the mineral nitrogen, and eventually organic nitrogen is returned to the soil as plant residues. Note also the pathways by which nitrogen is lost from the soil and the means by which it is replenished. The compartments represent various forms of nitrogen; the arrows represent processes by which one form is transformed into another.

Ammonium nitrogen is subject to five possible fates in the nitrogen cycle: (1) *immobilization* by microorganisms; (2) removal by *plant uptake;* (3) ammonium ions may be *fixed* in the interlayers of certain 2:1 clay minerals; (4) ammonium ions may be transformed into ammonia gas and lost to the atmosphere by *volatilization;* (5) ammonium ions may be oxidized to nitrite and subsequently to nitrate by a microbial process called *nitrification.*

Nitrogen in the nitrate form is highly mobile in the soil and in the environment. Whether added as fertilizer or produced in the soil by nitrification, nitrate may take any of four paths in the nitrogen cycle: (1) *immobilization* by microorganisms; (2) removal by *plant uptake;* (3) nitrate ions may be lost by *leaching* in drainage water; or (4) by *volatilization* to the atmosphere as several nitrogen-containing gases.

13.4   IMMOBILIZATION AND MINERALIZATION

At any one time, the great bulk (95–99%) of the nitrogen in a soil is in organic compounds that protect it from loss but leave it largely unavailable to higher plants. For this reason, much scientific effort has been devoted to the study of organic nitrogen—how it is stabilized and how it may be released to forms usable by plants.

Much of the nitrogen in soil organic matter is present as amine groups ($R - NH_2$), largely in proteins or as part of humic compounds. As organic compounds are attacked by soil microorganisms, the organic nitrogen combinations are first broken down into simple amino compounds[4] ($R - NH_2$). Then the amine groups are hydrolyzed and the nitrogen is released as ammonium ions (NH_4^+). Finally, some of the nitrogen appears in the nitrate (NO_3^-) form.

This conversion of organically bound nitrogen into inorganic mineral forms (NH_4^+ and NO_3^-) is termed **mineralization** (Figure 13.1). Although a whole series of reactions is involved, the net effect can be visualized quite simply. A wide variety of soil organisms simplifies and hydrolyzes the organic nitrogen compounds, ultimately producing the NH_4^+ and NO_3^- ions. The enzymatic process may be indicated as follows, using an amino compound ($R - NH_2$) as an example of the organic nitrogen source:

$$\xrightarrow{\hspace{3cm}} \text{Mineralization} \xrightarrow{\hspace{3cm}}$$

$$R{-}NH_2 \underset{-2H_2O}{\overset{+2H_2O}{\rightleftharpoons}} OH^- + R{-}OH + NH_4^+ \underset{-O_2}{\overset{+O_2}{\rightleftharpoons}}$$

$$4H^+ + \text{energy} + NO_2^- \underset{-1/2O_2}{\overset{+1/2O_2}{\rightleftharpoons}} \text{energy} + NO_3^-$$

$$\xleftarrow{\hspace{3cm}} \text{Immobilization} \xleftarrow{\hspace{3cm}}$$

Many studies have shown that about 1.5 to 3.5% of the organic nitrogen of a soil mineralizes annually. The exact rate of mineralization depends largely on the temperature, moisture, and aeration status of the soil. Maximum rates of annual mineralization occur with a year-long hot climate, well-aerated soils, and a fluctuating moisture content. Except in some desert soils, or very sandy soils low in organic matter, mineralization of organic nitrogen provides sufficient mineral nitrogen for normal growth of natural vegetation. Even for farmland that has been amended with synthetic nitrogen fertilizers, *isotope tracer studies* show that mineralized soil nitrogen constitutes a major part of the nitrogen taken up by a crop. If the organic matter content of a soil is known, one can made a rough estimate of the amount of nitrogen likely to be released by mineralization during a typical growing season (see Box 13.1).

The opposite of mineralization is **immobilization,** the conversion of inorganic nitrogen ions (NO_3^- and NH_4^+) into organic forms (Figure 13.2). As carbonaceous organic residues are decomposed in the soil, the growth of microbial colonies may require more nitrogen than is contained in the residues themselves. The microorganisms then have to incorporate mineral nitrogen ions to synthesize cellular components, such as proteins. When the organisms die, some of the organic nitrogen in their cells may be converted into forms that make up the humus complex, and some may be released as NO_3^- and NH_4^+ ions. Mineralization and immobilization occur simultaneously in the soil; whether the *net* effect is an increase or a decrease in the mineral nitrogen available depends primarily on the ratio of carbon to nitrogen in the organic residues undergoing decomposition (see Section 12.7).

13.5 AMMONIUM FIXATION BY CLAY MINERALS[5]

Like other positively charged ions, ammonium ions are attracted to the negatively charged surfaces of clay and humus where they are held in exchangeable form, available for plant uptake, but partially protected from leaching. However, because of the particular size of the ammonium ion (and potassium also), it can become entrapped within cavities in the crystal structure of certain clays (Figure 13.2). Several clay minerals with

---

[4]Amino acids such as lysine (CH_2NH_2COOH) and alanine (CH_3CHNH_2COOH) are examples of these simpler compounds. The R in these formulas represents the part of the organic molecule with which the amino group (NH_2) is associated. For example, for lysine, the R is CH_2COOH.

[5]This chemical fixation of ammonia is caused by the entrapment or other strong binding of NH_4^+ ions by organic matter and certain silicate clays. This type of fixation (by which K ions are similarly bound) is not to be confused with the very beneficial biological fixation of atmospheric nitrogen gas into compounds usable by plants (see Section 13.10).

BOX 13.1 CALCULATION OF NITROGEN MINERALIZATION

If the organic matter content of a soil, soil management practices, climate, and soil texture are known, it is possible to make a rough estimate of the amount of N likely to be mineralized each year. The following equation may be used:

$$\frac{\text{kg N mineralized}}{\text{ha 15 cm deep}} = \left(\frac{A \text{ kg SOM}}{100 \text{ kg soil}}\right)\left(\frac{B \text{ kg soil}}{\text{ha 15 cm deep}}\right)\left(\frac{C \text{ kg N}}{100 \text{ kg SOM}}\right)\left(\frac{D \text{ kg SOM mineralized}}{100 \text{ kg SOM}}\right)$$

where A = The amount of soil organic matter (SOM) in the A horizon soil, given in kg SOM per 100 kg soil. This value may range from close to zero to over 75% (in a Histosol) (see Section 12.13). Values between 0.5 and 5% are most common. ☞ Use a value of 2.5% (2.5 kg SOM/100 kg soil) for the example shown below.

  B = The weight of soil per hectare to the depth of the A horizon. Most nitrogen used by plants is likely to come from the A horizon. If the A horizon is 15 cm deep, 2.2×10^6 kg/ha is a reasonable estimate of its weight per hectare. See Section 12.3 to calculate the weight of the A horizon if bulk density of a soil is known. ☞ Use 2.2×10^6 kg soil/ha 15 cm deep in the example shown below.

  C = Indicates that typical SOM is approximately 5% nitrogen (5 kg N/100 kg soil organic matter; See Section 12.7) ☞ Assume that 5% N is a reasonable figure for the example shown below.

  D = Indicates the amount of SOM likely to be mineralized in one year for a given soil. This figure depends upon the soil texture, climate, and management practices. Values of around 2% are typical for a fine-textured soil, while values of around 3.5% are typical for coarse-textured soils. Slightly higher values are typical in warm climates; slightly lower values are typical in cool climates. ☞ Assume a value of 2.5 kg SOM mineralized/100 kg SOM for the example shown below.

The amount of nitrogen likely to be released by mineralization during a typical growing season may be calculated by substituting the example values indicated above into the equation:

$$\frac{\text{kg N mineralized}}{\text{ha 15 cm deep}} = \left(\frac{2.5 \text{ kg SOM}}{100 \text{ kg soil}}\right)\left(\frac{2.2 \times 10^6 \text{ kg soil}}{\text{ha 15 cm deep}}\right)\left(\frac{5 \text{ kg N}}{100 \text{ kg SOM}}\right)\left(\frac{2.5 \text{ kg SOM mineralized}}{100 \text{ kg SOM}}\right)$$

$$\frac{\text{kg N mineralized}}{\text{ha}} = \left(\frac{2.5}{100}\right)\left(\frac{2.2 \times 10^6}{1}\right)\left(\frac{5}{100}\right)\left(\frac{2.5}{100}\right) = 69 \text{ kg N/ha}$$

Contributions from the deeper layers of this soil might be expected to bring total nitrogen mineralized in the root zone of this soil during a growing season to over 120 kg N/ha. Most nitrogen mineralization occurs during the growing season when the soil is relatively moist and warm.

These calculations estimate the nitrogen mineralized annually from a soil that has not had large amounts of organic residues added to it. Animal manures, legume residues, or other nitrogen-rich organic soil amendments would mineralize much more rapidly than the native soil organic matter, and thus would substantially increase the amount of nitrogen available in the soil.

---

a 2:1-type structure have the capacity to *fix* both ammonium and potassium ions in this manner (Figure 14.25). Vermiculite has the greatest capacity, followed by fine-grained micas and some smectites. Ammonium and potassium ions fixed in the rigid part of a crystal structure are held in a nonexchangeable form, from which they are released only slowly to higher plants and microorganisms.

Ammonium fixation by clay minerals is generally greater in subsoil than in topsoil, due to the higher clay content of subsoils (Table 13.1). In soils with considerable 2:1 clay content, interlayer-fixed NH_4^+ typically accounts for 5 to 10% of the total nitrogen in the surface soil and up to 20 to 40% of the nitrogen in the subsoil. In highly weathered soils, on the other hand, ammonium fixation is minor because little 2:1 clay is present. While ammonium fixation may be considered an advantage because it provides a means of conserving nitrogen, the rate of release of the fixed ammonium is often too slow to be of much practical value in fulfilling plant needs.

TABLE 13.1 Total Nitrogen Levels of A and B Horizons of Four Cultivated Virginia Soils and the Percentage of the Nitrogen Present as Nonexchangeable or Fixed NH_4^+

Note the higher percentage of fixation in the B horizon.

Soil great group (series)	Total N (mg/kg)		Nitrogen fixed as NH_4^+ (%)	
	A Horizon	B Horizon	A Horizon	B Horizon
Hapludults (Bojac)	812	516	5	18
Paleudults (Dothan)	503	336	5	14
Hapludults (Groseclose)	1792	458	3	17
Hapludults (Elioak)	1110	383	6	26

From Baethgen and Alley (1987).

13.6 AMMONIA VOLATILIZATION

Ammonia gas (NH_3) can be produced in the soil–plant system and much nitrogen may be lost to the atmosphere in this form (Figure 13.2). The source of the ammonia gas may be animal manures (hence the familiar ammonia smell around barnyards and poultry houses), fertilizers (especially anhydrous ammonia and urea), decomposing plant residues (especially legume foliage that is rich in nitrogen), or even the foliage of living plants. In each case, ammonia gas is in equilibrium with ammonium ions according to the following reversible reaction:

$$NH_4^+ + OH^- \rightleftharpoons H_2O + NH_3 \uparrow$$

Dissolved ions   Gas

From the above reaction we can draw three conclusions. First, ammonia volatilization will be more pronounced as pH increases (i.e., as the supply of OH^- ions increases and drives the reaction to the right); second, ammonia-gas-producing amendments will have the immediate effect of rapidly raising the pH of the solution in which they are dissolved; and finally, as a moist soil dries, the water is removed from the right-hand side of the equation, again driving the equation to the right.

Soil colloids, both clay and humus, are capable of adsorbing ammonia gas, so ammonia losses are greatest where little of these colloids are present, or where the ammonia is not in close contact with the soil. For these reasons, ammonia losses can be quite large from sandy soils and from alkaline or calcareous soils, especially when the ammonia-producing materials are left at or near the soil surface and when the soil is drying out. High temperatures, as often occur on the surface of the soil, also favor the volatilization of ammonia.

Incorporation of manure and fertilizers into the top few centimeters of soil can reduce ammonia losses by 25 to 75% from those that occur when the materials are left on the soil surface. In natural grasslands and pastures, incorporation of animal wastes by earthworms and dung beetles is critical in maintaining a favorable nitrogen balance and a high animal-carrying capacity in these ecosystems.

VOLATILIZATION FROM WETLANDS. Gaseous ammonia loss from nitrogen fertilizers applied to the surface of fishponds and flooded rice paddies can also be appreciable, even on slightly acid soils. The applied fertilizer stimulates algae growing in the paddy water. As the algae photosynthize, they extract CO_2 from the water and reduce the amount of carbonic acid formed. As a result, the pH of the paddy water increases markedly, especially during daylight hours, to levels commonly above 9.0. At these pH levels, ammonia is released from ammonium compounds and goes directly into the atmosphere. As with upland soils, this loss can be reduced significantly if the fertilizer is placed below the soil surface. Natural wetlands lose ammonium by a similar daily cycle.

AMMONIA ABSORPTION. By the reverse of the ammonium loss mechanism just described, both soils and plants can absorb ammonia from the atmosphere. Thus the soil–plant system can help cleanse ammonia from the air, while deriving usable nitrogen for plants and soil microbes. Forests may receive a significant proportion of their nitrogen requirements as ammonia carried by wind from fertilized cropland and cattle feedlots located many kilometers away.

13.7 NITRIFICATION

If suitable conditions prevail, ammonium ions that begin to accumulate in the soil will be enzymatically oxidized by certain soil bacteria. These bacteria obtain their energy from ammonium oxidation rather than from organic matter oxidation. Bacterial oxidation of ammonium is termed **nitrification** (Figure 13.2). The process consists of two main sequential steps. The first step results in the conversion of ammonium to nitrite by a specific group of autotrophic bacteria (*Nitrosomonas*). The nitrite so formed is then immediately acted upon by a second group of autotrophs, *Nitrobacter*. The enzymatic oxidation releases energy, and may be represented very simply as follows:

Step 1

$$NH_4^+ + 1\tfrac{1}{2}O_2 \xrightarrow[\text{bacteria}]{Nitrosomonas} NO_2^- + 2H^+ + H_2O + 351 \text{ kJ energy}$$
Ammonium   Nitrite

Step 2

$$NO_2^- + \tfrac{1}{2}O_2 \xrightarrow[\text{bacteria}]{Nitrobacter} NO_3^- + 74 \text{ kJ energy}$$
Nitrite   Nitrate

So long as conditions are favorable for both reactions, the second transformation is thought to follow the first closely enough to prevent accumulation of nitrite. This is fortunate, because even at concentrations of just a few parts per million, nitrite is quite toxic to most plants and to mammals.

Regardless of the source of ammonium (i.e., ammonia-forming fertilizer, sewage sludge, animal manure, or any other organic nitrogen source), nitrification will significantly increase soil acidity by producing H^+ ions as shown in the above reaction. In humid regions, liming materials must be added to counteract acidity (see Chapter 9).

Soil Conditions Affecting Nitrification

The nitrifying bacteria are much more sensitive to environmental conditions than are the broad groups of heterotrophic organisms responsible for the release of ammonium from organic nitrogen compounds (**ammonification**). We will consider briefly some of the soil conditions that affect nitrification.

AMMONIA LEVEL. Nitrification can take place only if there is ammonium to be oxidized. Factors such as a high C/N ratio of residues, which prevents the release of ammonium, may prevent nitrification. However, too much ammonia gas (NH_3) can inhibit nitrification. High concentrations of urea or anhydrous ammonia fertilizers in alkaline soils may raise ammonia to levels toxic to *Nitrobacter*. Without active *Nitrobacter* bacteria, nitrite may accumulate to toxic levels.

AERATION. The nitrifying organisms are aerobic bacteria and require oxygen to make NO_2^- and NO_3^- ions. Therefore, soil aeration and good soil drainage promote nitrification, as does moderate tillage, which stirs air into the soil, and often results in a warmer soil as well. Nitrification is generally somewhat slower under minimum tillage than where plowing and some cultivation are practiced.

MOISTURE. Nitrification is retarded by both very low and very high moisture conditions (Figure 13.3). The optimum moisture for higher plants is also optimum for nitrification (about 60% of the soil pore space filled with water). However, appreciable nitrification does occur when soil moisture is at or near the wilting coefficient, which is too dry for plant growth.

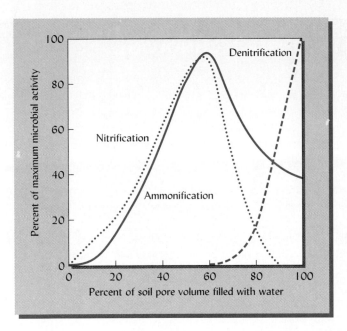

FIGURE 13.3 Because it depicts the balance of oxygen and water, percentage of water-filled pore space is closely related to rates of nitrification, ammonification, and denitrification. Note that ammonification can proceed in soils too waterlogged for active nitrification. There is very little overlap in the conditions suitable for nitrification and denitrification.

SOURCE OF CARBON. The nitrifiers use CO_2 and bicarbonate ions as sources of carbon to synthesize their cellular components. Being autotrophs, they do not require organic matter as either a carbon source or an energy source.

TEMPERATURE. The temperature most favorable for nitrification ranges from 25–35°C (77–95°F). Nitrification is slow when soils are cool, and virtually ceases below 5°C. If ammonium fertilizer is applied early in spring, nitrification and ready availability of nitrate will be delayed until the soil warms up later in the season. Nitrification rates also decline at temperatures above 35°C and essentially cease at temperatures above 50°C (122°F).

EXCHANGEABLE BASE-FORMING CATIONS AND pH. Nitrification proceeds most rapidly where there is an abundance of exchangeable base-forming cations. The need for base-forming cations accounts in part for the slow nitrification in acid mineral soils and the seeming sensitivity of the organisms to low pH. However, within reasonable limits, acidity itself seems to have less influence on nitrification when adequate base-forming cations are present. This is especially true of peat soils. Even at pH values below 5, peat soils may show remarkable accumulations of nitrates.

FERTILIZERS. Nitrifying organisms have mineral nutrient requirements not too different from those of higher plants. Consequently, nitrification may be stimulated by the presence of essential elements in adequate levels.

PESTICIDES. Nitrifying organisms are quite sensitive to some pesticides. If added at high rates, many of these chemicals inhibit nitrification to a large or small extent. Most studies suggest, however, that at ordinary field rates, the majority of the pesticides have only a minimal effect on nitrification. Chemicals intended to inhibit nitrification are discussed in Box 13.2, while unintentional inhibition by pesticides is discussed further in Chapter 19.

Provided that all the above conditions are favorable, nitrification is such a rapid process that nitrate is generally the predominant mineral form of nitrogen in most soils. Irrigation of an initially dry arid region soil, the first rains after a long dry season in the tropics, the thawing and rapid warming of frozen soils in spring, and sudden aeration by tillage are examples of environmental fluctuations that typically cause a flush of soil nitrate production (Figure 13.4). The growth patterns of natural vegetation and the optimum planting dates for crops are greatly influenced by such seasonal changes in nitrate levels.

BOX 13.2 REGULATION OF SOLUBLE NITROGEN WITH FERTILIZER TECHNOLOGY

Perhaps the most challenging aspect of nitrogen control is the regulation of the soluble forms of this element after it enters the soil. Availability at the proper time and in suitable amounts, with a minimum of loss, is the ideal. Even where commercial fertilizers are used to supply much of the nitrogen, maintaining an adequate but not excessive quantity of available nitrogen is not an easy task. Split applications of nitrogen fertilizer is one way to accomplish this. This technique involves splitting nitrogen application into several small doses applied as the growing crop develops, rather than applying the entire amount at or before planting time. In sandy soils or where rainfall or irrigation is high early in the season, applying all the nitrogen at planting time may result in much of the nitrogen leaching below the root zone before the crop has had a chance to use it. In regions with high rainfall during the winter, it is important to avoid leaving a large amount of soluble nitrogen in the soil profile after the crop has completed its growth.

NITRIFICATION INHIBITORS.* Because ammonium is much less susceptible than nitrate to losses by leaching or denitrification, nitrogen fertilizers might be used more efficiently if nitrification of ammonium from fertilizers could be slowed down until the crop is ready to make use of the nitrogen. To this end chemical companies have developed compounds called *nitrification inhibitors*, which inhibit the activity of the *Nitrosomonas* bacteria that convert ammonium to nitrate in the first step of nitrification. (Note that a chemical that inhibited *Nitrobacter* would *not* be useful as this would cause toxic nitrite to accumulate.) Three commercially available nitrification inhibitors are dicyandiamide (DCD), nitrapyrin (N-Serve®) and etridiazol (Dwell®). When mixed with nitrogen fertilizers, these compounds can temporarily prevent nitrate formation. The key word here is *temporarily*, for when conditions are favorable for nitrification, the inhibition usually lasts only a few weeks (less if soils are above 20°C). Research has shown that these materials may pay for themselves with improved fertilizer efficiency when conditions are very conducive to nitrogen loss by leaching or denitrification. Thus, nitrification inhibitors have been of value primarily in relatively wet years on sandy soils (reduced leaching) and on imperfectly drained soils (reduced denitrification).

SLOW-RELEASE NITROGEN FERTILIZERS. The use of slow-release fertilizer materials provides another means of reducing nitrogen losses from fertilized soils. Unlike most inorganic fertilizers, these materials release soluble nitrogen slowly in the soil so that nitrogen availability is better sychronized with plant uptake. Some stabilized organic materials such as compost and digested sewage sludge (e.g., Milorganite®) are used for this purpose. However, most slow-release nitrogen fertilizers are made by treating urea with materials that slow its dissolution or inhibit its hydrolysis to ammonium. Urea-formaldehyde, isobutylidene diurea (IDBU), resin-coated fertilizers (e.g., Osmocote®), and sulfur-coated urea are all examples of slow-release nitrogen fertilizers. In the case of the latter, the sulfur content (10 to 20%) may be a benefit where sulfur is in low supply (Section 13.23), or as a problem where the extra acidity generated by the sulfur (Section 13.21) would be undesirable. Because of additional manufacturing costs and lower concentrations of nitrogen, slow-release fertilizers cost from 1.5 (for sulfur-coated ureas) to 4 times as much as plain urea per unit of nitrogen. Nonetheless, slow-release materials are practical for high-value plant production where the cost of fertilizers is not critical (certain vegetables, turf, and ornamentals) and for certain situations in which nitrogen loss from urea would be very high (e.g., some rice paddy soils†). They are widely applied to turf grass because they are not likely to cause fertilizer burn, and because they provide the convenience of fewer applications. Turf fertilized by a single annual application of soluble nitrogen may lose 10 to 50% of the applied nitrogen by leaching, while 3% or less is lost if slow-release fertilizers are used.

---

*For a review of this subject, see Prasad and Power (1995).

†Sulfur-coated urea should not be applied in flooded rice paddies, as the reduced iron in the floodwater will combine with the sulfur coating to form insoluble FeS, which locks up the nitrogen in the fertilizer.

13.8 THE NITRATE LEACHING PROBLEM

In contrast to ammonium ions, which carry a positive charge, the negatively charged nitrate ions are not adsorbed by the negatively charged colloids that dominate most soils. Therefore, nitrate ions move downward freely with drainage water, and are thus readily leached from the soil. The loss of nitrogen in this manner is of concern for two basic reasons. Such loss represents an impoverishment of the ecosystem whether or not cultivated crops are grown, and leaching of nitrate causes several serious environmental

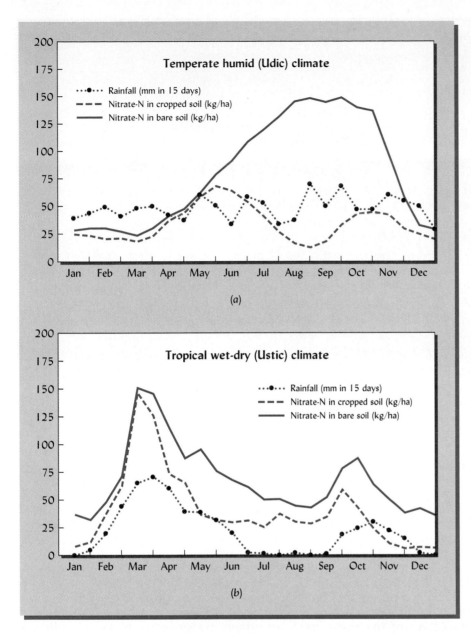

FIGURE 13.4 Typical seasonal patterns of nitrate concentration in representative surface soils with and without growing plants. The upper graph (a) represents a typical soil in a humid temperate region with cool winters and rainfall rather uniformly distributed throughout the year. Nitrates accumulate as the soil warms up in May and June, but are lost by leaching in the fall. The lower graph (b) represents a typical soil in a tropical region with a major and minor rainy season separated by several months of very dry, hot weather. The most notable feature here is the flush of nitrate that appears when the rain first moistens the soil after the long, dry season. This flush of nitrate is caused by the rapid decomposition and mineralization of the dead cells of microorganisms previously killed by the dry, hot conditions. Note that soil nitrate is lower in both climates when plants are grown, because much of the nitrate formed is removed by plant uptake.

problems. Productivity of the ecosystem suffers not only because of the loss of nitrogen, but also because the leaching of nitrate from acidic sources (nitrification or acid rain) facilitates the loss of calcium and other nutrient cations. Both types of nutrient losses will generally reduce the ecosystem productivity (see also Section 13.14). In the case of managed land, there is also an economic loss equal to the value of the lost nitrogen.

Environmental problems caused by nitrogen are mainly associated with its movement away from the soil. Nitrate lost by leaching is generally carried by drainage waters to the groundwater. It may reach domestic wells, and may also eventually flow underground to surface waters such as streams, lakes, and estuaries. The nitrate may contaminate drinking water and cause eutrophication and associated problems (Box 13.3).

The quantity of nitrate nitrogen lost in drainage water depends on two basic factors. First, nitrate leaching depends upon the amount of water draining through the soil profile, which is largely a function of the precipitation and, where used, irrigation. The potential for leaching is very low on nonirrigated semiarid and arid region soils in which there is rarely sufficient rain to moisten the entire profile, let alone a surplus of water that could carry nitrate to the groundwater. Where higher rainfall or irrigation provides a surplus of water, the texture and structure of the soil horizons will affect the rate at which this water enters the soil and percolates downward. Very sandy soils are particularly susceptible to nitrate leaching losses if cultivated and fertilized. Leaching may also be accentuated by use of conservation tillage systems, which may increase water infiltration and thereby increase leaching and concomitant loss of nitrates.

The second factor controlling nitrate leaching is the amount of soluble nitrate in the soil solution that is available for leaching. This second factor depends on the balance and timing of nitrogen inputs and outputs to and from the soil, and on the rates of nitrification and removal of nitrates from the soil solution. Some mature forest ecosystems achieve a very close balance between plant uptake of nitrogen and the return of nitrogen in litter fall and dead trees. Inputs of nitrogen from the atmosphere (e.g., nitrate in acid rain or from nitrogen fixation) can overload the forest ecosystem, resulting in annual leaching losses of up to 25 or 30 kg N/ha.

Even greater losses may come from agricultural systems in which inputs of nitrogen regularly exceed the amounts removed by plant uptake and harvest. Heavy nitrogen fertilization (especially common for vegetables and other cash crops) may exceed what the plants are able to utilize, and can be a major cause of excessive nitrate leaching. Ineffective management of manure from concentrated livestock production facilities is another common cause of nitrate contamination of ground- and surface waters.

Timing of nitrogen inputs is critical. The potential for groundwater contamination with nitrate is greatest where rainfall (or application of irrigation water) is high at the same time that rates of evaporation and plant use of both water and nitrates are low. Late fall, winter, and early spring are therefore the seasons during which most nitrate leaching occurs in humid temperate and Mediterranean climates. Soil managers should adopt practices that will minimize production of soil nitrate and maximize plant uptake of nitrogen during these critical periods. For example, where feasible, fall application of manure and nitrogen fertilizer should be avoided, while the planting of winter cover crops should be encouraged (see Section 16.2).

Even in regions of high leaching potential, careful soil management can prevent excessive nitrate losses. Fertilizer or manure should be applied at modest rates in keeping with plant requirements. The timing of these applications should allow nitrogen to be available when the plants need it, but not available in soluble form much before or after the period of active plant uptake. If these guidelines are followed, nitrogen leaching may be kept to less than 5 or 10% of the nitrogen applied. Limiting nitrogen leaching while ensuring that nitrogen is readily available when the crop is growing can be a challenge. The shallow groundwater under farmland often contains nitrate in excess of 50 mg/L (11 mg/L N).[6] In contrast, groundwater under undisturbed natural vegetation generally contains less than 10 (and often less than 1) mg/L nitrate. Except for certain mature forests, most natural ecosystems are nitrogen-limited and allow little nitrate to leach away. However, disturbances to these systems, such as timber harvest or drainage of wetlands, can result in large losses of nitrate.

13.9 GASEOUS LOSSES BY DENITRIFICATION

Nitrogen may be lost to the atmosphere when nitrate ions are converted to gaseous forms of nitrogen by a series of widely occurring biochemical reduction reactions termed **denitrification**.[7] The organisms that carry out this process are commonly present in large numbers, and are mostly facultative anaerobic bacteria in genera such as *Pseudomonas, Bacillus, Micrococcus,* and *Achromobacter*. These organisms are heterotrophs, which obtain their energy and carbon from the oxidation of organic compounds. Other denitrifying bacteria are autotrophs, such as *Thiobacillus denitrificans,* which obtain their energy from the oxidation of sulfide. The exact mechanisms vary depending on the conditions and organisms involved. In the reaction, NO_3^- (N^{5+}) is reduced in a series of steps to NO_2^- (N^{3+}), and then to nitrogen gases that include NO (N^{2+}), N_2O (N^+), and eventually N_2 (N^0):

$$2NO_3^- \xrightarrow{-2[O]} 2NO_2^- \xrightarrow{-2[O]} 2NO \uparrow \xrightarrow{-[O]} N_2O \uparrow \xrightarrow{-[O]} N_2 \uparrow$$

| Nitrate ions | Nitrite ions | Nitric oxide gas | Nitrous oxide gas | Dinitrogen gas | |
| (+5) | (+3) | (+2) | (+1) | (0) | ← Valence state of nitrogen |

Although not shown in the simplified reaction given here, the oxygen released at each step would be used to form CO_2 from organic carbon (or SO_4^{-2} from sulfide if *Thiobacillus* is the nitrifying organism).

Conditions necessary for significant denitrification to take place can be summarized as follows:

1. Nitrate must be available.

2. Readily decomposable organic compounds (or reduced sulfur compounds) must also be available to provide energy.

3. The soil air should contain less than 10% oxygen, or less than 0.2 mg/L of O_2 dissolved in the solution. Denitrification will proceed most rapidly if oxygen is completely absent. The entire soil need not be anaerobic, as localized microsites of low oxygen in the center of soil aggregates can provide a suitable environment in an otherwise well-aerated soil (see Section 7.5).

4. Temperature should be from 2 to over 50°C, with optimum temperatures between 25 and 35°C.

5. Very strong soil acidity (pH less than 5.0) inhibits rapid denitrification and tends to cause N_2O to be the dominant end product.

Generally when oxygen levels are very low the end product released from the overall denitrification process is dinitrogen gas (N_2). It should be noted, however, that NO and N_2O are commonly also released during denitrification under the fluctuating aeration conditions that often occur in the field (Figure 13.5). The proportion of the three main gaseous products seems to be dependent on the prevalent pH, temperature, degree of oxygen depletion, and concentration of nitrate and nitrite ions available. For example, the release of nitrous oxide (N_2O) is favored if the concentrations of nitrite and nitrate are high and the supply of oxygen is not too low. Under very acid conditions, almost all of the loss occurs in the form of N_2O. Nitric oxide (NO) loss is generally small and apparently occurs most readily under acid conditions.

---

[6]Nitrate concentrations are sometimes reported as "N in nitrate form." To convert mg/L nitrate (NO_3^-) to mg/L nitrate-N (N in nitrate form), multiply by 4.4.

[7]Nitrate can also be reduced to nitrite and to nitrous oxide gas by nonbiological chemical reactions, but these reactions are quite minor in comparison with biological denitrification.

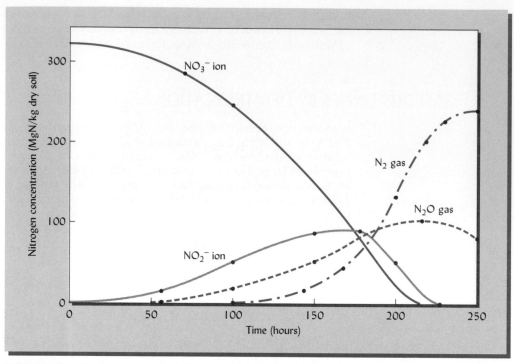

FIGURE 13.5 Changes in various forms of nitrogen during the process of denitrification in a moist soil incubated in the absence of atmospheric oxygen. [From Leffelaar and Wessel (1989)]

Atmospheric Pollution

The question of how much of each nitrogen-containing gas is produced is not merely of academic interest. Dinitrogen gas is quite inert and environmentally harmless, but the oxides of nitrogen are very reactive gases and have the potential to do serious environmental damage in at least four ways. First, NO and N_2O released into the atmosphere by denitrification can contribute to the formation of nitric acid, one of the principal components of acid rain. Second, the nitrogen oxide gases can react with volatile organic pollutants to form ground-level ozone, a major air pollutant in the photochemical smog that plagues many urban areas. Third, when NO rises into the upper atmosphere it contributes to the greenhouse effect by absorbing infrared radiation (see Section 12.2) that would otherwise escape into space.

Finally, and perhaps most significantly, as N_2O moves up into the stratosphere, it may participate in reactions that result in the destruction of ozone (O_3), a gas that helps shield the earth from harmful ultraviolet solar radiation. In recent years this protective ozone layer has been measurably depleted, probably by reaction with industrial CFC's as well as with N_2O and other gases. As this protective layer is further degraded, thousands of additional cases of skin cancer are likely to occur annually. While there are other important sources of N_2O, such as automobile exhaust fumes, a major contribution to the problem is being made by denitrification in soils, especially in rice paddies, wetlands, and heavily fertilized or manured agricultural soils.

QUANTITY OF NITROGEN LOST THROUGH DENITRIFICATION. As might be expected, the exact magnitude of the denitrification losses is difficult to predict and will depend on management practices and soil conditions. Studies of forest ecosystems have shown that during periods of adequate soil moisture, denitrification results in a slow, but relatively steady, loss of nitrogen from these undisturbed natural systems. In contrast, most field measurements of gaseous nitrogen loss from agricultural soils reveal that the losses are highly variable in both time and space. The greater part of the annual nitrogen loss often occurs during just a few days in summer, when heavy rain has temporarily caused the warm soils to become poorly aerated (Figure 13.6).

Low-lying, organic-rich areas and other hot spots may lose nitrogen 10 times as fast as the average rate for a typical field. Although as much as 10 kg/ha of nitrogen may be lost in a single day from the sudden wetting of a well-drained, humid region soil, such

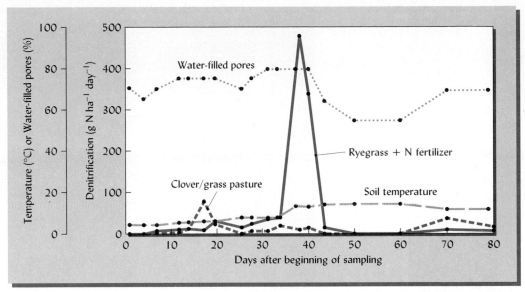

FIGURE 13.6 Changes in denitrification and soil conditions in spring on a well-drained, but somewhat fine-textured soil (East Keswick silty clay loam, Udalfs) in Wales, U.K. Two pasture systems were studied: ryegrass receiving 150 kg N ha⁻¹ annually and a clover-grass mixture. The fertilizer nitrogen applied to the ryegrass system was intended to approximately equal the nitrogen fixed from the atmosphere by the clover in the mixed system (see Section 13.11). Note the sporadic nature of the denitrification process, with most of the nitrogen loss occurring during a brief period when the soil was warm, wet, and high in nitrate. This episodic pattern is typical of disturbed agroecosystems and stands in contrast to the slow, steady pattern of denitrification usually found in undisturbed natural forests. [Data from Coulbourn (1993)]

soils rarely lose more than 5 to 15 kg N/ha annually by denitrification (Table 13.2). But where drainage is restricted and where large amounts of nitrogen fertilizer are applied, substantial losses might be expected. Losses of 30 to 60 kg N/ha/year of nitrogen have been observed in agricultural systems.

Denitrification in Saturated Soils and Groundwater

In flooded soils such as those found in natural wetlands or rice paddies (Figure 13.7a), losses by denitrification may be very high. Flooded soils have aerobic and anaerobic zones, allowing both nitrification and denitrification to take place simultaneously. Since the first process produces the substrate for the second, nitrogen losses can be very high when the two process are associated. As much as 60 to 70% of the fertilizer nitrogen applied to rice paddy water may be volatilized as oxides of nitrogen or elemental nitrogen. However, this loss can be dramatically reduced by deep placement of the fer-

TABLE 13.2 Fate of Applied Nitrogen in Two Experiments, One in Ohio and the Other in a Less Humid Area in Oklahoma

In the more humid area (Ohio) loss by leaching and volatilization was much higher than in the drier area (Oklahoma).

Fate of applied N	Sorghum-Sudan grass[a] (Oklahoma) (%)	Corn[b] (Ohio) (%)
Plant uptake	60	29
Soil organic matter	33	21
Leached	0	33
Gaseous loss[c]	7	17
	100	100

[a]Calculated from 3 years' data from eight soils (Smith et al., 1982).
[b]Calculated from 3 years' data (Chichester and Smith, 1978).
[c]Unaccounted for and presumed lost by volatilization.

(a)

(b)

FIGURE 13.7   Denitrification can be very efficient in removing nitrogen in flooded systems that combine aerobic and anaerobic zones and have high concentrations of available organic carbon. Some examples of such systems are (a) a flooded rice paddy, and (b) a tidal wetland. (Photos courtesy of R. Weil)

tilizer into the reduced zone of the soil. In this zone there is insufficient oxygen to allow nitrification to proceed, so the nitrogen remains in the ammonium form and is not susceptible to loss by denitrification (Figure 13.8).

The sequential combination of nitrification and denitrification also operates in natural and artificial wetlands. Tidal wetlands (Figure 13.7b), which become alternately anaerobic and aerated as the water level rises and falls, have particularly high potentials for converting nitrogen to gaseous forms. Often the resulting rapid loss of nitrogen is considered to be a beneficial function of wetlands in that the process protects estuaries and lakes from the eutrophying effects of too much nitrogen. In fact, wastewater high in organic carbon and nitrogen can be cleaned up quite efficiently by allowing it to flow slowly over a specially designed water-saturated soil system in a process known as *overland flow wastewater treatment* (Figure 13.7c).

Recent studies on the movement of nitrate in groundwater have documented the significance of denitrification taking place in the poorly drained soils under **riparian**

(c)

(d)

FIGURE 13.7 (*Continued*)   Denitrification can be very efficient in removing nitrogen in flooded systems that combine aerobic and anaerobic zones and have high concentrations of available organic carbon. Other examples of such systems are (c) a site for treating sewage effluent by overland flow (effluent applied by sprinklers, see arrows), and (d) a manure storage lagoon. (Photos courtesy of R. Weil)

vegetation (mainly woodlands adjacent to streams). In most cases studied in humid temperate regions, contaminated groundwater lost most of its nitrate load as it flowed through the riparian zone on its way to the stream. The apparent removal of nitrate may be quite dramatic, whether the nitrate source is septic drainfields or fertilized cropland (Figure 13.9). Most of nitrate is believed to be lost by denitrification, stimulated by organic compounds leached from the decomposing forest litter and by the anaerobic conditions that prevail in the wet riparian zone soils.

Whether nitrogen removal by denitrification is actually beneficial to the environment depends on which gases are produced. If significant amounts of N_2O or NO are produced, then wetlands and overland flow systems may merely trade water pollution for air pollution.

We have just discussed a number of biological processes that lead to losses of nitrogen from the soil system. We now turn our attention to the principal biological process by which soil nitrogen is replenished.

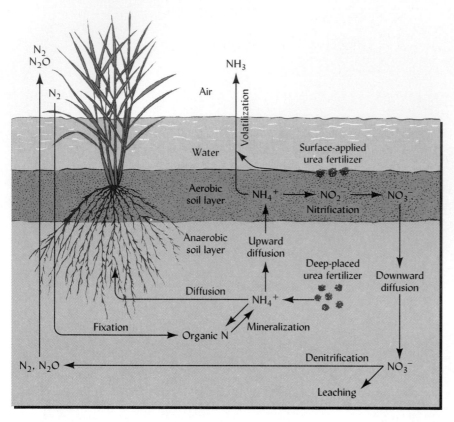

FIGURE 13.8 Nitrification–denitrification reactions and kinetics of the related processes controlling nitrogen loss from the aerobic–anaerobic layers of a flooded soil system. Nitrates, which form in the thin aerobic soil layer just below the soil–water interface, diffuse into the anaerobic (reduced) soil layer below and are denitrified to the N_2 and N_2O gaseous forms, which are lost to the atmosphere. Placing the urea or ammonium-containing fertilizers deep in the anaerobic layer prevents the oxidation of ammonium ions to nitrates, thereby greatly reducing the loss. [Modified from Patrick (1982)]

13.10   BIOLOGICAL NITROGEN FIXATION

Next to plant photosynthesis, **biological nitrogen fixation** is probably the most important biochemical reaction for life on earth. Through this process certain organisms convert the inert dinitrogen gas of the atmosphere (N_2) to nitrogen-containing organic compounds that become available to all forms of life through the nitrogen cycle. The process is carried out by a limited number of microorganisms, including several species of bacteria, a number of actinomycetes, and certain cyanobacteria (blue-green algae).

Globally, enormous amounts of nitrogen are biologically fixed each year. Terrestrial systems alone fix an estimated 139 million Mg (metric tons), about twice as much as is industrially fixed in the manufacture of fertilizers (Table 13.3).

TABLE 13.3   Global Nitrogen Fixation from Different Sources

		Nitrogen fixed per year	
Source of N fixation	Area (10^6 ha)	Rate (kg/ha)	Total fixed (10^6 Mg)
Biological Fixation			
Legume crops	250	140	35
Nonlegume crops	1,150	8	9
Meadows and grassland	3,000	15	45
Forest and woodland	4,100	10	40
Other vegetated land	4,900	2	10
Ice-covered land	1,500	0	0
Total Land	14,900		139
Sea	36,100	1	36
Total Biological	51,000		175
Lightning			8
Fertilizer industry			77
Grand Total			260

Calculated from Jenkinson (1990) and Burns and Hardy (1975).

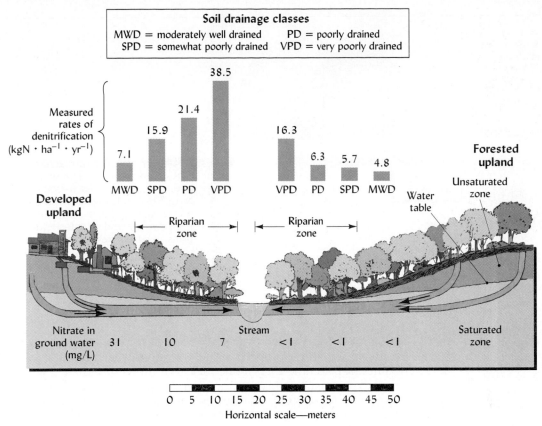

FIGURE 13.9 Denitrification in riparian wetlands receiving groundwater with high- and low-nitrate contents. The high-nitrate groundwater (left) came from sites with heavy development of houses using septic drainfields for sewage disposal. The low-nitrate groundwater (right side) came from an area of undeveloped forest. The riparian zones on both sides of the stream were covered with red-maple-dominated forest. Within a few meters after entering the riparian wetland, the nitrate content of the contaminated groundwater was reduced by 75%, from 31 ppm nitrate to less than 7 ppm nitrate. The soils in this study site are very sandy Inceptisols and Entisols. In other regions, riparian zones with finer textured soils have shown even more complete removal of nitrate from groundwater. [Data shown are from Hanson et al. (1994)]

THE MECHANISM. Regardless of the organisms involved, the key to biological nitrogen fixation is the enzyme nitrogenase, which catalyzes the reduction of dinitrogen gas to ammonia.

$$N_2 + 6H^+ + 6e^- \xrightarrow[\text{(Fe,Mo)}]{\text{(nitrogenase)}} 2NH_3$$

The ammonia, in turn, is combined with organic acids to form amino acids and, ultimately, proteins.

$$NH_3 + \text{organic acids} \longrightarrow \text{amino acids} \longrightarrow \text{proteins}$$

The site of N_2 reduction is the enzyme **nitrogenase**, a complex consisting of two proteins, the smaller of which contains iron while the larger component contains molybdenum and iron (Figure 13.10). Several salient facts about this enzyme and its function are worth noting, for nitrogenase is unique and its role in the nitrogen cycle is of great importance to humankind.

1. The reduction of N_2 to NH_3 by nitrogenase requires a great deal of energy to break the triple bond between the nitrogen atoms. Therefore the process is greatly enhanced by association with higher plants, which can supply this energy from photosynthesis.

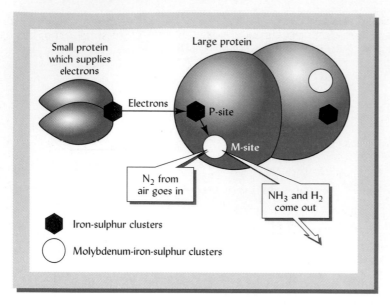

FIGURE 13.10 The nitrogenase complex consists of two proteins. The smaller protein supplies electrons needed to convert atmospheric N_2 to NH_3, while the larger protein converts the nitrogen. The M-sites capture nitrogen (N_2) from the air, while the P-sites receive the electrons provided by the small protein so that N_2 can be reduced to NH_3. [From Emsley (1991), reprinted with permission of *New Scientist*]

2. Nitrogenase is destroyed by free O_2, so organisms that fix nitrogen must protect the enzyme from exposure to oxygen. When nitrogen fixation takes place in root nodules (see Section 13.11), one means of protecting the enzyme from free oxygen is the formation of leghemoglobin.[8] This compound, which gives active nodules a red interior color, binds oxygen in such a way as to protect the nitrogenase while making oxygen available for respiration in other parts of the nodule tissue.

3. The reduction reaction is end-product inhibited—an accumulation of ammonia will inhibit nitrogen fixation. Also, too much nitrate in the soil will inhibit the formation of nodules (see Section 13.11).

4. Nitrogen-fixing organisms have a relatively high requirement for molybdenum, iron, phosphorus, and sulfur, because these nutrients are either part of the nitrogenase molecule or are needed for its synthesis and use.

FIXATION SYSTEMS. Biological nitrogen fixation occurs through a number of microorganism systems with or without direct association with higher plants (Table 13.4). Although the legume–bacteria symbiotic systems have received the most attention,

---

[8]Leghemoglobin is virtually the same molecule as the hemoglobin that gives human blood its red color when oxygenated. The use of hemoglobin to perform essentially similar functions in both legume root nodules and mammalian blood is a striking example of nature's conservative tendency and the unity of all life.

TABLE 13.4 **Information on Different Systems of Biological Nitrogen Fixation**

N-fixing systems	Organisms involved	Plants involved	Site of fixation
Symbiotic			
Obligatory			
Legumes	Bacteria *Rhizobia* and *Bradyrhizobia*	Legumes	Root nodules
Nonlegumes (angiosperms)	Actinomycetes (*Frankia*)	Nonlegumes (angiosperms)	Root nodules
Associative			
Morphological involvement	Cyanobacteria, bacteria	Various higher plants and microorganisms	Leaf and root nodules, lichens
Nonmorphological involvement	Cyanobacteria, bacteria	Various higher plants and microorganisms	Rhizosphere (root environment) Phyllosphere (leaf environment)
Nonsymbiotic	Cyanobacteria, bacteria	Not involved with plants	Soil, water independent of plants

recent findings suggest that the other systems involve many more families of plants worldwide, and may even rival the legume-associated systems as suppliers of biological nitrogen to the soil. Each major system will be discussed briefly.

13.11 SYMBIOTIC FIXATION WITH LEGUMES

The **symbiosis** (mutually beneficial relationship) of legumes and bacteria of the genera *Rhizobium* and *Bradyrhizobium* provide the major biological source of fixed nitrogen in agricultural soils. The genus *Rhizobium* contains fast-growing, acid-producing bacteria, while the *Bradyrhizobia* are slow growers that do not produce acid. Both will be considered together. These organisms infect the root hairs and the cortical cells, ultimately inducing the formation of **root nodules** that serve as the site of nitrogen fixation (Figure 13.11). In a mutually beneficial association, the host plant supplies the bacteria with carbohydrates for energy, and the bacteria reciprocate by supplying the plant with fixed-nitrogen compounds.

ORGANISMS INVOLVED. A given *Rhizobium* or *Bradyrhizobium* species will infect some legumes but not others. For example, *Rhizobium trifolii* inoculates *Trifolium* species (most clovers), but not sweet clover, which is in the genus *Melilotus*. Likewise, *Rhizobium phaseoli* inoculates *Phaseolus vulgaris* (dry beans), but not soybean, which is in the genus *Glycine*. This specificity of interaction is one basis for classifying rhizobia (see Table 13.5). Legumes that can be inoculated by a given *Rhizobium* species are included in the same cross-inoculation group.

In areas where a given legume has been grown for several years, the appropriate species of *Rhizobium* is probably present in the soil. Often, however, the natural *Rhizobium* population in the soil is too low or the strain of the *Rhizobium* species present is not effective (Figure 13.12). In such circumstances, special mixtures of the appropriate *Rhizobium* or *Bradyrhizobium* inoculant may be applied either by coating the legume seeds or by applying the inoculant directly to the soil. Effective and competitive strains of *Rhizobium,* which are available commercially, often give significant yield increases, but only if used on the proper crops. You may want to refer to Table 13.5 when planning legume rotations or purchasing commercial inoculant.

QUANTITY OF NITROGEN FIXED. The rate of biological fixation of nitrogen is greatly dependent on soil and climatic conditions. The legume–rhizobium associations generally function best on soils that are not too acid (although *Bradyrhizobium* associations gen-

(a)   (b)   (c)

FIGURE 13.11   Photos illustrating soybean nodules. In (a) the nodules are seen on the roots of the soybean plant, and a closeup (b) shows a few of the nodules associated with the roots. A scanning electron micrograph (c) shows a single plant cell within the nodule stuffed with the bacterium *Bradyrhizobium japonicum*. (Courtesy W. J. Brill, University of Wisconsin)

TABLE 13.5   Classification of Rhizobia Bacteria and Associated Legume Cross-Inoculation Groups

The genus Rhizobium *contains fast-growing, acid-producing bacteria, while those of* Bradyrhizobium *are slow growers that do not produce acid. A third genus,* Azorhizobium, *which is not shown, produces stem nodules on* Sesbania rostrata.

Bacteria		Host legume
Genus	*Species/subgroup*	
Rhizobium	R. leguminosarum	
	bv. *viceae*	*Vicia* (vetch), *Pisum* (peas), *Lens* (lentils), *Lathyrus* (sweet pea)
	bv. *trifolii*	*Trifolium* spp. (most clovers)
	bv. *phaseoli*	*Phaseolus* spp. (dry bean, runner bean, etc.)
	R. Meliloti	*Melilotus* (sweet clover, etc.), *Medicago* (alfalfa), *Trigonella*, (fenugreek)
	R. loti	*Lotus* (trefoils), *Lupinus* (lupins), *Cicer* (chickpea), *Anthyllis, Leucaena,* and many other tropical trees
	R. Fredii	*Glycine* spp. (e.g., soybean)
Bradyrhizobium	B. japonicum	*Glycine* spp. (e.g., soybean)
	B. sp.	*Vigna* (cowpeas), *Arachis* (peanut), *Cajanus* (pigeon pea), *Pueraria* (kudzu), *Crotolaria* (crotolaria), and many other tropical legumes

erally can tolerate considerable acidity) and that are well supplied with essential nutrients. However, high levels of available nitrogen, whether from the soil or added in fertilizers, tend to depress biological nitrogen fixation (Figure 13.13). Apparently, plants make the heavy energy investment required for symbiotic nitrogen fixation only when short supplies of mineral nitrogen make nitrogen fixation necessary.

Although quite variable from site to site, the amount of nitrogen biologically fixed can be quite high, especially for those systems involving nodules, which allow efficient supply of energy from photosynthates and protected conditions for the nitrogenase enzyme system (Table 13.6). Nonnodulating or nonsymbiotic systems generally fix relatively small amounts of nitrogen. Nonetheless, many natural plant communities and agricultural systems (generally involving legumes) derive the bulk of their nitrogen needs from biological fixation.

EFFECT ON SOIL NITROGEN LEVEL.   The data in Table 13.6 show that symbiotic nitrogen fixation is definitely beneficial to both forestry and agriculture. This source of nitrogen is

FIGURE 13.12 This soybean crop in East Africa was a total failure. The soybean seeds were not inoculated with the proper bacteria prior to planting in the newly cleared field, which had been cleared from forest vegetation and had never grown soybeans before. (Photo courtesy of R. Weil)

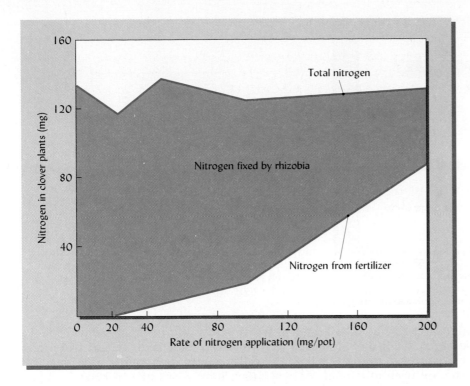

FIGURE 13.13 Influence of added inorganic nitrogen on the total nitrogen in clover plants, the proportion supplied by the fertilizer and that fixed by the rhizobium organisms associated with the clover roots. Increasing the rate of nitrogen application decreased the amount of nitrogen fixed by the organisms in this greenhouse experiment. [From Walker et al. (1956)]

not available to most coniferous and hardwood trees, grasses, cereal crops, and vegetables that lack symbiotic associations, unless they are grown in association with nodulated species. Over time, the presence of nitrogen-fixing species can significantly increase the nitrogen content of the soil and benefit nonfixing species grown in association with fixing species (see Figure 13.14).

TABLE 13.6 Typical Levels of Nitrogen Fixation from Different Systems

Crop or plant	Associated organism	Typical levels of nitrogen fixation kg N/ ha⁻¹ yr⁻¹
Symbiotic		
Legumes (nodulated)		
Ipil Ipil tree (*Leucena leucocephala*)	Bacteria (*Rhizobium*)	100–500
Locust tree (*Robina* spp.)		75–200
Alfalfa (*Medicago sativa*)		150–250
Clover (*Trifolium pratense L.*)		100–150
Lupine (*Lupinus*)		50–100
Vetch (*Vicia vilbosa*)		50–150
Bean (*Phaseolus vulgaris*)		30–50
Cowpea (*Vigna unguiculata*)	Bacteria (*Bradyrhizobium*)	50–100
Peanut (*Arachis*)		40–80
Soybean (*Glycine max L.*)		50–150
Pigeon pea (*Cajunus*)		150–280
Kudzu (*Pueraria*)		100–140
Nonlegumes (nodulated)		
Alders (*Alnus*)	Actinomycetes (*Frankia*)	50–150
Species of *Gunnera*	Cyanobacteria[a] (*Nostoc*)	10–20
Nonlegumes (nonnodulated)		
Pangola grass (*Degetaria decumbens*)	Bacteria (*Azospirillum*)	5–30
Bahia grass (*Pasalum notatum*)	Bacteria (*Azobactor*)	5–30
Azolla	Cyanobacteria[a] (*Anabena*)	150–300
Nonsymbiotic	Bacteria (*Azobactor, Clostridium*)	5–20
	Cyanobacteria[a] (various)	10–50

[a]Sometimes referred to as blue-green algae.

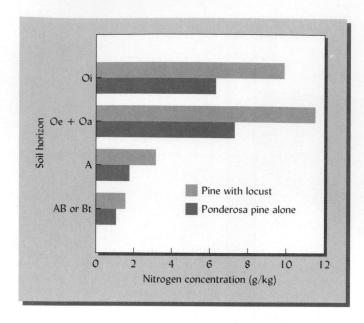

FIGURE 13.14 Nitrogen contents of forest soil horizons showing the effects of New Mexican locust trees (*Robinia neomexicana*) growing in association with ponderosa pine (*Pinus ponderosa*) in a region of Arizona receiving about 670 mm rainfall per year. The data are means from 20 stands of pondersosa pine, half of them with the nitrogen-fixing legume trees (locust) in the understory. The soils are Eutroboralfs and Argiborolls with loam and clay loam textures. [Data from Klemmedson (1994)]

As noted, nitrogen fixation is inhibited where mineral forms of N are available for plant uptake. Even where a nitrogen-fixing symbiosis is established, plants will preferentially take up nitrogen from the soil if it is available from such sources as mineralized organic matter, nitrogen in rainfall, and fertilizers. Some crops such as beans and peas are such weak nitrogen fixers that most of the nitrogen they absorb must come from the soil. Consequently, it should not be assumed that the symbiotic systems always increase soil nitrogen. Only in cases where the soil is low in available nitrogen and vegetation includes strong nitrogen fixers would this be likely to be true. In the case of legume crops harvested for seed or hay, most of the nitrogen fixed is removed from the field with the harvest. Nitrogen additions from such crops should be considered as nitrogen *savers* for the soil rather than nitrogen builders. On the other hand, considerable buildup of soil nitrogen can be achieved by perennial legumes (such as alfalfa) and by annual legumes (such as hairy vetch) whose entire growth is returned to the soil as **green manure**. Such nitrogen buildup should be taken into account when estimating nitrogen fertilizer needs for maximum plant production with minimal environmental pollution (see Sections 13.15 and 16.4).

FATE OF NITROGEN FIXED BY LEGUME BACTERIA.   The nitrogen fixed in root nodules goes in three directions. First, it is used directly by the host plant, which thereby benefits greatly from the symbiosis described in Section 13.10. A portion of the nitrogen-rich plant tissue may later be returned to the soil as residues.

Second, some of the fixed nitrogen may become available to nonfixing plants grown in association with nitrogen-fixing plants. Although some direct transfer may take place via mycorrhizal hyphae connecting the two plants, most of the transfer results from mineralization of nitrogen-rich compounds in root exudates and in sloughed-off root and nodule tissues. Ammonium and nitrate thus released into the soil are available to any plant growing in association with the legume. The vigorous development of a grass in a legume–grass mixture is evidence of this rapid release (Figure 13.15), as are the relatively high nitrate concentrations sometimes measured in groundwater under legume crops.

The third pathway for the fixed nitrogen is immobilization by heterotrophic microorganisms and eventual incorporation into the soil organic matter (see Section 13.4).

13.12   SYMBIOTIC FIXATION WITH NONLEGUMES

Nodule-Forming Nonlegumes

Nearly 200 species from more than a dozen genera of nonlegumes are known to develop nodules and to accommodate symbiotic nitrogen fixation. Included are several important groups of angiosperms, listed in Table 13.7. The roots of these plants, which are present in certain forested areas and wetlands, form distinctive nodules when their root hairs are invaded by soil actinomycetes of the genus *Frankia*.

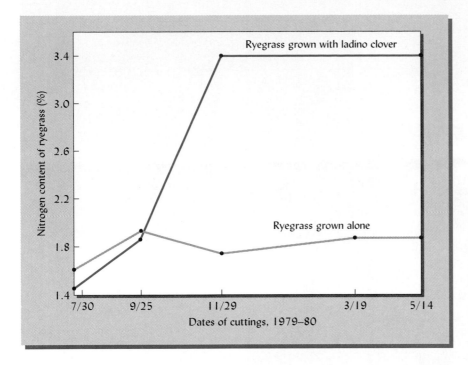

FIGURE 13.15 Nitrogen content of five field cuttings of ryegrass grown alone or with ladino clover. For the first two harvests, nitrogen fixed by the clover was not available to the ryegrass and the nitrogen content of the ryegrass forage was low. In subsequent harvests, the fixed nitrogen apparently was available and was taken up by the ryegrass. This was probably due to the mineralization of dead ladino clover root tissue. [From Broadbent et al. (1982)]

The rates of nitrogen fixation per hectare compare favorably with those of the legume–*Rhizobium* complexes (Table 13.6). On a worldwide basis, the total nitrogen fixed in this way may even exceed that fixed by agricultural legumes (Table 13.3). Because of their nitrogen-fixing ability, certain of the tree–actinomycete complexes are able to colonize infertile soils and newly forming soils on disturbed lands, which may have extremely low fertility as well as other conditions that limit plant growth (Figure 13.16). Once nitrogen-fixing plants become established and begin to build up the soil nitrogen supply through leaf litter and root exudation, the land becomes more hospitable for colonization by other species. *Frankia* thus play a very important role in the nitrogen economy of areas undergoing succession, as well as in established forests and marshes.

Certain cyanobacteria are known to develop nitrogen-fixing symbiotic relations with green plants. One involves nodule formation on the stems of *Gunnera*, an angiosperm common in marshy areas of the southern hemisphere. In this association, Cyanobacteria of the genus *Nostoc* fix 10 to 20 kg N/ha/year (Table 13.6).

Symbiotic Nitrogen Fixation Without Nodules

Among the most significant nonnodule nitrogen-fixing systems are those involving cyanobacteria. One system of considerable practical importance is the *Azolla/Anabaena* complex, which flourishes in certain rice paddies of tropical and semitropical areas. The

TABLE 13.7 **Number and Distribution of Major Actinomycete-Nodulated Nonlegume Angiosperms**

In comparison there are about 13,000 legume species.

Genus	Family	Species[a] nodulated	Geographic distribution
Alnus	Betulaceae	33/35	Cool regions of the northern hemisphere
Ceanothus	Rhamnaceae	31/35	North America
Myrica	Myricaceae	26/35	Many tropical, subtropical, and temperate regions
Casuarina	Casuarinaceae	24/25	Tropics and subtropics
Elaeagnus	Elaeagnaceae	16/45	Asia, Europe, N. America
Coriaria	Coriariaceae	13/15	Mediterranean to Japan, New Zealand, Chile to Mexico

Selected from Torrey (1978).
[a]Number of species nodulated/total number of species in genus.

FIGURE 13.16 Soil actinomycetes of the genus *Frankia* can nodulate the roots of certain woody plant species and form a nitrogen-fixing symbiosis that rivals the legume–rhizobia partnership in efficiency. The actinomycete-filled root nodule (a) is the site of nitrogen fixation. The red alder tree (b) is among the first pioneer tree species to revegetate disturbed or badly eroded sites in high-rainfall areas of the Pacific Northwest in North America. This young alder is thriving despite the nitrogen poor, eroded condition of the soil, because it is not dependent on soil nitrogen for its needs. (Photos courtesy of R. Weil)

Anabaena cyanobacteria inhabit cavities in the leaves of the floating fern *Azolla* and fix quantities of nitrogen comparable to those of the more efficient *Rhizobium*–legume complexes (Table 13.6).

A more widespread but less intense nitrogen-fixing phenomenon is that which occurs in the *rhizosphere* of certain grasses and other nonlegume plants. The organisms responsible are bacteria, especially those of the *Spirillum* and *Azotobacter* genera (Table 13.6). Plant root exudates supply these microorganisms with energy for their nitrogen-fixing activities.

Scientists have reported a wide range of rates for rhizosphere nitrogen fixation, with the highest values observed in association with certain tropical grasses. Even if typical rates are only 5 to 30 kg N/ha/year, the vast areas of tropical grasslands suggest that the total quantity of nitrogen fixed by rhizosphere organisms is likely very high (Table 13.6).

13.13   NONSYMBIOTIC NITROGEN FIXATION

Certain free-living microorganisms present in soils and water are able to fix nitrogen. Because these organisms are not directly associated with higher plants, the transformation is referred to as *nonsymbiotic* or *free-living*.

Fixation by Heterotrophs

Several different groups of bacteria and cyanobacteria are able to fix nitrogen nonsymbiotically. In upland mineral soils, the major fixation is brought about by species of two genera of heterotrophic aerobic bacteria: *Azotobacter* (in temperate zone soils) and *Beijerinckia* (in tropical soils). Certain anaerobic bacteria of the genus *Clostridium* are also able to fix nitrogen. Because pockets of low oxygen supply exist in soils even when they are in good tilth, aerobic and anaerobic bacteria probably work side by side in many well-drained soils.

The amount of nitrogen fixed by these heterotrophs varies greatly with the pH, soil nitrogen level, and sources of organic matter available. Because of their limited energy supply, under normal agricultural conditions the rates of nitrogen fixation by these organisms are thought to be in the range of 5 to 20 kg N/ha/year (Table 13.6)—only a small fraction of the nitrogen needed by crops. However, in some natural ecosystems this level of nitrogen fixation would make a significant contribution to meeting nitrogen needs.

Fixation by Autotrophs

Among the autotrophs able to fix nitrogen are certain photosynthetic bacteria and cyanobacteria. In the presence of light, these organisms are able to fix carbon dioxide and nitrogen simultaneously. The contribution of the photosynthetic bacteria is uncertain, but that of cyanobacteria is thought to be of some significance, especially in wetland areas and in rice paddies. In some cases, these algae have been found to fix sufficient nitrogen for moderate rice yields, but normal levels may be no more than 20 to 30 kg N/ha/year. Nitrogen fixation by cyanobacteria in upland soils also occurs, but the level is much lower than is found under wetland conditions.

13.14 ADDITION OF NITROGEN TO SOIL IN PRECIPITATION

The atmosphere contains ammonia gas and nitrates dissolved in water vapor, and other nitrogen compounds released from the soil and plants as well as from the combustion of coal and petroleum products. Nitrates also form in small quantities as a result of electrical discharges (lightning) in the atmosphere. Another source is the exhaust from automobile and truck engines, which contributes a considerable amount to the atmosphere, especially downwind from large cities. These atmosphere-borne nitrogen compounds are added to the soil through rain, snow, and dust. Although the rates of addition per hectare are typically small, the total quantity of nitrogen added annually is significant and may be tens of kg/ha near highly polluted areas.

The quantity of ammonia and nitrates in precipitation varies markedly with location and with season. The additions are greater in humid tropical regions than in humid temperate regions, and are larger in the latter than in semiarid, temperate climates. Rainfall additions of nitrogen are highest near cities and industrial areas and near large animal feedlots (Table 13.8). There is special concern for the deposition of

TABLE 13.8   Amounts of Nitrogen Brought Down in Precipitation Annually in Different Parts of the United States

| Areas in the United States | Range in annual deposition (kg/ha) | | Total nitrogen (kg/ha) |
	Nitrate nitrogen	Ammonium nitrogen	
Rural area of Ohio[a]	6.6–10.6	6.2–8.2	12.8–18.8
Industrialized Northeast[b]	4.3–7.4	8.6–14.8	12.9–22.2
Borders of NE industrialized areas[b]	2.8–4.1	5.6–8.2	8.4–12.3
Open areas in West[b]	0.4–0.6	0.8–1.2	1.2–1.8

[a]Owens et al., 1992.
[b]Ammonium nitrogen is calculated as twice the nitrate deposition, the approximate ratio from numerous measurements, although it may be incorrect at any specific site (U.S./Canada work group #2, 1982).

nitrates and other nitrogen oxides from these areas of concentration because they are associated with increased acidity. The environmental impact of acid rain and its influence on vegetation (forests and crops) and surface waters tends to overshadow the nutrient benefit of precipitation-supplied nitrates.

Typically, the nitrogen in precipitation is about two-thirds ammonium and one-third nitrate. The range of total nitrogen (nitrate + ammonium) added by precipitation annually is 1 to 25 kg N/ha. A figure of 5 to 8 kg N/ha would be typical for nonindustrial temperate regions. This modest annual acquisition of nitrogen is probably of more significance to the nitrogen budget in natural ecosystems than in agriculture.

Effects on Forest Ecosystems

In mature forests, net plant uptake and microbial immobilization are in such close balance with release of nitrogen by mineralization that many of these systems have very little ability to retain incoming nitrogen. Nitrogen from the atmosphere enters the forest largely as nitric acid (HNO_3) or as ammonium, which is converted to nitric acid upon microbial nitrification (see Section 13.7). As a result, in certain mature forests most of the nitrogen coming from polluted air seems to be lost as leached nitrate. The H^+ ions from these acids displace base-forming cations on the soil colloids. The displaced cations, especially calcium and magnesium, are lost from the soil as they leach downward with the nitrate (and sulfate; see Section 13.22). Thus the leaching of mobile anions (nitrate and sulfate) promotes the loss of calcium and magnesium, causing imbalances in the nutrition of the trees and a decline in forest productivity. Nitrogen in precipitation might be considered beneficial fertilizer when it falls on farmland, but it can be a serious pollutant when added to a soil supporting a mature forest. Chapter 9 should be consulted for a more detailed explanation of these reactions and soil differences.

13.15 REACTIONS OF NITROGEN FERTILIZERS

The worldwide use of nitrogen-containing fertilizers has expanded greatly during the past few decades. Most commercial fertilizers supply nitrogen in soluble forms, such as nitrate or ammonium, or as urea, which rapidly hydrolyzes to form ammonium. Ammonium and nitrate ions from fertilizer are taken up by plants and participate in the nitrogen cycle in exactly the same way as ammonium and nitrate derived from organic matter mineralization or other sources. The main difference is that nitrogen from heavy doses of fertilizer is far more concentrated in time (available all at once) and space (applied in narrow bands or in a thin layer in the soil) than is nitrogen from other sources. While judicious application of such concentrations may be very beneficial for plant growth, it may also have serious detrimental effects.

First, high concentrations of ammonia gas from ammonium-releasing fertilizers (such as urea and anhydrous ammonia) inhibit the activity of many of the soil flora and fauna, including nitrifying bacteria and earthworms. This partial sterilization is usually quite temporary, and is localized near the zone of fertilizer application. Such concentrations of ammonia also favor loss of this gas to the atmosphere, especially in alkaline soils. Also, because two moles of acidity are formed for every mole of ammonium nitrogen that undergoes nitrification to nitrates, use of ammonium fertilizers increases soil acidity (see Section 13.7). Soil acidity may also be increased by ammonia released from a heavy application of manure.

Concentration in time occurs when a year's supply of soluble nitrogen is applied all at once. Such single applications are low cost and convenient, so they are common in many agricultural and forestry enterprises. Unfortunately, the assimilatory processes of the nitrogen cycle (e.g., plant uptake and immobilization) may not be able to utilize the soluble fertilizer nitrogen fast enough to prevent major losses by leaching, surface runoff, denitrification, and ammonia volatilization. The environmental consequences of these losses have already been discussed. As a general rule, it is most environmentally benign to apply nitrogen fertilizer in multiple small doses that coincide with the plant demand.

13.16 PRACTICAL MANAGEMENT OF SOIL NITROGEN IN AGRICULTURE

The goals of nitrogen control are threefold: (1) the maintenance of an adequate nitrogen supply in the soil, (2) the regulation of the soluble forms of nitrogen to ensure that enough is readily available to provide all the nitrogen plants need for optimum growth, and (3) the minimization of environmentally damaging leakage from the soil–plant system.

To understand the problem of nitrogen control, it is useful to consider the nitrogen inputs and outputs for a farm or ecosystem. Figure 13.17 summarizes the inputs and outputs of nitrogen for a hypothetical single farm field. Most of the input and output pathways have equivalents in natural ecosystems. The features that make the cropland balance different from that for a natural ecosystem are mainly (1) the removal of nitrogen contained in the harvested product, which represents a major loss from the field and (2) the addition of nitrogen fertilizer, which represents a major input into the field.

A particular type of soil in any given combination of climate and farming system tends to assume what may be called a *normal* or *equilibrium content* of nitrogen. Consequently, under ordinary methods of cropping and manuring, any attempt to permanently raise the nitrogen content to a level higher than this will result in unnecessary waste as nitrogen leaches, volatilizes, or is otherwise lost before it may be used. There are some exceptions to the above general rule. If a soil that is naturally low in organic matter and nitrogen is heavily fertilized over a period of time, and if crop residues are returned to the soil, the soil nitrogen level will likely increase. Such a situation is common where soils of dry areas (Aridisols) are irrigated and cropped to high-fertilizer-requiring vegetables or alfalfa. If irrigation is discontinued and crop yields are lowered, however, the soil will eventually return to its previous level of nitrogen. The equilibrium level of soil nitrogen is often governed by the management practices employed.

Several basic strategies are proving useful in achieving a rational reduction of excessive nitrogen inputs while maintaining or improving production levels and profitability in agriculture (Box 13.4). These approaches include (1) taking into account the nitrogen contribution from *all* sources and reducing the amount of fertilizer applied accordingly, (2) improving the efficiency with which fertilizer as well as so-called waste products such as animal manure are used, (3) avoiding overly optimistic yield goals that lead to fertilizer application rates designed to meet crop needs that are much higher than actually occur in most years, and (4) improving crop response knowledge, which identifies the lowest nitrogen application that is likely to produce optimum profit. These strategies of nutrient management will be discussed further in Chapter 16.

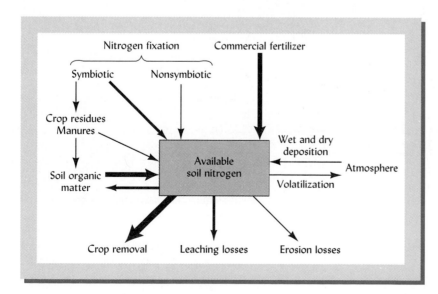

FIGURE 13.17 Major gains and losses of available soil nitrogen. The widths of the arrows indicate roughly the magnitude of the losses and the additions often encountered. It should be emphasized that the diagram represents average conditions only and that much variability is to be expected in the actual and relative quantities of nitrogen involved.

BOX 13.4   RATIONALIZING NITROGEN INPUTS IN AGRICULTURAL SYSTEMS

The relative importance and absolute magnitudes of the various nitrogen cycle pathways vary greatly with the particular crop grown and with variations in climate and management. A good crop of wheat or cotton may remove only 100 kg N/ha, nearly half of which may be returned to the soil in the stalks or straw. A bumper crop of silage corn, in contrast, may contain over 250 kg N/ha, and a good annual yield of alfalfa or well-fertilized grass hay removes more than 350 kg N/ha. Regardless of the crop, the goal should be to minimize the losses of nitrogen by all pathways except crop removal (this is, after all, what the farmer sells and what we all eat). The need for commercial nitrogen fertilizer will be governed by the degree to which the farmer is able to integrate symbiotic nitrogen fixation, manure and residue recycling, and loss minimization into the farming system.

If crops and livestock are combined in a farming system, then most of the nitrogen removed in crops can be returned in animal manures. Fertilizer should be considered as a supplement to the nitrogen made available from organic matter mineralization, biological nitrogen fixation, rainfall, animal manure, and crop residues. Nitrogen losses by leaching and denitrification generally become a problem only when nitrogen fertilization exceeds the amount needed to fill the gap between crop uptake needs and the supply from these other sources.

When we examine the nitrogen balance for cropland on a national scale, it is clear that more nitrogen is being applied than can be properly used (Table 13.9). In the United States, cropland receives approximately 20.9 million Mg of nitrogen annually, while it exports some 13.5 million Mg of nitrogen as crop harvests and residues. In other words, nitrogen inputs are more than 50% greater than intended outputs, leaving over 7 million Mg of nitrogen to either accumulate in soils (unlikely in most cases; see Section 13.16) or leak into the environment via leaching, erosion, and runoff, or via gaseous losses to the atmosphere. While some losses are inevitable from agroecosystems, it is becoming widely acknowledged that cropland in the United States and in other industrial countries has been receiving more nitrogen than plants can effectively use, and that this excess will have to be reduced if nitrogen pollution is to be controlled.

TABLE 13.9   Nitrogen Inputs and Outputs for All Cropland in the United States in 1987 and 1977

Only the major categories are considered. These are aggregate data for the entire country and should not be considered representative of any given farm field. The estimates for the two years were made independently of each other, so the close agreement lends validity to both.

Nitrogen output or input	Millions of Mg	
	1987[a]	1977[b]
Outputs		
Harvested crops	10.60	8.9
Crop residues	2.89	3.0
Total outputs	13.50	11.9
Inputs to cropland		
As commercial fertilizer	9.39	9.5
As legume N fixation	6.87	7.2
In crop residues returned	2.89	3.0
Recoverable manure	1.73	1.4
Total inputs	20.9	21.1
Balance = inputs – outputs	7.42	9.2

[a]Data from Table 6-3 in National Research Council (1993).
[b]Data abstracted from Power (1981) as cited in NRC (1993).

13.17 IMPORTANCE OF SULFUR[9]

Sulfur has long been recognized as indispensable for many reactions in living cells. In addition to its vital roles in plant and animal nutrition, sulfur is also responsible for several types of air, water, and soil pollution, and is therefore of increasing environmental interest. The environmental problems associated with sulfur include acid precipitation, certain types of forest decline, acid mine drainage, acid sulphate soils, and even some toxic effects in drinking water used by humans and livestock.

Roles of Sulfur in Plants and Animals

Sulfur is a constituent of the amino acids methionine, cysteine, and cystine, deficiencies of which result in serious human malnutrition. The vitamins biotin, thiamine, and B1 contain sulfur, as do many protein enzymes that regulate activities such as photosynthesis and nitrogen fixation. It is believed that sulfur-to-sulfur bonds link certain sites on long chains of amino acids, causing proteins to assume the specific three-dimensional shapes that are the key to their catalytic action. Sulfur is closely associated with nitrogen in the processes of protein and enzyme synthesis. Sulfur is also an essential ingredient of the aromatic oils that give the cabbage and onion families of plants their characteristic odors and flavors. It is not surprising that among the plants, the legume, cabbage, and onion families require especially large amounts of sulfur.

Deficiencies of Sulfur

Healthy plant foliage generally contains 0.15 to 0.45% sulfur, or approximately one-tenth as much sulfur as nitrogen. Plants deficient in sulfur tend to become spindly and to develop thin stems and petioles. Their growth is slow, and maturity may be delayed. They also have a chlorotic light green or yellow appearance. Symptoms of sulfur deficiency are similar to those associated with nitrogen deficiency (Section 13.1). However, unlike nitrogen, sulfur is relatively immobile in the plant, so the chlorosis develops first on the youngest leaves as sulfur supplies are depleted (in nitrogen-deficient plants, chlorosis develops first on the older leaves). Sulfur-deficient leaves on some plants show interveinal chlorosis or faint striping that distinguishes them from nitrogen-deficient leaves. Also unlike nitrogen-deficient plants, sulfur-deficient plants tend to have low sugar but high nitrate contents in their sap.

As a result of the three independent trends described below, sulfur deficiencies in agricultural plants have become increasing common during the past several decades:

1. Enforcement of clean air standards has led to reduction of sulfur dioxide (SO_2) emissions to the atmosphere from the burning of fossil coal and oil.

2. The fertilizer market is increasingly dominated by more concentrated, high-analysis products. These fertilizers supply their primary nutrients at lower cost, but they do not contain significant amounts of sulfur-bearing impurities as did the previously used materials. For example, the once widely used ammonium sulfate and single superphosphate fertilizers, which contain 24 and 12% sulfur, respectively, have been largely replaced by diammonium phosphate (DAP), urea, triple superphosphate (TSP), and other materials which have relatively high contents of nitrogen and phosphorus, but contain little or no sulfur. In addition, older sulfur-containing pesticides (such as the fungicidal copper sulfate) have been largely replaced by organic materials free of sulfur.

3. At the same time that reductions in the supply of sulfur to soils and plants have occurred, improved varieties and better management have resulted in higher crop yields. As harvests have grown larger, greater amounts of sulfur have been removed from soils. Thus, the need for sulfur has increased just as the supplies of this element have declined.

---

[9]For a discussion of sulfur and agriculture, see Tabatabai (1986).

AREAS OF DEFICIENCY.   Sulfur deficiencies have been reported in most areas of the world but are most prevalent in areas where soil parent materials are low in sulfur, where extreme weathering and leaching has removed this element, or where there is little replenishment of sulfur from the atmosphere. In many tropical countries, one or more of these conditions prevail and sulfur-deficient areas are common.

Burning of plant biomass results in a loss of sulfur to the atmosphere. In many parts of the world, crop residues and native vegetation are routinely burned as a means of clearing the land. Soils of the African savannas are particularly deficient in sulfur as a result of the annual burning of plant residues during the dry season. Fire converts much of the sulfur in the plant residues to sulfur gases such as sulfur dioxide. Sulfur in these gases, and in smoke particulates, is subsequently carried by the wind hundreds of kilometers away to areas covered by rain forest, where some of the sulfur dioxide is absorbed by moist soils and foliage, and some is deposited with rainfall. Thus the soils of the savannas tend to export their sulfur to those of the rain forest (e.g., Oxisols). Consequently, the latter often contain significant accumulations of sulfur in their profiles (see Figure 13.18).

In the United States, deficiencies of sulfur are most common in the Southeast, the Northwest, California, and the Great Plains. In the Northeast and in other areas with heavy industry and large cities, sulfur deficiencies are not yet widespread.

13.18   NATURAL SOURCES OF SULFUR

The three major natural sources of sulfur that can become available for plant uptake are organic matter, soil minerals, and sulfur gases in the atmosphere. In natural ecosystems where most of the sulfur taken up by plants is eventually returned to the same soil, these three sources combined are usually sufficient to supply the needs of growing plants (Figure 13.19). These three natural sources of sulfur will be considered in order.

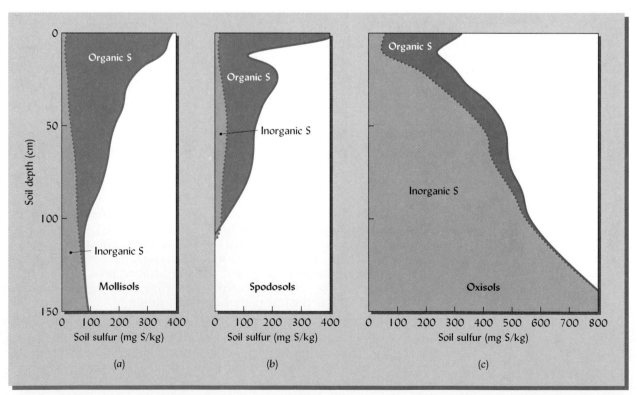

FIGURE 13.18   The distribution of organic and inorganic sulfur in representative soil profiles of the soil orders Mollisols, Spodosols, and Oxisols. In each, soil organic forms dominate the surface horizon. Considerable inorganic sulfur, both as adsorbed sulfate and calcium sulfate minerals, exists in the lower horizons of Mollisols. Relatively little inorganic sulfur exists in Spodosols. However, the bulk of the profile sulfur in the humid tropics (Oxisols) is present as sulfate adsorbed to colloidal surfaces in the subsoil.

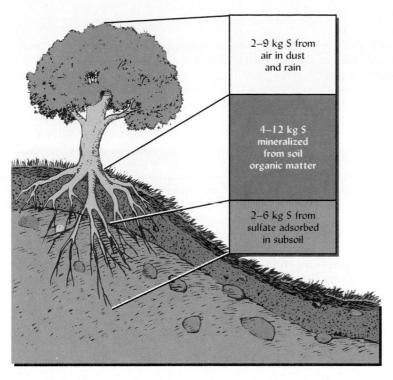

FIGURE 13.19 Plants take up sulfur primarily from three sources: sulfur in atmospheric gases and dust; sulfate mineralized from soil organic matter; and sulfate adsorbed on soil minerals. Typical ranges of sulfur uptake from these sources are shown. Where these three sources are insufficient for optimal growth, the application of sulfur-containing fertilizers may be warranted. In areas downwind of coal burning plants and metal smelters, the atmospheric contribution may be much larger than indicated here.

Organic Matter

In surface soils in temperate humid regions, 90 to 98% of the sulfur is usually present in organic forms (Figure 13.18). As is the case for nitrogen, the exact form of the sulfur in the organic matter is not known. Typically more than half of the sulfur is *carbon bonded* (C–S fraction), mostly in proteins and in amino acids such as cysteine, cystine, and methionine. These materials are bound with the humus and clay fractions, and are thereby protected from microbial attack. A somewhat more transitory pool of organic sulfur is in the *ester sulfate* form (C–O–S), in which S is bound to oxygen rather than directly to carbon. Examples of compounds in the two main fractions of organic sulfur are:

Ester sulfate (glucose sulfate)   Carbon-bonded sulfur (cysteine)

Over time, soil microorganisms break down these organic sulfur compounds into soluble inorganic forms, mainly sulfate. This mineralization of organic compounds to release sulfate is analogous to the release of ammonium and nitrate from organic matter discussed in Section 13.4.

In arid and semiarid regions, less organic matter is present in the surface soils. However, gypsum (CaSO$_4$ · 2H$_2$O), which supplies inorganic sulfur, is often present in the subsurface horizons. Therefore, the proportion of organic sulfur is not likely to be as high in arid and semiarid region soils as it is in humid region soils. This is especially true in the subsoils, where organic sulfur may comprise only a small fraction of the sulfur present.

The inorganic forms of sulfur are not as plentiful as the organic forms, but they include the soluble and available compounds on which plants and microbes depend.

Sulfur is held in several mineral forms in soils, with the sulfide and sulfate minerals being most common. The sulfate minerals are most easily solubilized, and the sulfate ion (SO_4^{2-}) is easily assimilated by plants. Sulfate minerals are most common in regions of low rainfall, where they accumulate in the lower horizons of some Mollisols and Aridisols (Figure 13.18). They may also accumulate as neutral salts in the surface horizons of saline soils in arid and semiarid regions.

Sulfides that are found in some humid region soils with restricted drainage must be oxidized to the sulfate form before the sulfur can be assimilated by plants. When these soils are drained, oxidation can occur, and ample available sulfur is released. In some cases, so much sulfur is oxidized that problems of extreme acidity result (see Section 13.21).

Another mineral source of sulfur is the clay fraction of some soils high in Fe,Al oxides and kaolinite. These clays are able to strongly adsorb sulfate from soil solution and subsequently release it slowly by anion exchange, especially at low pH. Oxisols and other highly weathered soils of the humid tropics and subtropics may contain large stores of sulfate, especially in their subsoil horizons (Figure 13.18). Considerable sulfate may also be bound by the metal oxides in the spodic horizons under certain temperate forests.

Atmospheric Sulfur[10]

The atmosphere contains varying quantities of carbonyl sulfide (COS), hydrogen sulfide (H_2S), sulfur dioxide (SO_2), and other sulfur gases, as well as sulfur-containing dust particles. These atmospheric forms of sulfur arise from volcanic eruptions, volatilization from soils, ocean spray, biomass fires, and industrial plants (such as electric-power-generation stations fired by high-sulfur coal and metal smelters). In recent decades, the contributions of the latter industrial sources have become the dominant sources in certain locations.

Some of these materials are oxidized in the atmosphere to sulfates, forming H_2SO_4 and sulfate salts such as $CaSO_4$ and $MgSO_4$. When these solids and gases return to the earth as dry particles and gases it is called *dry deposition;* when they are brought down with precipitation it is called *wet deposition.* Although the proportion of these two forms of deposition varies from one place to another, typically each supplies about half of the total.

The industrialized northeastern states have the highest deposition of sulfur in the United States, commonly ranging from 30 to 75 kg S/ha/yr downwind from industrial sites. Farther away from the industrial sources, deposition declines to about 8 to 15 kg S/ha/yr. In rural areas of the western United States, away from industrial cities and smelting plants, only 2 to 5 kg S/ha/yr is deposited. In rural Africa, 1 to 4 kg S/ha/yr is deposited. Thus, the effect of atmospheric deposition depends greatly on location, with the main effect of industrial emission occurring within a relatively small distance of the source (Figure 13.20). On the other hand, regional patterns of air quality and sulfur deposition are evident as the less acute effects of industrial emissions are felt hundreds of kilometers downwind (Figure 13.21).

Atmospheric sulfur becomes part of the soil–plant system in three ways. The wet deposition materials, which are usually high in H_2SO_4, are mostly absorbed by soils, with some also absorbed through plant foliage. Part of the dry deposition is also absorbed directly by soils, while some is absorbed directly by plants. The quantity plants can absorb directly is variable, but in some cases 25 to 35% of the plant sulfur can come from this source even if available soil sulfate is adequate. In sulfur-deficient soils, about half of the plant needs can come from the atmosphere (Figure 13.19).

Environmental concerns about high sulfur levels in the atmosphere and the resulting acid rain are of great practical significance to both forestry and agriculture. Efforts to reduce atmospheric sulfur are welcomed in areas near industrial plants, where levels of atmospheric forms of sulfur may be high enough to cause toxicity to trees and crops

---

[10]For a discussion of this topic, see NAS (1983).

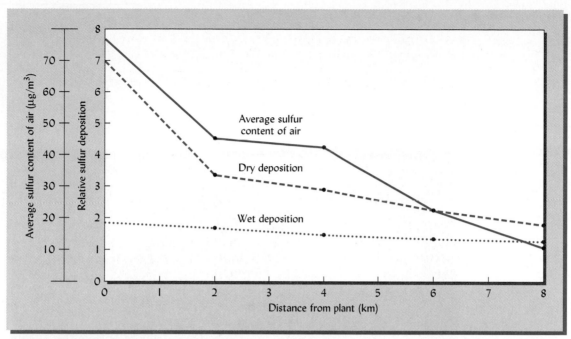

FIGURE 13.20 Industrial facilities such as coal- or oil-burning power plants or metal smelters can provide large inputs of sulfur to nearby soils. Note the rapid drop off in sulfur deposition at increasing distances from the source of sulfur emissions. Dry deposition of S from the atmosphere is dominant very close to the source, but at greater distances dry and wet deposition are nearly equal. Regulations in many countries would today require that such a power plant greatly reduce its sulfur emissions. [From Johannson (1960)]

(not to mention respiratory problems in people). Even beyond the immediate influence of industrial sources the elevated levels of atmospheric sulfur, and associated acid effects, are causing serious damage to certain types of forest ecosystems. On the other hand, continued reductions in atmospheric sulfur are resulting in plant deficiencies of this element in other areas, especially for high-yield-potential agricultural crops. In these cases, part of the crop's sulfur requirement may have to be supplied by increasing the sulfur content of the fertilizers used, thereby increasing crop production costs.

13.19 THE SULFUR CYCLE

The major transformations that sulfur undergoes in soils are shown in Figure 13.22. The inner circle shows the relationships among the four major forms of this element (*sulfides, sulfates, organic sulfur,* and *elemental sulfur*). The outer portions show the most important sources of sulfur and how this element is lost from the system.

Considerable similarity to the nitrogen cycle is evident (compare Figures 13.22 and 13.2). In each case, the atmosphere is an important source of the element in question. Both elements are held largely in the soil organic matter; both are subject to microbial oxidation and reduction; both can enter and leave the soil in gaseous forms; and both are subject to some degree of leaching in the anionic form. Microbial activities are responsible for many of the transformations that determine the fates of both nitrogen and sulfur.

Figure 13.22 should be referred to frequently in conjunction with the following more detailed examination of sulfur in plants and soils.

13.20 BEHAVIOR OF SULFUR COMPOUNDS IN SOILS

Mineralization

Sulfur behaves much like nitrogen as it is absorbed by plants and microorganisms and moves through the sulfur cycle. The organic forms of sulfur must be mineralized by soil

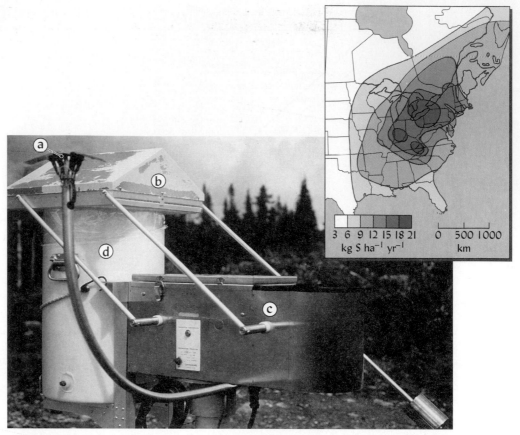

FIGURE 13.21 The apparatus shown in the picture collects both wet and dry sulfur deposition. A sensor (a) triggers the small roof (b) to move over and cover the dry deposition collection chamber (c) at the first sign of precipitation. The wet deposition chamber (d) is then exposed to collect precipitation. When the precipitation ceases, the sensor triggers the roof to move back over the wet deposition collection chamber so that dry deposition can again be collected. The map shows the typical geographic distribution of sulfur deposition from the atmosphere in eastern North America. The data are based on measurements of sulfur in wet deposition and the assumption that dry deposition contributes approximately an equal share of the total. [Photo courtesy of R. Weil; map modified from Olsen and Slavich (1986)]

organisms if the sulfur is to be used by plants. The rate at which this occurs depends on the same environmental factors that affect nitrogen mineralization, including moisture, aeration, temperature, and pH. When conditions are favorable for general microbial activity, sulfur mineralization occurs. Some of the more easily decomposed organic compounds in the soil are sulfate esters from which microorganisms release sulfate ions directly. However, in much of the soil organic matter, sulfur in the reduced state is bonded to carbon atoms in protein and amino acid compounds. In the latter case the mineralization reaction might be expressed as follows:

$$\text{Organic sulfur} \longrightarrow \text{decay products} \xrightarrow{O_2} SO_4^{2-} + 2H^+$$

| (proteins and other organic combinations) | (H₂S and other sulfides are simple examples) | Sulfates |

Because the release of available sulfate sulfur is mainly dependent on microbial processes, the supply of available sulfate in soils fluctuates with seasonal, and sometimes daily, changes in environmental conditions (Figure 13.23). These fluctuations lead to the same difficulties in predicting and measuring the amount of sulfur available to plants as were discussed in the case of nitrogen.

Immobilization

Immobilization of inorganic forms of sulfur occurs when low-sulfur, energy-rich organic materials are added to soils. The immobilization mechanism is thought to be the same as for nitrogen—the energy-rich material stimulates microbial growth, and the inorganic sulfate is assimilated into microbial tissue. When the microbial activity subsides, the inorganic sulfate reappears in the soil solution.

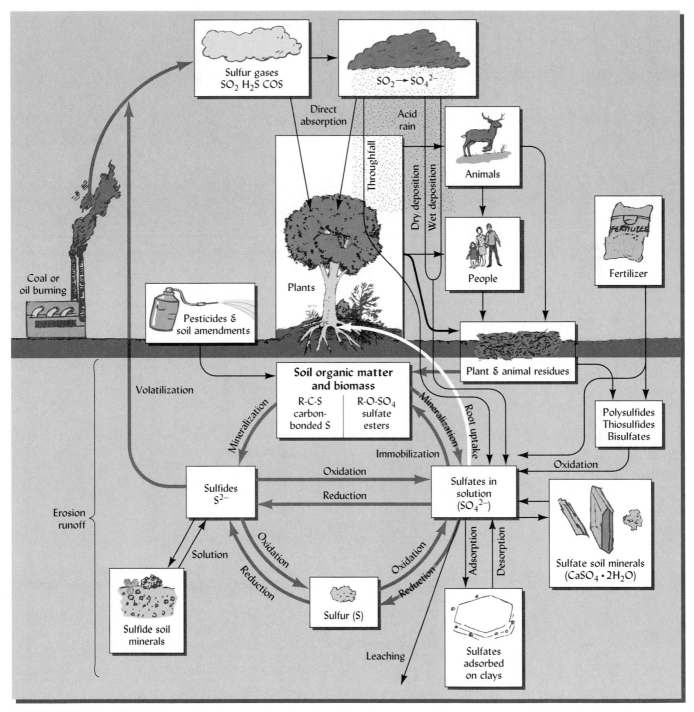

FIGURE 13.22 The sulfur cycle showing some of the transformations that occur as this element is cycled through the soil–plant–animal–atmosphere system. In the surface horizons of all but a few types of arid region soils, the great bulk of sulfur is in organic forms. However, in deeper horizons or in excavated soil materials, various inorganic forms may dominate. The oxidation and reduction reactions that transform sulfur from one form to another are mainly mediated by soil microorganisms.

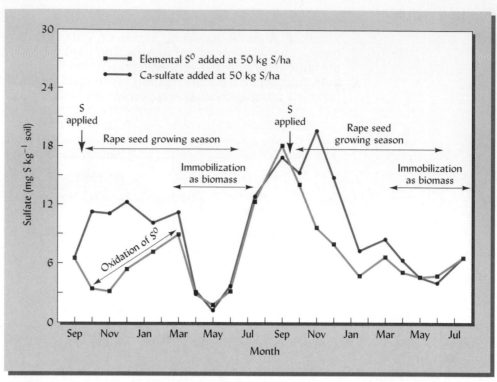

FIGURE 13.23 Seasonal changes in the sulfate form of sulfur available in the surface horizon of a soil (Argixeroll) in Oregon used to grow the oilseed crop, rape. This crop is sown in the fall, grows slowly during the winter and then rapidly during the cool spring months. Data are shown for plots which were fertilized with either elemental S or calcium sulfate. The vertical arrows indicate dates on which these amendments were applied. Note that sulfate concentration was greater in the calcium sulfate fertilized soils for the first few months after each application while the elemental S was slowly converted to sulfate by microbial oxidation. A distinct depression in sulfate concentration occurred each spring as the soil warmed up, stimulating both immobilization of sulfate into microbial biomass and uptake of sulfate into the rapeseed crop. Sulfate concentrations reached a peak in late summer and early fall when crop uptake ceased after harvest and microbial mineralization was rapidly occurring. Movement of dissolved sulfate from the lower horizons up into the surface soil may also have occurred during hot, dry weather. [Modified from Castellano and Dick (1991)]

The pattern of S immobilization in soils suggests that, like nitrogen, sulfur in soil organic matter may be associated with organic carbon in a reasonably constant ratio. The ratio among carbon, nitrogen, and sulfur for a number of soils on three different continents is given in Table 13.10. The C/N/S ratio of 100:8:1 is reasonably representative.

During the microbial breakdown of organic materials, several sulfur-containing gases are formed, including hydrogen sulfide (H_2S), carbon disulfide (CS_2), carbonyl sulfide (COS), and methyl mercaptan (CH_3SH). All are more prominent in anaerobic soils. Hydrogen sulfide is commonly produced in waterlogged soils by reduction of sulfates by anaerobic bacteria. Most of the others are formed from the microbial decomposition of sulfur-containing amino acids. Although these gases can be adsorbed by soil colloids, some escape to the atmosphere, where they undergo chemical changes and eventually return to the soil.

TABLE 13.10 Mean Carbon/Nitrogen/Sulfur Ratios in a Variety of Soils

Location	Description and number of soils	C/N/S ratio
North Scotland	Agricultural, noncalcareous (40)	104:7:1
Minnesota	Mollisols (6)	74:7:1
Minnesota	Spodosols (24)	108:8:1
Oregon	Agricultural, varied (16)	143:10:1
Eastern Australia	Acid soils (128)	126:8:1
Eastern Australia	Alkaline soils (27)	92:7:1

From Whitehead (1964).

13.21   SULFUR OXIDATION AND REDUCTION

The Oxidation Process

During the microbial decomposition of organic carbon-bonded sulfur compounds, sulfides are formed along with other incompletely oxidized substances such as elemental sulfur (S^0), thiosulfates ($S_2O_3^{2-}$), and polythionates ($S_{2x}O_{3x}^{2-}$). These reduced substances are subject to oxidation, just as are the ammonium compounds formed when nitrogenous materials are decomposed. The oxidation reactions may be illustrated as follows with hydrogen sulfide and elemental sulfur:

$$H_2S + 2O_2 \longrightarrow H_2SO_4 \longrightarrow 2H^+ + SO_4^{2-}$$

$$2S + 3O_2 + 2H_2O \longrightarrow 2H_2SO_4 \longrightarrow 4H^+ + 2SO_4^{2-}$$

The oxidation of some sulfur compounds, such as sulfites (SO_3^{2-}) and sulfides (S^{2-}), can occur by strictly chemical reactions. However, most sulfur oxidation in soils is *biochemical* in nature, carried out by a number of autotrophic bacteria, which include five species of the genus *Thiobacillus*. Since the environmental requirements and tolerances of these five species vary considerably, the process of sulfur oxidation occurs over a wide range of soil conditions. For example, sulfur oxidation may occur at pH values ranging from less than 2 to higher than 9. This flexibility is in contrast to the comparable nitrogen oxidation process, nitrification, which requires a rather narrow pH range closer to neutral.

The Reduction Process

Like nitrate ions, sulfate ions tend to be unstable in anaerobic environments. They are reduced to sulfide ions by a number of bacteria of two genera, *Desulfovibro* (five species) and *Desulfotomaculum* (three species). The organisms use the oxygen in sulfate to oxidize organic materials. A representative reaction showing the reduction of sulfur coupled with organic matter oxidation is:

$$2R\!-\!CH_2OH + SO_4^{2-} \longrightarrow 2R\!-\!COOH + 2H_2O + S^{2-}$$

Organic alcohol   Sulfate   Organic acid   Sulfide

In poorly drained soils, the sulfide ion reacts immediately with iron or manganese, which in anaerobic conditions are typically present in the reduced forms. By tying up the soluble reduced iron, the formation of iron sulfides helps prevent iron toxicity in rice paddies and marshes. This reaction may be expressed as follows:

$$Fe^{2+} + S^{2-} \longrightarrow FeS$$

Dissolved ferrous iron   Sulfide   Iron sulfide (solid)

$$Mn^{2+} + S^{2-} \longrightarrow MnS$$

Dissolved reduced manganese   Sulfide   Manganese sulfide (solid)

Sulfide ions will also undergo hydrolyses to form gaseous hydrogen sulfide, which causes the rotten egg smell of swampy or marshy areas. Sulfur reduction may take place with sulfur-containing ions other than sulfates. For example, sulfites (SO_3^{2-}), thiosulfates ($S_2O_3^{2-}$), and elemental sulfur (S^0) are readily reduced to the sulfide form by bacteria and other organisms.

The oxidation and reduction reactions of inorganic sulfur compounds play an important role in determining the quantity of sulfate (the plant-available nutrient form of sulfur) present in soils at any one time. Also, the state of sulfur oxidation is an important factor in the acidity of soil and of water draining from soils.

Sulfur Oxidation and Acidity

The reaction equations given for oxidation of elemental sulfur (S) and hydrogen sulfide (H_2S) show that, like nitrogen oxidation, sulfur oxidation is an acidifying process. For every sulfur atom oxidized, two hydrogen ions are formed. Because of this acidifying reaction, sulfur can be applied to extremely alkaline and sodic soils of arid regions to reduce the pH of these soils to a level more favorable for plant growth and disease control (Section 9.9). The acidity formed as sulfur oxidizes must also be considered when choosing a fertilizer, since some materials contain elemental sulfur and might lower the soil pH unfavorably.

The addition of atmospheric sulfur to soils through precipitation increases soil acidity.[11] The pH of this so-called *acid rain* may be 4 or even lower (see Section 9.6). In contrast, the pH of unpolluted rainwater varies from 5.6 in humid regions to over 7.0 in drier regions, where calcareous dust reacts with rainwater as it falls.

EXTREME SOIL ACIDITY.   The acidifying effect of sulfur oxidation can bring about extremely acid soil conditions. This may happen when coastal land inundated with brackish water or seawater is drained and put under cultivation. Ocean water and marine sediments contain relatively high amounts of sulfur. During periods when land is under water, sulfates are reduced to generally stable iron and manganese sulfides. If there are periods of partial drying, elemental sulfur can form by partial oxidation of the sulfides left in the soils that have been submerged. The sulfide and elemental sulfur content of such tidal marsh soils are hundreds of times greater than would be found in comparable upland soils.

If high-sulfide soils are drained, the sulfides and/or elemental sulfur quickly oxidize and form sulfuric acid.

$$2FeS + 4.5O_2 + 2H_2O \longrightarrow Fe_2O_3 + 4H^+ + 2SO_4^-$$

$$S + 1.5O_2 + H_2O \longrightarrow 2H^+ + SO_4^-$$

The soil pH may drop to as low as 1.5, a level not observed in normal upland soils. Obviously, plant growth cannot occur under these conditions. Furthermore, the quantity of limestone needed to neutralize the acidity is so high that it is impractical to remediate these soils by liming.

Sizable areas of high-sulfur wetland soils (classified as sulfaquepts and sulfaquents in *Soil Taxonomy* and commonly called ***acid sulfate soils*** or ***cat-clays***) are found in Southeast Asia and along the Atlantic coasts of South America and Africa. They also occur in tidal areas along the coasts of several other areas, including the Netherlands and both the Southeast and West Coast of the United States. So long as these soils are kept submerged, reduced sulfur is not oxidized and the soil pH does not drop prohibitively. Consequently, production of paddy rice is sometimes possible in these soils.

The occurrence of soils capable of producing extreme acidity by sulfur oxidation is not limited to the coastal wetlands just described. Marine sediments, possibly associated with coastal marshes millions of years ago, now form the rock and regolith covering large areas of land no longer submerged by oceans or marshes (see Section 2.8). Today's coastal plains are often underlain by such sediments. Also, many types of shales and other sedimentary rocks may contain reduced sulfur compounds, often in the form of the iron sulfide mineral, pyrite. These rocks are commonly associated with deposits of coal, a substance that also formed millions of years ago in marshy areas.

Excavations for road cuts and coal mining operations often expose long-buried pyrite-containing materials. When pyrite comes into contact with air and water the following net reaction, which is facilitated by sulfur oxidizing bacteria, occurs:

$$2FeS_2 + 7.5O_2 + 7H_2O \xrightarrow{Bacteria} 8H^+ + 4SO_4^{2-} + 2Fe(OH)_3$$

Pyrite   Oxygen   Water   Hydrogen ions   Sulfate   Iron hydroxide

The four moles of sulfuric acid (dissociated into hydrogen and sulfate ions) produced by the reaction above come from the oxidation of both the sulfate and the ferrous iron in pyrite in conjunction with the hydrolysis of water. The production of acid from this reaction can be an enormous problem, which prevents growth of vegetation on affected

---

[11]Note that some of this acidity results from nitrogen oxides as they are oxidized in the atmosphere to nitric and nitrous acids. Sulfur and nitrogen are jointly responsible for this problem.

areas (Figure 13.24) and makes it very difficult to restore certain land disturbed by mining operations. If allowed to proceed unchecked, the acids may wash into nearby streams. Thousands of kilometers of streams have been seriously polluted in this manner, the water and rocks in such streams often exhibiting orange colors from the iron compounds in the acid drainage.

13.22   SULFATE RETENTION AND EXCHANGE

The sulfate ion is the form in which plants absorb most of their sulfur from soils. Since many sulfate compounds are quite soluble, the sulfate would be readily leached from the soil, especially in humid regions, were it not for its adsorption by the soil colloids. As was pointed out in Chapter 8, most soils have some anion exchange capacity, which is associated with iron and aluminum oxide coatings and clays and, to a limited extent, with 1:1-type silicate clays. Sulfate ions are attracted by the positive charges that characterize acid soils containing these clays. They also react directly with hydroxy groups exposed on the surfaces of these clays. A general equation indicates how this may occur:

$$
\begin{array}{c}
\vert \\
Al \\
HO \diagdown \diagup OH \\
Al \\
\vert
\end{array}
+\ KHSO_4 \ \rightleftharpoons \
\begin{array}{c}
\vert \\
Al \\
HO \diagdown \diagup SO_4 - K \\
Al \\
\vert
\end{array}
+\ H_2O
$$

This reaction is driven to the right in acid soils. If the soil pH were increased, hydroxide ions would be added, the reaction would be driven to the left, and sulfate ions would be released. Accordingly, this is an ion exchange reaction, the hydroxide being exchanged for a KSO_4^- anion. Most soils will retain some sulfate, although the quantity held is generally small and its strength of retention is low compared to that of phosphate.

In warm humid regions, surface soils are typically quite low in sulfur. However, much sulfate may be held by the iron and aluminum oxides and 1:1-type silicate clays that tend to accumulate in the subsoil horizons of the Ultisols and Oxisols of these regions (Figure 13.17). Symptoms of sulfur deficiency commonly occur early in the growing season on Ultisols in the Southeast United States, especially those with sandy, low organic matter surface horizons. However, the symptoms may disappear as the crop matures and its roots reach the deeper horizons where sulfate is retained.

FIGURE 13.24   Construction of this interstate highway cut through several layers of sedimentary rock. One of these layers contained reduced sulfide materials. Now exposed to the air and water, this layer is producing copious quantities of sulfuric acid as the sulfide materials are oxidized. Note the failure of vegetation to grow below the zone from which the acid is draining. (Photo courtesy of R. Weil)

13.23 SULFUR AND SOIL FERTILITY MAINTENANCE

Figure 13.25 depicts the major gains and losses of sulfur from soils. The problem of maintaining adequate quantities of sulfur for mineral nutrition of plants is becoming increasingly important. Even though chances for widespread sulfur deficiencies are generally less than for nitrogen, phosphorus, and potassium, increasing crop removal of sulfur makes it essential that farmers be attentive to prevent deficiencies of this element. In some parts of the world (especially in certain semiarid grasslands), sulfur is already the next most limiting nutrient after nitrogen.

Crop residues and farmyard manures can help replenish the sulfur removed in crops, but these sources generally can help to recycle only those sulfur supplies that already exist within a farm. In regions with low-sulfur soils, greater dependence must be placed on fertilizer additions. Regular applications of sulfur-containing materials are now necessary for good crop yields in large areas far removed from industrial plants. There will certainly be an increased necessity for the use of sulfur in the future.

13.24 CONCLUSION

Sulfur and nitrogen have much in common concerning the manner in which they cycle in soils. Both are held by soil colloids in slowly available forms. Both are found in proteins and other organic forms as part of the soil organic matter. Their release to inorganic ions (SO_4^{2-}, NH_4^+, and NO_3^-), in which forms they are available to higher plants, is accomplished by soil microorganisms. Anaerobic soil organisms are able to change these elements into gaseous forms, which are then released to the atmosphere to be joined by similar gases released from industrial plants and from vehicle engines. These gases are then deposited on plants, soils, and other objects in forms that are popularly termed *acid rain*. This has serious consequences to forestry, agriculture, and to society in general.

There are some significant differences between nitrogen and sulfur. Large amounts of both soil nitrogen and soil sulfur are found in organic compounds. However, a higher proportion of soil sulfur is found in inorganic compounds, especially in some drier areas where gypsum ($CaSO_4 \cdot 2H_2O$) is abundant in the subsoil. Some soil organisms have the ability to fix elemental N_2 gas into compounds usable by plants. No analogous process occurs for sulfur. Also, nitrogen, which is removed in much larger quantities by plants, must be replenished regularly by organic residues, manure, or chemical fertilizers.

In rural areas away from cities and industrial plants, sulfur deficiencies in plants are increasingly common. As a result of increased crop removal and reductions in incidental applications of sulfur to soils, this element will likely join nitrogen as a regular component of soil fertility programs.

Several serious environmental problems are caused by excessive amounts of nitrogen or sulfur in certain forms. While water pollution by nitrates is probably the most

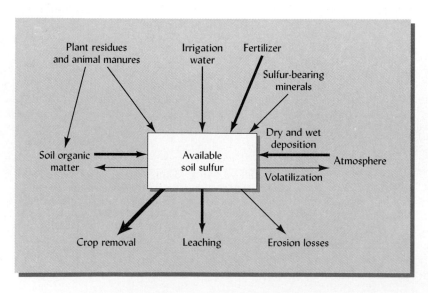

FIGURE 13.25 Major gains and losses of available soil sulfur. The thickness of the arrows indicates the relative amounts of sulfur involved in each process under average conditions. Considerable variation occurs in the field.

prominent of the nitrogen-related problems, acid deposition and drainage are the main environmental concerns associated with sulfur cycling in soils. Using our knowledge of the nitrogen and sulfur cycles, much can now be done to alleviate both sets of problems. However, much remains to be learned about the reactions of nitrogen and sulfur in soils and the environment.

REFERENCES

Addiscott, T. M., A. P. Whitmore, and D. S. Powlson. 1991. *Farming, Fertilizers and the Nitrate Problem* (Wallingford, Oxon, U.K.: C.A.B. International).

Baethgen, W. E., and M. M. Allen. 1987. "Nonexhanged ammonium nitrogen contributions to plant available nitrogen," *Soil Sci Soc Am J.,* **51**:110–115.

Broadbent, F. E., T. Nakashima, and G. Y. Chang. 1982. "Estimation of nitrogen fixation by isotope dilution in field and greenhouse experiments," *Agron. J.,* **74**:625–628.

Burns, R. C., and R. W. F. Hardy. 1975. *Nitrogen Fixation in Bacteria and Higher Plants* (Berlin: Springer-Verlag).

Castellano, S. D., and R. P. Dick. 1991. "Cropping and sulfur fertilization influence on sulfur transformations in soil," *Soil Sci Soc Am J.,* **54**:114–121.

Chichester, F. W., and S. J. Smith. 1978. "Disposition of ^{15}N-labeled fertilizer nitrate applied during corn culture in field lysimeters," *Journal of Environmental Quality,* **7**:227–233.

Colbourn, P. 1993. "Limits to denitrification in two pasture soils in a maritime temperate climate," *Agriculture, Ecosystems and Environment,* **43**:49–68.

Dutch, J., and P. Ineson. 1990. "Denitrification of an upland forest site," *Forestry,* **63**:363–376.

Emsley, J. 1991. "Metals trace the secrets of nitrogen fixation," *New Scientist,* **131(1784)**:19.

Hanson, G. C., P. M. Groffman, and A. J. Gold. 1994. "Denitrification in riparian wetlands receiving high and low groundwater nitrate inputs," *Journal of Environmental Quality,* **23**:917–922.

Hauck, R. D. (ed.). 1984. *Nitrogen in Crop Production* (Madison, Wis.: Amer. Soc. Agron., Crop Sci. Soc. Amer., Soil Sci. Soc. Amer.).

Jenkinson, D. S. 1990. "An introduction to the global nitrogen cycle," *Soil Use and Management,* **6**:56–61.

Johannson, O. 1960. "On sulfur problems in Swedish agriculture," *Kgl. Lanabr. Ann.,* **25**:57–169.

Klemmedson, J. O. 1994. "New Mexican locust and parent material: influence on forest floor and soil macronutrients," *Soil Sci Soc Am J.,* **58**:974–980.

Leffelaar, P. A. 1986. "Dynamics of partial anaerobiosis, denitrification and water in a soil aggregate," *Soil Science,* **142**:352–366.

National Research Council. 1993. *Soil and Water Quality: An Agenda for Agriculture* (Washington, D.C.: National Academy of Sciences), 516 pp.

National Academy of Sciences. 1983. *Acid Deposition: Atmospheric Processes in Eastern North America* (Washington, D.C.: National Academy Press).

Olsen, A. R., and A. L. Slavich. 1986. *Acid Precipitation in North America: 1984 Annual Data Summary* from Acid Deposition System Data Base. Environ. Protection Agency Rep EPA/600/4-86/033 (Washington, D.C.: U.S. Environmental Protection Agency).

Owens, L. B., W. M. Edwards, and R. W. Van Keuren. 1992. "Nitrate levels in shallow groundwater under pastures receiving ammonium nitrate or slow-release nitrogen fertilizer," *Journal of Environmental Quality,* **21**(4):607–613.

Patrick, W. H., Jr. 1982. "Nitrogen transformations in submerged soils," in F. J. Stevenson (ed.), *Nitrogen in Agricultural Soils,* Agronlomy Series No. 27 (Madison, Wis.: Amer. Soc. Agron., Crop Sci. Soc. Amer., Soil Sci. Soc. Amer.).

Power, J. F. 1981. "Nitrogen in the cultivated ecosystem," in *Terrestrial Nitrogen Cycles— Processes, Ecosystem Strategies and Management Impacts,* F. E. Clark and T. Rosswall (eds.), Ecological Bulletin No. 33. (Stockholm, Sweden: Swedish National Research Council), pp. 529–546.

Prasad, R., and J. F. Power. 1995. "Nitrification inhibitors for agriculture, health and the environment," *Advances in Agronomy,* **54**:233–280.

Smith, S. J., D. W. Dillow, and L. B. Young. 1982. "Disposition of fertilizer nitrates applied to sorghum–Sudan grass in the Southern Plains," *Journal of Environmental Quality,* **11**:341–344.

Stevenson, F. J. (ed.). 1982. *Nitrogen in Agricultural Soils,* Agronomy Series No. 22 (Madison, Wis.: Amer. Soc. Agron., Crop Sci. Soc. Amer., Soil Sci Soc. Amer.).

Tabatabai, S. J. 1986. *Sulfur in Agriculture.* Agronomy Series No. 27 (Madison, Wis.: Amer. Soc. Agron., Crop Sci. Soc. Amer., Soil Sci. Soc. Amer.).

Torrey, J. G. 1978. "Nitrogen fixation by actinomycete-induced angiosperms," *Bio-Science,* **28**:586–592.

U.S./Canada Work Group #2. 1982. *Atmospheric Sciences and Analysis,* Final Report, J. L. Ferguson and L. Machta, Co-chairmen (Washington, D.C.: U.S. Environmental Protection Agency).

Walker, T. W., et al. 1956. "Fate of labeled nitrate and ammonium nitrogen when applied to grass and clover grown separately and together," *Soil Sci.,* **81**:339–52.

Whitehead, D. C. 1964. "Soil and plant nutrition aspects of the sulfur cycle," *Soils and Fertilizers,* **27**:1–8.

14

SOIL PHOSPHORUS AND POTASSIUM

. . . residual effects of phosphorus fertilization are large and of long duration.
—*R. E. McCULLUM*

Phosphorus is at a premium in natural and agricultural ecosystems throughout the world. This macronutrient element is a key component of cellular compounds, and is vital to both plant and animal life. In world agriculture, management of phosphorus is second only to management of nitrogen in its importance for the production of healthy plants and profitable yields. The natural supply of phosphorus in most soils is small, and the availability of that which is present is very low. Inputs of phosphorus from the atmosphere and rainfall are negligible. Fortunately, most undisturbed natural ecosystems lose little of this nutrient because phosphorus does not form gases that can escape into the atmosphere, nor does it readily leach out of the soil with drainage water. Furthermore, natural ecosystems have evolved in ways that promote certain biological and chemical processes which allow plants to make relatively efficient use of the scant supplies of this element available to them.

Phosphorus is closely associated with animal (and human) activity. Bones and teeth contain large amounts of this element. Archaeologists study the phosphorus content of soil horizons because they know that unusually high concentrations of this element often accumulate where humans have congregated and have discarded the bones of wild or domesticated animals. Phosphorus is so scarce in most soils that high concentrations are often an indication of past animal or human activity in the area.

Phosphorus is also the culprit in two widespread and serious environmental problems. In industrialized countries liberal application of phosphorus to soils over many decades has significantly increased the level of this element in the surface layers of many soils. Where management practices are not sufficiently stringent, phosphorus lost from watersheds in surface runoff water and eroded sediments is severely upsetting the nutrient balance in streams, lakes, and estuaries. The added inputs of phosphorus are largely responsible for *cultural eutrophication*[1], which may jeopardize drinking water supplies and can severely restrict the use of these aquatic systems for fisheries, recreation, industry, and aesthetics. Phosphorus control is therefore a high priority of most national and regional water quality programs.

At the other extreme, lack of adequate available phosphorus is contributing to land degradation and subsequent water pollution in vast land areas, mostly in the lesser-

---

[1]The word *eutrophication* comes from the Greek word *eutrophos,* meaning well nourished or nourishing. Natural accumulation of nutrients over centuries causes lakes to fill in with plants and lose dissolved oxygen needed by fish—the process of eutrophication. Excessive inputs of nutrients under human influence tremendously speeds this process and is called *cultural eutrophication.*

developed countries of tropical and subtropical regions. Phosphorus deficiency often limits the growth of crops, and may even cause crop failure, which forces farmers to clear more land in order to survive. Without adequate phosphorus, regrowth of natural vegetation on disturbed forest and savanna sites is often too slow to prevent soil erosion and depletion of soil organic matter. Unless sources of available phosphorus can be added, growth of vegetation will be poor, and this may lead to even lower levels of soil productivity and a downward spiral of land degradation and water pollution.

While it does not play as direct a role in environmental quality as does phosphorus, low availability of soil potassium also commonly limits plant growth and reduces crop quality. Potassium works hand in hand with nitrogen and phosphorus to ensure healthy plants. Even though most soils have large total supplies of this nutrient, most of that present is tied up in the form of insoluble minerals and is unavailable for plant use. Also, plants require potassium in such large amounts that careful management practices are necessary in order to make this nutrient available rapidly enough to optimize plant growth.

Both phosphorus and potassium are used in large amounts as fertilizers, thereby earning them the title, along with nitrogen, of "the fertilizer elements." Ask most growers which nutrients they apply in the fertilizers they use and the answer will most likely be N-P-K (nitrogen, phosphorus, and potassium). Add to this soil fertility role the important environmental impacts of phosphorus, and it is clear that a good understanding of the soil processes that govern how these two macronutrients are cycled in nature is essential to anyone working with soils.

14.1 ROLE OF PHOSPHORUS IN PLANT NUTRITION AND SOIL FERTILITY[2]

Neither plants nor animals can grow without phosphorus. It is an essential component of the organic compound often called the *energy currency* of the living cell: **adenosine triphosphate** (ATP). Synthesized through both respiration and photosynthesis, ATP contains a high-energy phosphate group that drives most energy-requiring biochemical processes. For example, the uptake of nutrients and their transport within the plant, as well as their assimilation into different biomolecules, are energy-using plant processes that require ATP.

Phosphorus is an essential component of deoxyribonucleic acid (DNA), the seat of genetic inheritance, and of ribonucleic acid (RNA), which directs protein synthesis in both plants and animals. Phospholipids, which play critical roles in cellular membranes, are another class of universally important phosphorus-containing compounds. For most plant species the total phosphorus content of healthy leaf tissue is usually between 0.2 and 0.4% of the dry matter.

Phosphorus and Plant Growth

ASPECTS OF PLANT GROWTH ENHANCED. Adequate phosphorus nutrition enhances many aspects of plant physiology, including the fundamental processes of photosynthesis, nitrogen fixation, flowering, fruiting (including seed production), and maturation. Root growth, particularly development of lateral roots and fibrous rootlets, is encouraged by phosphorus. In cereal crops, good phosphorus nutrition strengthens structural tissues such as those found in straw or stalks (Table 14.1), thus helping to prevent lodging (falling over). Improvement of crop quality, especially in forages and vegetables, is another benefit attributed to this nutrient.

SYMPTOMS OF PHOSPHORUS DEFICIENCY IN PLANTS. Phosphorus deficiency is generally not as easy to recognize in plants as are deficiencies in many other nutrients. A phosphorus-deficient plant is usually stunted, thin-stemmed, and spindly, but its foliage is often dark, almost bluish, green. Thus, unless much larger, healthy plants are present to make a comparison, phosphorus-deficient plants often seem quite normal in appearance. In severe cases, phosphorus deficiency can cause yellowing and senescence of leaves. Some plants develop purple colors (Plate 16) in their leaves and stems as a result of phosphorus deficiency, though other related stresses such as cold temperatures can also cause

---

[2]For a review of the significance of this element, see Khasawneh et al. (1980).

TABLE 14.1 Influence of Phosphorus Nutrition on Cornstalk Rot

Breeding line of corn	P in ear leaf (mg P/kg dry matter)	Erect plants (% of all plants)	Crushing strength of stalk (kg/cm²)
High P accumulator	0.75	92	288
Low P accumulator	0.24	70	105

Data from Porter (1981).

purple pigmentation. Phosphorus is needed in especially large amounts in meristematic tissues, where cells are rapidly dividing and enlarging.

Phosphorus is very mobile within the plant, so when the supply is short, phosphorus in the older leaves is mobilized and transferred to the newer, rapidly growing leaves. Both the purpling and premature senescence associated with phosphorus deficiency are therefore most prominent on the older leaves. Phosphorus-deficient plants also are characterized by delayed maturity, sparse flowering, and poor seed quality.

The Phosphorus Problem in Soil Fertility

The phosphorus problem in soil fertility is threefold. First, the total phosphorus level of soils is low, usually no more than one-tenth to one-fourth that of nitrogen, and one-twentieth that of potassium. The phosphorus content of soils ranges from 200 to 2000 kg phosphorus per hectare–furrow slice (HFS), with an average of about 1000 kg P per HFS. Second, the phosphorus compounds commonly found in soils are mostly unavailable for plant uptake, often because they are highly insoluble. Third, when soluble sources of phosphorus, such as those in fertilizers and manures are added to soils, they are fixed[3] (changed to unavailable forms) and, in time, form highly insoluble compounds. We will examine these fixation reactions in some detail (Section 14.8) because they play an important role in determining how much and in what manner phosphorus should be added to soils.

Since fixation reactions in soils may allow only a small fraction (10 to 15%) of the phosphorus in fertilizer to be taken up by plants in the year of application, farmers who can afford to do so apply two to four times as much phosphorus as they expect to remove in the crop harvest. Repeated over many years, such practices have saturated the phosphorus-fixation capacity and built up the level of available phosphorus in many agricultural soils. Soils that have built up high levels of soil phosphorus no longer need to be fertilized with more than the amount of phosphorus removed in harvest. In fact, many agricultural soils in industrialized countries, especially those with long histories of phosphorus buildup from manure or fertilizer application, have accumulated so much available phosphorus that little or no additional phosphorus is needed until phosphorus is drawn down to more moderate levels over a period of years. The statistics on fertilizer use in the United States reflect the fact that farmers have recently begun to recognize that fertilizer applications can be reduced where soil phosphorus levels have been built up (Figure 14.1). The long-term buildup of phosphorus has improved soil fertility, but has also resulted in certain undesirable environmental consequences, discussed below.

14.2 EFFECTS OF PHOSPHORUS ON ENVIRONMENTAL QUALITY

Unlike certain nitrogen-containing compounds that are produced during the cycling of nitrogen (e.g., ammonia, nitrates, and nitrosoamines; see Chapter 13), phosphorus added to aquatic systems from soil is *not* toxic to fish, livestock, or humans. However, too much or too little phosphorus *can* have severe and widespread negative impacts on environmental quality. The principal environmental problems related to soil phosphorus are **land degradation** caused by too little available phosphorus and **accelerated eutrophication** caused by too much. Both problems are related to the role of phosphorus as a plant nutrient.

---

[3]Note that the term *fixation* as applied to phosphorus has the same general meaning as the chemical fixation of potassium or ammonium ions, that is, the chemical being fixed is bound, entrapped, or otherwise held tightly by soil solids in a form that is relatively unavailable to plants. In contrast, the fixation of gaseous nitrogen refers to the biological conversion of N_2 gas to combined forms that plants can use.

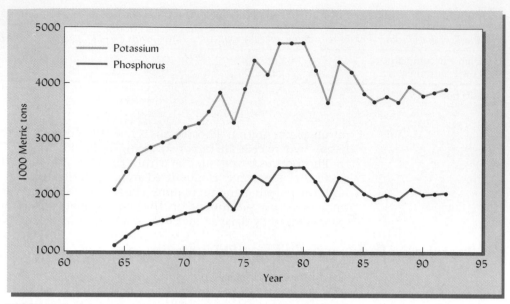

FIGURE 14.1 Estimated annual addition of fertilizer phosphorus and potassium to cropland in the United States since 1964. Note that annual applications of phosphorus increased steadily until about 1980, then declined somewhat to level off at about 2 million metric tons per year. The decline and leveling off of phosphorus application rates reflects the reduced need for phosphorus buildup in agricultural soils in which the phosphorus fixation capacity has been largely saturated. The trend toward lower rates of phosphorus application since 1980 was clearly driven by economic and agronomic influences, not environmental concerns, since the trends for potassium, a nonpolluting nutrient, are almost identical. As an example of economic effects, the oil embargo of 1973–1974 increased fertilizer prices dramatically, resulting in the notable dip in application rates in 1974. The sharp dip in 1983 reflected U.S. government policy of "payment in kind" to induce farmers to restrict their corn plantings. In other parts of the world, sustainable crop production will probably require increasing application of phosphorus and potassium for many years to come. (Based on data from the USDA Economic Research Service and the United Nations Food and Agriculture Organization)

Land Degradation

Many highly weathered soils in the warm, humid and subhumid regions of the world have very little capacity to supply phosphorus for plant growth. The low phosphorus availability is partly a result of extensive losses of phosphorus during long periods of relatively intense weathering and partly due to the low availability of phosphorus in the aluminum and iron combinations that are the dominant forms of phosphorus in these soils.

Undisturbed natural ecosystems in these regions usually contain enough phosphorus in the biomass and soil organic matter to maintain a substantial standing crop of trees or grasses. Most of the phosphorus taken up by the plants is that released from the decomposing residues of other plants. Very little is lost as long as the system remains undisturbed.

Once cleared for agricultural use (by timber harvest or by forest fires), the losses of phosphorus in eroded soil particles, in runoff water, and in biomass removals (harvests) can be substantial. Within just a few years the system may lose most of the phosphorus that had cycled between the plants and the soils. The remaining inorganic phosphorus in the soil is largely unavailable for plant uptake. In this manner, the phosphorus-supplying capacity of the disturbed soil rapidly becomes so low that regrowth of natural vegetation is sparse and, in land cleared for agricultural use, crops soon fail to produce useful yields (Figure 14.2).

Leguminous plants that might be expected to replenish soil nitrogen supplies are particularly hard hit by phosphorus deficiency because low phosphorus supply inhibits effective nodulation and retards the biological nitrogen-fixation process. The spindly plants, deficient in both phosphorus and nitrogen, can provide little vegetative cover to prevent heavy rains from washing away the surface soil. The resulting erosion will further reduce soil fertility and water-holding capacity. The increasingly impoverished soils can support less and less vegetative cover, and so the degradation accelerates.

Meanwhile, the soil particles lost by erosion become sediment farther down in the watershed, filling in reservoirs and increasing the turbidity of the rivers.

There are probably 1 to 2 billion hectares of land in the world where phosphorus deficiency limits growth of both crops and native vegetation. Much of this land lies in poor countries whose farmers have little money for fertilizers. Without properly managed inputs of phosphorus, there can be little hope of restoring these lands to productivity and their inhabitants to prosperity. Halting and reversing this type of land degradation will require managing the phosphorus cycle to make efficient use of scarce phosphorus resources.

FIGURE 14.2 On many highly weathered soils in the tropics and subtropics, low availability of phosphorus, as well as of other nutrients, may limit the regrowth of natural vegetation or the production of crops after natural vegetation is cleared. (a) Regrowth of natural vegetation is very sparse in the demarcated unfertilized plot in this disturbed area of Caribbean rain forest in Puerto Rico. (b) and (c) Phosphorus has been so depleted in this African woman's cornfield that there will be little or no grain to harvest from her severely phosphorus-deficient crop. In both cases, poor plant growth leaves the soil susceptible to accelerated erosion and further degradation. (Photos courtesy of R. Weil)

Accelerated or cultural eutrophication (see Box 14.1) is caused by phosphorus entering streams from both **point sources** and **nonpoint sources**. Point sources, such as sewage treatment plant outflows, industries, and the like, are relatively easy to identify, regulate, and clean up. During the past several decades many industrialized countries have made much progress in reducing phosphorus loading from point sources. Nonpoint sources, in contrast, are difficult to identify and control. Nonpoint sources of phosphorus are principally runoff water and eroded sediments from soils scattered throughout the affected watershed. These diffuse sources of phosphorus are now the main cause of eutrophication in many regions.

Phosphorus losses from a watershed can be increased by a variety of human activities, including timber harvest, intensive livestock grazing, soil tillage, and soil application of animal manures and phosphorus-containing fertilizers (Table 14.2). In many

BOX 14.1 EUTROPHICATION

Abundant growth of plants in terrestrial systems is usually considered beneficial; however, too much plant growth in aquatic systems can cause numerous water quality problems. The growth of algae (floating single-celled plants) and rooted aquatic vegetation depends on the same factors as those that affect any plants—namely, sunlight, carbon dioxide, nitrogen, phosphorus, and other mineral nutrients. Usually, one of these factors is in short supply and limits growth, keeping aquatic plants from completely filling up the lake or estuary. Most often the limiting factor is the supply of phosphorus or nitrogen dissolved in the water.

Oligotrophic lakes (those with low levels of nitrogen and phosphorus) generally have clear water and highly diverse communities of organisms. Some lakes are naturally eutrophic and have higher levels of nitrogen and phosphorus, are more turbid, and have less-diverse plant and animal communities. Eutrophic lakes eventually become so choked with plant growth that they slowly fill in to become bogs and form Histosols. Eutrophication is a natural process which usually proceeds slowly over a period of many centuries. Humans can cause **accelerated** or **cultural eutrophication** by adding nutrients to a naturally low-nutrient watershed.

Brackish and salty waters in estuaries are most often nitrogen-limited. Nitrogen is limiting in the saltier parts of estuaries partly because of the species of algae present and partly because denitrification in the water, sediments, and wetland soils keeps the native nitrogen level low. Therefore, pollution that adds nitrogen to such waters will likely trigger accelerated eutrophication.

Freshwater lakes and streams, on the other hand, are usually phosphorus-limited, partly because phosphorus levels in the water are typically very low and partly because certain cyanobacteria (blue-green algae) can prevent nitrogen limitation by supplying ample nitrogen through biological fixation of atmospheric nitrogen (see Section 13.10). Exceptions to the phosphorus limitation in freshwater do occur, especially in certain Rocky Mountain lakes, which are more limited by nitrogen. Lakes most likely to be sensitive to added phosphorus are those that contain a separate layer of cold water under the warm surface water during the summer months. Lakes in which the inflow of new water flushes out the old only slowly are also more susceptible to phosphorus pollution. Critical levels of phosphorus in water, above which eutrophication is likely to be triggered, are approximately 0.03 mg/L of dissolved phosphorus and 0.1 mg/L of total phosphorus.

When phosphorus is added to a phosphorus-limited lake, it stimulates a burst of algal growth (referred to as an *algal bloom*) and often a shift in the dominant algal species. The phosphorus-stimulated algae may cover the surface of the water with mats of algal scum. The lake may also become choked with higher plants that are also stimulated by the added phosphorus. When these aquatic weeds and algal mats die, they sink to the bottom, where their decomposition by microorganisms uses up much of the oxygen dissolved in the water. The decrease in oxygen (anoxic conditions) severely limits the growth of many aquatic organisms, especially the more desirable fish. Such eutrophic lakes often become turbid, limiting the growth of submerged aquatic vegetation and benthic (bottom-feeding) organisms that serve as food for much of the fish community. In extreme cases eutrophication can lead to massive fish kills (Figure 14.3).

(continued)

BOX 14.1 (Continued)   EUTROPHICATION

Excessive input of phosphorus can transform clear, oxygen-rich, good-tasting water that is home to a variety of sport fish into cloudy, oxygen-poor, foul smelling, bad-tasting, and possibly toxic water. Eutrophic conditions favor the growth of cyanobacteria (blue-green algae), most of which are not desirable as food for zooplankton (small, floating animals), thus reducing this major food source for fish. Certain cyanobacteria produce highly toxic substances and bad-tasting and -smelling compounds that make water unsuitable for drinking by humans or livestock. Some of the stimulated algae form long filaments that clog water treatment intake filters, further reducing the water's usability for drinking or industrial water supply. Dense growth of both algae and aquatic weeds may make the water useless for boating and swimming. Furthermore, eutrophic waters generally have a much lower level of biological diversity (fewer species) and fewer desirable species of fish. This may have severe consequences for many aquatic species, including those fished as recreational catch (i.e., trout).

FIGURE 14.3   *In extreme cases of eutrophication, massive fish kills can occur in sensitive lakes and rivers. The kills result from anoxic conditions, which are brought on by the decay of the masses of algae stimulated by elevated inputs of phosphorus (or sometimes of nitrogen). (Photo courtesy of R. Weil)*

areas application of phosphorus to soils has, over the years, dramatically increased the phosphorus content of surface soil, and thus the impact that runoff and sediment have on stream phosphorus-loading.

There are several reasons why surface soil phosphorus content has increased so much. First, farmers have long known that when soluble phosphorus fertilizer is applied to a low-phosphorus soil, most of the phosphorus rapidly becomes bound in forms plants cannot use (Sections 14.2 and 14.6). To compensate for this inefficiency, farmers in industrialized countries have traditionally applied more phosphorus to their soils than is removed in the harvest. Second, where animal manures or sewage sludges are used, the amount applied is most often calculated to meet the nitrogen needs of the crop. The phosphorus contained in that same amount of manure is likely to exceed by two to four times the amount of phosphorus taken up by the plants. Consequently, the phosphorus content of many intensively managed agricultural soils has been increased over the years; subsequently, the phosphorus content of the runoff water and sediment coming from these soils has also increased.

TABLE 14.2 Influence of Wheat Production and Tillage on Annual Losses of Phosphorus in Runoff Water and Eroded Sediments Coming from Soils in the Southern Great Plains

The total phosphorus lost includes the phosphorus dissolved in the runoff water and the phosphorus adsorbed to the eroded particles.[a] Although cattle grazing on the natural grasslands probably increased losses of phosphorus from these watersheds, the losses from the agricultural watersheds were about 10 times as great. The no-till wheat fields tended to lose much less particulate phosphorus, but more soluble phosphorus, than the conventionally tilled wheat fields.

Location and soil	Management	Soluble P	Particulate P	Total P
		kg P ha^{-1} yr^{-1}		
El Reno, OK. Paleustolls, 3% slope	Wheat with conventional plow and disk	0.21	3.51	3.72
	Wheat with no-till	1.04	0.43	1.42
	Native grass, heavily grazed	0.14	0.10	0.24
Woodward, OK. Ustochrepts, 8% slope	Wheat with conventional sweep plow and disk	0.23	5.44	5.67
	Wheat with no-till	0.49	0.70	1.19
	Native grass, moderately grazed	0.02	0.07	0.09

[a]Wheat was fertilized with up to 23 kg of P each fall. Data from Smith et al. (1991).

Studies in many regions have shown that soils devoted to crop production lose far more phosphorus to streams than do those covered by relatively undisturbed forests or natural grasslands (Figure 14.4). Streams draining watersheds with predominantly agricultural land use tend to carry much higher phosphorus loads because of the just-discussed phosphorus enrichment of the surface soils, and because many agricultural operations increase the level of surface runoff and erosion.

PHOSPHORUS LOSSES IN RUNOFF. Agricultural management that involves disturbing the soil surface with tillage generally increases the amount of phosphorus carried away on eroded sediment (i.e., **particulate P**). On the other hand, fertilizer or manure that is left unincorporated on the surface of cropland or pastures usually leads to increased losses of phosphorus dissolved in the runoff water (i.e., **dissolved P**). These trends can be seen in Table 14.2 by comparing phosphorus losses from no-till and conventionally tilled wheat.

The decision to use tillage to incorporate phosphorus-bearing soil amendments involves a trade-off between the advantages of incorporating phosphorus into the soil (less phosphorus dissolves in the runoff water and more is available for plant uptake) and the disadvantages associated with disturbing the surface soil (increased loss of soil particles by erosion) (Table 14.2). In no-tillage systems surface application of manure without incorporation may result in lower total loss of phosphorus, because this type of management achieves substantial reductions in soil erosion and total runoff. The effect of these reductions may outweigh the effect of the increased phosphorus concentration in the relatively small volume of water that does runoff. If the equipment is available to do so, the best option might be to inject the high-phosphorus amendment into the soil with a minimum of disturbance to the soil surface.

Disturbances to natural vegetation, such as timber harvest or wildfires, also increase the loss of phosphorus, primarily via eroded sediment. Erosion tends to transport predominantly the clay and organic matter fraction of the soil, which are relatively rich in phosphorus, leaving behind the coarser, lower phosphorus fractions. Thus, compared to the original soil, eroded sediment is often enriched in phosphorus by a ratio of 2 or more (Table 14.3).

The above examples make it clear that the control of eutrophication in many rivers and lakes will require improved management of the phosphorus cycle in soils. Specifically, it is critical to reduce the transport of phosphorus from soils to water.

14.3 THE PHOSPHORUS CYCLE[4]

In order to manage phosphorus for economic plant production and for environmental protection we will have to understand the nature of the different forms of phosphorus

[4]For a review of soil phosphorus, see Olsen and Khasawneh (1980) and Stevenson (1986).

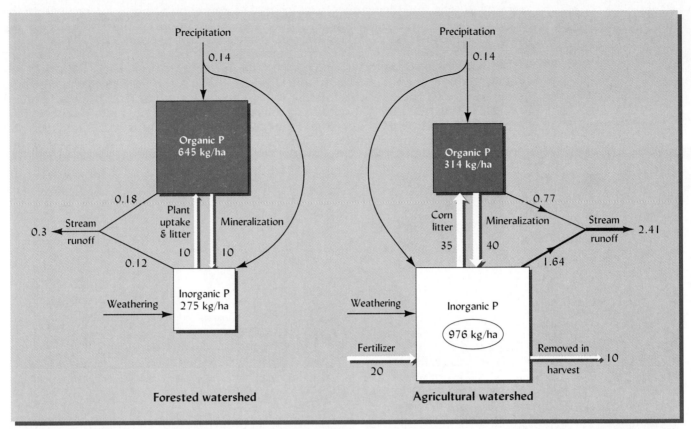

FIGURE 14.4 Phosphorus balance in surface soils (Ultisols) of adjacent forested and agricultural watersheds. The forest consisted primarily of mature hardwoods that had remained relatively undisturbed for 45 or more years. The agricultural land was producing row crops for more than 100 years. It appears that in the agricultural soil about half of the organic phosphorus has been converted into inorganic forms or lost from the system since cultivation began. At the same time, substantial amounts of inorganic phosphorus accumulated from fertilizer inputs. Compared to the forested soil, mineralization of organic phosphorus was about four times as great in the agricultural soil, and the amount of phosphorus lost to the stream was eight times as great. Flows of phosphorus, represented by arrows, are given as kg/ha/year. Although not shown in the diagram, it is interesting to note that nearly all (95%) of the phosphorus lost from the agricultural soil was in particulate form, while losses from the forest soil were 33% dissolved and 77% particulate. (Data from Vaithiyanathan and Correll, 1992)

TABLE 14.3   Effect of Forest Fire on Phosphorus Loss During the Following Year from a Scrub Forest in Northwestern Spain

The site was steeply sloped, with shallow sandy loam soils (lithic Haplumbrepts) relatively high in available P. Note that the more severe the burn, the greater were the losses of sediment and associated P. In all cases, the enrichment ratios were slightly greater than 2, indicating that the sediments eroded from the plots were more than twice as concentrated in phosphorus as were the surface soils themselves. In this, as in most cases, the soil fractions rich in phosphorus were the most susceptible to erosion. The levels of phosphorus lost from this particular forest are higher than levels typically lost from humid region forests on soils low in available P.

	Sediment lost (Mg/ha)	Total P lost (kg/ha)	Enrichment ratio[b]
Unburned control	2	1.4	2.6
Moderately burned by prescribed fire	5	4.3	2.1
Severely burned by wildfire	13+[a]	9.1+	2.2

Data from Saa et al. (1994).
[a]Values for the severely burned plots were measured for only 10 months, while the other plots were measured for 12 months.
[b]Enrichment ratio = (mg P/kg sediment)/(mg P/kg soil).

found in soils, and the manner in which these forms of phosphorus interact within the soil and in the larger environment. The cycling of phosphorus within the soil, from the soil to higher plants and back to the soil, is illustrated in Figure 14.5.

PHOSPHORUS IN SOIL SOLUTION.   Compared to other macronutrients such as sulfur and calcium, the concentration of phosphorus in the soil solution is very low, generally ranging from 0.001 mg/L in very infertile soils to about 1 mg/L in rich, heavily fertilized soils. Plant roots absorb phosphorus dissolved in the soil solution, mainly as phosphate ions (HPO_4^{-2} and $H_2PO_4^-$), but some soluble organic phosphorus compounds are also taken up. The chemical species of phosphorus present in the soil solution is determined by the solution pH as shown in Figure 14.6. In strongly acid soils (pH 4 to 5.5), the monovalent anion, $H_2PO_4^-$, dominates, while alkaline solutions are characterized by the

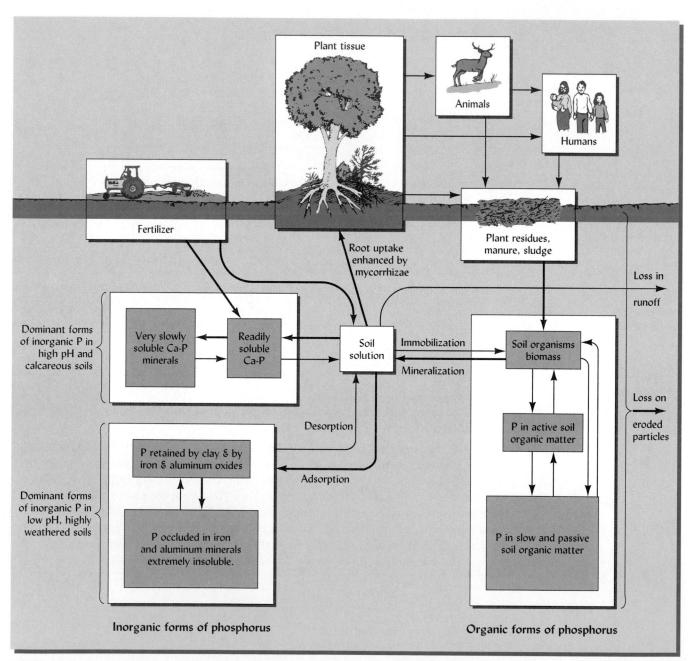

FIGURE 14.5   The phosphorus cycle in soils. The boxes represent pools of the various forms of phosphorus in the cycle, while the arrows represent translocations and transformations among these pools. The three largest white boxes indicate the principal groups of phosphorus-containing compounds found in soils. Within each of these groups, the less soluble, less available forms tend to dominate.

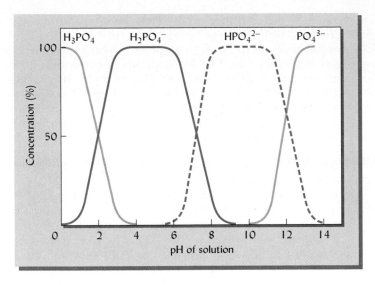

FIGURE 14.6 The effect of pH on the relative concentrations of the three species of phosphate ions. At lower pH values, more H^+ ions are available in the solution, and thus the phosphate ion species containing more hydrogen predominates. In near-neutral soils HPO_4^{-2} and $H_2PO_4^-$ are found in nearly equal amounts. Both of these species are readily available for plant uptake.

divalent anion HPO_4^{-2}. Both anions are important in near-neutral soils. Of the two anions, $H_2PO_4^-$ is thought to be slightly more available to plants, but effects of pH on phosphorus reactions with other soil constituents are more important than the particular phosphorus anion present.

UPTAKE BY ROOTS AND MYCORRHIZAE. Uptake by the plant root requires, not only that the phosphate ions be dissolved in the soil solution, but also that they move from the bulk soil to the surface of the root (Figure 14.7). This movement, usually over a distance of a few millimeters to a centimeter, takes place primarily by physical diffusion (see Section 1.16). However, because phosphate ions are very strongly adsorbed by soil particles, diffusion to the root may be so slow and intermittent as to limit the availability of phos-

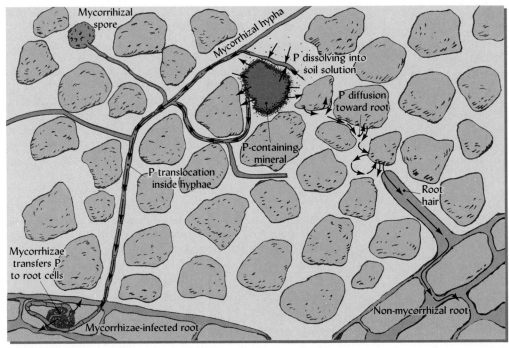

FIGURE 14.7 Roles of diffusion and mycorrhizal hyphae in the movement of phosphate ions to plant roots. In soils with low solution phosphorus concentration and high phosphorus fixation, slow diffusion may seriously limit the ability of roots to obtain sufficient phosphorus. The hyphae of symbiotic mycorrhizal fungi are particularly beneficial to the plant where phosphorus diffusion is slow, because phosphorus is transported inside the hyphae by cytoplasmic streaming, making the plant much less dependent on the diffusion of phosphate ions through the soil.

phorus to the plant. Root hairs growing into the soil can help to shorten the distance through which phosphate ions must diffuse before being taken up.

Note that Figure 14.7 shows, in addition to direct root uptake of phosphorus from the soil solution, a second pathway that can be used by many plants to obtain phosphorus from the soil. This second pathway involves symbiosis with mycorrhizal fungi (see Section 11.9). In soils low in available phosphorus many species of plants could barely survive without mycorrhizal assistance in obtaining phosphorus. The microscopic threadlike mycorrhizal hyphae extend out into the soil several centimeters from the root surface. The hyphae are able to absorb phosphorus ions as the ions enter the soil solution, and may even be able to access some strongly bound forms of phosphorus. The hyphae then bring the phosphate to the root by transporting it inside the hyphal cells where soil-retention mechanisms cannot interfere with phosphate movement (Figure 14.7).

Once in the plant, a portion of the phosphorus is translocated to the plant shoots where it becomes part of the plant tissues. As the plants shed leaves and roots die, or are eaten by people or animals, phosphorus returns to the soil in the form of plant residues, leaf litter, and wastes from animals and people. Microorganisms decompose the residues and temporarily tie up at least part of the phosphorus in their cells. Some of it becomes associated with the active and passive fractions of the soil organic matter (see Section 12.14) where it is subject to storage and future release. These organic forms are very slowly converted to the soluble forms that plant roots can absorb, thereby repeating the cycle.

CHEMICAL FORMS IN SOILS. In most soils, the amount of phosphorus available to plants from the soil solution at any one time is very low, seldom exceeding about 0.01% of the total phosphorus in the soil. The bulk of the soil phosphorus exists in three general groups of compounds—namely, *organic phosphorus, calcium-bound inorganic phosphorus* and *iron or aluminum-bound inorganic phosphorus* (Figure 14.5). Of the inorganic phosphorus, the calcium compounds predominate in most alkaline soils, while the iron and aluminum forms are most important in acidic soils. All three groups of compounds slowly contribute phosphorus to the soil solution, but most of the phosphorus in each group is of very low solubility and not available for plant uptake.

Unlike nitrogen and sulfur, phosphorus is not lost from the soil in gaseous form. Because soluble inorganic forms of phosphorus are strongly adsorbed by mineral surfaces (see Section 14.6), there is also no appreciable phosphorus lost by leaching (except in some organic soils or soils to which very high amounts of animal manures have been added).

GAINS AND LOSSES. The principal pathways by which phosphorus is lost from the soil system are plant removal (<1 to 30 kg/ha annually in harvested biomass), erosion of phosphorus-carrying soil particles (0.1 to 10 kg/ha annually on organic and mineral particles), and phosphorus dissolved in surface runoff water (0.01 to 3.0 kg/ha annually). For each pathway, the higher figures cited for annual phosphorus loss would most likely apply to cultivated soils.

The amount of phosphorus that enters the soil from the atmosphere (sorbed on dust particles) is quite small (0.05 to 0.5 kg/ha annually), but may nearly balance the losses from the soil in undisturbed forest and grassland ecosystems. As already discussed, in an agroecosystem the input from phosphorus fertilizer is likely to exceed the output in harvested crops. Figure 14.4 provides examples of phosphorus input-output balance in two adjacent watersheds, one devoted primarily to conventionally tilled row crop production and the other covered by mature deciduous forest.

14.4 ORGANIC PHOSPHORUS IN SOILS

Both inorganic and organic forms of phosphorus occur in soils, and both are important to plants as sources of this element. The relative amounts in the two forms vary greatly from soil to soil, but the data in Table 14.4 give some idea of their relative proportions in a range of mineral soils. The organic fraction generally constitutes 20 to 80% of the total phosphorus in surface soil horizons (Figure 14.8). The deeper horizons may hold large amounts of inorganic phosphorus, especially in soils from arid and semiarid regions.

TABLE 14.4 Total Phosphorus Content of Surface Soils from Different Locations and the Percentage of Total Phosphorus in the Organic Form

Soils	Number of samples	Total P (mg/kg)	Organic fraction (%)
Western Oregon			
Upland soils	4	357	66
Old valley-filling soils	4	1479	30
Recent valley soils	3	848	26
New York			
Histosols (cultivated)	8	1491	52
Iowa			
Mollisols and Alfisols	6	561	44
Arizona	19	703	36
Australia	3	422	75
Texas			
Ustolls	2	369	34
Hawaii			
Andisol	1	4700	37
Oxisol	1	1414	19
Zimbabwe			
Alfisols	22	899	56
Maryland			
Ultisols (silt loams)	6	650	59
Ultisols (forested sandy loams)	3	472	70
Ultisols (sandy loam, cropland)	4	647	25

Data for Oregon, Iowa, and Arizona from sources quoted by Brady (1974); Australia from Fares et al. (1974); New York from Cogger and Duxbury, 1984; Hawaii from Soltanpour et al. (1988); Zimbabwe courtesy R. Weil and F. Folle; Texas from Raven and Hossner (1993); Maryland (silt loams) from Weil et al. (1988); Maryland (sandy loams) from Vaithiyanathan and Correll (1992).

Organic Phosphorus Compounds

Until recently, scientists have focused more attention on the inorganic than on the organic phosphorus in soils, and our knowledge of the specific nature of most of the organic-bound phosphorus in soils is quite limited. However, three broad groups of organic phosphorus compounds are known to exist in soils. Although all three types of phosphorus compounds are also found in plants, most of the organic phosphorus compounds in soils are believed to have been synthesized by microorganisms. The three groups are (1) Inositol phosphates or phosphate esters of a sugarlike compound, inositol ($C_6H_6(OH)_6$); (2) nucleic acids, and (3) phospholipids. While other organic phosphorus compounds are present in soils, the identity and amounts present are less well understood.

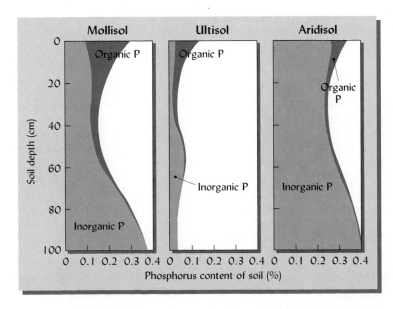

FIGURE 14.8 Phosphorus contents of representative soil profiles from three soil orders. All three soils contain a high proportion of organic phosphorus in their surface horizons. The Aridisol has a high inorganic phosphorus content throughout the profile because rainfall during soil formation was insufficient to leach much of the inorganic phosphorus compounds from the soil. The increased phosphorus in the subsoil of the Ultisol is due to adsorption of inorganic phosphorus by iron and aluminum oxides in the B horizon. In both the Mollisol and Aridisol, most of the subsoil phosphorus is in the form of inorganic calcium-phosphate compounds.

Inositol phosphates are the most abundant of the known organic phosphorus compounds, making up 10 to 50% of the total organic phosphorus. They tend to be quite stable in acid and alkaline conditions, and interact with the higher molecular weight humic compounds. These properties may account for their relative abundance in soils.

Nucleic acids are adsorbed by humic compounds as well as by silicate clays. Adsorption on these soil colloids probably helps protect the phosphorus in nucleic acids from microbial attack. Still, the nucleic acids and phospholipids together probably make up only 1 to 2% of the organic phosphorus in most soils.

The other chemical compounds that contain most of the soil organic phosphorus have not yet been identified, but much of the organic phosphorus appears to be associated with the fulvic acid fraction of the soil organic matter. Our ignorance of the specific compounds involved does not detract from the importance of these compounds as suppliers of phosphorus through microbial breakdown.[5]

Mineralization of Organic P

Phosphorus held in organic form can be mineralized and immobilized by the same general processes that release nitrogen and sulfur from soil organic matter (see Chapter 13):

$$\text{Organic P forms} \underset{\text{Microbes}}{\overset{\text{Microbes}}{\rightleftharpoons}} \underset{\text{Soluble phosphate}}{H_2PO_4^-} \underset{}{\overset{Fe^{3+},\ Al^{3+},\ Ca^{2+}}{\rightleftharpoons}} \underset{\text{Insoluble fixed P}}{\text{Fe, Al, Ca phosphates}}$$

(Immobilization arrow above pointing left; Mineralization arrow below pointing right)

Soluble phosphorus compounds are released when organic residues and humus decompose. The resulting soluble inorganic phosphate ion ($H_2PO_4^-$) is subject to uptake by plants or to fixation into insoluble forms by reaction with iron, aluminum, manganese, and calcium in soils. Should organic residues low in phosphorus but high in carbon and other nutrients be added to a soil, microbes would increase their activity and immobilize the phosphorus in their biomass. The available $H_2PO_4^-$ in the soil solution would temporarily disappear, just as was described for soluble NH_4^+, NO_3^-, and SO_4^{2-} ions (Sections 13.4 and 13.20). Net immobilization of soluble phosphorus is most likely to occur if residues added to the soil have a C/P ratio greater than about 300/1, while net mineralization is likely if the ratio is below 200/1.

When conditions are such that organic matter is increasing in a soil (for example, after conversion of cultivated land to a grass/legume pasture), organic phosphorus tends to accumulate. If inorganic phosphorus fertilizer is added to such a soil, much of the added phosphorus is eventually stored in the organic fraction after being assimilated by plants or microorganisms. The ratio of C/P in soil organic matter varies widely, commonly from 15/1 up to 75/1. This variability is in contrast to relatively stable ratios of C/N and C/S characteristic of soil organic matter, and results from the fact that the rates of phosphorus mineralization and immobilization do not always parallel those of N and S.

Contribution of Organic Phosphorus to Plant Needs

Recent evidence indicates that the readily decomposable or easily soluble fractions of soil *organic phosphorus* are often the most important factor in supplying phosphorus to plants in *highly weathered soils* (e.g., Ultisols and Oxisols), even though the total organic matter content of these soils may not be especially high. The inorganic phosphorus in the highly weathered soils is far too insoluble to contribute much to plant nutrition. Apparently plant roots and mycorrhizal hyphae are able to obtain some of the phosphorus released from organic forms before it forms inorganic compounds that quickly become insoluble. In contrast, it appears that the more soluble *inorganic forms* of phosphorus play the biggest role in phosphorus fertility of *less weathered soils* (e.g., Mollisols, Vertisols), even though these generally contain relatively high amounts of soil organic matter.

---

[5]Although plants are able to absorb some organic phosphorus compounds directly, the level of such absorption is thought to be very low compared to that of inorganic phosphates.

Mineralization of organic phosphorus in soils is subject to many of the same influences that control the general decomposition of soil organic matter—namely, temperature, moisture, tillage, etc. (Section 12.13). In temperate regions mineralization of organic phosphorus in soils typically releases 5 to 20 kg P/ha annually, most of which is readily absorbed by growing plants. These values can be compared to the annual uptake of phosphorus by most crops, trees, and grasses, which generally ranges from 5 to 30 kg P/ha. When forested soils are first brought under cultivation in tropical climates the amount of phosphorus released by mineralization may exceed 50 kg/ha annually, but unless phosphorus is added from outside sources, these high rates of mineralization will soon decline due to the depletion of readily decomposable soil organic matter. In Florida, rapid mineralization of organic matter in Histosols (Saprists) drained for agricultural use is estimated to release about 80 kg P/ha annually. Unlike most mineral soils, these organic soils possess little capacity to retain dissolved phosphorus, so water draining from them is quite concentrated in phosphorus (0.5 to 1.5 mg P/L) and is thought to be contributing to the degradation of the Everglades wetland system.

14.5 INORGANIC PHOSPHORUS IN SOILS

Of all the macronutrients found in soils, phosphorus has by far the smallest quantities in solution or in readily soluble forms in mineral soils. Likewise, the relative immobility of inorganic phosphorus in mineral soils is well known. Two phenomena tend to control the concentration of phosphorus in the soil solution and the movement of phosphorus in soils: (1) the solubility of phosphorus-containing minerals and (2) the fixation or adsorption of phosphate ions on the surface of soil particles. In practice it is difficult to separate the influence of these two types of reactions or even determine the exact nature of inorganic phosphorus compounds present in a particular soil.

FIXATION AND RETENTION. Dissolved phosphate ions in mineral soils are subject to many types of reactions that tend to remove the ions from the soil solution and produce phosphorus-containing compounds of very low solubility. These reactions are sometimes collectively referred to by the general terms *phosphorus fixation* and *phosphorus retention*. Phosphorus retention is a somewhat more general term that includes both precipitation and fixation reactions.

The tendency for soils to fix phosphorus in relatively insoluble, unavailable forms has far-reaching consequences for phosphorus management. For example, phosphorus fixation may be viewed as troublesome if it prevents plants from using all but a small fraction of fertilizer phosphorus applied. On the other hand, phosphorus fixation can be viewed as a benefit if it causes most of the dissolved phosphorus to be removed from phosphorus-rich wastewater applied to a soil (Box 14.2). The fixation reactions responsible in both situations will be discussed as they apply to the availability of phosphorus under acidic and alkaline soil conditions. We will begin by describing the various inorganic compounds and their solubility.

Inorganic Phosphorus Compounds

As indicated by Figure 14.5, most inorganic phosphorus compounds in soils fall into one of two groups: (1) those containing calcium and (2) those containing iron and aluminum (and, less frequently, manganese).

As a group, the calcium-phosphate compounds become more soluble as soil pH decreases; hence, they tend to dissolve and disappear from acid soils. On the other hand, the calcium phosphates are quite stable and very insoluble at higher pH, and so become the dominant forms of inorganic phosphorus present in neutral to alkaline soils.

Of the common calcium compounds containing phosphorus (Table 14.5), the **apatite** minerals are the least soluble, and therefore the least available source of phosphorus. Some apatite minerals (e.g., fluorapatite) are so insoluble that they persist even in weathered (acid) soils. The simpler mono- and dicalcium phosphates, are readily available for plant uptake. Except on recently fertilized soils, however, these compounds are present in only extremely small quantities because they easily revert to the more insoluble forms.

In contrast to calcium phosphates, the iron and aluminum hydroxy phosphate minerals, **strengite** ($FePO_4 \cdot 2H_2O$) and **variscite** ($AlPO_4 \cdot 2H_2O$), have very low solubilities

BOX 14.2 PHOSPHORUS REMOVAL FROM WASTEWATER

Environmental soil scientists and engineers remove phosphorus from wastewater by taking advantage of some of the same reactions that cause soluble phosphorus to precipitate as insoluble compounds in soils. Municipal sewage water is generally quite high in dissolved phosphorus derived largely from detergents and human wastes. Primary and secondary levels of sewage treatment are designed to reduce biological oxygen demand (BOD) created by the organic compounds that would deplete river water of oxygen as they decompose, but do not reduce phosphorus.

In order to deal with the problem of phosphorus-stimulated eutrophication, installation of tertiary sewage treatment processes is increasingly common in industrialized countries (Figure 14.9). In the tertiary treatment process, phosphorus is removed from the wastewater by precipitation reactions similar to those that occur in soils. In huge, specially designed tanks, aluminum or iron compounds are commonly added to the wastewater to cause the phosphorus to precipitate by reactions such as these:

$$Al_2(SO_4)_3 \cdot 14H_2O \; + \; 2PO_4^{3-} \; \longrightarrow \; 2AlPO_4 \; + \; 3SO_4^{2-} \; + \; 14H_2O$$

Alum   Soluble phosphate   Insoluble Al-P

$$Fe_2Cl_3 \; + \; 2PO_4^{3-} \; \longrightarrow \; 2FePO_4 \; + \; 3Cl^{1-}$$

Ferric chloride   Soluble phosphate   Insoluble Fe-P

The insoluble aluminum- or iron-phosphate compounds precipitate out of solution and settle to the bottom of the tank, while the now low-phosphorus water near the top of the tank flows out for further processing before being released into the river. The precipitated phosphate compounds are added to other solids removed from the wastewater during various earlier processes. These solids, containing much of the phosphorus (as well as N and C) originally in the sewage water, are concentrated and further processed to produce sewage sludge or *biosolids*.

A much less expensive approach to tertiary treatment is to spray the partially treated wastewater onto a specially designated area of vegetated soils and let natural soil and plant processes clean the phosphorus and other constituents out of the wastewater. *Infiltration systems* use relatively permeable soils and low rates of wastewater application to allow the water to infiltrate and percolate through the soil profile. Because of the iron, aluminum, or calcium compounds naturally occurring in soils, nearly all of the dissolved phosphorus is removed from the wastewater shortly after coming in contact with soil. Crops or forest trees on the site benefit from the wastewater irrigation and help remove some of the phosphorus (and other nutrients) from the wastewater. Other systems, termed *overland flow systems* use finer-textured, less-permeable soils and higher rates of wastewater application so that the water slowly flows over the soil surface, interacting with the upper few centimeters of

(continued)

FIGURE 14.9 *Increasingly, modern sewage treatment plants are required to include tertiary treatment facilities for phosphorus removal. The chemical reactions encouraged in the sewage treatment process are similar to those that affect phosphorus availability in soils. (Photo courtesy of R. Weil)*

BOX 14.2 (Continued) PHOSPHORUS REMOVAL FROM WASTEWATER

soil and with the vegetation to remove a large percentage of the dissolved phosphorus and other contaminants. The long-term effectiveness of both systems is dependent on the phosphorus-fixation capacity (Section 14.8) of the soil used.

As the following example illustrates, knowledge of soil properties and processes is essential for effective design of advanced land-based wastewater treatment systems. A large environmental engineering firm won a contract to build a new type of municipal wastewater treatment facility that would look more like a park than a sewage plant because it used constructed wetlands and soils to clean the wastewater. After flowing through a number of artificial marshes and filtering systems (Figure 14.10a), the wastewater was sprayed onto a large field covered with a layer of artificial soil several meters thick (Figure 14.10b).

(continued)

(a)

(b)

FIGURE 14.10   *After passing through an artificial wetland (a) to remove nitrogen by denitrification (see Chapter 13), the effluent goes through the final stage (b) of an innovative wastewater treatment facility designed to remove pollutant by "natural" processes. The dark colored "soil" that can be seen under the grass is a thick layer of peat through which the wastewater will percolate after being sprayed onto the field by an irrigation system (arrow). Peat has physical properties desirable for the construction of this "designer soil," but it had to be amended with iron before it developed the phosphorus-removal ability characteristic of most mineral soils. (Photos courtesy of R. Weil)*

in strongly acid soils, and become more soluble as soil pH rises. These minerals would therefore be quite unstable in alkaline soils, but are prominent in acid soils, in which they are quite insoluble and stable.

Other, similar compounds combining phosphorus with iron, aluminum, or manganese are also found in acid soils. Some are products of surface reactions between phosphate ions and the somewhat amorphous hydroxy polymers that often exist as coatings on soil particles. Evidence suggests that phosphate ions even react with aluminum near the edges of silicate clay crystals, forming insoluble products similar to the aluminum phosphates described in the preceding paragraph.

Effect of Aging on Inorganic Phosphate Availability

In both acid and alkaline soils, phosphorus tends to undergo sequential reactions that produce phosphorus-containing compounds of lower and lower solubility. Therefore, the longer that phosphorus remains in the soils, the less soluble, and therefore less plant-available, it tends to become. Usually when soluble phosphorus is added to a soil, a rapid reaction removes the phosphorus from solution (*fixes* the phosphorus) in the first few hours. Slower reactions then continue to gradually reduce phosphorus solubility for months or years as the phosphate compounds age. The freshly fixed phosphorus may be slightly soluble and of some value to plants. With time, the solubility of the fixed phosphorus tends to decrease to extremely low levels.

The effect of aging appears to be due to such factors as regularity and size of crystals in precipitated phosphates, more permanent bonding of adsorbed phosphate into the

TABLE 14.5 **Inorganic Phosphorus-Containing Compounds Commonly Found in Soils**

In each group, the compounds are listed in order of increasing solubility.

Compound	Formula
Iron and Aluminum Compounds	
Strengite	$FePO_4 \cdot 2H_2O$
Variscite	$AlPO_4 \cdot 2H_2O$
Calcium Compounds	
Fluorapatite	$[3Ca_3(PO_4)_2] \cdot CaF_2$
Carbonate apatite	$[3Ca_3(PO_4)_2] \cdot CaCO_3$
Hydroxy apatite	$[3Ca_3(PO_4)_2] \cdot Ca(OH)_2$
Oxy apatite	$[3Ca_3(PO_4)_2] \cdot CaO$
Tricalcium phosphate	$Ca_3(PO_4)_2$
Octacalcium phosphate	$Ca_8H_2(PO_4)_6 \cdot 5H_2O$
Dicalcium phosphate	$CaHPO_4 \cdot 2H_2O$
Monocalcium phosphate	$Ca(H_2PO_4)_2 \cdot H_2O$

calcium carbonate or metal oxide particles, and the extent to which sorbed phosphate is buried as surface precipitation reactions continue (Figure 14.11). The nature of these and other reactions of phosphorus in soils are discussed below.

14.6 SOLUBILITY OF INORGANIC PHOSPHORUS IN ACID SOILS

The particular types of reactions that fix phosphorus in relatively unavailable forms differ from soil to soil and are closely related to soil pH (Figure 14.12). In acid soils these reactions predominantly involve Al, Fe, or Mn, either as dissolved ions, as oxides, or as hydrous oxides. Many soils contain such hydrous oxides as coatings on soil particles and as interlayer precipitates in silicate clays. In alkaline and calcareous soils, the reactions primarily involve precipitation as various calcium-phosphate minerals (Table 14.5) or adsorption to the iron impurities on the surfaces of carbonates and clays. At moderate pH values, adsorption on the edges of kaolinite or on the iron oxide coating on kaolinite clays plays an important role.

Precipitation by Iron, Aluminum, and Manganese Ions

Probably the easiest type of phosphorus-fixation reaction to visualize is the simple reaction of $H_2PO_4^-$ ions with dissolved Fe^{3+}, Al^{3+}, and Mn^{3+} ions to form insoluble hydroxy phosphate precipitates (Figure 14.13a). In strongly acid soils, enough soluble Al, Fe, or Mn is usually present to cause the chemical precipitation of nearly all dissolved $H_2PO_4^-$ ions by reactions such as the following (using the aluminum cation as an example):

$$Al^{3+} + H_2PO_4^- + 2H_2O \rightleftharpoons 2H^+ + Al(OH)_2H_2PO_4$$

$$\text{(soluble)} \qquad\qquad\qquad\qquad\qquad \text{(insoluble)}$$

Freshly precipitated hydroxy phosphates are slightly soluble because they have a great deal of surface area exposed to the soil solution. Therefore, the phosphorus contained in them is, initially at least, somewhat available to plants. Over time, however, as the precipitated hydroxy phosphates age, they become less soluble and the phosphorus in them becomes almost completely unavailable to most plants.

Precipitation reactions similar to those just described are responsible for the rapid reduction in availability of phosphorus added to soil as soluble $Ca(H_2PO_4)_2 \cdot H_2O$ in fertilizers (Figure 14.14). This type of reaction can also be used to control the solubility of phosphorus in wastewater (see Box 14.2).

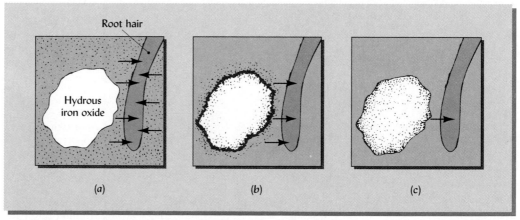

FIGURE 14.11 How relatively soluble phosphates are rendered unavailable by compounds such as hydrous oxides of Fe and Al. (a) The situation just after application of a soluble phosphate. The root hair and the hydrous iron oxide particle are surrounded by soluble phosphates. (b) Within a very short time most of the soluble phosphate has reacted with the surface of the iron oxide crystal. The phosphorus is still fairly readily available to the plant roots since most of it is located at the surface of the particle where exudates from the plant can encourage exchange. (c) In time the phosphorus penetrates the crystal and only a small portion is found near the surface. Under these conditions its availability is low.

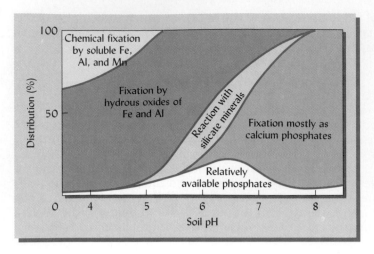

FIGURE 14.12 Inorganic fixation of added phosphates at various soil pH values. Average conditions are postulated, and it is not to be inferred that any particular soil would have exactly this distribution. The actual proportion of the phosphorus remaining in an available form will depend upon contact with the soil, time for reaction, and other factors. It should be kept in mind that some of the added phosphorus may be changed to an organic form in which it would be temporarily unavailable.

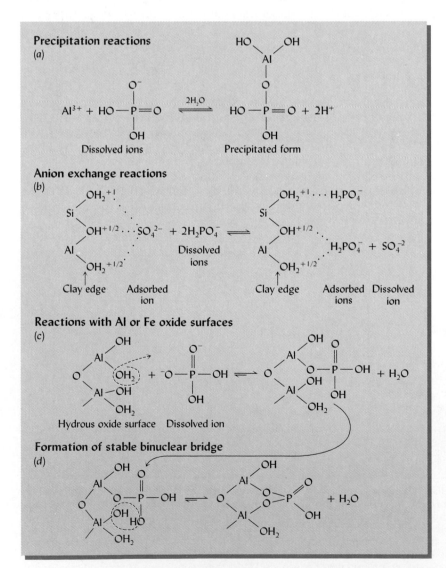

FIGURE 14.13 Several of the reactions by which phosphate ions are removed from soil solution and fixed by reaction with iron and aluminum in various hydrous oxides. Freshly precipitated aluminum, iron, and manganese phosphates (a) are relatively available, though over time they become increasingly unavailable. In (b) the phosphate is reversibly adsorbed by anion exchange. In reactions of the type shown in (c) a phosphate ion replaces an —OH$_2$ or an —OH group in the surface structure of Al or Fe hydrous oxide minerals. In (d) the phosphate further penetrates the mineral surface by forming a stable binuclear bridge. The adsorption reactions (b, c, d) are shown in order—from those that bind phosphate with the least tenacity (relatively reversible and somewhat more plant-available) to those that bind phosphate most tightly (almost irreversible and least plant-available). It is probable that, over time, phosphate ions added to a soil may undergo an entire sequence of these reactions, becoming increasingly unavailable.

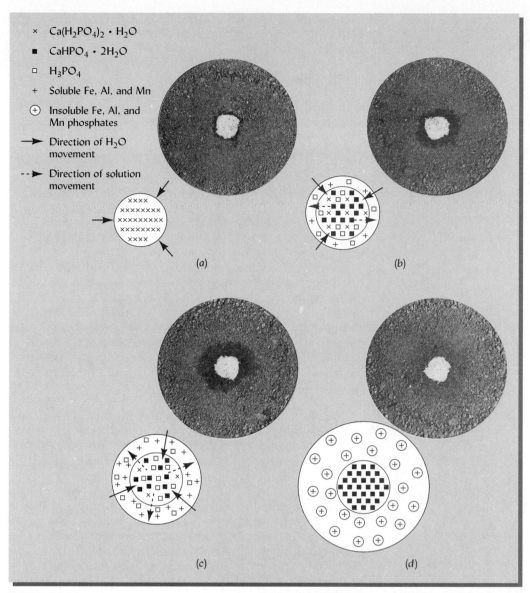

FIGURE 14.14 When a granule of soluble calcium monophosphate ($Ca(H_2PO_4)_2 \cdot H_2O$) fertilizer is added to a moist soil, the following series of reactions rapidly and dramatically reduce the availability of the added phosphorus: (a) The $Ca(H_2PO_4)_2 \cdot H_2O$ in the fertilizer granules attracts water from the soil. (b) In the moistened granule phosphoric acid is formed by the following reaction: $Ca(H_2PO_4)_2 \cdot H_2O + H_2O \rightarrow CaHPO_4 \cdot 2H_2O + H_3PO_4$. As more water is attracted, an H_3PO_4-laden solution with a pH of about 1.4 moves outward from the granule. (c) This solution is sufficiently acid to dissolve and displace large quantities of iron, aluminum, and manganese. These ions promptly react with the phosphate to form low-solubility compounds. (d) Later, these compounds revert to the hydroxy phosphates of iron, aluminum, and manganese in acid soils. In neutral to alkaline soils equally insoluble calcium phosphates are formed. In both cases insoluble dicalcium phosphate ($CaHPO_4 \cdot 2H_2O$) remains in the granule. Fortunately, the phosphorus in the freshly precipitated compounds is slightly available for plant uptake. But when these freshly precipitated compounds are allowed to age or to revert to more insoluble forms, the phosphorus becomes almost completely unavailable to plants in the short term. (Photos courtesy G. L. Terman and National Plant Food Institute, Washington, D.C.)

Reaction with Hydrous Oxides and Silicate Clays

Most of the phosphorus fixation in acid soils probably occurs when $H_2PO_4^-$ ions react with, or become adsorbed to, the surfaces of insoluble hydrous oxides of iron, aluminum, and manganese, such as gibbsite ($Al_2O_3 \cdot 3H_2O$) and goethite ($Fe_2O_3 \cdot 3H_2O$) (Figure 14.11). These hydrous oxides occur as crystalline and noncrystalline particles and as coatings on the interlayer and external surfaces of clay particles. Under some circumstances, structural Al at the edges of 1:1 silicate clays (e.g., kaolinite) can also react with phosphorus. Fixation of phosphorus by clays probably takes place over a relatively wide pH range (Figure 14.12). The large quantities of hydrous Fe, Al oxides, and 1:1 clays present in many soils make possible the fixation of extremely large amounts of phosphorus by these reactions.

Although all the exact mechanisms have not been identified, $H_2PO_4^-$ ions are known to react with iron and aluminum mineral surfaces in several different ways, resulting in different degrees of phosphorus fixation. Some of these reactions are shown diagrammatically in Figure 14.13.

The $H_2PO_4^-$ anion may be attracted to positive charges that develop under acid conditions on the surfaces of iron and aluminum oxides and broken edges of kaolinite clays (Figure 14.13b). The $H_2PO_4^-$ anions adsorbed by electrostatic attraction to these positively charged sites may be removed by certain other anions such as OH^-, SO_4^{2-}, MoO_4^{2-} or organic acids ($R\text{—}COO^-$) in the reversible process of anion exchange as discussed in Chapter 8. Since this type of adsorption of $H_2PO_4^-$ ions is reversible, the phosphorus may slowly become available to plants. Availability of such adsorbed $H_2PO_4^-$ may be increased by (1) liming the soil to increase the hydroxyl ions or (2) adding organic matter to increase organic acids (anions) capable of replacing $H_2PO_4^-$.

Alternatively, the phosphate ion may replace a structural hydroxyl to become chemically bound to the oxide (or clay) surface (Figure 14.13c). This reaction, while reversible, binds the phosphate too tightly to allow it to be replaced by other anions. The availability of phosphate bound in this manner is very low. Over time, a second oxygen of the phosphate ion may replace a second hydroxyl, so that the phosphate becomes chemically bound to two adjacent aluminum (or iron) atoms in the hydrous oxide surface (Figure 14.13d). With this step, the phosphate becomes an integral part of the oxide mineral and the likelihood of its release back to the soil solution is extremely small.

Finally, as more time passes, the precipitation of additional iron or aluminum hydrous oxide may bury the phosphate deep inside the oxide particle. Such phosphate is termed *occluded* and is the least available form of phosphorus in most soils.

Effect of Iron Reduction Under Wet Conditions

It should be noted that phosphorus bound to iron oxides by the mechanisms just discussed is very insoluble under well-aerated conditions. However, prolonged anaerobic conditions can reduce the iron in these complexes from Fe^{3+} to Fe^{2+}, making the iron-phosphate complex much more soluble and causing it to release phosphorus into solution. The release of phosphorus from iron-phosphates by means of the reduction and subsequent solubilization of iron improves the phosphorus availability in soils used for paddy rice.

These reactions are also of special relevance to concerns about water quality. Phosphorus bound to soil particles may accumulate in river- and lake-bottom sediments, along with organic matter and other debris. As the sediments become anoxic, the reducing environment may cause the gradual release of phosphorus held by hydrous iron oxides. Thus the phosphorus eroded from soils today may aggravate the problem of eutrophication for years to come, even after the erosion and loss of phosphorus from the land has been brought under control.

14.7 INORGANIC PHOSPHORUS AVAILABILITY AT HIGH pH VALUES[6]

The availability of phosphorus in alkaline soils is principally determined by the solubility of the various calcium-phosphate compounds present. In alkaline soils (e.g., pH = 8) soluble $H_2PO_4^-$ quickly reacts with calcium to form a sequence of products of decreasing solubility. For instance, highly soluble monocalcium phosphate ($Ca(H_2PO_4)_2 \cdot H_2O$) added as concentrated superphosphate fertilizer rapidly reacts with calcium carbonate in the soil to form first dicalcium phosphate and then tricalcium phosphate ($Ca_3(PO_4)_2$) as follows:

---

[6]See Sample et al. (1980) for a discussion of this subject.

$$Ca(H_2PO_4)_2 \cdot H_2O + 2CaCO_3 \longrightarrow \longrightarrow Ca_3(PO_4)_2 + 2CO_2\uparrow + 3H_2O$$

Monocalcium phosphate   Calcium carbonate   Tricalcium phosphate
(soluble)   (very low solubility)

The solubility of these compounds and, in turn, the plant availability of the phosphorus they contain, decrease as the phosphorus changes from the $H_2PO_4^-$ ion to tricalcium phosphate [$Ca_3(PO_4)_2$]. Although this compound is quite insoluble, it may undergo further reactions to form even more insoluble compounds such as the hydroxy-, oxy-, carbonate-, and fluorapatite compounds shown in Table 14.5. These compounds are thousands of times less soluble than freshly formed tricalcium phosphates. The extreme insolubility of apatites in neutral or alkaline soils generally makes powdered phosphate rock (which consists mainly of apatite minerals) virtually useless as a source of phosphorus for plants unless ground very fine (to increase weathering surface) and applied to relatively acidic soils.

Reversion of soluble fertilizer phosphorus to extremely insoluble calcium phosphate forms is most serious in the calcareous soils of low-rainfall regions (e.g., western United States). Iron and aluminum impurities in calcite particles may also adsorb considerable amounts of phosphate in these soils. Because of the various reactions with $CaCO_3$, phosphorus availability tends to be nearly as low in the Aridisols, Inceptisols, and Mollisols of arid regions as in the highly acid Spodosols and Ultisols of humid regions where iron, aluminum, and manganese limit phosphorus availability.

14.8   PHOSPHORUS-FIXATION CAPACITY OF SOILS

Soils may be characterized by their capacity to fix phosphorus in unavailable, insoluble forms. The phosphorus-fixation capacity of a soil may be conceptualized as the total number of sites on soil particle surfaces capable of reacting with phosphate ions. Phosphorus fixation may also be due to reactive, soluble iron, aluminum, or manganese. The different types of fixation mechanisms are illustrated schematically in Figure 14.15.

One way of determining the phosphorus-fixing capacity of a particular soil is to shake a known quantity of the soil in a phosphorus solution of known concentration. After about 24 hours an equilibrium will be approached and the concentration of phosphorus remaining in the solution (the **equilibrium phosphorus concentration** or **EPC**) can be determined. The difference between the initial and final (*equilibrium*) solution phosphorus concentrations represents the amount of phosphorus fixed by the soil. If this procedure is repeated using a series of solutions with different initial phosphorus concentrations, the results can be plotted as a phosphorus-fixation curve (Figure 14.16) and the maximum phosphorus-fixation capacity can be extrapolated from the value at which the curve levels off.

To a large degree, phosphorus fixation by soils is not easily reversible. However, if a portion of the fixed phosphorus is present in relatively soluble forms (see Section 14.6), and most of the fixation sites are already occupied by a phosphate ion, some release of phosphorus to solution is likely to occur when the soil is exposed to water with very low phosphorus concentration. This release (often called *desorption*) of phosphorus is indicated in Figure 14.16 where the curve for soil A crosses the zero fixation line and becomes negative (negative fixation = release). The solution concentration (x-axis) at which zero fixation occurs (phosphorus is neither released nor retained) is called the EPC_0. Such release of previously fixed phosphorus helps to resupply soil solution phosphorus depleted by plant uptake. Release of fixed phosphorus is also very important in determining losses of dissolved phosphorus in the surface runoff from a watershed. The EPC_0 is an important soil parameter because it indicates both the level of phosphorus fertility and the hazard of phosphorus loss by solution in runoff water.

QUANTITY–INTENSITY RELATIONSHIPS.   The relationship between phosphorus in solution and phosphorus in slowly soluble or fixed forms is an example of the balance between *quantity* factors and *intensity* factors in soil fertility. The intensity factor is the amount of a nutrient dissolved in the soil solution. The quantity factor is the amount of that nutrient associated with the solid framework of the soil and in equilibrium with the nutrient ions in solution. In Figure 14.16 the intensity factor would be represented on the x-axis and the quantity factor on the y-axis. The slope of the curve that defines the relation-

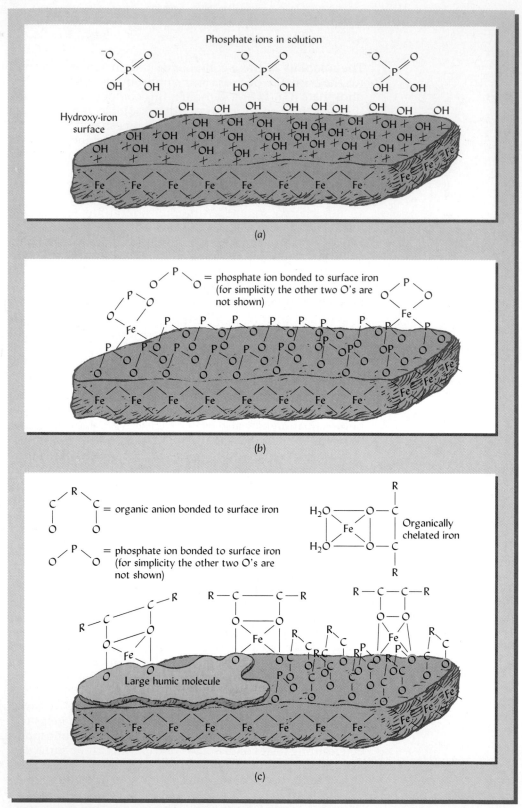

FIGURE 14.15 Schematic illustration of phosphorus-fixation sites on a soil particle surface showing hydrous iron oxide as the primary fixing agent. In part (a) the sites are shown as +'s indicating positive charges or hydrous metal oxide sites, each capable of fixing a phosphate ion. In part (b) the fixation sites are all occupied by phosphate ions (the soil's fixation capacity is satisfied). Part (c) illustrates how organic anions, larger organic molecules, and certain strongly fixed inorganic anions can reduce the sites available for fixing phosphorus. Such mechanisms account for the reduced phosphorus fixation and greater phosphorus availability brought about when organic matter is added to a soil.

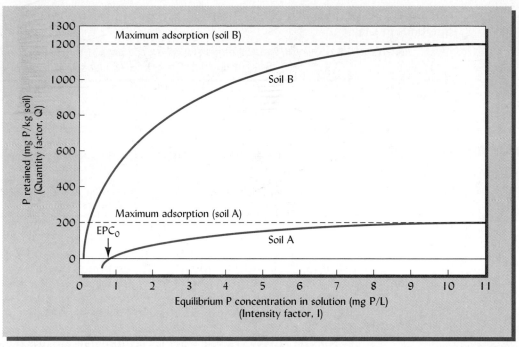

FIGURE 14.16 The relationship between phosphorus fixation and phosphorus in solution when two different soils (A and B) are shaken with solutions of various initial phosphorus concentrations. Initially, each soil removes nearly all of the phosphorus from solution, and as more and more concentrated solutions are used, the soil fixes greater amounts of phosphorus. However, eventually solutions are used that contain so much phosphorus that most of the phosphorus-fixation sites are satisfied, and much of the dissolved phosphorus remains in solution. The amount fixed by the soil levels off as the *maximum phosphorus fixing capacity* of the soil is reached (see horizontal dashed lines: for soil A, 200 mg P/kg; for soil B, 1200 mg P/kg soil). If the initial phosphorus concentration of a solution is equal to the *equilibrium phosphorus concentration (EPC)* for a particular soil, that soil will neither remove phosphorus from nor release phosphorus to the solution (i.e., phosphorus fixation = 0 and EPC = EPC_0). If the solution phosphorus concentration is less than the EPC_0, the soil will release some phosphorus (i.e., the fixation will be negative). In this example, soil B has a much *higher* phosphorus-fixing capacity and a much *lower* EPC than does soil A. It can also be said that soil B is highly *buffered* because much phosphorus must be added to this soil to achieve a small increase in the equilibrium solution phosphorus concentration. On the other hand, if a plant root were to remove a relatively large amount of total phosphorus from the soil, only a small change would occur in the equilibrium solution concentration.

ship for each soil represents the amount of change in the quantity factor Q that results from a given change in the intensity factor I; that is, slope = $\Delta Q/\Delta I$. This is another way of expressing the *potential buffering capacity* (PBC) of the soil:

$$PBC = \Delta Q/\Delta I$$

This general relationship applies not only to phosphorus, but to potassium and any other substance whose solution concentration is controlled by retention reactions with the soil solids. In Section 9.4, for example, buffering of pH was discussed.

Factors Affecting the Extent of Phosphorus Fixation in Soils

Soils that remove more than 350 mg P/kg of soil (i.e., a phosphorus-fixing capacity of about 770 kg P/ha) from solution are generally considered to be high phosphorus-fixing soils. High phosphorus-fixing soils tend to maintain low phosphorus concentrations in the soil solution and in runoff water. Table 14.6 lists values of maximum phosphorus-fixing capacity for a range of soils. The effects of clay content and type of clay are apparent. These and other factors will now be discussed.

The phosphorus-fixation capacity of a soil is positively related to certain soil properties such as calcium carbonate content, clay content, and contents of iron, aluminum, and manganese, especially the hydrous oxides of these metals.

TABLE 14.6 Maximum Phosphorus-Fixation Capacity of Several Soils of Varied Content and Kinds of Clays[a]

Fe, Al oxides (especially amorphous types) fix the largest quantities and silicate clays (especially 2:1 type) fix the least.

Soil Great Group (and series, if known)	Location	Clay		Maximum P fixation (mg P/kg soil)
		Percent	Type	
Evesboro (Quartzipsamment)	Maryland	6	Kaolinite, Fe, Al oxides	125
Kandiustalf	Zimbabwe	20	Kaolinite, Fe oxides	394
Kitsap (Xerochrept)	Washington	12	2:1 clays, allophane	453
Matapeake (Hapludult)	Maryland	15	Chlorite, kaolinite, Fe oxides	465
Rhodustalf	Zimbabwe	53	Kaolinite, Fe oxides	737
Newberg (Haploxeroll)	Washington	38	2:1 clays, Fe oxides	905
Tropohumults	Cameroon	46	Fe, Al oxides, kaolinite	2060

[a]Cameroon data courtesy of V. Ngachie; Washington data from Kuo, 1988.

AMOUNT OF CLAY PRESENT. Most of the compounds with which phosphorus reacts are in the finer soil fractions. Therefore, if soils with similar pH values and mineralogy are compared, phosphorus fixation tends to be more pronounced and phosphorus availability tends to be lowest in those soils with higher clay contents.

TYPE OF CLAY MINERALS PRESENT. Some clay minerals are much more effective at phosphorus fixation than others. Generally those clays that possess greater anion exchange capacity (due to positive surface charges) have a greater affinity for phosphate ions. For example, extremely high phosphorus fixation is characteristic of allophane clays typically found in Andisols and other soils associated with volcanic ash. Iron and aluminum hydrous oxides, such as gibbsite and goethite, also strongly attract and hold phosphorus ions. Among the layer silicate clays, kaolinite has a greater phosphorus-fixation capacity than most. The 2:1 clays of less weathered soils have relatively little capacity to bind phosphorus. Thus, the soil components responsible for phosphorus-fixing capacity are, in order of increasing extent and degree of fixation:

$$2\text{:}1 \text{ clays} \ll 1\text{:}1 \text{ clays} < \text{carbonate crystals} < \text{crystalline Al, Fe, Mn oxides}$$
$$< \text{amorphous Al, Fe, Mn oxides, allophane}$$

To some degree, the above phosphorus-fixing soil components are distributed among soils in relation to soil taxonomy. Vertisols and Mollisols generally are dominated by 2:1 clays and have low phosphorus-fixation capacities. Iron and aluminum oxides are prominent in Ultisols and Oxisols. Andisols, characterized by large quantities of amorphous oxides and allophane, have the greatest phosphorus-fixing capacity, and their productivity is often limited by this property (see Figure 14.17).

EFFECT OF SOIL pH. The greatest degree of phosphorus fixation occurs at very low and very high soil pH. As pH increases from below 5.0 to about 6.0, the iron and aluminum phosphates become somewhat more soluble. Also, as pH drops from greater than 8.0 to below 6.0, calcium phosphate compounds increase in solubility. Therefore, as a general rule in mineral soils, phosphate fixation is at its lowest (and plant availability is highest) when soil pH is maintained in the 6.0–7.0 range (Figure 14.12).

Even if pH ranges from 6.0 to 7.0, phosphate availability may still be very low, and added soluble phosphates will be readily fixed by soils. The low recovery by plants of phosphates added to field mineral soils in a given season is partially due to this fixation. A much higher recovery would be expected in organic soils (see Histosol, Figure 14.17) and in many potting mixes where calcium, iron, and aluminum concentrations are not as high as in mineral soils.

EFFECT OF ORGANIC MATTER. Organic matter generally has little capacity to strongly fix phosphate ions. Soils high in organic matter, especially active fractions of organic matter (see Section 12.14), generally exhibit relatively low levels of phosphorus fixation. Several mechanisms are responsible for the reduced phosphorus fixation associated with high soil organic matter. First, large humic molecules can adhere to the surfaces of

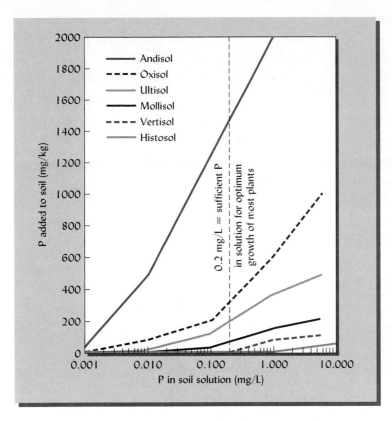

FIGURE 14.17 Phosphorus-fixing tendencies typical of several soil orders. The graph shows the concentrations of phosphorus that would be achieved in the soil solution for each type of soil if varying amounts of soluble phosphorus were added as fertilizer. Although some plants, especially if infected with mycorrhizal fungi, can grow well with lower-solution phosphorus concentrations, a concentration of 0.2 mg P/L in the soil solution is adequate for optimum growth of most crops (see Table 14.7). Therefore, the graph indicates that for optimum plant growth, nearly 1500 mg of phosphorus per kg of soil should be added to the Andisol, while somewhat less than 200 mg of phosphorus per kg of soil would be required for the same phosphorus availability in the Ultisol. Within a particular soil order, the phosphorus-fixing tendency of individual soils may vary considerably. The curves given are representative for soils in the taxonomic orders indicated, but it should be borne in mind that high rates of phosphorus application would eventually reduce the fixation capacity (reduce the slope of the curve) as phosphorus-fixation sites were used up. (Curves are representative of data from many sources.)

clays and metal hydrous oxide particles, masking the phosphorus-fixation sites and preventing them from interacting with phosphorus ions in solution. Second, organic acids produced by plant roots and microbial decay can serve as organic anions, which are attracted to positive charges and hydroxyls on the surfaces of clays and hydrous oxides. These organic anions may compete with phosphorus ions for fixation. Third, certain organic acids and similar compounds can entrap reactive Al and Fe in stable organic complexes called *chelates* (see Section 15.15). Once chelated, these metals are unavailable for reaction with phosphorus ions in solution. These three mechanisms are illustrated schematically in Figure 14.15c.

14.9   PRACTICAL CONTROL OF PHOSPHORUS AVAILABILITY

We have seen that most mineral soils have a large capacity to remove phosphorus ions from solution and fix them on particle surfaces, thus making it very difficult for plant roots to obtain sufficient amounts of this essential nutrient. The phosphorus in fertilizers is also subject to fixation on soil particle surfaces, and only 10 to 15% of the phosphorus added in fertilizer is likely to be taken up by plants in the year of application. Consequently, phosphorus availability is a limiting factor in many agroecosystems. Based on the principles of soil phosphorus behavior discussed in this chapter, a number of approaches can be suggested to ameliorate the phosphorus fertility problem.

1. SATURATION OF PHOSPHORUS-FIXING CAPACITY.   If enough phosphorus can be added, the phosphorus-fixation capacity of even high phosphorus-fixing soils can be largely saturated. This may be achieved all at once with one or two massive doses of phosphorus (usually as phosphorus fertilizer, phosphate rock, or high-phosphorus animal manure), or over a period of years or decades by annually adding more phosphorus than is removed in plant harvest. Either the rapid or slow buildup approach will eventually satisfy most of the phosphorus-fixation sites and lead to a soil so high in phosphorus that ample phosphorus is maintained in solution despite the initially high phosphorus-fixation capacity. These approaches, especially the rapid buildup approach, can be very expensive in terms of initial capital outlay for phosphorus fertilizer. For this reason, it is not practical for most of the world's farmers and foresters, but has been put into prac-

tice in some intensive agricultural systems. A second drawback to this approach is the potential for water pollution from large amounts of phosphorus desorbed to runoff water from phosphorus-saturated soils.

2. PLACEMENT OF PHOSPHORUS FERTILIZER. One strategy to maximize phosphorus fertilizer value is to minimize the opportunity for the fertilizer phosphorus to react with the soil before being taken up by root. Generally, if fertilizer is placed directly in a localized rooting zone (point placement) instead of thoroughly mixed with the soil (broadcast), one-half to one-third as much fertilizer may be used. Point placement is widely used where fertilizer is applied by hand, but new spike-wheel fertilizer injection machines (see Figure 16.25) are being developed that may make point placement feasible even in mechanized systems. Banded applications are standard practice (see Figure 16.23) for starter fertilizers. Trees are often fertilized with pellets. In untilled systems, a surface band is often created.

3. COMBINATION OF AMMONIUM AND PHOSPHORUS FERTILIZERS. Using ammonium in the same band with phosphorus fertilizers greatly increases root uptake of phosphorus, especially in alkaline soils. The increased phosphorus uptake is probably related to the acids that roots are known to produce when they take up most of their nitrogen in the ammonium form. Mono- and diammonium phosphate fertilizers offer this advantage.

4. CHOOSE PHOSPHORUS EFFICIENT PLANTS. Some plants are able to thrive in soils with far less available phosphorus than other plants that are less efficient at phosphorus uptake (see Table 14.7 for several examples). The mechanisms for the more efficient phosphorus uptake are only partially known. Some plants have fine root systems with very large root surface areas. Other plants are dependent on efficient mycorrhizal associations. Still others produce specific root secretions that solubilize either calcium phosphates (as is the case for buckwheat) or Fe phosphates (as is the case for pigeon pea).

5. INCREASED CYCLING OF ORGANIC PHOSPHORUS. Mineralization of organic materials (i.e., animal manures or other phosphorus-rich organic materials) added to soils may release phosphorus gradually and somewhat in synchrony with plant needs. This slow release allows plants to uptake phosphorus before the soil fixation reactions remove it from the available pool. Leaf fall from trees and residues from crops return phosphorus to the soil in organic, potentially mineralizable forms. Alternating phosphorus-efficient plants with less efficient ones may convert inorganic forms taken up by the efficient plants into organic forms in the residues of these plants. The subsequent decay and mineralization of these residues may enhance the supply of phosphorus for the less-efficient plants. In soils with relatively high total phosphorus content but very low phosphorus availability, crop rotation, residue management, and agroforestry all may help move more phosphorus into organic forms that are protected from phosphorus fixation. This phosphorus may be available in the future as the organic compounds are mineralized.

TABLE 14.7 Concentration of Phosphorus in Soil Solution Required for Near-Optimal Growth (95% of Maximum Yield) of Various Plants

Plant	Approximate P in soil solution mg/L
Cassava	0.005
Peanut	0.01
Corn	0.05
Sorghum	0.06
Cabbage	0.04
Soybean	0.20
Tomato	0.20
Head lettuce	0.30

Data from Fox (1981).

6. INCREASED CYCLING OF ORGANIC MATTER. In addition to providing organic phosphorus for release by mineralization, soil organic matter can improve phosphorus availability by reducing the tendency of the mineral fractions to fix phosphorus. The masking of phosphorus fixation sites by humus and organic acids and the chelation of reactive Al and Fe were discussed in Section 14.8. Return of crop residues, inclusion of green manure crops in rotations, mulching with various organic materials, and adding animal manures and other decomposable organic wastes can all increase available phosphorus.

7. CONTROL OF SOIL pH. Some control over phosphorus solubility can be obtained by maintaining soil pH between 6.0 and 7.0 (Figure 14.12). Maintenance of pH in this range is generally more practical in less weathered soils than in the highly weathered soils of warm, humid regions. Proper liming can contribute to improved phosphorus availability in many cases, but if a source of liming material is not locally available, the cost may be prohibitive.

8. ENHANCEMENT OF MYCORRHIZAL SYMBIOSIS. As already mentioned, mycorrhizal fungi (Section 14.3) can greatly enhance the ability of plants to obtain phosphorus from soils with low levels of phosphorus in solution. In the case of many shrubs and forest trees, inoculation of seedlings with efficient strains of mycorrhizal fungi is a practical measure when planting new trees (see Section 11.9). For most plants, however, the mycorrhizal symbiosis can best be enhanced by fostering appropriate soil conditions through crop rotations, organic matter additions, and minimum tillage.

In summary, the small pool of available phosphorus in most soils is depleted and replenished by a number of processes (Figure 14.18). Maintaining sufficient available phosphorus in a soil is basically a twofold program: (1) addition of phosphorus-containing fertilizers or amendments; and (2) regulation to some degree of soil processes that fix both added and native phosphates.

14.10 POTASSIUM: NATURE AND ECOLOGICAL ROLES[7]

Of all the essential elements, potassium is the third most likely, after nitrogen and phosphorus, to limit plant productivity. For this reason it is commonly applied to soils as fertilizer, and is a component of most mixed fertilizers.

The potassium story differs in many ways from that of phosphorus. Unlike phosphorus (or sulphur and, to a large extent, nitrogen), potassium is present in the soil solution only as a positively charged cation, K^+. Like phosphorus, potassium does not form any gases that could be lost to the atmosphere. Its behavior in the soil is influenced primarily by soil cation exchange properties (see Chapter 8) and mineral weathering (Chapter 2) rather than by microbiological processes. Unlike nitrogen and phosphorus, potassium causes no off-site environmental problems when it leaves the soil system. It is not toxic and does not cause eutrophication in aquatic systems.

14.11 POTASSIUM IN PLANT AND ANIMAL NUTRITION

Although potassium plays numerous roles in plant and animal nutrition, it is not actually incorporated into the structures of organic compounds. Instead, potassium remains in the ionic form (K^+) in solution in the cell, or acts as an activator for cellular enzymes. Potassium is known to activate over 80 different enzymes responsible for plant and animal processes such as energy metabolism, starch synthesis, nitrate reduction, photosynthesis, and sugar degradation. Certain plants, many of which evolved in sodium-rich semiarid environments, can substitute sodium or other monovalent ions to carry out some, but not all, of the functions of potassium.

As a component of the plant cytoplasmic solution, potassium plays a critical role in lowering cellular osmotic water potentials, thereby reducing the loss of water from leaf stomata and increasing the ability of root cells to take up water from the soil (see Section 5.3 for a discussion of osmotic water potential). Potassium is essential for photosynthesis, for protein synthesis, for nitrogen fixation in legumes, for starch formation,

---

[7]For further information on this topic, see Mengel and Kirkby (1987) and Munson (1985).

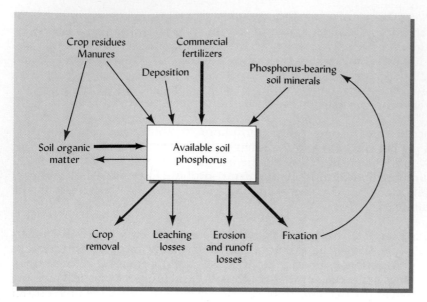

FIGURE 14.18 How the pool of available phosphorus in a soil is depleted and replenished in a typical agroecosystem. The width of the arrows is a relative indication of the quantity of phosphorus involved in each pathway. The two largest fluxes are the addition of phosphorus from soil amendments (usually commercial fertilizer) and the subsequent fixation of much of the added phosphorus in unavailable forms. The return of plant and animal wastes and the subsequent mineralization of soil organic matter can also play major roles in replenishing the available phosphorus. Note that leaching losses are usually negligible.

and for the translocation of sugars. As a result of several of these functions, a good supply of this element promotes the production of plump grains and large tubers. The potassium content of normal, healthy leaf tissue can be expected to be in the range of 1 to 4% in most plants, similar to that of nitrogen, but an order of magnitude greater than that of phosphorus.

Potassium is especially important in helping plants adapt to environmental stresses. Good potassium nutrition is linked to improved drought tolerance, improved winter-hardiness, better resistance to certain fungal diseases, and greater tolerance to insect pests (see Figure 14.19). In the latter role, potassium fertilization is often an important component of integrated pest-management programs designed to reduce the use of toxic pesticides. Potassium also enhances the quality of flowers, fruits, and vegetables by improving flavor and color and strengthening stems (thereby reducing lodging). In many of these respects, potassium seems to counteract some of the detrimental effects

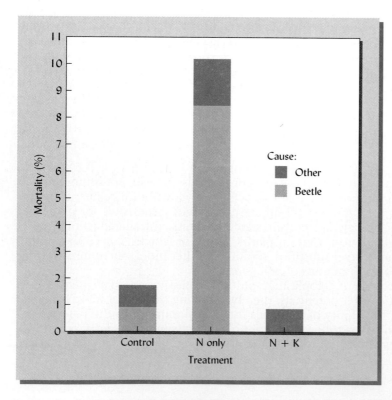

FIGURE 14.19 Influence of potassium and nitrogen fertilizer treatments on the percentage of ponderosa pine trees dying from beetle damage and other causes in the first four years after planting in western Montana. Nitrogen used alone (224 kg N/ ha) caused a large increase in tree mortality, but adding potassium (224 kg K/ha) completely counteracted this effect. Both fertilization treatments stimulated the growth of the surviving trees. [From Mandzak and Moore (1994)]

of excess nitrogen. Maintaining a balance between potassium and other nutrients (especially nitrogen, phosphorus, calcium, and magnesium) is an important goal in managing soil fertility.

In animals, including humans, potassium plays critical roles in regulating the nervous system and in maintenance of healthy blood vessels. Diets that include such high-potassium foods as bananas, potatoes, orange juice, and leafy green vegetables have been shown to lower human risk of stroke and heart disease. Maintaining a balance between potassium and sodium is especially important in human diets.

Deficiency Symptoms in Plants

Compared to deficiencies of phosphorus and many other nutrients, a deficiency of potassium is relatively easy to recognize in most plants. In addition to the characteristics mentioned above (reduced drought tolerance, increased lodging, etc.), specific foliar symptoms are associated with potassium deficiency. Because potassium is very mobile within the plant, it is translocated from older tissues to younger ones if the supply becomes inadequate. The symptoms of deficiency therefore usually occur earliest and most severely on the oldest leaves.

In general, when potassium is deficient the tips and edges of the oldest leaves begin to yellow (chlorosis) and then die (necrosis), so that the leaves appear to have been burned on the edges (Figure 14.20). On some plants the necrotic leaf edges may tear, giving the leaf a ragged appearance (Figure 14.20c). In several important forage and cover-crop legume species potassium deficiency produces small, white necrotic spots that form a unique pattern along the leaflet margins; this easily recognized symptom is one that people often mistake for insect damage (Figure 14.20d).

Potassium deficiency should not be confused with damage from excess salinity, which can also produce brown, necrotic leaf margins. Salinity damage is more likely to affect the newer leaves (see Figure 10.9).

14.12   THE POTASSIUM CYCLE

Figure 14.21 shows the major forms in which potassium is held in soils and the changes it undergoes as it is cycled through the soil–plant system. The original sources of potassium are the primary minerals such as micas (biotite and muscovite) and potassium feldspar (orthoclase and microcline). As these minerals weather, their rigid lattice structures become more pliable. For example, potassium held between the 2:1-type crystal layers of mica in time is made more available, first as nonexchangeable but slowly available forms near the weathered edges of minerals, and eventually as the readily exchangeable forms and the soil solution forms from which it is absorbed by plant roots.

Potassium is taken up by plants in large quantities. Depending on the type of ecosystem under consideration, a portion of this potassium is leached from plant foliage by rainwater (throughfall) and returned to the soil, and a portion is returned to the soil with the plant residues. In natural ecosystems, most of the potassium taken up by plants is returned in these ways, or as wastes (mainly urine) from animals feeding on the vegetation. Some potassium is lost with eroded soil particles and in runoff water, and some is lost to groundwater by leaching. In agroecosystems, from a fifth (e.g., in cereal grains) to nearly all (e.g., in hay crops) of the potassium taken up by plants may be exported to distant markets from which it is unlikely to return.

At any one time, most soil potassium is in primary minerals and nonexchangeable forms. In relatively fertile soils, the release of potassium from these forms to the exchangeable and soil solution forms (that plants can use directly) may be sufficiently rapid to keep plants supplied with enough potassium for optimum growth. On the other hand, where high yields of agricultural crops or timber are removed from the land, or where the content of weatherable potassium-containing minerals is low, the levels of exchangeable and solution potassium may have to be supplemented by outside sources such as chemical fertilizers, poultry manure, or wood ashes. Without these additions, the supply of available potassium will likely be depleted over a period of years and the productivity of the soil will likewise decline.

An example of depletion and restoration of available soil potassium is given in Figure 14.22. Farming without fertilizers for over a century depleted the exchangeable potassium in a sandy soil in New York. After the supply of available potassium was

(a)

(c)

(b)

(d)

FIGURE 14.20   Potassium deficiency often produces easily recognized foliar symptoms, mainly on older leaves: (a) chlorotic margins on boxwood leaves, (b) chlorotic leaf margins on soybean, (c) ragged, necrotic margins of older banana leaves, and (d) small, white necrotic spots on hairy vetch leaflets. [Photos (a), (c), and (d) courtesy of R. Weil; photo (b) courtesy of Potash & Phosphate Institute]

exhausted, the land was abandoned in the 1920s. In the 1920s and 1930s a red-pine forest was planted. The trees in some plots were fertilized with potassium, causing the expected rapid recovery of exchangeable potassium to its preagricultural level. Even where the trees were not fertilized, the level of exchangeable potassium in the surface soil was replenished, but slowly, over a period of about 80 years.

The replenishment of exchangeable potassium in the unfertilized forest plots provides an example of the ability of unharvested perennial plants such as forest trees and pasture legumes to ameliorate soil fertility over time. Deep-rooted perennials often act as "nutrient pumps," taking potassium from deep subsoil horizons into their root systems, translocating it to their leaves, and then recycling it back to the surface of the soil via leaf fall and leaching (see Section 16.4).

In most mature natural ecosystems the small (1 to 5 kg/ha) annual losses of potassium by leaching and erosion are more than balanced by weathering of potassium from primary minerals and nonexchangeable forms in the soil profile, followed by vegetative translocation to the surface of the soil. In most agricultural systems, leaching losses are far greater because much higher exchangeable K levels are generally maintained for crop production, but crop roots are active for only part of the year. The sections that follow give greater details on the reactions involved in the potassium cycle.

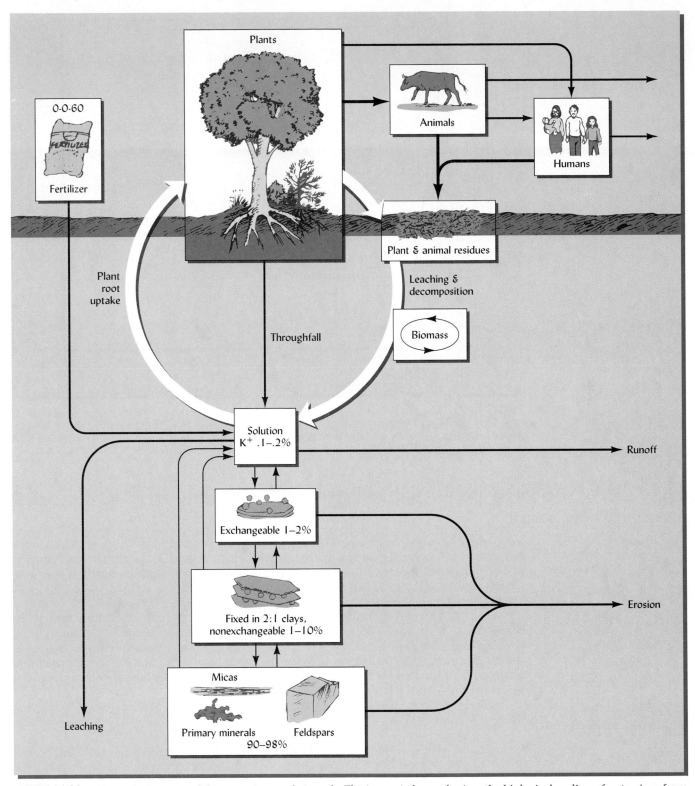

FIGURE 14.21 Major components of the potassium cycle in soils. The inner circle emphasizes the biological cycling of potassium from the soil solution to plants and back to the soil via plant residues or animal wastes. Primary and secondary minerals are the original sources of the element. Exchangeable potassium may include those ions held and released by both clay and humus colloids, but potassium is not a structural component of soil humus. The interactions among solution potassium, exchangeable, nonexchangeable, and structural potassium in primary minerals is shown. The bulk of soil potassium occurs in the primary and secondary minerals, and is released very slowly by weathering processes.

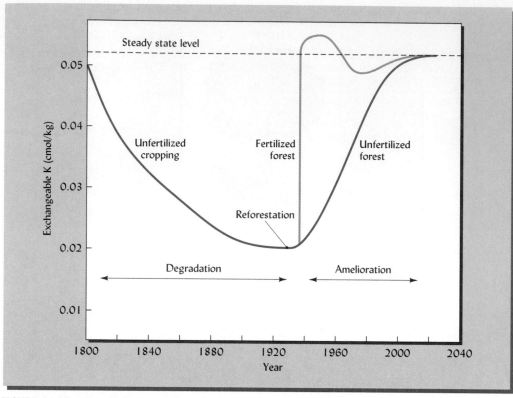

FIGURE 14.22 The general pattern of depletion of A-horizon exchangeable potassium by decades of exploitative farming, followed by its restoration under forest vegetation. The forest consisted of red-pine trees planted on a Plainfield loamy sand (Udipsamment) in New York. This soil has a very low cation exchange capacity and low levels of exchangeable K^+. [From C. A. Nowak, et al. (1991)]

14.13   THE POTASSIUM PROBLEM IN SOIL FERTILITY

Availability of Potassium

In contrast to phosphorus, potassium is found in comparatively high levels in most mineral soils, except those consisting mostly of quartz sand. In fact, the total quantity of this element is generally greater than that of any other major nutrient element. Amounts as great as 35,000–55,000 kg potassium per hectare–furrow slice (31,000–45,000 lb per acre–furrow slice) are not at all uncommon (see Table 1.3).

Yet the quantity of potassium held in an easily exchangeable condition at any one time often is very small. Most of this element is held rigidly as part of the primary minerals or is fixed in forms that are at best only moderately available to plants. Therefore, the situation with respect to potassium utilization parallels that of phosphorus and nitrogen in at least one way. A very large proportion of all three of these elements in the soil is insoluble and relatively unavailable to growing plants.

Leaching Losses

Potassium is much more readily lost by leaching than is phosphorus. Drainage waters from soils receiving liberal fertilizer applications usually contain considerable quantities of potassium. From representative humid region soils growing annual crops and receiving only moderate rates of fertilizer, the annual loss of potassium by leaching is usually about 25 to 50 kg/ha, the greater values being typical of acid, sandy soils.

Losses would undoubtedly be much larger were the leaching of potassium not slowed by the attraction of the positively charged potassium ions to the negatively charged cation exchange sites on clay and humus surfaces. Liming an acid soil to raise its pH can reduce the leaching losses of potassium because of the *complementary ion effect* (see Chapter 8 and Figure 14.23). The ease with which ions may be removed from the exchange complex varies among different elements. Typically, trivalent ions (Al^{3+}) are more tightly held than divalent ions (Ca^{2+} and Mg^{2+}). In a limed soil, where higher levels of exchangeable calcium and magnesium are present, monovalent potassium ions are better able to replace them on the exchange complex. Where higher levels of exchangeable aluminum saturate the exchange complex, potassium is less likely to be adsorbed. Thus, in a limed soil, K$^+$ can be more readily retained on the exchange complex, and leaching of this element is reduced.

Plant Uptake and Removal

Plants take up very large amounts of potassium, often five to ten times as much as for phosphorus and about the same amount as for nitrogen. If most or all of the above-ground plant parts are removed in harvest, the drain on the soil supply of potassium can be very large. For example, a 60-Mg/ha yield of corn silage may remove 160 kg/ha of potassium. Conventional bolewood timber harvest typically removes about 100 kg/ha of potassium. If the entire tree is chipped and removed, as for paper pulp, the removal of potassium may be twice as great. A high-yielding legume hay crop may remove 400 kg/ha of potassium each year.

Luxury Consumption

Moreover, this situation is made even more critical by the tendency of plants to take up soluble potassium far in excess of their needs if sufficiently large quantities are present. This tendency is termed *luxury consumption,* because the excess potassium absorbed does not increase crop yields to any extent.

The principles involved in luxury consumption are shown by the graph of Figure 14.24. For many plants there is a direct relationship between the available potassium (soil plus fertilizer) and the removal of this element by the plants. However, only a certain amount of this element is needed for optimum growth, and this is termed *required potassium*. Potassium taken up by the plant above this critical required level is considered a *luxury*. If plant residues are not returned to the soil, the removal of this excess potassium is decidedly wasteful.

In summary, then, the problem of potassium is at least threefold: (1) a very large proportion is relatively unavailable to higher plants; (2) it is subject to leaching losses; and

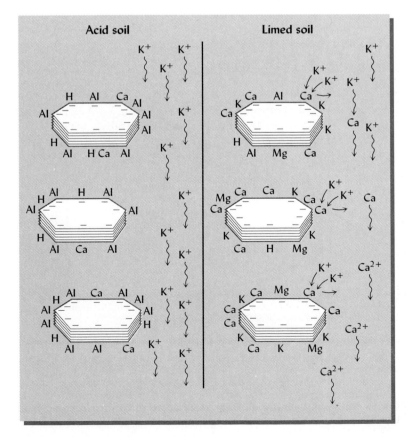

FIGURE 14.23 Diagrammatic illustration of how liming an acid soil can reduce leaching losses of potassium. The fact that the K$^+$ ions can more easily replace Ca^{2+} ions than they could replace Al^{3+} ions allows more of the K$^+$ ions to be removed from solution by cation exchange in the limed (high-calcium) soil. The removal of K$^+$ ions from solution by adsorption on the colloids will reduce their loss by leaching, but will also reduce their availability for plant uptake.

(3) the removal of potassium by plants is high, especially when luxury quantities of this element are supplied. With these ideas as a background, the various forms and availabilities of potassium in soils will now be considered.

14.14 FORMS AND AVAILABILITY OF POTASSIUM IN SOILS

Four forms of soil potassium are shown in the potassium-cycle diagram (Figure 14.21, from the bottom upward): (1) K in primary mineral crystal structures, (2) K in nonexchangeable positions in secondary minerals, (3) K in exchangeable form on soil colloid surfaces, and (4) potassium ions soluble in water. The total amount of potassium in a soil and the distribution of potassium among the four major pools shown in Figure 14.21 is largely a function of the kinds of clay minerals present in a soil. Generally, soils dominated by 2:1 clays contain the most potassium; those dominated by kaolinite contain the least (Table 14.8). In terms of availability for plant uptake, the following interpretation applies to the different forms of soil potassium:

K in primary mineral structure	*Unavailable*
Nonexchangeable K in secondary minerals	*Slowly available*
Exchangeable K on soil colloids ⎫ K soluble in water ⎭	*Readily available*

All plants can easily utilize the readily available forms, but the ability to obtain potassium held in the slowly available and unavailable forms differs greatly among plant species. Many grass plants with fine, fibrous root systems are able to exploit potassium held in clay interlayers and near the edges of mica and feldspar crystals of clay and silt size. A few plants, such as elephant grass (*Pennisetum purpureum* Schum), adapted to low-fertility sandy soils, have been shown to obtain potassium from even sand-sized primary minerals, a form of potassium usually considered to be unavailable.

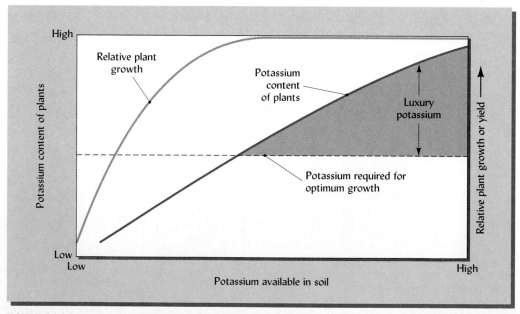

FIGURE 14.24 The general relationship between available potassium level in the soils, plant growth, and plant uptake of potassium. If available soil potassium is raised above the level needed for maximum plant growth, many plants will continue to increase their uptake of potassium without any corresponding increase in growth. The potassium taken up in excess of that needed for optimum growth is termed *luxury consumption.* Such luxury consumption may be wasteful, especially if the plants are completely removed from the soil. It may also cause dietary imbalance in grazing animals.

TABLE 14.8   The Influence of Dominant Clay Minerals on the Amounts of Water-Soluble, Exchangeable, Fixed (Nonexchangeable), and Total Potassium in Soils

The values given are means for many soils in 10 soil orders sampled in the United States and Puerto Rico.

Potassium pool	Dominant clay mineralogy of soils		
	Kaolinitic (26 soils)	Mixed (53 soils)	Smectitic (23 soils)
	mg K/kg soil		
Total potassium	3340	8920	15780
Exchangeable potassium	45	224	183
Water-soluble potassium	2	5	4

Data from Sharpley (1990).

Relatively Unavailable Forms

Some 90 to 98% of all soil potassium in a mineral soil is in relatively unavailable forms (Figure 14.21). The compounds containing most of this form of potassium are the feldspars and the micas. These minerals are quite resistant to weathering and supply relatively small quantities of potassium during a given growing season. However, their cumulative release of potassium over a period of years undoubtedly is of some importance. This release is enhanced by the solvent action of carbonic acid and of stronger organic and inorganic acids, as well as by the presence of acid clays and humus (see Section 2.3). As already mentioned, the roots of some plants can obtain a significant portion of their potassium supply from these minerals, apparently by depleting the potassium ions from the solution around the edges of these minerals, thereby favoring the dissolution of the mineral.

Readily Available Forms

Only 1 to 2% of the total soil potassium is readily available. Available potassium exists in soils in two forms: (1) in the soil solution and (2) exchangeable potassium adsorbed on the soil colloidal surfaces. Although most of this available potassium is in the exchangeable form (approximately 90%), soil solution potassium is most readily absorbed by higher plants but, unfortunately, is subject to considerable leaching loss.

As represented in Figure 14.21, these two forms of readily available potassium are in dynamic equilibrium. This equilibrium is of extreme practical importance. When plants absorb potassium from the soil solution, some of the exchangeable potassium immediately moves into the soil solution until the equilibrium is again established. When water-soluble fertilizers are added, the soil solution becomes potassium enriched and the reverse of the above adjustment occurs—potassium from soil solution moves onto the exchange complex. The exchangeable potassium can be seen as an important buffer mechanism for soil solution potassium.

Slowly Available Forms

In the presence of vermiculite, smectite, and other 2:1-type minerals, the K^+ ions (as well as the similarly sized NH_4^+ ions) in the soil solution (or added as fertilizers) not only become adsorbed but also may become definitely fixed by the soil colloids (Figure 14.25). Potassium (and ammonium) ions fit in between layers in the crystals of these normally expanding clays and become an integral part of the crystal. These ions cannot be replaced by ordinary exchange processes and consequently are referred to as *nonexchangeable ions*. As such, these ions are not readily available to most higher plants. This form is in equilibrium, however, with the more available forms and consequently acts as an extremely important reservoir of slowly available nutrients.

Release of Fixed Potassium

The quantity of nonexchangeable or *fixed* potassium in some soils is quite large. The fixed potassium in such soils is continually released to the exchangeable form in

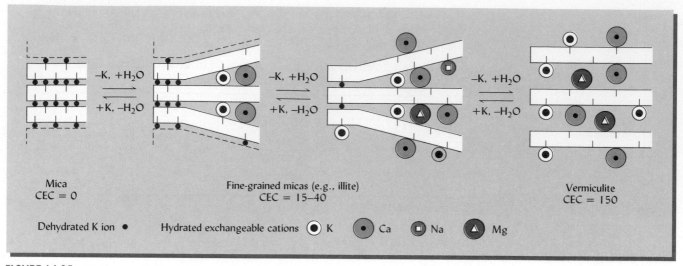

FIGURE 14.25 Diagrammatic illustration of the release and fixation of potassium between primary micas, fine-grained mica (illite clay), and vermiculite. In the diagram, the release of potassium proceeds to the right, while the fixation process proceeds to the left. Note that the dehydrated potassium ion is much smaller than the hydrated ions of Na^+, Ca^{2+}, Mg^{2+}, etc. Thus, when potassium is added to a soil containing 2:1 type minerals such as vermiculite, the reaction may go to the left and potassium ions will be tightly held (fixed) in between layers within the crystal, producing a fine-grained mica structure. Ammonium ions ($NH4^+$) are of a similar size and charge to potassium ions and may be fixed by similar reactions. [Modified from McLean (1978)]

amounts large enough to be of great practical importance. The data in Table 14.9 indicates the magnitude of the release of nonexchangeable potassium from certain soils. In these soils, the potassium removed by plants was supplied largely from nonexchangeable forms. The entire equilibrium may be represented for potassium as follows.

$$\text{Nonexchangeable K} \underset{}{\overset{\text{Slow}}{\rightleftharpoons}} \text{Exchangeable K} \underset{}{\overset{\text{Rapid}}{\rightleftharpoons}} \text{Soil solution K}$$

As a result of these relationships, very sandy soils with low CEC are poorly buffered with respect to potassium. In them, the potassium ion concentration may be quite high at the beginning of a growing season or just after fertilization, but the soils have little capacity to maintain the potassium concentration, as plants remove the dissolved potassium from the soil solution during the growing season. Late-season potassium deficiency may result. In finer-textured soils with a greater CEC (and therefore greater buffering capacity), the initial solution concentration of potassium may be somewhat lower, but the soil is capable of maintaining a fairly constant supply of solution potassium ions throughout the growing season.

TABLE 14.9 Potassium Removal by Very Intensive Cropping and the Amount of This Element Coming from the Nonexchangeable Form

Soil	Total K used by crops (kg/ha)	Total K used by crops (lb/acre)	Percent derived from nonexchangeable form
Wisconsin soils[a]			
Carrington silt loam	133	119	75
Spencer silt loam	66	59	80
Plainfield sand	99	88	25
Mississippi soils[b]			
Robinsonville fine silty loam	121	108	33
Houston clay	64	57	47
Ruston sandy loam	47	42	24

[a]Average of six consecutive cuttings of ladino clover, from Evans and Attoe (1948).
[b]Average of eight consecutive crops of millet, from Gholston and Hoover (1948).

14.15 FACTORS AFFECTING POTASSIUM FIXATION IN SOILS

Four soil conditions markedly influence the amounts of potassium fixed: (1) the nature of the soil colloids, (2) wetting and drying, (3) freezing and thawing, and (4) the presence of excess lime.

Effects of Type of Clay and Moisture

The ability of the various soil colloids to fix potassium varies widely. Kaolinite and other 1:1-type clays fix little potassium. On the other hand, clays of the 2:1 type, such as vermiculite, fine-grained mica (illite), and smectite, fix potassium very readily and in large quantities. Even silt-sized fractions of some micaceous minerals fix and subsequently release potassium.

The potassium and ammonium ions are attracted between layers in the negatively charged clay crystals. The tendency for fixation is greatest in minerals where the major source of negative charge is in the silica (tetrahedral) sheet. Consequently, vermiculite has a greater fixing capacity than montmorillonite (see Table 8.5 for formulas for these minerals).

Alternate wetting/drying and freezing/thawing has been shown to enhance both the fixation of potassium in nonexchangeable forms as well as the release of previously fixed potassium to the soil solution. Although the practical importance of this is recognized, its mechanism is not well understood.

Influence of pH

Applications of lime sometimes result in an increase in potassium fixation of soils (Figure 14.26). This is not surprising since in strongly acid soils the tightly held H^+ and hydroxy aluminum ions prevent the potassium ions from being closely associated with the colloidal surfaces, which reduces their susceptibility to fixation. As the pH increases, the H^+ and hydroxyl aluminum ions are removed or neutralized, and it is easier for potassium ions to move closer to the colloidal surfaces where they are more susceptible to fixation in 2:1 clays.

In soils where the negative charge is pH-dependent, liming increases the cation exchange capacity, which results in an increased potassium adsorption by the soil colloids and a decrease in the potassium level in the soil solution (see Figure 14.27).

Furthermore, high calcium and magnesium levels in the soil solution may reduce potassium uptake by the plant because cations tend to compete against one another for

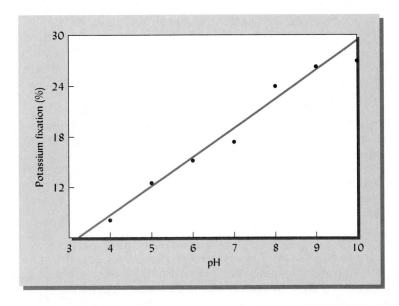

FIGURE 14.26 The effect of pH on the fixation of potassium soils in India. [From Grewal and Kanwar (1976)]

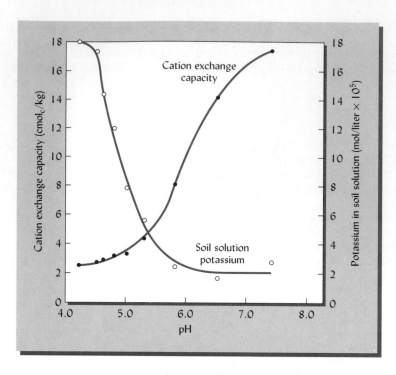

FIGURE 14.27 The influence of increased pH resulting from lime additions on the pH-dependent cation exchange capacity of a soil and the level of potassium in the soil solution. As the cation exchange capacity increases, some of the soil solution potassium is attracted to the adsorbing colloids. [Data from Magdoff and Bartlett (1980)]

uptake by roots. Since the absorption of the potassium ion by plant roots is affected by the activity of other ions in the soil solution, some authorities prefer to use the ratio

$$\frac{[K^+]}{\sqrt{[Ca^{2+}] + [Mg^{2+}]}}$$

rather than the potassium concentration to indicate the available potassium level in solution. Finally, potassium deficiency frequently occurs in calcareous soils even when the amount of exchangeable potassium present would be adequate for plant nutrition on other soils. Potassium fixation as well as cation ratios may be responsible for these adverse effects on calcium-carbonate-rich soil.

14.16 PRACTICAL ASPECTS OF POTASSIUM MANAGEMENT

Except in very sandy soils, the problem of potassium fertility is rarely one of total supply, but rather one of adequate *rate* of transformation from nonavailable to available forms. Where little plant material is removed (e.g., in forests, range, and some ornamental systems), cycling between plant and soil may be adequate for continued plant growth. However, where crops are removed, especially if little plant residue is returned, then the plant–soil cycle must be supplemented by release of potassium from less available mineral forms and, to some degree, by fertilization.

Vigorous growth of high-potassium-content plants places great demands on the soil supply of available potassium. Moreover, the rate of potassium uptake is not constant, but varies with plant growth stage and season. If high yields of forage legumes such as alfalfa are to be produced, the soil may have to be capable of supplying potassium for very high uptake rates during certain periods, resulting in the need for high levels of fertilization even on soils well supplied with weatherable minerals.

Frequency of Application

Although a heavy dressing applied every few years may be most convenient, more frequent light applications of potassium may offer the advantages of reduced luxury consumption of potassium by some plants, reduced losses of this element by leaching, and reduced opportunity for fixation in unavailable forms before plants have had a chance to use the potassium applied. Although such fixation has definite conserving features, these in most cases tend to be outweighed by the disadvantages of leaching and luxury consumption.

Potassium-Supplying Power of Soils

Full advantage should be taken of the potassium-supplying power of soils. The idea that each kilogram of potassium removed by plants or through leaching must be returned in fertilizers may not always be correct. In many soils the large quantities of moderately available forms already present can be utilized so that only a part of the total amount removed by harvest need be replaced by fertilizer. Moreover, the importance of lime in reducing leaching losses of potassium should not be overlooked as a means of effectively utilizing the power of soils to furnish this element.

Soils of arid regions are often well supplied with weatherable potassium-containing minerals and can therefore supply adequate potassium for many years, even under irrigation where leaching is more important and plant removal is great. However, continued crop removal can deplete the available potassium pools even in these soils. Also, deep-rooted plants such as cotton and fruit trees may depend on the subsoil for much of their potassium. Increasing the availability of this element at depths below the plow layer is difficult.

Potassium Losses and Gains

The problem of maintaining soil potassium is outlined diagrammatically in Figure 14.28. Plant removal of potassium generally exceeds that of the other essential elements, with the possible exception of nitrogen. Annual losses from plant removal as great as 200 kg/ha or more of potassium are not uncommon, particularly if the plant is a legume and is cut several times for hay. As might be expected, therefore, the return of plant residues and manures is very important in maintaining soil potassium.

The annual losses of available potassium by leaching and erosion greatly exceed those of nitrogen and phosphorus. They are generally not as great, however, as the corresponding losses of available calcium and magnesium. Such losses of soil minerals have serious implications for sustainable soil productivity.

Increased Use of Potassium Fertilizers

The global use of potassium fertilizers will continue to increase as the pools of available potassium are depleted and increasing crop yields place demands on the soil that cannot be met by mineral weathering alone. This is especially true in cash-crop areas and in regions where sandy soils or highly weathered soils are prominent. Figure 14.1 shows that farmers in the United States increased their use of potassium fertilizer for many years until about 1980, by which time the potassium-supplying power of their soils (on

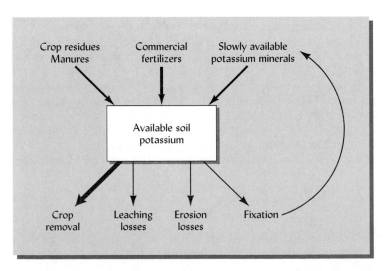

FIGURE 14.28 Gains and losses of *available* soil potassium under average field conditions. The approximate magnitude of the changes is represented by the width of the arrows. For any specific case the actual amounts of potassium added or lost undoubtedly may vary considerably from this representation. As was the case with nitrogen and phosphorus, commercial fertilizers are important in meeting crop demands.

average) had largely reached the level at which only maintenance additions were needed. This maintenance level in many regions is about half the level of average potassium removal (the other half of removals being replaced by potassium released by mineral weathering). In less developed agricultural regions of the world potassium fertilizer use will have to increase for many years to come if yields are to be increased or even maintained.

14.17   CONCLUSION

Soil phosphorus presents us with a double-edged sword; its management is crucially important from both an environmental standpoint and for soil fertility. On one hand, too little phosphorus commonly limits the productivity of natural and cultivated plants and is the cause of widespread soil and environmental degradation. On the other hand, industrialized agriculture has concentrated too much phosphorus in some cases, resulting in losses from soil that causes egregious eutrophication in surface waters. Some of these situations have been caused by excessive buildup of soil phosphorus with fertilizer. Others are the result of concentration of animal production so that phosphorus in the manure produced at many livestock facilities far exceeds that required by the crops grown on the surrounding land.

Except in cases of extreme buildup, the availability of phosphorus to plant roots has a double constraint: the low total phosphorus level in soils and the small percentage of this level that is present in available forms. Furthermore, even when soluble phosphates are added to soils, they are quickly fixed into insoluble forms that in time become quite unavailable to growing plants. In acid soils, the phosphorus is fixed primarily by iron, aluminum, and manganese; in alkaline soils, by calcium and magnesium. This fixation greatly reduces the efficiency of phosphate fertilizers, with little of the added phosphorus being taken up by plants. In time, however, this unused phosphorus can build up and serve as a reserve pool for plant absorption.

Potassium is generally abundant in soils, but it, too, is present mostly in forms that are quite unavailable for plant absorption. Fortunately, however, some soils contain considerable nonexchangeable but slowly available forms of this element. Over time this potassium can be released to exchangeable and soil solution forms that can be quickly absorbed by plant roots. This is fortunate since the plant requirements for potassium are high—five to ten times that of phosphorus and similar to that of nitrogen.

REFERENCES

Brady, N. C. 1974. *The Nature and Properties of Soils,* 8th ed. (New York: Macmillan).

Cogger, C., and J. M. Duxbury. 1984. "Factors affecting phosphorus loss from cultivated organic soils," *Journal of Environmental Quality,* 13:111–114.

Evans, C. E., and O. J. Attoe. 1948. "Potassium supplying power of virgin and cropped soils," *Soil Sci.,* 66:323–34.

Fares, F., et al. 1974. "Quantitative survey of organic phosphorus in different soil types," *Phosphorus in Agriculture* (Madison, Wis.: Amer. Soc. Agron.), pp. 25–40.

Fox, R. L. 1981. "External phosphorus requirements of crops," in *Chemistry in the Soil Environment,* ASA Special Publication No. 40 (Madison, Wis.: Amer. Soc. Agron. and Soil Sci. Soc. Amer.), pp. 223–239.

Gholston, L. E., and C. D. Hoover. 1948. "The release of exchangeable and nonexchangeable potassium from several Mississippi and Alabama soils upon continuous cropping," *Soil Sci. Soc. Amer. Proc.,* 13:116–21.

Grewal, J. S., and J. S. Kanwar. 1976. *Potassium and Ammonium Fixation in Indian Soils* (review) (New Delhi, India: Indian Council for Agricultural Research).

Khasawneh, F. E., et al. (eds). 1980. *The Role of Phosphorus in Agriculture* (Madison, Wis.: Amer. Soc. Agron.).

Kuo, S. 1988. "Application of modified Langmuir isotherm to phosphate sorption by some acid soils," *Soil Sci. Soc. Am. J.,* 52:97–102.

Magdoff, F. R., and R. J. Bartlett. 1980. "Effect of liming acid soils on potassium availability," *Soil Sci.,* 129:12–14.

Mandzak, J. M., and J. A. Moore. 1994. "The role of nutrition in the health of inland Western forests," *J. Sustainable Forestry,* **2**:191–210.

McLean, E. O. 1978. "Influence of clay content and clay composition on potassium availability," in G. S. Sekhon (ed.), *Potassium in Soils and Crops* (New Delhi, India: Potash Research Institute of India), pp. 1–19.

Mengel, K., and E. A. Kirkby. 1987. *Principles of Plant Nutrition* (4th edition) (Bern, Switzerland: International Potash Institute).

Munson, R. D. (ed.). 1985. *Potassium in Agriculture* (Madison, Wis.: Amer. Soc. Agron.).

Nowak, C. A., R. B. Downard, Jr., and E. H. White. 1991. "Potassium trends in red pine plantations at Pack Forest, New York," *Soil Sci. Soc. J.,* **55**:847–850.

Olsen, S. R., and F. E. Khasawneh. 1980. "Use and limitations of physical-chemical criteria for assessing the status of phosphorus in soils," in F. E. Khasawneh et al. (eds.), *The Role of Phosphorus in Agriculture* (Madison, Wis.: American Society of Agronomy, Crop Science Society of America, Soil Science Society of America).

Porter, R. M., J. E. Ayers, M. W. Johnson, Jr., and P. E. Nelson. 1981. "Influence of differential phosphorus accumulation on corn stalk rot," *Agron. J.,* **73**:283–287.

Raven, K. P., and L. R. Hossner. 1993. "Phosphorus desorption quantity-intensity relationships in soils," *Soil Sci. Soc. Am. J.,* **57**:1501–1508.

Saa, A., M. C. Trasar-Cepeda, B. Soto, F. Gil-Sotres, and F. Diaz-Fierros. 1994. "Forms of phosphorus in sediments eroded from burnt soils," *J. Environ. Quality,* **23**:739–746.

Sample, E. C., et al. 1980. "Reactions of phosphate fertilizers in soils," in F. E. Khasawneh, et al. (eds), *The Role of Phosphorus in Agriculture* (Madison, Wis.: Amer. Soc. Agron.).

Sharpley, A. N. 1990. "Reaction of fertilizer potassium in soils of differing mineralogy," *Soil Sci.,* **49**:44–51.

Smith, S. J., A. N. Sharpley, J. W. Naney, W. A. Berg, and O. R. Jones. 1991. "Water quality impacts associated with wheat culture in the Southern Plains," *J. Environ. Qual.,* **20**:244–249

Soltanpour, P. N., R. L. Fox, and R. C. Jones. 1988. "A quick method to extract organic phosphorus from soils," *Soil Science Soc. of America Journal,* **51**:255–256.

Stevenson, F. J. 1986. *Cycles of Soil Carbon, Nitrogen, Phosphorus, Sulfur, and Micronutrients* (New York: John Wiley and Sons).

Vaithiyanathan, P., and D. L. Correll. 1992. "The Rhode River watershed: phosphorus distribution and export in forest and agricultural soils," *J. Environ. Quality,* **21**:280–288.

Weil, R. R., P. W. Benedetto, L. J. Sikora, and V. A. Bandell. 1988. "Influence of tillage practices on phosphorus distribution and forms in three Ultisols," *Agronomy Journal,* **80**:503–509.

15

MICRONUTRIENT ELEMENTS

Of the eighteen elements known to be essential for plant growth, nine are required in such small quantities that they are called *micronutrients*[1] or *trace elements*. These elements are iron, manganese, zinc, copper, boron, molybdenum, nickel, cobalt, and chlorine.

Other elements, such as silicon, vanadium, and sodium, appear to improve the growth of at least certain plant species. Animals, including humans, also require most of these elements in their diets. Some elements, such as selenium, chromium, tin, iodine, and fluorine, have been shown to be essential for animal growth but are apparently not required by plants.

The terms, *micronutrient* and *trace element,* must not be construed to imply that these nutrients are somehow less important than macronutrients. To the contrary, the effects of micronutrient deficiency can be very severe in terms of stunted growth, low yields, dieback, and even plant death. By the same token, where they are needed, very small applications of micronutrients may produce dramatic results.

Increasing attention is being directed toward micronutrient deficiencies for several reasons:

1. Intensive plant production practices have increased crop yields, resulting in greater removal of micronutrients from soils.

2. The trend toward high-analysis fertilizers has reduced the use of impure salts and organic manures, which formerly supplied significant amounts of micronutrients.

3. Increased knowledge of plant nutrition and improved methods of analysis in the laboratory are helping in the diagnosis of micronutrient deficiencies that might formerly have gone unnoticed.

4. Increasing evidence indicates that food grown on soils with low levels of trace elements may provide insufficient human dietary levels of certain elements, even though the crop plants show no signs of deficiency themselves.

Interest in trace elements has also been sparked by problems of toxicity resulting from an oversupply of these elements. Levels of trace elements toxic to plants or to animals may result from natural soil conditions, from pollution, or from soil management practices.

---

[1]For review articles on this subject, see Welch (1995) and Stevenson (1986).

This chapter will provide some of the background and tools needed to recognize and deal effectively with the varied deficiencies and toxicities encountered in a wide range of soil–plant systems.

15.1 DEFICIENCY VERSUS TOXICITY

It is axiomatic that anything can be toxic if taken in large enough amounts. At low levels of a nutrient, deficiency and reduced plant growth may occur (*deficiency range*). As the level of nutrient is increased, plants respond by taking up more of the nutrient and increasing their growth. If a level of nutrient availability has been reached that is sufficient to meet the plant's needs (*sufficiency range*), raising the level further will have little effect on plant growth, although the concentration of the nutrient may continue to increase in the plant tissue. At some level of availability, the plant will take up too much of the nutrient for its own good (*toxicity range*), causing adverse physiological reactions to take place. The relationship among deficient, sufficient, and toxic levels of nutrient availability is described in Figure 15.1.

For macronutrients the sufficiency range is very broad and toxicity occurs rarely, if ever. However, for micronutrients the difference between deficient and toxic levels may be very narrow, making the possibility of toxicity quite real. For example, in the cases of boron and molybdenum, severe toxicity may result from applying as little as 3 to 4 kg/ha of available nutrient to a soil initially deficient in these elements. While the sufficiency range for other micronutrients is much wider, and toxicities are not as likely from overfertilization, toxicities of copper and zinc have been observed on soils contaminated by industrial sludges, metal smelter wastes, and long-term application of copper sulfate fungicide.

High levels of molybdenum may occur naturally in certain poorly drained alkaline soils. In some cases, enough of this element may be taken up by plants to cause toxicity, not only to susceptible plants, but also to livestock grazing forages on these soils. Boron, too, may occur naturally in alkaline soils at levels high enough to cause plant toxicity. Although somewhat larger amounts of most other micronutrients are required and can be tolerated by plants, great care needs to be exercised in applying micronutrients, especially for maintaining nutrient balance.

In addition, irrigation water in dry regions may contain enough dissolved boron, molybdenum, or selenium (a trace element not thought to be a plant nutrient) to damage sensitive crops, even if the original levels of these trace elements in the soil were not very high. Therefore it is prudent to monitor the content of these elements in water used for irrigation (see Section 10.5).

Micronutrients are required in very small quantities, their concentrations in plant tissue being one or more orders of magnitude lower than for the macronutrients (Figure 15.2). The ranges of plant tissue concentrations considered deficient, adequate, and toxic for several micronutrients are illustrated in Figure 15.3.

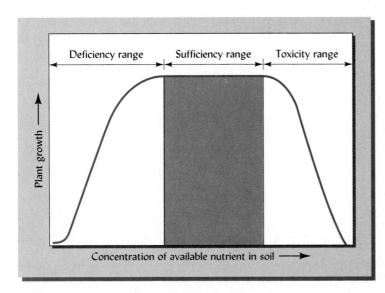

FIGURE 15.1 The relationship between the amount of a micronutrient available for plant uptake and the growth of the plant. Within the deficiency range, as nutrient availability increases so does plant growth (and uptake, which is not shown here). Within the sufficiency range, plants can get all of the nutrient they need, and so their growth is little affected by changes within this range. At higher levels of availability a threshold is crossed into the toxicity range, in which the amount of nutrient present is excessive and causes adverse physiological reactions leading to reduced growth and even death of the plant. A similar relationship exists between plant growth and the concentration of nutrients in plant tissue (see Figure 16.26).

15.2 ROLE OF THE MICRONUTRIENTS[2]

Physiological Roles

Micronutrients play many complex roles in plant nutrition. While most of the micronutrients participate in the functioning of a number of enzyme systems (Table 15.1), there is considerable variation in the specific functions of the various micronutrients in plant and microbial growth processes. For example, copper, iron, and molybdenum are capable of acting as electron carriers in enzyme systems that bring about oxidation-reduction reactions in plants. Such reactions are essential steps in photosynthesis and many other metabolic processes. Zinc and manganese function in many plant enzyme systems as bridges to connect the enzyme with the substrate upon which it is meant to act.

Molybdenum and manganese are essential for certain nitrogen transformations in microorganisms as well as in plants. Molybdenum and iron are components of the enzyme *nitrogenase,* which is essential for the processes of symbiotic and nonsymbiotic nitrogen fixation. Molybdenum is also present in the enzyme *nitrate reductase,* which is responsible for the reduction of nitrates in soils and plants.

Nickel has only recently been added to the list of elements shown to be essential to higher plants. It is essential for the function of several enzymes, including urease, the enzyme that breaks down urea into ammonia and carbon dioxide. Nickel-deficient legumes accumulate toxic levels of urea in their leaves; seeds of cereal plants deficient in nickel are not viable, and fail to germinate.

Zinc plays a role in protein synthesis, in the formation of some growth hormones, and in the reproductive process of certain plants. Copper is involved in both photosynthesis and respiration, and in the use of iron. It also stimulates lignification of cell walls. The roles of boron have yet to be clearly defined, but boron appears to be involved with cell division, water uptake, and sugar translocation in plants. Iron is involved in chlorophyll formation and degradation, and in the synthesis of proteins and nucleic acids. Manganese seems to be essential for photosynthesis, respiration, and nitrogen metabolism.

The role of chlorine is still somewhat obscure; however, it is known to influence photosynthesis and root growth. Cobalt is essential for the symbiotic fixation of nitrogen. In addition, legumes and some other plants have a cobalt requirement independent of nitrogen fixation, although the amount required is small compared to that for the nitrogen-fixation process.

Deficiency Symptoms

Insufficient supply of a nutrient is often expressed by certain visible plant symptoms, some of which are quite useful as indicators of particular micronutrient deficiencies. Most of the micronutrients are relatively immobile in the plant. That is, the plant cannot efficiently transfer the nutrient from older leaves to newer ones. Therefore the concentration of the nutrient tends to be lowest, and the symptoms of deficiency most pronounced, in the younger leaves, which develop after the supply of the nutrient has run low.[3] The iron-deficient sorghum in Plate 19 and the zinc-deficient corn and peach leaves in Plates 17 and 20 illustrate the pattern of pronounced deficiency symptoms on the younger leaves.

---

[2]For a review of the role of individual micronutrients, see Mengel and Kirkby (1987) and Marschner (1986). For nickel see Brown et al. (1987).

[3]This pattern is in contrast to most of the macronutrients (except sulfur), which are more easily translocated by the plants and so become most deficient in the older leaves.

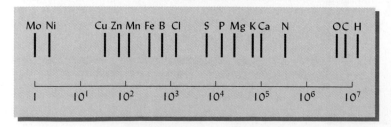

FIGURE 15.2 Relative numbers of atoms of the essential elements in alfalfa at bloom stage, expressed logarithmically. Note that there are more than 10 million hydrogen atoms for each molybdenum atom. Even so, normal plant growth would not occur without molybdenum. [Modified from Viets (1965)]

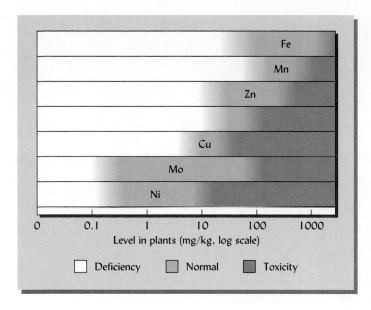

FIGURE 15.3 Deficiency, normal, and toxicity levels in plants for seven micronutrients. Note that the range is shown on a logarithm base and that the upper limit for manganese is about 10,000 times the lower range for molybdenum and nickel. (Based on data from many sources)

In the case of zinc deficiency in corn, broad white bands on both sides of the midrib are typical in young corn plants, but the symptoms may disappear as the soil warms up and the maturing plant's root system expands into a larger volume of soil. Interveinal chlorosis (darker green veins and light yellowish areas between the veins) on the younger leaves, and the whirl of tiny leaves at the terminal end of the branch (a symptom known as *little leaf*) are characteristic of tree foliage suffering from too little zinc.

Boron deficiency generally affects the growing points of plants, such as their buds, fruits, flowers, and root tips. Plants low in boron may produce deformed flowers (Figure 15.4), aborted seeds, thickened, brittle, puckered leaves, or dead growing points (see Figure 15.5). In all of these cases, a small dose of boron applied at the correct time could make the difference between a marketable or unmarketable plant product.

15.3 SOURCE OF MICRONUTRIENTS

Deficiencies of micronutrients may be related to low contents of these elements in the parent rocks or transported parent material. Similarly, toxic quantities are sometimes related to abnormally large amounts in the soil-forming rocks and minerals. More

TABLE 15.1 **Functions of Several Micronutrients in Higher Plants**

Micronutrient	Functions in higher plants
Zinc	Present in several dehydrogenase, proteinase, and peptidase enzymes; promotes growth hormones and starch formation; promotes seed maturation and production.
Iron	Present in several peroxidase, catalase, and cytochrome oxidase enzymes; found in ferredoxin, which participates in oxidation-reduction reactions (e.g., NO_3^- and SO_4^{2-} reduction, N fixation); important in chlorophyll formation.
Copper	Present in laccase and several other oxidase enzymes; important in photosynthesis, protein and carbohydrate metabolism, and probably nitrogen fixation.
Manganese	Activates decarboxylase, dehydrogenase, and oxidase enzymes; important in photosynthesis, nitrogen metabolism, and nitrogen assimilation.
Nickel	Essential for urease, hydrogenases, and methyl reductase; needed for grain filling, seed viability, iron absorption, and urea and ureide metabolism (to avoid toxic levels of these nitrogen-fixation products in legumes).
Boron	Activates certain dehydrogenase enzymes; facilitates sugar translocation and synthesis of nucleic acids and plant hormones; essential for cell division and development.
Molybdenum	Present in nitrogenase (nitrogen fixation) and nitrate reductase enzymes; essential for nitrogen fixation and nitrogen assimilation.
Cobalt	Essential for nitrogen fixation; found in vitamin B_{12}.

(a)   (b)

FIGURE 15.4 A misshapen rose, or *bullhead* on a boron-deficient tea rose (a). A well-formed bloom on the same plant after a solution containing boron was applied to the leaves and young buds (b). (Photos courtesy of R. Weil)

FIGURE 15.5 The importance of boron in plant growth is clearly illustrated by comparing the normal bean plant on the left to the boron-deficient plant on the right. Once the bean plant on the right used all the boron from its seed, development of new leaves from the terminal bud (see arrow) was completely arrested. (Photo courtesy of R. Weil)

often, however, deficiency and toxicity of these elements result from certain soil conditions which either increase or decrease the solubility and availability of these elements to plants.

Inorganic Forms

Sources of the nine micronutrients vary markedly from area to area. The wide variability in the content of these elements in soils and suggested contents in a representative soil is shown on Table 15.2.

All of the micronutrients have been found in varying quantities in igneous rocks. Two of them, iron and manganese, have prominent structural positions in primary silicate minerals, such as biotite and hornblende. Others, such as cobalt and zinc, also may occupy structural positions as minor replacements for the major constituents of silicate minerals, including clays.

The mineral forms of micronutrients are altered as mineral decomposition and soil formation occur. Oxides and, in some cases, sulfides of elements such as iron, manganese, and zinc are formed (Table 15.2). Secondary silicates, including the clay minerals, may contain considerable quantities of iron and manganese and smaller quantities of zinc and cobalt. Ultramafic rocks, especially serpentinite, are high in nickel. The micronutrient cations released as weathering occurs are subject to colloidal adsorption, just as are the calcium or aluminum ions (see Section 8.3).

Anions such as borate and molybdate in soils may undergo adsorption or reactions similar to those of the phosphates. For boron, the adsorption may be represented as follows:

Chlorine, by far the most soluble of the group, is added to soils in considerable quantities each year through rainwater. Its incidental addition to soils in fertilizers and in other ways helps prevent the deficiency of chlorine under field conditions. Chlorine is only very weakly adsorbed to soil colloids.

Organic Forms

Organic matter is an important secondary source of some of the trace elements. Several of them tend to be held as complex combinations by organic (humus) colloids. Copper is especially tightly held by organic matter, so much so that its availability can be very low in organic soils (Histosols). In uncultivated profiles, there is a somewhat greater concentration of micronutrients in the surface soil, much of it presumably in the organic fraction. Correlations between soil organic matter and contents of copper,

TABLE 15.2 Major Natural Sources of the Nine Micronutrients and Their Suggested Contents in a Representative Humid Region Surface Soil

Element	Major forms in nature	Analyses of soils	
		Range (mg/kg)[a]	Representative surface soil (mg/kg)
Iron	Oxides, sulfides, and silicates	10,000–100,000	25,000
Manganese	Oxides, silicates, and carbonates	20–4,000	1,000
Zinc	Sulfides, carbonates, and silicates	10–300	50
Copper	Sulfides, hydroxy carbonates, and oxides	2–100	20
Nickel	Silicates (especially serpentine)	5–10,000	20
Boron	Borosilicates, borates	2–100	10
Molybdenum	Sulfides, oxides, and molybdates	0.2–5	2
Chlorine[b]	Chlorides	7–50	10
Cobalt	Silicates	1–40	8

[a]Equivalent to parts per million.
[b]Much higher in saline and alkaline soils.

molybdenum, and zinc have been noted. Although the elements thus held are not always readily available to plants, their release through decomposition is undoubtedly an important fertility factor. Animal manures are a good source of micronutrients, much of it present in organic forms.

Forms in Soil Solution

The dominant forms of micronutrients that occur in the soil solution are listed in Table 15.3. The specific forms present are determined largely by the pH and by soil aeration (i.e., redox potential). Note that the cations are present in the form of either simple cations or hydroxy metal cations. The simple cations tend to be dominant under highly acid conditions. The more complex hydroxy metal cations become more prominent as the soil pH is increased.

Molybdenum is present mainly as MoO_4^{2-}, an anionic form that reacts at low pH in ways similar to those of phosphorus (see Chapter 14). Although boron also may be present in anionic form at high pH levels, research suggests that undissociated boric acid (H_3BO_3) is the form that is dominant in the soil solution and is absorbed by higher plants.

The cycling of micronutrients through the soil–plant–animal system is illustrated in a generalized way by Figure 15.6. Although not every micronutrient will participate in every pathway shown in Figure 15.6, it can be seen that organic chelates, soil colloids, soil organic matter, and soil minerals all contribute micronutrients to the soil solution and, in turn, to growing plants. As we turn our attention to micronutrient availability, it will be helpful to refer to this figure to see the relationships among the processes involved.

15.4 GENERAL CONDITIONS CONDUCIVE TO MICRONUTRIENT DEFICIENCY

Micronutrients are most apt to limit crop growth in (1) highly leached, acid, sandy soils; (2) organic soils; (3) soils of very high pH; and (4) soils that have been very intensively cropped and heavily fertilized with macronutrients only. Zinc deficiency is quite common where low organic matter subsoil is exposed by terracing, leveling, or grading operations.

Strongly leached, acid, sandy soils are low in most micronutrients for the same reasons they are deficient in most of the macronutrients—their parent materials were initially low in the elements, and acid leaching has removed much of the small quantity of micronutrients originally present. In the case of molybdenum, acid soil conditions markedly decrease availability.

The micronutrient contents of organic soils depend on the extent of the washing or leaching of these elements into the bog area as the soils were formed. In most cases, this rate of movement is too slow to produce deposits as high in micronutrients as are the surrounding mineral soils. The ability of organic soils to bind certain elements, notably copper, also accentuates micronutrient deficiencies.

Intensive cropping, in which large amounts of plant nutrients are removed in the harvest, accelerates the depletion of micronutrient reserves in the soil and increases the

TABLE 15.3 **Forms of Micronutrients Dominant in the Soil Solution**

Micronutrient	Dominant soil solution forms
Iron	Fe^{2+}, $Fe(OH)_2^+$, $Fe(OH)^{2+}$, Fe^{3+}
Manganese	Mn^{2+}
Zinc	Zn^{2+}, $Zn(OH)^+$
Copper	Cu^{2+}, $Cu(OH)^+$
Molybdenum	MoO_4^{2-}, $HMoO_4^-$
Boron	H_3BO_3, $H_2BO_3^-$
Cobalt	Co^{2+}
Chlorine	Cl^-
Nickel	Ni^{2+}, Ni^{3+}

From data in Lindsay (1972).

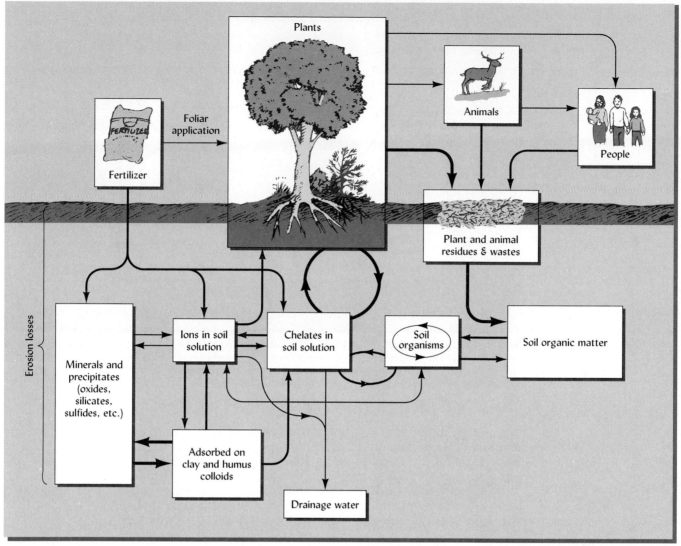

FIGURE 15.6 Cycling and transformations of micronutrients in the soil–plant–animal system. Although all micronutrients may not follow each of the pathways shown, most are involved in the major components of the cycle. The formation of chelates, which keep most of these elements in soluble forms, is a unique feature of this cycle.

likelihood of micronutrient deficiencies. Such depletion is most common where high yields are produced with the aid of chemical fertilizers that supply some of the macronutrients, but not the micronutrients.

Soil pH, especially in well-aerated soils, has a decided influence on the availability of all the micronutrients except chlorine. Under very acid conditions, molybdenum is rendered unavailable; at high pH values, availability of all the micronutrient cations is unfavorably affected.

15.5 FACTORS INFLUENCING THE AVAILABILITY OF THE MICRONUTRIENT CATIONS

Each of the six micronutrient cations (iron, manganese, zinc, copper, nickel, and cobalt) is influenced in a characteristic way by the soil environment. However, certain soil factors have the same general effects on the availability of all six cations.

Soil pH

The micronutrient cations are most soluble and available under acid conditions. In very acid soils, there is a relative abundance of the ions of iron, manganese, zinc, and copper (see Figure 15.7). Nickel also follows a general pattern of increased availability at low

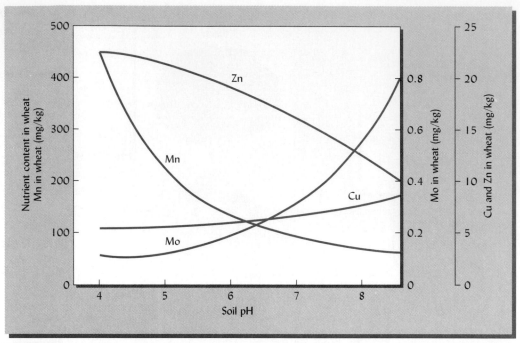

FIGURE 15.7 Effect of soil pH on the concentrations of manganese, zinc, copper, and molybdenum in wheat plants. The soils were from different countries around the world. The molybdenum levels are extremely low, but increase with increasing pH. Manganese and zinc levels decrease as the pH rises, while copper is little affected. [Redrawn from Sillanpaa (1982)]

pH. In fact, under these conditions, the soil solution concentrations or activities of one or more of these elements (most commonly manganese) is often sufficiently high to be toxic to common plants. As indicated in Chapter 9, one of the primary reasons for liming acid soils is to reduce the solubility of manganese and aluminum.

As the pH is increased, the ionic forms of the micronutrient cations are changed first to the hydroxy ions and, finally, to the insoluble hydroxides or oxides of the elements. The following example uses the ferric ion as typical of the group:

$$Fe^{3+} \xrightarrow{OH^-} Fe(OH)^{2+} \xrightarrow{OH^-} Fe(OH)_2^{+} \xrightarrow{OH^-} Fe(OH)_3$$

Simple cation   Hydroxy metal cations   Hydroxide
(soluble)   (soluble)   (insoluble)

All of the hydroxides of the micronutrient cations are relatively insoluble, some more so than others. The exact pH at which precipitation occurs varies from element to element and between oxidation states of a given element. For example, the higher-valence states of iron and manganese form hydroxides that are much more insoluble than their lower-valence counterparts. In any case, the principle is the same—at low pH values, the solubility of micronutrient cations is high, and as the pH is raised, their solubility and availability to plants decrease. Overliming of a soil with a naturally low pH often leads to deficiencies of iron, manganese, zinc, copper, and sometimes boron. Such deficiencies associated with high pH occur naturally in many of the calcareous soils of arid regions.

The general desirability of a slightly acid soil (with a pH between 6 and 7) largely stems from the fact that for most plants this pH condition allows micronutrient cations to be soluble enough to satisfy plant needs without becoming so soluble as to be toxic (see Figure 9.18). Certain plants, especially those which are native to very acid soils, have only a poor ability to take up iron and other micronutrients unless the soil is quite acid (pH about 5). Such acid-loving plants therefore become deficient in these elements when the soil pH is such that iron solubility is lowered (usually above pH 5.5) (see Figure 15.8). Section 9.7 gives more specific information on the pH preferences of various plants.

Zinc availability, like that of iron, is reduced when soil pH is raised by liming. However, in addition to the pH effects, addition of high-magnesium liming materials (e.g.,

FIGURE 15.8 The chlorotic foliage of these azaleas is a sign of iron deficiency inadvertently induced by a too-high soil pH. In this case the iron deficiency was caused by the use of marble gravel as a decorative mulch. Marble consists mostly of calcium carbonate. Rainwater percolating through the gravel mulch dissolved enough of the calcium to lime the soil below, raising the pH gradually from 5.2 to 6.0. At pH 6.0 the solubility of iron is too low for azaleas to obtain what they need. Calcium leaching from concrete walkways can have a similar effect on acid-loving vegetation growing in adjacent soil. (Photo courtesy of R. Weil)

dolomitic limestone) further decreases zinc availability because zinc is tightly adsorbed to dolomite and magnesium carbonate crystal surfaces. Zinc deficiency may also be aggravated by interaction between zinc and magnesium in the plant.

Oxidation State and pH

The trace element cations, iron, manganese, nickel, and copper, occur in soils in more than one valence state. In the lower-valence states, the elements are considered *reduced,* in the higher-valence state they are *oxidized.* Metallic cations generally become reduced when the oxygen supply is low, as occurs in wet soils containing decomposable organic matter. Reduction can also be brought about by organic metabolic reducing agents such a NADPH or caffeic acid, produced by plants and microorganisms in the soil. Oxidation and reduction reactions in relation to soil drainage are discussed in Section 7.5.

The changes from one valence state to another are, in most cases, brought about by microorganisms and organic matter. In some cases, the organisms may obtain their energy directly from the inorganic reaction. For example, the oxidation of manganese from Mn^{2+} in manganous oxides (MnO) to Mn^{4+} in manganic oxides (MnO_2) can be carried out by certain bacteria and fungi. In other cases, organic compounds formed by microbes or plant roots may be responsible for the oxidation or reduction. In general, high pH favors oxidation, whereas acid conditions are more conducive to reduction.

INTERACTION OF SOIL REACTION AND AERATION. At pH values common in soils, the oxidized states of iron, manganese, and copper are generally much less soluble than are the reduced states. The hydroxides (or hydrous oxides) of these high-valent forms precipitate even at low pH values and are extremely insoluble. For example, the hydroxide of trivalent ferric iron precipitates at pH values of 3.0 to 4.0, whereas ferrous hydroxide does not precipitate until a pH of 6.0 or higher is reached.

The interaction of soil acidity and aeration in determining micronutrient availability is of great practical importance. Iron, manganese, and copper are generally more available under conditions of restricted drainage or in flooded soils (Figure 15.9). Very acid soils that are poorly drained may supply toxic quantities of iron and manganese. Manganese toxicity has been reported to occur when certain high-manganese acid soils are thoroughly wetted during irrigation. Andisols (volcanic soils) with high organic matter melanic epipedons are known to cause manganese toxicity problems when they are wet by heavy rains. Iron toxicity is common in flooded rice paddies. Such toxicity is much less apt to occur under well-drained conditions, unless soil pH is very low.

At the high end of the soil-pH range, good drainage and aeration often have the opposite effect. Well-oxidized calcareous soils are sometimes deficient in available iron, zinc, or manganese even though adequate total quantities of these trace elements are present. The hydroxides of the high-valence forms of these elements are too insoluble to supply the ions needed for plant growth. In contrast, at high soil pH values, molybdenum availability may be excessively high. *Molybdenosis* is a potentially fatal disorder caused by excessive molybdenum in the diet of livestock grazing plants grown on certain very high pH soils.

There are marked differences in the sensitivity of different plant varieties to iron deficiency in soils with high pH. This is apparently caused by differences in their ability to solubilize iron immediately around the roots. Efficient varieties respond to iron stress by acidifying the immediate vicinity of the roots and by excreting compounds capable of reducing the iron to a more soluble form, with a resultant increase in its availability (Figure 15.10). It appears that most of the iron-reducing activity is concentrated in the root hairs of actively growing young roots (Figure 15.11).

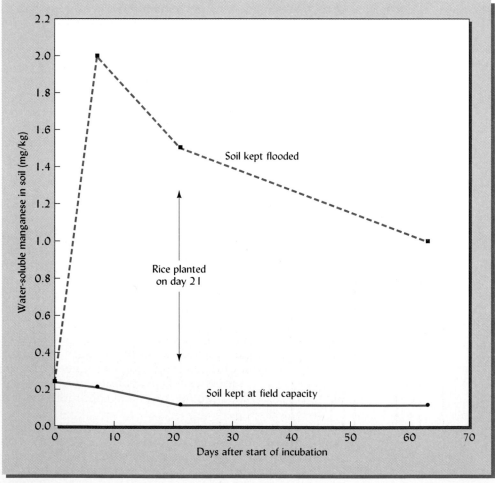

FIGURE 15.9 Effect of flooding on the amount of water-soluble manganese in soils. The data are the averages for 13 unamended Ultisol horizons with initial pH values ranging from 3.9 to 7.1. [Data from Weil and Holah (1989)]

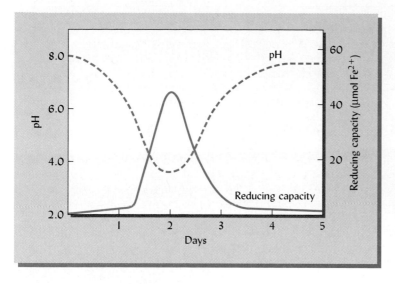

FIGURE 15.10 Response of one variety of sunflowers to iron deficiency. When the plant became stressed owing to iron deficiency, plant exudates lowered the pH and increased the reducing capacity immediately around the roots. Iron is solubilized and taken up by the plant, the stress is alleviated, and conditions return to normal. [From Marschner et al. (1974) as reported by Olsen et al. (1981); used with permission of *American Scientist*]

Other Inorganic Reactions

Micronutrient cations interact with silicate clays in two ways. First, they may be involved in cation exchange reactions much like those of calcium or aluminum. Second, they may be tightly bound or fixed to certain silicate clays, especially the 2:1 type. Zinc, manganese, cobalt, and iron ions sometimes occur in the crystal structure of these clays. Depending on conditions, they may be released from the clays or fixed by them in a manner similar to that by which potassium is fixed (see Section 14.14). The fixation may cause serious deficiency in the case of cobalt, and sometimes zinc, because these elements are present in soil in such small quantities (Table 15.2).

The application of large quantities of phosphate fertilizers can adversely affect the supply of some of the micronutrients. The uptake of both iron and zinc may be reduced in the presence of excess phosphates. For environmental quality reasons, as well as from a practical standpoint, phosphate fertilizers should be used in only those quantities required for good plant growth.

Lime-Induced Chlorosis

Iron deficiency in fruit trees and many other plants is encouraged by the presence of the bicarbonate ion. Bicarbonate-containing irrigation waters increase the level of this ion in some soils. In other soils, especially poorly buffered sandy soils, the problem stems from application of more lime than needed to reach the proper pH. The chlorosis apparently results from iron deficiency in soils with high pH because the bicarbonate ion interferes in some way with iron metabolism. This interference is coupled with the fact that iron solubility in the soil is greatly reduced with higher pH.

Role of Mycorrhizae

A symbiosis between most higher plants and certain soil fungi produces mycorrhizae (fungus roots) which are far more efficient than normal plant roots in several respects. The nature of the mycorrhizal symbiosis and its importance in phosphorus nutrition were described in Sections 11.9 and 14.3, but it is worth mentioning here that mycorrhizae have also been shown to increase plant uptake of micronutrients (Table 11.4). Crop rotations and other practices that encourage a diversity of mycorrhizal fungi may thereby improve micronutrient nutrition of plants (Figure 15.12).

Surprisingly, mycorrhizae also appear to protect plants from excessive uptake of micronutrients and other trace elements where these elements are present in potentially toxic concentrations. Seedlings of trees such as birch, pine, and spruce are able to grow well on sites contaminated with high levels of zinc, copper, nickel, and aluminum only if their roots are sheathed by ectomycorrhizae. The mycorrhizae apparently help exclude these metallic cations from the root stele and prevent long-distance transport of metal cations within the plant.

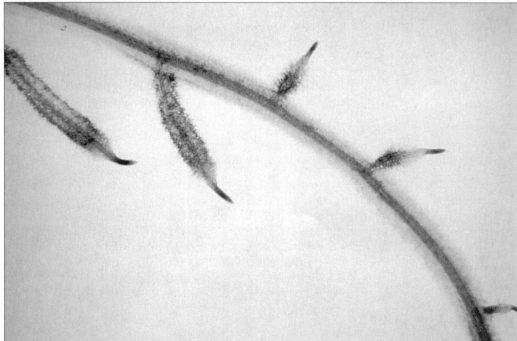

FIGURE 15.11 Reduction of iron by tomato roots is made visible by soaking live roots in a solution containing a stain that turns a dark blue color when the iron in the stain is reduced to the ferrous form. Note that the reduction reaction occurred only in the mature part of young branch rootlets (not in the zone just behind the root tip). The expanding root tips had not yet developed the reducing mechanisms. The close-up (top) shows that the reduction reaction occurs only in the root hairs. The main part of the root remains undarkened. (Photos courtesy of Paul Bell, Louisiana State University Agricultural Center)

15.6 ORGANIC COMPOUNDS AS CHELATES

The cationic micronutrients react with certain organic molecules to form organometallic complexes called *chelates.* If these complexes are soluble they increase the availability of the micronutrient and protect it from precipitation reactions. Conversely, formation of an insoluble complex will decrease the availability of the micronutrient.

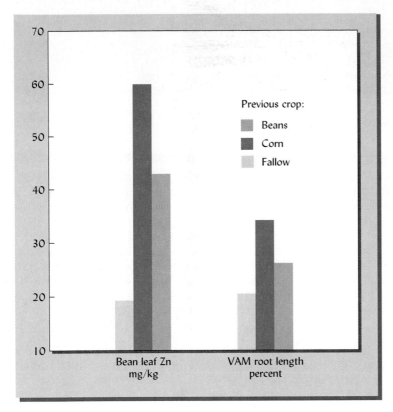

FIGURE 15.12 Effect of crop rotation on uptake of zinc and formation of mycorrhizae by beans. The bean crop, grown with furrow irrigation on an Aridisol (Calciorthid) in Idaho, was preceded by either a corn crop, a bean crop, or a year of bare fallow. The corn-followed-by-bean rotation favored both vesicular arbuscular mycorrhizae (VAM) formation and zinc uptake by the second-year bean crop. [Data from Hamilton et al. (1993)]

A chelate (from a Greek word *chele*, claw) is an organic compound in which two or more atoms are capable of bonding to the same metal atom, thus forming a ring. These organic molecules may be synthesized by plant roots and released to the surrounding soil, may be present in the soil humus, or may be synthetic compounds added to the soil to enhance micronutrient availability. In complexed form, the cations are protected from reaction with inorganic soil constituents that would make them unavailable for uptake by plants. Iron, zinc, copper, and manganese are among the cations that form chelate complexes. Two examples of an iron chelate ring structure are shown in Figure 15.13.

The effect of chelation can be illustrated with iron. In the absence of chelation, when an inorganic iron salt such as ferric sulfate is added to a calcareous soil, most of the iron is quickly rendered unavailable by reaction with hydroxide, as follows:

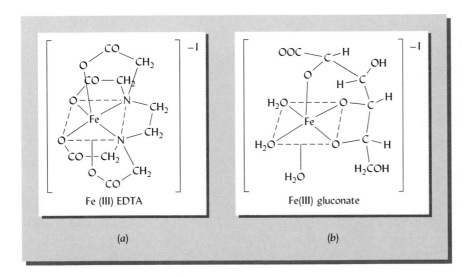

FIGURE 15.13 Structural formula for two common iron chelates, ferric ethylenedi-aminetetra-acetate (Fe-EDTA) (a) and ferric gluconate (b). In both chelates, the iron is protected, and yet can be used by plants. [Diagrams from Clemens et al. (1990); reprinted by permission of Kluwer Academic Publishers]

$$Fe^{3+} + 3OH^- \rightleftharpoons FeOOH + H_2O$$

(available)   (unavailable)

In contrast, if the iron is chelated, it largely remains in the chelate form, which is available for uptake by plants. In this reaction, the available iron chelate reactant is favored:

$$Fe \; chelate + 3OH^- \rightleftharpoons FeOOH + chelate^{3-} + H_2O$$

(available)   (unavailable)

The mechanism by which micronutrients from chelates are absorbed by plants is different for different plants. Many dicots appear to remove the metallic cation from the chelate at the root surface, reducing (in the case of iron) and taking up the cation while releasing the organic chelating agent in the soil solution. Roots of certain grasses have been shown to take in the entire chelate-metal complex, reducing and removing the metallic cation inside the root cell, then releasing the organic chelate back to the soil solution (Figure 15.14). In both cases it appears that the primary role of the chelate is to allow metallic cations to remain in solution so they can diffuse through the soil to the root. Once the micronutrient cations are inside the plant, other organic chelates (such as citrates) may be carriers of these cations to different parts of the plant.

Stability of Chelates

Some of the major synthetic chelating agents are listed in Table 15.4. Many similar chelating compounds occur naturally in the soil.

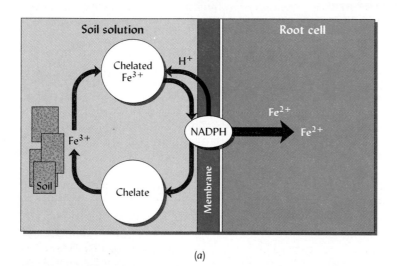

(a)

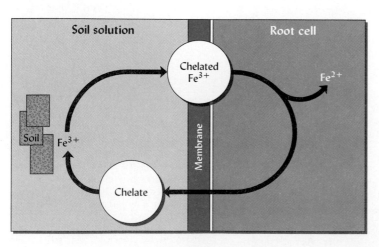

(b)

FIGURE 15.14 Two ways in which plants utilize micronutrients held in chelated form. (Top) Dicotyledonous plants such as cucumber and peanuts produce strong reducing agents (NADPH) that reduce iron at the outer surface of the root membrane. They then take in only the reduced iron, leaving the organic chelate in the soil solution where it can complex another iron atom. (Bottom) Grass plants such as wheat or corn apparently take the entire chelate-metal complex into their root cells. They then remove the iron, reduce it, and return the chelate to the soil solution.

TABLE 15.4 Stability Constants (K) for Selected Chelating Agents and Nutrient Cations

The stability constants are given as logarithms, so a difference of 1.0 represents a 10-fold difference in stability. Calcium is included because, although it is not a micronutrient, it is usually the metallic cation with by far the greatest activity in the soil solution. As such, calcium competes with micronutrient cations for binding sites in chelating agents. The relative cost of chelating agents is also given.

Chelating agent	Log K[a]						Relative cost[b]
	Fe^{3+}	Fe^{2+}	Zn^{2+}	Cu^{2+}	Mn^{2+}	Ca^{2+}	
EDTA	25.0	14.27	14.87	18.70	13.81	11.0	4.4
EDDHA	33.9	14.3	16.8	23.94	—	7.2	43
HEEDTA	19.6	12.2	14.5	17.4	10.7	8.0	5.5
Citrate	11.2	4.8	4.86	5.9	3.7	4.68	—
Gluconate	37.2	1.0	1.7	36.6	—	1.21	1.0

$$^aK = \frac{[\text{metal-chelate}]}{[\text{metal}]\,[\text{chelate}]}$$

[b]Cost per kg of iron relative to the cost of iron gluconate. Based on 1989 prices (from Clemens et al., 1990).

Chelates vary in their stability and therefore in their suitability as sources of micronutrients. The stability of a chelate is measured by its stability constant, which is related to the tenacity with which a metal ion is bound in the chelate. If the binding and release of a metal (M) by chelating agent is represented by the following reaction,

$$\text{Metal-chelate}^{-1} \rightleftharpoons \text{metal}^{2+} + \text{chelate}^{-3}$$

then the stability constant (K) for the metal-chelate complex is calculated as follows, where the values in brackets are concentrations in solution.

$$K = \frac{[\text{metal-chelate}]}{[\text{metal}]\,[\text{chelate}]}$$

The larger the stability constant, the greater the tendency for the metal to remain chelated. The stability constant (K) for each metal-chelate complex is different and is usually expressed as the logarithm of K (see Table 15.4).

The stability constant is useful in predicting which chelate is best for supplying which micronutrient. An added metal chelate must be reasonably stable within the soil if it is to have lasting advantage. For example, the stability constant for EDDHA-Fe^{3+} is 33.9, but that for EDDHA-Zn^{2+} is only 16.8. We can therefore predict that if EDDHA-Zn were added to a soil, the Zn in the chelate would be rapidly and almost completely replaced by Fe^{3+} from the soil, leaving the Zn in the unchelated form and subject to precipitation:

$$\text{Zn chelate}^{-1} + Fe^{3+} \rightleftharpoons \text{Fe chelate} + Zn^{2+}$$

Since the iron chelate is more stable than its zinc counterpart, the reaction goes to the right, and the released zinc ion is subject to reaction with the soil. Similarly, calcium can replace micronutrients from chelates. Even though the stability constant for Ca chelates is generally low, calcium often replaces micronutrients from chelates in practice because the concentration of Ca^{2+} in the soil solution is far greater than the concentrations of micronutrients.

It should not be inferred that only iron chelates are effective. The chelates of other micronutrients, including zinc, manganese, and copper, have been used successfully to supply these nutrients (see Figure 15.15). Apparently, replacement of other micronutrients in the chelates by iron from the soil is sufficiently slow to permit absorption by plants of the other added micronutrients. Also, because foliar spray and banded appli-

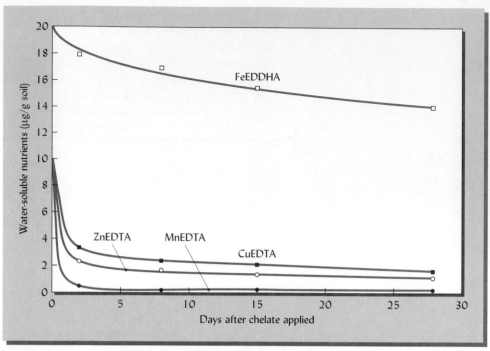

FIGURE 15.15 Average reduction in water solubility of four chelated micronutrients when incubated with four calcareous soils with pH higher than 8. The stability constants listed in Table 15.4 explain the behavior of the four micronutrient chelates shown here. The iron chelate was most stable in these soils, that with manganese the least. Because of the small quantities needed by plants, even the copper and zinc chelates would likely provide adequate nutrients for plant absorption. [Drawn using data from Ryan and Hariq (1983)]

cations are often used to supply zinc and manganese, the possibility of reaction of these elements in chelates with iron and calcium in the soil can be reduced or eliminated.

The use of synthetic chelates in industrial countries is substantial in spite of the fact that they are quite expensive. They are used primarily to ameliorate micronutrient deficiencies of fruit trees and ornamentals. Although chelates may not replace the more conventional methods of supplying most micronutrients, they offer possibilities in special cases. Some of the chelators used in micronutrient fertilizers, such as gluconate, are naturally occurring and can supply certain micronutrients much more economically than can the more expensive aminopolycarboxylate compounds (e.g., EDDHA) listed in Table 15.4. Agricultural and chemical research will likely continue to increase the opportunities for effective use of chelates.

In addition, selection of plant varieties whose roots produce their own chelating agents and practices that encourage the production of natural chelating agents from decomposing organic matter may take increasing advantage of the chelation phenomena to improve micronutrient fertility of soils.

15.7 FACTORS INFLUENCING THE AVAILABILITY OF THE MICRONUTRIENT ANIONS

Unlike the cations needed in trace quantities by plants, the anion micronutrients seem to have relatively little in common with each other. Chlorine, molybdenum, and boron are quite different chemically, so little similarity would be expected in their reaction in soils.

Chlorine

Chlorine is absorbed in larger quantities by most crop plants than any of the micronutrients except iron. Most of the chlorine in soils is in the form of the chloride ion, which leaches rather freely from humid region soils. In semiarid and arid regions, a higher concentration might be expected, with the amount reaching the point of salt toxicity in some of the poorly drained saline soils. In most well-drained areas, however, one would not expect a high chlorine content in the surface of arid region soils.

Except where toxic quantities of chlorine are found in saline soils, there are no common soil conditions that reduce the availability and use of this element. Accretions of chlorine from the atmosphere along with those from fertilizer salts such as potassium chloride are sufficient to meet most crop needs. However, crop responses to chloride additions have been noted for winter wheat, and are thought to be due to reduction of diseases such as "take all." Tropical palms, adapted to growth in coastal soils where ocean spray contributes much chlorine, sometimes show chlorine deficiency if they are grown on inland soils with relatively low chlorine levels.

Boron

Boron is one of the most commonly deficient of all the micronutrients. The availability of boron is related to the soil pH, this element being most available in acid soils. While it is most available at low pH, boron is also rather easily leached from acid, sandy soils. Therefore, although deficiency of boron is relatively common on acid, sandy soils, it occurs because of the low supply of total boron rather than because of low availability of the boron present.

When soil pH is increased by liming, boron availability decreases, being lowest between pH 7 and 9. Apparently this element is fixed or bound by soil colloids as the pH increases, resulting in lime-induced boron deficiency in some cases. Note that the mechanism is likely different from that of phosphate fixation, which is high in acid soils (see Section 14.8).

Boron is also adsorbed by humus, the strength of binding being even greater than for inorganic colloids. Boron is also a component of soil organic matter that is released by microbial mineralization. Consequently, organic matter serves as a major reservoir for boron in many soils and exerts considerable control over the availability of this nutrient. The adsorption of boron to organic and inorganic colloids increases with increasing pH (Figure 15.16).

Boron availability is impaired by long dry spells, especially following periods of optimum moisture conditions. Boron deficiencies are also common in calcareous Aridisols and in neutral to alkaline soils with a high pH.

Soluble boron is present in soils mostly in the form of boric acid (H_3BO_3). A second soluble form ($H_2BO_3^-$) is present in smaller quantities, its activity increasing as soil pH increases to neutral and alkaline conditions.

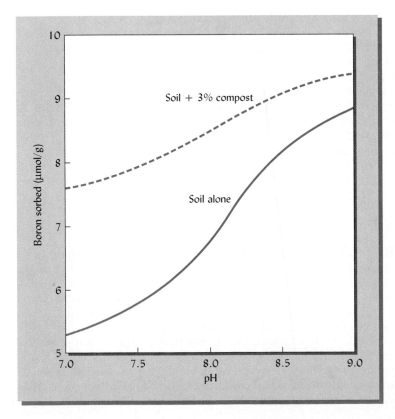

FIGURE 15.16 Effect of organic matter (added as compost in this experiment) and soil pH on adsorption of boron. Along with clays, humic compounds in decomposed organic matter adsorb boron. The adsorption capacities from both clays and humic compounds increase with rising pH in the neutral to alkaline range. While adsorption lowers the boron concentration in the soil solution, and hence its availability for plant uptake, it also protects boron from leaching loss. The soil in this study was a Calcic Haploxeralf. [Redrawn from Yermiyahu et al. (1995)]

Molybdenum

Soil pH is the most important factor influencing the availability and plant uptake of molybdenum. The following equation shows the forms of this element present at low and high soil pH:

$$H_2MoO_4 \underset{+H^+}{\overset{+OH^-}{\rightleftharpoons}} HMoO_4^- + H_2O \underset{+H^+}{\overset{+OH^-}{\rightleftharpoons}} MoO_4^{2-} + H_2O$$

At low pH values, the relatively unavailable H_2MoO_4 and $HMoO_4^-$ forms are prevalent, whereas the more readily available MoO_4^{2-} anion is dominant at pH values above 5 or 6. The MoO_4^{2-} ion is subject to adsorption by oxides of iron and possibly aluminum just as is phosphate, but calcium molybdate is much more soluble than its phosphorus counterpart.

The liming of acid soils will usually increase the availability of molybdenum (Figure 15.7). The effect is so striking that some researchers, especially those in Australia and New Zealand, argue that the primary reason for liming very acid soils is to supply molybdenum. Furthermore, in some instances 30 grams or so of molybdenum added to acid soils has given about the same increase in the yield of legumes as has the application of several megagrams of lime.

The phosphate anion seems to improve the availability of molybdenum by competing with the latter for sorption sites on soil surfaces. For this reason, molybdate salts are often applied along with phosphate carriers to molybdenum-deficient soils. This practice apparently encourages the uptake of both elements and is a convenient way to add the extremely small quantities of molybdenum required. Legume seeds coated in a mixture of superphosphate and sodium molybdate have also been used successfully to improve the grazing quality of acid soil range and savanna lands.

A second common anion, the sulfate ion, seems to have the opposite effect on plant utilization of molybdenum. Sulfate reduces molybdenum uptake, and seems to compete with molybdenum at functional sites on plant metabolic compounds.

15.8 NEED FOR NUTRIENT BALANCE

Nutrient balance among the trace elements is as essential as, but even more difficult to maintain than, macronutrient balance. Some of the plant enzyme systems that depend on micronutrients require more than one element. For example, both manganese and molybdenum are needed for the assimilation of nitrates by plants. The beneficial effects of combinations of phosphates and molybdenum have already been discussed. Apparently, some plants need zinc and phosphorus for optimum use of manganese. The use of boron and calcium depends on the proper balance between these two nutrients. A similar relationship exists between potassium and copper and between potassium and iron in the production of good-quality potatoes. Copper utilization is favored by adequate manganese, which in some plants is assimilated only if zinc is present in sufficient amounts. Of course, the effects of these and other nutrients will depend on the specific plant being grown, but the complexity of the situation can be seen from the examples cited.

Antagonism and Synergism

Some enzymatic and other biochemical reactions requiring a given micronutrient may be poisoned by the presence of a second trace element in toxic quantities. Other negative effects occur because one element competes with or otherwise reduces uptake of a second element by the plant root. On the other hand, a good supply of a certain nutrient may enhance the utilization of a second nutrient element in what is termed a *synergistic effect*. Some of these interactions are summarized in Table 15.5.

Some of the antagonistic effects may be used effectively in reducing toxicities of certain of the micronutrients. For example, copper toxicity of citrus groves caused by residual copper from fungicidal sprays may be reduced by adding iron and phosphate fertilizers. Sulfur additions to calcareous soils containing toxic quantities of soluble molybdenum may reduce the availability, and hence the toxicity, of molybdenum.

TABLE 15.5 Some Antagonistic (Negative) and Synergistic (Positive) Effects of Other Nutrients on Micronutrient Utilization by Plants[a]

The occurrence of so many interactions emphasizes the need for balance among all nutrients and avoidance of excess application of any particular nutrient.

Micronutrient	Elements decreasing utilization		Elements increasing utilization	
	Soil and root surface reactions	Plant metabolic reactions	Soil and root surface reactions	Plant metabolic reactions
Fe	B, Cu, Zn, Mo, Mn	Mn, Mo, P, S, Zn	B, Mo	
Mn	Fe, B	Fe	B	
Zn	Mg, Cu, B, Fe, P	Fe, N	N, B	Fe, Mg
Cu	B, Zn, Mo	P, N	B	
B	Ca, K			N
Mo	S, Cu	S	P	P
Ni	Ca, Fe	Fe, Zn		

[a]Summarized from many sources.

These examples of nutrient interactions, both beneficial and detrimental, emphasize the highly complicated nature of the biological transformations in which micronutrients are involved. The total land area on which unfavorable nutrient balances require special micronutrient treatment is increasing as soils are subjected to more intensive cropping methods.

15.9 SOIL MANAGEMENT AND MICRONUTRIENT NEEDS

Although the characteristics of each micronutrient are quite specific, some generalizations with respect to management practices are possible.

In seeking the cause of plant abnormalities, one should keep in mind the conditions under which micronutrient deficiencies or toxicities are likely to occur. Sandy soils, mucks, and soils having very high or very low pH values are prone to micronutrient deficiencies. Areas of intensive cropping and heavy macronutrient fertilization may be deficient in the micronutrients.

Changes in Soil Acidity

In very acid soils, one might expect toxicities of iron and manganese and deficiencies of phosphorus and molybdenum. These can be corrected by liming and by appropriate fertilizer additions. Calcareous soils may have deficiencies of iron, manganese, zinc, and copper and, in a few cases, a toxicity of molybdenum.

No specific statement can be made concerning the pH value most suitable for all the elements. However, medium-textured soils generally supply adequate quantities of micronutrients when the soil pH is held between 6 and 7. In sandy soils, a somewhat lower pH may be justified because the total quantity of micronutrients is low, and even at pH 6.0, some cation deficiencies may occur. It is important to be on guard to recognize inadvertent increases in soil pH, such as those occurring from application of lime-stabilized sewage sludge (see Chapter 16) or leaching of calcareous gravel and pavements (see Figure 15.8).

Soil Moisture

Drainage and moisture control can influence micronutrient solubility in soils. Improving the drainage of acid soils will encourage the formation of the oxidized forms of iron and manganese. These are less soluble and, under acid conditions, less toxic than the reduced forms.

Flooding a soil will favor the reduced forms of iron and manganese, which are more available to growing plants (see Figure 15.17). Excessive moisture at high pH values can have the opposite effect since poorly drained soils have a high carbon dioxide concentration, encouraging the formation of bicarbonate ions which reduce iron availability. Poor drainage also increases the availability of molybdenum in some alkaline soils to the point of producing plants with toxic levels of this element.

FIGURE 15.17 Manganese toxicity in young cotton plants as affected by soil moisture during the three weeks prior to planting the cotton seed. The crinkled-leaf symptom and stunted size resulted from toxic levels of manganese in the soil that had been flooded prior to planting. Both pots were maintained at field-capacity water content from sowing to harvest. The Jackland soil (Aquic Hapludult) used is high in manganese-containing minerals and has a pH of approximately 5.8. (Photo courtesy of R. Weil)

Fertilizer Applications

Micronutrient deficiencies are relatively unlikely to be a problem in most soils to which plant residues are returned, and organic amendments such as animal manure or sewage sludge are regularly applied. Animal manures applied at normal rates sufficient to supply macronutrient needs carry enough copper, zinc, manganese, and iron to supply a major portion of micronutrient needs as well (see Table 16.7). In addition, the chelates produced from manure enhance the availability of these micronutrients.

Nonetheless, the most common management practice to overcome micronutrient deficiencies (and some toxicities) is the application of commercial fertilizers. Examples of fertilizer materials applied to supply micronutrients are shown in Table 15.6. The materials are most commonly applied to the soil, although foliar sprays and even seed treatments can be used. Foliar sprays of dilute inorganic salts or organic chelates are more effective than soil treatments where high soil pH and other factors render the soil-applied nutrients unavailable.

For soil applications, about one-half to one-fourth as much fertilizer is needed if the application is banded rather than broadcast (see Figure 16.23). About one-fifth to one-tenth as much fertilizer is needed if the material is sprayed on the plant foliage. Foliar application may, however, require repeated applications in a single year. Treating seeds with small dosages (20–40 g/ha) of molybdenum has had satisfactory results on molybdenum-deficient acid soils. Typical rates of application to soil are given in Table 15.7.

The micronutrients can be applied to the soil either as separate materials or incorporated in standard macronutrient carriers. Unfortunately, the solubilities of copper, iron, manganese, and zinc can be reduced by such incorporation, but boron and molybdenum remain in reasonably soluble condition. Liquid macronutrient fertilizers containing polyphosphates encourage the formation of complexes that protect added micronutrients from adverse chemical reactions. In effect, polyphosphates in fertilizer solutions act as chelating agents for micronutrients. High-surface-area pitted glasslike beads, called *frits,* are manufactured with boron, copper, zinc, and other micronutrients incorporated into the glass. These fritted materials slowly release their micronutrients as

TABLE 15.6 A Few Commonly Used Fertilizer Materials That Supply Micronutrients

Micronutrient	Commonly used fertilizers		Nutrient content (%)
Boron	Borax	$Na_2B_4O_7 \cdot 10H_2O$	11
	Sodium pentaborate	$Na_2B_{10}O_{16} \cdot H_2O$	18
Copper	Copper sulfate	$CuSO_4 \cdot 5H_2O$	25
Iron	Ferrous sulfate	$FeSO_4 \cdot 7H_2O$	19
	Iron chelates	NaFeEDDHA	6
Manganese	Manganese sulfate	$MnSO_4 \cdot 3H_2O$	26–28
	Manganese oxide	MnO	41–68
Molybdenum	Sodium molybdate	$Na_2MoO_4 \cdot 2H_2O$	39
	Ammonium molybdate	$(NH_4)_6Mo_7O_{24} \cdot 4H_2O$	54
Zinc	Zinc sulfate	$ZnSO_4 \cdot H_2O$	35
	Zinc oxide	ZnO	78
	Zinc chelate	$Na_2ZnEDTA$	14

Selected from Murphy and Walsh (1972).

the glass weathers in the soil. This slow release avoids some of the problems of precipitation and sorption that might otherwise occur, particularly in an alkaline soil.

Economic responses to micronutrients are becoming more widespread as intensity of plant production increases. For example, responses of fruits, vegetables, and field crops to zinc and iron applications are common in areas with neutral to alkaline soils. Even on acid soils, deficiencies of these elements are increasingly encountered. Molybdenum, which has been used for some time for forage crops and for cauliflower and other vegetables, has received attention in recent years for forest nurseries and soybeans, especially on acid soils. These examples, along with those from muck areas and sandy soils where micronutrients have been used for decades, illustrate the need for these elements if optimum yields are to be maintained.

A soil will often produce a micronutrient deficiency in some plants but not in others. Plant species, and varieties within species, differ widely in their susceptibility to micronutrient deficiency or toxicity. Table 15.7 provides examples of plant species known to be particularly susceptible or tolerant to several micronutrient deficiencies.

TABLE 15.7 Plants Known to Be Especially Susceptible or Tolerant to, and Soil Conditions Conducive to, Micronutrient Deficiencies

Plants which are most susceptible to deficiency of a micronutrient often have a relatively high requirement for that nutrient and may be relatively tolerant to levels of that nutrient that would be high enough to cause toxicity to other plants.

Micronutrient	Common range in rates recommmended for soil application[a] (kg/ha)	Plants most commonly deficient (high requirement or low efficiency of uptake)	Plants rarely deficient (low requirement or high efficiency of uptake)	Soil conditions commonly associated with deficiency
Iron	0.5–10.0	Blueberries, azaleas, roses, holly, grapes, nut trees, maple, bean, sorghum, oaks	Wheat, alfalfa, sunflower, cotton	Calcareous, high pH, waterlogged alkaline soils
Manganese	2–20	Peas, oats, apple, sugar beet, raspberry, citrus	Cotton, soybean, rice, wheat	Calcareous, high pH, drained wetlands, low organic matter, sandy soils
Zinc	0.5–20	Corn, onion, pines, soybeans, beans, pecans, rice, peach, grapes	Carrots, asparagus, safflower, peas, oats, crucifers, grasses	Calcareous soils, acid, sandy soils, high phosphorus
Copper	0.5–15	Wheat, corn, onions, citrus, lettuce, carrots	Beans, potato, peas, pasture grasses, pines	Histosols, very acid, sandy soils
Boron	0.5–5	Alfalfa, cauliflower, celery, grapes, conifers, apples, peanut, beets, rapeseed, pines	Barley, corn, onion, turf grass, blueberry, potato, soybean	Low organic matter, acid, sandy soils, recently limed soils, droughty soils, soils high in 2:1 clays
Molybdenum	0.05–0.5	Alfalfa, sweet clover, crucifers (broccoli, cabbage, etc.), citrus, most legumes	Most grasses	Acid sandy soils, highly weathered soils with amorphous Fe and Al

[a]The lower end of each range is typical for banded application; the higher end is typical for broadcast applications.

Marked differences in crop needs for micronutrients make fertilization a problem where rotations are being followed. On general-crop farms, vegetables are sometimes grown in rotation with small grains and forages. If the boron fertilization is adequate for a vegetable crop such as red beets or even for alfalfa, the small-grain crop grown in the rotation may show toxicity damage. These facts emphasize the need for specificity in determining crop nutrient requirements and for care in meeting these needs.

Micronutrient Availability

Major sources of micronutrients and the general reactions that make them available to higher plants and microorganisms are summarized in Figure 15.18. Original and secondary minerals are the primary sources of these elements, while the breakdown of organic forms releases ions to the soil solution. The micronutrients are used by higher plants and microorganisms in important life-supporting processes. Removal of nutrients in crop or timber harvested reduces the soluble ion pool, and it may need to be replenished with manures or chemical fertilizers to avoid nutrient deficiencies.

Worldwide Management Problems

Micronutrient deficiencies have been diagnosed in most areas of crop production and some forest areas of the United States and Europe. However, in some developing countries, particularly in the tropics, the extent of these deficiencies is much less well known. Limited research suggests that there may be large areas with deficiencies or toxicities of one or more of these elements. Irrigation schemes that bring calcareous soils in desert areas under cultivation are often plagued with deficiencies of iron, zinc, copper, and manganese. Copper and zinc deficiency have been noted in cereal crops on highly weathered soils in the humid and subhumid tropics. As macronutrient deficiencies are addressed and yields are increased, more micronutrient deficiencies will undoubtedly come to the fore. One encouraging factor is the small quantity of micronutrients usually needed, making transportation of fertilizer materials to remote parts of the world much less of an expense than is the case for macronutrient fertilizers. The management principles established for the economically developed countries should also be helpful in alleviating micronutrient deficiencies in less developed countries.

15.10   CONCLUSION

Micronutrients are becoming increasingly important to world agriculture as crop removal of these essential elements increases. Soil and plant tissue tests confirm that these elements are limiting crop production over wide areas and suggest that attention to them will likely increase in the future.

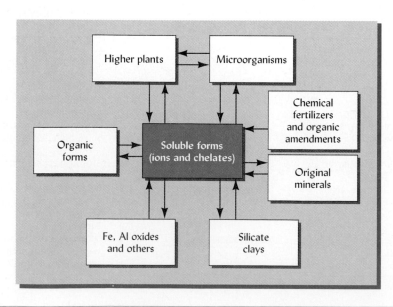

FIGURE 15.18 Diagram of soil sources of soluble forms of micronutrients and their utilization by plants and microorganisms.

In most cases, soil management practices that avoid extremes in soil pH, that optimize the return of plant residues and animal manures, and that promote chelate production by actively decomposing organic matter will minimize the risk of micronutrient deficiencies or toxicities. But increasingly noted are situations where micronutrient problems can be most practically solved by the application of micronutrient fertilizers. Such materials are becoming common components of fertilizers for field and garden use and will likely become even more so in the future.

REFERENCES

Brown, P. H., R. M. Welch, and E. E. Cary. 1987. "Nickel: a micronutrient essential for higher plants," *Plant Physiol.,* **85**:801–803.

Clemens, D. F., B. M. Whitehurst, and G. B. Whitehurst. 1990. "Chelates in agriculture," *Fertilizer Research,* **25**:127–131.

Elliott, H. A., M. R. Liberati, and C. P. Huang. 1986. "Competitive adsorption of heavy metals by soils," *J. Environ. Qual.,* **15**:214–219.

Gupta, U. C., and J. Lipsett. 1981. "Molybdenum in soils, plants, and animals," *Adv. in Agron.,* **34**:73–116.

Hamilton, M. A., D. T. Westermann, and D. W. James. 1993. "Factors affecting zinc uptake in cropping systems," *Soil Sci. Soc. Am. J.,* **57**:1310–1315.

Lindsay, W. L. 1972. "Inorganic phase equilibria of micronutrients in soils," in J. J. Mortvedt, P. M. Giordano, and W. L. Lindsay (eds.), *Micronutrients in Agriculture* (Madison, Wis.: Soil Sci. Soc. Amer.).

Marschner, H., A. Kalisch, and V. Romheld. 1974. "Mechanism of iron uptake in different plant species," *Proc. 7th Int. Colloquium on Plant Analysis and Fertilizer Problems,* Hanover, West Germany.

Marschner, H. 1986. *Mineral Nutrition of Higher Plants* (New York: Academic Press).

Mengel, K., and E. A. Kirkby. 1987. *Principles of Plant Nutrition* (4th ed.) (Bern, Switzerland: Int. Potash Institute).

Mortvedt, J. J., P. M. Giordano, and W. L. Lindsay (eds.) 1975. *Micronutrients in Agriculture* (Madison, Wis.: Soil Sci. Soc. Amer.).

Murphy, L. S., and L. M. Walsh. 1972. "Correction of micronutrient deficiencies with fertilizers," in J. J. Mortvedt, P. M. Giordano, and W. L. Lindsay (eds.), *Micronutrients in Agriculture* (Madison, Wis.: Soil Sci. Soc. Amer.).

Norvell, W. A. 1972. "Equilibria of metal chelates in soil solution," in J. J. Mortvedt, P. M. Giordano, and W. L. Lindsay (eds.), *Micronutrients in Agriculture* (Madison, Wis.: Soil Sci. Soc. Amer.).

Olsen, R. A., R. B. Clark, and J. H. Bennett. 1981. "The enhancement of soil fertility by plant roots," *Amer. Scientist,* **69**:378–384.

Ryan, J., and S. N. Hariq. 1983. "Transformation of incubated micronutrients in calcareous soils," *Soil Sci. Soc. Amer. Jour.,* **47**:806–810.

Sillanpaa, M. 1982. *Micronutrients and the Nutrient Status of Soils: A Global Study* (Rome, Italy: U.N. Food and Agricultural Organization).

Stevenson, F. J. 1986. *Cycles of Soil Carbon, Nitrogen, Sulfur and Micronutrients* (New York: John Wiley and Sons).

Viets, F. J., Jr. 1965. "The plants' need for and use of nitrogen," in *Soil Nitrogen* (*Agronomy,* 10) (Madison, Wis.: Amer. Soc. Agron.).

Weil, R. R., and Sh. Sh. Holah. 1989. "Effects of submergence on the availability of micronutrients in three Ultisols," *Plant and Soil,* **114**:147–157.

Welch, R. M. 1995. "Micronutrient nutrition of plants," *Critical Reviews in Plant Science,* **14**(1):49–82.

Yermiyahu, U., R. Keren, and Y. Chen. 1995. "Boron sorption by soil in the presence of composted organic matter," *Soil Science Society of America Journal,* **59**:405–409.

16

PRACTICAL NUTRIENT MANAGEMENT

Using scientific knowledge and ecological wisdom we can manage the earth . . .
—RENE J. DUBOS, HUMANIZING THE EARTH

As stewards of the land, soil managers must keep nutrient cycles in balance. By doing so they maintain the soil's capacity to supply the nutritional needs of plants and, indirectly, of us all. While undisturbed ecosystems generally need no intervention from soil managers or from anyone else, few ecosystems are so undisturbed. More often, human hands have directed the output of the ecosystem for human ends. Forests, farms, fairways, and flower gardens are ecosystems modified to provide us with paper, food, recreational opportunities, and aesthetic satisfaction. By their very nature, managed ecosystems need management.

In managed ecosystems, nutrient cycles become unbalanced through increased removals (e.g., harvest of timber and crops), through increased system leakage (e.g., leaching and runoff), through simplification (e.g., monoculture, be it of pine tree or sugarcane), and through increased demands for rapid plant growth (whether the soil is naturally fertile or not). Hence, the land manager is, of necessity, also a nutrient manager.

In this chapter we will discuss the goals of nutrient management and describe the tools that may help achieve those goals. These tools include methods of enhancing nutrient recycling as well as sources of additional nutrients that can be applied to soils or plants. We will learn how to diagnose nutritional disorders of plants and correct soil fertility problems. Building on the foundation of principles set out in earlier parts of this book, this chapter contains much practical information on the profitable use and conservation of organic and inorganic nutrient resources in producing abundant, high-quality plant products, and in maintaining the quality of both the soil and the rest of the environment.

16.1 GOALS OF NUTRIENT MANAGEMENT[1]

Nutrient management is one aspect of a holistic approach to managing soils in the larger environment. The goals for managing nutrients will influence the specific practices adopted. In general, nutrient management aims to achieve four broad, interrelated goals: (1) cost-effective production of high-quality plants; (2) efficient use and conservation of

---

[1]For an overview of issues and advances in nutrient management for agriculture and environmental quality, see the special supplement to the March–April 1994 issue of *Journal of Soil and Water Conservation*, Volume 49(2). A standard textbook on management of agricultural soil fertility and fertilizers is Tisdale, et al. (1993).

nutrient resources; (3) maintenance or enhancement of soil quality; and (4) protection of the environment beyond the soil. Each of these goals will now be briefly considered.

Plant Production

People nurture plants for many purposes. To simplify our discussion we can generalize that people engage in three types of plant production: agriculture, forestry, and ornamental landscaping. In the case of small-scale farming and gardening the products may be consumed directly by the grower's family. For commercial agriculture, the principal goal is to produce plant products that can be sold at a profit (or used as animal feed). The main objective of nutrient management is to increase the yield, that is, the amount of marketable product. From this point of view, the expense of nutrient management must not exceed the market value of the resulting extra yield. In other words, nutrient management is meant to increase profits. Although the time frame for fully evaluating the results of soil management may span several human generations, and improvements in biological nutrient cycling often require a decade or so for full effect, the time frame for profitability in agriculture tends to be rather short, a matter of a few months to a few years in most cases.

In forestry, the principal plant product may be measured in terms of volume of lumber or paper produced. In these instances, nutrient management aims to enhance rate of growth so that the time between investment and payoff can be minimized. The survival rate of tree seedlings is also important. Wildlife habitat and recreational values may also be primary or secondary products. The time frame in forestry, measured in decades or even centuries, tends to limit the intensity of nutrient management interventions that can be profitably undertaken.

When soils are used for ornamental landscaping purposes the principal objective is to produce quality, aesthetically pleasing plants. Whether the plants are produced for sale or not, relatively little attention is paid to yield of biomass produced. In fact, high yield may be undesirable, as in the case of turf grass for which high yield requires more frequent mowing. Hardiness, resistance to pests, color, and abundance of blooms are much more important. Labor costs and convenience are generally of more concern than fertilizer costs; hence, expensive, slow-release fertilizers are widely used.

Conservation of Nutrient Resources

Two concepts that are key to the goal of conserving nutrient resources are (1) renewal or reuse of the resources and (2) nutrient budgeting that reflects a balance between system inputs and outputs.

The first law of thermodynamics suggests that all material resources are ultimately renewable since the elements are not destroyed by use, but are merely recombined and moved about in space. In practical terms, however, once a nutrient has been removed from a plot of land and dispersed into the larger environment, it may be difficult if not impossible to use it again as a soil nutrient. For instance, phosphorus deposited in a lake bottom with eroded sediment and nitrogen buried in a landfill as a component of garbage are not available for reuse. In contrast, land application of composted municipal garbage and irrigation with sewage effluent are examples of practices that treat nutrients as *reusable* resources.

Recycling is a form of reuse in which nutrients are returned to the *same* land from which they were previously removed. Leaving crop residues in the field and spreading barnyard manure onto the land from which the cattle-feed grains were harvested are both examples of nutrient recycling. The term *renewable resource* best applies to soil nitrogen, which can be replenished from the atmosphere by biological nitrogen fixation (see Sections 13.10–13.12). Manufacture of nitrogen fertilizers also fixes atmospheric nitrogen, but at a large cost in nonrenewable fossil fuel energy.

Other fertilizer nutrients such as potassium and phosphorus are mined or extracted from nonrenewable mineral deposits, or from mineral-laden seas and oceans. The size of known global reserves varies according to the nutrient, but for most nutrients the mineral reserves are estimated to be large enough to last for at least several centuries. However, should third-world farmers increase their use of phosphorus fertilizer to levels now common in Europe and the United States, the supplies of this crucial element might become depleted within as little as a single century. Therefore, we must not let the lack of immediate scarcity lull us into ignoring conservation and efficiency. One reason is that as the best, most concentrated, and most accessible sources of these nutrients are depleted, the cost of producing fertilizer will likely rise in terms of money, energy, and environmental disruption. Second, the need for these resources will increase, not decrease, over the years. By its very nature, an essential element cannot be replaced by substitute substances. Therefore it would be wise to make careful husbandry of nutrient resources a part of any long-term nutrient management program.

A very useful step in planning nutrient management is to conceptualize the nutrient flows for the particular system under consideration. Such a flowchart should attempt to account for all the major inputs and outputs of nutrients. Simplified examples of such a conceptual budget are shown in Figure 16.1.

Nutrient Imbalances

In many cases addressing nutrient imbalances, shortages, and surpluses calls for analysis of nutrient flows, not just on a single farm or enterprise, but on a watershed, regional, or even national scale. For example, studies have shown that many countries in Africa are net exporters of nutrients. That is, exports of agricultural and forest products carry with them more nutrients than are imported into the country as fertilizers, food, or animal feed. This imbalance appears to be a contributing factor in the impoverishment of African soils, reduction of agricultural productivity, and stagnation or decline of national economies (see Section 20.7).

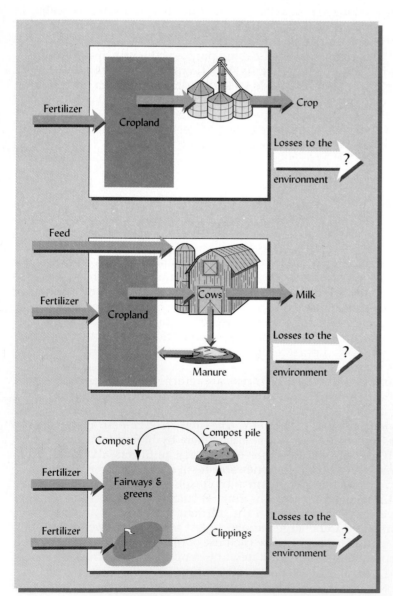

FIGURE 16.1 Representative conceptual nutrient flows for a cash-grain farm, a dairy farm, and a golf course. Only the intentional inputs and outputs are shown. Other, unmanaged inputs (such as nutrient deposition in rainfall) and outputs (such as leaching or runoff losses) are not shown. Although information on unmanaged inputs and outputs is not always readily available, it must also be taken into consideration in developing a complete nutrient management plan. Flowcharts such as those shown are a starting point to help to identify imbalances between inputs and outputs that could lead to wasted resources, reduced profitability, and environmental damage.

In industrialized countries, changes in the structure of agricultural production have led to regional nutrient imbalances and concomitant serious water pollution problems in many areas. The concentration of livestock production away from the place of feed production is a case in point. As long as a local area produces all the feed required by its livestock, it should be feasible to return the nutrients in the animal manure to the soils from which they came, allowing productivity to be maintained with only enough fertilizer importation to make up for the export of nutrients in livestock products. Excessive concentration of livestock in one area requires the importation of large quantities of nutrients as feed, resulting in more manure production than can be locally utilized in an efficient and environmentally safe manner (Figure 16.2). The concentrated poultry industry in Delaware is one example for which the nutrient imbalance has been documented (see Table 16.1).

High

Medium

Low

FIGURE 16.2 Map of the United States showing regions in which the ratio of livestock manure produced to cropland available for manure application ranges from high to low. In some of the high-ratio areas as much as 4 Mg of dry manure (about 10 to 20 Mg of fresh manure) is produced for each hectare of available cropland or improved pasture land. A high or medium ratio usually indicates that many farms produce far more manure than can be properly used on land close enough for economical transportation. Even in regions with a low ratio, animal manure may not be applied to enough of the theoretically available cropland. [Source: USDA Natural Resources Conservation Service (1991)]

TABLE 16.1 Statewide Nitrogen and Phosphorus Balance for Poultry-Based Agriculture in Delaware

It is apparent that each year about 10,000 more Mg of N is applied to the soils of Delaware's farms than is needed by the crops grown. This is nearly 50 kg of nitrogen per hectare. Much of this excess nitrogen leaves these farms as nitrates in surface runoff and groundwater. Phosphorus is proportionally even further out of balance, resulting in the buildup of this element in surface soils and increased losses in runoff.

Nitrogen requirement or Source	Land area (hectares)	Amount of phosphorus (Mg/year)	Amount of nitrogen (Mg/year)
Requirements:			
Corn	69,600	940	9,800
Soybeans	80,600	1085	0
Wheat	24,300	330	2,200
Barley	11,000	150	1,000
Other crops	32,400	435	3,600
Total	217,900	2,940	16,600
Nutrient Sources:			
Poultry manure	—	3,495	7,865
Fertilizer sales	—	2,955	19,275
Other wastes	—	unknown	unknown
Total Sources of Nutrient	—	6,450+	27,140+
Balance (excess N or P)	—	3,510+	10,580+

From Sims and Wolf (1994).

The concept of using nutrient management to enhance soil quality goes far beyond simply supplying nutrients for the current year's plant growth. Rather, it includes the long-term nutrient-supplying and -cycling capacity of the soil, improvement of soil physical properties or tilth, maintenance of above- and belowground biological functions and diversity, and the avoidance of chemical toxicities. Likewise, the management tools employed go far beyond the application of various fertilizers (although this may be an important component of nutrient management). Nutrient management requires the integrated management of physical, chemical, and biological processes. The effects of tillage on organic matter accumulation (Chapter 12), the increase in nutrient availability brought about by earthworm activity (Chapter 11), the role of mycorrhizal fungi in phosphorus uptake by plants (Chapter 14), and the impact of fire on soil nutrient and water supplies (Chapter 7) are all examples of components of integrated nutrient management.

16.2   ENVIRONMENTAL QUALITY

Nutrient management impacts the environment most directly in the area of water quality, and that is the aspect which will be emphasized in this chapter. The principal water pollutants to be considered are nitrogen, phosphorus, and sediment. You may therefore wish to consult Chapters 13 and 14 to review the many pathways by which nitrogen and phosphorus can be lost from soils. Clearly, the primary means to avoid damaging surface waters with excess nitrogen or phosphorus is to keep the rate at which these nutrients are supplied in balance with the rate with which they are used by plants. In addition, a number of strategies can be used to reduce the transport of nutrients (and other pollutants) from soils to groundwater and surface waters. In the United States, practices officially sanctioned to implement these strategies are known as *best management practices* or BMPs. Four general practices will now be briefly considered: buffer strips, cover crops, selective timber cutting, and conservation tillage.

Buffer Strips[2]

Buffer strips of dense vegetation are a simple and generally cost-effective method of protecting a body of water from the polluting effects of a nutrient-generating land use. Fertilized cropland, poultry or livestock operations, farmland which has been amended with organic wastes, forest harvest operations, and urban development are examples of land uses that have the potential to generate nutrient or sediment loadings. The vegetation in the buffer strips may consist of natural vegetation or planted vegetation, and may include grasses or trees (see Figure 16.3).

Water running off the surface of the nutrient-rich land passes through the buffer strip before it reaches the stream. Trees and the litter layer they form, or grass plants and the thatch layer they form, reduce the water velocity and increase the tortuosity of the water's travel paths. Under these conditions, most of the sediment and attached nutrients will settle out of the slowly flowing water. Also, dissolved nutrients are adsorbed by the organic mulch and mineral soil, or are taken up by the buffer strip plants. The decreased flow velocity also increases the retention time—the length of time during which microbial action can work to break down pesticides before they reach the stream. Under some circumstances, buffer strips along streams can also reduce the nitrate levels in the groundwater flowing under them (although emission of nitrous oxide to the atmosphere may be a consequence of nitrate reduction) (see Figure 13.8).

Buffer strips can work amazingly well, provided they are densely vegetated and not overly trampled by cattle or people. A width of 10 meters or so is usually sufficient for significant removal of sediment and nutrients (see Figure 16.4). However, the width needed for optimum cleanup of the runoff water may vary from as little as 6 meters to as much as 60 meters, depending on the slope, runoff volume, and nutrient load. A practical compromise must be reached regarding the width of buffer strip used, since a wider buffer means less land is left for forest production, farming, or development. On the other hand, a buffer strip may be useful to the landowner for purposes such as access or

---

[2]For a discussion of buffer strips in forestry, see Belt and O'Laughlin (1994).

FIGURE 16.3 The zone of grassy vegetation maintained along the banks of this stream is an example of a buffer strip installed in an agricultural watershed. A large proportion of the nutrients and other pollutants carried off the crop fields by surface runoff is trapped in these strips before the runoff water can reach the stream. In the forested areas in the background, a similar strip of land along the stream will be left undisturbed to act as a buffer during logging operations. (Photo courtesy of R. Weil)

turnaround space for equipment, hay production, enhanced recreational value of the stream banks, and better fish habitat. Buffer strips can often be a win-win proposition.

Cover Crops

A cover crop is not harvested, but rather is allowed to provide vegetative cover for the soil and then either killed and left on the surface as a mulch or tilled into the soil as a green manure. Growing a cover crop provides numerous benefits compared to leaving the soil bare for the off season. The plants, if leguminous, may increase the available nitrogen in

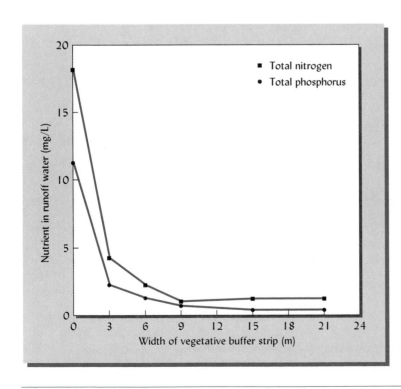

FIGURE 16.4 Removal of phosphorus from runoff by vegetated buffer strips of various widths. Swine manure had been applied to the field upslope from the buffer strips. During a simulated rainfall, surface runoff was sampled after it had traveled through different widths of grassed buffer strip. As is typical, most of the reduction in nutrient loading was achieved by the first few meters of buffer strip. In this case, a width of 9 meters would seem to be justified. Depending on the type of vegetation, steepness of slope, and type of soils, the width of buffers needed to achieve near-optimum nutrient removal may vary from as narrow as 6 meters to as wide as 60 meters. [Data from Chauby et al. (1994)]

the soil (see Section 16.4); they may provide habitat for wildlife and for beneficial insects; they may protect the soil from the erosive forces of wind and rain (see Section 17.8); and they add to the soil organic matter (see Section 12.14). More relevant to the topic of this chapter, cover crops may reduce the loss of nutrients and sediment in surface runoff, and they may conserve nutrients that would otherwise leach below the root zone.

A cover crop reduces nutrient losses in runoff for two reasons. First, the vegetative cover reduces the formation of a crust at the soil surface, thus maintaining a high rate of infiltration. In any given storm, more infiltration leaves less water to run off on the soil surface. Second, for the runoff that does occur, the cover crop helps remove both sediment and nutrients by the same mechanisms that operate in a buffer strip, as described above.

COVER CROPS REDUCE LEACHING LOSSES. Perhaps the most important use of cover crops in nutrient management is to reduce the leaching losses of nutrients, principally nitrogen. In many temperate humid and subhumid regions, the greatest potential for leaching of nitrate from cropland occurs during fall, winter, and spring, after harvest and before planting of the main crop. During the main growing season, crop roots take up both water and nutrients, resulting in little or no movement of water below the root zone and relatively low concentrations of nitrate in the soil water. However, if the soil is bare after harvest of the main crop, nitrate will readily leach downward. High leaching potential will persist (unless the soil is frozen) until the next year's crop is up and growing.

During this time of vulnerability, the presence of an actively growing cover crop will slow percolation and remove much of the nitrate from the soil solution, incorporating this nutrient into plant tissue that will largely be decomposed in the soil during the following year. For this purpose, an ideal cover crop should produce as extensive a root system as possible, as quickly as possible, once the main crop has ceased growth. Grass plants such as winter annual cereals (rye, wheat, oats, etc.) have proven to be more efficient than legumes (vetch, clover, etc.) at mopping up leftover soluble nitrogen (Figure 16.5).

It also should be noted that a cover crop, like any plant, requires a balanced nutrient supply, and will not be capable of effectively reducing nitrate leaching if the soil is poorly supplied with other nutrients (Figure 16.6). In this regard P and K are especially important for cover crops growing during periods of cool temperatures.

Selective Timber Cutting

Most forests, especially those with young, rapidly growing trees, are characterized by relatively slow rates of nutrient release (mainly by organic matter mineralization) matched rather closely by comparable rates of uptake by the trees. If environmental costs are not accounted for, many forest companies find that tree harvest by clear-cut or even-age management methods is most profitable (Figure 16.7). When these methods of harvest are used, all the trees in a large block of forest land are removed, and the nutrient balance is severely upset for three primary reasons.

First, the removal of all the trees leaves the soil with few active roots to take up nutrients dissolved in the soil solution. Second, organic matter decomposition and mineral nutrient release is greatly accelerated by the increased temperature of the no-longer-shaded forest floor and by the physical disturbance of the O and A horizons by log skidding and clearing operations. Third, the slash (branches and leaves) left behind when the logs are removed presents an enormous additional source of nutrients to be released either by slash burning or decomposition.

As a result of these factors, streams draining clear-cut watersheds have been shown to carry greatly elevated loads of nitrate and other nutrients for some time after the timber harvest has occurred (Figure 16.8). The impact of clear-cutting is greatest where it is combined with herbicide treatment of the harvested site to reduce the level of weed competition experienced by newly planted tree seedlings.

Selective cutting practices, in contrast, remove only a small percentage of the trees present in any one year. The soil remains permeated with nutrient-scavenging tree roots, and the forest floor suffers only minimum disturbance. Streams are less polluted and the forest soils are less depleted of essential nutrients.

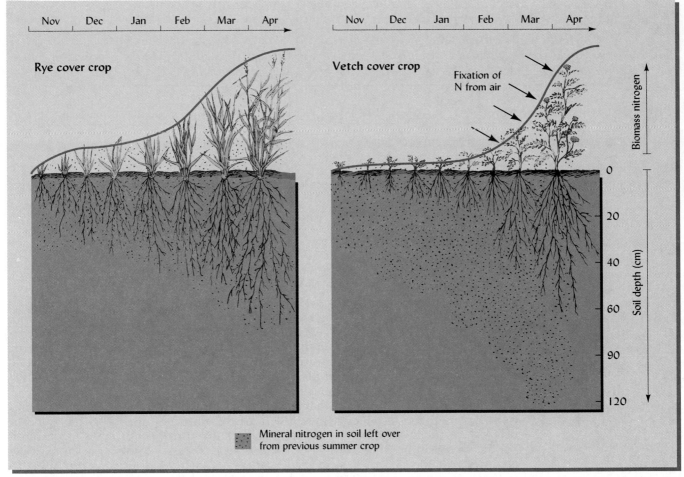

FIGURE 16.5 Relative effectiveness of a nonlegume (rye) and a legume (vetch) winter cover crop in mopping up the nitrate remaining in the soil after the harvest of a heavily fertilized summer crop. Rye grows very rapidly during mild fall weather just after harvest, while the legume grows very little until the soil warms again the following spring. These traits, and the facts that, unlike the legume, rye is dependent on soil nitrogen and has a fibrous root system, combine to make rye an excellent choice of cover crop to reduce winter leaching of soil nitrate.

Conservation Tillage

The term *conservation tillage* applies to agricultural tillage practices which keep at least 30% of the soil surface covered by plant residues. The effects of conservation tillage on soil properties and on the prevention of soil erosion are discussed elsewhere in this textbook (Section 17.12). Here, we emphasize the effects on nutrient losses.

Studies in many regions of the world make it clear that, compared to plowed fields with little or no residue cover, conservation tillage reduces the total amount of water running off the land surface, and reduces even more the total load of nutrients and sediment carried by that runoff (Table 16.2). When combined with a cover crop, the effect is greater still. There has been some concern that fertilizers and organic wastes applied to conservation-tillage fields are not well (if at all) incorporated into the soil, and therefore are more likely to be washed downhill. In most situations, however, the less the soil surface is disturbed by tillage, even when manure or sewage sludge is spread on it, the smaller are the losses of nutrients to runoff water (although leaching may be enhanced).

In the remainder of this chapter we will concentrate on the properties and uses of various nutrient sources, measures designed to meet the goals of nutrient management outlined above, and methods of assessing plant and soil nutrient status in order to determine which nutrients need to be supplemented and in what quantities.

TABLE 16.2 Effect of Land Use and Tillage on the Loss of Nutrients in Runoff from a Silty Clay Loam Hapludalf in Ohio

The forest was a mixed stand of oak and pine, the alfalfa was a three-year-old hayfield, and the cultivated plots were planted to corn after having been in grass for 10 years. Ridge tillage is a form of conservation tillage that leaves the soil covered by residues in winter and disturbs only part of the soil in spring.

	Tillage system			
Water/nutrient loss	Forest	Untilled alfalfa	Ridge-tilled corn	Conventionally tilled corn
Runoff (% of rainfall)	5	18	33	40
Nutrient loss[a] (kg ha^{-1} year^{-1})				
Nitrogen	19	13	49	315
Phosphorus	0.26	0.21	1.12	2.65

Data from Thomas et al. (1992).
[a]Sediment and runoff.

16.3 NUTRIENT RESOURCES

The principal resources from which the pool of available soil nutrients is resupplied will vary with the type of ecosystem, the rates of productivity and nutrient removals, and the kinds of management practices applied. Internal nutrient resources resupply the soil from within the ecosystem, be it a forest, a watershed, a farm, or a home in the suburbs. Generally, both financial and environmental costs are minimized when internal resources are fully used. These resources include the process of mineral weathering within the soil profile, biological nitrogen fixation, acquisition of nutrients from atmospheric deposition, and various forms of internal recycling, such as animal manure application and plant residue management.

When internal resources prove inadequate, as will often be the case for at least some nutrients in highly productive systems, nutrients must be imported from resources external to the system. Such external resources are usually purchased inorganic or organic fertilizers. Some of the organic residues and wastes available in the United States are listed in Table 16.3 along with the percentage of each that is used on the land.

We will now turn our attention to specific internal and external nutrient resources.

FIGURE 16.6 As for any plants, vigorous growth of a cover crop requires adequate levels of available nutrients in the soil. The thin, slow-growing oat cover crop on the low-P and -K soil in the foreground is not able to help very much in preventing nitrate leaching on this coarse-textured Ultisol. The soil in the background with much more vigorously growing oats received phosphorus and potassium during the previous year at the rate of 36 and 70 kg/ha. (Photo courtesy of R. Weil)

TABLE 16.3 Estimated Quantities of Major Organic Wastes Generated Annually in the United States, and Percentages of These Materials That Are Used on the Land

Organic waste	Annual production (millions of dry metric tons)	Used on land (%)
Crop residues	450	75
Animal manures	175	90
Municipal refuse	145	10
Logging and wood manufacture	35	10
Industrial organics	9	5
Sewage sludge and septage	6	40
Food processing	3	15

Estimates from USDA (1980) and U.S. Federal Registry (1988) and other sources.

16.4 SOIL–PLANT–ATMOSPHERE NUTRIENT CYCLES

Weathering of Parent Materials: An Internal Resource

Depending on the parent materials and climate, weathering of parent materials (see Section 2.3) may release significant quantities of nutrients (Table 16.4). For timber production, most nutrients are released fast enough from either parent materials or decaying organic matter to supply adequate nutrients. In agricultural systems, however, some nutrients generally must be added. Negligible amounts of nitrogen are released by weathering of mineral parent materials, so other mechanisms (including biological nitrogen fixation) are needed to resupply this important nutrient. Release of nutrients by mineralization of soil organic matter is important in short-term nutrient cycling, but in the long run, the organic matter and the nutrients it contains must be replenished or soil fertility will be depleted.

FIGURE 16.7 Clear-cut harvest of loblolly pine on Coastal Plain soils. A few mature trees were left standing as seed trees. The forest floor is covered with slash of all sizes, but little living vegetation. (Photo courtesy of R. Weil)

TABLE 16.4 Amounts of Selected Nutrients Released by Mineral Weathering in a Representative Humid, Temperate Climate, Compared with Amounts Removed by Silvicultural and Agricultural Harvests

Additional losses by leaching are not shown.

	P	K	Ca	Mg
	(kg/ha)			
Weathered from igneous parent material over 50 years	5–25	250–1000	150–1500	50–500
Removed in harvest of 50-year-old deciduous bole wood	10–20	60–150	175–250	25–100
Removed by 50 annual harvests of a corn–wheat–soybean rotation	1,200	2,000	150	500

Estimated from many sources.

Nutrient Economy of Trees[3]

Forests have evolved various mechanisms to conserve nutrients in nutrient-poor sites. For instance, conifers in cold regions (e.g., black spruce) may retain their needles for several decades rather than shed them more frequently, enabling these trees to grow on sites with very low levels of nitrogen availability (the needles are much higher in nitrogen than the woody tissues). Forests generally produce more aboveground biomass (100–200 kg) per kg of nitrogen taken up than other ecosystems (e.g., corn produces about 60–70 kg biomass per kg nitrogen taken up). Forests of cold climates, in which slow rates of organic matter decomposition result in the immobilization of the system's nitrogen in the forest floor, are more efficient in this regard than are temperate or tropical forests (Table 16.5).

Although trees, like agricultural crops, may obtain most of their nutrients from the surface horizons, certain nutrient acquisition patterns are unique to trees. Because of their perennial growth habit and deep root systems, trees are well adapted to gathering nutrients from deep in the soil profile where much of the nutrient release from parent material takes place. Overall nutrient-use efficiency can sometimes be increased by combining trees and agricultural crops into what are known as *agroforestry systems* (Figures 16.9 and 20.7).

---

[3]For an in-depth treatment of this topic see Attiwill and Leeper (1987).

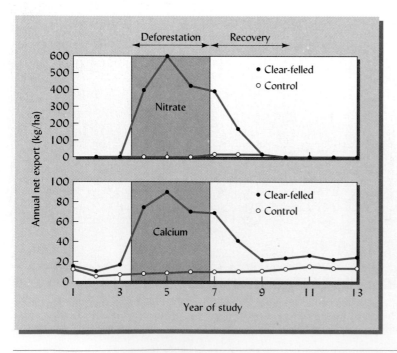

FIGURE 16.8 The effect of clear-cutting on the net export of two nutrients from a forested watershed in the White Mountains of New Hampshire. Clear-cut harvesting took place during the fourth year of the study and regrowth was suppressed with herbicides during the three years labeled deforestation. Losses of nutrients would have been less dramatic had vegetation been allowed to regrow immediately after cutting. [Data from Bormann and Likens (1979) © Springer-Verlag]

TABLE 16.5   Representative Nutrient Distribution in Several Types of Forest Ecosystems[a]

	In vegetation (kg/ha)	In forest floor (kg/ha)	Residence time (years)[b]
	Nitrogen		
Boreal coniferous	300–500	600–1,100	200
Temperate deciduous	100–1,200	200–1,000	6
Tropical rain forest	1,000–4,000	30–50	0.6
	Phosphorus		
Boreal coniferous	30–60	75–150	300
Temperate deciduous	60–80	20–100	6
Tropical rain forest	200–300	1–5	0.6
	Potassium		
Boreal coniferous	150–350	300–750	100
Temperate deciduous	300–600	50–150	1
Tropical rain forest	2,000–3,500	20–40	0.2
	Calcium		
Boreal coniferous	200–600	150–500	150
Temperate deciduous	1,000–1,200	200–400	3
Tropical rain forest	3,500–5,000	100–200	0.3

[a]Data derived from many sources.
[b]Residence time in forest floor (O and A horizons).

Trees may improve soil fertility in several ways. In Chapter 2 (Figure 2.23) we saw that trees can act as nutrient pumps, taking up nutrients that occur deep in the profile because of weathering or leaching, and depositing them at the soil surface as litter that will decompose and release the nutrients where they can be of use to relatively shallow-rooted agricultural crops.

Nitrogen-fixing trees (mostly legumes) can also add nitrogen to the surface soil with their nitrogen-rich leaf litter. Trees may also enhance the fertility of the soil in their vicinity by trapping windblown dust, thus increasing the deposition of nutrients such as calcium, phosphorus, and sulfur.

(a)

(b)

FIGURE 16.9 Two examples of agroforestry systems. (a) The deep-rooted *Acacia albida* enriches the soil under its spreading branches (arrow points to a man). Conveniently, these trees leaf out in the dry season and lose their leaves during the rainy season when crops are grown. It is an African tradition to leave these trees standing when land is cleared for crop production. Crops growing under the trees yield more and have higher contents of sulfur, nitrogen, and other nutrients. (b) Branches pruned from widely spaced rows of leguminous trees are spread as a mulch on the soil surface in the alleys between the tree rows, thus enriching the alleys with nutrients from the leaves as well as conserving soil moisture. The crops grown in this alley-cropping system may yield better than crops grown alone, but only if competition between trees and crop plants for light and water can be kept to a minimum. (Photos courtesy of R. Weil)

Leguminous Cover Crops to Supply Nitrogen

We have already mentioned many of the benefits of cover crops. Here we will consider the value of legume cover crops in supplying nitrogen for uptake by a subsequent non-legume crop.

Winter annual legumes such as vetch, clovers, and peas can be sown in fall after the main crop harvest or, if the growing season is short, they can be aerially seeded while the main crop is still in the field. The objective is to get the legume established before the winter weather becomes too cold for the cover crop to grow. Then, in spring, the cover crop will resume growth and biological nitrogen fixation, often fixing as much as 3 kg/ha of nitrogen daily during the warmer spring weather. The cover is then killed and left on the surface as a mulch in a no-till system or plowed under in a conventional tillage system. A vigorous winter cover crop can provide a significant amount of nitrogen to the main crop that follows (Figure 16.10).

A cover crop system may be able to replace part or all of the nitrogen fertilizer normally used to grow the main crop. Depending on the cost of fertilizer and cover crop seed, a net financial savings may accrue. The use of winter legume cover crops to provide nitrogen for nonlegume crops is becoming increasingly popular as more sustainable agricultural systems are developed. Such cover crop systems have been adapted for use in orchards, rice paddies (drained during the dry season), corn fields, vegetable fields, and gardens.

Crop Rotations

Growing one particular crop year after year on the same land generally produces lower yields of that crop and engenders more negative impacts on the soil and environment than does growing a sequence of different crops. The improved plant productivity noted when crops are grown in sequences or rotations may result from the effect that switching crops has on interrupting weed, disease, and insect pest cycles; the soil fertility effect of different rooting patterns; different types of residues and nutrient require-

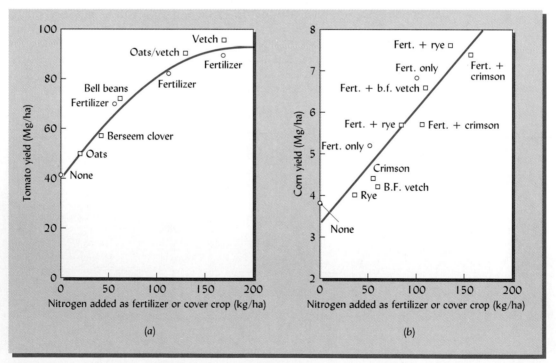

FIGURE 16.10 Comparative effects of winter cover crops and inorganic fertilizer on the yield of the following main crop. (a) Yield of processing tomatoes grown on a Xeralf soil in California. (b) Yield of corn grown on a Udalf in Kentucky. The amount of nitrogen shown on the x-axis was added either as inorganic fertilizer or as aboveground residues of the cover crops, or a combination of both. Note that nitrogen from either source seemed to be equally effective. For the tomato crop, which requires nitrogen over a long period of time, vetch alone or vetch mixed with oats produced enough nitrogen for near-optimal yields. [Data (left) abstracted from Stivers et al. (1993) © Lewis Publishers, an imprint of CRC Press, Boca Raton, Florida and (right) Ebelhar, et al. (1984)]

ments; positive allelopathic effects (see Section 12.10); and, possibly, positive effects on mycorrhizal diversity (see Section 11.9). These and other phenomena may explain why, even where pests and nutrient supply are optimally controlled, crops consistently yield 10 to 20% more in rotations than in continuous culture.

Although a detailed consideration of different cropping systems and rotations is beyond the scope of this book, the nutrient management aspects of rotating legumes with nonlegumes deserve mention here. Similar to the cover crop effects just discussed, a legume main crop may substantially reduce the amount of nitrogen that needs to be added to grow a subsequent nonleguminous crop, even though much of the biomass and accumulated nitrogen in the legume are removed with the crop harvest. Perennial forage legumes, such as alfalfa, tend to produce the greatest effects in this regard, but the nitrogen contributions of grain legumes such as soybeans and peanuts should also be taken into account when planning nitrogen application to nonlegumes such as cereal grains (Figure 16.11).

Protection from Wildfires

Fires convert a great deal of nitrogen and sulfur, and some phosphorus, to gaseous forms in which they are lost from the site. Burned-over land also loses more phosphorus in runoff for several years after a high-intensity burn (see Table 14.3). Ashes, both from prescribed burns and from wildfires, contain high levels of soluble K, Mg, Ca, and P, increasing the short-term availability of these nutrients, but also increasing the rate of loss of the cations from the forest ecosystem. Wildfires result in much greater nutrient losses because the high heat associated with these fires destroys some of the soil organic matter as well as the aboveground biomass.

Another nutrient aspect of forest fires is the fact that fire retardants used to fight wildfires are mostly fertilizer-type materials. Diammonium phosphate (DAP), monoammonium phosphate (MAP), and ammonium sulfate (AS) are widely used as fire retardants. The nitrogen and phosphorus in these chemicals may assist rapid regrowth in burned-over areas. However, they may also pose a water quality threat when lost to streams because of the low rate of plant uptake and increased runoff that occur immediately after most of the vegetation has been destroyed by an intense fire.

Borates are also used as fire suppressants. Levels of boron could be phytotoxic (see Section 15.1) for several years, possibly retarding the growth of plants attempting to revegetate the burned-over, borate-treated area.

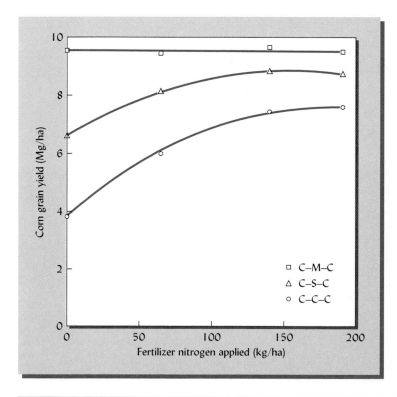

FIGURE 16.11 The nitrogen fertilizer requirements of corn are usually reduced by growing corn in rotation with legumes. This is one reason why the cropping history of a field is a consideration in deciding the optimum amount of fertilizer nitrogen to apply to cereal crops. The cropping sequences represented are: corn-corn-corn (C-C-C), corn-meadow-corn (C-M-C) and corn-soybean-corn (C-S-C). The meadow in the rotations consisted of several years of mixed alfalfa/brome grass. The data represent results on a variety of Iowa soils, but are typical of observations in many parts of the world. Less nitrogen should be applied to corn grown after soybeans than corn grown after corn. Corn following several years of mixed legume meadow is unlikely to respond to any nitrogen application. Rotation benefits other than nitrogen supply apparently account for the fact that, regardless of the amount of nitrogen fertilizer applied, corn grown after corn produced less yield than corn grown in rotation. [Based on data of A. M. Blackmer and J. R. Webb as reported in National Research Council (1993)]

16.5 RECYCLING NUTRIENTS THROUGH ANIMAL MANURES

For centuries the use of farm manure has been synonymous with a successful and stable agriculture. Not only does manure supply organic matter and plant nutrients to the soil, but it is associated with animal agriculture and with forage crops, both of which are soil protecting and conserving. A high proportion of the solar energy captured by growing plants ultimately is embodied in farm manure (see Figure 16.12). Crop and animal production and soil conservation are enhanced by its use.

Huge quantities of farm manure are available each year for possible return to the land. For each kilogram liveweight of farm animals about 4 kg dry weight of manure is produced in a year. In the United States the farm animal population voids about 10 times as much manure solids as does the human population.

More than 1 billion Mg (fresh weight) of this manure is produced in feedlots or large poultry or swine complexes where manure *disposal* as a problem tends to overshadow manure *utilization* as an opportunity (Figure 16.13). A 50,000-head beef feedlot operation, for example, produces about 90,000 Mg of manure solids annually, after some decomposition. A conventional application rate of 15 Mg per hectare (7 tons per acre) would supply approximately 150 kg N per hectare. Approximately 6000 hectares (1500 acres) of land would be required to properly utilize this manure. To find this much cropland, the manure would likely have to be hauled 6 to 10 km (3.7 to 6 miles) from the feedlot. Until it can be spread, the manure is usually kept in huge piles. Groundwater under such feedlots is often polluted with nitrate and unfit to drink (see Section 13.8). Application of the manure to nearby cropland at higher-than-needed rates would save

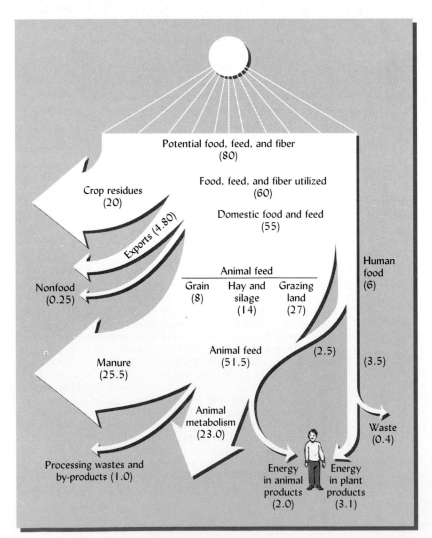

FIGURE 16.12 Estimated energy flow for the human food chain in the United States (data in billions of joules per year) showing the high proportion of the energy ultimately found in animal manures. An even larger proportion of the mineral nutrients (not shown) in the food chain end up in animal manure. [From Stickler et al. (1975); courtesy Deere & Company, Moline, Ill.]

on costs of transporting the manure to more distant fields, but can result in salinity damage to crops and soils and excess nitrogen and phosphorus loss to surface and groundwater.

Nutrient Composition of Animal Manures

Since manure as it is applied in the field is a combination of feces and urine, along with bedding (litter) and spilled feed, its composition is quite variable (Table 16.6). For a particular type of animal, the actual water and nutrient content of a load of manure will depend on the nutritional quality of the animals' feed, how the manure was handled,

FIGURE 16.13 Aerial view of a large feedlot in Colorado where 100,000 cattle are fed on grain imported from distant farms. The inset is a closeup of a similar feedlot. It is unlikely that the feedlot managers will be able to recycle the nutrients in the manure back to the land on which the cattle feed was grown. Instead of being a valued resource, the manure in this situation may be considered to be a waste requiring disposal. The challenge is to find ways to structure agriculture in a more ecologically balanced manner. (Large photo courtesy Monfort Feed Lots, Greeley, Colo.; inset courtesy of R. Weil)

TABLE 16.6 Example of the Variable Composition of Farm Manure

The data are based on 28 samples of horse manure sent in to one lab over a period of five years.

	Total N	Soluble N	Phosphorus	Potassium	Water
			Percent of fresh weight		
Lowest analysis	0.21	0.0	0.04	0.07	39
Highest analysis	0.85	0.14	0.75	1.0	80
Average analysis	0.51	0.03	0.16	0.35	63

Courtesy of V. A. Bandel, University of Maryland.

and the conditions under which it has been stored. The variability from one type of animal to another (e.g., broiler chicken manure compared to horse manure) is even greater. Therefore one has to be cautious in interpreting general statements about the value and use of manure.

Animals excrete large portions of the nutrient elements consumed by them in their feeds. Generally, about three-fourths of the nitrogen, four-fifths of the phosphorus, and nine-tenths of the potassium ingested is voided by the animals and appears in the manures. For this reason, animal manures are valuable sources of both macro- and micronutrients (Table 16.7).

Both the urine (except for poultry, which produce solid uric acid instead of urine) and feces are valuable components of animal manure. On the average, a little more than *one-half of the nitrogen,* almost *all of the phosphorus,* and about *two-fifths of the potassium* are found in the solid manure. Nevertheless, this higher nutrient content of the solid manure is offset by the ready availability of the constituents carried by the urine. Care must be taken in handling and storing the manure to minimize the loss of the liquid portion.

The data in Table 16.7 show that manures have a relatively low nutrient content in comparison with commercial fertilizer, and a nutrient ratio that is considerably lower in phosphorus than in nitrogen and potassium. On a *dry weight basis* animal manures contain from 2 to 5% N, 0.5 to 2% P, and 1 to 3% K. These values are one-half to one-tenth as great as are typical for modern commercial fertilizers.

However, unless it has been specially processed, manure is not spread in the dry form, but contains a great deal of water. As it comes from the animal, manure has a high water content, commonly varying from 30 or 40% for poultry to 70 or 80% for dairy cattle (see Table 16.7). If the fresh manure is handled as a solid and spread directly on the land (Figure 16.14), the high water content is a nuisance and adds to the expense of hauling. But if the manure is handled and digested in a liquid form or slurry and applied to the land as such, even more water must be added. In any case, all this water dilutes the nutrient content of manure, as normally spread in the field, to values much lower than those cited for dry manure. The large water content makes it difficult to transport the manure to distant fields where it might do the most good.

16.6 STORAGE, TREATMENT, AND MANAGEMENT OF ANIMAL MANURES

Where animal and crop production are integrated on a farm, manure handling is not too much of a problem. The use of pasture can be maximized so that the animals themselves spread much of the manure while grazing. The manure from confined animals is produced in small enough quantities to be hauled daily to the fields, or stored under cover during periods when soil conditions are not favorable for spreading manure. The total amount of nutrients in the manure produced on the farm is likely to be somewhat less than that needed to grow the crops; thus modest amounts of inorganic fertilizers may be needed to make up the difference.

Where animals are raised in confined quarters in large numbers, the large concentration of manure production leads to problems of offensive odors, along with the possibilities of pollution of streams and drinking water with pathogens, organic materials, nitrates, and phosphates. The problem of disposal of manure then takes precedence over its utilization, bringing about major changes in manure management.

On farms with concentrated animal production, four general management systems are now being used to handle farm manures: (1) collection and spreading of the fresh manure daily; (2) storage and packing in piles and allowing the manure to ferment before spreading; (3) aerobic liquid (air stirred in) storage and treatment of the manure prior to application; and (4) anaerobic liquid storage and treatment prior to application. Each of these methods of handling affects the nutrient value of the manure (Table 16.8). The greatest loss of nutrients, especially nitrogen, occurs either from open-lot storage or from lagoon storage. In open-lot storage, much nitrogen is volatilized as ammonia gas (see Section 13.6), and considerable quantities of nutrients may be washed away when it rains. Proper choice of site and design are necessary to prevent pollution of the groundwater and surface water. When manures are stored anaerobically (as in a large, densely packed pile or in an unaerated lagoon), much nitrogen loss occurs through denitrification (see Section 13.9).

TABLE 16.7 Commonly Used Organic Nutrient Sources: Their Nutrient Contents and Other Characteristics

Material	Water[a] %	Total N	P	K	Ca	Mg	S	Fe	Mn	Zn	Cu	B	Mo	
		Percent of dry weight						g/Mg of dry weight						
Poultry (broiler) manure[b]	35	4.4	2.1	2.6	2.3	1.0	0.6	1,000	413	480	172	40	0.7	May contain high C bedding, high soluble salts, and ammonia.
Dairy cow manure	75	2.4	0.7	2.1	1.4	0.8	0.3	1,800	165	165	30	20	—	May contain high C bedding.
Swine manure[c]	72	2.1	0.8	1.2	1.6	0.3	0.3	1,100	182	390	150	75	0.6	May contain elevated Cu levels.
Sheep manure	68	3.5	0.6	1.0	0.5	0.2	0.2	—	150	175	30	30	—	
Horse manure	63	1.4	0.4	1.0	1.6	0.6	0.3	—	200	125	25	—	—	May contain high C bedding.
Feedlot cattle manure[d]	80	1.9	0.7	2.0	1.3	0.7	0.5	5,000	40	8	2	14	1	May contain high (15%) soluble salts.
Young rye green manure	85	2.5	0.2	2.1	0.1	0.05	0.04	100	50	40	5	5	.05	Nutrient content decreases with advanced growth stage.
Spoiled legume hay	40	2.5	0.2	1.8	0.2	0.2	0.2	100	100	50	10	1,500	3	
Municipal solid waste compost[e]	40	1.2	0.3	0.4	3.1	0.3	0.2[f]	14,000	500	650	280	60	7	May have high C/N and contain heavy metals, plastic, glass.
Sewage sludge	80	4.5	2.0	0.3	1.5[g]	0.2	0.2	16,000[g]	200	700	500	100	15	May contain high soluble salts and toxic heavy metals.
Wood wastes	—	—	0.2	0.2	0.2	1.1	0.2	2,000	8,000	500	50	30	—	Very high C/N ratio, must be supplemented by other N.

Data derived from many sources.

[a]Water content given for fresh materials. Processing and storage methods may alter alter water content to less than 5% (heat-dried) or to more than 93% (slurry).

[b]Broiler and dairy manure composition estimated from means of approximately 800 and 400 samples analyzed by the University of Maryland manure analysis program from 1985–1990.

[c]Composition of swine, sheep, and horse manure calculated from North Carolina Cooperative Extension Service Soil Fact Sheets prepared by Zublena et al. (1993).

[d]Feedlot manure composition is based on average analysis reported in Eghball and Power (1994).

[e]Composition of municipal solid waste compost based on mean values for the products of 10 composting facilities in the United States as reported by He et al. (1995).

[f]Sulfate-S.

[g]Sludge contents of Ca and Fe may vary 10-fold depending on the wastewater treatment processes used.

*The values are ranges of percent loss of nutrient from the time the manure is excreted
until it is applied to the land.*

	Percentage lost		
Manure handling and storage method	*N*	*P*	*K*
Solid Systems			
Daily scrape and haul	15–35	10–20	20–30
Manure pack or compost	20–40	5–10	5–10
Liquid Systems			
Aerobic tank storage	5–25	5–10[a]	0–5
Lagoon (anaerobic) storage	70–80	50–80	50–80

From Sutton (1994).
[a]Most of the P and K in a lagoon system settles to the bottom of the lagoon and is recoverable only when the sludge in the lagoon is dredged out.

Methods of manure handling that both prevent pollution and preserve nutrients in a form that can be easily transported and sold commercially would make a major contribution to ameliorating the manure problem for concentrated animal production enterprises. Several options are currently being developed, three of which are described below.

Heat-Dry and Pelletize

This technology dries the manure with heat and then compresses the dried product into small pellets that handle like commercial fertilizer. This manure-processing method requires considerable energy use and capital investment, but the product has generated a large demand, especially in the landscaping and lawn industries where much money is already being spent on slow-release fertilizers.

FIGURE 16.14   Spreading solid dairy manure on cropland recycles nutrients efficiently but is labor-intensive and time-consuming for the farmer. Many loads of manure will have to be hauled to fertilize this field. The manure should be incorporated as soon as possible after spreading, and spreading should not be done on frozen soils. Calibration of spreaders is important to prevent unintentional overapplication of nutrients. (Photo courtesy of R. Weil)

Commercial Composting

A second method under development is improved **composting** of manure (see Section 12.17). In the case of poultry manure, dead birds are composted along with the manure, and the final product is a very stable, nonoffensive, relatively high analysis, slow-release fertilizer that is also easy to handle and has been enthusiastically received in test markets. Composting is a natural aerobic decomposition process and is much less energy- and capital-intensive than heat-drying and pelleting.

Anaerobic Digestion with Biogas Production

In this method the manure is made into a liquid slurry and allowed to become anaerobic (air is not stirred in). As discussed in Section 12.6, methane is a major gaseous product of anaerobic decomposition. The biogas produced by anaerobic manure digestion contains about 80% methane and 20% carbon dioxide. This gas can be burned much like commercial natural gas. Developing countries have attempted to use small-scale manure digesters to supply cooking and heating fuel for remote villages.

Recently, several commercial ventures have been established in the United States for large-scale production of biogas from manure. The gas is most commonly used to generate electricity, which is sold to local utility companies. The slurry remaining after digestion still contains most of the nutrients from the manure (though most of the organic carbon has been transformed to gases) and can be pumped to nearby fields or further processed (see above methods) to allow for wider sale and distribution.

In the future, such technologies, along with more integrated animal farming systems, may help to redress the serious nutrient imbalances that have developed with regard to manure production in modern agriculture.

16.7   INDUSTRIAL AND MUNICIPAL BY-PRODUCTS

Farm animals are not alone in generating large quantities of nutrient-containing wastes. People, and their industrial activities, do likewise, mostly in the relatively concentrated setting of metropolitan urban areas. Four major types of organic wastes are of significance in land application: (1) municipal garbage, (2) sewage effluents and sludges, (3) food-processing wastes, and (4) wastes of the lumber industry. Because of their uncertain content of toxic chemicals, other industrial wastes may or may not be acceptable sources of organic matter and plant nutrients for land application.

Society's concern for environmental quality has forced waste generators to seek non-polluting, but still affordable, ways of disposing of these materials. Although the primary reason for adding these wastes to land may be to safely dispose of them, they also serve as valuable soil amendments for agriculture, forestry, and disturbed-land reclamation. Once seen as mere waste products to be flushed into rivers and out to sea, these materials are increasingly seen as sources of nutrients and organic matter that can be beneficially used to promote soil productivity.

Garbage

Municipal garbage has traditionally been used extensively to enhance crop production in China and other Asian countries. In a growing number of locations in the United States and Europe, the inorganic glass and metal, etc., is first removed and municipal organic wastes are composted, sometimes in conjunction with sewage sludge, poultry manure, or other nutrient-rich materials, to produce municipal solid waste (MSW) compost that is then applied to the land. Most municipal garbage in the United States is incinerated or landfilled (see Section 18.10), but growing concerns about air quality and scarcity of landfill space are raising the level of interest in using soil application as a means of disposal. Because its nutrient content is even lower than that of most animal manures (see Table 16.7), and its source is even farther from the fields where it can be used as a soil amendment, MSW compost is subject to even greater economic limitations related to high transportation costs. On the other hand, municipalities have to dispose of the material somehow, and the cost of hauling MSW compost to land-application sites is becoming economically more competitive with alternative methods of disposal. As with manures, there is a trend in some regions to produce an attractive product that can be sold to landscaping and specialty users as a soil amendment or planting medium.

Food-Processing Wastes

Land application of food-processing wastes is being practiced in selected locations, but the practice is focused almost entirely on pollution abatement and not on crop production. Liquid wastes are commonly applied through sprinkle irrigation to permanent grass. Plant-processing schedules dictate the timing and rate of application, which may not be suitable for optimum crop production.

Sawdust

Sawdust, wood chips, and shredded bark from the lumber industry have long been sources of soil amendments and mulches, especially for home gardeners and landscapers. These wastes are high in lignin and related materials, have very high C/N ratios, and therefore decompose very slowly—desirable properties in a mulch material. However, they do not readily supply plant nutrients. In fact, sawdust incorporated into soils to improve soil physical properties may cause plants to become nitrogen deficient unless an additional source of nitrogen is applied with the sawdust (see Section 12.8).

16.8   SEWAGE EFFLUENT AND SLUDGE

Sewage treatment has evolved over the past century in response to society's desire (and resultant regulations) to avoid polluting our rivers and oceans with pathogens, oxygen-demanding organic debris, and eutrophying nutrients. The ever more stringent efforts to clean up the wastewater before returning it to natural waters has two basic consequences: (1) The amount of material *removed* from the wastewater during the treatment process has increased tremendously. This solid material, known as *sewage* **sludge** or **biosolids,** must also be disposed of safely. (2) A goal of advanced wastewater treatment is to remove nutrients (mainly phosphorus, but increasingly also nitrogen) from the **effluent** (the treated water that is returned to the stream), a job that can be accomplished safely and economically by allowing the partially treated effluent to interact with a soil–plant system. Therefore there is a growing interest in using soils to assist with the sewage problem in two ways: (1) as a system of assimilating, recycling, or disposing of the solid sludge and (2) as a means of carrying out the final removal of nutrients and organics from the liquid effluent.

Sewage Effluent

Sewage effluents have been applied to land for decades in Europe and in selected sites in the United States. Some cities operate sewage farms on which they produce crops, usually animal feeds and forages, that offset part of the expense of effluent disposal. The City of Muskegon, Michigan, has operated such a farm for decades, as has Paris, France. Sewage effluent may become more important in the future as a source of organic matter, nutrients, and water for crop production.

One very beneficial use of nutrient-rich sewage effluent is the irrigation of forest land (Figure 16.15). This method of advanced wastewater treatment is used by a number of cities around the world. Forest irrigation is a cost-effective method of final effluent cleanup and produces enhanced tree growth as a bonus. The rate of wood production is greatly increased as a result of both the additional water and the additional nutrients supplied therewith.

In a carefully planned and managed effluent irrigation system, the combination of (1) nutrient uptake by the plants, (2) adsorption of inorganic and organic constituents by soil colloids, and (3) degradation of organic compounds by soil microorganisms results in the purification of the wastewater. Percolation of the purified water eventually replenishes the groundwater supply.

FIGURE 16.15 Final treatment of sewage effluent and recharge of groundwater are being accomplished by natural soil and plant processes in this effluent-irrigated forest on Ultisols near Atlanta, Georgia. Nutrient flows, groundwater quality, and tree growth are carefully monitored. Some of the greatly increased production of wood is used as an energy source to run the sewage treatment plant. (Photo courtesy of R. Weil)

Sewage Sludge

Sewage sludge is the solid by-product of domestic and/or industrial wastewater treatment plants (see Figure 16.16). It has been spread on the land for decades, and its use will likely increase in the future. The product Milorganite®, a dried sludge sold by the Milwaukee Sewerage Commission, has been used widely in North America since 1927, especially on turf grass. Numerous other cities market composted sludge products to landscaping and other specialty users. However, the great bulk of sewage sludge used on land is applied as a liquid slurry or as a partially dried cake. The cake is less costly to transport because it carries only 40 to 70% water, while the slurry is generally 80 to 90% water. Further drying and processing decreases the cost of transportation, an important consideration since large cities have difficulty finding sufficient nearby land to receive their biosolids. One program *daily* ships over 600,000 kg of dried, treated sludge by rail from New York City to western Texas, a distance of over 2000 km, for use in restoring fertility to 7000 hectares of degraded range land.

COMPOSITION OF SEWAGE SLUDGE. As might be expected, the composition of the sludge varies from one sewage treatment plant to another, depending on the nature of treatment the sewage receives, especially the degree to which the organic material is allowed to digest. Representative values for plant nutrients are given in Table 16.7. Like manure and other organic waste products, sewage sludge contributes micronutrients as well as macronutrients. Levels of plant micronutrient metals (zinc, copper, iron, manganese, and nickel) as well as other heavy metals (cadmium, chromium, lead, etc.) are determined largely by the degree to which industrial wastes have been mixed in with domestic wastes. In the United States, the levels of metals in sewage are far lower than they were in the past, because of source-reduction programs that require industrial facilities to remove pollutants *before* sending their sewage to municipal treatment plants (see Section 18.7).

In comparison with inorganic fertilizers, sludges are generally low in nutrients, especially potassium. Representative levels of N, P, and K are 4, 2, and 0.4%, respectively (Table 16.7). The potassium concentration in sewage sludge is rather low because most of the potassium in sewage is soluble and remains with the effluent during treatment. The phosphorus content is higher because advanced sewage treatment is designed to

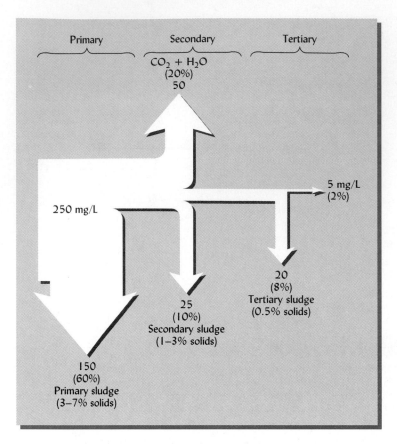

FIGURE 16.16 Diagram showing the removal of suspended solids from wastes. Primary treatment permits the separation of most of the solids from raw sewage. Secondary treatment encourages oxidation of much of the organic matter and separation of more solids. Tertiary treatment usually involves the use of calcium, aluminum, or iron compounds to remove phosphorus from the sewage water. [Redrawn from Loehr et al. (1979); used with permission of Van Nostrand Reinhold Company, New York]

remove phosphorus from the effluent and deposit it in the sludge (see Box 14.2). If the sewage treatment precipitates phosphorus by reactions with iron or aluminum compounds, the phosphorus in the sludge will likely have a very low availability to plants.

16.9 PRACTICAL UTILIZATION OF ORGANIC NUTRIENT SOURCES

In Section 12.11 we discussed the many beneficial effects on soil physical and chemical properties that can result from amendment of soils with decomposable organic materials such as manure or sludge. Here we will focus on the nutrient management aspects of organic waste utilization. Whether the material is sewage sludge, farm manure, or MSW compost, several general principles apply to the ecologically sound application of the material to soils:

1. The rate of application is generally governed by the amount of nitrogen that the organic material will make available to plants. This is the first criterion because nitrogen is needed in the largest quantity by most plants, and because excess nitrogen can present a pollution problem (see Section 13.8). It should be noted, however, that supply of plant nitrogen needs from most organic sources will include more phosphorus than the plants take up, and will thus eventually lead to excessive levels of soil phosphorus (see Section 14.2).

2. Most of the nitrogen in organic sources is not immediately available to plants, as it is with most inorganic commercial fertilizers (Section 16.11). A small fraction of the nitrogen in manure or sludge may be soluble (ammonium) forms and immediately available, but the bulk of the nitrogen must be released by microbial mineralization of organic compounds. Table 16.9 lists the percentage of the organic nitrogen that is likely to be released in the first, second, and third years after application. Note that those materials that have been partially decomposed during treatment and handling (e.g., composts and digested sludge) release a lower percentage of the nitrogen. The release rates of other nutrients are less critical because the availability from the organic sources is generally comparable (or even higher in the case of micronutrients) to commercial fertilizers.

TABLE 16.9   Rates of Release of Mineral Nitrogen from Various Sources of Organic Nitrogen Applied to Soils

The values given are percent of the organic nitrogen present in the original material. For example, if 10 Mg of poultry litter initially contains 300 kg (3.0%) nitrogen in organic forms, 50% or 150 kg of nitrogen would be mineralized in the first year. Another 15% (0.15 × 300) or 45 kg of nitrogen would be released in the second year. These values are approximate and may need to be increased for warm climates or sandy soils.

Organic nitrogen source	First year	Second year	Third year	Fourth year
Poultry floor litter	50	15	8	3
Dairy manure (fresh solid)	35	18	9	4
Swine manure lagoon liquid	50	15	8	3
Lime-stabilized, aerobically digested sewage sludge	30	11	5	2
Anaerobically digested sewage sludge	20	8	4	1
Composted sewage sludge	10	5	3	2
Activated, unstabilized sewage sludge	40	12	4	2

3. If a field is treated annually with an organic material, the application rate needed will become progressively smaller because, after the first year, the amount of nitrogen released from material applied in previous years must be subtracted from the total to be applied afresh (see Box 16.1, calculations for organic nitrogen application rates).

4. The nutrient and moisture contents of organic amendments vary widely among sources, and even from batch to batch, depending on how the material has been stored and treated. Therefore, general values such as those in Table 16.7 should not be relied upon when calculating rates of material to apply. Instead, representative samples of the material should be analyzed in a laboratory. Analyses are generally required by laws regulating land application of sewage sludge, but other materials are not as widely regulated in this regard.

5. Nutrients from organic sources will produce the highest return and cause the least environmental damage if applied to fields that are relatively poorly supplied with nitrogen and phosphorus. Cost of transport and considerations of convenience unfortunately provide incentives to apply organic materials on land close to their source, often leading to the nutrient imbalances discussed in Section 16.5.

Special Uses

There are a number of special uses of organic nutrient sources for which the organic matter component plays a special role. Examples include applications to denuded soil areas resulting from erosion, from land-leveling for irrigation, or from mining operations. Improvements in water-holding capacity and soil structure brought about by organic material application may be just as important as the long-term, slow-release, nutrient-supplying features of the organic wastes. Initial applications of 50 to 100 Mg/ha may be worked into the soil in the affected areas. These rates may be justified to supply organic matter as well as nutrients, provided the material is not so high in nitrogen as to create the potential for nitrate leaching.

Special cases of micronutrient deficiency can be ameliorated with manure application. Such treatments are sometimes used when there is some uncertainty about which specific nutrient is lacking. Manure applications can be made with little concern for adding toxic quantities of the micronutrients.

16.10   INTEGRATED RECYCLING OF WASTES

For most of the industrialized countries, widespread recycling of organic wastes other than animal manures is a relatively recent phenomenon. In heavily populated areas of Asia, however, and particularly in China and Japan, such recycling has long been practiced. Figure 16.17 illustrates the many ways in which organic wastes are used in China.

This example is based on the principle that the amount of nitrogen made available in any year should meet, but not exceed, the amount of nitrogen that plants can use for optimum growth. Therefore, the rate of release of available nitrogen usually determines the proper amount of manure, sludge, or other organic nutrient source to apply. Our example is a field producing corn two years in a row.

Year 1

Assumptions: Crop to be planted is corn. The goal is to produce 7000 kg/ha of grain, a yield that normally requires the application of 120 kg/ha of available nitrogen.

Available nitrogen needed = 120 kg.

Organic material to be applied is lime stabilized sewage sludge.

Analysis of the sludge showed 0.2% mineral N (ammonium and nitrate) and 4.5% total nitrogen.

Calculation of amount of sludge to apply:

% organic N in sludge   = total N − mineral N = 4.5% − 0.2% = 4.3%.

Organic N in 1 Mg of sludge   = 0.043 × 1,000 kg = 43 kg N.

Mineral N in 1 Mg of sludge   = 0.002 × 1,000 kg = 2 kg N.

Mineralization rate for lime stabilized sludge (Table 16.9) = 30% of organic N.

Available N mineralized from 1 Mg sludge in first year = 0.30 × 43 kg N = 12.9 kg N.

Total available N from 1 Mg sludge = mineral N + mineralized N

= 2.0 + 12.9 = 14.9 kg N.

Amount of (dry) sludge needed = 120 kg N/(14.9 kg available N/Mg dry sludge)

= 120/14.9 Mg dry sludge = 8.05 Mg dry sludge.

Adjust for moisture content of sludge (e.g., assume sludge has 25% solids and 75% water).

Amount of wet sludge to apply: 8.05 Mg dry sludge/(0.25 Mg dry sludge/Mg wet sludge) = 8.05/0.25 = 32.2 Mg wet sludge.

Year 2

Assumptions: Crop to be planted is again corn. The goal again is to produce 7000 kg/ha of grain, a yield that normally requires the application of 120 kg/ha of available nitrogen.

Available nitrogen needed = 120 kg/ha.

Organic material to be applied is dairy manure.

Analysis of the manure showed 0.2% mineral N and 2.4% total N.

Calculation of amount of manure to apply:

% organic N in manure = total N − mineral N = 2.4% − 0.2% = 2.2%.

Organic N in 1 Mg of manure   = 0.022 × 1,000 kg = 22 kg N.

Mineral N in 1 Mg of manure   = 0.002 × 1,000 kg = 2 kg N.

Mineralization rate for dairy manure is (Table 16.9) = 35% of organic N.

Available N mineralized from 1 Mg manure in first year = 0.35 × 22 kg N = 7.7 kg N.

Total available N from 1 Mg manure = mineral N + mineralized N = 2.0 + 7.7 = 9.7 kg N.

(continued)

Agriculture is only one of the recipients of these wastes. Much is used for biogas production, as food for fish, and as a source of heat from compost piles to help warm homes, greenhouses, and household water. Most important, the plant nutrients and organic matter are recycled and returned to the soil for future plant utilization. Such conservation is likely to be practiced more widely in the future by other countries, including the United States.

Calculate amount of N mineralized in year 2 from sludge applied in year 1:

Second year mineralization rate (Table 16.9) = 11% of original organic N.

N mineralized from sludge in year 2 = 0.11 × 43 Kg N/Mg × 8.05 Mg dry sludge

= 38 kg N from sludge in second year.

Amount of N needed from manure in year 2 = N needed by corn − N from sludge

= 120 kg − 38 kg = 82 kg N needed/ha.

Amount of (dry) manure needed per ha = 82 kg N/(9.7 kg available N/Mg dry manure) = 82/9.7 = 8.5 Mg dry manure/ha.

Adjust for moisture content of manure (e.g., assume manure has 25% solids and 75% water).

Mg of wet manure to apply: 8.5 Mg dry manure/(0.25 Mg dry manure/Mg wet manure) = 8.9/0.25 = 34.0 Mg wet manure.

Note that the 8.09 Mg sludge and 8.5 Mg manure also provided plenty of P: 161 kg P/ha from the sludge plus 58 kg P/ha from the manure (assuming 2 and 0.65% P in the two materials, see Table 16.7).

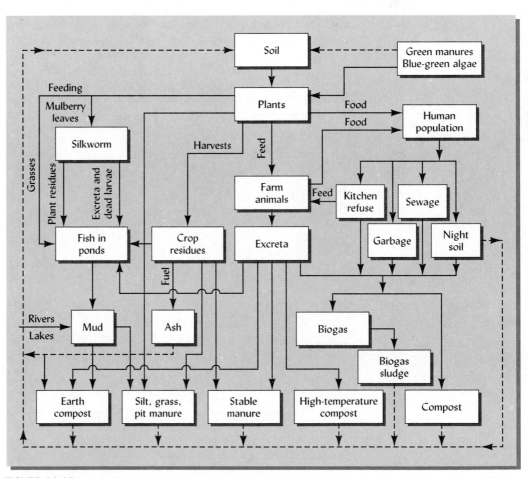

FIGURE 16.17 Recycling of organic wastes and nutrient elements in the People's Republic of China. Note the degree to which the soil is involved in the recycling processes. [Modified from FAO (1977)]

16.11 INORGANIC COMMERCIAL FERTILIZERS

The worldwide use of fertilizers increased dramatically during the latter half of the twentieth century (Figure 16.18), accounting for a significant part of the equally dramatic increases in crop yields during the same period. Improved soil fertility through the application of fertilizer nutrients is an essential factor in enabling the world to feed the billions of people that are being added to its population (see Section 20.1).

The need to supplement forest soil fertility is increasing as the demands for forest products increase the removal of nutrients, and competing uses of land leave forestry with more infertile, marginal sites. Currently, most fertilization in forestry is concentrated on tree nurseries and on seed trees, where the benefits of fertilization are relatively short term and of high value and the logistics of fertilizer application are not so difficult and expensive as for extensive forest stands. In the warmer, more humid regions of Japan,[4] about 50% of new forest plantings are fertilized. Forest fertilization is also increasingly practiced in the Southeast United States, the Scandinavian countries, and Australia.

Properties of Inorganic Fertilizers

The term *fertilizer* usually refers to inorganic salts that contain plant nutrient elements. Most, but not all, readily dissolve in water and make their nutrients available for plant uptake as soon as they are applied. Some fertilizers are simply purified products mined from natural sedimentary deposits (e.g., potassium chloride), some are chemically altered geologic materials (e.g., superphosphates are produced by treating apatite-bearing rocks with strong acids), and some are produced by complete chemical synthesis in chemical fertilizer factories (e.g., urea and ammonium phosphates). Fertilizers take many physical forms. While most are solid powders, granules, or pellets, others are liquids (e.g., phosphoric acid) or aqueous solutions (e.g., urea–ammonium nitrate solution, UAN), and still others are sold as a gas (e.g., anhydrous ammonia). Special machines have been designed to handle and apply each type.

---

[4]Where the climate is capable of supporting rapid growth of the more valuable timber species (such as "sugi"—*Cryptomeria japonica* and *Quercus acutissima*—whose logs are in demand to grow shiitake mushrooms).

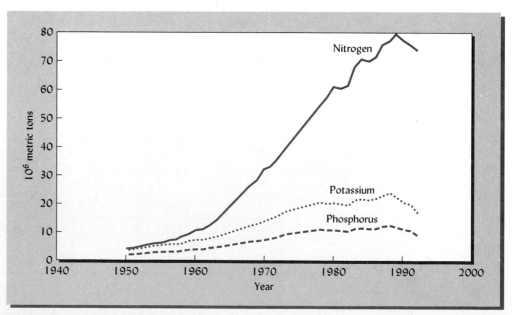

FIGURE 16.18 World fertilizer use since 1950 by major nutrient element. Fertilizer use leveled off during the 1990s as industrialized countries cut back somewhat from supraoptimal rates of use while developing countries continued to increase their use, striving to reach optimal levels. Compare this figure to Figure 14.1. (Data compiled from United Nations Food and Agriculture Organization reports.)

The composition of inorganic commercial fertilizers is much more precisely defined than is the case for the organic material discussed above. Table 16.10 lists the nutrient element contents and other properties of some of the more commonly used inorganic fertilizers. In most cases, fertilizers are used to supply plants with the macronutrients nitrogen, phosphorus, and/or potassium—sometimes called the *primary fertilizer elements*. Fertilizers that supply sulfur, magnesium, and the micronutrients are also manufactured.

It can be seen from the data in Table 16.10 that a particular nutrient (say, nitrogen) can be supplied by many different *carriers* or fertilizer compounds. Decisions as to which fertilizers to use must take into account not only the nutrients they contain, but also a number of other characteristics of the individual carriers. Table 16.10 provides information about some of these characteristics, such as the salt hazard (see also Section 10.5), acid-forming tendency (see also Section 13.7), tendency to volatilize, ease of solubility, and content of nutrients other than the principal one. Of the nitrogen carriers, anhydrous ammonia, nitrogen solutions, and ureas are the most widely used. Diammonium phosphate and potassium chloride supply the bulk of the phosphorus and potassium used in the United States.

Fertilizer Grade

Commercial fertilizers were first manufactured in the late 19th and early 20th centuries. Contemporary analytical procedures and laws passed to regulate the new industry resulted in certain labeling and marketing conventions that are still in common use today.

Of these conventions it is most important to be familiar with the *fertilizer grade*. Every fertilizer label states the grade as a three-number code, such as 10-5-10 or 6-24-24. These numbers stand for percentages indicating the *total* nitrogen (N) content, the *available* phosphoric acid (P_2O_5) content, and the *soluble* potash (K_2O) content. Plants do not take up phosphorus and potassium in these chemical forms, nor do any fertilizers actually contain P_2O_5 or K_2O.

These forms of expression are relics of the days when geochemists reported the contents of rocks in terms of the oxides formed upon heating. The fertilizer industry is slowly moving away from this archaic method of expressing fertilizer content. In all scientific work and in this textbook the simple elemental contents are used (P and K rather than P_2O_5 and K_2O). Box 16.2 explains how to convert between the elemental and oxide forms of expression.

The grade is important from an economic standpoint because it conveys the analysis or concentration of the nutrient in a carrier. When properly applied, most fertilizer carriers give equally good results for a given amount of nutrient element. The more concentrated carriers are usually the most economical to use because less weight of fertilizer must be transported to supply the needed quantity of a given nutrient. Hence, economic comparisons among different equally suitable fertilizers should be based on the price per kilogram of nutrient, not the price per kilogram of fertilizer.

Fate of Fertilizer Nutrients

A common *myth* about fertilizers suggests that inorganic fertilizers applied to soil directly feed the plant, and that therefore the biological cycling of nutrients, such as described by Figures 13.2 for nitrogen and 14.5 for phosphorus, are of little consequence where inorganic fertilizers are used. The reality is that nutrients added by normal application of fertilizers, whether organic or inorganic, are incorporated into the complex soil nutrient cycles, and that relatively little of the fertilizer nutrient (from 10 to 60%) actually winds up in the plant being fertilized during the year of application. Even when the application of fertilizers greatly increases both plant growth and plant uptake of a particular nutrient, the fertilizer acts partially by stimulating increased cycling of the nutrient, and the actual atoms taken up by the plant come largely from various pools in the soil and not directly from the fertilizer. This knowledge has been obtained by careful analysis of dozens of nutrient studies that used fertilizer with isotopically tagged nutrients. Results from such a study are summarized in Table 16.11, which shows somewhat more N uptake from fertilizer than is typically reported. Generally, as fertilizer rates are increased, the efficiency of fertilizer nutrient use decreases, leaving behind in the soil an increasing proportion of the added nutrient.

Fertilizer	Percent by weight							Salt hazard	Acid-forming tendency[b] (kg CaCO$_3$ kg^{-1})	Comments
	N	P	K	S	Ca	Mg	Micronutrients			
Primarily sources of nitrogen										
Anhydrous ammonia	82							Low	−148	Pressurized equipment needed; toxic gas.
Urea	45							Moderate	−84	Soluble; urease enzyme rapidly hydrolyses urea to ammonium forms.
Ammonium nitrate	33							High	−59	Absorbs moisture from air; can explode.
Sulfur-coated urea	30–40			13–16				Low	−110	Variable slow rate of release.
UF (ureaformaldehyde)	30–40							Very low	−68	Slowly soluble.
UAN solution	30							Moderate	−52	Most commonly used liquid N.
IBDU (isobutylidene diurea)	30							Very low	—	Slowly soluble.
Ammonium sulfate	21			24				High	−110	Rapidly lowers soil pH; very easy to handle.
Sodium nitrate	16						About 0.6% Cl	Very high	+29	Hardens, disperses soil structure.
Potassium nitrate	13		36	0.2	0.4	0.3	1.2% Cl	Very high	+26	Very rapid plant response.
Primarily sources of phosphorus										
Monoammonium phosphate	11	21–23	1–2					Low	−65	Best as starter.
Diammonium phosphate	18–21	20–23	0–1					Moderate	−70	Best as starter.
Triple superphosphate		19–22	1–3	15				Low	0	
Phosphate rock		8–18[a]		30				Very low	Variable	Low to extremely low availability. Best as fine powder on acid soils. Contains some Cd, F, etc.
Single superphosphate		7–9	11	20				Low	0	Nonburning, can place with seed.
Bonemeal	1–3[a]	10[a]	0.4	20				Very low	—	Slow availability of N, P as for phosphate rock.
Colloidal phosphate		8[a]		20				Very low	—	P availability as for phosphate rock.

TABLE 16.10 (Continued)

Fertilizer	Percent by weight							Salt hazard	Acid-forming tendency[b] (kg CaCO₃ kg⁻¹)	Comments
	N	P	K	S	Ca	Mg	Micronutrients			
Primarily sources of potassium										
Potassium chloride			50				47% Cl	High	0	Cl may reduce some diseases.
Potassium sulfate			42	17				Moderate	0	Use where Cl not desirable.
Wood ashes		1	4–8		5–10	3–6	0.2% Fe, 0.8% MN, 0.05% Zn, 0.005% Cu, 0.03% B	Moderate to high	+30	About ⅔ the liming value of limestone; caustic.
Greensand		0.6	6					Very low	0	Very low availability.
Granite dust			4					Very low		Very slow availability.
Primarily sources of other nutrients										
Basic slag		1–7			3–30	3	10% Fe, 2% Mn	Low	+70	Industrial by-product, slow availability, best on acid soils.
Gypsum				19	23			Low	0	Stabilizes soil structure, no effect on pH, Ca and S readily available.
Calcitic limestone					36			Very low	+95	Slow availability; raises pH.
Dolomitic limestone					24	12		Very low	+95	Very slow availability; raises pH.
Epsom salts (magnesium sulfate)				13		10		Moderate	0	No effect on pH; water soluble.
Sulfur, flowers				95				—	−300	Irritates eyes, very acidifying, slow acting, requires microbial oxidation.
Solubor							20.5% B	Moderate	—	Very soluble, compatible with foliar sprays.
EDTA chelate							13% Cu or 10% Fe or 12% Mn or 12% Zn	—	—	One or more micronutrients, see label.

Data derived from many sources.
[a]Highly variable contents.
[b]A negative number indicates that acidity is produced, a positive number indicates that alkalinity is produced.

BOX 16.2 HOW MUCH NITROGEN, PHOSPHORUS, AND POTASSIUM IS IN A BAG OF 6-24-24?

Conventional labeling of fertilizer products reports percentage N, P_2O_5, and K_2O. Thus, a fertilizer package (Figure 16.19) labeled as 6-24-24 (6% nitrogen, 24% P_2O_5, 24% K_2O) actually contains 6% N, 10.5% P, and 19.9% K (see calculations below).

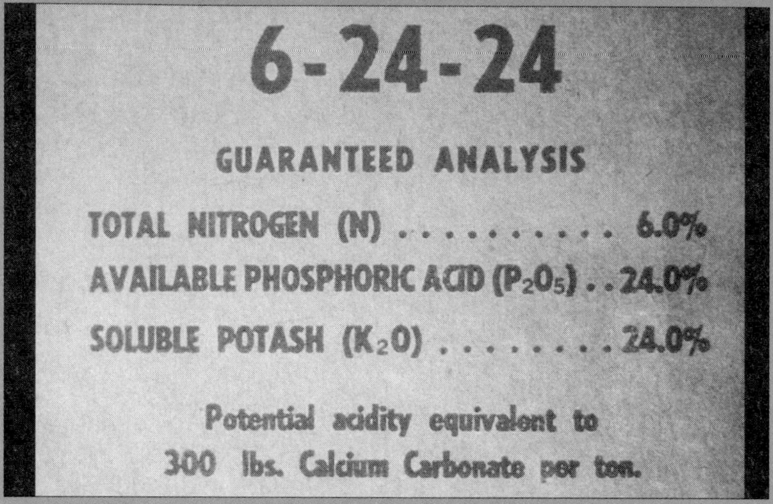
6-24-24
GUARANTEED ANALYSIS
TOTAL NITROGEN (N) 6.0%
AVAILABLE PHOSPHORIC ACID (P2O5) . . 24.0%
SOLUBLE POTASH (K2O) 24.0%
Potential acidity equivalent to
300 lbs. Calcium Carbonate per ton.

FIGURE 16.19 A typical commercial fertilizer label. Note that a calculation must be performed to determine the percentage of the nutrient elements P and K in the fertilizer since the contents are expressed as if the nutrients were in the forms of P_2O_5 and K_2O. Also note that after interacting with the plant and soil, this material would cause an increase in soil acidity that could be neutralized by 300 units of $CaCO_3$ per 2000 units (1 ton = 2000 lbs) of fertilizer material.

In order to determine the amount of fertilizer needed to supply the recommended amount of a given nutrient, it is necessary to convert percent P_2O_5 and percent K_2O to percent P and K. This may be done by calculating the proportion of P_2O_5 that is P and the proportion of K_2O that is K. The following caluclations may be used:

Given that the molecular weights of P, K, and O are as follows:
P 30.97 g/mol
K 39.10 g/mol
O 16.00 g/mol

Molecular weight of $P_2O_5 = 2(30.97) + 5(16.00) = 141.94$ g/mol
Proportion P in $P_2O_5 = (2(30.97))/141.94 = 0.4364$
To convert $P_2O_5 \rightarrow P$ multiply percent P_2O_5 by 0.44

Molecular weight of $K_2O = 2(39.10) + 16.00 = 94.20$
Proportion K in $K_2O = (2(39.10))/94.20 = 0.8301$
To convert $K_2O \rightarrow K$ multiply percent K_2O by 0.83

Thus, if the bag in Figure 16.19 contains 25 kg of the 6-24-24 fertilizer, it will supply the following amounts of N, P, and K:

Fertilizer anlaysis	*Conversion to percent element*	*Element in 25 kg (kg)*
6% N	No conversion $\rightarrow$ 6%	0.06×25 kg = 1.5 kg of N
24% P_2O_5	$(24\% \times 0.44) = 10.5\%$	0.105×25 kg = 2.6 kg of P
24% K_2O	$(24\% \times 0.83) = 19.9\%$	0.199×25 kg = 5.0 kg of K

TABLE 16.11   Source of Nitrogen in Corn Plants Grown in North Carolina on an Enon Sandy Loam Soil (Ultic Hapludalf) Fertilized with Three Rates of Nitrogen as Ammonium Nitrate

The source of the nitrogen in the corn plant was determined by using fertilizer tagged with the isotope ^{15}N. Note that moderate use of nitrogen fertilizer increased the uptake of N already in the soil system as well as that derived from the fertilizer.

Fertilizer nitrogen applied	Corn grain yield	Total N in corn plant	Fertilizer-derived N in corn	Soil-derived N in corn	Fertilizer-derived N in corn as percent of total N in corn	Fertilizer-derived N in corn as percent of N applied
kg/ha	Mg/ha		kg/ha		%	%
50	3.9	85	28	60	33	56
100	4.6	146	55	91	38	55
200	5.5	157	86	71	55	43

Calculated from Reddy and Reddy (1993).

16.12   THE CONCEPT OF THE LIMITING FACTOR

A famous German chemist named Justus von Liebig is credited with first publishing the concept that *plant production can be no greater than that level allowed by the growth factor present in the lowest amount relative to the optimum amount for that factor.* This growth factor, be it temperature, nitrogen, or water supply, will *limit* the amount of growth that can occur and is therefore called the *limiting factor* (see Figure 16.20).

If a factor is not the limiting one, increasing it will do little or nothing to enhance plant growth. In fact, increasing a nonlimiting factor may actually reduce plant growth by throwing the system further out of balance. For example, if a plant is limited by lack of phosphorus, adding more nitrogen may only aggravate the phosphorus deficiency.

Looked at another way, applying available phosphorus (the first limiting nutrient in this example) may allow the plant to respond positively to a subsequent addition of

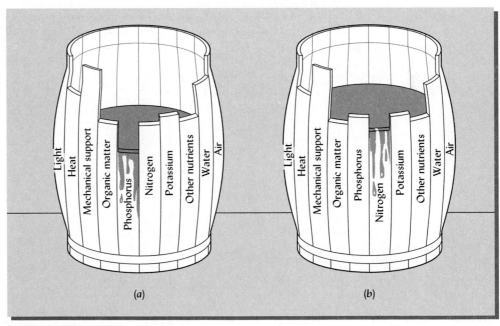

FIGURE 16.20   An illustration of the law of the minimum and the concept of the limiting factor. Plant growth is constrained by the essential element (or other factor) that is most limiting. The level of water in the barrel represents the level of plant production. (a) Phosphorus is represented as being the factor that is most limiting. Even though the other elements are present in more than adequate amounts, plant growth can be no greater than that allowed by the level of phosphorus available. (b) When phosphorus is added, the level of plant production is raised until another factor becomes most limiting—in this case nitrogen.

nitrogen. Thus, the increased growth obtained by applying two nutrients together often is much greater than the sum of the growth increases obtained by applying each of the two nutrients individually. Such an *interaction* or *synergy* between two nutrients can be seen in the data of Figure 16.21.

16.13   FERTILIZER APPLICATION METHODS

Wise, effective fertilizer use involves making correct decisions regarding *which nutrient element(s)* to apply, *how much* of each needed nutrient to apply, *what type* of material or carrier to use (Tables 16.7 and 16.10 list some of the choices), *in what manner* to apply the material, and finally *when* to apply it. We will leave information on the first two decisions until Sections 16.14 and 16.15. Here we will discuss the alternatives available with regard to the last two decisions.

There are three general approaches to applying fertilizers: broadcast application, localized placement, and foliar application. Each method has some advantages and disadvantages and may be particularly suitable for different situations. Often some combination of the three methods is used.

Broadcasting

In many instances fertilizer is spread evenly over the entire field or area to be fertilized. This method is called *broadcasting*. Often the broadcast fertilizer is mixed into the plow layer by means of tillage, but in some situations it is left on the soil surface and allowed to be carried into the root zone by percolating rain or irrigation water. This method is most appropriate when a large amount of fertilizer is being applied with the aim of raising the fertility level of the soil over a long period of time. Broadcasting is the most economical way to spread large amounts of fertilizer over wide areas (Figure 16.22).

For close-growing vegetation, broadcasting provides an appropriate distribution of the nutrients. It is therefore the most commonly used method for rangeland, pastures, small grains, turf grass, and forests. Fertilizers are also broadcast on some row cropland, especially in the fall when it is most convenient, although certainly not most efficient. The broadcast fertilizer may or may not be incorporated into the soil (see Figure 16.23b–c).

For phosphorus, zinc, manganese, and other nutrients that tend to be strongly retained by the soil, broadcast applications are usually much less efficient than localized

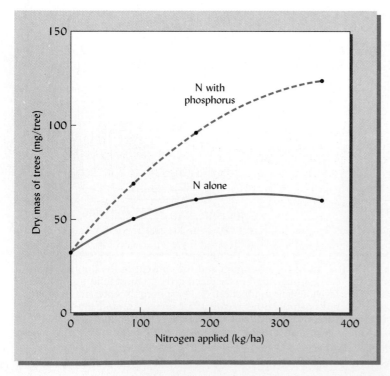

FIGURE 16.21   Increase in biomass of one-year-old white-pine seedlings in response to nitrogen fertilizer given with or without phosphorus on a sandy soil in southern Ontario, Canada. Apparently, phosphorus was the most limiting nutrient, and therefore little response was obtained from adding nitrogen until the phosphorus level was raised. This is an example of a nitrogen-by-phosphorus interaction. [Redrawn from Teng and Timmer (1994)]

(a)

(b)

FIGURE 16.22 (a) Broadcasting granular phosphorus and potassium fertilizer on a hayfield in Maryland; (b) Broadcasting a nitrogen solution (mixed with fungicide) on a wheat field in France. (Photos courtesy of R. Weil)

placement. Often 2 to 3 kg of fertilizer must be broadcast to achieve the same response as from 1 kg that is placed in a localized area.

A heavy one-time application of phosphorus and potassium fertilizer, broadcast and worked into the soil, is a good preparation for establishing perennial lawns and pastures. It may be necessary to broadcast a top dressing in subsequent years, being careful not to allow fertilizers with high-salt hazards (see Table 16.10) to remain in contact with the foliage long enough to cause salt burn. Because of its mobility in the soil, nitrogen does not suffer from reduced availability when broadcast, but if left on the soil surface much may be lost by volatilization. Volatilization losses are especially troublesome for urea and ammonium fertilizers applied to soils with a high pH. Nitrogen is commonly broadcast (sprayed) in a liquid form, often in a solution also containing other nutrients or chemicals (Figure 16.22b).

Another disadvantage of surface broadcasting is that the fertilizer is easily washed away with runoff during heavy rains. In fact, many runoff studies have shown that most of the annual loss of nutrients (or herbicides, if surface broadcast) usually occurs during the first one or two heavy-rainfall events after the broadcast application.

APPLICATION IN IRRIGATION WATER. Where irrigation is practiced, liquid fertilizers can be applied in the irrigation water, a practice sometimes called "fertigation." Liquid ammonia, nitrogen solutions, phosphoric acid, and even complete fertilizers are dissolved in the irrigation stream or the overhead sprinkler system. The nutrients are thus broadcast onto the soil in solution and carried down with the infiltrating water. Not only are appli-

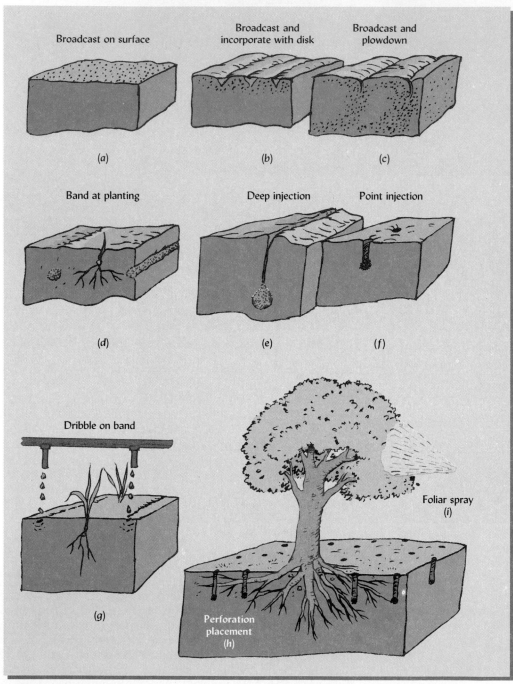

FIGURE 16.23 Fertilizers may be applied by many different methods, depending on the situation. Methods (a)–(c) represent broadcast fertilizer, with or without incorporation. Methods (d)–(h) are variations of localized placement. Method (i) is foliar application and has special advantages, but also limitations. Commonly, two or three of these methods may be used in sequence. For example, a field may be prepared with (c) before planting, (d) may be used during the planting operation, (g) may be used as a side dressing early in the growing season, and finally (i) may used to correct a micronutrient deficiency that shows up in the middle of the season.

cation costs reduced, but also relatively inexpensive nitrogen carriers can be used. Some care must be taken, however, to prevent ammonia loss by evaporation and to avoid certain fertilizer compounds which can clog the irrigation system with precipitates.

Finally, it should be pointed out that for crops with wide row spacing or young tree seedlings in forest plantings, broadcasting places the fertilizer where it is just as accessible to the *weeds* as to the target plants.

Localized Placement

Although it is commonly thought that nutrients must be thoroughly mixed throughout the root zone if plants are to be able to readily satisfy their needs, research has clearly shown that a plant can easily obtain its entire supply of a nutrient from a concentrated localized source in contact with only a small fraction of its root system. In fact, a small portion of a plant's root system can grow and proliferate in a band of fertilizer even though the salinity level caused by the fertilizer would be fatal to a germinating seed or to a mature plant were a large part of its root system exposed. This finding allowed the development of techniques for localized fertilizer placement.

There are at least two reasons why fertilizer is often more effectively used by plants if it is placed in a localized concentration rather than mixed with soil throughout the root zone. First, localized placement reduces the amount of contact between soil particles and the fertilizer nutrient, thus minimizing the opportunity for adverse fixation reactions. Second, the concentration of the nutrient in the soil solution at the root surface will be very high, resulting in greatly enhanced uptake by the roots.

STARTER APPLICATION.   Localized placement is especially effective for young seedlings, in cool soils in early spring, and for plants that grow rapidly with a big demand for nutrients early in the season. For these reasons *starter* fertilizer is often applied in *bands* on either side of the seed as the crop is planted. Since germinating seeds can be injured by fertilizer salts, and since these salts tend to move upward as water evaporates from the soil surface, the best placement for starter fertilizer is approximately 5 cm below and 5 cm off to the side from the seed row (see Figure 16.23d and Figure 16.24).

LIQUID FERTILIZERS.   Liquid fertilizers and slurries of manure and sewage sludge can also be applied in bands rather than broadcast. Bands of these liquids are placed 10 to 30 cm deep in the soil by a process known as *knife injection* (Figure 16.25). In addition to the advantages mentioned for banding fertilizer, injection of these organic slurries reduces runoff losses and odor problems.

Anhydrous ammonia and pressurized nitrogen solutions must be injected into the soil to prevent losses by volatilization. Injecting bands at depths of 15 and 5 cm, respectively, are considered adequate for these two materials.

FIGURE 16.24 Many modern planters are equipped with a fertilizer-banding attachment that places starter fertilizer for row crops slightly below and slightly to the side of the seed. This placement eliminates the danger of fertilizer burn, yet concentrates the nutrients near the seed where the crop roots will encounter them shortly after the seed germinates. (Courtesy National Plant Food Institute, Washington, D.C.)

FIGURE 16.25 (Upper) Sewage-sludge slurry being knife-injected into the soil before planting crops on the land. This injection method reduces runoff losses and objectionable odors. Lighter knives (not shown) are used to inject liquid fertilizers. (center) A spoke-wheel applicator designed for point injection of liquid fertilizer. (inset) Close-up view of spoke-wheel applicator. [Photo (top) courtesy of R. Weil; photos (center and inset) courtesy of Fluid Fertilizer Foundation]

DRIBBLE APPLICATION. Another approach to banding liquids (though not slurries) is to dribble a narrow stream of liquid fertilizer alongside of the crop row as a side dressing. The use of a stream instead of a fine spray changes the application from broadcast to banding and results in enough liquid in a narrow zone to cause the fertilizer to soak into the soil. This action greatly reduces volatilization loss of nitrogen.

POINT INJECTION. Localized placement of fertilizer can be carried one step beyond banding with a new system called *point injection*. With this system small portions of liquid fertilizer can be applied next to every individual plant without significantly disturbing either the plant root or the surface residue cover left by conservation tillage. The point injection implement shown in Figure 16.25 (center) is a modern version of the age-old dibble stick with which peasant farmers in the third world plant their seeds and later apply a portion of fertilizer in the soil next to each plant, all with a minimum of disturbance of the surface mulch.

DRIP IRRIGATION. The use of drip irrigation systems (see Section 6.10) has greatly facilitated the localized application of nutrients in irrigation water. Because drip fertigation is applied at frequent intervals, the plants are essentially spoon-fed, and the efficiency of nutrient use is quite high.

PERFORATION METHOD FOR TREES. Trees in orchards and ornamental plantings are best treated individually, the fertilizer being applied around each tree within the spread of the branches but beginning approximately 1 meter from the trunk (Figure 16.23h). The fertilizer is best applied by what is called the *perforation* method. Numerous small holes are dug around each tree within the outer half of the branch-spread zone and extending well into the upper subsoil. Into these holes, which are afterward filled up, is placed a suitable amount of an appropriate fertilizer. Special large fertilizer pellets are available for this purpose. This method of application places the nutrients within the tree root zone and avoids an undesirable stimulation of the grass or cover that may be growing around the trees. If the cover crop or lawn around the trees needs fertilization, it is treated separately, the fertilizer being drilled in at the time of seeding or broadcast later.

Foliar Application

Plants are capable of absorbing nutrients through their leaves in limited quantities. Under certain circumstances the best way to apply a nutrient is *foliar application*—spraying a dilute nutrient solution directly onto the plant leaves. Diluted NPK fertilizers, micronutrients, or small quantities of urea can be used as foliar sprays, although care must be taken to avoid significant concentrations of Cl^- or NO_3^-, which can be toxic to some plants. Foliar fertilization may conveniently fit in with other field operations for horticultural crops because the fertilizer is often applied simultaneously with pesticides.

The amount of nutrients that can be sprayed on leaves in a single application is quite limited. Therefore, while a few spray applications may deliver the entire season's requirement for a micronutrient, only a small portion of the macronutrient needs can be supplied in this manner. The danger of leaf injury is especially high during dry, hot weather, when the solution quickly evaporates from the leaf surface, leaving behind the fertilizer salts. Spraying on cool, overcast days or during early morning or late evening hours reduces the risk of injury, as does the use of a dilute solution containing, for example, only 1 or 2% nitrogen.

16.14 TIMING OF FERTILIZER APPLICATION

The timing of nutrient applications in the field is governed by several basic considerations: (1) making the nutrient available when the plant needs it; (2) avoiding excess availability, especially of nitrogen, before and after the principal period of plant uptake; (3) making nutrients available when they will strengthen, not weaken, long-season and perennial plants; (4) conducting field operations when conditions make them practical and feasible.

Availability When the Plants Need It

For mobile nutrients such as nitrogen (and to some degree potassium) the general rule is to make applications as close as possible to the period of rapid plant nutrient uptake. For a rapid-growing summer annuals such as corn, this means making only a small starter application at planting time, and applying most of the needed nitrogen as a side dressing just before the plants enter the rapid nutrient accumulation phase, usually about four to six weeks after planting. For cool-season plants such as winter wheat or certain turf grasses, most of the nitrogen should be applied about the time of spring "green-up" when the plants resume a rapid growth rate. For trees, the best time is when new leaves are forming. With slow-release organic sources, some time should be allowed for mineralization to take place prior to the plants' period of maximum uptake.

Environmentally Sensitive Periods

In temperate (**Udic** and **xeric**) climates, most leaching takes place in the winter and early spring when precipitation is high and evapotranspiration is low. Nitrates left over after plant uptake has ceased have the potential for leaching during this period. In this regard it should be noted that, for grain crops, the rate of nutrient uptake begins to decline during grain-filling stages and has virtually ceased long before the crop is ready for harvest. With inorganic nitrogen fertilizers, avoiding leftover nitrates is largely a matter of limiting the amount applied to what the plants are expected to take up. However, for slow-release organic sources applied in late spring or early summer, mineralization is likely to continue to release nitrates after the crop has matured and ceased taking them up. To the extent that this timing of nitrate release is unavoidable, nonleguminous cover crops should be planted in the fall to absorb the excess nitrate being released.

SPLIT APPLICATIONS. In high-rainfall conditions and on permeable soils, dividing a large dose of fertilizer into two or more split applications may avoid leaching losses prior to the crop's establishment of a deep root system. In cold climates, another environmentally sensitive period occurs during early spring, when snowmelt over frozen or saturated soils results in torrents of runoff water that may carry to rivers and streams soluble nutrients from manure or fertilizer that may be at or near the soil surface.

Application of nitrogen fertilizers to mature forests is usually carried out when rains can be expected to wash the nutrients into the soil and minimize volatilization losses. However, observation of nitrate content in stream flow often reveals that such applications result in a pulse of nitrate leaving the watershed for several weeks following the fertilizer application.

Physiologically Appropriate Timing

It is important to make nutrients available when they will strengthen plants and improve their quality. For example, too much nitrogen in the summer may stress a cool-season turf grass, while high nitrogen late in the season will reduce the sugar content of sugar crops. A good supply of potassium is particularly important in the fall to enable plants to improve their winter-hardiness. Broadcast fertilization of trees soon after the seedlings are planted may benefit fast-growing weeds more than the desired trees. Later in the development of a forest stand, when the tree canopy has matured, application of fertilizer, usually from a helicopter, can be quite beneficial.

Practical Field Limitations

Sometimes it is simply not possible to apply fertilizers at the ideal time of the year. For example, although a crop may respond to a late-season side dressing, if the plants are too tall to drive over without damaging them, such an application will be difficult. Application by airplane may increase the flexibility possible. Early spring applications may be limited by fields too wet to support tractors. Economic costs or the time demands of other activities may also require that compromises be made in the timing of nutrient application.

16.15 DIAGNOSTIC TOOLS AND METHODS[5]

Three basic tools are available for diagnosing soil fertility problems: (1) field observations, (2) plant tissue analysis, and (3) soil analysis (soil testing). To effectively guide the application of nutrients as well as to diagnose problems as they arise in the field, all three approaches should be integrated. There is no substitute for careful observation and *recording* of circumstantial evidence and symptoms in the field. Effective observation and interpretation requires skill and experience as well as an open mind. It is not uncommon for a supposed soil fertility problem to actually be caused by soil compaction, weather conditions, pest damage, or human error. The task of the diagnostician is to use all the tools available in order to identify the factor that is limiting plant growth, and then devise a course of action to alleviate the limitation.

16.16 PLANT SYMPTOMS AND FIELD OBSERVATIONS

This detective-like work can be one of the more exciting and challenging aspects of nutrient management. To be an effective soil fertility diagnostician, several general guidelines are helpful.

First, develop an organized way to *record* your observations. The information you collect may be needed to properly interpret soil and plant analytical results obtained at a later date.

Second, look for spatial patterns—how the problem seems to be distributed in the landscape and in individual plants. Linear patterns across a field may indicate a problem related to tillage, drain tiles, or the incorrect spreading of lime or fertilizer. Poor growth concentrated in low-lying areas may relate to the effects of soil aeration. Poor growth on the high spots in a field may reflect the effects of erosion and possibly exposure of subsoil material with an unfavorable pH.

Third, closely examine individual plant leaves to characterize any foliar symptoms. Nutrient deficiencies can produce characteristic symptoms on leaves and other plant parts. Examples of such symptoms are shown in several figures in Chapters 13–15 and in Plates 16, 17, 19, and 20. Determine if the symptoms are most pronounced on the younger leaves (as is the case for most of the micronutrient cations) or on the older leaves (as is the case for nitrogen, potassium, and magnesium). Some nutrient deficiencies are quite reliably identified from foliar symptoms, while others produce symptoms that may be confused with herbicide damage, insect damage, or damage from poor aeration.

Fourth, observe and *measure* differences in plant growth and crop yield that may reflect different levels of soil fertility, even though no leaf symptoms are apparent. Check *both* aboveground and belowground growth. Are mycorrhizae associated with tree roots? Are legumes well nodulated? Is root growth restricted in any way?

Fifth, obtain records on plant growth or crop yield from previous years, and ascertain the history of management for the site for as many years as possible. It is often useful to sketch a map of the site showing features you have observed and the distribution of symptoms.

16.17 PLANT ANALYSIS AND TISSUE TESTING

The concentration of essential elements in plant tissue is related to plant growth or crop yield as shown in Figure 16.26. The *sufficiency range* and the *critical range* have been well characterized for many plants, especially for agronomic and major horticultural crops. Less is known about forest trees and ornamentals. Sufficiency ranges for 11 essential elements in a variety of plants are listed in Table 16.12.[6] Plants with tissue concentrations lower than the smaller value given in the sufficiency range for a particular nutrient are likely to respond to additions of that nutrient if no other factor is more limiting.

---

[5]For a detailed discussion of both plant analysis and soil testing, see Westerman (1990).

[6]Detailed information on tissue analyses for a large number of plant species can be found in Reuter and Robinson (1986).

TABLE 16.12 A Guide to Tissue Analysis for Selected Plant Species

The values given are for sufficiency ranges.

Plant species	Part to sample	N	P	K	Ca	Mg	S	Fe	Mn	Zn	B	Cu
		%						µg/g				
Pine trees (*Pinus spp.*)	Current-year needles near terminal	1.2–1.4	0.10–0.18	0.30–0.45	0.13–0.16	0.05–0.09	0.08–0.12	20–100	50–600	20–50	3–9	2–6
Turf grasses	Clippings	2.75–3.5	0.30–0.55	1.0–2.5	0.50–1.2	0.20–0.60	0.20–0.45	35–1000	25–150	20–55	10–60	5–20
Corn (*Zea mays*)	Ear-leaf at tasseling	2.5–3.5	0.20–0.50	1.5–3.0	0.2–1.0	0.16–0.40	0.16–0.50	25–300	20–200	20–70	6–40	6–40
Soybean (*Glycine max*)	Recently matured trifoliate, flowering stage	4.0–5.0	0.31–0.50	2.0–3.0	0.45–2.0	0.25–0.55	0.25–0.55	50–250	30–200	25–50	25–60	8–20
Apple (*Malus spp.*)	Leaf from base of nonfruiting shoots	1.8–2.4	0.15–0.30	1.2–2.0	1.0–1.5	0.25–0.50	0.13–0.30	50–250	35–100	20–50	20–50	5–20
Wheat (*Triticum spp.*)	First leaf blade from top of plant	2.2–3.3	0.24–.36	2.0–3.0	0.28–0.42	0.19–0.30	0.20–0.30	35–55	30–50	20–35	5–10	6–10
Tomato (*Lycopersicon esculentum*)	Most recently matured leaf at early bloom	3.2–4.8	0.32–0.48	2.5–4.2	1.7–4.0	0.45–0.70	0.60–1.0	120–200	80–180	30–50	35–55	8–12
Alfalfa (*Medicago sativa*)	Upper third of plant at first flower	3.0–4.5	0.25–0.50	2.5–3.8	1.0–2.5	0.3–0.8	0.3–0.5	50–250	25–100	25–70	6–20	30–80

Data derived from many sources.

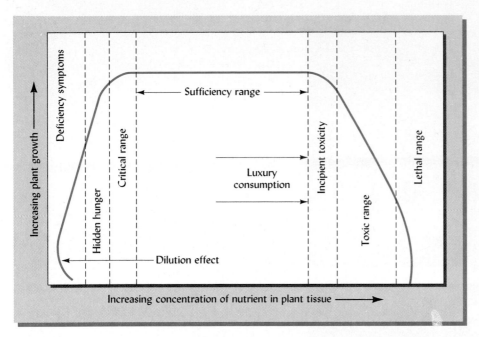

FIGURE 16.26 The relationship between plant growth or yield and the concentration of an essential element in the plant tissue. For most nutrients there is a relatively wide range of values associated with normal, healthy plants (the sufficiency range). Beyond this range, plant growth suffers from either too little or too much of the nutrient. The critical range (CR) is commonly used for the diagnosis of nutrient deficiency. Nutrient concentrations below the CR are likely to reduce plant growth even if no deficiency symptoms are visible. This moderate level of deficiency is sometimes called *hidden hunger*. The odd hook at the lower left of the curve is the result of the so-called dilution effect that is often observed when extremely stunted, deficient plants are given a small dose of the limiting nutrient. The growth reponse may be so great that even though somewhat more of the element is taken up, it is diluted in a much greater plant mass.

Tissue analysis can be a powerful tool for identifying plant nutrient problems if several simple precautions are taken. First, it is critical that the correct plant part be sampled. Second, the plant part must be sampled at the specified stage of growth because the concentrations of most nutrients decrease considerably as the plant matures. Third, it must be recognized that the concentration of one nutrient may be affected by that of another nutrient, and that sometimes the ratio of one nutrient to another (e.g., Mg/K, S/N, or Fe/Mn) may be the most reliable guide to plant nutritional status. In fact, several elaborate mathematical systems for assessing the ratios or balance among nutrients have proven useful for certain plant species.[7] Because of the uncertainties and complexities in interpreting tissue concentration data, it is wise to sample plants from the best and worst areas in a field or stand. The differences between such samples may provide valuable clues concerning the nature of the nutrient problem.

16.18   SOIL ANALYSIS

Since the total amount of an element in a soil tells us very little about the ability of that soil to supply that element to plants, more meaningful *partial soil analyses* have been developed. *Soil testing* is the routine partial analysis of soils for the purpose of guiding nutrient management.

The soil testing process consists of three critical phases: (1) sampling the soil, (2) chemically analyzing the sample, and (3) interpreting the analytical result to make a recommendation on the kind and amounts of nutrients to apply.

Sampling the Soil

Soil sampling is widely acknowledged to be one of the weakest links in the soil testing process. Part of the problem is that about a teaspoonful of soil (see Figure 16.27) is eventually used to represent millions of kilograms of soil in the field. Since soils are highly variable, both horizontally and vertically, it is essential to carefully follow the sampling instructions from the soil testing laboratory.

---

[7]The most well developed of the multinutrient ratio systems is known as the Diagnostic Recommendation Integrated System (DRIS). For details on the DRIS, see Walworth and Sumner (1987).

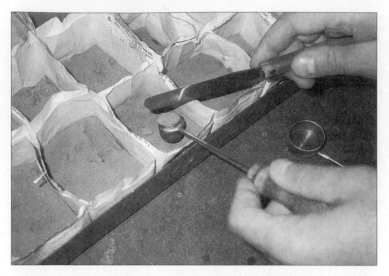

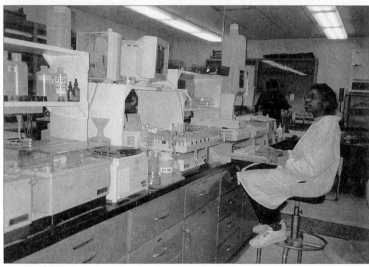

FIGURE 16.27 (Top) After a soil sample is received by the soil testing lab, the soil is ground and screened to make a homogenous powder. Then a small scooped or weighed sample is analyzed chemically. This small amount of soil must represent thousands of metric tons of soil in the field. (Bottom) After a portion of the nutrients have been extracted from the soil sample, the solution containing these nutrients undergoes elemental analysis. Since soil test labs must run hundreds or thousands of samples each day, the analysis is generally automated and the results are recorded and interpreted by computer. (Photos courtesy of R. Weil)

Usually, a soil probe is used to remove a thin cylindrical core of soil from at least 15 to 20 randomly scattered places within the land area to be represented (see Figure 16.28). The 15 to 20 subsamples are thoroughly mixed in a plastic bucket, and about 0.5 L of the soil is placed in a labeled container and sent to the lab. If the soil is moist, it should be air-dried without sun or heat prior to packaging for routine soil tests. Heating the sample might cause a falsely high result for certain nutrients.[8]

Two questions must be addressed when sampling a soil: (1) To what depth should the sample be taken? (2) What time of year should the soil be sampled?

The standard depth of sampling for a plowed soil is the depth of the plowed layer, about 15 to 20 cm, but various other depths are also used (Figure 16.28). Because in many unplowed soils nutrients are stratified in contrasting layers, the depth of sampling can greatly alter the results obtained.

Seasonal changes are often observed in soil test results for a given field. For example, the potassium level is usually highest in early spring after freezing and thawing has released some "fixed" K ions from clay interlayers, and lowest in late summer after plants have removed much of the readily available supply. The time of sampling is especially important if year-to-year comparisons are to be made. A good practice is to sample each field every year or two (always at the same time of year), so that the soil test levels can be tracked over the years to determine whether nutrient levels are being maintained, increased, or depleted.

---

[8]Samples intended for analysis of soil nitrates are an exception. These should be rapidly dried under a fan or in an oven at about 65°C.

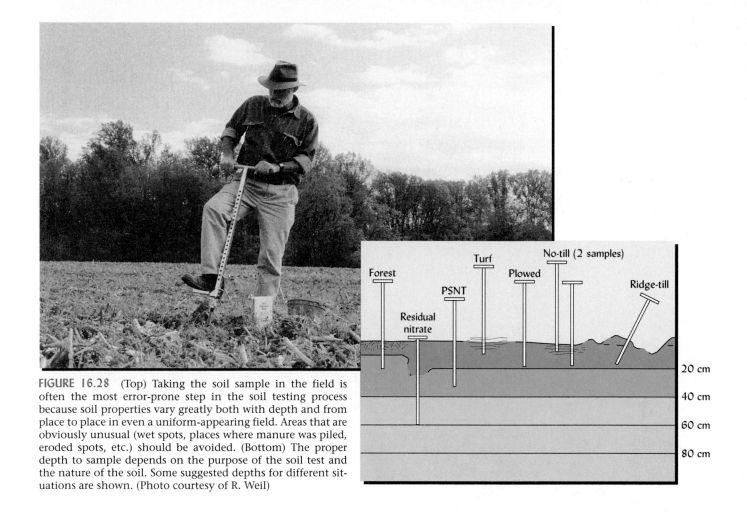

FIGURE 16.28 (Top) Taking the soil sample in the field is often the most error-prone step in the soil testing process because soil properties vary greatly both with depth and from place to place in even a uniform-appearing field. Areas that are obviously unusual (wet spots, places where manure was piled, eroded spots, etc.) should be avoided. (Bottom) The proper depth to sample depends on the purpose of the soil test and the nature of the soil. Some suggested depths for different situations are shown. (Photo courtesy of R. Weil)

TIMING FOR SPECIAL NITROGEN TESTS.   Timing is especially critical to determine the amount of mineralized nitrogen in the root zone. In relatively dry, cold regions (e.g., the Great Plains), the residual nitrate test is done on 60-cm-deep samples obtained sometime between fall and before planting in spring. In humid regions, where nitrate leaching is more pronounced, a special test has been developed to determine whether a soil will mineralize enough nitrogen for a corn crop. The samples for this Pre-Sidedress Nitrate Test (PSNT) are taken from the upper 30 cm when the corn is about 30 cm tall, just in time to determine how much nitrogen to apply as the crop enters its period of most rapid nitrogen uptake. In this case the soil must be sampled during the narrow window of time when spring mineralization has peaked but plant uptake has not yet begun to deplete the nitrate produced.

Chemical Analysis of the Sample

In general, soil tests attempt to extract from the soil amounts of essential elements that are correlated with the nutrients taken up by plants. Different extraction solutions are employed by various laboratories. Buffered salt solutions such as sodium or ammonium acetate, or mixtures of dilute acids and chelating agents are the extracting agents most commonly used. The extractions are accomplished by placing a small measured quantity of soil in a bottle with the extracting agent and shaking the mixture for a certain number of minutes. The amount of the various nutrient elements brought into solution is then determined. The whole process is usually automated so that a modern laboratory can handle hundreds of samples each day (Figure 16.29).

The most common and reliable tests are those for soil pH, potassium, phosphorus, and magnesium. Micronutrients are sometimes extracted using chelating agents, especially for calcareous soils in the more arid regions. Predicting the availability of nitrogen and sulfur is considerably more difficult because of the many biological factors involved, but the nitrate and sulfate present in the soil at the time of sampling can be measured.

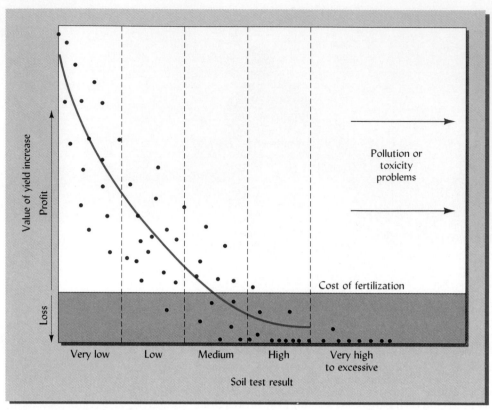

FIGURE 16.29 The relationship between soil test results for a nutrient and the extra yield obtained by fertilizing with that nutrient. Each data point represents the *difference* in plant yield between the fertilized and the unfertilized soil. Because many factors affect yield and because soil tests can only approximately predict nutrient availability, the relationship is not precise, but the data points are scattered about the trend line. If the point falls above the fertilizer cost line, the extra yield was worth more than the cost of the fertilizer and a profit would be made. For a soil testing in the very low and low categories, a profitable response to fertilizer is very likely. For a medium testing soil, a profitable response is a 50:50 proposition. For soils testing in the high category, a profitable response is unlikely.

Because the methods used by different labs may be appropriate for different types of soils, it is advisable to send a soil sample to a lab in the same region from which the soil originated. Such a lab is likely to use procedures appropriate for the soils of the region, and should have access to data correlating the analytical results to plant responses on soils of a similar nature.

Soil tests designed for use on soils or soil-based potting media generally do not give meaningful results when used on artificial peat-based soilless potting media. Special extraction procedures must be used for the latter, and the results then correlated to nutrient uptake and growth of plants grown in similar media.

These examples should further emphasize the importance of providing the soil testing lab with complete information concerning the nature of your soil, its management history, and your plans for its future use.

Interpreting the Results to Make a Recommendation

This is, perhaps, the most controversial aspect of soil testing. The soil test values themselves are merely *indices* of nutrient-supplying power. They do not indicate the actual amount of nutrient that will be supplied. For this reason, it is best to think of soil test reports as more qualitative than quantitative.

Many years of field experimentation at many sites are needed to determine which soil test level indicates a low, medium, or high capacity to supply the nutrient tested. These categories are used to predict the likelihood of obtaining a profitable response from the application of the nutrient tested (see Figure 16.29).

Recommendations for nutrient applications take into consideration practical knowledge of the plants to be grown, the characteristics of the soil under study, and other environmental conditions. Management history and field observations can help relate soil test data to fertilizer needs.

The interpretation of soil test data is best accomplished by experienced and technically trained personnel, who fully understand the scientific principles underlying the common field procedures. In modern laboratories, the factors to be considered in making fertilizer recommendations are programmed into a computer, and the interpretation is printed out for the farmer's or gardener's use (see Figure 16.30).

Merits of Soil Testing

It must not be inferred from the preceding discussion that the limitations of soil testing outweigh its advantages. When the precautions already described are observed, soil testing is an invaluable tool in making fertilizer recommendations. These tests are most useful when they are correlated with the results of field fertilizer experiments (see Figure 16.31).

Report on soil tests
Auburn University
Soil testing laboratory
Auburn University, AL 36849

Name — Alabama resident
Address — 118 Main Street
City — Hometown, AL 36830

County — Lee
District — 2

Lab. no.	Sender's sample designation and	Crop to be grown	Soil* group	pH**	Phosphorus P***		Potassium K***		Magnesium Mg***		Limestone Tons/acre	N	P_2O_5	K_2O
												Pounds per acre		
23887	1	Soybeans	2	5.3	L	70	M	70	H	160	2.0	0	80	40

Comment 224—soil acidity (low pH) can be corrected with either dolomitic or calcitic lime.

| 23888 | 2 | Corn | 1 | 5.6 | L | 70 | M | 70 | H | 160 | 1.0 | 120 | 80 | 40 |

See comment 224 above.
Comment 15—corn on sandy soils may respond to nitrogen rates up to 150 lbs. per acre. On sandy soils apply 3 lbs. zinc (Zn) per acre in fertilizer after liming or where pH is above 6.0.

| 23889 | 3 | Bahia | 1 | 6.0 | M | 100 | H | 140 | H | 240 | 0.0 | 60 | 40 | 0 |

| 23890 | Garden | Vegetables | 3 | 5.2 | M | 90 | M | 70 | H | 160 | 3.0 | 120 | 120 | 120 |

See comment 224 above.
Comment 82—per 100 ft. of row apply 6 lbs. 8-8-8 (3 quarts) at planting and sidedress with 4 lbs. 8-8-8 (2 quarts).

***On summergrass pastures apply P and K as recommended and 60 lbs. of N before growth starts. Up to September 1 repeat the N applications when more growth is desired.

*** 1.0 ton limestone per acre is approximately equivalent to 50 lbs. per 1,000 sq. ft.

***For cauliflower, broccoli and root crops, apply 1.0 lb. of boron (B) per acre. (For home gardens, 1 tablespoon borax per 100 ft. of row.)

The number of samples processed in this report is 4.

*1. Sandy soils
2. Loams & light clays
3. Heavy clays (excluding Blackbelt)
4. Heavy clays of the Blackbelt

**7.4 or higher   Alkaline   6.5 or lower   Acid
6.6–7.3   Neutral   5.5 or lower   Very acid

*** Rating & fertility (percent sufficiency)

- - - - - - - - - - - - - - -
Approved

Soil testing Form B

FIGURE 16.30 Example of a soil test report giving the soil test levels and the recommendations for fertilizers and lime. [From Cope et al. (1981)]

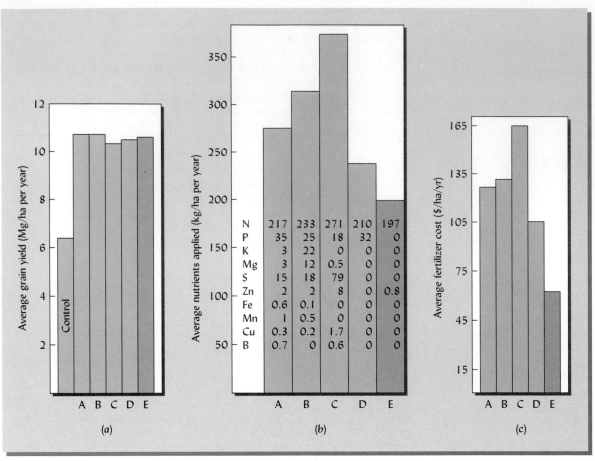

FIGURE 16.31 (a) The yield of corn from plots near North Platte, Nebraska, receiving five different rates of fertilizer as recommended by five soil testing laboratories (A–E). Data averaged over a six-year period. The soil was a Cozad silt loam (Fluventic Haplustoll). (b) The nutrient levels recommended by these laboratories and (c) the cost of the fertilizer applied each year. Note that the yields from all fertilized plots were about the same, even though rates of fertilizer application differed markedly. The fertilizer rates recommended by laboratory E utilized a sufficiency-level concept that was based on the calibration of soil tests with field yield responses. Those recommended by laboratories A–D were based either on a maintenance concept that required replacement of all nutrients removed by a crop or on the supposed need to maintain set cation ratios (Ca/Mg, Ca/K, and Mg/K). Obviously, the sufficiency-level concept provided more economical results in this trial. As a result of this and many other similar comparison studies, most soil test labs have adopted the sufficiency-level approach. [From Olson et al. (1982); used with permission of American Society of Agronomy]

Generally, soil testing has been most relied upon in agricultural systems, while foliar analysis has proved more widely useful in forestry. The limited use of soil testing in forestry is probably a result of the complex stratification in forested soils that creates a great deal of uncertainty about how to obtain a representative sample of soil for analysis. Also, because of the comparably long time frame in forestry, little information is available on the correlation of soil test levels with timber yield in the sense that such information is widely available for agronomic crops (Figure 16.29). Even though the relationship between tree growth and soil test level is not well known for most forest systems, standard agronomic soil testing can still be useful in distinguishing those soils whose ability to supply P or K is adequate from those with very low supplying power for these nutrients.

16.19 BROADER ASPECTS OF FERTILIZER PRACTICE

Fertilizer practice involves many intricate details regarding soils, plants, and fertilizers. Because of the great variability from place to place in each of these three factors, it is difficult to arrive at generalizations for fertilizer use. However, because of the prominence of the response by most nonlegume plants, and because of its implications for environmental quality, the initial focus is generally on nitrogen in most fertilizer schemes.

Applications of phosphorus and potassium are made to balance and supplement the nitrogen supply whether it be from the soil, crop residues (especially legumes), organic wastes, or added fertilizers.

Because it is very difficult to predict the nitrogen-supplying ability of a soil from chemical tests, nitrogen fertilizer recommendations are usually based on field experiments that define the relationship between added nitrogen and plant growth or crop yield. Generally, these field studies are carried out on a variety of soils and under a range of different weather conditions (which cause the response to nitrogen to vary greatly from one year to the next).

From the shape of the response curve and economic considerations (see below), the optimal level of nitrogen fertilizer is determined. This optimal fertilizer rate should be adjusted by the amount of any additional gains or losses not taken into account in the standard response curve. For example, nitrogen contributions from previous or current manure applications (see Box 16.1), legume cover crop (Figure 6.10), or previous legume in the rotation (Figure 16.11) should be subtracted from the amount of fertilizer recommended.

A second aspect relates to economics. Farmers do not use fertilizers just to grow big crops or to increase the nutrient content of their soils. They do so to make a living. As a result, any fertilizer practice that does not give a fair economic return will not stand the test of time. In crop production, the most profitable rate is determined by the ratio of the value of the extra yield expected to the cost of the fertilizer applied. The law of diminishing returns applies. Therefore, the most profitable fertilizer rate will be somewhat less than the rate which would produce the very highest yield (see Figure 16.32).

Traditionally, economic analysis of optimum fertilizer rates has assumed that the plant response to fertilizer inputs was represented by a smooth curve following a quadratic function. In fact, actual data obtained can be just as well described by a number of other mathematical functions (Figure 16.33). This seemingly esoteric observation can have a great effect on the amount of fertilizer recommended and, in turn, on the likelihood of environmental harm from excessive fertilizer use (see Figure 16.34).

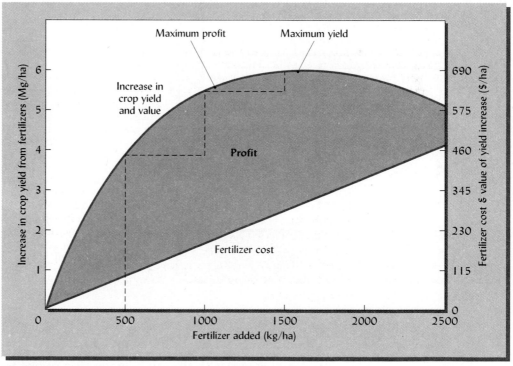

FIGURE 16.32 Relationships among the rate of fertilizer addition, crop yield increase, fertilizer costs, and profit from adding fertilizers. Note that the yield increase (and profit) from the first 500 kg of fertilizer was much greater than from the second and third 500 kg. Also note that the maximum profit is obtained at a lower fertilizer rate than that needed to give a maximum yield. The calculations assume that the yield response to added fertilizer takes the form of a smooth quadratic curve.

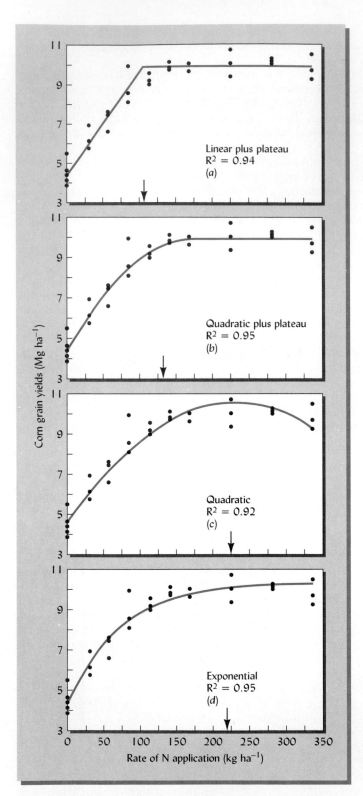

FIGURE 16.33 An example of how the mathematical function chosen to represent fertilizer-response data can effect the amount of fertilizer recommended. The data in all five graphs are exactly the same and represent the response of corn yields in Iowa to increasing levels of nitrogen fertilization. The vertical arrows indicate the recommended, most profitable rate of nitrogen to apply according to each mathematical model. Note that all models fit the data equally well (as indicated by the very similar R^2 values), but that the linear-plus-plateau model predicts an optimum nitrogen rate of 104 kg/ha, while the standard quadratic model suggests that 222 kg/ha is the optimum. Apparently the extra 116 kg/ha of nitrogen may have no effect on crop yield, but may greatly increase the risk of environmental damage. [Data from Cerrato and Blackmer (1990)]

Anyone involved with the actual production of plants can testify to the enormous improvements that accrue from judicious use of organic and inorganic nutrient supplements. The preceding discussion should not imply that such supplements are not needed, but only that their optimum levels are difficult to determine with precision.

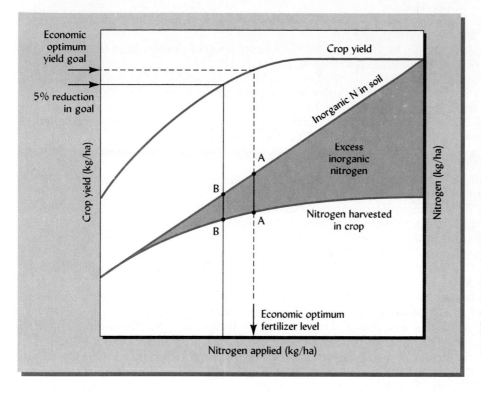

FIGURE 16.34 Influence of rate of nitrogen fertilization on crop yield, nitrogen removed in harvest, and the amount of excess inorganic soil nitrogen potentially available for loss into the environment. The fine lines show that a relatively small decrease (5%) in the yield goal for which nitrogen was applied would result in a relatively large reduction (about 30%) in the potential for nitrogen pollution. Line A-A represents the excess nitrogen present when fertilizer is applied at a rate designed to reach the economic optimum yield goal. Line B-B represents the excess nitrogen if the yield goal is 5% less than the economic optimum. Unfortunately, in many cases a yield reduction of 5% may reduce profits by a much larger percentage. [Adapted from National Research Council (1993)]

16.20 CONCLUSION

The continuous availability of plant nutrients is critical for the sustainability of most ecosystems. The challenge of nutrient management is threefold: (1) to provide adequate nutrients for plants in the system, and simultaneously (2) to ensure that inputs are in balance with plant utilization of nutrients, thereby conserving nutrient resources and (3) preventing contamination of the environment with unutilized nutrients.

The recycling of plant nutrients must receive primary attention in any ecologically sound management system. This can be accomplished in part by returning plant residues to the soil. These residues can be supplemented by judicious application of organic wastes that are produced in abundance by municipal, industrial, and agricultural operations worldwide. The use of cover crops grown specifically to be returned to the soil is an additional organic means of recycling nutrients.

For sites from which crops or forest products are removed, nutrient losses commonly exceed the inputs from recycling. Inorganic fertilizers will continue to supplement natural and managed recycling to replace these losses and to increase the level of soil fertility so as to enable humankind to not only survive, but to flourish on this planet. In extensive areas of the world, fertilizer use will have to be increased above current levels to avoid soil and ecosystem degradation and to enable profitable food production.

The use of fertilizers, both inorganic and organic, should not be done in a simply habitual manner or for so-called insurance purposes. Rather, soil testing and other diagnostic tools should be used to determine the true need for added nutrients. Where soils are low in available nutrients, inorganic fertilizers often return several dollars' worth of improved yield for every dollar invested. However, where the nutrient-supplying power of the soil is already sufficient, adding fertilizers is likely to be damaging to both the bottom line and to the environment.

REFERENCES

Attiwill, P. M., and G. W. Leeper. 1987. *Forest Soils and Nutrient Cycles* (Melbourne, Australia: Melbourne University Press).

Belt, G. H., and J. O'Laughlin. 1994. "Buffer strip design for protecting water quality and fish habitat," *Western Journal of Applied Forestry,* **9**(2):41–45.

Bormann, F. H., and G. E. Likens. 1979. *Pattern and Process in a Forested Ecosystem* (New York: Springer-Verlag), 253 pp.

Cerrato, M. E., and A. M. Blackmer. 1990. "Comparison of models from describing corn yield response to nitrogen fertilizer," *Agron. J.*, 138–143.

Chauby, I., D. R. Edwards, T. C. Daniel, P. A. Moore, Jr., and D. J. Nichols. 1994. "Effectiveness of vegetative filter strips in retaining surface-applied swine manure constituents," *Trans. Am. Soc. Agric. Engineers*, 37:845–850.

Cope, J. T., C. E. Evans, and H. C. Williams. 1981. *Soil Test Fertilizer Recommendations for Alabama Crops*, Circular 251, Agr. Exp. Sta., Auburn Univ.

Ebelhar, S. A., W. W. Frye, and R. L. Blevins. 1984. "Nitrogen from legume cover crops for no-tillage corn," *Agron. J.*, 76:51–55.

Eghball, B., and J. F. Power. 1994. "Beef cattle feedlot manure management," *J. Soil and Water Conservation*, 49:113–122.

FAO. 1977. *China: Recycling of Organic Wastes in Agriculture*, FAO Soils Bull. 40 (Rome: Food and Agriculture Organization of the United Nations).

He, Xin-Tao, Terry Logan, and Samuel Traina. 1995. "Physical and chemical characteristics of selected U.S. municipal solid waste composts," *J. Environ. Quality*, 24:543–552.

Loehr, R. C. et al. 1979. *Land Application of Wastes*, Vol. I (New York: Van Nostrand Reinhold Co.).

National Research Council. 1993. *Soil and Water Quality: An Agenda for Agriculture* (Washington, D.C.: National Academy of Sciences), 516 p.

Olson, R. A., K. D. Frank, P. H. Graboushi, and G. W. Rehm. 1982. "Economic and agronomic impacts of varied philosophies of soil testing," *Agron. J.*, 74:492–499.

Reddy, G. B., and K. R. Reddy. 1993. "Fate of nitrogen-15 enriched ammonium nitrate applied to corn," *Soil Sci. Soc. Am. Journ.*, 57:111–115.

Reuter, D. J., and J. B. Robinson. 1986. *Plant Analysis: An Interpretation Manual* (Melbourne, Australia: Inkata Press).

Sims, J. T., and D. C. Wolf. 1994. "Poultry waste management: agricultural and environmental issues," *Advances in Agronomy*, 52:1–83.

Stickler, F. C., et al. 1975. *Energy from Sun to Plant to Man* (Moline, Ill.: Deere and Company).

Stivers, L. J., C. Shennen, L. E. Jackson, K. Groody, C. J. Griffin, and P. R. Miller. 1993. "Winter cover cropping in vegetable production systems in California," In M. G. Paoletti, W. Foissner, and D. Coleman, eds., *Soil Biota, Nutrient Cycling and Farming Systems* (Boca Raton, Fla.: Lewis Publishers).

Sutton, A. L. 1994. "Proper animal manure utilization," in *Nutrient Management*, Supplement to *J. Soil Water Conserv.*, 49(2), pp. 65–70.

Teng, Y., and V. R. Timmer. 1994. "Nitrogen and phosphorus in an intensely managed nursery soil–plant system," *Soil Sci. Soc. Am. Journ.*, 58:232–238.

Thomas, M. L., R. Lal, T. Logan, and N. R. Fausey. 1992. "Land use and management effects on non-point loading from Miamian soil," *Soil Sci. Soc. Am. Journ.*, 56:1871–1875.

Tisdale, S. L., W. L. Nelson, J. D. Beaton, and J. L. Havlin. 1993. *Soil Fertility and Fertilizers*, 5th ed. (New York: Macmillan).

USDA. 1980. Appraisal, 1980 Soil and Water Resources Conservation Act, Review Draft, Part I (Washington, D.C.: U.S. Dept. of Agric.).

USDA. Natural Resources Conservation Service. 1991. *Water Quality Indicator Guide: Surface Waters*, Report No. SCS-TP-161 (Washington, D.C.: USDA).

Walworth, J. L., and M. E. Sumner. 1987. "The diagnosis and recommendation integrated system (DRIS)," *Advances in Soil Science*, 6:149–187.

Westerman, R. L. (ed.). 1990. *Soil Testing and Plant Analaysis*, 3d ed. (Madison, Wis.: Soil Sci. Soc. Amer.), 784 pp.

Wilkinson, D. M., and N. M. Dickinson. 1995. "Metal resistance in trees: the role of mycorrhizae," *Oikos*, 72:298–299.

Zublena, J. P., J. C. Barker, and T. A. Carter. 1993. "Poultry manure as a fertilizer source," *Soil Facts* (Raleigh, N.C.: North Carolina Cooperative Extension Service, North Carolina State University).

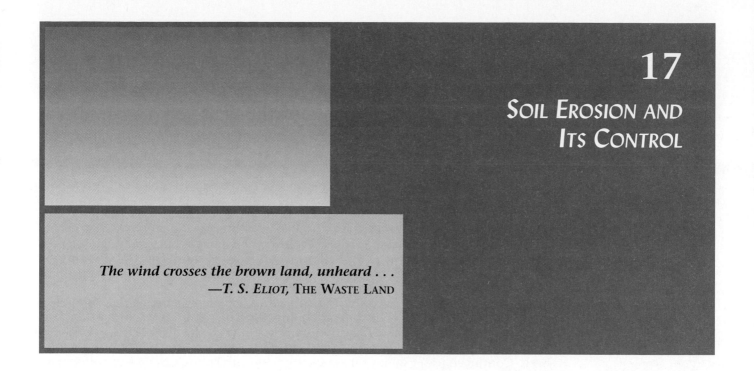

17

SOIL EROSION AND
ITS CONTROL

The wind crosses the brown land, unheard . . .
—T. S. ELIOT, THE WASTE LAND

No soil phenomenon is more destructive worldwide than soil erosion. This degradation process has always had ill effects on human welfare. Past civilizations have disintegrated as their soils, once deep and productive, have been washed away, leaving only thin, infertile, droughty, and rocky relics of the past. The hill lands in parts of Greece and in central India are sites of once flourishing and productive agricultural communities that suffered serious soil erosion and now depend on others for their food supplies.

The threat of soil erosion today is more ominous than at any time in history. In our generation, farmers have had to more than double world food output to meet needs of unprecedented increases in human population. Most of the increases are in the low-income countries where the ratio of human population to available cropland is already very high. To produce the extra food, nations have had to expand the area of land under cultivation, clearing and burning steep, forested slopes and plowing grasslands. Population pressures have also led to overgrazing rangelands and overexploiting timber resources. All these activities degrade or remove natural vegetation, causing these areas to be much more prone to erosion loss of soil. Even more deforestation would have occurred had it not been for the intensified cultivation of the more level lands that has increased yields and relieved somewhat the food shortages.

The effects of erosion on farmers' fields, logged-over forests, and degraded range-lands are only part of the story. The fine particles washed from these areas are subsequently deposited elsewhere—in nearby low-lying landscape sites; in river floodplains; or in downstream reservoirs, lakes, and river harbors. The economic and social costs to society to remove and manage the sediments from erosion are often greater than the on-site upstream costs of lost soil productivity.

17.1 SIGNIFICANCE OF RUNOFF AND SOIL EROSION[1]

Effects of Runoff

As discussed in Chapter 6, a primary principle of soil water management is to encourage water movement into rather than off of the soil. The soil can serve as a reservoir for

---

[1]For interesting reviews of this subject, see Batie (1983) and Follett and Stewart (1985).

future plant uptake. Surface runoff reduces or prevents penetration. Water that could have been retained in the plant root zone for subsequent absorption is lost.

Runoff levels vary greatly from region to region and from soil to soil. In some humid regions, losses as high as 50 to 60% of the annual precipitation occur. Annual runoff losses may be much lower in semiarid and arid regions, but high losses are not unusual during heavy storms, which are common in these regions. In any case, runoff losses seriously constrain plant productivity, and attempts to prevent them must receive high priority.

Soil Erosion

Soil erosion is a worldwide problem, as suggested by the data in Table 17.1. It is a common phenomenon in the United States, but is even more prevalent in many other countries. Preliminary investigations in the tropical areas suggest that erosion rates for parts of Africa and Latin America may be at least as high as those found for Asia. The high sediment losses shown in Table 17.2 from eight major river systems, including the Mississippi-Missouri, illustrate the seriousness of soil erosion. The Yellow (Yangtze) River in China and Kori River in Nepal/India carry very heavy loads of sediment—four to six times that of the Mississippi.

Soil erosion in the United States accelerated when the first European settlers chopped down trees and began to farm sloping to hilly lands of the humid eastern part of the country. Soil erosion was a factor in the declining productivity of these lands that in time led to their abandonment and the westward migration in search of new farmlands. Continued use of some of these soils for cultivated row crops such as cotton, particularly on the sloping lands of the piedmont areas of the Southeast, resulted in the further loss of topsoil and the reduction of soil organic matter. Had it not been for the significant use of chemical fertilizers to compensate for the loss of natural soil fertility, crop yields would have declined even further.

It took the worldwide depression and widespread droughts of the early 1930s to call the nation's attention to the rapid deterioration of its soils. In 1930, Dr. H. H. Bennett and associates recognized the damage being done and obtained federal government support for erosion-control efforts. Figure 17.1 indicates the extent of soil erosion found in a 1977 survey by the U.S. Department of Agriculture. In that survey, soil erosion was found to still be a prominent problem in about half the total cropland area of the country.

Subsequent surveys have confirmed the seriousness of the soil erosion problems but also suggested that wise conservation practices can reduce somewhat both water and wind erosion. Figure 17.2 illustrates the reduction that has taken place in the rate of erosion on cropland between 1982 and 1992. This is a remarkable achievement, but we must be reminded that even such evidence of progress illustrates continuing problems. Thus, more than 80% of the nation's cultivated cropland is still losing through water and wind erosion more than 11 Mg/ha/yr, the maximum loss tolerated for sustainable development. The loss of soil from both agricultural and nonagricultural lands continues to be a serious problem in the United States. About 4 billion tons (Mg) of soil are moved annually by soil erosion, some two-thirds by water and one-third by wind. More

TABLE 17.1 Comparative Soil Loss by Erosion in the United States and Other Countries of Europe and Asia

Unfortunately, little comprehensive data are available for Africa and Latin America, but studies suggest that losses comparable for those in Asia are likely.

Country	Total cropland (million ha)	Excessive soil loss (millions mg)	(mg/ha)
China	99	3628	36.6
India	140	4716	33.7
United States	167	1524	9.1
Former Soviet Union	251	2268	9.0
Other	607	11201	18.5
Total	1265	23337	18.4

From Brown and Wolf (1984).

TABLE 17.2 Annual Sediment Loads for Nine of the World's Major Rivers Including the Mississippi River

River	Countries	Annual sediment load (million Mg)	Erosion (Mg/ha drained)
Yangtze	China	1600	479
Ganges	India, Nepal	1455	270
Amazon	Brazil, Peru, etc.	363	13
Mississippi	United States	300	93
Irrawaddy	Burma	299	139
Kosi	India, Nepal	172	555
Mekong	Vietnam, Thailand, etc.	170	43
Red	China, Vietnam	130	217
Nile	Sudan, Egypt, etc.	111	8

Data from different sources compiled by El-Swaify and Dangler (1982).

than half of the water erosion and about 60% of the wind erosion are from croplands that produce most of the country's food. Such huge losses are simply not acceptable for long-term sustainability and must be further reduced.

In addition to the sediment carried downstream by rivers, another 1.5 billion Mg of sediment is deposited each year in the nation's reservoirs. Because of sedimentation, these reservoirs can hold less water for irrigation, domestic, and industrial purposes. The turbidity caused by eroded sediments is one of the most serious forms of water pollution.

Nutrient Losses

The quantity of essential nutrients lost from the soil by erosion is quite high, as shown in Table 17.3. The total quantity of nitrogen, phosphorus, and potassium removed by erosion from soils nationwide is estimated at more than 38 million Mg. Such losses can be counterbalanced only in part by adding fertilizers. Much of the nitrogen and phosphorus lost in the soil is in the organic matter, finer particles of which are included in the eroded sediments. Experiments have shown organic matter and nitrogen in the eroded material to be five times as high as in the original topsoil. Comparable figures for phosphorus (see Table 14.3) and potassium are three and two, respectively.

Total potassium losses shown in Table 17.3 are higher than those for nitrogen or phosphorus because this element is commonly present in larger total quantities in soil

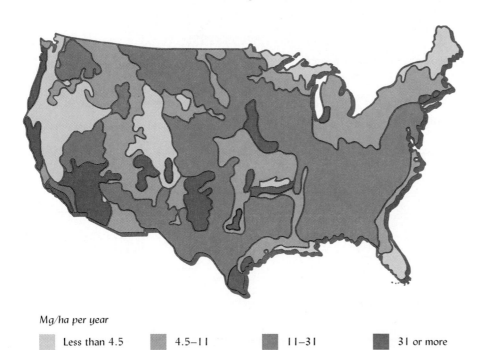

Mg/ha per year

　Less than 4.5　　　4.5–11　　　11–31　　　31 or more

FIGURE 17.1 Extent of water (sheet and rill) and wind erosion on cropland in continental United States in 1977. Gully erosion is not included. For the average soil, erosion rates greater than 11 Mg/ha (5 tons/acre) per year will result in soil productivity decline. [From USDA (1984)]

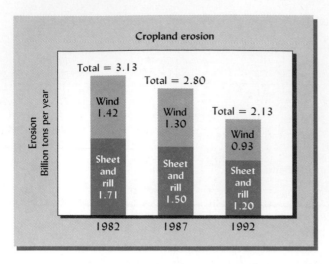

Cropland erosion

Total = 3.13

Total = 2.80

Total = 2.13

Erosion Billion tons per year

Wind 1.42

Wind 1.30

Wind 0.93

Sheet and rill 1.71

Sheet and rill 1.50

Sheet and rill 1.20

1982　　　1987　　　1992

FIGURE 17.2 Soil erosion from cropland in the United States declined significantly in the 10-year period 1982–1992, largely as a result of the Conservation Reserve Program which encouraged farmers to plant grass or trees on their most erodable croplands. The rate of reduction was about the same for wind erosion and water erosion (sheet and rill). [From the 1992 National Resources Inventory—USDA (1994)]

solids. But the amount of available nitrogen lost through erosion is higher than for the other two elements. This may be because of increasing nitrogen fertilizer applications to the soil surface where the nitrogen may be more subject to loss in runoff and erosion.

On-Site versus Off-Site Costs[2]

Erosion damage involves two kinds of costs. The on-site costs are incurred by the land owner at the site of the erosion. The off-site costs relate to the management and removal of sediment and excess water downstream. While either or both of these costs may not be apparent today, in time they will grow and eventually must be covered by landowners and/or by society as a whole.

It is not easy to estimate the costs of erosion, but some attempts have been made to do so. One such recent estimate for the United States suggests extremely high total costs. Using average wind and water erosion values for the entire country, and assigning appropriate costs for nutrient and water losses and for yield reductions due to erosion, the total annual on-site costs are estimated at $27 billion.

Using similar techniques, these researchers estimated the off-site costs for items such as cleaning up and treating domestic water supplies, reduced water storage in reservoirs, dredging harbors and waterways, loss of wildlife habitat, widespread flooding and health costs. The total of these annual off-site costs is estimated at $17 billion. This brings the grand total to $44 billion as the annual cost of erosion to this country. Even if these estimates are somewhat high, they are sobering, and remind us that soil erosion is a burden on everyone.

---

[2]This discussion is based on the work of Pimentel et al. (1995).

TABLE 17.3　**Estimated Losses of Total and Available Nitrogen, Phosphorus, and Potassium in Eroded Sediments**

In thousands of metric tons.

Region	Nitrogen		Phosphorus		Potassium	
	Total	Available	Total	Available	Total	Available
Pacific	100	18	29	0.6	1,154	23
Mountain	176	32	64	1.3	2,550	51
Southern Plains	512	94	101	2.0	3,043	61
Northern Plains	2,068	380	293	5.9	11,711	234
Lake states	622	114	107	2.1	3,643	73
Corn belt	4,360	802	624	12.5	24,959	499
Delta states	478	88	141	2.8	4,220	84
Southeastern states	202	37	101	2.0	1,007	20
Appalachian states	676	124	169	3.4	3,381	67
Northeastern states	300	55	75	1.5	2,252	45
Total	9,494	1,744	1,704	34.1	57,920	1,158

Data computed by Larson et al. (1983).

17.2 ACCELERATED EROSION: MECHANICS

Water Erosion

Water erosion is one of the most common geologic phenomena. It accounts in large part for the leveling of mountains and the development of plains, plateaus, valleys, river flats, and deltas. The vast deposits that now appear as sedimentary rocks originated in this way. This normal erosion amounts annually to about 0.2–0.5 Mg/ha (0.1–0.2 tons/acre). It operates slowly, yet inexorably. Erosion that exceeds this normal rate becomes unusually destructive and is referred to as *accelerated erosion.* This is usually caused by human activities that make the soil more susceptible to the action of water and wind.

Two steps are recognized in accelerated erosion—the **detachment** of soil particles, which is a preparatory action, and **transportation** of these particles by floating, rolling, dragging, and splashing. Freezing and thawing, flowing water, and raindrop impact are the major detaching agencies. Raindrop splash and especially running water facilitate the carrying away of the loosened soil. In channels cut into the soil surface, most of the loosening and cutting is due to water flow; on comparatively smooth soil surfaces, the beating of raindrops causes most of the detachment.

Influence of Raindrops

Raindrop impact exerts three important effects: (1) it detaches soil; (2) its beating tends to destroy granulation; and (3) its splash, under certain conditions, causes an appreciable transportation of soil (see Figure 17.3). So great is the force exerted by raindrops that soil granules are not only loosened and detached but may even be beaten to pieces. The dispersed material may develop into a hard crust on drying that will prevent the emergence of seedlings of large-seeded crops such as beans and will encourage runoff from subsequent precipitation (see Section 4.11).

Transportation of Soil: Splash Effects

In soil translocation, runoff water plays the major role. So familiar is the power of water to cut and carry that little more need be said regarding these capacities. In fact, so much publicity has been given to runoff that the public generally ascribes to it all of the damage done by heavy rainfall.

Under certain conditions, however, splash transportation is of considerable importance (see Figure 17.3). On a soil subject to easy detachment, a very heavy rain may

FIGURE 17.3 A raindrop (left) and the splash that results when the drop strikes a wet bare soil (right). Such rainfall impact not only tends to destroy soil granulation and encourage sheet and rill erosion but also effects considerable transportation by splashing. A ground cover, such as sod, will largely prevent this type of erosion. (Courtesy USDA Natural Resources Conservation Service)

splash as much as 225 Mg/ha of soil, some of the particles rising as high as 0.7 m and moving horizontally perhaps 1–2 m. On a slope or if the wind is blowing, splashing greatly aids and enhances runoff translocations of soil; the two together account for the total soil movement that finally occurs.

17.3   TYPES OF WATER EROSION

Three types of water erosion are generally recognized: *sheet, rill,* and *gully* (see Figure 17.4). In **sheet erosion,** soil is removed more or less uniformly from every part of the slope. However, sheet erosion is often accompanied by tiny channels (rills) irregularly dispersed, especially on bare land newly planted or in fallow. This is **rill erosion.** The rills can be smoothed over by tillage, but the damage is already done—the soil is lost.

Where the volume of runoff water is further concentrated, the formation of larger channels or gullies occurs by undermining and downward cutting. This is **gully erosion.** The gullies are obstacles to tillage and cannot be removed by ordinary tillage practices. While all types may be serious, the losses due to sheet and rill erosion, although less noticeable, are responsible for most of the field soil deterioration. The cropland area in the United States classed by its erodibility by sheet and rill erosion is shown in Figure 17.5.

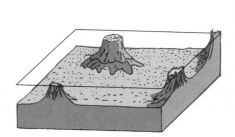

(a) Sheet erosion

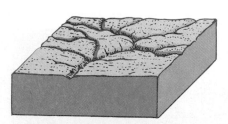

(b) Rill erosion

FIGURE 17.4   Three major types of soil erosion. Sheet erosion is relatively uniform erosion from the entire soil surface. Note the perched stones and pebbles have protected the soil underneath from sheet erosion. Rill erosion is initiated when the water concentrates in small channels (rills) as it runs off the soil. Subsequent cultivation may erase rills but it does not replace the lost soil. Gully erosion creates deep channels that cannot be erased by cultivation. Although gully erosion looks more catastrophic, far more total soil is lost by the less obvious sheet and rill erosion. [Drawings from FAO (1987); photos courtesy USDA Natural Resources Conservation Service]

(c) Gully erosion

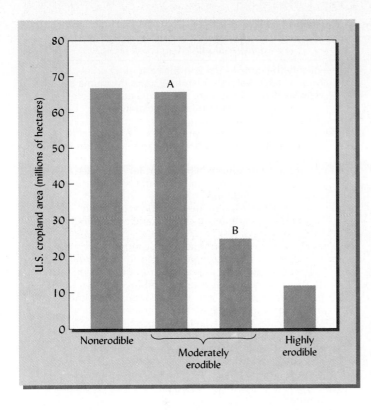

FIGURE 17.5 Hectares of United States cropland in 1982 classed by its erodibility by water. The nonerodible land will lose less than 11 Mg/ha (5 tons/acre) per year (the allowable limit) regardless of treatment. The moderately erodible land A will lose less than 11 Mg/ha per year if appropriately managed, while land B probably will lose somewhat more than 11 Mg/ha per year even if well managed. Erosion from the highly erodible land is very high, and permanent vegetation should be established on this land. [Data from Heimlich and Bills (1986)]

17.4 FACTORS AFFECTING ACCELERATED EROSION: REVISED UNIVERSAL SOIL-LOSS EQUATION[3]

Decades of agricultural research coupled with centuries of farmers' experience have clearly identified the major factors affecting accelerated erosion. These factors are included in the *universal soil-loss equation* (USLE).

$$A = RKLSCP$$

A, the predicted soil loss, is the product of

R = climatic erosivity (rainfall and runoff)
K = soil erodibility
L = slope length
S = slope gradient or steepness
C = cover and management
P = erosion-control practice

Working together, these factors determine how much water enters the soil, how much runs off, and the manner and rate of its removal.

This equation has been used widely since the 1970s. It now serves as a basis for a modern, computerized, erosion-prediction tool called the *revised universal soil-loss equation* (RUSLE). RUSLE uses the same basic factors of USLE shown above, although some are better defined and their interrelationship is recognized to improve the accuracy of soil-loss prediction. RUSLE is a readily available computer software package that can be constantly improved and modified as experience is gained from its use. This new tool makes possible greater accuracy of soil-loss prediction at specific locations around the world. The major differences between USLE and RUSLE are shown in Table 17.4. These

---

[3]For discussions of this topic, see Wischmeier and Smith (1978) and Renard et al. (1994). Since the factors used in the soil-loss equation are calculated in English units rather than SI units, the practice will be continued in this text. The soil loss (*A*) calculated by the USLE is expressed in tons per acre, which can easily be converted to Mg/ha by multiplying by 2.24.

TABLE 17.4 Summary of Major Differences Between USLE and RUSLE

Factor	Universal soil-loss equation (USLE)	Revised universal soil-loss equation (RUSLE)
R	Based on long-term average rainfall conditions for specific geographic areas in the United States	Generally the same as USLE in the eastern United States. Values for western states (Montana to New Mexico and west) are based on data from more weather stations and thus are more precise for any given location. RUSLE computes a correction to R to reflect, for flat land, the effect of raindrop impact on water ponded on the surface.
K	Based on soil texture, organic matter content, permeability, and other factors inherent to soil type.	Same as USLE but adjusted to account for seasonal changes such as freezing and thawing, soil moisture, and soil consolidation.
LS	Based on length and steepness of slope, regardless of land use.	Refines USLE by assigning new equations based on the ratio of rill to interrill erosion, and accommodates complex slopes.
C	Based on cropping sequence, surface residue, surface roughness, and canopy cover, which are weighted by the percentage of erosive rainfall during the six crop stages. Lumps these factors into a table of soil-loss ratios, by crop and tillage scheme.	Uses these subfactors: prior land use, canopy cover, surface cover, surface roughness, and soil moisture. Refines USLE by dividing each year in the rotation into 15-day intervals, calculating the soil-loss ratio for each period. Recalculates a new soil-loss ratio every time a tillage operation changes one of the subfactors. RUSLE provides improved estimates of soil-loss changes as they occur throughout the year, especially relating to surface and near-surface residue and the effects of climate on residue decomposition.
P	Based on installation of practices that slow runoff and thus reduce soil movement. P factor values change according to slope ranges with some distinction for various ridge heights.	P factor values are based on hydrologic soil groups, slope, row grade, ridge height, and the 10-year single storm erosion index value. RUSLE computes the effect of strip-cropping based on the transport capacity of flow in dense strips relative to the amount of sediment reaching the strip. The P factor for conservation planning considers the amount and location of deposition.

From Renard et al. (1994).

differences will be referred to as we consider the various factors influencing water-induced erosion.

17.5 RAINFALL AND RUNOFF FACTOR

The rainfall and runoff erosive factor, *R,* represents the driving force for sheet and rill erosion. It takes into consideration the **total rainfall** and, more important, the **rainfall intensity** and **seasonal distribution** of the rainfall. Gentle rains of low intensity may cause little erosion, even if the total annual precipitation is high. In contrast, a few torrential downpours may result in severe damage, even in areas of low annual rainfall. Likewise, soil losses are heavy if the rain falls when there is little vegetative cover on the soil, as is the case when seedbeds have been prepared with conventional tillage.

The *R* factor, sometimes called the *rainfall erosion index,* takes into account the erosive effects of storms. The total kinetic energy of each storm (related to intensity and total rainfall) plus the average rainfall during the 30-minute period of greatest intensity is considered. The sum of the indexes for all storms occurring during a year provides an annual index. An average of such indexes for several years is used in the universal soil-loss equation. The RUSLE includes more precise values, especially for the western part of the United States. It also gives special consideration to raindrops striking water ponded on the surface of flat slopes and to runoff from the thawing of frozen soils, especially in the Northwest.

Rainfall indexes computed for the United States are shown in Figure 17.6. Note that they vary from less than 20 in areas of the west to more than 550 along the coasts of the gulf states.

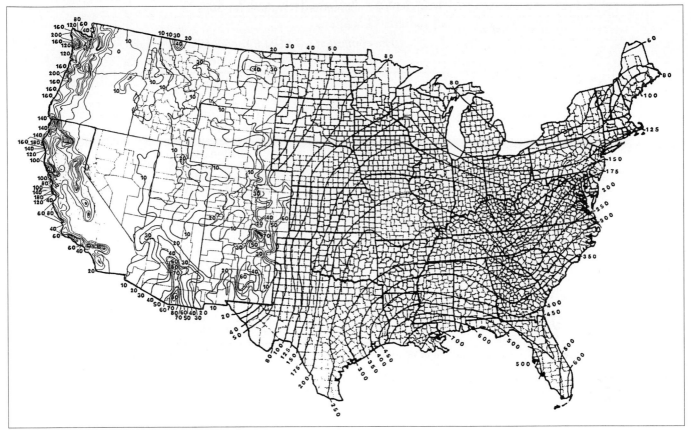

FIGURE 17.6 The *R*-values contour map. Note the relatively high values in the humid East, low values in the western Great Plains and Rocky Mountain regions, and intermediate values in the agricultural areas of California and the Northwest. [Redrawn from Handbook No. 703, U.S. Dept. of Agriculture (1995)]

17.6 SOIL ERODIBILITY FACTOR

The soil erodibility factor, *K*, indicates soil's inherent susceptibility to erosion. *K* gives an indication of the soil loss from a unit plot 22 m (72 ft) long with a 9% slope kept continuously bare by tillage. The two most significant and closely related soil characteristics influencing erosion are (1) infiltration capacity, and (2) structural stability. In turn, these are affected by soil properties such as organic matter content, soil texture, the kind and amount of swelling clays, soil depth, tendency to crust, and the presence of impervious soil layers.

Stable soil aggregates resist the beating action of rain, thereby saving soil even though runoff does occur. The marked granule stability of certain tropical clay soils high in hydrous oxides of iron and aluminum accounts for the resistance of these soils to the action of torrential rains. Downpours of a similar magnitude on swelling-type clays would be disastrous.

The RUSLE also takes into account the fact that *K* values vary seasonally, being highest in the spring in temperate regions when the soil is more "fluffy" from freeze-thaw interactions. Also, the protection provided by rock fragmentation on the soil surface lowers slightly the *K* factor. Approximate *K* values for soils at selected locations are shown in Table 17.5. Computer-ascertained site-specific values will likely vary somewhat from these values. Note that the soil erodibility, or *K* factor, normally varies from near zero to about 0.6. *K* is low for soils into which water readily infiltrates, such as well-drained sandy soils or friable tropical clays high in hydrous oxides of iron and aluminum or kaolinite. Erodibility indexes of less than 0.2 are normal for these readily infiltrated soils. Soils with intermediate infiltration capacities and moderate soil structural stability generally have a *K* factor of 0.2–0.3, while the more easily eroded soils with low infiltration capacities will have a *K* factor of 0.3 or higher.

TABLE 17.5   Computed K Values for Soils at Different Locations

The computerized RUSLE model may give some more accurate values for some locations

Soil	Location	Computed K
Udalf (Dunkirk silt loam)	Geneva, N.Y.	0.69
Udalt (Keene silt loam)	Zanesville, Ohio	0.48
Udult (Lodi loam)	Blacksburg, Va.	0.39
Udult (Cecil sandy clay loam)	Watkinsville, Ga.	0.36
Udoll (Marshall silt loam)	Clarinda, Iowa	0.33
Udalf (Hagerstown silty clay loam)	State College, Pa.	0.31
Ustoll (Austin silt)	Temple, Tex.	0.29
Aqualf (Mexico silt loam)	McCredie, Mo.	0.28
Udult (Cecil sandy loam)	Clemson, S.C.	0.28
Udult (Cecil sand loam)	Watkinsville, Ga.	0.23
Udult (Tifton loamy sand)	Tifton, Ga.	0.10
Ochrept (Bath flaggy silt loam)	Arnot, N.Y.	0.05
Alfisols	Indonesia	0.14
Alfisols	Benin	0.10
Ultisols	Hawaii	0.09
Ultisols	Nigeria	0.04
Alfisols	Nigeria	0.06
Oxisols	Puerto Rico	0.01
Oxisols	Ivory Coast	0.10

From Wischmeier and Smith (1978); data for tropical soils cited by Cassel and Lal (1992).

17.7   TOPOGRAPHIC FACTOR

The topographic factor, *LS*, reflects the influence of **length** and **steepness** of slope on soil erosion. *LS* is the ratio of soil loss from the area in question to that of a unit plot with 9% slope, 22 m (72 ft) long, and continuously fallowed. The longer the slope, the greater is the opportunity for concentration of the runoff water. However, soil loss is generally less sensitive to slope length than to steepness of slope. Also, the complexity of the slope (its unevenness) influences soil erosion, as does the susceptibility of the soil to rill erosion relative to interrill erosion.

The widespread availability of computer technology now makes it easy, using RUSLE, to make *LS* calculations for specific field situations that were not possible when USLE was first established in 1978. Consequently, generalized *LS* factor values are not used for RUSLE since the computer program calculates the more location-specific values. However, the *LS* factors in Table 17.6 illustrate the increases that occur as slope length and steepness increase.

17.8   COVER AND MANAGEMENT FACTOR

The cover and management factor, *C*, indicates the influence of cropping systems and management variables on soil loss. *C* is the factor over which the soil manager has the most control. Undisturbed forests and grass provide the best natural protection known for soil and are about equal in their effectiveness, but forage crops (both legumes and grasses) are next in effectiveness because of their relatively dense cover. Small grains such as wheat and oats are intermediate and offer considerable obstruction to surface wash. Row crops such as corn, soybeans, and potatoes offer relatively little cover during the early growth stages and thereby encourage erosion unless crop residues cover the soil surface (Section 17.12). Most subject to erosion are bare areas where no living plants or plant residues cover the soil. The effect on erosion of leaving wheat stubble as surface mulch is illustrated in Figure 17.7.

The marked differences among crops in their ability to maintain soil cover emphasize the value of appropriate crop rotation to reduce soil erosion. The inclusion of a close-growing forage crop in rotation with row crops will help control both erosion and runoff. Likewise, the use of conservation tillage systems, which leave most of the crop residues on the surface, greatly decreases erosion hazards.

TABLE 17.6 Values for the Topographic Factor (LS) for Different
Slope Length and Steepness
on Soils with Low and High Ratios
of Rill-to-Interrill Erosion

Slope	Approximate slope length (m)			
	25	50	75	100
Low ratio of rill to interrill erosion				
2	0.21	0.23	0.25	0.26
4	0.36	0.43	0.46	0.50
8	0.64	0.79	0.90	0.99
12	1.01	1.31	1.52	1.69
16	1.38	1.85	2.18	2.46
High ratio of rill to interrill erosion				
2	0.16	0.21	0.25	0.28
4	0.26	0.38	0.47	0.55
8	0.45	0.70	0.91	1.10
12	0.71	1.15	1.54	1.88
16	0.98	1.64	2.21	2.73

Selected from *USDA Agriculture Handbook 703*
(1995).

Technically, the *C* value is the ratio of soil loss under the conditions found in the field in question to that which would occur under clean-tilled, continuously fallow conditions. This ratio (*C*) will be high (approaching 1.0 where there is little soil cover) (e.g., bare soil in the spring before a crop canopy develops). It will be low (e.g., <0.10) where large amounts of crop residues are left on the land or in areas of dense forests.

The computerized RUSLE divides the *C* factor into a number of subfactors that take into consideration prior land use, surface cover, crop canopy, surface roughness, and probable residue decomposition. Each year is divided into 15-day intervals and the soil-loss prediction is calculated for each period. The total expected loss for the years can then be calculated by adding up the 15-day-interval losses.

Examples of *C* values for cropping systems in northern Illinois are given in Table 17.7. While more specific values would likely be used in RUSLE, these examples illustrate the effect of cropping and tillage systems on *C* values.

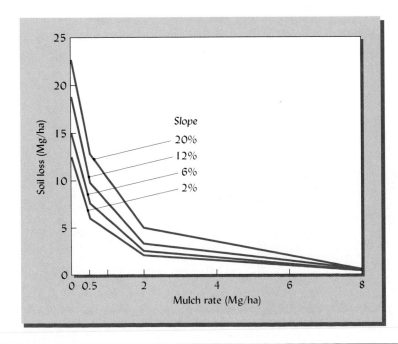

FIGURE 17.7 The effect of wheat straw mulch rate and soil slope on soil loss by erosion. Note the striking reduction in erosion with increased coverage of the soil surface (mulching) even on the steep slopes. A prime objective of conservation tillage is to keep the cover on the soil. [From Lattanzi et al. (1979); used with permission of the Soil Science Society of America]

TABLE 17.7 Crop Management or C Values for Different Crop Sequences in Northern Illinois

Note the dominating effect of tillage systems and of the maintenance of soil cover. Values would differ slightly in other areas but the principles illustrated would pertain.

Crop sequence[a]	Conventional tillage[b]		Minimum tillage Residue level (kg/ha)		No tillage Residue level (kg/ha)	
	Residue left	Residue removed	450–900	900–1800	450–900	900–1800
Continuous soybeans (Sb)	0.49	—	0.33	—	0.29	—
Continuous corn (C)	0.37	0.47	0.31	0.07	0.29	0.06
C–Sb	0.43	0.49	0.32	0.12	0.29	0.06
C–C–Sb	0.40	0.47	0.31	0.12	0.29	0.06
C–C–Sb–G–M	0.20	0.24	0.18	0.09	0.14	0.05
C–Sb–G–M	0.16	0.18	0.15	0.09	0.11	0.05
C–C–C–M	0.12	0.16	0.13	0.08	0.09	0.04

Selected data from Walker (1980).
[a]Crop abbreviations: C = corn; Sb = soybeans; G = small grain (wheat or oats); M = meadow.
[b]Spring plowed; assumes high crop yields.

17.9 SUPPORT PRACTICE FACTOR

The support practice factor, P, reflects the benefits of contouring, strip-cropping, and other supporting factors. P is the ratio of soil loss with a given support practice to the corresponding loss when crop culture is up and down the slope. If there are no support practices, the P factor is 1.0. On sloping fields, the protection offered by surface coverage and crop management must be supported by other practices that help slow the runoff water and, in turn, reduce soil erosion. These support practices (factor P) include tillage on the contour, contour strip-cropping, terrace systems, and grassed waterways, all of which will tend to reduce the P factor.

Contour Tillage

Except on fields with only modest slopes, it is important that cultivation be across the slope rather than with it. This is called **contour tillage.** Crop rows may be relied upon to slow the flow of runoff water, or ridges may be built on the contour with special tillage equipment. However, ridges must be designed to carry heavy runoff safely from the field (see Figure 17.8). On long slopes subject to sheet and rill erosion, the fields may be laid out in narrow strips across the incline, alternating the tilled crops, such as corn and potatoes, with hay and grain. Water cannot achieve an undue velocity on the narrow strips of cultivated land, and the hay and grain crops check the rate of runoff. Such a layout is called **strip-cropping** and is the basis for much of the erosion control now advocated.

When the cross strips are laid out rather definitely on the contours, the system is called **contour strip-cropping.** The width of the strips will depend primarily on the

FIGURE 17.8 Contour ridges must be carefully laid out with sufficient height to hold back water from even heavy rainfall. Here, surface retention is fast becoming surface runoff. (Photo courtesy of R. Weil)

degree of slope, the permeability of the land, and its erodibility. Widths of 30–125 m (100–300 ft) are common. Contour strip-cropping often is guarded by diversion ditches and waterways between fields. When their grade is high, these waterways should be sown with grass (grassed waterways) to prevent underwashing (see Figure 17.9).

Terraces

There are a variety of terraces in use around the world today (see Figure 17.10). Bench terraces are used where rather complete control of the water runoff must be achieved, such as in rice paddies. Broad-based terraces are more common where large-scale farm machinery is used and where the entire land surface is to be cropped. The terrace channel has a grade of from 50–65 cm per 100 m (6–8 in./100 ft) to permit gentle but controlled runoff. The runoff water is commonly carried in the terrace channel to a broader-based grassed waterway through which the water moves to a nearby canal, stream, or river.

Since some terraces are low and broad, they may be cropped without difficulty, and offer no obstacle to cultivating and harvesting machinery. Such terraces waste little or no land and are quite effective if properly maintained. Where necessary, waterways can be sodded. Terracing of this kind is really a more or less elaborate type of contour strip-cropping.

Examples of P values for contour tillage and strip-cropping at different slope gradients are shown in Table 17.8. Note that P values increase with slope and that they are low for strip-cropping, illustrating the importance of this practice for erosion control. Note that terracing also reduces the P values.

Vegetative Barriers

In recent years, considerable interest and experience have developed in the use of "living terraces" that are stimulated by a row of grasses or shrubs planted on the contour across the slope. Tropical grasses (e.g., a deep-rooted, drought-tolerant species called *vetiver grass*) have shown some promise. The vetiver plants are quite deep-rooted and have a matted root system that tends to serve as a sort of sieve to hold soil particles that may be moving down the slope. Such particles build up on the upslope side of the row of plants and, in time, actually create a terrace that may be 15–50 cm above the downslope side of the plants (see Figure 17.11).

FIGURE 17.9 Aerial photograph of fields in Kentucky where contour strip-cropping is being practiced. (Courtesy USDA Natural Resources Conservation Service)

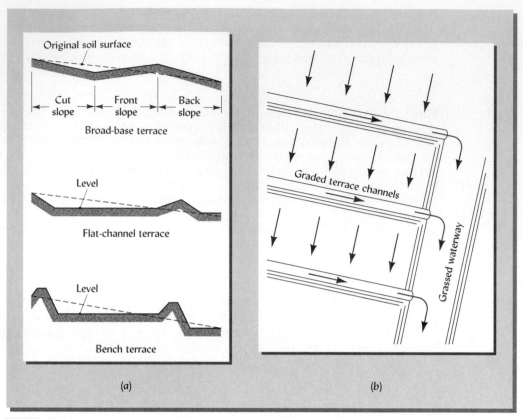

FIGURE 17.10 (a) Three types of terraces in use around the world. Broad-based terraces permit the entire surface to be cropped and are widely used in the United States. Flat-channel terraces allow large volumes of water to move off the soil without erosion. Bench terraces can keep water on the land and are used widely in rice production. (b) Diagram of controlled flow of runoff water from a field to terrace channels and onto a grassed waterway from which it can move to a canal or stream channel.

The deep and matted root system of vetiver grass permits it to forage well for water, but sometimes at the expense of food crops that may be growing beside it. Attempts are being made to evaluate other grasses that may not be quite so aggressive as vetiver. In the United States grasses such as perennial fall wheatgrass are being evaluated for a combination of wind erosion control and capturing of winter snows in the Northern Plains states. In developing countries the "live terrace" models have some appeal because of their relatively low cost and probable utilization by peasant farmers who have no machinery with which to build and maintain regular terraces. The benefits of reduced erosion and increased crop yields resulting from several vegetation erosion control practices are summarized in Table 17.9.

TABLE 17.8   P Factors for Contour and Strip-Cropping at Different Slopes and the Terrace Subfactor at Different Terrace Intervals

The product of the contour or strip-cropping factors and the terrace subfactor gives the P value for terraced fields.

Slope (%)	Contour P factor	Strip cropping P factor	Terrace interval (m)	Terrace subfactor Closed outlets	Open outlets
1–2	0.60	0.30	33	0.5	0.7
3–8	0.50	0.25	33–34	0.6	0.8
9–12	0.60	0.30	43–54	0.7	0.8
13–16	0.70	0.35	55–68	0.8	0.9
17–20	0.80	0.40	69–60	0.9	0.9
21–25	0.90	0.45	90	1.0	1.0

Contour and strip-cropping factors from Wischmeier and Smith (1978); terrace subfactor from Foster and Highfill (1983).

FIGURE 17.11 The use of vegetative barriers to create natural terraces. (Above) A tropical grass (vetiver) has been planted vegetatively on the contour in a cassava field by pushing root sprigs into the soil. (Below left) The root cuttings are planted perpendicular to the slope direction. In a year or so the grass will be well established and its dense root and shoot growth will serve as a barrier to hold soil particles while permitting some water to pass on through. Note the buildup of soil above the grass, basically forming a terrace wall. Perennial tall wheatgrass is being used experimentally in the Northern Plains area of the United States to serve as a barrier against snow movement and later against wind erosion. (Photo courtesy of Centro Internacional Agricultura Tropical in Cali, Colombia)

RUSLE is able to compute more accurately than USLE the effects of contouring and strip-cropping on erosion loss. The interaction of the conservation practices with other subfactors such as slope and soil water infiltration are also considered. As is the case for the other factors, *P* values for RUSLE are based on many more studies than was the case for USLE.

TABLE 17.9 **Range in the Effects of Mulching, Contour Cultivation, and Grass Contour Hedges on Soil Erosion and Crop Yields**

Practice	Reduction in soil erosion (%)	Increase in crop yield (%)
Mulching	78–98	7–188
Contour cultivation	50–86	6–66
Grass contour hedges	40–70	38–73

From a review of more than 200 studies, from Doolette and Smyle (1990).

17.10 CALCULATING EXPECTED SOIL LOSSES

The RUSLE computer software is designed to calculate the expected soil loss from specific cropping systems at a given location. The Soil and Water Conservation Society in Ankeny, Iowa, has been given primary responsibility for transferring the technology to users, and has arranged training courses as well as books and videos on its use. In the future, the system will likely become widely used for predicting soil erosion losses.

The principles involved in both USLE and RUSLE can be verified by making calculations using USLE and its associated factors. In Box 17.1, calculations are made for a field with no conservation practices and for the same field using some easily applied conservation practices. The reduction in soil erosion losses predicted is most impressive.

17.11 SHEET AND RILL EROSION CONTROL

From the preceding sections it is obvious that three objectives must be attained if sheet and rill erosion are to be held to a reasonable level. First, we must encourage as much of the precipitation as possible to run *into* the soil rather than *off* it. Second, the **sediment burden** (soil solids) of the runoff water should be kept low. And third, we must direct the runoff water into channels, thereby minimizing its cutting action and the sediment burden it can carry.

The key to control of sheet and rill erosion is found in factors included in the universal soil-loss equation. The control measures needed will be determined by the characteristics of the rainfall, by fundamental properties of the soil, and by the slope gradient and length. Care must be exercised to keep vegetative cover on or near the soil surface. Close-growing crops such as grasses and small grains provide good cover, especially if adequate nutrients have been provided from fertilizer, lime, and manure. Furthermore, mechanical support practices such as **contour tillage** and **strip-cropping** should continue to receive high priority on sloping fields. But perhaps the most significant and relatively inexpensive means of reducing sheet and rill erosion is the use of so-called conservation tillage practices. These will now be considered.

BOX 17.1   CALCULATIONS OF EXPECTED SOIL LOSS USING USLE

Assume, for example, a location in Iowa on a Marshall silt loam with an average slope of 4% and an average slope length of 50 m (160 ft). Assume further that the land is clean-tilled and fallowed.

Figure 17.6 shows that the R factor is about 150 for this location. The K factor for a Marshall silt loam in central Iowa is 0.33 (Table 17.5) and the topographic factor (LS) from Table 17.6 is 0.43. The C factor is 1.0, since there is no cover or other management practice to discourage erosion. If we assume the tillage is up and down the hill, the P value is also 1.0. Thus, the anticipated soil loss can be calculated by the USLE ($A = RKLSCP$):

$$A = (150)(0.33)(0.43)(1.0)(1.0) = 21.3 \text{ tons/acre or } 47.8 \text{ Mg/ha}$$

If the crop rotation involved wheat–hay–corn–corn and conservation tillage practices (e.g., minimum-tillage) were used, a reasonable amount of residue would be left on the soil surface. Under these conditions, the C factor may be reduced to about 0.2. Likewise, if the tillage and planting were done on the contour, the P value would drop to about 0.5 (Table 17.8). P could be reduced further to $0.4(0.5 \times 0.8)$ if terraces with outlets were installed about 40 m apart. With these figures the soil loss becomes:

$$A = (150)(0.33)(0.43)(0.2)(0.5)(0.8) = 1.7 \text{ tons/acre or } 3.8 \text{ Mg/ha}$$

The benefits of good cover and management and support practices are obvious. The figures cited were chosen to provide an example of the utility of the universal soil-loss equation, but calculations can be made for any specific location. In the United States, pertinent factor values that can be used for erosion prediction in specific locations are generally available from state offices of the USDA Natural Resource Conservation Service.

17.12 CONSERVATION TILLAGE PRACTICES[4]

Until fairly recently, conventional agricultural practice encouraged extensive soil tillage. Most of the crop residues were incorporated into the soil using the moldboard plow. The soil surface was then tilled further using a disc harrow to provide a seedbed devoid of clods. Once row crops were planted, a cultivator was used, often several times, to control weeds. Thus the soil was tilled repeatedly at great cost in terms of time and energy. More important, however, the soil was usually left bare immediately after plowing until later in the year when crop growth was sufficient to provide some ground cover. This means that the soil was left unprotected during the spring of the year when runoff and erosion pressures were greatest.

Three developments of the past two decades have allowed drastic changes in tillage practices. First, the use of herbicides to control major weeds has reduced the need for cultivating and even for plowing in some cases. Second, high fuel costs have encouraged tractor-dependent farmers to reduce their tillage operations. Third, public awareness of the environmental consequences of soil erosion has increased. These three developments have stimulated farmers to adopt reduced-tillage methods that are both energy-saving and environmentally sound. Reduced-tillage practices are collectively known as *conservation tillage,* in contrast to the **conventional tillage** system described above.

Conservation Tillage Systems

There are several conservation tillage systems in use today, and they tend to have two characteristics in common. They leave significant amounts of organic residues on the soil surface, and they require less energy than conventional systems. Some conservation tillage systems are described in Table 17.10. In studying them, keep in mind that conventional tillage involves plowing (from one to three passes with a harrow), crop planting, and sometimes subsequent tillage with a cultivator.

The conservation tillage systems vary from those that merely reduce excess tillage to the **no-tillage system,** which permits direct planting in the residue of the previous crop and uses only that localized tillage necessary to plant the seed. Figure 17.12 shows a moldboard plow in action and also illustrates a no-tillage system, where one crop is planted into the residues of another and organic cover is kept on the land.

---

[4]For recent reviews of conservation tillage practices, see Carter (1994) and Blevins and Frye (1992).

TABLE 17.10   **General Classification of Different Conservation Tillage Systems**

All systems maintain at least 30% of the crop residues on the surface.

Tillage system	Operation involved
No-till	Soil undisturbed prior to planting, which occurs in narrow seedbed, 2.5–7.5 cm wide. Weed control primarily by herbicides.
Ridge till (till, plant)	Soil undisturbed prior to planting, which is done on ridges 10–15 cm higher than row middles. Residues moved aside or incorporated on about one-third of soil surface. Herbicides and cultivation to control weeds.
Strip till	Soil undisturbed prior to planting. Narrow and shallow tillage in row using rotary tiller, in-row chisel, etc. Up to one-third of soil surface is tilled at planting time. Herbicides and cultivation to control weeds.
Mulch till	Soil surface disturbed by tillage prior to planting, but at least 30% of residues left on or near soil surface. Tools such as chisels, field cultivators, disks, and sweeps are used (e.g., stubble mulch). Herbicides and cultivation to control weeds.
Reduced till	Any other tillage and planting system that keeps at least 30% of residues on the surface.

Definitions used by Conservation Technology Information Center, West Lafayette, Ind.

(a)   (b)   (c)

FIGURE 17.12 Photographs showing action of the moldboard plow and two conservation tillage practices. The moldboard plow (a) essentially turns the top 15–20 cm of soil, covers the crop residues, and leaves bare soil on the surface. In contrast, the no-till systems leave all the residue on the surface, the new crop being planted in the previous crop's residue. (b) Wheat is being planted; disks (coulters) cut through the sorghum residues and open slits in which wheat is planted. (c) On an Illinois field, corn was planted in wheat stubble and herbicides were used to control the weeds. [(a) Courtesy USDA National Tillage Laboratory; (b) courtesy John Deere, Inc., Moline, Ill.; and (c) courtesy USDA Natural Resources Conservation Service]

It is obvious from Figure 17.12 that a prime objective of conservation tillage is to keep some (generally more than 30%) of the soil surface covered with plant residues. Extensive field research shows that the conventional moldboard plow systems leave only 1–5% of the soil covered with crop residues (Table 17.11). Conservation tillage systems (e.g., chiseling or disking) commonly leave more than 30% soil coverage, while with no-tillage systems we can expect 50–100% of the land to remain covered.

Runoff and Erosion Control

Since conservation tillage systems were initiated, hundreds of field trials have demonstrated that these tillage systems result in much less soil erosion than that experienced with conventional tillage methods. Surface runoff is also decreased, although the differences are not as pronounced as with soil erosion. Typical of the results obtained are those presented in Figure 17.13.

TABLE 17.11 **The Effect of Tillage Systems in Nebraska on the Percentage of Land Surface Covered by Crop Residues**

Note that the conventional moldboard plow system provided essentially no cover while no till and planting on a ridge (ridge till) provided best cover.

		Percentage of fields with residue cover greater than			
Tillage system	Number of fields	15%	20%	25%	30%
Moldboard	33	3	0	0	0
Chisel	20	40	15	5	0
Disk	165	40	20	9	4
Field cultivate	13	46	23	0	0
Ridge till (till, plant)	2	100	50	50	0
No-till	3	100	100	100	100
All systems	236	36	18	9	4

From Dickey et al. (1987).

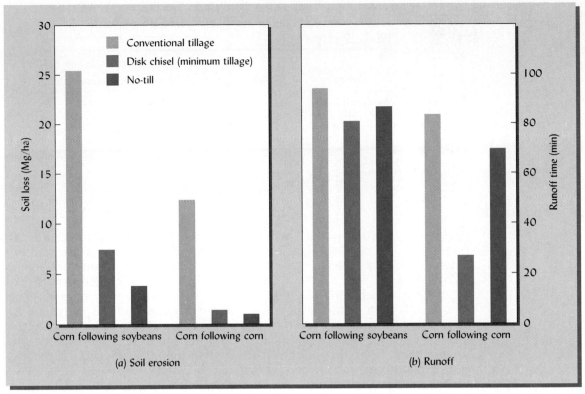

FIGURE 17.13 Effect of tillage systems on soil erosion and runoff from corn plots in Illinois following corn and following soybeans. Soil loss by erosion was dramatically reduced by the conservation tillage practices. The period of runoff was reduced most by the disk chisel system where corn was grown after corn. The soil was a Typic Argiudoll (Catlin silt loam), 5% slope, planted up-and-down slope, tested in early spring. [Data from Oschwald and Siemens (1976)]

The need for conservation tillage is determined in part by the erosion potential of the soil on which crops are to be grown. Estimates shown in Table 17.12 illustrate the point. Note that on soils with high and very high erosion potentials, estimated erosion losses from the conventional plow system are two to four times that considered tolerable if current soil productivity levels are to be maintained (11.2 Mg/ha). In contrast, when the no-till system is used, soil losses, even from soils with high erosion potential, are estimated to be far below the tolerable limit. Conservation tillage also can significantly reduce losses of nutrients (see Tables 14.2 and 16.2).

TABLE 17.12 Estimates of Annual Soil Losses as Affected by Tillage System and Soil Erosion Potential

Three tillage systems were used on soils with low, medium, high, or very high erosion potential. Corn was grown following soybeans.

	Tillage system		
Soil erosion potential[a]	Plow[b] (2% cover)	Chisel (14% cover)	No-till (42% cover)
	Soil loss by erosion (Mg/ha)		
Low (½ T)	6.2	3.4	0.8
Medium (1 T)	12.4	6.8	1.7
High (2 T)	24.8	13.6	3.4
Very high (4 T)	49.5	27.2	6.7

Estimated by Baker and Laflen (1983).
[a]T = loss soil tolerance of 11.2 Mg/ha (5 tons/acre), thought to be the maximum allowable to maintain productivity of most soils.
[b]Plow = plow, disk, and plant; chisel = chisel, disk, and plant; no-till = plant in untilled land.

TABLE 17.13   Average Corn Grain Yields from Well-Drained Soils in Kentucky Under No-Tillage and Conventional Tillage Systems

		Grain yields (kg/ha)	
Soil type	Years tested	No-tillage	Conventional tillage
Maury silt loam	8	9,136	8,932
Crider silt loam	5	9,886	8,318
Tilsit silt loam	5	7,705	7,705
Allegheny silt loam	3	10,977	10,909

From Phillips et al. (1980).

Effect on Crop Yields

Conservation tillage systems generally provide yields equal to or greater than those from conventional tillage, provided the soil is not poorly drained and can be kept free of weeds without cultivation. Typical of the results obtained from numerous field trials on well-drained soils are those shown in Table 17.13. Long-term trials show higher yields from conservation tillage systems than those where conventional tillage is used.

On soils with restricted drainage, yields with conservation tillage are commonly inferior in temperate regions to those from conventionally tilled areas (Figure 17.14). Reasons for these lower yields include higher moisture contents and consequent lower soil temperatures in the spring on the conservation tillage plots. Also, the incidence of certain plant diseases may be higher under the somewhat higher moisture conditions that characterize these plots.

The use of the ridge till system (Table 17.10), which involves planting on 10–15-cm ridges, at least partially alleviates early season moisture and soil temperature problems.

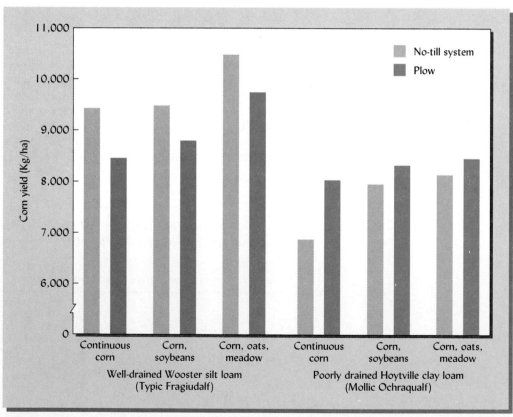

FIGURE 17.14   Effect of tillage systems on the yield of corn grown in different rotations on a well-drained and a poorly drained soil. The no-till system was superior on the well-drained soil but inferior on the soil with poor drainage. Data averages of five years. [Data from Van Doren et al. (1976)]

The moisture level in the ridge is lower and the soil temperature higher than in the surrounding soil—permitting earlier seed germination and enhancing seedling growth. In any case, however, the yield constraints for crops grown on poorly drained soils have been a reason for low adoption of conservation tillage on these soils.

Effect on Soil Properties

Conservation tillage has variable effects on soil properties, depending on the particular system chosen. No-tillage systems, which maintain high surface soil coverage, have resulted in significant changes in soil properties, especially in the upper few centimeters.

PHYSICAL PROPERTIES. Conservation tillage has significant and generally positive effects on soil physical properties. Figure 17.15 illustrates the effects of two such systems compared to the conventional moldboard plow system after 28 years of continuous corn production. The effects of tillage systems on bulk density vary, but are generally not very significant. However, aggregate stability is generally much higher in no-tillage plots than in conventionally tilled plots. This is to be expected, since no-tillage plots are commonly higher in organic matter. Likewise, saturated hydraulic conductivity and available water capacity are generally higher on the no-tillage plots. Again, these characteristics are linked to the abundance of surface residues and higher soil organic matter. All of these important properties increase the probability of increased infiltration of water, less runoff, and less erosion from no-tillage systems.

Effects of conservation tillage on soil properties are related to the degree of surface cover or mulching. Consequently, conservation tillage practices such as stubble mulch farming, till plant, and ridge planting, which leave part of the soil unprotected, would likely be intermediate between no-tillage and conventional tillage in their effect on soil properties.

CHEMICAL PROPERTIES. No-tillage systems significantly increase the organic matter content of the upper few centimeters of soil (Figure 17.15). The higher soil microbe population needed to break down the surface residues also encourages the immobilization rather

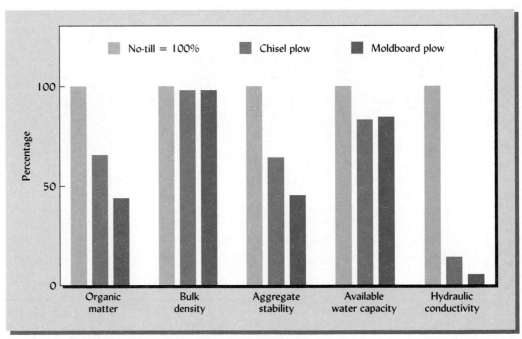

FIGURE 17.15 The comparative effects of 28 years of three tillage systems on soil organic matter content and a number of soil physical properties of an Alfisol in Ohio. Values for the no-till system were taken as 100, and the others are shown in comparison. Bulk density was about the same for each tillage system, but for all other properties the no-till system was decidedly more beneficial than either of the other two systems. The saturated hydraulic conductivity was especially high with no-till. [From Mahboubi et al. (1993); used with permission of the Soil Science Society of America]

than mineralization of soil nutrients, especially nitrogen. High moisture and low oxygen levels are known also to stimulate denitrification at times. Coupled with greater leaching losses of nitrate ions sometimes associated with no-tilled areas, these gaseous losses can be of some concern.

Response to nitrogen fertilizer also may be affected by soil nitrogen reactions stimulated by the tillage system used. In some cases, yield increases from low nitrogen fertilizer applications are smaller under no-tillage systems than under conventional systems. This is likely due to the immobilization of nitrogen in the no-till surface layers as microbes decompose the organic residues. But when higher levels of nitrogen are applied, the response under no-tillage is fully as great if not greater than where the conventional plow-and-cultivate systems are used.

No-tillage in humid areas has been found to significantly reduce soil pH, especially in the upper horizon (see Figure 17.16). Although some pH reduction is apparently due to acids that form when crop residues are broken down, most of it results from the acidifying effect of surface-applied nitrogen fertilizers. This acidity must be countered by application of liming materials to the soil.

In no-tillage systems, the essential nutrient elements tend to accumulate in the upper few centimeters of soil. These elements are added to the soil surface in crop residues, animal manures, and chemical fertilizers and lime; and if the elements are rather immobile (such as phosphorus) surface concentrations can increase significantly. To prevent undesirable buildup of nutrients in the surface layers, occasional conventional plowing may be necessary.

The effects of no-tillage on both physical and chemical properties of soils are considerably less marked in areas where crop residues do not cover a high percentage of the soil surface. Likewise, some effects are not as noticeable in subhumid and semiarid areas as in more humid regions. The effects of no-tillage also vary somewhat depending on the cropping sequence. Crops with high residues, such as corn, produce more pronounced effects than crops with little residue, such as soybeans, cotton, potatoes, and sugar beets.

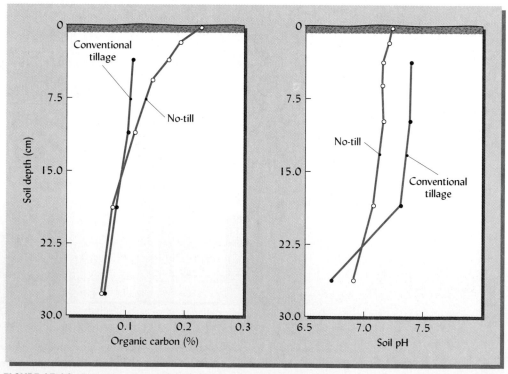

FIGURE 17.16 Comparative effects of 19 years of no-tillage and conventional tillage on the organic carbon content and pH of a Typic Fragiudult (Wooster silt loam). Keeping organic residues near the surface (no-tillage) increased the organic content of the upper 10 cm but reduced the soil pH in the upper 20 cm. The lower pH was likely due to increased nitrification where soil organic matter was higher. [From Dick (1982); used with permission of the Soil Science Society of America]

Biological Effects

Conservation tillage practices, which leave most of the crop residues on or near the surface, tend to encourage higher microbial populations in the few centimeters of surface soil. Also, numbers of soil fauna are commonly higher in minimum-tillage plots compared with conventionally tilled plots (Table 17.14). Higher numbers of earthworms on no-till plots may account for the greater ease with which water can infiltrate the soil even though tillage has not been used to loosen the soil.

Labor and Energy Requirements

A primary reason for a farmer to adopt conservation tillage practices is their lower labor and energy requirements. Although the time and energy savings vary depending on the conservation tillage system chosen, a number of studies suggest labor requirements are cut significantly when some conservation tillage practices are used. Since the number of passes over a field is reduced by adopting conservation practices, lowered labor costs are not surprising. Likewise, fewer machines are required in conservation tillage, which also tends to reduce costs.

Fuel consumption for conservation tillage is also markedly lower, being only one-third to one-half those for the conventional systems. Even when the energy needed to manufacture the greater quantity of herbicides used to control weeds in the conservation tillage systems is taken into consideration, in most cases the conservation tillage systems require less total energy than do their conventional counterparts.

Probable Future Expansion

Conservation tillage practices have spread rapidly in the United States in the last 20 years. Whereas in 1965 these practices were used on only 2–3% of the harvested cropland, by 1979 this figure had risen to about 20% and in 1994 to 35%. Regional use of different conservation tillage systems in the United States is shown in Table 17.15 and specific county adoption rates in Figure 17.17.

Future projections suggest that this trend will continue, and that soon after the turn of the century, about 75% of the country's cropland will be under conservation tillage. The expansion in area covered by conservation tillage practices is encouraging because of its probable constraint on soil erosion.

Up to this point, consideration has been given to sheet and rill water erosion. It is appropriate to have done so since most soil loss occurs from one of these processes. But in local areas, gully erosion is of greatest importance. This destructive process will now receive our attention.

17.13   GULLY EROSION

Small channels (rills), though at first insignificant, may soon expand into large gullies that quickly eat into the land, exposing its subsoil and increasing the already detrimental sheet and rill erosion (see Figure 17.4). If small enough, such gullies can be plowed in and seeded to grass.

TABLE 17.14   **Effect of Conventional and No-Tillage Systems on the Number and Weight of Soil Fauna**

Except for nematodes, all organisms were more abundant in no-till plots.

Organism	Numbers (per m^2)		Dry weight (mg/m^2)	
	Conv.	No-till	Conv.	No-till
Nematodes	3,008	2,473	344	252
Microarthropods	49,430	95,488	135	343
Macroarthropods	14	78	7	44
Earthworms	149	967	3,129	20,307

From Hendrix et al. (1986).

TABLE 17.15 The Total Cropland on Which Certain Conservation Tillage Practices Are Followed and the Percent of Land on Which These Practices Are Used in the Different Geographic Regions of the United States

| Region | Total cropland (Millions of ha) | Percent using different tillage systems | | | | 15–30% soil coverage |
| | | 30% soil coverage | | | | |
		No-till	Ridge till	Mulch till	Total	Various systems
Appalachian	5.6	32.9	0.1	11.8	44.8	9.2
Corn belt	31.0	26.7	0.7	18.1	45.4	21.3
Delta	6.3	9.3	1.6	7.9	18.9	19.3
Lake states	13.2	8.6	1.7	18.7	29.1	26.7
Mountain	8.8	4.2	0.3	25.5	29.9	35.4
Northeast	3.3	20.3	0.1	17.3	37.7	14.9
Northern Plains	27.2	8.7	3.0	20.6	38.2	33.1
Southern Plains	11.4	1.7	0.4	22.9	24.9	31.5
Pacific	4.8	1.7	0.1	18.5	20.3	21.4
Southeast	3.7	7.3	0.4	8.1	15.9	18.4
	115.3	13.7	1.3	20.0	35.0	25.8

Data for 1994 from Conservation Technology Information Center, West Lafayette, Indiana.

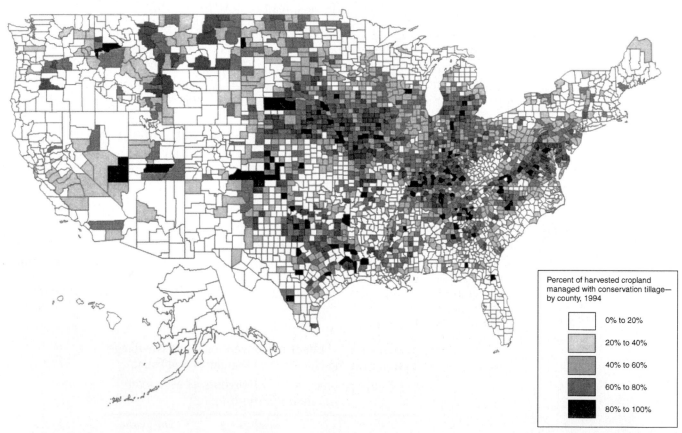

Percent of harvested cropland managed with conservation tillage—by county, 1994

- 0% to 20%
- 20% to 40%
- 40% to 60%
- 60% to 80%
- 80% to 100%

FIGURE 17.17 The percentage use of conservation tillage on different areas of the United States in 1994. Note that in some areas about two-thirds of the farmers are using these soil-saving systems. (Courtesy Conservation Technology Information Center, West Lafayette, Ind.)

When the gully erosion is too active to be thus checked and the ditch is still small, dams of rotted manure or straw at intervals of 5–7 m are very effective. Such dams may be made more secure with strips of wire netting below them. After a time, the ditch may be plowed in and the site of the gully seeded and kept in sod permanently. The gully then becomes a **grassed waterway**, an important feature of most successful erosion-control systems (see Figure 17.9).

With very large gullies, dams of earth, concrete, or stone are often used successfully. Most of the sediment is deposited above the dam, and the gully is slowly filled. The use of semipermanent check dams, flumes, and **rip-rap**-lined channels are also recommended on occasion, especially for construction sites. Unfortunately, such engineering features are generally too expensive for extensive use on agricultural land.

17.14   WIND EROSION: IMPORTANCE AND CONTROL

Wind erosion is most common in arid and semiarid regions, but it occurs to some extent in humid climates as well. It is virtually limited to situations in which the soil surface is relatively dry. All kinds of soils and soil materials are affected, and at times their more finely divided portions are carried to great heights and for hundreds of kilometers.

In the great dust storm of May 1934, which originated in western Kansas, Texas, Oklahoma, and contiguous portions of Colorado and New Mexico, clouds of powdery debris, silt, clay, and organic matter were carried eastward to the Atlantic seaboard and even hundreds of kilometers out over the ocean. Such activity is not a new phenomenon but has been common in all geologic ages. Strong winds gave rise to the wind-blown loess deposits now so important agriculturally in the United States (see Figure 2.18) and other countries.

Wind erosion is often very destructive. Not only is the land robbed of its richest soil material, but crops are either blown away, left to die with roots exposed (see Figure 17.18), or they are covered by the drifting debris. Wind moves close to half of the soil transported by all types of erosion in the United States. About 12% of the continental United States is somewhat affected by wind erosion—8% moderately so and perhaps 2–3% greatly. Figure 17.19 shows relative wind erosion forces at four locations in the United States.

In six of the Great Plains states, wind erosion exceeds water erosion. Wind erosion in these states is shown in Figure 17.20. Here mismanagement of plowed lands and overgrazing of range grasses have greatly increased the susceptibility of the soils to wind action. In dry years, the results have been most deplorable.

FIGURE 17.18   The devastating effects of wind erosion. The roots of this desert grass in Niger were exposed when the soil was blown away. [From IUCN (1986)]

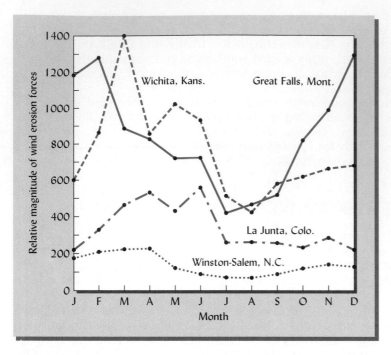

FIGURE 17.19 Monthly variation in the relative magnitude of wind erosion forces at four locations in the United States. One would expect little soil movement in Winston-Salem, N.C., not only because of the low wind erosion forces but also because of the probable high soil moisture content. [From Skidmore and Woodruff (1968)]

Although most of the damage is confined to regions of low rainfall, some serious wind erosion occurs in humid sections. Sand dune movement is a good example. More important agriculturally, sandy soils and peat soils cultivated for many years are often affected by the wind. The drying out of the finely divided surface layers of these soils leaves them extremely susceptible to wind erosion.

Wind erosion is a worldwide problem. Large areas of the former Soviet Union and sub-Saharan Africa have been badly damaged by wind erosion. Overgrazing and other misuses of the fragile lands of arid and semiarid areas have depressed soil productivity and brought starvation and human misery to millions of people in these areas. Wind erosion together with water erosion threatens humankind's capacity to sustain food and fiber production.

Desertification

In the Sahel region of Africa, long drought periods in the 1970s coupled with overgrazing of common lands and expanded cultivation on marginal lands brought about serious wind erosion. The soil was left bare over wide areas and the hot, dry winds picked up the soil and moved it toward the Saharan desert. This resulted in an apparent advance southward of the desert, the process being called **desertification.** Similar advances of desert areas have been noted elsewhere, and wind erosion is a major consequence of the process. More favorable rainfall in the Sahel during the 1980s and 1990s has stopped the southward advance of the desert, and even caused it to recede as plants have become reestablished in some areas.

There are differences of opinion about the steps needed to prevent and reverse desertification, encouraging the desert to recede to its earlier boundaries. However, there seems to be agreement on the critical role that plants play in stabilizing the soil—a stability that is lost when the land is overgrazed and insufficient water is available to maintain reasonable plant coverage of the soil.

Mechanics of Wind Erosion

As was the case for water erosion, the loss of soil by wind movement involves two processes: (1) *detachment* and (2) *transportation*. The lifting and abrasive action of the wind results in some detachment of tiny soil grains from the granules or clods of which they are a part. When the wind is laden with soil particles, however, its abrasive action is greatly increased. The impact of these rapidly moving grains dislodges other particles from soil clods and aggregates. These dislodged particles are now ready for movement.

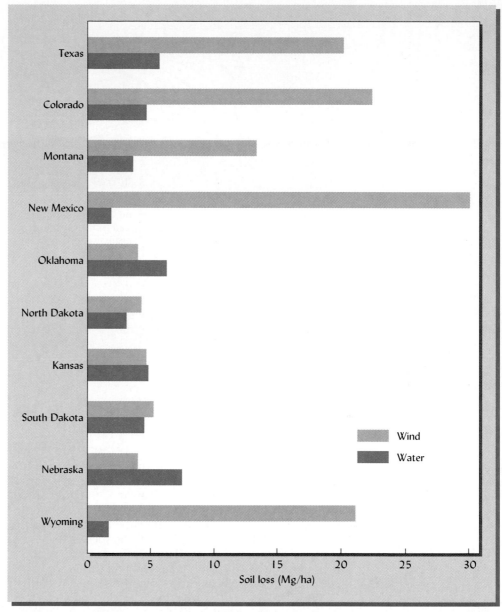

FIGURE 17.20 Average wind and water erosion rates on cropland in 10 Great Plains states. In eight states wind erosion exceeds water erosion, especially in Texas, Colorado, Montana, New Mexico, and Wyoming. [From USDA (1994)]

The transportation of the particles once they are dislodged takes place in several ways. The first and most important is that of **saltation**, or movement of soil by a short series of bounces along the surface of the ground (Figure 17.21). The particles remain fairly close to the ground as they bounce, seldom rising more than 30 cm or so. Depending on conditions, this process may account for 50–70% of the total movement of soil.

Saltation also encourages **soil creep**, or the rolling and sliding along the surface of the larger particles. The bouncing particles carried by saltation strike the large aggregates and speed up their movement along the surface. Soil creep accounts for the movement of particles up to about 1.0 mm in diameter, which may amount to 5–25% of the total movement.

The most spectacular method of transporting soil particles is by movement in **suspension**. Here, dust particles of a fine-sand size and smaller are moved parallel to the ground surface and upward. Although some of them are carried at a height no greater than a few meters, the turbulent action of the wind results in others being carried kilometers upward into the atmosphere and many hundreds of kilometers horizontally.

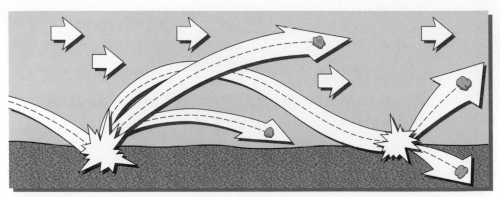

FIGURE 17.21 The process of saltation. Medium-sized particles (0.05 to 0.5 mm in diameter) bounce along the soil surface, striking and dislodging other particles as they move. They are too large to be carried long distances suspended in the air but small enough to be transported by the wind. [From Hughes (1980); used with permission of Deere & Company, Moline, Ill.]

These particles return to the earth only when the wind subsides and/or when precipitation washes them down. Although it is the most striking manner of transportation, suspension seldom accounts for more than 40% of the total and is generally no more than about 15%.

Factors Affecting Wind Erosion

Susceptibility to wind erosion is related to the moisture content of soils. Wet soils do not blow. The moisture content is generally lowered by hot, dry winds to the wilting point and lower before wind erosion takes place.

Other factors that influence wind erosion are (1) wind velocity and turbulence, (2) soil surface conditions, (3) soil characteristics, and (4) the nature and orientation of the vegetation. Obviously, the rate of wind movement, especially gusts having greater than average velocity, will influence erosion. Tests have shown that wind speeds of about 20 km/hr (12 mph) are required to initiate soil movement. At higher wind speeds, the soil movement is proportional to the cube of the wind velocity. Thus, the quantity of soil carried by wind goes up very rapidly as speeds above 30 km/hr are reached.

Wind turbulence also influences the capacity of the atmosphere to transport matter. Although the wind itself has some direct influence in picking up fine soil, the impact of wind-carried particles as they strike the soil is probably more important.

Wind erosion is less severe where the soil surface is rough. This roughness can be obtained by proper tillage methods, which leave large clods or ridges on the soil surface. Leaving stubble mulch (see Section 6.9) is probably an even more effective way of reducing wind-borne soil losses.

In addition to moisture content, several other soil characteristics influencing wind erosion are (1) mechanical stability of soil clods and aggregates, (2) stability of soil crust, and (3) bulk density and size of erodible soil fractions. The clods must be resistant to the abrasive action of wind-carried particles. If a soil crust resulting from a previous rain is present, it too must be able to withstand the wind's erosive power without deteriorating. The importance of clay, organic matter, and other cementing agents here is quite apparent. Sandy soils, which are low in such agents, are easily eroded.

Soil particles or aggregates about 0.1 mm in diameter are most erodible, those larger or smaller in size being less susceptible to movement. Thus, fine sandy soils are quite susceptible to wind erosion. Particles or aggregates about 0.1 mm in size are also responsible to a degree for the movement of larger or smaller particles. By saltation, most of the erodible particles bounce against larger particles, causing surface movement (creep), and against smaller soil particles, resulting in movement in suspension.

Vegetation or a stubble mulch will reduce wind erosion hazards, especially if the rows run perpendicular to the prevailing wind direction. This effectively presents a barrier to wind movement. In addition, plant roots help bind the soil and make it less susceptible to wind damage.

Predicting Wind Erosion

A wind erosion prediction equation (WEQ) has been in use since the late 1960s:

$$E = f(ICKLV)$$

The predicted wind erosion (E) is a function (f) of:

I = soil erodibility factor
C = climate factor
K = soil-ridge-roughness factor
L = width of field factor
V = vegetative cover factor

The WEQ involves the major factors that determine the severity of the erosion, but it also considers how these factors interact with each other. Consequently, it is not as simple as RUSLE for water erosion. Evidence of the interaction among factors is seen in Figure 17.22.

The **soil erodibility factor** (I) relates to the properties of the soil and to the degree of slope of the site in question. The **soil-ridge-roughness factor** (K) takes into consideration the cloddiness of the soil surface, vegetative cover (V) and ridges on the soil surface. Notice the degree to which the various factors interact.

The **climatic factor** (C) involves wind velocity, soil temperature, and precipitation (which helps control soil moisture). The **width of field factor** (L) is the width of a field in the downwind direction. Naturally the width changes as the direction of the wind changes, so the prevailing wind direction is generally used. The **vegetative cover** (V) relates not only the degree of soil surface covered with residues but the nature of the cover, whether it is living or dead, still standing or flat on the ground.

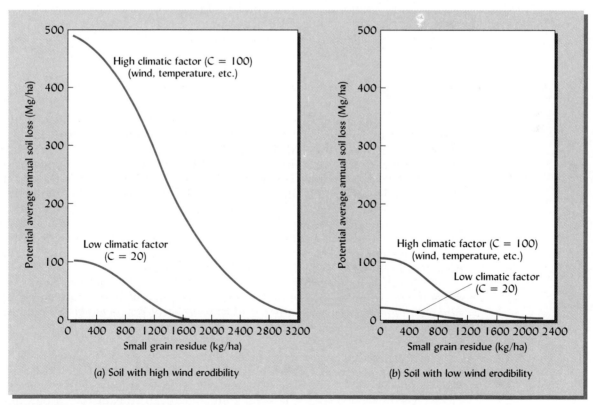

FIGURE 17.22 Effect of small grain residues on the potential wind erosion of soils with high (a) and low (b) erodibility. Strong, dry winds and high temperatures encourage erosion, especially on the soil with high wind-erodibility characteristics. Surface residues can be used to help control this wind erosion. [From Skidmore and Siddoway (1978); used with permission of the American Society of Agronomy]

Because of the many factors and subfactors influencing wind erosion, specific calculations are needed to provide valid predictions of erosion. For this reason, attempts are being made to develop computer models that will properly address the interaction among the various factors and increase the accuracy of the predictions. Accordingly, a Wind Erosion Prediction System (WEPS) program has been developed by the USDA and is being tested. It, too, will soon be available for more widespread use.

Control of Wind Erosion

The factors of the wind erosion equation give clues to methods of reducing wind erosion. Obviously, if the soil can be kept moist, there is little danger of wind erosion. A vegetative cover also discourages soil blowing, especially if the plant roots are well established. In dry-farming areas, however, sound moisture-conserving practices require summer fallow on some of the land, and hot, dry winds reduce the moisture in the soil surface. Consequently, other means must be employed on cultivated lands of these areas.

Conservation tillage practices described in Section 17.12 were used for wind erosion control long before they became popular as water erosion control practices. Keeping the soil surface rough and maintaining some vegetative cover is accomplished by using appropriate tillage practices. However, the vegetation should be well lodged into the soil to prevent it from blowing away. Stubble mulch has proved to be effective for this purpose (see Section 6.9).

Tillage to provide for a cloddy surface condition should be at right angles to the prevailing winds. Likewise, strip-cropping and alternate strips of cropped and fallowed land should be perpendicular to the wind. Barriers such as shelter belts (Figure 17.23) are effective in reducing wind velocities for short distances and for trapping drifting soil.

Various devices are used to control blowing of sands, sandy loams, and cultivated peat soils (even in humid regions). Windbreaks and tenacious grasses and shrubs are especially effective. Picket fences and burlap screens, though less efficient as windbreaks than trees such as willows, are often preferred because they can be moved from place to place as crops and cropping practices are varied. Rye, planted in narrow strips across the field, is sometimes used on peat lands. All of these devices for wind erosion control, whether applied in arid or humid regions and whether vegetative or purely mechanical, are closely associated with the broad problem of soil moisture control.

17.15   SOIL-LOSS TOLERANCE

The loss of any amount of soil by erosion generally is not considered beneficial, but years of field experience as well as scientific research indicate that some loss can be tolerated.[5] Scientists of the USDA Natural Resources Conservation Service working in cooperation with field personnel throughout the country have developed tentative soil-loss tolerance limits, known as *T-values,* for most cultivated soils of the country.

A tolerable soil loss (*T*-value) is considered as the maximum combined water and wind erosion that can take place on a given soil without degrading that soil's long-term productivity. As might be expected, there is insufficient research data to ascertain accurately the *T*-values for all soils. However, the soil-loss equations coupled with years of practical experience and the judgment of knowledgeable soil scientists have made it possible to set these values for the more widely used soils.

The *T*-values for soils in the United States commonly range from 5–11 Mg/ha. They depend on a number of soil-quality and management factors, including soil depth, organic matter content, and the use of water-control practices. At the present time, 11 Mg/ha is the maximum *T*-value assigned for most soils in this country; many have lower values.

The soil tolerance values set in the United States should be used with caution in other regions and particularly in the tropics. In some tropical soils, the nutrient-supplying power of the subsoil horizons is low. Since the subsoil becomes part of the plow layer in eroded soils, the *T*-value in some deep soils of the tropics may well be lower than that in the United States.

---

[5]For a discussion of this subject, see Schertz (1983).

There is considerable controversy as to whether the current *T*-values should be increased or lowered. For soils with deep, favorable rooting zones, the current 11-Mg/ha *T*-value may be too low. Some scientists contend, for example, that for a soil with a rooting depth of 200 cm, a *T*-value of 15 or even 20 Mg/ha would be appropriate. Others are concerned, however, not only about soil productivity but about the sediment from eroded fields as it affects environmental quality, including the quality of water for downstream users. These scientists contend that the *T*-values currently in use may be too high. In any case, the concept of *T*-values is useful if for no other reason than to focus attention on practical means of reducing soil erosion.

Maintenance of Soil Productivity

The soil-loss tolerance (*T*) concept suggests that the productive capacity of a soil is reduced only when the erosion loss exceeds the tolerable level. The 1992 U.S. National

(a)

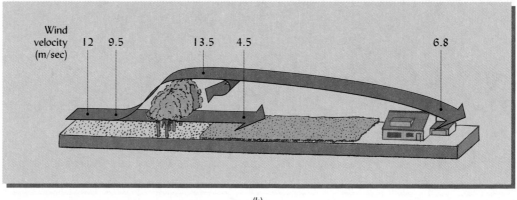

(b)

FIGURE 17.23 (a) Shrubs and trees make good windbreaks and add beauty to a North Dakota farm homestead. (b) Effect of a windbreak on wind velocity. The wind is deflected upward by the trees and is slowed down even before reaching them. On the leeward side further reduction occurs, the effect being felt as far as 20 times the height of the trees. [(a) Courtesy USDA Natural Resources Conservation Service; (b) from FAO (1987)]

Resource Inventory suggests that under current land use and management, soil productive capacity is going down on about 102 million hectares (252 million acres) of nonfederal land in the United States. Thirty-three percent of all cropland is eroding at rates in excess of T, and about 15% at twice the tolerable rate or more. It is obvious that progress must continue in order to bring these excessive losses under control.

The ultimate influence of soil loss by erosion on soil productivity is determined by such soil properties as soil depth and permeability. As shown in Figure 17.24, a deep, well-managed, and well-drained soil may not lose its crop productivity even though it suffers some erosion. In contrast, erosion from a shallow, poorly drained soil may bring about rapid decline in soil productivity.

17.16 LAND CAPABILITY CLASSIFICATION

The U.S. Department of Agriculture has developed a land capability classification system. While the system is useful in ascertaining the appropriate use of soils for many purposes, it is especially helpful in identifying practices that can minimize soil erosion.

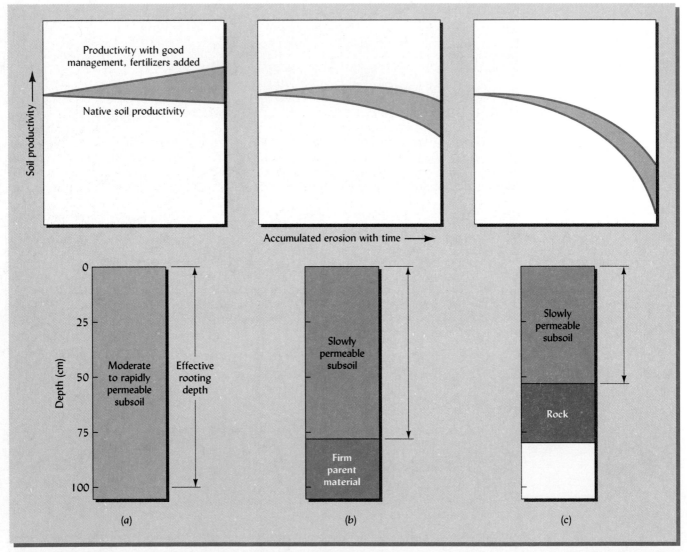

FIGURE 17.24 Effect of erosion over time on the productivity of three soils differing in depth and permeability. Productivity on soil A actually increases with time because of good management practices and fertilizer additions, even though the native soil productivity declines as a result of erosion. Because soil C is shallow and has restricted permeability, its productivity declines rapidly as a result of erosion, a decline that good management and fertilizers cannot prevent. Soil B, which is intermediate in both depth and permeability, suffers only a slight decline in productivity due to erosion. Obviously, the effect of erosion on soil productivity is influenced by soil characteristics.

Eight land capability classes are recognized in this classification system. Class I is least susceptible to erosion and Class VIII the most susceptible. Figure 17.25 shows the appropriate intensity of use allowable for each of these land capability classes if damage from erosion is to be avoided. A brief description of the characteristics and safe use of each land capability class follows.

Class I

Soils in this land class have few limitations on their use. They can be cropped intensively, used for pasture, range, woodlands, or wildlife preserves. The soils are deep and well drained, and the land is nearly level (Figure 17.26). Class I lands are either naturally fertile or have characteristics that encourage good response of crops to applications of fertilizer.

The soils in Class I need only prudent crop management practices to maintain their productivity. Good management entails the use of fertilizer and lime and the return of manure and crop residues, including green manures. Less than 3% of the nonfederal rural lands in the United States are in Class I.

Class II

Soils in Class II have some limitations that reduce the choice of uses or require moderate conservation practices. Less intensive cropping systems must be used on Class II than on Class I land, or, if the same cropping systems are used, Class II lands require some conservation practices. About 21% of the rural lands are in Class II.

The use of soils in Class II may be limited by one or more factors, such as (1) gentle slopes, (2) moderate erosion hazards, (3) inadequate soil depth, (4) less than ideal soil structure and workability, (5) slight to moderate alkaline or saline conditions and, (6) somewhat restricted drainage.

The management practices that may be required for soils in Class II include terracing, strip-cropping, contour tillage, rotations involving grasses and legumes, and grassed waterways (see Section 17.8). In addition, those prudent management practices that were suggested for Class I land are also generally required for soils in Class II.

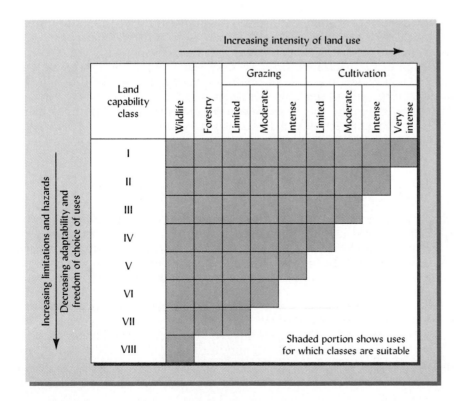

FIGURE 17.25 Intensity with which each land capability class can be used with safety. Note the increasing limitations on the safe uses of the land as one moves from Class I to Class VIII. [Modified from Hockensmith and Steele (1949)]

FIGURE 17.26 Several land capability classes in San Mateo County, California. A range is shown from the nearly level land in the foreground (Class I), which can be cropped intensively, to the badly eroded hillsides (Classes VII and VIII). Although topography and erosion hazards are emphasized here, it should be remembered that other factors—drainage, stoniness, droughtiness—also limit soil usage and help determine the land capability class. (Courtesy USDA Natural Resources Conservation Service)

Class III

About 20% of the rural land in the United States is in Class III. Soils in this class have severe limitations that reduce the choice of plants or require special conservation practices or both. The same crops may be grown on Class III land as on Classes I and II land, but crops that provide soil cover such as grasses and legumes must be more prominent in the rotations used. The amount of land used for row crops is severely restricted.

Limitations in the use of soils in Class III result from factors such as (1) moderately steep slopes, (2) high erosion hazards, (3) very slow water permeability, (4) shallow depth and restricted root zone, (5) low water-holding capacity, (f) low fertility, (6) moderate alkalinity or salinity, and (h) unstable soil structure.

Soils in Class III often require special conservation practices. These methods mentioned for Class II land must be employed, frequently in combination with restrictions in kinds of crops. Tile or other drainage systems also may be needed in poorly drained areas.

Class IV

Soils in this class can be used for cultivation, but there are severe limitations on the choice of crops. Also, careful management may be required. The alternative uses of Class IV soils are more limited than for Class III. Close-growing crops must be used extensively, and row crops cannot be grown safely in most cases. The choice of crops may be limited by excess moisture as well as by erosion hazards. Some 14% of the rural lands are in this class.

Limiting factors on Class IV soils may include (1) steep slopes, (2) severe erosion susceptibility, (3) severe past erosion, (4) shallow soils, (5) low water-holding capacity, (6) poor drainage, and (7) severe alkalinity or salinity. Soil conservation practices must be used more frequently than for soils in Class III and must be combined with strict limitations in choice of crop.

Class V

Soils in Classes V–VIII are generally not suited to cultivation. Those in Class V are limited in their safe use by factors other than erosion hazards. Such limitations include (1) frequent stream overflow, (2) growing season too short for crop plants, (3) stony or rocky soils, and (4) ponded areas where drainage is not feasible. Often, pastures can be improved on this class of land. Only about 2% of rural lands in the United States are in this class.

Class VI

Nineteen percent of United States rural lands fall in Class VI. Soils in this class have extreme limitations that restrict use largely to pasture, range, woodland, or wildlife. The limitations are the same as those for Class V land, but they are more rigid.

Class VII

Soils in Class VII have severe limitations that restrict their use to grazing, woodland, or wildlife. The physical limitations are the same as for Class VI except they are so strict that pasture improvement is impractical. Nearly 20% of nonfederal rural lands are in this category.

Class VIII

In this land class are soils that should not be used for any kind of commercial plant production. Land use is restricted to recreation, wildlife, water supply, or aesthetic purposes. Examples of kinds of soils or landforms included in Class VIII are sandy beaches, river wash, and rock outcrop. Fortunately, they comprise only about 2% of the nonfederal lands in the United States.

Subclasses

In each of the land capability classes are **subclasses** that have the same kind of dominant limitations for agricultural uses. The four kinds of limitations recognized in these subclasses (along with their letter designations) are risks of erosion (e); wetness, drainage, or overflow (w); root-zone limitations (s); and climatic limitations (c). Thus, a soil may be found in Class III(e), indicating that it is in Class III because of risks of erosion.

Land Use Capability in the United States

In 1992, the U.S. Department of Agriculture made a national inventory of soil and water conservation needs for the United States (USDA, 1994). This inventory included information on land capability classification. Figure 17.27 presents summary information from this inventory.

About 43% of the land in the United States (242 million hectares) is suitable for regular cultivation (Classes I, II, and III). Another 14% (78 million ha) is marginal for growing cultivated crops. The remainder is suited primarily for grasslands and forests, and is used mostly for this purpose.

Table 17.16 provides a summary of areas with major limitations on land use in the United States. Overall, for nearly 60% of the land, erosion and sedimentation problems are the most serious limitations, although excessive wetness and shallowness are each problems on about 20% of the land.

This land classification scheme illustrates the practical use that can be made of soil surveys (see Section 19.7). The many soils delineated on a map by the soil surveyor are viewed in terms of their safest and best longtime use. The eight land capability classes have become the starting point in the development of farm plans so useful to thousands of American farmers.

17.17 CONSERVATION TREATMENT IN THE UNITED STATES

Some progress in controlling soil erosion has been made in the last decade in the United States. A major part of this reduction is due to government support programs that have

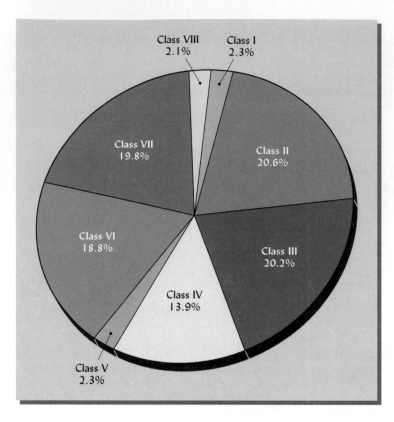

FIGURE 17.27 Percentage of total rural nonfederal land in each of the eight land use capability classes of the USDA. Note the very small percentage of land that can be managed with few limitations (Class I), while nearly one-fifth of the land falls in each of Classes II, III, IV, V, and VI. [From USDA (1994)]

reduced the area of cropland, shifting some land use to pastures and forests. Between 1982 and 1992, 13 million hectares of cropland were diverted to other noncultivated uses through what is termed the *Conservation Reserve Program* (CRP). Nearly 7 million hectares of these diverted lands had been highly erodible croplands. Changing the use of these lands from cropland to pasture and forests reduced the sheet and rill erosion from an average of 19.3 to 1.3 Mg/ha and the wind erosion from 24 to 2.9 Mg/ha. It is estimated that some 60% of the reduction in soil erosion is related to the CRP program.

The growing acceptance of conservation tillage is also helping to reduce soil erosion in the United States. Conservation tillage practices that provide at least 30% vegetative cover of the soil are now being used on 35% of the nation's croplands. Tillage practices on another 26% have 15 to 30% of the soil covered with crop residues. As we have seen, conservation tillage systems greatly decrease soil erosion and likely contribute significantly to the progress the country is making to better control this destructive process.

As satisfying as is the progress of the past decade, soil losses by erosion are still much too high. Continued efforts must be made to protect the soil and to hold it in place. In the United States nearly 34 million hectares of highly erodible cropland continues to

TABLE 17.16 **Percentage of Land with Major Limitations in the Different Land Use Capability Classes**

Erosion and sedimentation are the most significant limitations, but wetness and shallowness are very prominent.

	% of land with indicated major limitation			
LUC class	Erosion and sedimentation (%)	Wetness (%)	Shallowness (%)	Climate hazards (%)
II	51	34	8	7
III	67	22	9	2
IV	69	16	14	1
VI	62	8	26	3
VII	47	6	45	2
Total	59	18	21	3

Calculated from USDA (1994).

lose an average of more than 15 Mg/ha of soil each year from water erosion and an equal amount from wind erosion. In spite of remarkable progress, conservation tillage systems have *not* been adopted for 65% of the nation's cropland. Obviously, the battle to bring erosion under control has just begun—not only in the United States but throughout the world.

17.18   CONCLUSION

Runoff and erosion losses, particularly on prime farmland, have profound significance for crop production and for long-term soil productivity. Erosion losses occur worldwide, and damage not only the soils in upstream areas but also cause equal if not greater damage to downstream reservoirs, waterways, and municipal supply facilities.

Soil loss by sheet and rill water erosion is most critical in subhumid and humid areas, whereas wind erosion exacts a higher toll in semiarid and arid areas. Maintenance of soil surface cover by living plants or plant residues offers the most effective means of controlling both types of erosion. Practices known as *conservation tillage* have become increasingly popular in the United States and are used on about one-third of the harvested cropland. Conservation tillage systems are undoubtedly the most significant soil and water conservation practices developed in modern times.

Conservation tillage must be complemented by a number of other practices. In the first place, highly erodible land should simply not be cultivated. It should be used for controlled grazing or be allowed to reseed with forest trees or other natural vegetation. Lands that erode more than the critical T (11 Mg/ha/yr) must be managed as suggested in the USDA Land Use Capability System, using appropriate practices such as crop and rotation selection, strip-cropping, contour tillage, and cropping-terrace creation and maintenance. Every effort must be made to apply the principles of soil science and good land management to preserve our natural heritage of soil resources for the use of future generations.

REFERENCES

Baker, J. L., and J. M. Laflen. 1983. "Water quality consequences of conservation tillage," *J. Soil Water Cons.,* **38**:186–194.

Batie, S. S. 1983. *Soil Erosion—Crisis in America's Cropland* (Washington, D.C.: The Conservation Foundation).

Blevins, R. L., and W. W. Frye. 1992. "Conservation tillage: an ecological approach to soil management," *Adv. in Agron.,* **51**:33–78.

Brown, L. R., and E. C. Wolf. 1984. *Soil Erosion: Quiet Crisis in the World Economy,* Worldwatch Paper 60 (Washington, D.C.: Worldwatch Institute).

Carter, M. R. 1994. *Conservation Tillage in Temperate Agroecosystems* (Boca Raton, Fla.: Lewis Publishers).

Cassel, D. K., and R. Lal. 1992. "Soil physical properties of the tropics: common beliefs and management constraints," in R. Lal and P. A. Sanchez (eds.), *Myths and Science of Soils of the Tropics,* SSA Special Publication No. 29 (Madison, Wis.: Soil Sci. of Amer.), pp. 61–89.

Dick, W. A. 1982. "Organic carbon, nitrogen and phosphorus concentrations and pH in soil profiles as affected by tillage intensity," *Soil Sci. Soc. Amer. J.,* **47**:102–107.

Dickey, E. C., et al. 1987. "Conservation tillage: perceived and actual use," *J. Soil Water Cons.,* **42**:431–34.

Doolette, J. B., and J. W. Smyle. 1990. "Soil and moisture conservation technologies: review of literature," in J. B. Doolette and W. B. Magrath (eds.), *Watershed Development in Asia: Strategies and Technologies* (Washington, D.C.: World Book Technical Paper 127).

El-Swaify, and E. W. Dangler. 1982. "Rainfall erosion in the tropics: a state-of-the-art," *Soil Erosion and Conservation in the Tropics,* ASA Special Publication No. 43 (Madison, Wis.: Amer. Soc. of Agron.).

FAO. 1987. *Protect and Produce* (Rome, Italy: UN Food and Agriculture Organization).

Follet, R. F., and B. F. Stewart (eds.). 1985. *Soil Erosion and Crop Productivity* (Madison, Wis.: Amer. Soc. of Agron.).

Foster, G. R., and R. E. Highfill. 1983. "Effect of terraces on soil loss: USLE *P* factor values for terraces," *J. Soil and Water Cons.,* 38:48–51.

Heimlich, R. E., and N. L. Bills. 1986. "An improved soil erosion classification: update, comparison and extension," *Soil Conservation, Assessing the National Resources Inventory,* Vol 2, (Washington, D.C.: National Academy Press).

Hendrix, P. F., R. W. Parmalee, D. A. Crosley, Jr., D. C. Coleman, E. P. Odum, and P. M. Goffman. 1986. "Detritus food webs in conventional and no-tillage agroecosystems," *BioScience,* 36:374–379, 1986.

Hockensmith, R. D., and J. G. Steel. 1949. "Recent trends in the use of the land-capability classification," *Soil Sci. Soc. Amer. Proc.,* 14:383–388.

Hughes, H. A. 1980. *Conservation Farming* (Moline, Ill.: Deere and Company).

IUCN. 1986. *The IUCN Sabel Report.* (Gland, Switzerland: Intl. Union Cons. Nature).

Kern, J. S., and M. G. Johnson. 1993. "Conservation tillage impacts on national soil and atmospheric carbon levels," *Soil Sci. Soc. Amer. J.,* 57:200–210.

Larson, W. E., F. J. Pierce, and R. H. Dowdy. 1983. "The threat of soil erosion to long-term crop productivity," *Science,* 219:458–465.

Lattanzi, A. R., L. D. Meyer, and M. F. Baumgardner. 1979. "Influence of mulch rate and slope steepness in interrill erosion," *Soil Sci. Soc. Amer. Proc.,* 38:946–950.

Mahboubi, A. A., R. Lal, and N. R. Faussey. 1993. "Twenty-eight years of tillage effects on two soils in Ohio," *Soil Sci. Soc. Am. J.,* 57:506–512.

Oschwald, W. R., and J. C. Siemens. 1976. "Conservation tillage: a perspective," SM-30, *Agron. Facts* (Urbana, Ill.: Univ. of Illinois).

Phillips, R. E., R. L. Blevins, G. W. Thomas, W. W. Frye, and S. H. Phillips. 1980. "No-tillage agriculture," *Science,* 208:1108–1113.

Pimental, D., et al. 1995. "Environmental and economic costs of soil erosion and conservation benefits," *Science,* 267:1117–1122.

Renard, K. G., G. Foster, D. Yoder, and D. McCool. 1994. "RUSLE revisted: status, questions answers and the future," *J. Soil and Water Cons.,* 49:213–220.

Schertz, D. L. 1983. "The basis for soil loss tolerance," *J. Soil Water Cons.,* 30:10–14.

Skidmore, E. L., and F. H. Siddoway. 1978. "Crop residue requirements to control wind erosion," in W. R. Oschwalk (ed.), *Crop Residue Management Systems,* ASA Special Publication No. 31 (Madison, Wis.: Amer. Soc. Agron.; Crop Sci. Soc. Amer.; and Soil Sci. Soc. Amer.).

Skidmore, E. L., and N. P. Woodruff. 1968. *Wind Erosion Forces in the United States and Their Use in Predicting Soil Loss,* Agr. Handbook 346 (Washington, D.C.: U.S. Department of Agriculture).

USDA. 1987. *The Second RCA Appraisal, Analysis of Conditions and Trends,* Review Draft (Washington, D.C.: U.S. Department of Agriculture).

USDA. 1994. *Summary Report: 1992 National Resources Inventory* (Washington, D.C.: USDA Natural Resources Conservation Service).

USDA. 1995. *Agricultural Handbook,* No. 703 (Washington, D.C.: U.S. Department of Agricultural).

Van Doren, D. M., Jr., G. B. Tripett, Jr., and J. E. Henry. 1976. "Influence of long-term tillage, crop rotation and soil type combinations on corn yields," *Soil Sci. Soc. Amer. J.,* 40:100–105.

Walker, R. D. 1980. "USLE, a quick way to estimate your erosion losses," *Crops and Soils,* 33:10–13.

Wischmeier, W. J., and D. D. Smith. 1978. *Predicting Rainfall Erosion Loss—A Guide to Conservation Planning,* Agr. Handbook No. 537 (Washington, D.C.: U.S. Department of Agriculture).

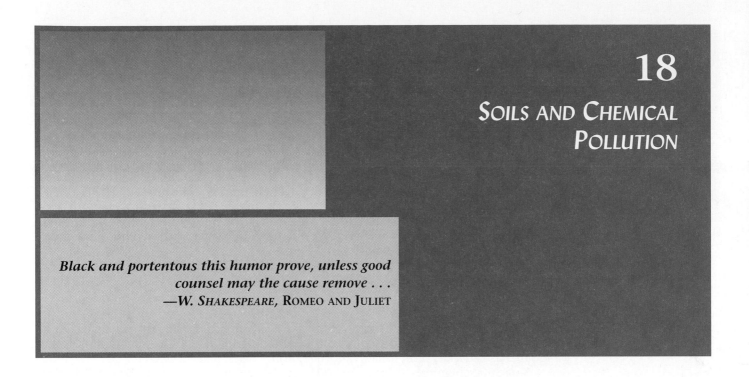

Black and portentous this humor prove, unless good
counsel may the cause remove . . .
—W. SHAKESPEARE, ROMEO AND JULIET

Soil science is increasingly called upon to apply basic biological, chemical, and physical principles to understanding environmental problems. Examples of such applications discussed in this textbook include the production and sequestering of various greenhouse gases (e.g., Sections 12.2 and 13.9), the contamination of ground- and surface waters with excess nitrogen (Section 13.8) and phosphorus (Section 14.2), and the reactions of acids impinging on soils from atmospheric deposition (e.g., Section 9.6). Many of these issues involve effects of soil management on water and air pollution.

In this chapter we will focus on chemicals that contaminate and degrade soils, although in many cases the damages also extend to water, air, and living things. The soil is a primary recipient, by design or by accident, of the waste products and chemicals used in modern society. Once these materials enter the soil, they become part of a cycle that affects all forms of life. The brief review of soil pollution in this chapter is intended as an introduction to the nature of the major pollutants, their reactions in soils, and alternative means of managing, destroying, or inactivating them.

18.1 TOXIC ORGANIC CHEMICALS

Modern industrialized societies have developed thousands of synthetic organic compounds for thousands of uses. An enormous quantity of organic chemicals is manufactured every year—about 60 million metric tons in the United States alone. Included are plastics and plasticizers, lubricants and refrigerants, fuels and solvents, pesticides and preservatives. Some are extremely toxic to humans and other life. Through accidental leakage and spills or through planned spraying or other treatments, synthetic organic chemicals can be found in virtually every corner of our environment—in the soil, in the groundwater, in the plants, and in our own bodies.

Environmental Damage from Organic Chemicals

Some of these organic compounds are relatively inert and harmless, but others are biologically damaging even in very small concentrations. Those that find their way into soils may produce several undesirable results: (1) Certain soil organisms may be inhibited or killed, undermining the balance of the soil community (see Section 11.14). (2) The chemical may be transported from the soil to the air or water or vegetation, where it may be contacted, inhaled, or ingested, over a wide area by any number of organisms. It is imperative, therefore, that we control the release of organic chemicals and that we learn of their fate and effects once they enter the soil.

Organic chemicals may enter the soil as contaminants in industrial and municipal organic wastes applied to or spilled on soils, as components of discarded machinery, in large or small lubricant and fuel leaks, and as pesticides applied to terrestrial ecosystems.

While some pesticides are meant to be applied to soils, most reach the soil because they have missed the insect or plant leaf that was the target of the application. When pesticides are sprayed in the field, most of the chemical misses the target organism. For pesticides aerially applied to forests, about 25% reaches the tree foliage and far less than 1% reaches a target insect. About 30% may reach the soil, while about half of the chemical applied is likely to be lost into the atmosphere or in runoff water.

Pesticides are probably the most widespread organic pollutant associated with soils. In the United States, pesticides are used on some 150 million hectares of land, three-fourths of which is agricultural land. We will therefore focus most of our discussion on the pesticide problem.

The Nature of Pesticides

Pesticides are chemicals that are designed to kill pests (that is, any organism that the pesticide user perceives to be damaging). Since the offending organisms may be plants (weeds), insects, fungi, nematodes, or rats, pesticides include different compounds designated as *herbicides, insecticides, fungicides, nematocides, rodenticides,* etc.

Some 600 chemicals in about 50,000 formulations are used to control pests. They are used extensively in all parts of the world. About 350 million kg of organic pesticide chemicals are used annually in the United States, with similar amounts used in western Europe and Asia. Although the total amount of pesticides used has remained relatively constant or even dropped since the 1980s, formulations in use today are generally more potent, so that smaller quantities are applied per hectare to achieve toxicity to the pest.

BENEFITS OF PESTICIDES. Pesticides have provided many benefits to society. They have helped control mosquitos and other vectors of human diseases such as yellow fever and malaria. They have protected crops and livestock against insects and diseases. Without the control of weeds by chemicals called *herbicides,* conservation tillage (especially no-tillage) would be much more difficult to adopt; much of the progress made in controlling soil erosion probably would not have come about without herbicides. Also, pesticides reduce the spoilage of food as it moves from farm fields to often distant dinner tables.

COSTS OF PESTICIDES. However, pesticides should not be seen as a panacea or as indispensable. Some farmers produce profitable yields without the use of pesticides. Even with the almost universal use of pesticides, insects, diseases, and weeds still cause the loss of one-third of the crop production in the United States, about the same proportion of crops lost to these pests in the 1940s, before many pesticides were in use. And while the benefits to society from pesticides are great, so are the costs (Table 18.1).

Designed to kill living things, many of these chemicals are potentially toxic to organisms other than the pests for which they are intended. Some are detrimental to nontarget organisms such as beneficial insects and certain soil organisms. Those chemicals that do not quickly break down may be biologically magnified as they move up the food chain. For example, as earthworms ingest contaminated soil, the chemicals tend to concentrate in the earthworm bodies. When birds and fish eat the earthworms, the pesticides can build up further to lethal levels. The near extinction of certain birds of prey (including the bald eagle) during the 1960s and 1970s called public attention to the sometimes devastating environmental consequences of pesticide use.

18.2 KINDS OF PESTICIDES

Pesticides are commonly classified according to the target group of pest organisms: (1) insecticides, (2) fungicides, (3) herbicides (weed killers), (4) rodenticides, and (5) nematocides. In practice, all find their way into soils. Since the first three are used in the

TABLE 18.1 Total Estimated Environmental and Social Cost from Pesticide Use in the United States

The death of an estimated 60 million wild birds may represent an additional substantial cost in lost revenues from hunters, bird watchers, etc.

Type of impact	Cost ($ million/year)
Public health impacts	787
Domestic animal deaths and contamination	30
Loss of natural enemies	520
Cost of pesticide resistance	1,400
Honeybee and pollination losses	320
Crop losses	942
Fishery losses	24
Groundwater contamination and cleanup costs	1,800
Cost of government regulations to prevent damage	200
Total	6,023

From Pimental et al. (1992). © Amer. Instit. of Biological Sciences.

largest quantities and are therefore more likely to contaminate soils, they will be given primary consideration. Figure 18.1 shows that most pesticides contain aromatic rings of some kind, but that there is great variability in pesticide chemical structures.

Insecticides

Most of these chemicals are included in three general groups. The *chlorinated hydrocarbons,* such as DDT, were the most extensively used until the early 1970s, when their use was banned or severely restricted in many countries due to their low biodegradability and persistence, as well as toxicity to birds and fish.

The *organophosphate* pesticides are generally biodegradable, and thus less likely to build up in soils and water. However, they are extremely toxic to humans, so great care must be used in handling and applying them. The *carbamates* are considered least dangerous because of their ready biodegradability and relatively low mammalian toxicity. However, they are highly toxic to honeybees and other beneficial insects and to earthworms.

Fungicides

Fungicides are used mainly to control diseases of fruit and vegetable crops and as seed coatings to protect against seed rots. Some are also used to protect harvested fruits and vegetables from decay, to prevent wood decay, and to protect clothing from mildew. Organic materials such as the thiocarbamates and triazoles are currently in use.

Herbicides

The quantity of herbicides used in the United States exceeds that of the other two types of pesticides combined. Starting with 2,4-D (a chlorinated *phenoxyalkanoic acid*), dozens of chemicals in literally hundreds of formulations have been placed on the market (see Figure 18.1). These include the *triazines* (used mainly for weed control in corn); *substituted ureas;* some *carbamates;* the relatively new *sulfonylureas,* which are potent at very low rates; *dinitroanilines;* and *acetanilides,* which have proved to be quite mobile in the environment. As one might expect, this wide variation in chemical makeup provides an equally wide variation in properties. Most herbicides are biodegradable, and most of them are relatively low in mammalian toxicity. However, some are quite toxic to fish and perhaps to other wildlife. They can also have deleterious effects on beneficial aquatic vegetation that provides food and habitat for fish and shellfish.

Nematocides

Although nematocides are not widely used, some of them are known to contaminate soils and the water draining from treated soils. For example, some carbamates used as nematocides are quite soluble in water, are not adsorbed by the soil, and consequently leach downward and into the groundwater.

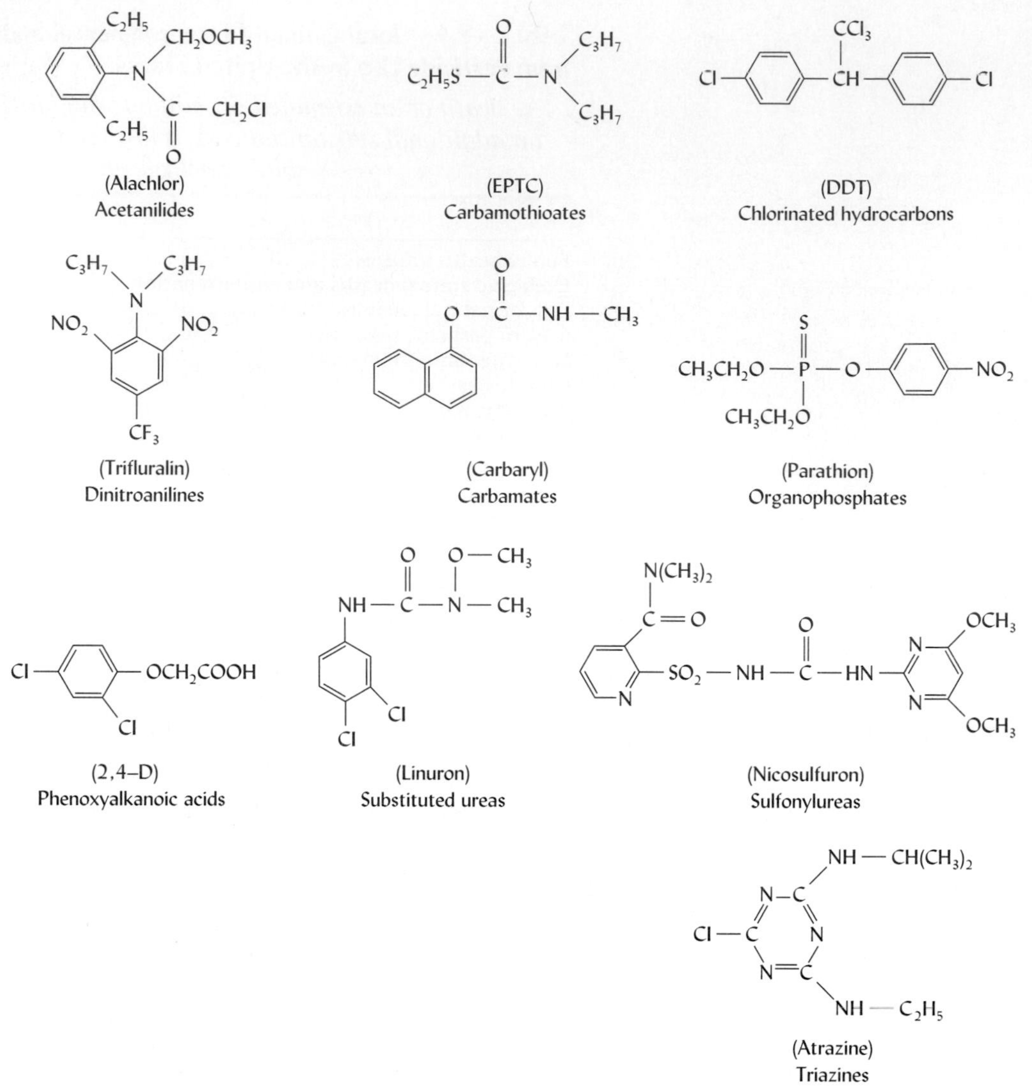

FIGURE 18.1 Structural formulae of representative compounds in 10 classes of widely used pesticides. Carbaryl, DDT, and parathion are insecticides; the other compounds shown are herbicides. The widely differing structures result in a great variety of toxicological properties and reactions in the soil.

18.3 BEHAVIOR OF ORGANIC CHEMICALS IN SOILS[1]

Once they reach the soil, organic chemicals such as pesticides or hydrocarbons move in one or more of seven directions (Figure 18.2): (1) They may vaporize into the atmosphere without chemical change; (2) they may be absorbed by soils; (3) they may move downward through the soil in liquid or solution form and be lost from the soil by leaching; (4) they may undergo chemical reactions within or on the surface of the soil; (5) they may be broken down by soil microorganisms; (6) they may wash into streams and rivers in surface runoff; and (7) they may be taken up by plants or soil animals and move up the food chain. The specific fate of these chemicals will be determined at least in part by their chemical structures, which are highly variable.

Volatility

Organic chemicals vary greatly in their volatility and subsequent susceptibility to atmospheric loss. Some soil fumigants such as methyl bromide (now banned from most uses) were selected because of their very high vapor pressure, which permits them to penetrate soil pores to contact the target organisms. This same characteristic encourages

[1]For reviews on organic chemicals in the soil environment, see Sawhney and Brown (eds.) (1989) and Pierzynski et al. (1994); for pesticides, see Cheng (ed.) (1990).

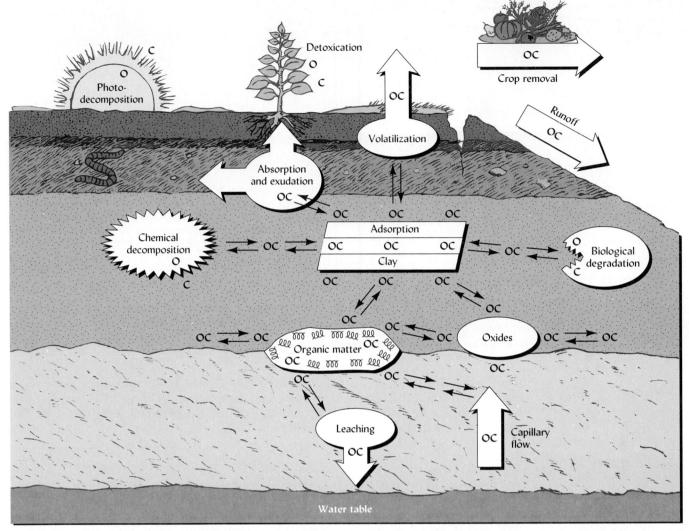

FIGURE 18.2 Processes affecting the dissipation of organic chemicals (OC) in soils. Note that the OC symbol is split up by decomposition (both by light and chemical reaction) and degradation by microorganisms, indicating that these processes alter or destroy the organic chemical. In transfer processes the OC remains intact. [From Weber and Miller (1989)]

rapid loss to the atmosphere after treatment, unless the soil is covered or sealed. A few herbicides (e.g., trifluralin) and fungicides (e.g., PCNB) are sufficiently volatile to make vaporization a primary means of their loss from soil. The lighter fractions of crude oil and many solvents vaporize to a large degree when spilled on the soil.

The assumption that disappearance of pesticides from soils is evidence of their breakdown is questionable. Some chemicals lost to the atmosphere are known to return to the soil or surface waters with the rain.

Adsorption

The adsorption of organic chemicals by soil is determined largely by the characteristics of the compound and of the soils to which they are added. Soil organic matter and high-surface-area clays tend to be the strongest adsorbents for some compounds (Figure 18.3), while oxide coatings on soil particles strongly adsorb others. The presence of certain functional groups such as —OH, —NH_2, —NHR, —$CONH_2$, —COOR, and —$^+NR_3$ in the chemical structure encourages adsorption, especially on the soil humus. Hydrogen bonding (see Section 8.3) and protonation (adding of H^+ to a group such as an —NH_2 (amino) group) probably promotes some of the adsorption. Everything else being equal, larger organic molecules with many charged sites are more strongly adsorbed.

Some organic chemicals with positively charged groups, such as the herbicides diquat and paraquat, are strongly adsorbed by silicate clays. Adsorption by clays of

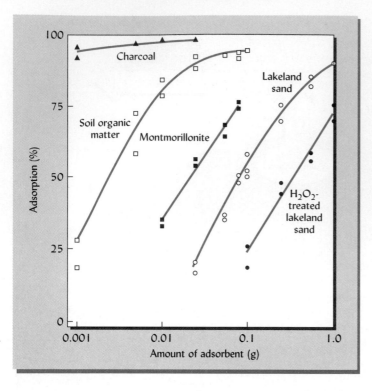

FIGURE 18.3 Adsorption of polychlorinated biphenyl (PCB) by different soil materials. The Lakeland sand (Typic Quartzipsamments) lost much of its adsorption capacity when treated with hydrogen peroxide to remove its organic matter. The amount of soil material required to adsorb 50% of the PCB was approximately 10 times as great for montmorillonite (a 2:1 clay mineral) as for soil organic matter, and 10 times again as great for H_2O_2-treated Lakeland sand. Later tests showed that once the PCB was adsorbed, it was no longer available for uptake by plants. Note that the amount of soil material added is shown on a log scale. [From Strek and Weber (1982)]

some pesticides tends to be pH-dependent (see Figure 18.4), maximum adsorption occurring at low pH levels, which encourages protonation. Adding an H^+ ion to functional groups (e.g., —NH_2) yields a positive charge on the herbicide, resulting in greater attraction to negatively charged soil colloids.

Leaching and Runoff

The tendency of organic chemicals to leach from soils is closely related to their solubility in water and their potential for adsorption. Some compounds, such as chloroform and phenoxyacetic acid, are a million times more water-soluble than others, such as DDT and PCBs, which are quite soluble in oil but not in water. High water solubility favors leaching losses.

Strongly adsorbed molecules are not likely to move down the profile (see Table 18.2). Likewise, conditions that encourage such adsorption will discourage leaching. Leaching is apt to be favored by water movement, the greatest leaching hazard occur-

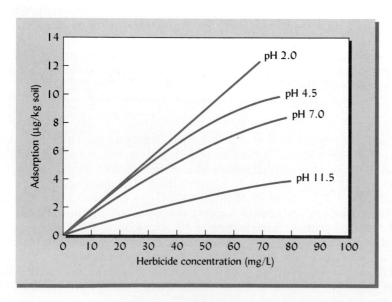

FIGURE 18.4 The effect of pH of kaolinite on the adsorption of glyphosate, a widely used herbicide. [Reprinted with permission from J. S. McConnell and L. R. Hossner, *J. Agric. Food Chem.* **33**:1075–78 (1985); copyright 1985 American Chemical Society]

TABLE 18.2 The Degree of Adsorption of Selected Herbicides

Weakly adsorbed herbicides are more susceptible to movement in the soil than those that are more tightly adsorbed.

Common name or designation	Trade name	Adsorptivity to soil colloids
Dalapon	Dowpon	None
Chloramben	Amiben	Weak
Bentazon	Basagran	Weak
2,4-D	Several	Moderate
Propachlor	Ramrod	Moderate
Atrazine	AAtrex	Strong
Alachlor	Lasso	Strong
EPTC	Eptam	Strong
Diuron	Karmex	Strong
Paraquat	Paraquat	Very strong
Trifluralin	Treflan	Very strong
DCPA	Dacthal	Very strong

Selected data from DMI (1981).

ring in high-permeability, sandy soils that are also low in organic matter. Periods of high rainfall around the time of application of the chemical promote both leaching and runoff losses (Table 18.3). With some notable exceptions, herbicides seem to be somewhat more mobile than most fungicides or insecticides, and therefore more likely to find their way to groundwater supplies and streams.

Contamination of Groundwater

Experts once maintained that contamination of groundwater by pesticides occurred only from accidents such as spills; but it is now known that many pesticides reach the groundwater from normal agricultural use. Since many people (e.g., 40% of Americans) depend on groundwater for their drinking supply, leaching of pesticides is of wide concern. Table 18.4 lists some of the 46 pesticides found in a national survey of well waters in the United States. The concentrations are given in parts per billion (see Box 18.1). In some cases the amount of pesticide found in the drinking water has been high enough to raise long-term health concerns.

Chemical Reactions

Upon contacting the soil, some pesticides undergo chemical modification independent of soil organisms. For example, iron-cyanide compounds decompose within hours or days if exposed to bright sunlight. DDT, diquat, and the triazines are subject to slower photodecomposition in sunlight. The triazine herbicides (e.g., atrazine) and organophosphate insecticides (e.g., malathion) are subject to hydrolysis and subsequent degradation. While the complexities of molecular structure of the pesticides suggest different mechanisms of

TABLE 18.3 Surface Runoff and Leaching Losses (Through Drain Tiles) of the Herbicide, Atrazine, from a Clay Loam Lacustrine Soil (Alfisols) in Ontario Canada

The herbicide was applied at 1700 g/ha in late May. The data are the average of three tillage methods. Note that the rainfall for May and June is related to the amount of herbicide lost by both pathways.

Year of study	Grams of atrazine / ha			Percent of total applied (%)	Rainfall, May–June (mm)
	Surface runoff loss	Drainage water loss	Total dissolved loss		
1	18	9	27	1.6	170
2	1	2	3	0.2	30
3	51	61	113	6.6	255
4	13	32	45	2.6	165

Data abstracted from Gaynor et al. (1995).

TABLE 18.4   Pesticides Present in Groundwater from Normal Agricultural Use

Note the wide range in concentrations which are considered to be risky to health. The great majority of wells sampled were uncontaminated, but when pesticides were detected they were often near or above the health-advisory level.

Pesticide	Use[a]	Median level found	Maximum level found	Health-advisory level[b]
		(parts per billion)		
Alachlor	H	0.90	113.00	
Aldicarb	I	9.00	315.00	10.00
Atrazine	H	0.50	40.00	3.00
Bromacil	H	9.00	22.00	90.00
Carbofuran	I	5.30	176.00	40.00
Cyanazine	H	0.40	7.00	10.00
2,4-D (1,2-Dichlorophenoxyacetic acid)	F	1.40	49.50	70.00
DBCP*	FUM	0.01	0.02	—
DCPA	H	109.00	1,039.99	4,000.00
Dinoceb*	H, I, F	0.70	36.70	7.00
EDB	F	0.90	14.00	—
Fonofos	I	0.10	0.90	10.00
Malathion	I	41.50	53.00	200.00
Metolachlor	H	0.40	32.30	100.00
Metribuzin	H	0.60	6.80	200.00
Oxamyl	I	4.30	395.00	200.00
Trifluran	H	0.40	2.20	5.0

Data from General Accounting Office (1991).
[a]H = herbicide; I = insecticide; F = fungicide; FUM = fumigant.
[b]Health-advisory level is the concentration that is suspected of causing health problems over a 70-year lifetime. Blank means no advisory level has been set.
*Most uses of this pesticide have been banned in the United States.

breakdown, it is important to realize that degradation independent of soil organisms does in fact occur.

Microbial Metabolism

Biochemical degradation by soil organisms is the single most important method by which pesticides are removed from soils. Certain polar groups such as —OH, —COO⁻, and —NH$_2$ on the pesticide molecules provide points of attack for the organisms.

DDT and other chlorinated hydrocarbons such as aldrin, dieldrin, and heptachlor are very slowly broken down, persisting in soils for 20 or more years. In contrast, the organophosphate insecticides such as parathion are degraded quite rapidly in soils, apparently by a variety of organisms (see Figure 18.5). Likewise, the most widely used herbicides, such as 2,4-D, the phenylureas, the aliphatic acids, and the carbamates, are readily attacked by a host of organisms. Exceptions are the triazines, which are slowly degraded, primarily by chemical action. Most organic fungicides are also subject to microbial decomposition, although the rate of breakdown of some is slow, causing troublesome residue problems.

Plant Absorption

Pesticides are commonly absorbed by higher plants. This is especially true for those pesticides (e.g., systemic insecticides and most herbicides) which must be taken up in order to perform their intended function. The absorbed chemicals may remain intact inside the plant or they may be degraded. Some degradation products are harmless, but others are even more toxic to humans than the original chemical that was absorbed. Understandably, society is quite concerned about pesticide residues found in the parts of plants that people eat, whether as fresh fruits and vegetables or as processed foods. The use of pesticides and the amount of pesticide residues in food are strictly regulated by law to ensure human safety. Despite widespread concerns, there is little evidence that the small amounts of residues permissible in foods by law have had any ill effects on public health. However, routine testing by regulatory agencies has shown that about 1 to 2% of the food samples tested contained pesticide residues above the levels permissible.

Persistence in Soils

The persistence of chemicals in soils is a summation of all the reactions, movements, and degradations affecting these chemicals. Marked differences in persistence are the rule (see Figure 18.5). For example, organophosphate insecticides may last only a few days in soils. The widely used herbicide 2,4-D persists in soils for only two to four weeks. PCBs, DDT, and other chlorinated hydrocarbons may persist for from 3 to 20 years or longer (Table 18.5). The persistence times of other pesticides and industrial organics fall generally between the extremes cited. The majority of pesticides degrade rapidly enough to prevent buildup in soils having normal annual applications. Those that resist degradation have a greater potential to cause environmental damage.

Continued use of the same pesticide on the same land can increase the rate of microbial breakdown of that pesticide. Apparently, having a constant food source allows the population to build up among those microbes equipped with the enzymes needed to break down the compound. This is an advantage in relation to environmental quality and is a principle sometimes applied in environmental cleanup of toxic organic compounds, but the breakdown may become sufficiently rapid to reduce a pesticide's effectiveness.

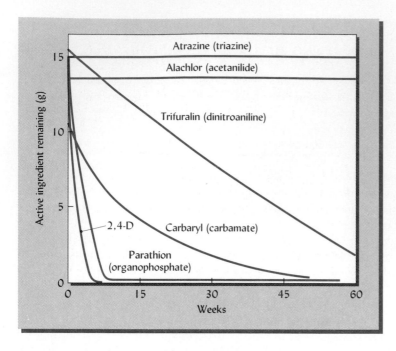

FIGURE 18.5 Degradation of four herbicides (alachlor, atrazine, 2,4-D, and trifuralin) and two insecticides (parathion and carbaryl), all of which are used extensively in the Midwest of the United States. Note that atrazine and alachlor are quite slowly degraded, whereas parathion and 2,4-D are quickly broken down. [Reprinted with permission from R. G. Krueger and J. N. Seiber, *Treatment and Disposal of Pesticide Wastes,* Symposium Series 259; copyright 1984 American Chemical Society]

18.4 EFFECTS OF PESTICIDES ON SOIL ORGANISMS

Since pesticides are formulated to kill organisms, it is not surprising that some of these compounds are toxic to specific soil organisms. At the same time, the diversity of the soil organism population is so great that, excepting a few fumigants, most pesticides do not kill a broad spectrum of soil organisms.

Fumigants

Fumigants are compounds used to free a soil of a given pest, such as nematodes. These compounds have a more drastic effect on both the soil fauna and flora than do other pesticides. For example, 99% of the microarthropod population is usually killed by the fumigants DD and vampam, and it may take as long as two years for the population to fully recover. Fortunately, the recovery time for the microflora is generally much less.

Fumigation reduces the number of species of both flora and fauna, especially if the treatment is repeated, as is often the case where nematode control is attempted. At the same time, the total number of bacteria is frequently much greater following fumigation than before. This increase is probably due to the relative absence of competitors and predators following fumigation.

TABLE 18.5   **Common Range of Persistence of a Number of Organic Compounds**

Risks of environmental pollution are highest with those chemicals with greatest persistence.

Organic chemical	Persistence
Chlorinated hydrocarbon insecticides (e.g., DDT, chlordane, dieldrin)	3–20 yr
PCBs	2–10 yr
Triazine herbicides (e.g., atrazine, simazine)	1–2 yr
Benzoic acid herbicides (e.g., amiben, dicamba)	2–12 mo
Urea herbicides (e.g., monuron, diuron)	2–10 mo
Vinyl chloride	1–5 mo
Phenoxy herbicides (2,4-D, 2,4,5-T)	1–5 mo
Organophosphate insecticides (e.g., malathion, diazinon)	1–12 wk
Carbamate insecticides	1–8 wk
Carbamate herbicides (e.g., barban, CIPC)	2–8 wk

Effects on Soil Fauna

The effects of pesticides on soil animals varies greatly from chemical to chemical and from organism to organism. Nematodes are not generally affected except by specific fumigants. Mites are generally sensitive to most organophosphates and to the chlorinated hydrocarbons, with the exception of aldrin. Springtails vary in their sensitivity to both chlorinated hydrocarbons and organophosphates, some chemicals being quite toxic to these organisms.

EARTHWORMS.   Fortunately, many pesticides have only mildly depressing effects on earthworm numbers, but there are exceptions. Among insecticides, most of the carbamates (carbaryl, carbofuran, aldicarb, etc.) are highly toxic to earthworms. Among the herbicides, simazine is more toxic than most. Among the fungicides, benomyl is unusually toxic to earthworms. The concentration of pesticides in the bodies of the earthworms are closely related to the levels found in the soil (see Figure 18.6). Thus earthworms can magnify the pesticide exposure of birds, rodents, and other creatures that prey upon them.

Pesticides have significant effects on the numbers of certain predators and, in turn, on the numbers of prey organisms. For example, an insecticide that reduces the numbers of predatory mites may stimulate numbers of springtails which serve as prey for the mites (see Figure 18.7). Such organism interaction is normal in most soils.

Effects on Soil Microorganisms

The overall levels of bacteria in the soil are generally not too seriously affected by pesticides. However, the organisms responsible for nitrification and nitrogen fixation are sometimes adversely affected. Insecticides and fungicides affect both processes more than do most herbicides, although some of the latter can reduce the numbers of organisms carrying out these two reactions. Recent evidence suggests that some pesticides can enhance biological nitrogen fixation by reducing the activity of protozoa and other organisms that are competitors or predators of the nitrogen-fixing bacteria. These findings illustrate the complexity of life in the soil.

Fungicides, especially those used as fumigants, can have marked adverse effects on soil fungi and actinomycetes, thereby slowing down the humus formation in soils. Interestingly, however, the process of ammonification is often stimulated by pesticide use.

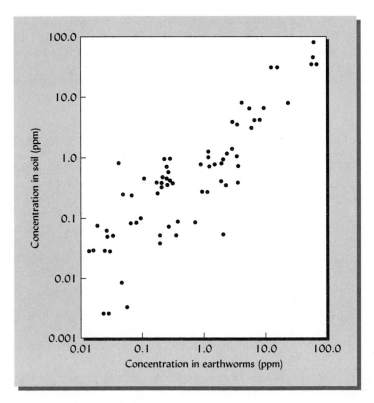

FIGURE 18.6 Effect of concentration of pesticides in soil on their concentration in earthworms. Birds eating the earthworms at any level of concentration would further concentrate the pesticides. [Data from several sources gathered by Thompson and Edwards (1974); used with permission of Soil Science Society of America]

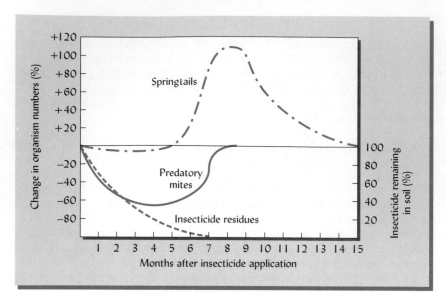

FIGURE 18.7 The direct effect of insecticide on predatory mites in a soil and the indirect effect of reducing mite numbers on the population of springtails (tiny insects) that serve as prey for the mites. [Replotted from Edwards (1978); used with permission of Academic Press, Inc., London]

The negative effects of most pesticides on soil microorganisms are temporary, and after a few days or weeks, organism numbers generally recover. But exceptions are common enough to dictate caution in the use of the chemicals. Care must be taken to apply them only as recommended on their labels.

This brief review of the behavior of organic chemicals in soils reemphasizes the complexity of the changes that take place when new and exotic substances are added to our environment. Our knowledge of the soil processes involved certainly reaffirms the necessity for a thorough evaluation of potential environmental impacts prior to approval of new chemicals for extensive use on the land.

18.5 BIOREMEDIATION OF SOILS CONTAMINATED WITH ORGANIC CHEMICALS[2]

Organic chemical contamination of soils presents serious public health and ecological hazards that must be addressed. The widespread but less acute contamination of soils by intentional pesticide application is probably best addressed by reducing the amounts of pesticides used and by using less toxic, less mobile, and more rapidly degraded compounds. In most cases the soil ecosystem will recover its function and diversity (self-remediation) within a reasonably short time after the offending chemicals cease to be applied at damaging rates. The recovery process can be accelerated by incorporating plenty of organic matter into the soil, using the practices outlined in Section 12.14.

SOIL WASHING AND INCINERATION. *Remediation* of *acute* contamination, such as in industrial waste sites and from accidental leaks and spills of organic chemicals, is a more complicated matter. Industry and government are spending billions of dollars annually to clean up contaminated soils. To date, most of the methods used take an engineering approach that virtually "kills the soil in order to save it." Typical off-site remediation involves excavating huge quantities of contaminated soil, then treating the excavated material in a "soil washer" that removes the toxic organic chemical by means of solvents. Finally, the pollutants are recovered and destroyed by incineration or placed in a secured toxic-waste landfill (see Section 18.9). Other systems essentially incinerate the contaminated soil in order to destroy the contaminant. Sometimes the contaminated soil is landfilled without attempting to cleanse it. These approaches to remediation are extremely expensive and rarely result in a truly recovered soil that resembles a productive natural soil.

BIOREMEDIATION. For many heavily contaminated soils there is a biological alternative to incineration, soil washing, and landfilling—namely, *bioremediation.* Simply put, this technology uses enhanced microbial action to degrade organic contaminants into

---

[2]For a comprehensive treatment of this topic see Alexander (1994).

harmless metabolic products. The process is sometimes carried on in situ (on-site), thus avoiding the high cost and disruptive effects of excavation and of hauling large quantities of soil.

NUTRIENT SUPPLEMENTATION. Bioremediation technology assists natural chemical breakdown in several ways. Usually, the soil naturally contains some bacteria or other microorganisms that can degrade the specific contaminant. But the rate of natural degradation may be far too slow to be very effective. Both growth and metabolic rate of organisms capable of using the contaminant as a carbon source are often limited by insufficient mineral nutrients, especially nitrogen and phosphorus (see Section 12.8 for a discussion of the C/N ratio in organic decomposition). Special fertilizers have been formulated and used successfully to greatly speed up the degradation process.

OIL SPILL CLEANUP. The 30 or more different genera of bacteria and fungi known to degrade hydrocarbons are found in almost any soil or aquatic environment. But they may need help. The cleanup of crude oil contamination from the 1989 Exxon Valdez oil spill in Alaskan waters was a spectacular case of successful bioremediation by fertilization (Figure 18.8). A special fertilizer was sprayed on the oil-soaked beaches (Entisols). The fertilizer was formulated to be oliophilic (soluble in oil but not in water) so that it would stay with the oil and not contribute to eutrophication of Prince William Sound. Within a few weeks, and despite the cold temperatures, most of the oil was degraded. The success of bioremediation was greatest where nitrogen was most available to the microorganisms.

IN SITU TECHNIQUES. Other situations call for other bioremediation techniques. In some cases, low soil porosity causes oxygen availability to be limiting. Techniques are being developed for the use of bioremediation to clean up soils and groundwater contamination in situ. For example, organic-solvent-contaminated soils have been bioremediated (Figure 18.9) by piping in a mixture of air (for oxygen), methane (to act as a carbon source to stimulate specific bacteria), and phosphorus (a nutrient that limits bacteria growth).

Some success has been achieved by inoculating contaminated soils with improved organisms that can degrade the pollutant more readily than can the native population. Although genetic engineering may prove useful in making "superbacteria" in the future, most inoculation has been achieved with naturally occurring organisms. Organ-

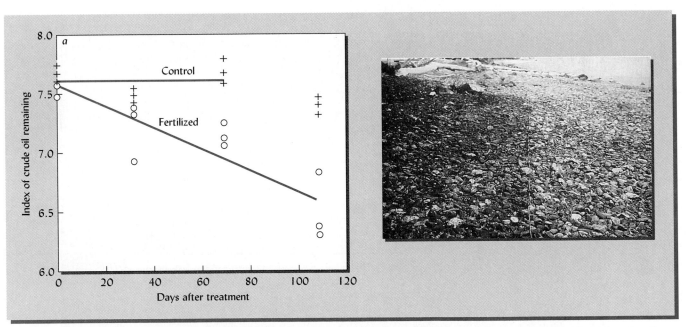

FIGURE 18.8 Bioremediation of crude oil from the Exxon Valdez oil spill off the coast of Alaska. The oil contaminating the beach soils was degraded by indigenous bacteria when an oil-soluble fertilizer containing nitrogen and phosphorus was sprayed on the beach (o data points). The control sections of the beach (+ data points) were left unfertilized for 70 days. By then the effect of the fertilization was so dramatic that a decision was made to treat the control sections as well. The index of oil remaining is based on natural logarithms, so each whole number indicates more than doubling of oil remaining. The photo inset shows the clear delineation between the oil-covered control section and the fertilized parts of the beach. [Data from Bragg (1994); reprinted with permission from *Nature*, © 1994 Macmillan Magazines Limited; photo P. H. Pritchard, USEPA, Gulf Breeze, Fla.; courtesy of Pritchard et al. (1992); reprinted by permission of Kluwer Academic Publishers]

isms isolated from sites with a long history of the specific contamination or grown in lab culture on a diet rich in the pollutant in question tend to become acclimated to metabolizing the target chemical.

18.6 CONTAMINATION WITH TOXIC INORGANIC SUBSTANCES[3]

The toxicity of inorganic contaminants released into the environment every year is now estimated to exceed that from organic and radioactive sources combined. A fair share of these inorganic substances ends up contaminating soils. The greatest problems most likely involve mercury, cadmium, lead, arsenic, nickel, copper, zinc, chromium, molybdenum, manganese, fluorine, and boron. To a greater or lesser degree, all of these elements are toxic to humans and other animals. Cadmium and arsenic are extremely poisonous; mercury, lead, nickel, and fluorine are moderately so; boron, copper, manganese, and zinc are relatively lower in mammalian toxicity. Although the metallic elements (which exclude fluorine and boron) are not all, strictly speaking, "heavy" metals, for the sake of simplicity this term is often used in referring to them. Table 18.6 provides background information on the uses, sources, and effects of some of these elements.

Sources and Accumulation

There are many sources of the inorganic chemical contaminants which can accumulate in soils. The burning of fossil fuels, smelting, and other processing techniques release into the atmosphere tons of these elements, which can be carried for miles and later deposited on the vegetation and soil. Lead, nickel, and boron are gasoline additives that are released into the atmosphere and carried to the soil through rain and snow.

---

[3]For a review of this subject, see Kabata-Pendias and Pendias (1992).

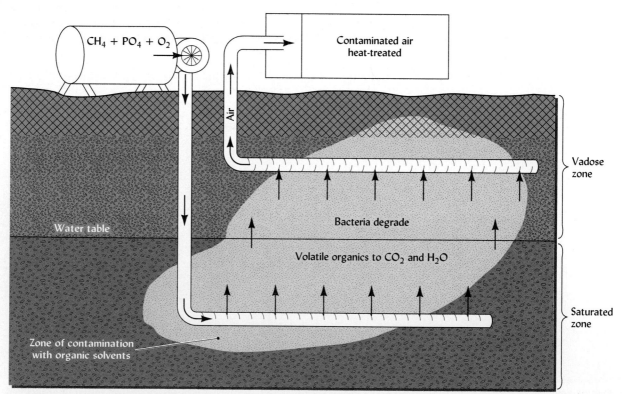

FIGURE 18.9 In situ bioremediation of soil and groundwater contaminated with volatile organic solvents at an industrial site in Georgia. The methane-air-phosphorus mixture was pumped *intermittently* into the soil through slotted pipes, while another pipe "vacuumed" air out of the soil. The air and nutrients stimulated the growth of certain bacteria which, when the methane was taken away, turned to the solvent for a carbon source. It was estimated that the bioremediation technology would cut the time to clean up the site from more than 10 years to less than 4, saving $1.6 million in costs. [Redrawn from Hazen (1995)]

TABLE 18.6　Sources of Selected Inorganic Soil Pollutants

Chemical	Major uses and sources of soil contamination	Organisms principally harmed[a]	Human health effects
Arsenic	Pesticides, plant desiccants, animal feed additives, coal and petroleum; mine tailings and detergents	H, A, F, B	Cumulative poison, possibly cancer
Cadmium	Electroplating, pigments for plastics and paints, plastic stabilizers, and batteries, phosphate fertilizers	H, A, F, B, P	Heart and kidney disease, bone embrittlement
Chromium	Stainless steel, chrome-plated metals, pigments, and refractory brick manufacture, leather tanning	H, A, F, B	Mutagenic, also essential nutrient
Copper	Mine tailings, fly ash, fertilizers, windblown copper-containing dust, water pipes	F, P	Rare, essential nutrient
Lead	Combustion of oil, gasoline, and coal; iron and steel production, solder on water pipe joints	H, A, F, B	Brain damage, convulsions
Mercury	Pesticides, catalysts for synthetic polymers, metallurgy, thermometers	H, A, F, B	Nerve damage
Nickel	Combustion of coal, gasoline, and oil; alloy manufacture, electroplating, batteries, mining	F, P	Lung cancer
Zinc	Galvanized iron and steel, alloys, batteries, brass, rubber manufacture, mining, old tires	F, P	Rare, essential nutrient

Data selected from Moore and Ramamoorthy (1984) and numerous other sources.
[a]H = humans, A = animals, F = fish, B = birds, P = plants.

Borax is used in detergents, fertilizers, and forest fire retardants, all of which commonly reach the soil. Superphosphate and limestone usually contain small quantities of cadmium, copper, manganese, nickel, and zinc. Cadmium is used in plating metals and in the manufacture of batteries. Arsenic was for many years used as an insecticide on cotton, tobacco, fruit crops, lawns, and as a defoliant or vine killer. Some of the above-mentioned elements are found as constituents in specific organic pesticides and in domestic and industrial sewage sludge. Additional localized contamination of soils with metals results from ore-smelting fumes, industrial wastes, and air pollution.

Some of the toxic metals are being released to the environment in increasing amounts, while others (most notably lead, because of changes in gasoline formulation) are decreasing. All are daily ingested by humans either through the air or through food, water, and—yes—soil.

Concentration in Organism Tissue

Irrespective of their sources, toxic elements can and do reach the soil, where they become part of the food chain: soil → plant → animal → human (see Figure 18.10). Unfortunately, once the elements become part of this cycle, they may accumulate in animal and human body tissue to toxic levels. This situation is especially critical for fish and other

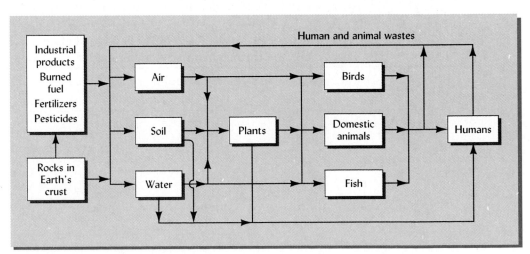

FIGURE 18.10　Sources of heavy metals and their cycling in the soil–water–air–organism ecosystem. It should be noted that the content of metals in tissue generally builds up from left to right, indicating the vulnerability of humans to heavy metal toxicity.

wildlife and for humans at the top of the food chain. It has already resulted in restrictions on the use of certain fish and wildlife for human consumption. Also, it has become necessary to curtail the release of these toxic elements in the form of industrial wastes.

18.7 POTENTIAL HAZARDS OF CHEMICALS IN SEWAGE SLUDGE

The domestic and industrial sewage sludges considered in Chapter 16 can be important sources of potentially toxic chemicals. Nearly half of the municipal sewage sludge produced in the United States is being applied to the soil, either on agricultural land or to remediate land disturbed by mining and industrial activities. Industrial sludges commonly carry significant quantities of inorganic as well as organic chemicals that can have harmful environmental effects.

SOURCE REDUCTION PROGRAMS. A great deal was learned during the 1970s and 1980s about the contents, behavior, and toxicity of metals in municipal sewage sludges. As a result of the research, source reduction programs were implemented which required industries to clean pollutants out of their wastewater *before* sending it to municipal wastewater treatment plants. In many cases, the recovery of valuable metal pollutants was actually profitable for industries. Because of these programs, municipal sewage sludges are much cleaner than in the past (see Table 18.7). Note that the median levels of the most toxic industrial pollutants (Cd, Cr, Pb, and PCB) declined dramatically between the 1976 survey and the 1990 survey. Since much of the copper comes from the plumbing in homes (metallic copper is slightly solubilized in areas with acidic water supplies), that metal has been less affected by the source reduction regulations.

REGULATION OF SLUDGE APPLICATION TO LAND. The lower levels of metals (and of organic pollutants) make municipal sewage sludges much more suitable for application to soils than in the past. Today, the amount of sludge that can be applied to agricultural land is more often limited by the potential for nitrate pollution from nitrogen it contains, rather than by the metal content of the sludge. Nonetheless, application of sewage sludge to farmland is closely regulated to ensure that the metal concentrations in the sludge do not exceed the standards and that the total amount of metal applied to the soil over the years does not exceed the maximum accumulative loading limit listed in Table 18.8. The fact that metal loading standards differ considerably between the United States and Europe (Table 18.8) is an indication that the nature of the metal contamination threat is still somewhat controversial.

TOXIC EFFECTS FROM SLUDGE. The uncertainties as to the nature of many of the organic chemicals found in the sludge, as well as the cumulative nature of the metals problem, dictate continued caution in the regulations governing application of sludge to croplands. The effect of application of a high-metal sludge on heavy metal content of soils and of earthworms living in the soil is illustrated in Table 18.9. The sludge-treated soil areas, as well as the bodies of earthworms living in these soils, were higher in some of

TABLE 18.7 **Median Pollutant Concentrations Reported in Sewage Sludges Surveyed Across the United States in 1976 and 1990 and Uncontaminated Agricultural Soils and Cow Manure**

Pollutant	Sludges surveyed in 1990[a]	Sludges surveyed in 1976[b]	Agricultural soils[d]	Typical values for cow manure
	(mg/kg dry weight)			
As	6	10	—	4
Cd	7	260	0.20	1
Cr	40	890	—	56
Cu	463	850	18.5	62
Hg	4	5	—	0.2
Mo	11	—	—	14
Ni	29	82	18.2	29
Pb	106	500	11.0	16
Zn	725	1,740	53.0	71
PCB	0.21	9[c]	—	0

[a]Data from Chaney (1990).
[b]Data from Sommers (1977).
[c]1976 PCB value is median of cities in New York; from Furr et al. (1976).
[d]Median of 3045 surface soils reported by Holmgren et al. (1993).

TABLE 18.8   Regulatory Limits on Inorganic Pollutants (Heavy Metals) in Organic Wastes Applied to Agricultural Land

Element	Maximum concentration in sludge, USEPA[a] (mg/kg)	Annual pollutant loading rates, USEPA (kg ha^{-1} y^{-1})	Cumulative pollutant loading rates (in kg/ha) USEPA	Germany	Ontario
As	75	2.0	41		14
Cd	85	1.9	39	6	1.6
Cr	3000	150.0	3000	200	210
Cu	4300	75.0	1500	200	150
Hg	840	15.0	300	4	0.8
Mo	57	0.85	17	—	4
Ni	75	0.90	18	100	32
Pb	420	21.0	420	200	90
Se	100	5.0	100	—	2.4
Zn	7500	140	2800	600	330

[a]U.S. Environmental Protection Agency (1993).

these elements than in areas where sludge had not been applied. One would expect further concentration to take place in the tissue of birds and fish, many of which consume the earthworms.

Farmers must be assured that the levels of inorganic chemicals in sludge are not sufficiently high to be toxic to plants (a possibility mainly for zinc and copper) or to humans and other animals who consume the plants (a serious consideration for Cd, Cr, and Pb). For relatively low-metal municipal sludges, application at rates just high enough to supply needed nitrogen seems to be quite safe (Table 18.10).

Direct ingestion of soils and sludge is also an important pathway for human and animal exposure. Animals should not be grazed on sludge-treated pastures until rain or irrigation has washed the sludge from the forage. Children may eat soil while they play, and a considerable amount of soil eventually becomes dust in many households.

18.8   REACTIONS OF INORGANIC CONTAMINANTS IN SOILS

Heavy Metals in Sewage Sludge

Concern over the possible buildup of heavy metals in soils, resulting from large land applications of sewage sludges, has prompted research on the fate of these chemicals in soils. Most attention has been given to zinc, copper, nickel, cadmium, and lead, which are commonly present in significant levels in these sludges. Studies have shown that these elements are generally bound by soil constituents, that they do not easily leach from the soil, and that they are not readily available to plants. Only in moderately to strongly acid soils is there significant movement down the profile from the layer of application of the sludge. Monitoring soil acidity and using judicious applications of lime can prevent leaching into groundwaters and can minimize uptake by plants.

TABLE 18.9   The Effect of Sewage Sludge Treatment on the Content of Heavy Metals in Soil and in Earthworms Living in the Soil

Note the high concentration of cadmium and zinc in the earthworms.

Metal	Concentration of metal (mg/kg) Soil Control	Sludge-treated	Earthworms Control	Sludge-treated
Cd	0.1	2.7	4.8	57
Zn	56	132	228	452
Cu	12	39	13	31
Ni	14	19	14	14
Pb	22	31	17	20

From Beyer et al. (1982).

TABLE 18.10 Uptake of Metals by Corn After 19 Years of Fertilizing a Minnesota Soil (Typic Hapludoll) with Lime-Stabilized Municipal Sewage Sludge

Note that the metals show the typical pattern of less accumulation in the grain than in the leaves and stalks (stover). The annual sludge rate of about 10.5 Mg was designed to supply the nitrogen needs of the corn. The sludge had little effect on the metal content of the plants, except in the case of zinc (which increased, but not beyond the normal range for corn).

Treatment	Zn	Cu	Cd	Pb	Ni	Cr
			(mg/kg)			
Stover						
Fertilizer	18	8.4	0.16	0.9	0.7	0.9
Sludge	46.5	7.0	0.18	0.8	0.6	1.4
Grain						
Fertilizer	20	3.2	0.29	0.4	0.4	0.2
Sludge	26	3.2	0.31	0.5	0.3	0.2
kg/ha						
Cumulative metal applied	175	135	1.2	49	4.9	1045

Data abstracted from Dowdy et al. (1994).

FORMS FOUND IN SOILS TREATED WITH SLUDGE. By using chemical extractants, researchers have found that heavy metals are associated with soil solids in four major ways (Table 18.11). *First,* a very small proportion is held in adsorbed or exchangeable forms which are available for plant uptake. *Second,* the elements are bound by the soil organic matter and by the organic materials in the sludge. A high proportion of the copper is commonly found in this form, while lead is not so highly attracted. Organically bound elements are not readily available to plants, but can be released over a period of time.

The *third* association of heavy metals in soils is with carbonates and with oxides of iron and manganese. These forms are less available to plants than either the exchangeable or organically bound forms, especially if the soils are not allowed to become too acid. The *fourth* association is commonly known as the *residual form,* which consists of sulfides and other very insoluble compounds that are less available to plants than any of the other forms.

It is fortunate that soil-applied heavy metals are not readily absorbed by plants and that they are not easily leached from the soil. However, the immobility of the metals means that they will accumulate in soils if repeated sludge applications are made. Care must be taken not to add such large quantities that the capacity of the soil to react with a given element is exceeded. It is for this reason that regulations set maximum cumulative loading limits for each metal (Table 18.8).

TABLE 18.11 Forms of Six Heavy Metals Found in a Greenfield Sandy Loam (Coarse Loamy, Mixed, Thermic Typic Haloxeralf) That Had Received 45 Mg/ha Sewage Sludge Annually for 5 Years

Forms	Percentage of elements in each form					
	Cd	Cr	Cu	Ni	Pb	Zn
Exchangeable/adsorbed	1	1	2	5	1	2
Organically bound	20	5	34	24	3	28
Carbonate/iron oxides	64	19	36	33	85	39
Residual[a]	16	77	29	40	12	31

From Chang et al. (1984).
[a]Sulfides and other very insoluble forms.

Chemicals from Other Sources

Arsenic has accumulated in orchard soils following years of application of arsenic-containing pesticides. Being present in an anionic form (e.g., $H_2AsO_4^-$), this element is absorbed (as are phosphates) by hydrous iron and aluminum oxides. In spite of the capacity of most soils to tie up arsenates, long-term additions of arsenical sprays can lead to toxicities for sensitive plants and earthworms. The arsenic toxicity can be reduced by applications of sulfates of zinc, iron, and aluminum, which tie up the arsenic in insoluble forms.

Contamination of soils with *lead* comes primarily from airborne lead from automobile exhausts; therefore, it is most concentrated within 100 m of major roadways and near urban centers. Some lead is deposited on the vegetation and some reaches the soil directly. In any case, most of the lead is tied up in the soil as insoluble carbonates, sulfides, and in combination with iron, aluminum, and manganese oxides (Table 18.11). Consequently the lead is largely unavailable to plants.

Soil contamination by *boron* can occur from irrigation water high in this element or by excessive fertilizer application. The boron can be adsorbed by organic matter and clays but is still available to plants, except at high soil pH. Boron is relatively soluble in soils, toxic quantities being leachable especially from acid sandy soils. Boron toxicity is usually considered a localized problem and is probably much less important than a deficiency of the element.

Fluorine toxicity is also generally localized. Drinking water for animals and fluoride fumes from industrial processes often contain toxic amounts of fluorine. The fumes can be ingested directly by animals or deposited on nearby plants. If the fluorides are adsorbed by the soil, their uptake by plants is restricted. The fluorides formed in soils are highly insoluble, the solubility being least if the soil is well supplied with lime.

Mercury contamination of lake beds and of swampy areas has resulted in toxic levels of mercury among certain species of fish. Insoluble forms of mercury in soils, not normally available to plants or (in turn) to animals, are converted by microorganisms to an organic form (methylmercury), in which it is more soluble and available for plant and animal absorption. The methylmercury is concentrated in fatty tissue as it moves up the food chain, until it accumulates in some fish to levels that may be toxic to humans. This series of transformations illustrates how reactions in soil can influence human toxicities.

18.9 PREVENTION AND ELIMINATION OF INORGANIC CHEMICAL CONTAMINATION

Three primary methods of alleviating soil contamination by toxic inorganic compounds are (1) to eliminate or drastically reduce the soil application of the toxins; (2) to immobilize the toxin by means of soil management, to prevent it from moving into food or water supplies; and (3) in the case of severe contamination, to remove the toxin by chemical, physical, or biological remediation.

Reducing Soil Application

The first method requires action to reduce unintentional aerial contamination from industrial operations and from automobile, truck, and bus exhausts. Decision makers must recognize the soil as an important natural resource that can be seriously damaged if its contamination by accidental addition of inorganic toxins is not curtailed. Also, there must be judicious reductions in intended applications to soil of the toxins through pesticides, fertilizers, irrigation water, and solid wastes.

Immobilizing the Toxins

Soil and crop management can help reduce the continued cycling of these inorganic chemicals. This is done primarily by keeping the chemicals in the soil rather than encouraging their uptake by plants. The soil becomes a "sink" for the toxins, and thereby breaks the soil–plant–animal (humans) cycle through which the toxin exerts its effect. The soil breaks the cycle by immobilizing the toxins. For example, most of these elements are rendered less mobile and less available if the pH is kept near neutral or above (Figure 18.11). Liming of acid soils reduces metal mobility; hence, regulations require that the pH of sludge-treated land be maintained at 6.5 or higher.

Draining wet soils should be beneficial since the oxidized forms of the several toxic elements are generally less soluble and less available for plant uptake than are the

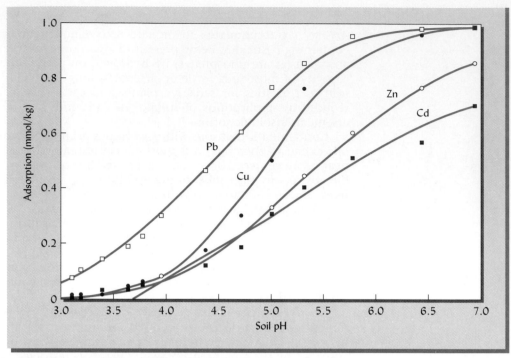

FIGURE 18.11 The effect of soil pH on the adsorption of four heavy metals. Maintaining the soil near neutral provides the highest adsorption of each of these metals and especially of lead and copper. The soil was a Typic Paleudult (Christiana silty clay loam). [From Elliot et al. (1986)]

reduced forms. However, the opposite is true for chromium, which occurs principally in two forms, Cr^{3+} and Cr^{6+}. Hexavalent Cr forms compounds that are mobile under a wide range of pH conditions and that are highly toxic to humans. Trivalent Cr, on the other hand, forms oxides and hydroxides that are quite immobile except in very acid soils. Therefore, it is desirable to reduce Cr^{6+} to Cr^{3+} in chromium-contaminated soils. Fortunately, active soil organic matter is quite effective at reducing chromium, and the Cr^{3+}, once formed, does not tend to reoxidize (see Figure 18.12).

Heavy phosphate applications reduce the availability of some metal cations but may have the opposite effect on arsenic, which is found in the anionic form. Leaching may be effective in removing excess boron, although moving the toxin from the soil to water may not be of any real benefit.

Care should be taken in selecting plants to be grown on metal-contaminated soil. Generally, plants translocate much larger quantities of metals to their leaves than to their fruits or seeds (Table 18.10). The greatest risk for food chain contamination with metals is therefore through leafy vegetables, such as lettuce and spinach, or through forage crops eaten by livestock.

Bioremediation by Metal Hyperaccumulating Plants

Certain plants that have evolved in soils naturally very high in metals are able to take up and accumulate extremely high concentrations of metals without suffering from toxicity. Plants have been found that accumulate more than 20,000 mg/kg nickel, 40,000 mg/kg zinc, and 1000 mg/kg cadmium. While such *hyperaccumulating* plants would pose a serious health hazard if eaten by animals or people, they may facilitate a new kind of bioremediation for metal-contaminated soils.

If sufficiently vigorously growing genotypes of such plants can be found, it may be possible to use them to remove metals from contaminated soils. For example, several plants in the genus *Thalaspi* have been grown in soils contaminated by smelter fumes (Figure 18.13). These soils are so contaminated that they are virtually barren. Accumulating more than 30,000 mg/kg (about 3%) zinc in their tissues, the *Thalaspi* plants grown on this site could be harvested to remove large quantities of the metals from the soil. The plant tissue is so concentrated that it could be used as an "ore" for smelting new metal. This and other bioremediation technologies for metals (e.g., the bioreduction of

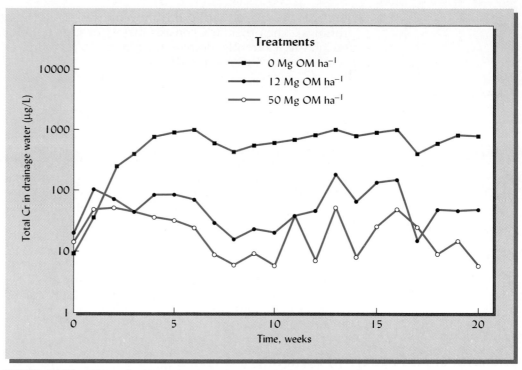

FIGURE 18.12 Effect of adding dried cattle manure (OM) on the concentration of chromium in drainage water from a chromium-contaminated soil. Oxidation of the manure caused the reduction of toxic, mobile Cr^{6+} to relatively immobile Cr^{3+}. Note the log scale for Cr in the water. The coarse-textured soil was a Typic Torripsamment in California. [Adapted from Losi, et al. (1994)]

chromium discussed earlier) hold promise for cleaning up badly contaminated soils without resorting to expensive and destructive excavation and soil-washing methods.

18.10 LANDFILLS[4]

A visit to the local landfill would convince anyone of the wastefulness of modern societies. Roughly 250 million metric tons of municipal wastes are generated each year by people in the United States. Most (about 70%) of this waste material is organic in

---

[4]For a good technical review of landfill design and function, see Baghchi (1994).

FIGURE 18.13 *Thalaspi caerulescens,* a zinc and cadmium hyperaccumulator plant growing in smelter-contaminated soil near Palmerton, Pa. This plant has been reported to accumulate up to 4% zinc in its tissue (dry weight basis). Research with such plants aims at developing technology to biologically remove and recover metals from heavily contaminated soils. (Photo by H. Witham; courtesy of R. Chaney, USDA)

nature, largely paper, cardboard, and yard wastes (e.g., grass clippings, leaves, and tree prunings). The other 30% consists mainly of such nonbiodegradables as glass, metals, and plastic. Currently, despite an upsurge in recycling efforts, the great majority of these materials are buried in the earth.

The Solid Waste Problem

We know that the entire waste disposal problem could be greatly reduced by creating less waste in the first place. Second, it is possible to eliminate most problems associated with waste disposal by two simple measures: (1) keeping the metals, glass, plastics, and paper separate in the household for easy recycling and (2) composting the yard wastes, food wastes, and some of the paper products. The composted product from a number of municipalities is successfully used as a beneficial soil amendment (see Section 16.7). The small fraction of more hazardous wastes remaining can then be detoxified or concentrated and immobilized.

The present reality is that most municipal solid wastes are buried in the earth and will probably continue to be disposed of in this manner for some time to come. In the past, wastes were merely placed in open dumps and often set afire. The term *landfill* came into use because wastes were often dumped in swampy lowland areas where, eventually, their accumulation filled up the lowland, creating upland areas for such uses as city parks and other facilities. Locating landfills on wetlands is no longer an acceptable practice.

Environmental Impacts of Landfills

Today, regulations require that wastes be buried in carefully located and designed sanitary landfills. A major concern with regard to landfills is the potential water pollution from the rainwater that percolates through the wastes, dissolving and carrying away all manner of organic and inorganic contaminants (Table 18.12). In addition to the oxygen-demanding general dissolved organics, many of the contaminants in landfill leachate are highly toxic and would create a serious pollution problem if they reached the groundwater under the landfill.

The production of methane gas by the anaerobic decomposition of organic wastes in a landfill can present a very serious explosion hazard if this gas is not collected (and possibly burned as an energy source). Where the soil is rather permeable, the gas may dif-

TABLE 18.12 Partial List of Organic and Inorganic Contaminants in Untreated Leachate from the City of Guelph (Ontario, Canada) Municipal Solid Waste Landfill

Typical sources of these contaminant in landfills and mechanisms by which soils can attenuate the contaminants are also given. Although leachates vary greatly among landfills, the contaminants in this list are fairly typical.

Chemical	Concentration (µg/L)	Common sources	Mechanisms of attenuation
		Organics	
Chemical oxygen demand (COD)	14,300	Rotting yard wastes, paper, garbage	Biological degradation
Benzene	20	Adhesives, deodorants, oven cleaner, solvents, paint thinner, medicines	Filtration, biodegradation, methanogenesis
Dichloroethane	406	Adhesives, degreasers	Biodegradation, dilution
Toluene	165	Glues, paint cleaners and strippers, adhesives, paints, dandruff shampoo, carburetor cleaners	Biodegradation, dilution
Xylene	212	Oil and fuel additives, paints, carburetor cleaners	Biodegradation, dilution
		Metals	
Nickel	0.38	Batteries, electrodes, spark plugs	Adsorption, precipitation
Chromium	0.14	Cleaners, paint, linoleum, batteries	Precipitation, adsorption, exchange
Cadmium	0.03	Paint, plastics	Precipitation, adsorption

Leachate data from Cureton et al. (1991).

fuse into basements up to several hundred meters away from the landfill. A number of fatal explosions have occurred by this process. Anaerobic decomposition in landfills also emits other harmful gases, the effects of which are less well known.

Although landfill designs vary with the characteristics of both the site and the wastes, two basic types of landfills (see Figure 18.14) can be distinguished: (1) the natural attenuation or unsecured landfill and (2) containment or secured landfill. We will briefly discuss the main features of each.

Natural Attenuation Landfills

The purpose of a natural attenuation landfill is to contain nonhazardous municipal wastes in a sanitary manner, protect them from animals and wind dispersal, and finally to cover them sufficiently to allow revegetation and possible reuse of the site. Some rainwater is allowed to percolate through the waste and down to the groundwater. Natural attenuation processes are relied upon to attenuate the leachate contaminants before the leachate reaches the groundwater. Soils play a major role in these natural attenuation processes through physical filtering, adsorption, biodegradation, and chemical precipitation (see Table 18.12).

SOIL REQUIREMENTS. Finding a site with suitable soil characteristics is critical for this type of landfill. There must be at least 1.5 meters of soil material between the bottom of the landfill and the highest groundwater level. This layer of soil should be only moderately permeable. If too permeable (sandy or gravelly or highly structured), it will allow the leachate to pass through so quickly that little attenuation of contaminants will take place. The soil must have sufficient cation exchange capacity to adsorb the cations (NH^{4+}, K^+, Na^+, and other metallic cations) that the wastes are expected to release. If too slowly permeable, the leachate will build up, flood the landfill, and seep out laterally.

The site for a natural attenuation landfill should also provide soils suitable for daily and final cover materials (Figure 18.14). At the end of every workday the waste must be covered by a layer of relatively impermeable soil material. The final cover for the landfill is much thicker than the daily covers, and includes a thick layer of low-permeability, fine-textured material underneath a thinner layer of loamy "topsoil." The impermeable cover is meant to minimize percolation of water into the landfill, and the topsoil is meant to support a vigorous plant cover that will prevent erosion and use up water by evapotranspiration. The whole system is designed to limit the amount of water percolating through the waste, so that the amounts of contaminated leachate generated will not overwhelm the attenuating capacity of the soil between the landfill bottom and the groundwater.

Containment or Secured Landfills

The second main type of landfill is much more complex and expensive to construct, but its construction is much less dependent on finding a site with suitable soils. The design is intended to contain all leachate from the landfill, rather than depending on the soil for cleansing before the leachate enters the groundwater. To accomplish the containment, one or more impermeable liners are set in place around the sides and bottom of the landfill. These are often made of expanding clays (e.g., bentonite) that swell to a very low permeability when wet. Plastic and tough, nonwoven, synthetic fabric (geotextiles) are also used in making the liners. A layer of gravel or sand is used to protect the liner from accidental punctures, and a system of slotted pipes and pumps is installed to collect all the leachate from the bottom of the landfill. The collected leachate is then treated on or off the site. The principal soil-related concern is the requirement for suitable sources of sand and gravel, for clayey material to form the final cover, and for topsoil to support protective vegetation (see Figure 18.14).

18.11 SOILS AS ORGANIC WASTE DISPOSAL SITES

In the United States nearly 250 million Mg of domestic wastes are generated each year. To this must be added the nearly 2 billion Mg of farm animal wastes, as well as millions of megagrams of organic wastes from food and fiber processing plants and industrial operations.

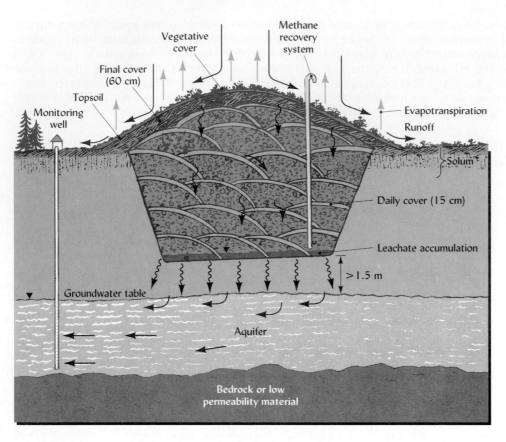

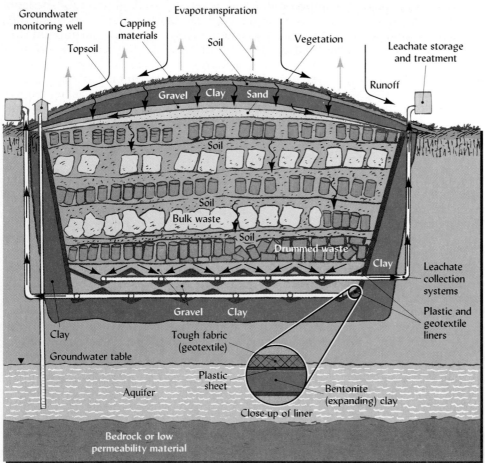

FIGURE 18.14 Two types of landfills. The natural attenuation landfill (top) depends largely on soil processes to attenuate the contaminants in the leachate before they reach the groundwater. The containment-type landfill (bottom) is used for more hazardous wastes or when soil conditions on the site are unsuitable for natural attenuation. It is designed to collect all the leachate and pump it out for storage and treatment.

It is no longer environmentally acceptable to dispose of these wastes by dumping them into waterways or the oceans or by burning, thus releasing reaction products into the atmosphere. The soil offers an alternative disposal sink which is being more and more widely used. These organic wastes can improve soil physical and chemical properties and can provide nutrients for increased plant growth. Such positive effects will likely encourage continued land application of these wastes. At the same time, when wastes are applied in excess quantities, soil productivity may be depressed by salts or soil, and water pollution may occur.

18.12   SOIL SALINITY

Contamination of soils with salts is primarily an agricultural pollution problem. More than 4 million hectares of land are affected in the western United States and many times that amount in other arid regions of the world. Whole civilizations in both the New and Old Worlds are known to have crumbled due to salt buildup in their irrigated soils. The same principles govern the management of irrigated soils today; and the same dangers—salt buildup and concomitant soil deterioration—continue to exist (see Section 10.2).

When more salt moves into the plant rooting zone than moves out, salt accumulation ensues. Salt buildup is often caused by irrigating poorly drained soils or by applying too much irrigation water without providing appropriate artificial drainage to prevent a rise in the water table. The mistakes of the ancients are today still being repeated in large and small irrigation projects the world over (see Figure 19.11 for space satellite documentation of this problem).

The standard solution to the problem of salt buildup is to apply extra irrigation water to promote leaching of the salts from the profile. Unfortunately, as with all such pollution problems, this apparent solution may save the farmers' soils but merely shift the problem somewhere else—namely, to the rivers draining the irrigated areas, causing damage and great expense elsewhere in the environment (Section 10.7). A relatively new approach is to apply a carefully controlled amount of leaching water to move the salts to a depth just below the root zone, where some of the salts will precipitate as relatively insoluble carbonate compounds. This, too, is only a temporary and a partial solution to the problem of salt pollution of soils.

Other sources of soluble salt pollution include manures (especially those from confined animals fed high-salt diets) and sewage sludge when these materials are applied for disposal purposes at levels far higher than needed to supply plant nutrients.

18.13   RADIONUCLIDES IN SOIL

Nuclear fission in connection with atomic weapons testing and nuclear power generation provides another source of soil contamination. To the naturally occurring radionuclides in soil (e.g., ^{40}K, ^{87}Rb, and ^{14}C), a number of fission products have been added. However, only two of these are sufficiently long-lived to be of significance in soils: strontium 90 (half-life = 28 years) and cesium 137 (half-life = 30 years). The average levels of these nuclides in soil in the United States are about 388 millicuries (mC)/km^2 for ^{90}Sr and 620 mC/km^2 for ^{137}Cs. A comparable figure for the naturally occurring ^{40}K is 51,800 mC/km^2. These normal soil levels of the fission radionuclides are not high enough to be hazardous. Even during the peak periods of weapons testing, soils did not contribute significantly to the level of these nuclides in plants. Atmospheric fallout on the vegetation was the primary source of radionuclides in the food chain. Consequently, only in the event of a catastrophic supply of fission products could toxic soil levels of ^{90}Sr and ^{137}Cs be expected. Fortunately, considerable research has been accomplished on the behavior of these two nuclides in the soil–plant system.

Strontium 90

Strontium 90 behaves in soil in much the same manner as does calcium, to which it is closely related chemically. It enters soil from the atmosphere in soluble forms and is quickly adsorbed by the colloidal fraction, both organic and inorganic. It undergoes cation exchange and is available to plants much as is calcium. Contamination of forages and ultimately of milk by this radionuclide is of concern, as the ^{90}Sr could potentially be assimilated into the bones of the human body. The possibility that strontium is involved in the same plant reactions as calcium probably accounts for the fact that high soil calcium tends to decrease the uptake of ^{90}Sr.

Cesium 137

Although chemically related to potassium, cesium tends to be less readily available in many soils. Apparently ^{137}Cs is firmly fixed by vermiculite and related interstratified minerals. The fixed nuclide is nonexchangeable, much as is fixed potassium in some interlayers of clay (see Section 14.15). Plant uptake of ^{137}Cs from such soils is very limited. Where vermiculite and related clays are absent, as in some tropical soils, ^{137}Cs uptake is more rapid. In any case, the soil tends to dampen the movement of ^{137}Cs into the food chain of animals, including humans.

Radioactive Wastes[5]

In addition to radionuclides added to soils as a result of weapons testing and accidents (such as that which occurred at Chernobyl, Ukraine), soils may interact with low-level radioactive waste materials which have been buried for disposal. Even though the materials may be in solid form when placed in shallow land burial pits, some dissolution and subsequent movement in the soil are possible. Plutonium, uranium, americium, neptunium, curium, and cesium are among the elements whose nuclides occur in radioactive wastes.

Nuclides in wastes vary greatly in water solubility, uranium compounds being quite soluble, compounds of plutonium and americium being relatively insoluble, and cesium compounds intermediate in solubility. Cesium, a positively charged ion, is adsorbed by soil colloids. Uranium is thought to occur as a UO_2^{2+} ion that is also adsorbed by soil. The charge on plutonium and americium appears to vary, depending on the nature of the complexes these elements form in the soil.

There is considerable variability in the actual uptake by plants of these nuclides from soils, depending on properties such as pH and organic matter content. The uptake from soils by plants is generally lowest for plutonium, highest for neptunium, and intermediate for americium and curium. Fruits and seeds are generally much lower in these nuclides than are leaves, suggesting that human foods may be less contaminated by nuclides than forage crops.

Since soils are being used as burial sites for low-level radioactive wastes, care should be taken that soils are chosen whose properties discourage leaching or significant plant uptake of the chemicals. Data in Table 18.13 illustrate differences in the ability of different soils to hold breakdown products of two radionuclides. It is evident that monitoring of nuclear waste sites will likely be needed to assure minimum transfer of the nuclides to other parts of the environment.

---

[5]This summary is based largely on papers on this subject in *Soil Science*, **132** (July 1981).

TABLE 18.13   **Concentrations of Several Breakdown Products of Uranium 238 and Thorium 232 (Nucleotides) in Six Different Soil Suborders in Louisiana**

Note marked differences among levels in the different soils.

Soil suborder	No. of samples	^{238}U breakdown products			^{232}Th breakdown products		
		^{226}Ra	^{214}Pb	^{214}Bi	^{212}Pb	^{137}Cs	^{40}K
Udults	22	37.3	27.7	28.9	27.4	16.7	136
Aquults	24	30.4	36.7	38.1	50.0	10.9	100
Aqualfs	37	51.1	38.3	36.6	59.7	13.5	263
Aquepts	93	92.2	47.6	45.2	63.8	16.1	636
Aquolls	57	90.4	45.8	44.7	59.5	8.7	608
Hemists	18	136.3	49.4	49.0	74.9	19.4	783

From Meriwether et al. (1988).

18.14 RADON GAS FROM SOILS

The soil is the primary source of the colorless and odorless radioactive gas *radon*, which has been shown to cause lung cancer. Radon is formed from the radioactive decay of radium, a breakdown product of uranium found in minute quantities in most soils (see Figure 18.15). The health hazard from this gas stems from the transformation of its radioactive decay products into alpha rays, which can penetrate the lung tissue and cause cancer. Radon enters homes and other buildings from the surrounding soil. Since modern airtight buildings permit little exchange of air with the outside, radon can accumulate to harmful levels.

Radon usually moves into the buildings through cracks in the basement walls and floors and around openings where utility pipes enter the basement. If radon tests show that undesirable levels are present, the simplest remedial action is to seal all cracks and points of entry. Additional steps include constructing more elaborate systems to better ventilate the basement with outside air, in order to prevent a buildup of unhealthful levels of radon gas.

Since radon is an inert gas, it does not react with the soil, which merely serves as a channel through which the gas moves. Coarse, gravelly soils are more likely to transfer radon rapidly to basements than would finer-textured soils.

18.15 CONCLUSION

Three major conclusions may be drawn about soils in relation to environmental quality. First, since soils are valuable resources, they should be protected from environmental contamination, especially that which does permanent damage. Second, because of their vastness and remarkable capacities to absorb, bind, and break down added materials, soils offer promising mechanisms for disposal and utilization of many wastes that otherwise may contaminate the environment. Third, soil contaminants and products of their breakdown in the soil reactions can be toxic to humans and other animals if they move from the soil into plant, the air, and particularly into water supplies.

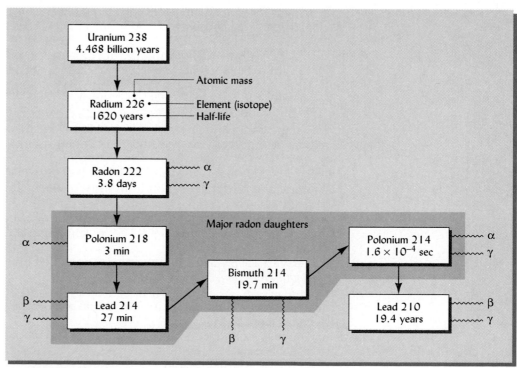

FIGURE 18.15 Radioactive decay of uranium 238 in soils that results in the formation of inert but radioactive radon. This gas emits alpha (α) particles and gamma (γ) rays and forms *radon daughters* that are capable of emitting alpha (α) and beta (β) particles and gamma (γ) rays. The alpha particles damage lung tissue and cause cancer. Radon gas may account for about 10,000 deaths annually in the United States. [Modified from Boyle (1988)]

To gain a better understanding of how soils might be used and yet protected in waste management efforts, soil scientists devote a considerable share of their research effort to environmental-quality problems. Furthermore, soil scientists have much to contribute to the research teams that search for better ways to clean up environmental contamination. Finding appropriate sites where soils can be safely used to clean up or store hazardous wastes involves geographic information about soils, the topic of the next chapter.

REFERENCES

Alexander, Martin. 1994. *Biodegradation and Bioremediation.* (San Diego: Academic Press), 302 pp.

Bagchi, A. 1994. *Design, Construction and Monitoring of Landfills,* 2d ed. (New York: John Wiley & Sons), 361 pp.

Beyer, W. N., R. L. Chaney, and B. M. Mulhern. 1982. "Heavy metal concentration in earthworms from soil amended with sewage sludge," *J. Envir. Qual.,* 11:381–385.

Boyle, M. 1988. "Radon testing of soils," *Env. Sci. Tech.,* 22:1397–1399.

Bragg, J. R., R. C. Prince, E. J. Harner, and R. M. Atlas. 1994. "Effectiveness of bioremediation for the Exxon Valdez oil spill," *Nature,* 368:413–418.

Chaney, R. L. 1990. "Public health and sludge utilization," Part II, *Biocycle* 31(10): 68–73.

Chang, A. C., A. L. Page, J. E. Warneke, and E. Grgurevic. 1984. "Sequential extraction of soil heavy metals following a sludge application," *J. Envir. Qual.,* 13:33–38.

Cheng, H. H. (ed.). 1990. *Pesticides in the Soil Environment: Processes, Impacts, and Modeling* (Madison, Wis.: Soil Science Society of America).

Cureton, P. M., P. H. Groenevelt, and R. A. McBride. 1991. "Landfill leachate recirculation: effects on vegetation vigor and clay surface cover infiltration," *Journ. Environ. Qual.,* 20:17–24.

DMI. 1981. *Farming in the Profit Zone Through Plant Nutrition and Conservation Tillage* (Goodfield, Ill.: DMC Inc.).

Dowdy, R. H., C. E. Clapp, D. R. Linden, W. E. Larson, T. R. Halbach, and R. C. Polta. 1994. "Twenty years of trace metal partitioning on the Rosemount sewage sludge watershed," in C. E. Clapp, W. E. Larson, and R. H. Dowdy (eds.), *Sewage Sludge: Land Utilization and the Environment* (Madison, Wis.: Soil Science Society of America), pp. 149–155.

Edwards, C. A. 1978. "Pesticides and the micro-fauna of soil and water," in I. R. Hill and S. J. Wright (eds.), *Pesticide Microbiology* (London: Academic Press), pp. 603–622.

Elliott, H. A., M. R. Liberati, and C. P. Huang. 1986. "Competitive adsorption of heavy metals in soils," *J. Environ. Qual.,* 15:214–19.

Furr, A. K., A. W. Lawerence, S. S. C. Tong, M. C. Grandolfo, R. A. Hofstader, C. A. Bache, W. H. Gutemann, and D. J. Lisk. 1976. "Multielement and chlorinated hydrocarbon analysis of municipal sewage sludges of American cities," *Envir. Sci. and Tech.,* 10: 683–687.

Gaynor, J. D., D. C. MacTavish, and W. I. Findlay. 1995. "Atrazine and metolachlor loss in surface and subsurface runoff from three tillage treatments in corn," *Jour. Environ. Qual.,* 24:246–256.

General Accounting Office. 1991. *Pesticides—EPA Could Do More to Minimize Groundwater Contamination,* GAO/RCED-91-75 (Washington, D.C.: U.S. General Accounting Office).

Hazen, Terry C. 1995. "Savannah river site—a test bed for cleanup technologies," *Environmental Protection,* April, pp. 10–16.

Holmgren, G. G. S., M. W. Meyer, R. L. Chaney, and R. B. Daniels. 1993. "Cadmium, lead, zinc, copper, and nickel in agricultural soils of the United States of America," *J. Environ. Qual.,* 22:335–348.

Kabata-Pendias, A., and H. Pendias. 1992. *Trace Elements in Soils and Plants* (Boca Raton, Fla.: CRC Press).

Losi, M. E., C. Amrheim, and W. T. Frankenberger, Jr. 1994. "Bioremediation of chromate-contaminated groundwater by reduction and precipitation in surface soils," *Jour. Environ. Qual.,* 23:1141–1150.

Meriwether, J. R., J. N. Beck, D. F. Keeley, M. P. Langley, R. N. Thompson, and J. C. Young, 1988. "Radionuclides in Louisiana soils," *J. Environ. Qual.,* 17:562–568.

Moore, J. W., and S. Ramamoorthy, 1984. *Heavy Metals in Natural Waters* (New York: Springer-Verlag).

Pierzynski, G. M., J. T. Sims, and G. F. Vance. 1994. *Soils and Environmental Quality* (Boca Raton, Florida: CRC Press/Lewis Publishers), 313 pp.

Pimental, D., H. Acquay, M. Biltonen, P. Rice, M. Silva, J. Nelson, V. Lipner, S. Giordano, A. Horowitz, and M. D'Amore. 1992. "Environmental and economic costs of pesticide use," *Bioscience,* **42**:750–760.

Pritchard, P. H., J. G. Mueller, J. C. Rogers, F. V. Kremer, and J. A. Glaser. 1992. "Oil spill bioremediation: experiences, lessons and results from the Exxon Valdez oil spill in Alaska," *Biodegradation,* **3**:315–335.

Sawhney, B. L., and K. Brown (eds.). 1989. "Reactions and movement of organic chemicals in soils," (Madison, Wis.: Soil Sci. Soc. of America) 305 pp.

Sommers, L. E. 1977. "Chemical composition of sewage sludges and analysis of their potential use as fertilizer," J. Environ. Qual. **6**:225–232.

Strek, H. J., and J. B. Weber. 1982. "Adsorption and reduction in bioactivity of polychlorinated biphenyl (Aroclor 1254) to redroot pigweed by soil organic matter and montmorillonite clay," *Soil Sci. Soc. Amer. J.,* **46**:318–322.

Thompson, A. R., and C. A. Edwards. 1974. "Effects of pesticides on nontarget invertebrates in freshwater and soil," in W. D. Guenzi (ed.), *Pesticides in Soil and Water* (Madison, Wis.: Soil Sci. Soc. Amer.), pp. 341–386.

U.S. Environmental Protection Agency. 1993. *Clean Water Act,* Section 503, Vol. 58, No. 32 (Washington, D.C.: USEPA).

Weber, J. B., and C. T. Miller. 1989. "Organic chemical movement over and through soil," pp. 305–34 in B. L. Sawhney and K. Prown (eds.), *Reactions and Movement of Organic Chemicals in Soils,* SSSA Spec. Pub. 22 (Madison, Wis.: Soil Sci. Soc. Amer.).

19

GEOGRAPHIC SOILS INFORMATION

The soil/landscape portrait thus evolved is an artwork of the soil scientist . . .

—L. P. WILDING

The one great constant concerning soils is their variability. Anyone who works intimately with soils soon realizes that soils are anything but uniform. In our discussion of soil profile development (Chapter 2) we focused on vertical variability in soils—the differences among soil horizons. The subject of this chapter is horizontal variability—how soils differ from place to place across the landscape.

In order to make practical use of soil science principles, the land resource manager must know not only the "what" and "why" of soils, but must also know the "where." If builders of an airport runway are to avoid the hazards of swelling clay soils, they must know *where* these troublesome soils are located. An irrigation expert probably knows what soil properties will be necessary for cost-efficient irrigation; but for the project to be a success, he or she must also know *where* soils with these properties can be found. Almost any project involving soils, from planning a game park to fertilizing a farm field, can take advantage of geographic information about soils and soil properties. This chapter provides an introduction to some of the tools that tell us what is where.

19.1 SOIL SPATIAL VARIABILITY IN THE FIELD

A quick review of Figure 7.8 (changing aeration from the surface of a soil granule to its center) and Figure 12.16 (changing soil organic matter from Texas to Minnesota) reminds us that soil properties are variable at all scales. In this chapter we will consider soil variations occurring across distances having geographic meaning for land management—from a few meters to many kilometers.

When attempting to understand geographic variation of soils and how best to use each part of the land, it is often useful to analyze each site in terms of the five factors responsible for soil formation: climate, parent material, organisms, topography, and time.

Small-Scale Soil Variability

Soil properties are likely to change markedly across small distances: within a few hectares of farmland, within a suburban house lot, and even within a single soil individual (as defined in Section 3.1). At this scale, variations are most often due to small changes in topography and thickness of parent material layers or to the effects of organisms (e.g., the effects of individual trees or past human management).

This small-scale variability may be difficult to measure and not readily apparent to the casual observer. In some cases the height and vigor of vegetation reflects the subsurface variability. In other cases the changes in soil properties are detected by analyzing soil samples taken from many evenly spaced borings made throughout the plot of land in question (Figure 19.1). This and other techniques will be discussed in Section 19.2 with regard to the practical use of such observations in making soil maps.

Variability in soil fertility often reflects past soil management practices (see Figure 19.2) as well as differences in soil profile characteristics. Analysis of small-scale variability has practical uses in managing soil fertility for a given field or nursery. As described in Section 16.17, soil tests for fertility management are traditionally performed on one composite sample (a mixture of small cores from 15 to 20 randomly scattered spots) that represents an entire field, or an area as large as 10 to 20 ha.

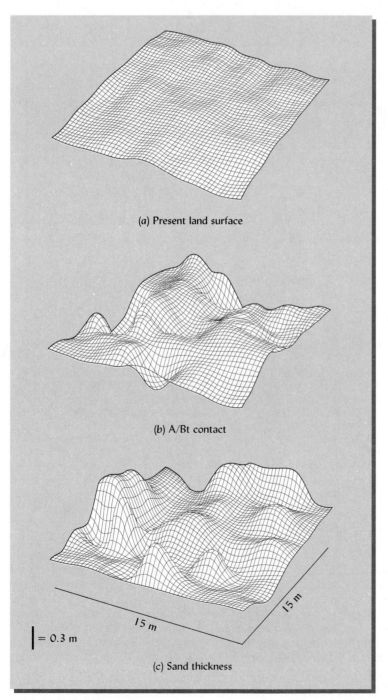

(a) Present land surface

(b) A/Bt contact

= 0.3 m

15 m

15 m

(c) Sand thickness

FIGURE 19.1 Three-dimensional computer drawings of the variability in a small plot of land (15 × 15 m) in southern Texas. Here, an important feature in the soil profile is a layer of very sandy material, probably a relic of an ancient and now buried land surface. (a) The relatively flat, uniform land surface as it would be seen by a person standing on this soil. (b) The wavy boundary between the A and Bt horizons, indicating that the A horizon is two to three times as thick in some spots as in others. (c) The variation of the sand layer below, from virtually nonexistent to about 1 m thick. The graphs were generated from measurements taken at 1-meter intervals in a grid pattern across the small plot of land. [From Wilding (1985)]

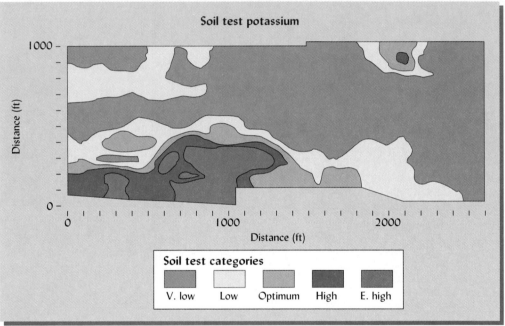

FIGURE 19.2 An oblique air photo (top) of a farm in central Wisconsin. The 22-ha field outlined by the dashed line was studied in detail to make a map (bottom) of the spatial variability of available (soil test) potassium. The map was computer-generated using soil test values from 199 samples (each made up of five subsamples) taken at 32-m intervals in a grid pattern across the entire field. Note that the very high potassium levels correspond with a section of the field closest to the farmstead (lower left in photo). In the past, the farmer had managed that section as a separate field and had used it to dispose of manure from the nearby barnyard. The high-potassium spot in the upper right of the field marks the location of a manure pile that existed a number of years prior to the study. [From Wollenhaupt et al. (1994)]

Where small-scale variability exists, the actual level of fertility at most spots in the field is likely to be either considerably higher or lower than the average soil test value for the field. To visualize the effect of the unaccounted-for variability, consider an analogy with shoe sizes. Fertilizer recommendations based on a field average may fit the needs of the soil about as well as buying everyone in a family size 6 shoes because it happens to be the average shoe size for that family. Fortunately, several technological developments (see Section 19.3) have made it much easier to detect and manage small-scale soil variability.

Medium-Scale Soil Variability

For many soil properties, variability across a landscape is related primarily to differences in a particular soil-forming factor such as soil topography (drainage) or parent material. If one understands the influences of these soil-forming factors in a landscape, it is often possible to define sets of individual soils that tend to occur together in sequence across a landscape. Identifying one member of the set often makes it possible to predict soil properties in the landscape positions occupied by other members of the set. Such sets of soils include *lithosequences* (occurring across a sequence of parent materials), *chronosequences* (occurring across similar parent materials of varying age), and *toposequences* (with soils arranged according to changes in relief).

Well-drained, moderately well drained, and somewhat poorly drained soils are often found together in a characteristic toposequence that defines the patterns of different soils found across a landscape. Where all the soils in a toposequence have developed from the same parent materials, the set of soils that differ on the basis of drainage or due to differences in relief is known as a *catena.* The concept of a soil catena is helpful in relating the soils to the landscape in a given region.

The relationship can be seen by referring to Figure 19.3, where the Bath-Mardin-Volusia-Alden catena is shown. Although all four or five members of a catena are not always found together in a given area, the diagram illustrates the spatial relationship among the soils with respect to their drainage status. As shown, the drainage status of each catena member gives rise to distinct profile characteristics that affect plant rooting depth and species adaptation.

A *soil association* is a more general grouping of individual soils that occur together in a landscape. Soil associations are named after the two or three dominant soils in the group, but may contain several additional, less extensive soils. The soils may be from the same soil order or they may be from different orders (see Figure 19.4a and b). They may have formed in the same or in different parent materials. The only requirement is that the soils occur together in the same area. Identifying soil associations is of practical importance, since they enable us to characterize landscapes over large areas and they assist in planning general patterns of land use. A given soil association represents a defined range of soil properties and landscape relationships, even though the range of conditions included may be quite large (Figure 19.4b).

Large-Scale Soil Variability

At a very large scale, soil patterns are principally the result of climatic and vegetation patterns and secondarily related to parent material differences. Although it is often useful to refer to general regional soil characteristics, it must be remembered that much localized variation exists within each regional grouping. A study of the World Soils Map printed on the endpapers of this textbook and the U.S. Soils Map in Appendix A will reveal important regional patterns. These patterns are also highlighted in the series of small soil order maps shown in Chapter 3. From these maps it can be seen that highly weathered Oxisols are located principally in the hot, humid regions of South America and Africa drained by the Amazon and Congo rivers, respectively. Mollisols can be seen to characterize the semiarid grasslands of the world; Aridisols are located in the desert regions. The great expanse of loess deposits in the central United States (see Figure 2.15) is an example of the regional influence of parent materials. Soils information on this scale can make an important contribution to inventorying the natural resources of a state, province, or nation.

19.2 TECHNIQUES AND TOOLS FOR MAPPING SOILS

Geographic information about soils is often best communicated to land managers by means of a soils map. Soils maps are in great demand as tools for practical land planning and management. Many soil scientists therefore specialize in mapping soils. Their basic task is threefold: (1) to define each soil unit to be mapped (see Section 19.3); (2) to compile information about the nature of each soil; and (3) to delineate the boundaries where each soil unit occurs in the landscape. We will now discuss some of the tools that soil scientists use to delineate soils in the field.

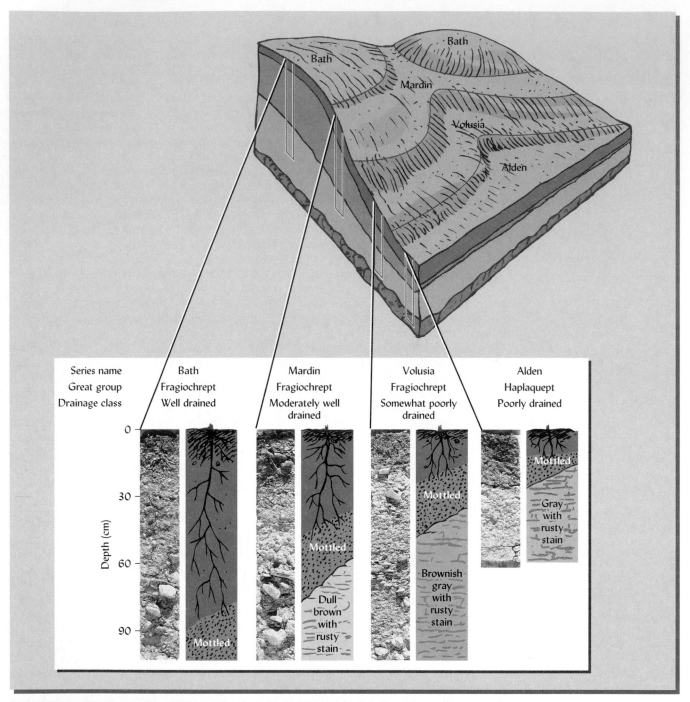

FIGURE 19.3 Profile monoliths of four soils of a drainage catena (below) and a block diagram showing their topographic association in a landscape (above). Note the decrease in the depth of the well-aerated zone (above the mottled layers) from the Bath (well-drained, upslope) to the Alden (poorly drained, downslope). The Alden soil remains wet throughout the growing season. These soils are all developed from the same parent material and differ only in drainage and topography. All four soils belong to the Inceptisols order. With the exception of the cultivated Volusia, the monoliths shown were taken from forested sites. [Based on Cline and Marshall (1977)]

Soil Description[1]

Soil scientists may use computers and satellites, but they also use spades and augers. Despite all the technological advances of recent years, the heart of soil mapping is still the soil pit. A soil pit, whether dug by hand or with a backhoe, is basically a rectangular hole large enough and deep enough to allow one or more soil scientists to enter and

[1]For details on procedures for making soil descriptions, see Soil Survey Staff (1993a) and FAO (1977).

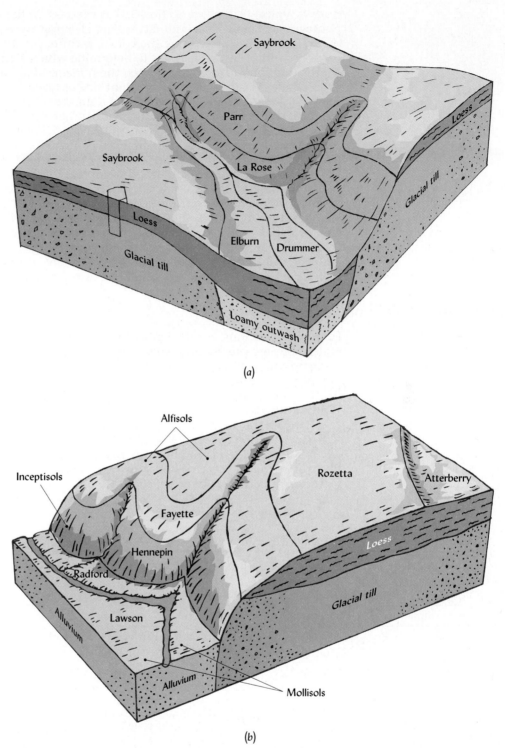

FIGURE 19.4 Two soil associations from Bureau County, Illinois. The soils of the Saybrook–Parr–La Rosa association (a) are all Mollisols. They differ principally with regard to topography and parent material (Parr and La Rosa developed in glacial till, the others mainly in loess). See Table 19.1 for a description of the pedon indicated in the Saybrook soil. The Rozetta–Fayette–Hennepin association (b) includes a wider range of soil conditions and includes soils from three soil orders. [Based on Zwicker (1992)]

study a typical pedon (see Section 3.1) as exposed on the pit face. Color Plates 1–11 are photographs taken of such pit faces. After cleaning away loose debris from the pit face, the soil scientist will examine the colors, texture, consistency, structure, plant rooting patterns, and other soil features to determine which horizons are present and at what depths their boundaries occur. Often the horizon boundaries are marked with a trowel or soil knife, as can be seen on the right side of Color Plate 8.

A soil description is then written in a standard format (see Table 19.1 for an example) that facilitates communication with other soil scientists and comparison with other soils. Sometimes the soil scientist will use field kits to make chemical tests such as those to indicate free carbonate minerals (which effervesce with carbon dioxide if wetted with 10% hydrochloric acid) or soil pH (see Figure 9.20).

As far as possible at this stage, the soil horizons will be given master (A, E, B, etc.) and subordinate (2Bt, Ap, etc.) designations (see Table 2.4). Finally, samples of soil material will be obtained from each horizon. These will be used for detailed laboratory analyses and for archiving. The laboratory analyses will provide information for the chemical, physical, and mineralogical *characterization* of each soil.

In this manner, the soil scientists assigned to map an area will familiarize themselves with the soils they expect to find, learning certain unique characteristics that they can look for to quickly identify each soil and distinguish it from other soils in the area.

Delineating Soil Boundaries

For obvious reasons, a soil scientist cannot dig pits all over the landscape to determine which soils are present and where their boundaries are located. Instead, he or she will bring up soil material from numerous small boreholes made with a hand auger (see Figure 19.5a). The texture, color, and other properties of the soil material from various depths can be compared mentally to characteristics of the known soils in the region. With hundreds of different soils in many regions, this might seem to be a hopeless task.

However, the job is not as daunting as one might suppose, for the soil scientist is not blindly or randomly boring holes. Rather, he or she is working from an understanding of the soil associations and how the five soil-forming factors determine which soils are likely to be found in which landscape positions. Usually there are only a few soils likely to occupy a particular location, so only a few characteristics must be checked. The soil

TABLE 19.1 **Typical Soil Profile Description**

Soil classification		Typic Argiudolls, Saybrook Series
Location of pedon described		79 m west and 375 m south of the northeast corner of section 9, Township 17N, Range 7E[a]

Horizon designation	Horizon boundaries	Description
Ap	0–20 cm	Very dark brown (10 YR 2/2) silt loam, dark grayish brown (10 YR 4/2) dry; moderate medium granular structure; friable; common fine roots; medium acid; abrupt smooth boundary
A	20–36 cm	Very dark brown (10 YR 2/2) silt loam, dark grayish brown (10 YR 4/2) dry; weak medium subangular blocky structure parting to moderate medium granular; friable; common fine roots; slightly acid; clear smooth boundary
Bt1	36–56 cm	Dark yellowish brown (10 YR 4/4) silty clay loam; moderate medium subangular blocky structure; friable; few fine roots; many distinct very dark brown (10 YR 2/2) clay films on faces of peds; slightly acid; clear smooth boundary
Bt2	56–76 cm	Dark yellowish brown (10 YR 4/4) silty clay loam; moderate medium subangular blocky structure; friable; few fine roots; many distinct dark brown (10 YR 3/3) clay films on faces of peds; slightly acid; clear smooth boundary
2Bt3	76–91 cm	Dark brown (7.5 YR 4/4) clay loam; weak medium subangular blocky structure; friable; few fine roots; few distinct dark brown (7.5 YR 4/2) clay films on faces of peds; few pebbles; neutral; clear smooth boundary
2C	91–152 cm	Brown (7.5 YR 5/4) loam; massive; friable; few fine rounded dark accumulations of iron and manganese oxide; few pebbles; violent effervescence; moderately alkaline

Adapted from Zwicker (1992).
[a]The township-range system is used to describe the location of parcels of land in most of the United States.

(a)

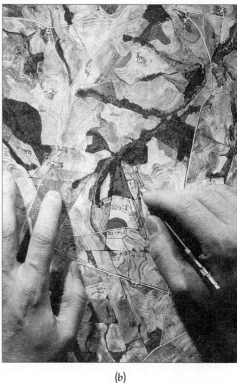

(b)

FIGURE 19.5 Soil maps are prepared by soil scientists who examine the soils in the field using a soil auger and other diagnostic tools (a). Most soil maps are initially prepared by outlining the boundaries of soil mapping units on an air photo (b). A permanent map is then made by superimposing soil boundaries, roads, and other features on an aerial photo map base (c). Each soil mapping unit outlined is identified by a three-part code. For example, in the two small areas (arrows) labeled "145B2," the "145" is the code for the soil series and surface texture phase, in this case a Saybrook silt loam. (On many soil survey maps a two-letter code is used in place of this three-digit one.) The capital letter "B" indicates slopes of 2 to 5%, and the final "2" indicates that the soil is moderately eroded. Thus the areas so marked contain the 2–5% slope, slightly eroded, silt loam surface soil phase of the Saybrook soil series. [Photo (a) courtesy of Diane Shields, USDA/NRSC]

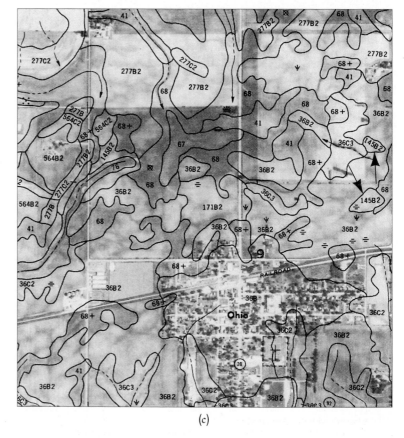

(c)

auger is primarily used to confirm that the type of soil predicted to occur in a particular landscape position is the type actually there.

The nature of soil units and the locations of the boundary lines surrounding them are inferred from information obtained by auger borings at numerous locations across a landscape. A simple, but very laborious and time-consuming approach to obtaining soils information is to make auger borings at regular intervals (say, every 50 m) in a grid pattern across the landscape (Figure 19.6a). Points with similar properties can then be connected to form soil boundaries. This approach is sometimes used in developing countries where labor to survey the sampling points and auger the soils is very inexpensive.

Knowledge of the interplay of the soil-forming factors in a landscape can greatly expedite the soil scientist's work. Making auger borings is time-consuming (and therefore expensive); so an efficient soil mapper will use clues from changes in topography, vegetation, and soil surface colors as a guide in locating sites to make borings. A typical approach is to traverse the landscape along selected transects (straight-line paths), augering only at enough points to confirm expected soil properties and boundaries. In order to pinpoint soil boundaries, more frequent borings are made near the places

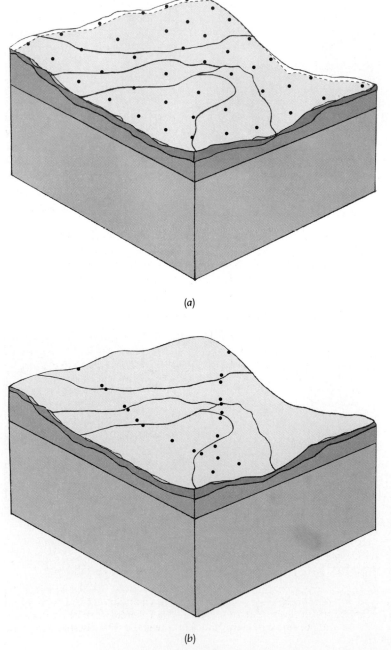

(a)

(b)

FIGURE 19.6 Two approaches to collecting information on soil properties and boundaries by soil auger borings (indicated by dots). The regularly spaced grid pattern (a) of borings is simple in concept, but very labor-intensive to carry out. In a much more efficient approach (b), the soil scientist traverses the landscape along selected transects (straight-line paths), augering only at enough points to confirm soil properties and boundaries predicted on the basis of soil–landscape relationships. Note that in order to pinpoint soil boundaries, extra borings are made near the places these boundaries are expected to occur.

where breaks in slope or other landscape clues suggest that soil boundaries will occur (see Figure 19.6b).

19.3   MODERN TECHNOLOGY FOR SOIL INVESTIGATIONS

While it is still the mainstay of soil investigations and mapping, soil augering is intrusive (i.e., it makes holes) and labor-intensive. Several nonintrusive methods of soil investigation are finding increasing use in helping to identify subsurface features and locate soil boundaries. These technologies include (1) ground-penetrating radar, (2) electromagnetic induction, (3) geopositioning systems, and (4) remote sensing of surface features.

Ground-Penetrating Radar

Ground-penetrating radar (GPR) has come into limited use to increase the quality and reduce the cost of making soil maps. High-frequency impulses of energy are transmitted into the soil. Energy is reflected back to the surface when the impulse strikes an interface between soil particles. This reflected energy is measured and displayed on a recorder. An illustration of the field equipment used and of the resulting graphic profile is shown in Figure 19.7. This technique is not suitable for all soils, since the reflectance is influenced by factors such as moisture, salt content, and type of clay. Where it is effective, however, the cost is only about one-third the cost of detecting soil boundaries by multiple borings with a soil auger.

Electromagnetic Induction[2]

Electromagnetic induction (EM) techniques offer another noninvasive, rapid method of investigating subsurface features. Using a handheld instrument approximately the size and shape of a carpenter's level, this method measures the apparent conductivity of the soil for electromagnetic energy. The conductivity measured is influenced by the moisture content, salinity of the soil (see Section 5.5), and the amount and type of clays in the soil. This technique has been successfully used to map out the depth and thickness of claypan horizons in humid regions and to investigate groundwater contamination and salinity in arid regions.

It should be noted that these electronic methods of soil investigation are not yet available as off-the-shelf technology, but must be adapted to each situation by the user. The user must initially quantify the relationship between the soil properties of interest and the electronic signals recorded by the instrument. A suitable computer program

---

[2]For an example of the use of this technique see Doolittle, et al. (1994).

 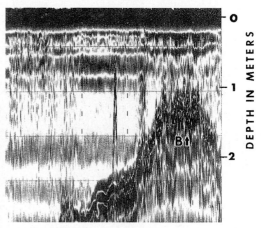

FIGURE 19.7   Ground-penetrating radar in use to investigate Florida soils (left) and the resulting graphic profile (right) showing the depth to the Bt horizon. On the left side of the GPR profile are the deep sands of the Chandler soil, a Quartzipsamment without an argillic horizon in the upper 2 meters. On the right side of the graph, the Bt horizon of an Apopka soil (a Paleudult) can be clearly seen. The boundary between the two soils is gradual but could be delineated at approximately the middle of the graph. The distance across the graph is approximately 140 meters. [Photos from Doolittle (1987)]

may be needed to analyze the data. Once so adapted, these technologies can provide detailed information about subsurface features and can be a great aid in determining the location of soil boundaries.

Global Positioning Systems

An obvious prerequisite for delineating the location of soil bodies in the field is that the soil mappers know where they themselves are located as they traverse a landscape. Traditionally, soil mappers have used large-scale base maps or air photos (see Section 19.5) to ascertain location. However, in nearly featureless terrain or heavily vegetated areas these aids are of limited use. Fortunately, soil mappers can now take advantage of satellite technology to identify precise locations anywhere in the world.

When the United States Department of Defense recently made its Global Positioning System (GPS) available for civilian use, it opened opportunities for many applications in mapping and other soil investigations in the field (see Box 19.1). The GPS

BOX 19.1 USING THE GLOBAL POSITIONING SYSTEM FOR "PRECISION FARMING"

The network of Global Positioning System (GPS) satellites developed by the United States Department of Defense can be used by civilians anywhere in the world to determine their precise locations almost instantaneously. The GPS satellites are constantly orbiting the earth, and their positions relative to a particular ground location are constantly changing. The group of satellites in communication with the receiver at any one time is called the satellite *constellation*.

If the constellation includes three or more satellites, the best civilian receivers currently available can determine a position within 10 to 15 meters. An electronic clock measures the time it takes for radio signals to travel from each satellite to the GPS receiver. Given that the signals travel at the velocity of light, the receiver can calculate the distance to each satellite (distance = velocity × time). For greater accuracy, a stationary receiver can be installed at a known location, such as a fire tower or farm silo. In addition, commercially available *differential corrections* broadcast from FM radio towers can be used by the receiver to correct errors and distortions in the satellite signals, making the system accurate to within approximately 1 to 3 meters.

As discussed in Section 19.3, a GPS capability can improve the speed and accuracy of making geographic soils investigations, including soil maps. The GPS is also being integrated with several other computer-based technologies to improve the precision of fertilizer and pesticide applications on large agricultural fields. The resulting *precision farming* systems attempt to treat each small soil unit according to its particular properties. Where considerable spatial variability exists for the relevant soil properties, precision farming should be more efficient than the traditional method, which averages the properties of all sections of a large field and treats the entire field according to the average properties.

Figure 19.8b shows a "map" of a 22-hectare farm field that was divided into 21 cells of roughly 1 hectare each. A composite soil sample was obtained from the plow-layer soil in the center of each cell, according to standard sampling techniques. A small commercial GPS receiver, mounted on an all-terrain soil-sampling vehicle, was used to determine the location (geographic coordinates) of each soil sample as it was collected (Figure 19.8a). The 21 samples were then analyzed for soil organic matter, pH, and nutrient availability indices by a soil testing lab (see Section 16.16). The results of each type of test were used by a special computer program to produce maps such as the one shown in Figure 19.8 for soil-test phosphorus level. In this case, if a single composite soil sample had been used to determine an average phosphorus soil test level, cells 8 and 9 would probably have been overfertilized (wasting money and nutrients and increasing the potential for environmental damages), while cells 4 and 5, as well as cells 19–21, would probably have been underfertilized (possibly resulting in less than the most profitable yields). For certain herbicides, the optimum rate of application could be similarly affected by variations in such soil properties as amount of clay and organic matter in the soil.

Geographic soils information such as that displayed in Figure 19.8b can be further integrated into a system that will apply fertilizer or other materials (such as pesticides or crop seeds) at rates that vary as the soil varies across a field. Figure 19.8c shows a computerized tractor cab which is equipped with a GPS receiver that continuously indicates the tractor's location as it travels across the field. A computer plots the location of the tractor with respect to the soil properties map in Figure 19.8b. Signals are then sent to automated controllers that can adjust the application rate and mix of different fertilizers to match the varying needs of the soil (Figure 19.8d).

(continued)

---

consists of a network of earth-orbiting satellites that constantly transmit signals that can be used to determine longitude (east-west coordinate) and latitude (distance north or south from the equator) on the ground.

To use signals from the satellite network, one carries a civilian GPS receiver, an instrument not much bigger than a TV remote control. At least two, but preferably three or four, of the GPS satellites must be "in view" of the receiver for it to be able to calculate its position (see Figure 19.8). The coordinates given by the GPS receiver are then transferred to the base map for each soil observation made.

19.4 REMOTE SENSING TOOLS FOR SOILS INVESTIGATIONS[3]

Remote sensing describes the gathering of information from a distance. In this general sense of the term, we are remote-sensing any time we use our eyes to see an object from a distance rather than picking up the object in our hands and feeling it. When we perceive an object, our brains are forming a mental image in response to light energy that has reflected off the object into our eyes. In an analogous manner, a photographic or digital image can be formed by sensors (e.g., cameras) mounted on a platform (e.g., an airplane or space satellite), providing a suitable vantage point for observing a particular area of land. While our eyes respond only to reflected energy with wavelengths in the visible range, other sensors can form images from additional wavelengths of energy, such as infrared energy. We will briefly describe several types of imagery and their uses in making geographic soils investigations. Air photos and other imagery covering most locations in the United States and the world are available from a number of government agencies and private companies (see Table 19.2).

19.5 AIR PHOTOS

Most air photos are made with panchromatic black-and-white film, which "sees" a range of light wavelengths that closely corresponds to the spectrum visible to our eyes. The photograph produced is black and white or, more correctly, many shades of gray. Black-and-white air photos can reveal a wealth of information about landforms, vegetation, human influences, and yes, soils. But experience is required to recognize the various gray tones and patterns as different types of vegetation, drainage patterns, and soil bodies.

Other films, such as natural color or infrared film, are also used for aerial photography. Black-and-white prints from infrared film are commonly used by forest managers because conifer needles absorb infrared energy much more completely than do hardwood leaves, allowing the conifers to be easily distinguished by their darker gray tones on the photograph.

Air photos have been used since 1935 to increase the speed and accuracy of making soil maps. They can be used to assist soil investigations in at least three ways: in providing a base map; as a source of proxy information; and in directly sensing soil properties.

Air Photographs as Base Maps

A sufficiently detailed air photo allows the soil scientist to determine his or her location in the field in relation to features such as buildings, roads, and streams, visible on both

---

[3]For a general discussion of remote sensing, see Lillesand and Kiefer (1994).

FIGURE 19.8 Part of a complex system designed to use the Global Positioning System to aid in detecting and responding to spatial variations in agriculturally significant soil properties. (a) A GPS receiver and computer mounted on an all-terrain vehicle used to collect soil samples. (b) A map of soil test phosphorus levels for 21 1-hectare cells in a 22-hectare farm field. (c) Computer equipment in a tractor cab, designed to integrate positioning information with prior soil test map information and control of fertilizer application rates. (d) Tractor with GPS receiver pulling a fertilizer spreader with computer-controlled variable-rate capability. (Photographs courtesy of Applications Mapping, Inc., Roswell, Georgia)

TABLE 19.2 Partial Listing of Sources for Remote Sensing Imagery

Many other reputable sources exist and may provide equally good products and services.[a]

Type of imagery	Source	Address
Air photos	Bureau of Land Management	BLM-Denver Service Center Division of Technical Services, Bldg. 46 P.O. Box 25047 Denver, CO 80225
	WAC Corporation	520 Conger Street Eugene, OR 97402-2795
	ASCS/USDA	Aerial Photography Field Office P.O. Box 3010 Salt Lake City, UT 84130
Landsat Imagery	EOSAT, Inc.	4300 Forbes Boulevard Lanham, MD 20706
	Earth Satellite Corp.	6011 Executive Blvd., Suite 400 Rockville, MD 20852
SPOT Imagery	SPOT Image Corporation	1897 Preston White Drive Reston, VA 22091
Radar Imagery	Radarsat International	Building D, Suite 200 3851 Shell Road Richmond, BC V6X 2W2, Canada

[a]Neither the authors nor the publishers promote the listed examples of sources in preference to other imagery providers.

the photograph and on the ground. Rather than using surveying equipment and a plane table to make a map from blank paper, the soil scientist can draw soil boundaries directly on the air photo. In this way the photograph serves as a *base map*.

It should be noted that uncorrected air photos are severely distorted because the land areas shown near the edges of the photograph were considerably farther from the camera than were the areas directly beneath the passing airplane. Also, in hilly or mountainous terrain the hilltops would have been closer to the camera than the valleys. An ortho photograph is one that has been corrected for both types of distortion. In most modern published soil surveys ortho photographs are used as base maps on which soil boundaries are drawn.

Air Photos for Proxy Information

Soil investigations are usually concerned with features of soil profiles and other subsurface information. However, an air photo records radiant energy reflected off surfaces above ground or, at best, a few centimeters below ground. Nonetheless the surface tones, patterns, and features shown on the photograph often are related to conditions below ground. Once the soil scientist has learned what these relationships are, air photos can be used as a source of *proxy* information about soil conditions.

For example, the photographs do not directly record the depth to the water table; but dark tones indicate moist, high organic matter surface soil that may correlate with a seasonally shallow water table. If the soil scientist knows that a certain type of soil occurs in the drainageways, then drainageways visible by patterns of dark gray tones on the photograph serve as a proxy for digging a soil pit or boring an auger hole. Stereo pairs of photographs (see Figure 19.9) give the viewer a three-dimensional (but vertically exaggerated) view of the land surface, and are especially useful in locating drainageways, slope breaks, and other features of relief. The features, in turn, help to locate the soils that are members of known soil associations and drainage catenas.

Vegetation often provides clues about the underlying soils. For example, a certain type of vegetation, recognizable on air photos, may grow only in areas of sodic soils

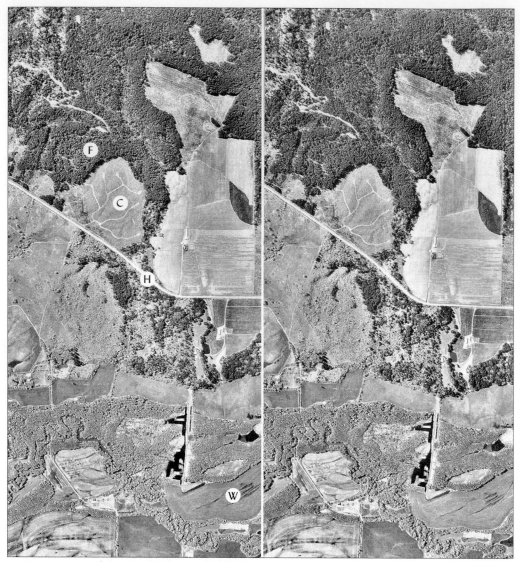

FIGURE 19.9 A stereo pair of air photos providing overlapping coverage of a scene in the Willamette valley of western Oregon. The soils are mostly Humults and Xerolls. The left- and right-hand photos show the same ground area, but were taken when the airplane was in two different positions. If you place a stereoscope over these photos, the two views of the same scene will trick your eyes into seeing the photos three-dimensionally. Note the coniferous forest (F) in the steeper terrain of the upper third of the scene above the highway (H), including a roundish clear-cut area (medium-gray zone, C) with a network of logging roads (white lines with "nodes" indicating leveled areas where logs were loaded onto trucks). In the lower third of the scene, the land is a nearly level river valley, and waterlogged spots (W) appear in the agricultural fields. Soil differences are visible as different gray tones in fields where plowing has exposed bare soils. (Photos courtesy WAC Corporation, Eugene, Oreg.)

with a natric horizon. The patches of this vegetation seen on the photograph then indicate the locations of the sodic soils.

Drainage patterns visible on air photos usually reflect the nature of soils and parent materials. For instance, many closely spaced (less than 1 cm apart on a 1:20,000 air photo) gullies and streams indicate the presence of relatively impervious bedrock and clayey, low-permeability soils. As another example, silty soils developed in loess usually produce a pinnate drainage pattern in which many small gullies and streams branch off from fairly straight larger streams at angles only slightly less than 90°.

Such interpretations can greatly speed the soil investigation by eliminating the need to investigate every mapping unit on the ground. However, the relationships between patterns on the photos and soil properties must be established afresh by ground investigation for each soil association or landscape type. The relationships so determined are likely to hold true only for landscapes in a limited area, typically 500 to 1500 km².

Certain properties of the upper 2 to 20 mm of soil alter the manner in which the soil surface reflects various wavelengths of radiant energy (Figure 19.10). Thus, depending on the wavelengths being monitored, a number of soil properties can be directly sensed and recorded. Soil properties that have been successfully determined by remote sensing include: mineral content; texture; soil water content; soil organic matter content; iron oxide content; albedo (overall reflectivity); temperature; and soil structure. Direct sensing of soil properties usually requires a remote sensor tuned in to specific bands of wavelengths and a computer program designed to interpret the complex data.

An example of computer-assisted interpretation is video image analysis (VIA), by which air photos are examined to distinguish up to 256 shades of gray (compared to only 32 for the human eye). These gray shades or tones are related to soil and vegetation variations. Using appropriate computer techniques, delineation of soil differences can be accomplished. Again, such interpretations of remotely sensed data must be checked by going to the site with an auger and verifying soil types and boundaries (*ground truth*).

19.6   SATELLITE IMAGERY

Many of the principles and techniques just mentioned with regard to air photos apply to satellite imagery as well. However, the imagery available from the sophisticated satellites now in orbit may be more complex and versatile than the types of air photos just discussed. Most satellite images are produced from computer-analyzed digital data obtained by multispectral scanners rather than by film cameras. The images are usually computer enhanced by classification of each cell (pixel) of the image with regard to the type of vegetation, soil cover, land use, or similar theme. This classification is based on computer identification of each land cover spectral fingerprint or pattern of reflectance over different wavelengths (as shown in Figure 19.10). An image may combine data sensed by different instruments, sometimes on different dates. In this manner, combinations of spectra can be used to identify vegetation types, soil properties, cultural features, water, etc.

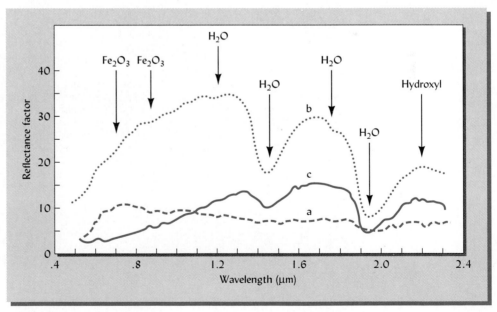

FIGURE 19.10 Spectral curves showing reflection of different wavelengths of energy by three soils. Prominent water and iron absorption bands (low reflection) are indicated by arrows. High iron content correlates with low reflectance in the iron oxide bands, while high organic matter results in low reflectance in the hydroxyl band. The soils represented are: (a) a Typic Haplaquoll with 5.6% organic matter, 0.8% iron oxide, and 41% water; (b) a Typic Calciorthid with 0.6% organic matter, 0.03% iron oxide, and 17% water; and (c) a Typic Haplorthox with 2.3% organic matter, 26% iron oxide, and 33% water. [Adapted from Baumgardner et al. (1985)]

The *resolution*[4] of satellite imagery available today is not as high as that obtained with low-altitude air photos, but it has steadily improved over the years. Imagery with a resolution as high as 10 meters is now available from the French satellite Système Probatoire de la Observation de la Terre (SPOT). Figure 19.11 illustrates the improved resolution between the early and current Landsat satellite technologies. The older image (made in 1973) has a relatively low resolution of 80 meters, while the newer image (from 1990) can resolve objects as small as 30 meters. Note that the higher resolution

---

[4]The resolution of an image is said to be high if very small objects can be discerned.

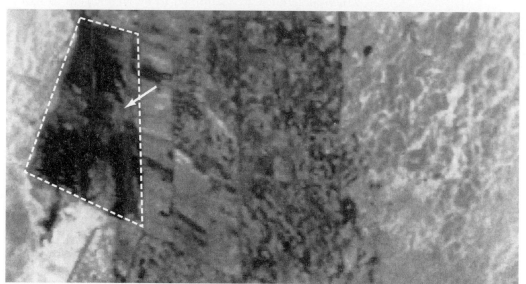

Landsat Multispectral Image
Bands 1-2-4   28 June 73

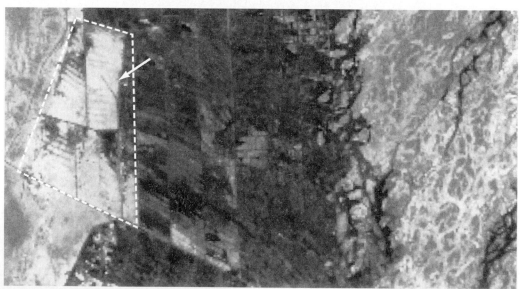

Landsat Thematic Mapper Image
Bands 2-3-4   27 April 90

FIGURE 19.11   A black-and-white reproduction of two false-color Landsat satellite images showing a portion of an irrigation project in western Marja, Afghanistan. The upper photo is a Landsat multispectral image obtained in 1973; the lower photo is a Landsat Thematic Mapper digital image obtained in 1990. Each image used three different spectral bands. The highly reflective white tones indicate barren, salinized land. Black areas indicate irrigated cropland that has become waterlogged. Medium-gray tones (red on the original image) indicate well-drained, irrigated cropland. In 1973 an area of approximately 1000 hectares (outlined) appeared to be cultivated but waterlogged, probably as a result of blocked drainage outlets. By 1990 the same area appears to be abandoned and salinized. (Photos courtesy of Earth Satellite Corporation, Rockville, Md.)

results in a clearer, more detailed image. The difference in resolution is most obvious in distinguishing the drainage canals (arrow).

This pair of Landsat images highlights the capability of multispectral imagery to portray soil conditions. The soil parent materials in the area are naturally high in soluble salts. Where irrigation without proper drainage raised the water table (evidenced by the dark tones of waterlogged soil in 1973), evaporation from groundwater caused salts to accumulate in the surface soils (as seen by the white, barren soils in 1990). The lack of proper maintenance of the drainage channels may have been related to the political upheaval and war experienced in the country of Afghanistan.

The capability of satellite imagery to clearly portray landforms and vegetation over a wide area is illustrated by Figure 19.12. Examine this image closely. It is a black-and-white reproduction of a false-color Landsat Thematic Mapper image of the Palo Verde valley where the Colorado River forms the border between California and Arizona. The image is a composite made from three different spectral bands. Rugged mountainous terrain and alluvial fans are visible surrounding a nearly level, irrigated valley. The arid landscape is virtually bare of vegetation, except in rectangular fields where irrigation water has been applied. The small circles in the upper left are center-pivot irrigation systems, which result in lush green vegetation (red on the original image) in circular fields almost 1 km in diameter. Note the grid of streets that is the town of Palo Verde, near the center of the irrigation project.

The outlined rectangular portion of the Palo Verde valley image is enlarged as Color Plate 14. Much can be learned by closely examining this color plate. Notice the different blue-gray colors indicating soil differences visible in bare fields. Bright red areas are indicative of dense, green crops. The wavy lines of red to the east of the Colorado River indicate natural vegetation and weeds growing where gullies and drainageways have collected the little rainwater that falls in this arid region. Many bare fields are highlighted by yellow outlines which were added to the image by a GIS program (see Section 19.9). The yellow-outlined fields are those that the City of Los Angeles paid the owners to keep bare, so that the city could obtain the water that would otherwise have been used for irrigating crops. The satellite image was used to verify that the owners of these fields were complying with this agreement.

A final example of satellite imagery is shown in Color Plate 15, which is a Landsat Thematic Mapper image of the Potomac River basin from Washington, D.C., south, including parts of southern Maryland and northern Virginia. Major land use patterns are clearly visible. The urbanized land of Washington and its suburbs is shown in blue-gray tones (the United States Capitol building grounds are visible as a white spot surrounded by a dark square of lawn). An enormous amount of sediment (yellow and brown colors) appears to be entering the Potomac River from streams draining the southern suburbs, possibly because of poor sediment control at construction sites. Forests appear green, most farm fields appear light yellow (mature corn and soybeans). The red-colored areas indicate Histosols and other hydric soils of the tidal marshes along the Potomac River and the Chesapeake Bay estuary. The bay itself is seen as the dark blue body of water at the far upper right. An image such as this is very useful in resource inventory and general regional land planning.

Clearly the range of imagery and technology available for soil investigations is rapidly growing, presenting challenges and opportunities for those soil scientists specializing in geographic information about soils.

19.7 SOIL SURVEYS

A **soil survey** is more than simply a soils map. The glossary describes a *soil survey* as "a systematic examination, description, classification, and mapping of the soils in a given area." Under some circumstances, soil scientists make special-purpose soil surveys which do not attempt to delineate and describe natural soil bodies, but merely aim to map the geographic distribution of selected soil properties such as suitability for an irrigation project or conformation to a legal definition of wetlands.

However, soil surveys are most useful if they characterize and delineate natural soil bodies. Once the natural bodies are delineated and their properties described, the soil survey will be useful in making interpretations for all kinds of soil uses, not just those uses which were intended at the time the soil survey was conducted.

FIGURE 19.12 A black-and-white reproduction of a Landsat Thematic Mapper image of Palo Verde valley on the border between California and Arizona. The image is a composite made from bands 2 (green), 3 (red), and 4 (near infrared). The outlined rectangle is reproduced in Color Plate 14. (Photo courtesy of Earth Satellite Corporation, Rockville, Md.)

The basic steps in making a soil survey are:

1. *Mapping of the soils.* Described in Sections 19.1–19.3.
2. *Characterization of the mapping units.* See Section 19.2.
3. *Classification of the mapping units.* See Chapter 3 and Section 19.2.
4. *Correlation with other soil surveys.* Once a team of soil scientists completes a soil map, the map is reviewed by other soil scientists to verify that the soil boundaries match those mapped for adjacent areas and that the characterization and classification of mapping units are consistent with other soil surveys.
5. *Interpretation of soil suitability for various land uses.* A report is written to accompany the soils map in order to describe the suitability of each mapping unit for various land uses. The interpretative tables in the report often reflect many person-years of experience and observation, as well as standard interpretations of measured soil properties.

Soil surveys may be conducted at different *orders* or levels of detail, ranging from very detailed surveys of small parcels of land (first order) to general reconnaissance surveys of very large regions or entire continents (fifth order). Different kinds of *mapping units* and remotely sensed data sources are used to produce different orders of soil surveys (see Table 19.3).

Map Scales

The *scale* of a map is the ratio of length on the map to actual length on the ground. A scale of 1:20,000 is commonly used for detailed soil maps and indicates that 1 cm on the map represents a distance of 20,000 cm (or 0.2 km) on the ground. The ratio is unitless, therefore 1 inch on the same map represents 20,000 inches (0.316 miles) on the ground.

The terms *large scale* and *small scale* sometimes confuse people. A *small-scale* map is one with a small *scale ratio* (e.g., 1:1,000,000 = 0.000001), on which a given object, such as a 100-ha lake, occupies only a tiny spot on the map. By contrast, a *large-scale* map has a large *scale ratio* (e.g., 1:10,000 = 0.0001), and a 100-ha lake would occupy a relatively large part of the map.

TABLE 19.3 **Different Orders of Soil Surveys**

Soil surveys may be conducted at various scales or orders, ranging from very detailed surveys of small parcels of land to general surveys of very large regions. Different kinds of mapping units and remotely sensed data sources are used in producing different orders of soil surveys. The guidelines given in this table should be considered flexible and approximate.

	Order of Soil Survey				
	5th order	4th order	3d order	2d order	1st order
Type of survey	Reconnaissance		Semidetailed	Detailed	Intensive
Survey scale	1:250,000–1:10,000,000	1:50,000–1:300,000	1:20,000–1:65,000	1:12,000–1:32,000	1:000–1:15,000
Size of mapping unit	2.5–500 km²	15–250 ha	1.5–15 ha	0.5–4 ha	Smaller than 0.5 ha
Typical components of map units	Orders, suborders, great groups	Great groups, subgroups, families	Families, series, phases of series	Soil series, phases of series	Phases of soil series
Kind of map unit	Associations, some consociations, and undifferentiated groups	Associations, complexes, some consociations	Associations or complexes, some consociations	Consociations and complexes, few associations	Mostly consociations, some complexes
Remote sensing sources	←————————Landsat Thematic Mapper digitized data————————→				
	←————————SPOT image digital data————————→				
			←————High-altitude aerial photography————→		
				←————Low-altitude aerial photography————→	
Use of soil survey in land planning	←————Resource inventory————→				
		←————Project location————→			
			←————Feasibility surveys————→		
				←————Management surveys————→	

Adapted from Baumgardner et al. (1985) and Soil Survey Staff (1993b).

Soil Taxonomy (or some other classification system; see Chapter 3) is usually the basis for preparing a soil survey. Because local features and requirements will dictate the nature of the soil maps and, in turn, the specific soil units that are mapped, the field *mapping units* may be somewhat different from the *classification units* found in *Soil Taxonomy*. The mapping units may represent some further differentiation below the soil series level—namely, *phases* of soil series; or the soil mappers may choose to group together similar or associated soils into conglomerate mapping units. Examples of such soil mapping units follow.

SOIL PHASE. Although technically not included as a class in *Soil Taxonomy,* a *phase* is a subdivision based on some important deviation that influences the use of the soil, such as surface texture, erosion, slope, stoniness, or soluble salt content. Thus, a Cecil sandy loam, 3–5% slope, and a Hagerstown silt loam, stony phase, are examples of phases of soil series.

CONSOCIATIONS.   The smallest practical mapping unit for most detailed soil surveys is an area that contains primarily one soil series and usually only one phase of that soil series. For example, a mapping unit may be labeled as the consociation "Saybrook silt loam, 2 to 5% slopes, moderately eroded." Quality-control standards for county soil surveys indicate that a consociation mapping unit should be 50% "pure" and that the "impurities" should be so similar to the named phase that the differences do not affect land management. That is to say, if you walk out on the land to an area mapped as the above-mentioned consociation and make 20 auger borings, at least 10 of the borings should reveal properties within the range defined for the Saybrook silt loam. However, it would be expected that a few of the borings would indicate a similar soil, such as the Catlina silt loam, in which the loess layer is somewhat thicker than for the Saybrook soil but for which land use interpretations are the same. Inclusions of *contrasting* soils should occupy less than 15% of the consociation.

COMPLEXES OR SOIL ASSOCIATIONS.   Sometimes *contrasting* soils occur adjacent to each other in a pattern so intricate that the delineation of each kind of soil on a soil map becomes difficult, if not impossible. In such cases, a soil *complex* is indicated on a soil map, and an explanation of the soils present in the complex is contained in the soil survey report. A complex often contains two or three distinctly different soil series. As described in Section 19.1, relatively large-scale soil maps (e.g., third order) may display only soil associations—general groupings of soils that typically occur together in a landscape.

UNDIFFERENTIATED GROUPS.   These units consist of soils that are *not* consistently found together, but are grouped together because their suitabilities and management are very similar for common land uses.

19.8   THE COUNTY SOIL SURVEY REPORT AND ITS UTILIZATION

In the United States an effort has been under way for more than 50 years to complete a detailed soil survey of the entire country, on a county-by-county basis. In some states, many counties have yet to be mapped. Other counties were mapped decades ago and their soil surveys are being updated. This soil survey, known as *The National Cooperative Soil Survey,* is an ongoing joint effort of federal, state, and local governments. The principal federal role is played by U.S. Department of Agriculture Natural Resources Conservation Service (formerly the Soil Conservation Service). Sometimes, other federal government agencies such as the Bureau of Land Management or the Forest Service are also involved. The state-level cooperating agency is usually the Agricultural Experiment Station associated with the land grant university of that state. We will now consider the typical U.S. county soil survey report and its uses.

A modern county soil survey report consists of two major parts: the *soil map* and accompanying *descriptive information* on the soil mapping units and their suitability for various land uses. In effect, the descriptive part of the report serves as a very elaborate key to explain the soil map.

The soil map usually consists of two parts. The first part is a fourth-order *General Soil Map* of the entire county, printed in color at a scale of approximately 1:200,000 (Figure

19.13a). This map shows the soil associations grouped into the main physiographic regions of the county. The General Soil Map is useful in providing an overview of the county land resources, but is too general to be used for site planning. Associated with the General Soil Map is an *index map* for the detailed map sheets. This index map is printed at the same scale as the General Soil Map and is divided into numerous rectangular sections, each representing a detailed map sheet. The index map shows enough cultural features, such as roads and towns, to enable a user to locate the map sheet covering the area of interest.

The second part of the county soils map is a detailed (second-order) soil map consisting of many individual map sheets (Figure 19.13b), usually folded and bound into the back of the Soil Survey Report. Each map sheet covers an area of approximately 20 to 30 km², typically at a scale of 1:15,840 (4 inches = 1 mile), 1:20,000 (1 cm = 0.2 km), or 1:24,000 (1 cm = 0.24 km or 1 inch = 2000 feet).[5] At these scales, individual trees and houses are discernable on the air photo map base, and areas of soil as small as 1 hectare can be delineated (Figure 19.13c). The mapping units are mostly consociations consisting of soil phases. As such, these detailed maps are extremely useful for site planning.

Use of the county soil survey for land use or site planning requires that the geographic information on the map be integrated with the descriptive information in the report. The report usually contains detailed soil profile descriptions for all of the mapping units, as well as tables that provide soil characterization data and interpretive rankings for the mapping units. Older soil surveys offer interpretive information on yield potentials for various locally important crops, on the suitability of soils for different irrigation methods, drainage requirements for soils, land capability classification of each mapping unit, and other information useful in making farm plans. Newer soil surveys also provide interpretive information on many nonagricultural land uses, including wildlife habitat, forestry, landscaping, waste disposal, building construction, and source of roadbed materials.

As an example of how the soil survey report may be used, consider the parcel of land shown in the photograph in Figure 19.13d. This area is also included in the air photo stereo pair shown in Figure 19.14. Study Figure 19.14, preferably with a stereoscope, and compare the topography observed with that shown in the photograph in Figure 19.13d. Locating this scene on the detailed map sheet (Figure 19.13c), we can determine that the parcel contains five mapping units encoded as Cu, LeB2, LeD2, LfE, and NeB2 (see Figure 19.5 for an explanation of this coding system). These codes correspond to the five mapping units described in Table 19.4, which also lists a few of the many interpre-

---

[5]To facilitate digitization using feet and inches, most soil survey updates are being produced at scales of 1:12,000 or 1:24,000.

TABLE 19.4 A Sampling of Some of the Interpretive Information Provided by the Soil Survey of Harford County Area, Maryland

The four mapping units described are those in the site shown in Figure 19.13d. The interpretations included are some of those that would be of interest for developing the site as a county park.

| Mapping unit code | Name of mapping unit | Land use Capability Classification[a] | Limitations for use of soil | | | Suitability for open land wildlife |
			Camp areas	Septic filter fields	Paths and trails	
Cu	Codorus silt loam	IIw-7	Severe—flood hazard	Severe—high water table, flood hazard	Moderate—flood hazard	Good
LeB2	Legore silt loam, 3 to 8% slopes, moderately eroded	IIe-10	Slight	Slight	Slight	Good
LeD2	Legore silt loam, 15 to 25% slopes, moderately eroded	IVe-10	Severe—slope	Severe—slope	Moderate—slope	Fair
LfE	Legore very stony silt loam, 25 to 45% slopes	VIIs-3	Severe—slope	Severe—slope	Severe—slope	Poor
NeB2	Neshaminy silt loam, 3 to 8% slopes, moderately eroded	IIe-4	Slight	Moderate—mod. permeability	Slight	Good

Data abstracted from Smith et al. (1975).
[a]See Section 17.16 for an explanation of the USDA Land Capability Classification System.

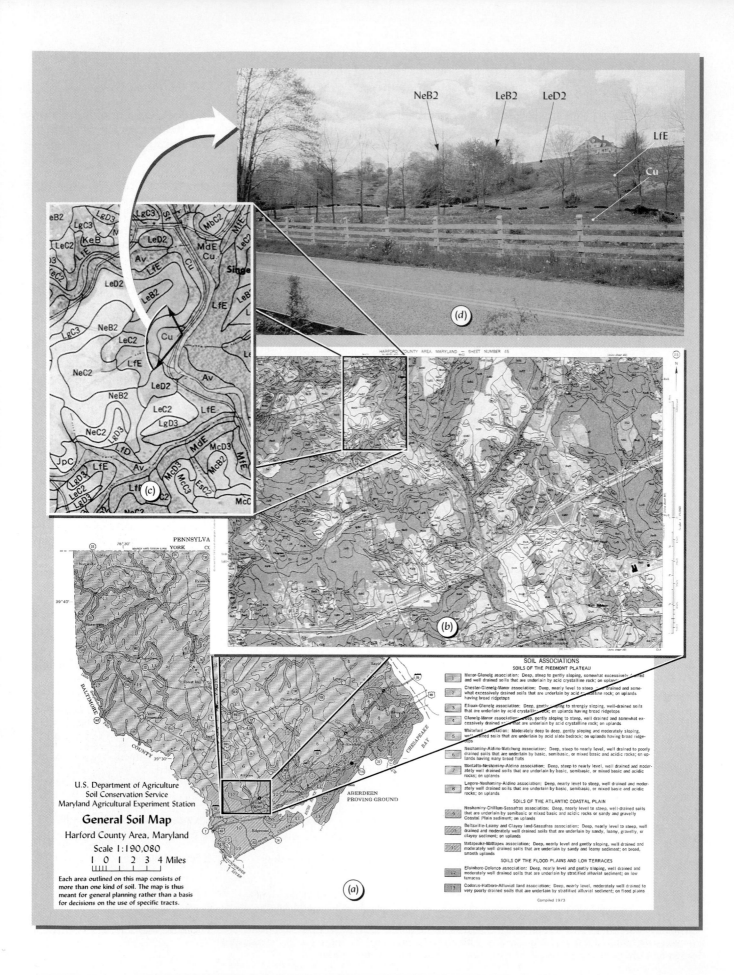

SOIL ASSOCIATIONS

SOILS OF THE PIEDMONT PLATEAU

1   Manor-Glenelg association: Deep, steep to gently sloping, somewhat excessively drained and well drained soils that are underlain by acid crystalline rock; on uplands

2   Chester-Glenelg-Manor association: Deep, nearly level to steep, well drained and somewhat excessively drained soils that are underlain by acid crystalline rock; on uplands having broad ridgetops

3   Elioak-Glenelg association: Deep, gently sloping to strongly sloping, well-drained soils that are underlain by acid crystalline rock; on uplands having broad ridgetops

4   Glenelg-Manor association: Deep, gently sloping to steep, well drained and somewhat excessively drained soils that are underlain by acid crystalline rock; on uplands

5   Whiteford association: Moderately deep to deep, gently sloping and moderately sloping, well drained soils that are underlain by acid slate bedrock; on uplands having broad ridgetops

6   Neshaminy-Aldino-Watchung association: Deep, steep to nearly level, well drained to poorly drained soils that are underlain by basic, semibasic, or mixed basic and acidic rocks; on uplands having many broad flats

7   Montalto-Neshaminy-Aldino association: Deep, steep to nearly level, well drained and moderately well drained soils that are underlain by basic, semibasic, or mixed basic and acidic rocks; on uplands

8   Legore-Neshaminy-Aldino association: Deep, nearly level to steep, well drained and moderately well drained soils that are underlain by basic, semibasic, or mixed basic and acidic rocks; on uplands

SOILS OF THE ATLANTIC COASTAL PLAIN

9   Neshaminy-Chillum-Sassafras association: Deep, nearly level to steep, well-drained soils that are underlain by semibasic or mixed basic and acidic rocks or sandy and gravelly Coastal Plain sediment; on uplands

10   Beltsville-Loamy and Clayey land-Sassafras association: Deep, nearly level to steep, well drained and moderately well drained soils that are underlain by sandy, loamy, gravelly, or clayey sediment; on uplands

W   Matapeake-Mattapex association: Deep, nearly level and gently sloping, well drained and moderately well drained soils that are underlain by sandy and loamy sediment; on broad, smooth uplands

SOILS OF THE FLOOD PLAINS AND LOW TERRACES

12   Elsinboro-Delance association: Deep, nearly level and gently sloping, well drained and moderately well drained soils that are underlain by stratified alluvial sediment; on low terraces

13   Codorus-Hatboro-Alluvial land association: Deep, nearly level, moderately well drained to very poorly drained soils that are underlain by stratified alluvial sediment; on flood plains

Compiled 1973

U.S. Department of Agriculture
Soil Conservation Service
Maryland Agricultural Experiment Station

General Soil Map

Harford County Area, Maryland

Scale 1:190,080

1 0 1 2 3 4 Miles

Each area outlined on this map consists of more than one kind of soil. The map is thus meant for general planning rather than a basis for decisions on the use of specific tracts.

FIGURE 19.14 A stereo pair of air photos including the area shown on the soil map in part (c) of Figure 19.13. Major features to notice are the dam and reservoir in the upper left, the small housing development in the middle left, the stream (Winters Run) flowing below the reservoir, and the hill and floodplain in the center of the photo. The latter features are the same as those shown from ground level in Figure 19.13d. (Air photos courtesy of USDA Agricultural Stabilization and Conservation Service)

tive ratings given in the soil survey report. If you were planning to develop a park on this site, the information in Table 19.4 would suggest that the Cu mapping unit would be suitable for a nature trail, but not for a visitor center with rest rooms.

For complex land planning analyses, the enormous quantity of information stored in a county soil survey report can perhaps be used to best advantage with the help of a computerized geographic information system (GIS). We will now briefly consider how a GIS works.

19.9 GEOGRAPHIC INFORMATION SYSTEMS[6]

An information system is a series of organized steps used to handle information or data, including the plan for the collection of data, actual collection of the data, manipulation of the data, and finally use of the data in making a product such as a report or map. In a geographic information system (GIS), the data is tied to spatial locations. For example the properties of a soil profile are associated with the location of the appropriate soil mapping units. A GIS includes the following five steps: (1) data acquisition, (2) preprocessing, (3) data management, (4) manipulation and analysis of data, and (5) product generation.

The soil mapping process described in Section 19.4 could be part of the data acquisition step, as could the collection of existing soil maps, reports, and zoning ordinances.

[6]For a comprehensive introduction to GIS, see Star and Estes (1990).

◀ FIGURE 19.13   The soil map of Harford County, Maryland, an example of the soil map in a modern county soil survey. (a) One part of the soil map is the small-scale General Soil Map (originally at a scale of 1:190,080, in this case) reproduced here at much-reduced scale. (b) The detailed soil map consists of the collection of many large-scale map sheets (originally at a scale of 1:15,840, in this case), one of which (map sheet 45) is shown here at reduced scale. (c) A rectangular area of approximately 1.1 km² is outlined on the detailed map sheet and reproduced at the original scale, so that the mapping units and features on the air photo map base are visible. (d) The scene showing the road in the foreground and the house on the hill in the background was photographed from the location indicated where the two arrows originate in (c). (Maps from Smith and Mathews (1975); photo courtesy of R. Weil)

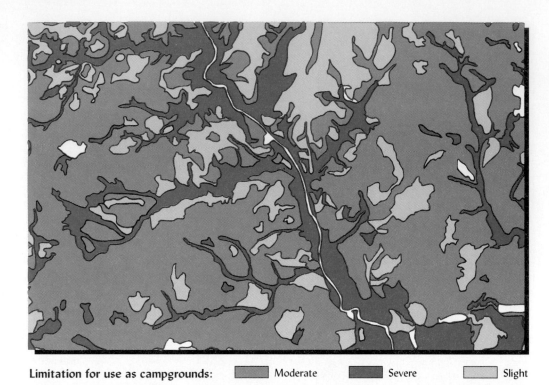

Limitation for use as campgrounds: ▨ Moderate ▨ Severe ▨ Slight

FIGURE 19.15 A simple interpretive soil map covering the same area as map sheet 45 from Harford County, Maryland. The ratings of limitations from tables in the soil survey report have been associated with each of the thousands of soil polygons on the map sheet. The dark gray areas along the major streams are either too steep or too wet to be favorable for campgrounds. This map is a single layer in that it contains only information on spatial distribution of campground limitations. It does not show roads, streams, or other data (nonsoil areas are left blank). (Courtesy of Margaret Mayers, University of Maryland)

This data will have to be preprocessed to be put into a form that can be integrated with other data. Data acquisition and preprocessing can be very time-consuming and expensive. An example of preprocessing is the digitization of an existing soils map, entering all the soil boundaries and other spatial features into a computer database as *points, lines,* and geometrically defined areas called *polygons*. This process may be accomplished using a scanner, but often must be done by hand using a digitizing palette. The descriptive information, such as that illustrated in Table 19.4, also must be entered into the database. Currently, only a few county soil surveys are available in digital form, but as soil surveys are updated many are being digitized.

Once all the necessary information has been entered into the database, it can be managed with computer programs and manipulated to create new types of information and insights. Finally, a product such as a map or report is produced. A number of complex computer programs have been designed to carry out the last three steps in a coordinated manner. Figure 19.15 shows a simple example of a product of such a system. The map in this figure is based on information taken from the detailed soil map sheet shown in Figure 19.13b. Although digitizing and entering the descriptive and geographic information for this single map sheet required more than 100 hours, subsequent computer-generation of the map of campground limitations took only a few minutes. The map in Figure 19.15 has simplified a set of more than 90 different mapping units by grouping them into three simple, spatially displayed categories. The same information is already available in the soil survey report tables, but not in a spatially displayed form.

The above-mentioned map (Figure 19.15) comprises a single *layer* of information. A GIS can be used to integrate many layers of soils information from a soil survey report. As a hypothetical example of the use of a GIS, consider the following scenario. A local agency wished to find a suitable site for a public park in one of 20 large tracts of land delineated in the area covered by detailed map sheet 45 (from Figure 19.13b). The planners decided that the site for the park must meet the following four criteria:

1. Only slight limitations for picnic areas
2. Only slight limitations for paths and trails
3. Land capability class greater than II (to avoid using prime agricultural land that should be preserved for food production)
4. Woodland subclass 20, indicating a site highly productive for hardwood trees

The boundaries of every mapping unit polygon on map sheet 45 were entered into the GIS database, and each polygon was associated with the appropriate interpretive and descriptive data from the soil survey report. The GIS computer program searched for mapping unit polygons that met all four of the above criteria and printed a map with shading used only for the qualifying polygons. Finally, the boundaries of the 20 tracts of land, as well as the locations of streams, were plotted on the map. The resulting map is shown in Figure 19.16. The uppermost tract just right of center appears to contain the largest area of land meeting the criteria for the proposed park.

In many land planning projects, soil properties comprise just one of several types of geographic information that must be integrated in order to take best advantage of a given site or to find the site best suited for the proposed land use. Planning the best use of different farm fields, finding a site suitable for a sanitary landfill, and designing a wildlife refuge are some examples of projects that could make good use of both soils and

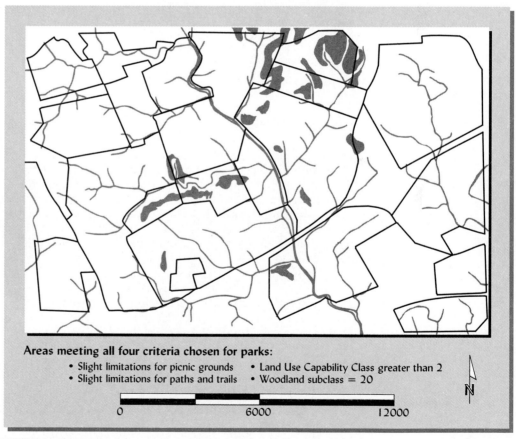

Areas meeting all four criteria chosen for parks:
- Slight limitations for picnic grounds
- Slight limitations for paths and trails
- Land Use Capability Class greater than 2
- Woodland subclass = 20

0   6000   12000

FIGURE 19.16   A map of a portion of Harford County, Maryland (map sheet 45), showing areas determined to be suitable sites for a new park according to the four criteria listed. The land use capability class restriction (LUCC > 2) avoids taking prime farmland out of production (see Section 17.16). The property of "park suitability" is a new attribute created by the geographic information system. In addition to "park suitability," two other layers of geographic information are displayed: (a) streams (light, branched lines) and (b) boundaries of 20 hypothetical tracts of land (heavy, straighter lines). (Courtesy of Margaret Mayers, University of Maryland)

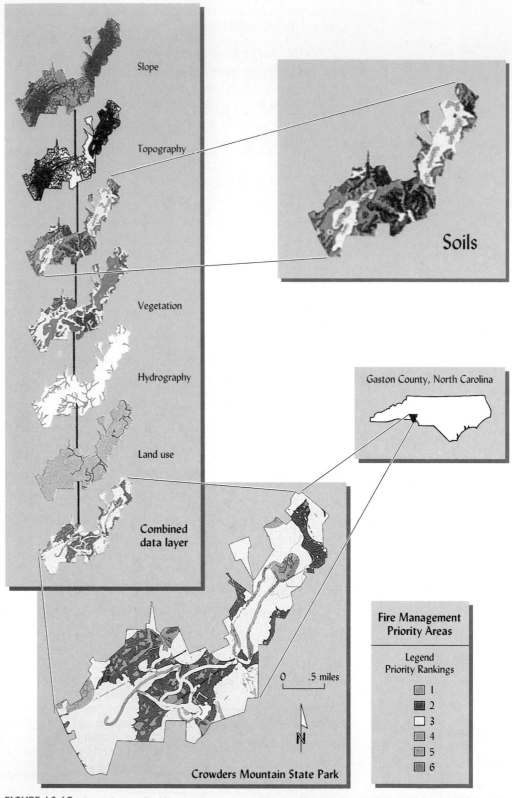

Slope

Topography

Soils

Vegetation

Hydrography

Gaston County, North Carolina

Land use

Combined
data layer

Fire Management
Priority Areas

Legend
Priority Rankings

1
2
3
4
5
6

0   .5 miles

N

Crowders Mountain State Park

FIGURE 19.17 In order to develop management plans for fighting forest fires in Crowders Mountain State Park in North Carolina, spatial information about soil properties was combined with other types of spatial information using a Geographic Information System computer program designed to overlay various types of maps. In this way soils information is integrated into the final plan as one of many components. [Modified from Gronlund et al. (1994) with permission of GIS World, Inc.]

nonsoils geographic information. Nonsoils information to be considered might include topography, streams, vegetation, and present land use. A GIS can be used to combine all these types of information, giving greatest weight to those factors that the planner deems most critical.

An example of such an integrated GIS model is shown in Figure 19.17. Here, the objective was to develop a plan for fighting forest fires in a recreational area. In order to predict where fires are most likely to break out, where different types of equipment and fire-fighting techniques are appropriate, and where a wildfire would likely cause the greatest damage, a number of factors must be considered. The nature of the land determines most of these factors. A number of maps showing the spatial distribution of soils and nonsoils factors were overlaid in order to produce a fire management priorities map. The final map differentiated zones within the park according to their priority in a fire management plan.

As more soils information becomes available in digital form, its use in making land planning decisions based on GIS analyses will undoubtedly become increasingly common. The quality of those decisions will continue to depend on the quality of the geographic soils information provided by soil scientists and the quality of the decision criteria developed by planners and their clients.

19.10 CONCLUSION

Making soil surveys is both a science and an art by which many soil scientists apply their understanding of soils and landscapes to the real world. Mapping soils is not only a profession; many would say that it is a way of life. Working alone outdoors in all kinds of terrain and carrying all the necessary equipment, the soil scientist collects ground truth to be integrated with data from satellites and laboratories. The resulting soil maps and descriptive information in the soil survey reports are used in countless practical ways by soil scientists and nonscientists alike. The soil survey, combined with powerful geographic information systems, enables planners to make rational decisions about what should go where. The challenge for soil scientists and concerned citizens is to develop the foresight and fortitude to use criteria in the GIS planning process that will help preserve our most valuable soils—not hasten their destruction under shopping malls and landfills.

REFERENCES

Baumgardner, M., L. Silva, L. Biehl, and E. Stoner. 1985. "Reflective properties of soils," *Advances in Agronomy,* **38**:1–44.

Cline, M. G., and R. L. Marshall. 1977. "Soils of New York Landscapes," Inf. Bull. 119, Physical Sciences, Agronomy 6, New York State College of Agriculture and Life Sciences, Cornell University.

Doolittle, J. A. 1987. "Using ground-penetrating radar to increase the quality and efficiency of soil surveys," in W. U. Reybold and G. W. Peterson (eds.), *Soil Survey Techniques,* SSSA Spec. Publ. No. 20 (Madison, Wis.: Soil Sci. Soc. Amer.).

Doolittle, J. A., K. A. Sudduth, N. R. Kitchen, and S. J. Indorante. 1994. "Estimating depths to claypans using electromagnetic induction methods," *Journ. Soil Water Conserv.,* **49**:572–575.

FAO 1977. *Guidelines for Soil Profile Description* (2d ed.) (Rome, Italy: Land and Water Development Div., Food and Agriculture Organization of the United Nations), 66 pp.

Gronlund, A. G., W. N. Xiang, and J. Sox. 1994. "GIS, expert systems technologies improve forest fire management techniques," *GIS World,* **7(2)**:32–36.

Lillesand, T. M., and R. W. Kiefer. 1994. *Remote Sensing and Image Interpretation,* 3d ed. (New York: John Wiley and Sons).

Smith, H., and E. Matthews, 1975. *Soil Survey of Harford County Area, Maryland* (Washington, D.C.: U.S. Soil Conservation Service).

Soil Survey Staff. 1993a. *National Soil Survey Handbook,* Title 430 VI (Washington, D.C.: U.S. Department of Agriculture).

Soil Survey Staff. 1993b. *Soil Survey Manual,* Handbook No. 18 (Washington, D.C.: U.S. Department of Agriculture).

Star, J., and J. Estes. 1990. *Geographic Information Systems: An Introduction* (Englewood Cliffs, N.J.: Prentice Hall).

Wilding, L. P. 1985. "Spatial variability: its documentation, accommodation and implication to soil surveys," in D. R. Nielsen and J. Bouma (eds.), *Soil Spatial Variability*, Pudoc, Wageningen, The Netherlands, pp. 166–194.

Wollenhaupt, N. C., R. P. Wolkowski, and M. K. Clayton. 1994. "Mapping soil test phosphorus and potassium for variable rate fertilizer application," *Jour. Prod. Agric.*, 7:441–448.

Zwicker, S. E. 1992. *Soil Survey of Bureau County, Illinois* (Washington, D.C.: U.S. Natural Resources Conservation Service).

20

SOILS AND THE WORLD'S FOOD SUPPLY

The basic problem is that God stopped making land some time ago, but He still is making people . . .
—ANTHONY BAILEY

Hunger is not new to the world. It has always been a threat to human survival. Through the centuries at some place on earth scarcity of food has brought human misery, disease, and death. But never in recorded history has the threat of widespread hunger been greater than it is today. This threat is not due to the reduced capacity of the world to supply food. Indeed, this capacity is greater today than it has ever been, and is continuing to grow at a reasonable rate. The problem lies in the even more rapid rate at which world population is increasing. World food production *per person* is barely holding its own. In some regions, it is declining.

Rapid increases in population have been matched during the past 30 years with unprecedented increases in food production, especially in many of the low-income countries. Except for sub-Saharan Africa, per capita food production has increased throughout the developing world. While some of this increase occurred because of expansion of arable land areas, most was due to improved technologies. Improved crop varieties, coupled with large increases in irrigation and chemical fertilizer use stimulated food production. The so-called green revolution proved that developing-country farmers would accept and use improved technologies if it were to their advantage to do so.

Unfortunately during the past few years this rapid *rate of increase* of food production has declined, now barely equaling the rate of population growth worldwide, and being decidedly below population growth rates in some countries. This slowdown is due to several factors. For economic and environmental reasons, the cutting of forests and burning of natural grasslands to make room for agriculture has slowed. Likewise, new irrigated projects are more expensive, and they face opposition from those who may be displaced to accommodate the projects. While scientists continue to produce new and improved crop varieties, the quantum jump in yields experienced in the 1960s has already been reached, and further increases are more difficult to achieve. Last, except for sub-Saharan Africa, fertilizer use is generally already sufficiently high that investments in more fertilizer are not as profitable as has been the case in the past.

These factors influencing food supplies are balanced, and in some cases overbalanced, by factors influencing the consumption of food—the most important of which is human population numbers, which we will now consider briefly.

20.1 EXPANSION OF WORLD POPULATION[1]

Science is in part responsible for the marked expansion in world population growth, especially that which has occurred in the developing nations. Until about the midpoint of the 20th century, high birth rates in most of South America, Africa, and Asia were largely negated by equally high death rates. High infant mortality, poor health facilities, inadequate medical personnel, and disease-spreading insects took their toll. Population expansion was held in check.

After World War II, advances in medical science and their application throughout the world changed this situation. Death rates were drastically reduced, especially among the young. Medical services became available in remote areas of the developing nations. The result was unprecedented population growth. The population is now doubling every 20 to 30 years in many developing areas, where two-thirds of the world's population now live (Figure 20.1). Experts predict the world's population will be between 8 and 9 billion by the year 2025—nearly 25% higher than it is today. Furthermore, *about 90% of this increase will occur in developing nations* where food supplies are already critical and where the technologies for increased food production are wholly inadequate. The world food supply continues to be among humanity's most serious problems. As one political scientist (Paarlberg, 1994) recently described the situation, "Farmers in poor countries will have to produce more food over the first 20 years of the next century than they have produced during the entire 10,000 years since sedentary farming first began."

For hungry, low-income people, the implications of the balance between food production and population numbers are quite sobering. During the 25 years of remarkable success (1969–1985), the *proportion* of the populations in the developing countries who were malnourished dropped from about 27 to 21.5%. However, the *absolute number* of malnourished increased from 460 to 512 million. The need to improve that situation is a pressing challenge.

[1]A stimulating discussion of the relationship between population growth and food production is given in FAO (1984).

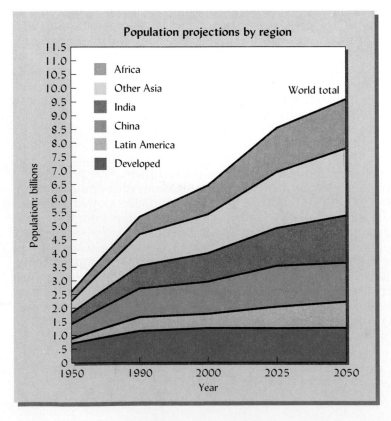

Population projections by region

Africa
Other Asia
India
China
Latin America
Developed

World total

FIGURE 20.1   From the beginning of the human race until 1960 the world's population increased to about 3 billion. Less than 40 years will be needed to provide the second 3 billion. The total is expected to rise to 8.5 billion by the year 2025. Note that essentially all the growth is in the developing countries and regions that are already pressed to provide food for their growing populations. [From UNFPA (1992)]

20.2 FACTORS AFFECTING WORLD FOOD SUPPLIES

The ability of a nation to produce food is determined by many social, economic, and political factors that affect the farmer's incentive and ability to produce. Food production is also affected by physical and biological factors such as the following:

1. The natural resources available, especially soil and water
2. Available technologies, including the knowledge of proper management of plants, animals, and soils
3. Improved plant varieties and animal breeds that respond to proper management
4. Supplies of production inputs such as manures, lime, fertilizers, and irrigation water

How each of these factors apply to food production depends on the area and quality of soils, their natural productivity, and response to management. There is good reason to place satisfactory soil properties high on the list of requisites for an adequate world food supply.

20.3 THE WORLD'S LAND RESOURCES

There is a total of more than 13 billion ha of land on the major continents (Table 20.1), but most of it is not suited for cultivation. About half the land is completely nonarable. It is mountainous, too cold, or too steep for tillage; it is swampland; or it is desert that is too dry for any but the sparsest vegetation.

About one-fourth of the land areas support enough vegetation to provide grazing for animals, but for various reasons cannot be cultivated. This leaves only about 25% of the land with the physical potential for cultivation (Table 20.1), and less than half of this potentially arable land is actually being cropped.

Continental Differences

There is great variation from continent to continent in the availability of potentially arable land (Table 20.1). In Asia and Europe, where population pressures have been strong for years, most of the potentially arable land is already under cultivation. In contrast, only 17% of the potentially arable land is cultivated in South America, 25% in Africa, and 31% in Oceania, including Australia and New Zealand. In these last three areas the physical potential for greater utilization of arable land is high.

Caution on Use of Potentially Arable Lands

The abundance of uncultivated but potentially arable lands will likely have relatively less than the desired impact on world food supplies. In the first place, the land areas are

TABLE 20.1 Population and Cropped Land on Each Continent, Along with Cropland Per Person and Percent of Potentially Arable Land That Was Cropped in 1991

Area	Population in 1990 (millions)	Area (million ha)			Cropland per person (ha)	Potentially arable land cropped (%)
		Total	Potentially arable	1991 cropland		
Africa	642	2964	733	182	0.27	25
Asia	3113	2679	627	458	0.14	73
Europe	498	473	174	138	0.27	79
North America	427	2138	465	272	0.69	59
South America	297	1753	680	115	0.38	17
Former U.S.S.R.	289	2190	356	228	0.81	65
Oceania	27	845	154	48	1.79	31
Total	5295	13,042	3189	1442	0.27	45

Data for potentially arable land from President's Science Advisory Committee Panel on World Food Supply (1967); all other from *World Resources* 1994–1995.

not always located where the people are. The vast areas in the Amazon River basin in South America and the Congo River basin in Africa, for example, are far removed from the nearest regions with heavy population pressures. The great distances and expense involved in transporting food produced in these areas to cities and towns where people live limit the practical potential of these areas for food production and utilization.

Second, these river basins are the homes of tropical forests that are so vital in the maintenance of biological diversity. These forested areas contain millions of species of plants and animals that would be destroyed if the forests were removed and the land cultivated for agriculture. Consequently there are sound environmental as well as economic reasons for conserving the natural forests (and natural grasslands) and looking to lands already under cultivation to provide most of the future world food needs.

Third, removal of the natural vegetation in humid regions invariably leads to increased runoff and erosion. Aside from the resultant destruction of soils, downstream floods and the pollution of domestic water supplies are the consequences. In large areas of the tropics, the clearing of sloping and hilly lands has already resulted in serious erosion damage. In some mountainous areas, land now used for crops should be allowed to return to forests and other natural vegetation—a sequence that has been followed in much of Europe and the eastern part of the United States.

A last reason for using caution before destroying forests for agricultural purposes relates to global warming. The destruction of forests releases vast quantities of carbon dioxide into the atmosphere, enhancing the so-called greenhouse effect that could result in increased global warming (see Section 12.2). Since there are viable alternatives to agricultural land expansions as means of providing future world food needs, the clearing of tropical forests should be an alternative of last resort.

Special constraints seem to apply to food production in Africa. The distances between areas with high food-producing potential and population centers are long, and the transportation systems to bring food to markets and to transport needed inputs such as fertilizers are poorly developed. Furthermore, erratic rainfall distribution and frequent droughts make it difficult to fully realize the potential of soils to produce food. Unfortunately, population numbers are rising so rapidly that the per capita food production in Africa is declining (Figure 20.2). Were it not for outside supplies of purchased or donated food, human hunger that is already all-too-prevalent would be truly rampant.

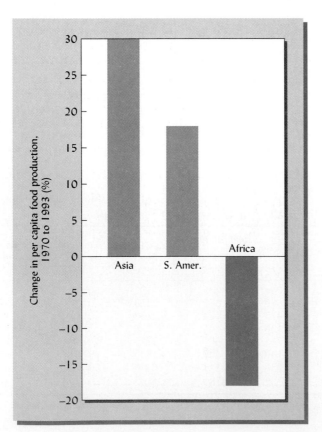

FIGURE 20.2 Percentage change in per capita food production in several regions (1970 to 1993). Note that Africa's per capita food production fell dramatically over this period. [Drawn from FAO data in *World Resources* (1994–1995)]

Choices of Action

There are three routes that nations may follow to increase their food production: (1) clear and cultivate arable land that has heretofore not been tilled, (2) increase cropping intensity (number of crops per year), or (3) intensify production on lands already under cultivation. During the past 30 years most of the increase in production came from increases in yields per hectare and increased cropping intensities (Table 20.2). However, for some developing countries in Africa and Latin America a significant amount came from expansion in land area.

Some nations, notably those in Europe and Asia, have only the choice of increasing yields per hectare. They have little opportunity to expand land under cultivation since most of their arable land is already in use and they are growing as many crops annually as feasible. Only by increasing annual yields per hectare can they produce more food.

In areas outside Asia and Europe, the physical potential for increasing land under cultivation is great. For example, in Africa and South America more than 1 billion ha of arable land are not now being cultivated. For environmental and economic reasons, however, most of these lands should probably not be cleared for agricultural purposes. They include the tropical rain forests with their rich biological diversity that should be conserved. Furthermore, the economic viability of agricultural land expansion in most of these areas is reduced by the high costs of clearing land and building roads for transporting fertilizers and other production inputs into the areas and farm products from the areas to outside markets.

In most areas of the world, increasing yield per hectare of land already under cultivation is the preferred method of increasing food production. In time, however, as economic development, transportation, and knowledge of soil management progress, selective expansion of land under cultivation is almost certain to occur. Table 20.3 shows estimates of future increases in crop production from growth in arable land compared to cropping intensity and increased yields per hectare. Note that more than two-thirds of the increase is expected to come from increased yields per hectare and increased cropping intensity.

20.4 POTENTIAL OF DIFFERENT SOIL ORDERS

The percentage distribution of 10 soil orders worldwide and in the humid and semiarid tropics is shown in Table 20.4. Note the high proportion of the humid tropical areas of Africa and Latin America that are dominated by Oxisols or Ultisols. These are highly leached acid soils relatively low in available plant nutrients. But with proper management, including the application of lime and fertilizers, these soils can be made quite productive. Unfortunately, large areas of Oxisols and Ultisols are located far from supplies of either lime or fertilizer, and crop production is thereby constrained. The huge Amazon basin of South America and the Congo River basin of Central Africa have a

TABLE 20.2 Percent of Increase in Food Production in Different Regions Between 1961–1963 and 1989–1990 Attributable to Increases in Area Cropped and to Increases in Yields Per Hectare

Region	Increased area (%)	Increased yields[a] (%)
Low-income countries		
sub-Sarahan Africa	47	52
Latin America	30	71
Middle East/North Africa	23	77
South Asia	14	86
East Asia	6	94
High-income countries	2	98
World	8	92

Data from the Food and Agriculture Organization (FAO).
[a]Includes both increasing the number of crops per year and increased yields per hectare.

TABLE 20.3   Percentage Contribution to Increased Food Production from Increases in the Area of Arable Land, Cropping Intensity, and Yields Per Hectare in Selected Developing Country Regions from 1982–1984 to 2000

Note that, except for Latin America, the majority of the expected production increase is from increased yields per hectare.

Region	Contribution to crop output growth (%)		
	Increased area of arable land	Increased cropping intensity	Increased yield per hectare
Sub-Saharan Africa	26	17	57
Near East/North Africa	0	22	77
Asia (excl. China)	11	20	69
Latin America	39	12	49

From FAO (1987).

high proportion of these highly weathered soils. While in time these soil areas may be more fully utilized for agriculture, the cost of making them suitable for crops and the environmental consequences of deforestation make this choice of action extremely questionable.

Soils developed on alluvial parent materials commonly have high crop-production potential. Many of these soils are classified as Entisols or Inceptisols and are found in the river valleys of Asia, Africa, and Latin America. They are commonly used for paddy rice production, for which they are well suited. Although rice yields are high in some areas, average yields of this crop in the humid tropics are commonly about half those obtained in temperate areas such as Japan, Korea, Italy, and the United States. Obviously these soils are not being utilized to their full potential.

In the semiarid tropics about one-third of the soil areas are classified as Alfisols. This suggests they have not been as highly leached as have the Oxisols and Ultisols of humid tropical areas. With proper soil and water management, these soils can be made quite productive. Unfortunately, the semiarid tropics are subject to considerable variability in annual precipitation rates. This means that periods of drought are quite common. The widespread droughts in sub-Saharan Africa in recent years indicate the inherent difficulty of maintaining sustainable crop productivity in these semiarid areas. Lack of reliable moisture is a major constraint on crop production in the African continent.

The highly productive, base-rich Mollisols so common in the midwestern part of the United States and of Russia are not found extensively in the tropics.

The crop-production potential of Aridisols is largely dependent on their access to irrigation water. Such soils are very important agriculturally in irrigated arid areas of India, Pakistan, the Middle East, and the United States. Irrigated Aridisols can produce

TABLE 20.4   Percentages of the Land Area of the World and of the Humid and Semiarid Tropics of Three Continents Classified as the Different Soil Orders of *Soil Taxonomy*

Soil order	In the world	In humid tropics			In semiarid tropics		
		Latin America	Africa	Asia	Latin America	Africa	Asia
Oxisols	9	50	40	4	—	—	—
Ultisols	8	32	15	35	3	2	6
Inceptisols	16	9	17	24	—	3	9
Entisols	11	5	20	24	5	17	—
Alfisols	14	3	5	4	34	32	38
Histisols	1	—	1	6	—	—	—
Spodosols	4	2	1	2	—	—	—
Mollisols	4	—	—	2	—	—	—
Vertisols	2	<1	<1	<1			
Aridisols	23	—	<1	<1	11	30	15
Others[a]	6	—	—	—	47	16	32
Total area (million ha)	10,504	666	445	379	313	1462	319

World distribution from U.S. Department of Agriculture (excludes miscellaneous lands); other humid tropics data calculated from NAS (1982); semiarid tropics data calculated from USDA (1986).
[a]For semiarid tropics, others category includes areas of soil orders that are not specified.

bumper crops when their nutrients are supplemented by fertilizers. The high solar radiation that typifies these areas helps stimulate high crop productivity.

Europe and Asia are already utilizing most of their arable land. North America and Australia are areas of excess food production. For these reasons, attention is focused appropriately on Africa, where food shortages are occurring and where the soil resources are not being effectively utilized or managed.

Major Production Constraints

Table 20.5 illustrates the principal limitation of soils for agriculture in major geographic areas of the world. A deficiency of soil moisture is probably the most significant worldwide constraint. This constraint is most notable in Australia, Africa, and South Asia (India, Pakistan).

Mineral deficiencies provide serious limitations in all areas but are most severe in Southeast Asia and South America. Shallow soils are problems in north, central, and south Asia. Excess water is a more serious problem in Southeast Asia than in any other region but the rice plants that grow on these wet areas are not affected by this constraint. The heavy monsoon rains in this area account for the wetness.

Production constraints are somewhat greater in developing areas of the world than in areas with greater industrialization, such as Europe and North America. Perhaps of greater importance, however, is the comparative difficulty of removing these constraints. For example, while the portion of Africa's soils characterized as being nutrient-deficient is even less than the world average, it is not easy to remove these constraints in Africa. The remoteness of many cultivated areas in Africa from sources of chemical fertilizers makes it difficult and expensive to supply the needed nutrients. Also, the prevalence of certain animal diseases makes it impossible to produce cattle over large areas of Africa, thereby denying farmers sources of animal manures.

Europe has the highest proportion of soils without serious limitation (36%). The comparable figure for Central and North America is slightly more than 20%.

20.5 PROBLEMS AND OPPORTUNITIES IN THE TROPICS

One cannot help but wonder why soils of the humid tropics have not been more effectively utilized. They seemingly have many advantages over their temperate zone counterparts. The crop-growing season is usually year-round. In the humid areas ample moisture is available throughout the growing season. And some of the soils have physical characteristics far superior to the soils of the temperate zones. The hydrous oxide and kaolinitic types of clays that dominate in these areas permit cultivation under very high rainfall conditions.

A comparison of the distribution of some major soil groupings in the humid tropics and in other nondryland areas of the world is shown in Table 20.6. The high proportion of the soils that are acid and infertile (Oxisols and Ultisols), especially in South America

TABLE 20.5 **World Soil Resources and Their Major Limitations for Agriculture**

Percent of total land area primarily affected by each limitation.

Region	Limitation					
	Drought	Mineral stress[a]	Shallow depth	Water excess	Permafrost	No serious limitation
North America	20	22	10	10	16	22
Central America	32	16	17	10	—	25
South America	17	47	11	10	—	15
Europe	8	33	12	8	3	36
Africa	44	18	13	9	—	16
South Asia	43	5	23	11	—	18
North and Central Asia	17	9	38	13	13	10
Southeast Asia	2	59	6	19	—	14
Australia	55	6	8	16	—	15
World	28	23	22	10	6	11

Data compiled from FAO/UNESCO soil map of the world by Dent (1980).
[a]Nutritional deficiencies or toxicities related to chemical composition or mode of origin.

TABLE 20.6 Distribution of General Soil Groups Found in the Humid Tropics Compared to Other Nondryland Soil Areas (%)

Note the high proportion of acid, infertile soils, especially in South America and Africa.

General soil grouping	Humid tropical soil areas (%)				Other nondryland soil areas (%)
	Central and South America	Africa	Asia	World	
Acid, infertile soils (Oxisols and Ultisols)	82	56	38	63	10
Moderately fertile, well-drained soils (Alfisols, Vertisols, Mollisols, Andepts, Tropepts, Fluvents)	7	12	33	15	32
Poorly drained soils (Aquepts)	6	12	6	8	10
Very infertile sandy soils (Psamments, Spodosols)	2	16	6	7	13
Others	3	4	16	17	35[a]
Total	100	100	100	100	100

From National Research Council (1982) and U.S. Department of Agriculture data.
[a]Mostly mountainous areas.

and Africa, helps explain why soils of these continents have not been more extensively utilized. These soils are low in nutrients and often are far from sources of limestone or fertilizers, making it difficult to increase their fertility.

Limiting Factors

In Table 20.7 are shown the main soil chemical constraints in five agroecological regions of the tropics. Note that aluminum toxicity and phosphorus fixation are serious problems, especially in the humid tropics and acid savannas. Such limitations are not greatly different from those found in humid areas of temperate regions, but the greater difficulty and expense of providing lime and fertilizers in the tropics makes these limitations more constraining to food production than in temperate areas.

It is interesting to note in Table 20.7 that low cation exchange capacities are not serious limitations for most tropical soils. This is somewhat unexpected if we consider only the mineral colloids as cation-adsorbing constituents. The oxides of iron, aluminum, and kaolinite that are prominent in Oxisols and Ultisols that dominate some humid tropical areas have low cation capacities. This means that much of the cation exchange in some tropical soils is associated with the organic fraction. In tropical forested areas the litter falling from trees is many times higher than in temperate areas, rates of 20–25 Mg/ha/yr having been recorded. Even though the rate of decomposition of these residues is also high, the equilibrium levels of soil organic matter in natural forested areas in the tropics is as high if not higher than that in soils of temperate forested areas.

Special problems relate to tropical rain forest areas, which are characterized by high rainfall and by dense forest. Canopies made up of trees and other woody species vary in height from 40 m to essentially ground level. This vegetative cover is extremely rich in organic materials (biomass) and in mineral nutrients. The quantities of these con-

TABLE 20.7 Percentage of Soil Areas in Five Agroecological Regions of Tropics Having Different Chemical Constraints to Plant Growth

Soil constraint	Humid tropics (%)	Acid savannas (%)	Semiarid tropics (%)	Tropical steeplands (%)	Tropical wetlands (%)	Total (%)
Aluminum toxicity	56	50	13	29	4	32
Acidity without Al toxicity	18	50	29	16	29	25
High P fixation by Fe Oxides	37	32	9	20	0	22
Low CEC	11	4	6	—	—	5
Salinity	1	0	2	0	7	1

From Sanchez and Logan (1992).

TABLE 20.8 Distribution of Major Nutrient Elements and Organic Materials in the Soil and in the Living and Dead Plant Species (Biomass) in a Tropical Forested Area in Brazil

Element	Percentage distribution in	
	Biomass	Soil
Nitrogen	26.9	73.1
Phosphorus	31.9	68.1
Potassium	89.6	10.3
Calcium	100.0	0.0[a]
Magnesium	92.3	7.7
Organic matter	68.4	31.6

From Salati and Vose (1984).
[a] This sampling showed essentially all the calcium in the biomass.

stituents in the biomass are often greater than can be found in the soil profile. Note in Table 20.8 the comparative distribution of chemical nutrients and organic matter in the biomass and soil in a rain forest area in Brazil. Obviously, the biomass is the dominant component of a nutrient cycle that is largely responsible for life in this area. If the forests are harvested or cut down and burned, and the land used for agriculture, much of the organic matter is subject to rapid decomposition and the nutrients are readily leached from the soil.

As a continent, Africa has a problem of low water availability. As shown in Table 20.9 inadequate water is a serious limiting factor for much of Africa. Water-conserving efforts must be paramount for crop production on this continent. At the same time, vast areas of arable soils in Africa receive adequate rainfall but are low in productivity. Soil and crop scientists are challenged to devise systems that will change this situation.

Research on Tropical Soils

Unfortunately, all too little is known about tropical soils and their management. Research on these soils, their properties, and their potential for crop production is insignificant compared to that on temperate region soils. As research results emerge, however, it is apparent that the basic principles involved in managing tropical soils are not greatly different from those relating to soils in the temperate regions.

What we have learned about some tropical soils is encouraging. For example, some of the soils (Oxisols) of Hawaii and the Philippines are excellent for pineapple and sugarcane production. They respond well to modern management and mechanization. In contrast, modern mechanized farming was a dramatic failure for the "Groundnut Scheme" carried out by the British in Tanganyika (Tanzania) following World War II. In that case, exposing the cleared soil to tropical rains resulted in catastrophic erosion. Knowledge of the nature of the soil and the adoption of management practices to keep vegetative cover on the soil might well have prevented much of the project's $100 million loss.

There is a good likelihood that more intensive study will identify other complexities among tropical soils similar to those known for temperate regions. We already know of

TABLE 20.9 Water Characteristics of Africa Compared to Those of North and South America and the World

Note that Africa is dramatically drier than the other areas.

Water characteristics	Africa	North America	South America	World
Precipitation (cm/yr)	70.7	64.7	151.6	70.2
Evaporation (cm/yr)	58.0	38.9	86.8	45.3
Precip. minus evap. (cm/yr)	12.7	25.8	64.8	24.9

Calculated from Mather (1984).

much variability among soils of tropical areas. Some are deep and friable, easily manipulated and tilled. At the other extreme are soils that, when denuded of their upper horizon, expose layers that harden into a surface resembling a pavement (Figure 20.3). Such soils when eroded are essentially worthless from an agricultural point of view. Fortunately, these soil areas are not very extensive.

The chemical characteristics of Oxisols differ drastically from those of soils of temperate regions. The high hydrous oxide content dictates enormous phosphate-fixing capacities. The low cation exchange capacities and heavy rainfall result in removal of not only macronutrients but micronutrients as well. Indeed, the level of technology needed to manage Oxisols is fully as high as that required for temperate zone soils.

Plantation Management Systems in the Tropics

The plantation system of agriculture has been successful in raising crops such as bananas, sugarcane, pineapples, rubber, coffee, and cacao. These crops are usually grown for export and require considerable skill and financial inputs for their production. The plantation system generally imports the best available technology from developed countries and sometimes has associated with it sizable research staffs to gain new knowledge for improved technology. Although it has been generally successful in producing and marketing crops and animals, it provides little benefit for the indigenous small-scale producer. Consequently social and political problems have plagued this system.

20.6 SHIFTING CULTIVATION

At the opposite extreme from the plantation approach are indigenous systems that require little from the outside and have evolved mostly by trial and error by the native cultivators. One of the most widespread of these systems—that of **shifting cultivation**—will be described briefly to illustrate what indigenous peoples have learned from centuries of experience (Figure 20.4).

While there are variations in the practice of shifting cultivation, in general it involves three major steps:

1. The cutting and burning of trees or other native plants with their ashes left on the soil. Sometimes only the vegetation in the immediate area is burned. In other cases, this is supplemented with plants brought in from nearby areas.

2. Growing crops on the cleared area for a period of 1–5 yr, thereby utilizing nutrients left from the clearing and burning of native plants. Vegetation from outside the area may be brought in and burned between plantings.

FIGURE 20.3 Area in central India with a pavement-like surface where little crop growth occurs. The cementation results from removal of the upper layers by erosion and the exposure at the surface of iron-rich materials called *plinthite* (Greek *plinthis,* brick) that harden irreversibly when repeatedly wetted and dried. Sizeable areas with such soils are found in India, Africa, South America, and Australia.

FIGURE 20.4 Expansion of unsustainable forms of slash-and-burn agriculture presents serious environmental and agricultural challenges in some tropical and subtropical areas. The natural vegetation (or regrowth from previous clearing of forests) is slashed by the farmer (left), allowed to dry, and then burned. After the burning, food crops are grown for two to three years using the ashes from the burned vegetation for plant nutrients (right). Yields soon decline, however, and the land is abandoned, the farmer moving on to slash and burn at a new site. (Photos taken in Peru)

3. Fallowing the areas for a period of 5–12 yr, thereby permitting regrowth of the native trees and other plants that rejuvenate the soil. Nutrients are accumulated in the native plants and some, such as nitrogen, are released to the soil. The cutting and burning are repeated and the cycle starts again.

Advantages of the System

Shifting cultivation is primarily a system of nutrient conservation, accumulation, and recycling. The native plants absorb available nutrients from the soil and from the atmosphere. Some are nitrogen fixers. Others are deep rooted and bring nutrients from lower horizons to the surface. All help protect the soil from the devastating effects of rain and sunshine. Also, the cropped area is generally small in size and is surrounded by native vegetation. This reduces the chances of gully formation or severe sheet erosion from runoff water.

There are benefits of the system other than those relating to nutrients and erosion control. The burning is likely to destroy some weed seeds and even some unwanted insects and disease organisms. The short period of actual cropping (1–5 yr out of 10–20 yr) discourages the buildup of weeds, insects, and diseases harmful to the cultivated crops. Although shifting cultivation may seem primitive, it deserves careful study. In some tropical areas it is more successful than the seemingly more efficient temperate zone systems.

Soil Deterioration

Unparalleled increases in human populations have placed great strain on the traditional shifting cultivation systems in some areas of the tropics. The more than 300 million people who depend upon the shifting cultivation for most of their food represent a dramatic increase in numbers during the past two decades. To provide food for their growing families, farmers have been forced to shorten the period of fallow—to recrop a given area after only 3–6 yr of fallow compared to 10 yr or longer in the past. As a consequence, insufficient time is allowed for soil rejuvenation between cropping periods. This leads to lower crop yields and greater exposure to erosion. The mechanism by which shortened fallow periods have reduced soil fertility is illustrated in Figures 20.5 and 20.6. Today's shorter fallow period is one of the reasons for a decline in per capita food production in Africa during the past decade.

Promising Alternatives

Some soils in the tropics must have continuous vegetative cover to remain productive. If they dry out, especially if erosion removes the surface layers, the doughy subsoil layers

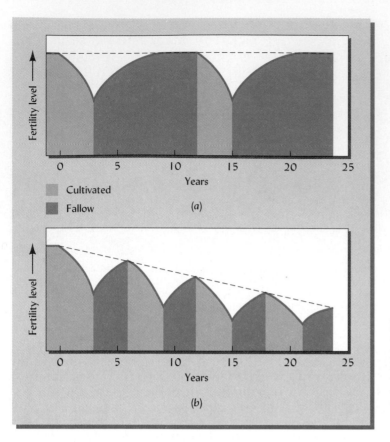

Cultivated

Fallow

(a)

(b)

FIGURE 20.5 Changes in fertility levels under two shifting-cultivation systems. (a) In the system with a long fallow period, the natural vegetation is able to help regenerate the fertility level after cropping. (b) With a shorter fallow period so common in areas of high population pressure, insufficient regeneration time is available and the fertility level declines rapidly.

harden irreversibly, making plant growth impossible. This turn of events may be prevented by planting the cultivated crop among native plants or other crop plants without completely removing the latter. Since the plants involved are in most cases trees, this system has been termed the *mixed tree crop* system. The desired crop or crops are introduced by removing some of the existing plants and replacing them with crop plants. In time, a given area may be planted entirely to a number of crop species. The essential feature of the system, however, is that the soil will at no time be free of vegetation. The mixed tree crop system succeeds where indigenous farmers possess detailed knowledge of the many native and crop plant species involved. Up to now science has not been able to develop more successful alternatives for the management of some of these tropical soils.

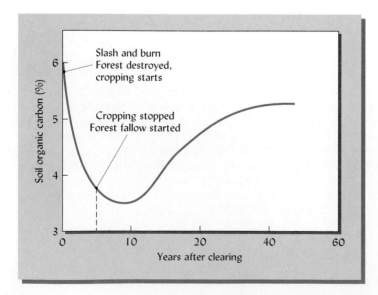

FIGURE 20.6 Effect of slash and burn, cropping for five years to pineapple, and then returning the land to forest (fallow) on the organic carbon content of Paleudults soils of Mexico. Reduction in organic matter continues even after cropping is stopped until forest species residues are sufficient to begin to replenish the soil organic matter. [Redrawn from Wadsworth et al. (1988)]

Attempts are being made to develop viable alternatives to the shifting cultivation system—alternatives that would permit crop production without slashing and burning. Some research has shown that modest inputs of fertilizer and lime, the selection of acid-tolerant crop cultivars, and the inclusion of legumes in crop rotations help sustain crop yields. Data in Table 20.10 from research in Peru illustrate this point. Small quantities of fertilizer and lime along with better means of weed control were effective in stabilizing yields and maintaining reasonable soil cover. Small-scale farmers near the research plots have adopted the scheme and are pleased with its greater stability.

A second alternative closely related to the mixed tree crop system already discussed is that of *alley cropping*. This is an agroforestry system that involves growing food crops in alleys, the borders of which are formed by fast-growing trees or shrubs (see Figure 16.9b). These woody species are usually leguminous and can provide fixed nitrogen to the system. The hedgerows are pruned to prevent shading of the food crop. The cuttings obtained by pruning are used as mulch and as a source of nutrients for the food crop. The mulch helps control weeds, prevents runoff and erosion, and reduces evaporation from the soil's surface.

Figure 20.7 provides an example of the functioning of an alley cropping system. Note that during the dry season the border trees are permitted to grow and then are cut back at crop-planting time. In some cases the hedgerow woody-plant leaves are fed to livestock and the manure is then applied to the land.

In some areas, this system appears to function effectively (Table 20.11). The yield of corn was sustained at a reasonable level over a period of five years when the prunings of the woody legume were used as a mulch. In contrast, yields dropped off sharply where no mulch material was returned to the soil. This system is being tried on farmers' fields; should it prove to be acceptable, a major step will have been taken toward a viable alternative to the slash-and-burn system.

Alley cropping systems not only maintain crop yields but also maintain soil organic matter and nutrient element levels (Table 20.12). Apparently the deep-rooted woody species used as hedgerows cycle mineral nutrients from the lower soil horizons to leaves and stems and eventually to the soil surface. These residues are incorporated into the soil by soil organisms, thereby enhancing organic matter and mineral element contents while simultaneously maintaining crop yields. Also, the leguminous woody species fix nitrogen annually in quantities that compare favorably with annual amounts fixed by some forage legumes in temperate regions.

The effectiveness of alley cropping systems is constrained in some areas due to the competition for water and nutrients between the border woody species and food crops grown in the alleys. Also, on very acid soils the leguminous woody species often perform poorly and provide few residues to help reduce erosion and to increase crop yields. Searches are under way for alternative legume shrubs or trees that will perform better. Alley cropping will likely be beneficial in some areas, but it is not a panacea alternative for all areas.

Potential of Tropical Agriculture

One cannot help but be optimistic about the future of agriculture in the tropics. The basic requirements for maximum year-round production appear to be higher in the humid tropics than anywhere else. Total annual solar radiation and warm to hot climates provide unmatched photosynthetic potential. Several crops a year can be grown on a given field in the tropics. Already researchers have developed new crop varieties

TABLE 20.10 Average Yields of Four Crops with and Without Modest Fertilizer Applications Following Burning in a Slash-and-Burn System in Peru

		Average harvest yield (Mg/ha)	
Crop	No. of harvests	No fertilization or lime	Fertilizer and lime added
Rice	37	0.99	2.71
Corn	17	0.21	2.81
Soybeans	24	0.24	2.30
Peanut	10	0.69	3.46

From Sanchez et al. (1983).

TABLE 20.11   Yield of Corn Over a Period of Six Years in an Alley Cropping
Experiment Using *Leucaena leucocephala* as the Hedgerow Component
on a Psammentic Ustorthent Soil in Nigeria

Nitrogen rate[a] (kg/ha)	Leucaena prunings	Yield of corn[a] (Mg/ha)						
		1979	1980	1981	1982	1983	1984	1986
0	Removed	—	1.04	0.48	0.61	0.26	0.69	0.66
0	Used as mulch	2.15	1.91	1.21	2.10	1.92	1.99	2.10
80	Used as mulch	3.40	3.26	1.89	2.91	3.16	3.67	3.00

From Kang et al. (1989).
[a]All plots received P, K, Mg, and Zn. Crop affected by drought in 1983; land fallowed in 1985.

especially adapted to warm climates. Appropriate means of controlling pests and use of fertilizer are becoming more common in tropical areas. The potential for food production is enormous, and it must be realized if disastrous food shortages are to be avoided.

20.7   REQUISITES FOR THE FUTURE

The world's ability to feed itself depends upon many factors, not the least of which is improved agricultural technology in the developing nations of this world. This technology depends largely on research and education, which must have direct relevance to the developing countries and not be a mere transplant of what is available in the more

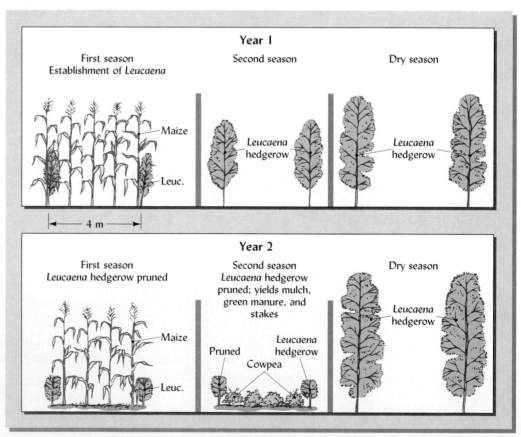

FIGURE 20.7   Cropping sequence diagram for establishing *Leucaena leucocephala* hedgerows (spaced 4 m) for alley cropping with sequentially cropped maize and cowpeas. Note that the hedgerow is never allowed to shade the food crop and that the clippings from the *Leucaena* are placed as a mulch on the food crop rows. [From Kang et al. (1984)]

TABLE 20.12 Effect of Six Years of Alley Cropping Using *Leucaena* in the Hedgerows on Selected Chemical Properties of the Upper 15 cm of a Psammentic Ustorthent

Nitrogen rate (kg/ha)	Leucaena prunings	pH	Organic matter (g/kg)	Exchangeable cations (cmol/kg)		
				K	Ca	Mg
0	Removed	6.0	6.5	0.19	2.90	0.35
0	Retained	6.0	10.7	0.28	3.45	0.50
80	Retained	5.8	11.9	0.26	2.80	0.45

From Kang et al. (1989).

developed nations. We cannot assume that the technology of western Europe or the United States can be transferred directly to the underdeveloped countries. Disastrous failures have shown the fallacy of this approach.

Sustainability

One requisite for future progress in world food production is that the farming systems chosen must be *sustainable*. In the past, the degradation of some of our natural resources was tolerated since these overall resources were so vast that the percentage degraded was considered insignificant. Truly, this is no longer the case. The world's soils are being degraded by erosion, chemical pollution, and salinization at rates that are no longer tolerable.

We recognize that extraordinary increases in food production must be achieved in the decades ahead. These increases must come primarily from land already in cultivation, thereby minimizing the need to cut down more forests and plow more native grasslands. But increases must be achieved while simultaneously reducing the rate of soil degradation. These are the challenges of the century for world agriculture.

Two Requisites

Past accomplishments suggest that there are two major requirements for increased agricultural production in the tropics. *First,* with the aid of farmers in the tropics, improved and sustainable technologies focused specifically on tropical agriculture must be developed and put into place. *Second,* the political, social, and economic climate in the developing countries must be such as to make it profitable for the cultivator to adopt and use the new technologies. While the second of these requirements is fully as important as the first, it is beyond the purview of this text. Consequently, we will focus on some of the technologies needed.

Genetic Development

New crop varieties adapted to conditions in the developing areas, along with methods of pest control, are among the first requirements. For example, the dwarf wheats and rices that are highly responsive to fertilizer applications were the twin engines of the green revolution that helped feed Asia and Latin America during the past quarter century. The average yield of wheat in Mexico doubled as a result of introducing new varieties, especially in irrigated areas. The production of wheat in India has tripled since the mid-1960s. Similarly, rice production in Indonesia more than doubled in the past 20 years. Luckily the high-yielding rice and wheat varieties are adapted to these and other food-deficit countries.

The creation and use of improved cultivars that will tolerate or resist insect pests and diseases are essential since they can enhance pest management without excessive use of chemical pesticides. The breeding of varieties that will tolerate high soil acidity and exchangeable aluminum levels would revolutionize crop production and soil vegetative coverage in large areas, especially in Latin America and Africa. The creation of drought and salinity-tolerant varieties would be most helpful to peasant farmers who may not be able to afford the irrigation and drainage facilities needed to remove water and salt constraints. Changing the plant characteristics may be less expensive than changing the environment in which it grows. Consequently, soil scientists must team up with plant breeders, geneticists, and biotechnologists in achieving joint objectives.

Crop and Residue Management

The maintenance of vegetative cover of soils is perhaps the most important means of protecting the soil and of reducing soil erosion. This principle has proved to be just as effective in tropical countries as it has in the United States and elsewhere. Unfortunately, however, techniques so common in industrialized countries are not widely practiced in the tropics. For example, in most areas in India, crop residues are not left on the soil surface. They are more badly needed as a source of fuel to cook food or as fodder for milk animals than they are as mulch for the soil. In the farmer's view, soil conservation takes second place to family health. In this situation the creation of a nearby community fuelwood plantation may be essential before sustainable soil management practices can be realized.

Crop residue management also requires a reexamination of the monoculture cropping systems so widely used in mechanized, temperate region agriculture. Mixed crop culture with or without rows is in many cases better suited to efficient use of water and nutrients and to soil conservation. It may be necessary to consider the advantages of some indigenous practices to better achieve sustainable agriculture goals.

In monoculture cropping systems research trials, conservation tillage has proven to be effective in reducing soil erosion in the tropics (Table 20.13). The challenge is to find mechanisms of keeping residues on the surface, mechanisms that must fit in with the needs of the farmer and his or her family.

Water Management

Remarkable increases in the amount of irrigated land have taken place in the past 35 years, especially in Asia. Worldwide, more than 275 million ha of land are now being irrigated, 185 million of which are in Asia. Irrigation coupled with high-yielding varieties and fertilizers has stimulated unparalleled increases in food production in Asia and will likely continue to do so.

Unfortunately, in Africa, which is the major dry continent of the world, only 13 million hectares are irrigated. This is one reason for Africa's slow agricultural growth. While there is considerable potential for increased irrigation in Africa, a number of economic and managerial factors have limited achievements.

There is growing recognition that irrigation without proper soil drainage can be disastrous. Salt buildup in some irrigated arid region soils results from poor drainage. Salinization seriously affects crop production from more than 20 million ha, or about 7% of the world's irrigated land. Drainage simply must be considered as an integral part of any irrigation scheme.

The efficiency of irrigation water use is generally abysmal—only about 37% on a global basis and considerably lower in developing countries. Concrete-lined irrigation canals are expensive, as is land-leveling to ensure better field water delivery. Microirrigation techniques (e.g., trickle, drip irrigation) have only begun to be used, but the viability of such systems to provide vegetables to nearby towns and cities should be explored in developing countries.

TABLE 20.13 **The Influence of Conservation Tillage and Alley Cropping Using Two Woody Legumes as Hedgerows on Runoff and Soil Erosion Losses in Nigeria**

System	Runoff (% of rainfall)		Soil erosion (Mg/ha)	
	Corn	Cowpeas	Corn	Cowpeas
No hedgerows				
Plow-till	17.0	4.3	4.3	0.63
No-till	1.3	0.8	0.1	0.03
Hedgerows				
Leucaena	4.9	1.1	0.6	0.13
Gliricidia	4.3	0.7	0.6	0.07

From Lal (1989).

Low water-use efficiency is also a critical factor in the management of rain-fed agriculture. Management practices must be used to increase that efficiency. For example, crop residue cover must be enhanced to reduce water losses by evaporation and by surface runoff. The quantity of residues available for such coverage could be increased by ensuring adequate plant nutrient levels, using either manures or chemical fertilizers to supply the chemicals.

Crop management systems also affect the efficiency of soil water utilization. For example, multiple cropping systems using crops of different height, growth patterns, and rooting systems to affect the infiltration and evaporative losses must receive attention.

Appropriate Technologies for Soil Erosion Control

Low-cost and easily maintained systems for contour farming must be developed and thoroughly tested. The machinery-intensive contour tillage and farming systems characteristic of industrialized countries are too expensive and time-consuming. Vegetative barriers (such as vetiver grass) should be tried in areas where competition for soil moisture is not a serious problem. Other such live barriers should be found and their effectiveness thoroughly tested.

Nutrient Cycling

The notoriously low efficiency of use of plant nutrients that plague agriculture worldwide has similar negative consequences for farmers in the tropics. Practices must be followed that will enhance the recycling of nutrients from the soil to the plant and through plant residues to the soil again. Chemical fertilizers and farm manures can be used to supplement the nutrient supply in the soil. The inclusion of nitrogen-fixing legumes in the cropping system with modest phosphorus applications to stimulate the legumes have proven successful. Such combinations of organic and inorganic sources of nutrients must be used to the extent feasible. The place of cover crops to provide additional organic residues must also be explored (Figure 20.8).

The role of organic matter in nutrient cycling is critical. The same principles governing the storage and slow release of mineral elements in organic matter pertinent in

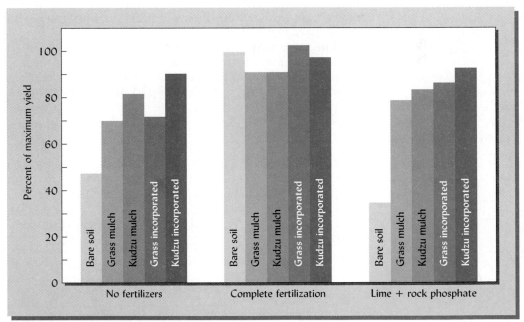

FIGURE 20.8   Effect of crop residues with and without fertilizers and lime on the yield of five consecutive crops grown in the Amazon basin. Yields are expressed as percent of that obtained from complete fertilization on bare soil. [From Wade and Sanchez (1983); used with permission of the American Society of Agronomy]

temperate regions holds for tropical conditions as well. While the higher soil temperatures in the tropics accelerate the breakdown of the organic residues, the conversion of some of these materials into more long-lasting humus still occurs. It is possible that organic matter could play an even more critical role in the recycling of nutrients in the tropics than it does in temperate regions.

In some soils the plant biomass production is simply not high enough to accommodate reasonable cycling levels needed for good crop production. Chemical fertilizers must be utilized to provide the nutrients needed. The dramatic increases in food production of the past three decades have been in part stimulated by dramatic increases in supplies of chemical fertilizers. From the 1960s to 1979 fertilizer use rose eightfold in the developing countries. The total N-P-K consumption in these countries was about 60 million Mg in 1992. This has resulted in a demand for information on soil characteristics to determine the kinds of fertilizers that are needed. In highly leached areas, evaluations are being made of micronutrient deficiencies as well as those of nitrogen, phosphorus, potassium, and lime. Considerable attention is being given to fertilizer that resists rapid reaction with the soil or volatilization by microbial action.

Africa lags far behind the other areas in rates of fertilizer application. In 1992 an average of only 20 kg of fertilizer per hectare was consumed in Africa, compared to about 44 kg/ha in Latin America and more than 123 kg/ha in Asia. Fertilizer trials have demonstrated the response that can be obtained from fertilizers in Africa, but economic, political, and supply issues have prevented more extensive use. As a consequence, removal of plant nutrients in Africa far exceeds the nutrients supplied in fertilizers. This means that the soils are being mined of their chemical nutrients. Table 20.14 gives some idea of the extent of this nutrient loss from African soils.

Soil Characterization and Survey

Agricultural development activities have emphasized the critical need for the characterization of soils. For example, the salinity and alkalinity status can well determine the probable success of an irrigation project. Nutrient deficiencies can be identified, as can the potential for erosion and drainage problems. The time and effort being devoted to soil characterization are insignificant compared to the need for this kind of information.

Reliable soil survey information is inadequate or unavailable in most of the developing areas. This is due to lack of information about the soil characteristics upon which a classification scheme can be based and to ignorance of the significance and value of the soil survey. Planners sometimes look upon soils without regard to the vast differences that exist among them—differences that could affect markedly the success of plans that are made to use the soils.

Soil surveys are of special significance in two ways. First, they make possible the extrapolation of research results from a given area to other areas where the same kinds of soils are found. Second, they provide one of the criteria to determine the economic feasibility of clearing and preparing for tillage lands that have as yet not been used for agriculture.

Human Resources

A final requisite for increased food production is human resources—people who will find out what to do and then to do it. The range needed goes from basic scientists (from

TABLE 20.14 The Mining of Plant Nutrients in Sub-Saharan Africa

Nutrients removed by plants are not being replaced in residues, manures, or chemical fertilizers.

| Year | Nutrients not being replaced 10^3 Mg | | | |
	N	P	K	Total
1983	4,400	520	3,100	8,020
2000	6,100	740	4,600	11,440

From Stoorvogel and Smaling (1990).

whose test tubes and field plots new technologies and food products can come) to the cultivators and their families who can help frame the constraints, work with the scientists to seek solutions, and put the products/systems to work in the field. Researchers whose interest relates directly to the solution of the world food problem are needed, but a new paradigm of cooperation between researchers and farmers is essential. Such a collaborative system will better assure that the technologies developed are appropriate, sustainable, and usable.

The fight to feed the world is not yet lost. But to win it will require technological and scientific inputs of a magnitude not yet realized. Among the most important of these inputs are those relating to soils and soil science. The struggle will also require involvement of the farm family in every step of the way. Soil scientists will be among those who participate in this exciting and rewarding experience.

20.8 CONSEQUENCES OF FOOD PRODUCTION SUCCESS/FAILURE

If all who are concerned with food production work together as a team, it should be possible not only for the world as a whole to feed itself but for most developing countries to do likewise. The consequences of their not doing so will be tragic, not only from the standpoint of the hundreds of millions of hungry people, but from the point of view of natural resource conservation as well. The food production successes of the past 30 years have had mostly favorable effects on the quality of the environment. In India alone, for example, it has been estimated that were it not for the yield gains from the green revolution (1964–1993) the country would have been forced to plow an additional 40 million hectares of fragile lands, mostly in forests (Figure 20.9). Similar benefits are pertinent elsewhere where high-yield farming was successful. More fully utilizing the flatter lands to reduce the need for the slopes was a strategy that paid off.

In spite of these successes, however, production intensification of the past three generations has brought some problems. The pumping of groundwater for irrigation in some cases far exceeds the recharge rates. Some irrigation project areas have become salinized, yields have been reduced, and the soils have become physically puddled. Excessive applications of fertilizers in some instances have increased nitrates in drainage waters to undesirable levels. Monoculture on some sloping lands has resulted in erosion beyond sustainable levels. The desired sustainability of high-yield agriculture is yet to be achieved.

As we look to the next three decades, the dual challenge of further production intensifications and improved natural resource conservation faces all concerned with food security around the world. The management and effective use of soils will be one of the key factors that determines whether that challenge will be met.

20.9 CONCLUSION

The world population explosion of this century presents humankind with one of its most complex challenges. This challenge is greatest in the low-income countries of the world where the bulk of the 90 million annual population increase is occurring. To feed an increase of population equivalent every three years to the entire population of the United States will tax the world's ingenuity and determination.

It is in the developing countries and especially Africa where the challenge is most severe. The conservation and improvement of soils is one of the keys to meeting this challenge. Soil and crop management systems must be developed and adapted to increase food production and simultaneously prevent soil deterioration. Soil scientists will play a key role in developing these systems.

Soil scientists must also gain greater knowledge of the characteristics and potential of arable but uncultivated lands. This knowledge will be used in determining the extent to which new lands should be brought into agricultural production to help meet the world's food demands.

There is potential for the world to produce adequate food for current and future generations. Whether this potential will be reached will depend in part on the wisdom that is used in managing and conserving soils. This wisdom will be based on a partnership involving soil scientists and their science colleagues working in cooperation with the ultimate users of improved farming systems—the farmer and his or her family.

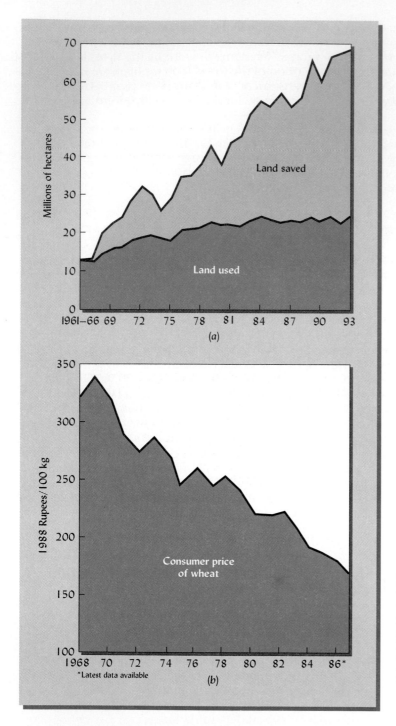

FIGURE 20.9 (a) If India had to produce today's wheat harvest with technologies and varieties of the early 1960s it would be necessary to cultivate 40 million more hectares of farmland than is currently being used. Most of this extra farmland would have to come from easily erodible forested lands that are characterized by steep slopes. (b) The increased production intensity simultaneously reduced the price that consumers (mostly poor people) had to pay for the wheat. [From CIMMYT (1995)]

REFERENCES

CIMMYT. 1995. *CIMMYT 1994 Annual Report* (Mexico, DF Mexico: International Maize and Wheat Improvement Center).

Dent, F. J. 1980. "Major production systems and soil-related constraints in Southeast Asia," in *Priorities for Alleviating Soil-Related Constraints to Food Production in the Tropics* (Manila: International Rice Research Institute of Ithaca; N.Y.: Cornell University), pp. 78–106.

FAO. 1984. *Land, Food, and People* (Rome, Italy: UN Food and Agriculture Organization).

FAO. 1987. *Agriculture: Toward 2000* (Rome, Italy: UN Food and Agriculture Organization).

Kang, B. T., G. F. Wilson, and T. L. Lawson. 1984. *Alley Cropping: A Stable Alternative to Shifting Cultivation* (Ibadan, Nigeria: Int. Inst. of Tropical Agric.).

Kang, B. T., L. Reynolds, and A. N. Atta-Krah. 1990. "Alley Farming," *Adv. Agron.*, **43**:316–360.

Lal, R. 1989. "AgroForestry systems and soil surface management of a tropical Alfisol: II: Water runoff, soil erosion and nutrient loss," *AgroForestry Systems*, **8**:97–111.

Mather, J. R. 1984. *Water Resources: Distribution, Use and Management* (New York: Wiley).

NAS. 1982. *Ecological Aspects of Development in the Humid Tropics* (Washington, D.C.: National Academy Press).

National Research Council. 1982. *Ecological Aspects of Development in the Humid Tropics*, report of the Committee on Selected Biological Problems in the Humid Tropics (Washington, D.C.: National Academy of Science Press).

Paarlberg, R. L. 1994. "The politics of agricultural resource abuse," *Environment*, **36**:6–41.

Salati, E., and P. B. Vose. 1984. "Amazon basin: a system in equilibrium," *Science*, **225**:129–38.

Sanchez, P. A., and T. J. Logan. 1992. "Myths and science about the chemistry and fertility of soils in the tropics," in P. A. Sanchez and R. Lal (eds.), *Myths and Science of Soils of the Tropics*, SSSA Special Publ. No. 29 (Madison Wis.: Soil Sci. Soc. Amer.), pp. 35–46.

Sanchez, P. A., J. H. Villachica, and D. E. Bandy. 1983. "Soil fertility dynamics after clearing a tropical rainforest in Peru," *Soil Sci. Soc. Amer. J.*, **47**:1171–78.

Stoorvogel, J. J., and E. M. A. Smaling. 1990. "Assessment of soil nutrient depletion in SubSaharan Africa: 1983–2000" (Wageningen, Netherlands: Rept #28 Winand Staring Centre for Integrated Land, Soil and Water Research).

The President's Science Advisory Committee Panel on World Food Supply. 1967. *The World Food Problem*, Vols. I–III (Washington, D.C.: The White House).

UNFPA. 1992. *The State of World Population* (New York: UN Population Fund).

USDA. 1986. *Symposium on Low Activity Clay (LAC) Soils*, SMSS Technical Monograph No. 14, U.S. Department of Agriculture and U.S. Agency for International Development.

Wade, M. K., and P. A. Sanchez. 1983. "Mulching and green manure applications for continuous crop production in the Amazon basin," *Agron. J.*, **75**:39–45.

Wadsworth, G. R., R. J. Southhard, and M. J. Singer. 1988. "Effects of fallow length on organic carbon and soil fabric of some tropical Udults," *Soil Sci. Soc. Amer. J.*, **52**:1424–30.

World Resources 1994–1995 (New York: Oxford University Press).

Appendix A
U.S. Soil Taxonomy Map and Simplified Key

Soil surveys have been prepared for most counties of the continental United States. The Soil Survey Staff of the U.S. Natural Resources Conservation Service, in cooperation with scientists from other countries, have used information from these surveys and analyses to develop generalized soils maps based on *Soil Taxonomy*. Areas dominated by specific soil orders and suborders are delineated on the map using map symbols (A.3a, D.1s, etc.). The soil orders and suborders to which these symbols refer are shown on the page facing the map.

Keys have been developed to help scientists determine which soil order, suborder, great group, etc., is present in a given field site. A simplified key is shown in Figure A.2. This key is based strictly on soil properties that can be measured by physical and chemical means. Use of this and more complex keys assures that two or more independent soil surveyors will arrive at the same classification of a given soil area.

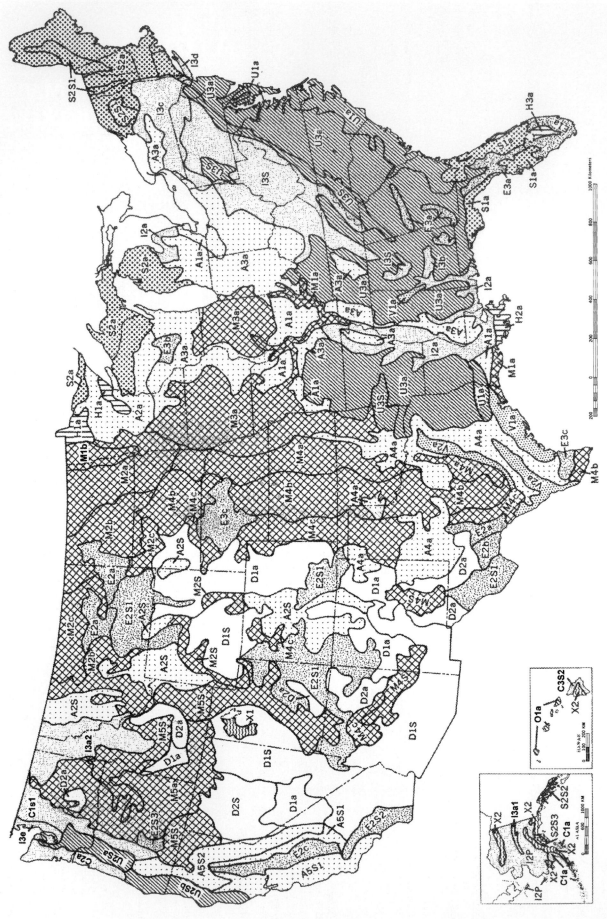

FIGURE A.1 General soil map of the United States showing patterns of soil orders and suborders based on *Soil Taxonomy*. Explanations of symbols follow. (Courtesy USDA Natural Resources Conservation Service, National Soil Survey Center, 1995)

ALFISOLS

Aqualfs
A1a Aqualfs with Udalfs, Aquepts, Udolls, Aquolls; gently sloping.

Boralfs
A2a Boralfs with Udipsamments and Histosols; gently and moderately sloping.
A2S Cryoboralfs with Borolls, Cryochrepts, Haplocryods, and Rock outcrops; steep.

Udalfs
A3a Udalfs with Aqualfs, Aquolls, Rendolls, Udolls, and Udults; gently or moderately sloping.

Ustalfs
A4a Ustalfs with Ustochrepts, Ustolls, Usterts, Ustipsamments, and Ustorthents; gently or moderately sloping.
A4b Ustalfs with Ustolls, Ustochrepts, Ustolls, Torriorthents, and Usterts; gently or moderately sloping.
A4c Aridic subgroups of Ustalfs with Ustolls, Cambids, Ustipsamments, Ustorthents, Ustochrepts, Ustorthents, Torriorthents, and Usterts; gently or moderately sloping.
A4S Ustalfs with Argids and Torriorthents; moderately sloping or steep.

Xeralfs
A5S1 Xeralfs with Xerolls, Xerorthents, and Xererts; moderately sloping to steep.
A5S2 Ultic and lithic subgroups of Haploxeralfs with Xerands, Xerults, Xerolls, and Xerochrepts; steep.

ANDISOLS

Cryands
C1a Cryands, Cryaquepts, Cryaquands, Histosols, and Rock outcrop; gently or moderately sloping.
C1S1 Cryands with Cryochrepts, Cryumbrepts, and Cryods; steep.

Udands
C2a Udands with Haplumbrepts, Dystrochrepts, Spodosols, and Aquepts.

Ustands
C3S2 Ustands with Tropepts, Ustolls, Udands, and Tropofolists; moderately sloping to steep.

ARIDISOLS

Argids
D1a Argids with Cambids, Orthents, Psamments, and Ustolls; gently and moderately sloping.
D1S Argids with Cambids, gently sloping; and Torriorthents, gently sloping to steep.

Cambids
D2a Cambids with Argids, Orthents, Psamments, and Xerolls; gently or moderately sloping.
D2S Cambids, gently sloping to steep, with Argids, gently sloping; lithic subgroups of Torriorthents and Xerorthents, both steep.

ENTISOLS

Aquents
E1a Aquents with Quartzipsamments, Aquepts, Aquolls, and Aquods; gently sloping.

Orthents
E2a Torriorthents, steep, with xerollic and ustic subgroups of Aridisols; Usterts and aridic and vertic subgroups of Borolls; gently or moderately sloping.
E2b Torriorthents with torrerts; gently or moderately sloping.
E2c Xerorthents with Xeralfs, Cambids, and Argids; gently sloping.
E2S1 Torriorthents; steep, and Argids, Torrifluvents, Ustolls, and Borolls; gently sloping.
E2S2 Xerorthents with Xeralfs and Xerolls; steep.
E2S3 Cryorthents with Cryopsamments and Cryands; gently sloping to steep.

Psamments
E3a Quartzipsamments with Aquults and Aquolls; gently or moderately sloping.
E3b Udipsamments with Aquolls and Udalfs; gently or moderately sloping.
E3c Ustipsamments with Ustalfs and Aquolls; gently sloping.

HISTOSOLS

Histosols
H1a Borohemists with Psammaquents and Udipsamments; gently sloping.
H2a Medihemists and Medisaprists with Fluvaquents, Hydraquents, and Aquepts; gently sloping.
H3a Medifibrists, Medihemists, and Medisaprists with Psammaquents; gently sloping.

INCEPTISOLS

Aquepts
I2a Epi and endo great groups of Aquepts with Aqualfs, Aquerts, Aquolls, Udalfs, and Fluvaquents; gently sloping.
I2P Cryaquepts with Cryorthents, Cryochrepts, and cryic great groups of Histosols; gently sloping to steep.

Ochrepts
I3a1 Cryochrepts with Cryaquepts, Histosols, and Cryods; gently or moderately sloping.
I3a2 Cryochrepts, Cryands with Xerochrepts, Xerolls, and Cryods; moderately sloping to steep.
I3b Eutrochrepts with Uderts; gently sloping.
I3c Fragiochrepts with Fragiaquepts, gently or moderately sloping; and Dystrochrepts, steep.
I3d Dystrochrepts with Udipsamments and Haplorthods; gently sloping.
I3e Xerochrepts, Xerands with Aquepts and Orthods; gently or moderately sloping.
I3S Dystrochrepts, steep, with Udalfs and Udults; gently or moderately sloping.

MOLLISOLS

Aquolls
M1a Aquolls with Udalfs, Fluvents, Udipsamments, Aquepts, and Eutrochrepts; gently sloping.
M1b Aquolls and Aquerts with Borolls and Udipsamments; gently sloping.

Borolls
M2a Udic subgroups of Borolls with Aquolls and Udorthents; gently sloping.
M2b Typic subgroups of Borolls with Ustipsamments, Ustorthents, and Boralfs; gently sloping.
M2c Aridic subgroups of Borolls and Ustic subgroups of Argids, Cambids, and Torriorthents; gently sloping.
M2S Borolls with Boralfs, Argids, Torriorthents, and Ustolls; moderately sloping or steep.

Udolls
M3a Udolls, with Aquolls, Udalfs, Udalfs, Aqualfs, Fluvents, Udipsamments, Udorthents, Aquepts, and Albolls; gently or moderately sloping.

Ustolls
M4a Udic subgroups of Ustolls with Ustorthents, Ustochrepts, Usterts, Aquents, Fluvents, and Udolls; gently or moderately sloping.
M4b Typic subgroups of Ustolls with Ustalfs, Ustipsamments, Ustorthents, Ustochrepts, Aquolls, and Usterts; gently or moderately sloping.
M4c Aridic subgroups of Ustolls with Ustalfs, Cambids, Ustipsamments, Ustorthents, Ustochrepts, Torriorthents, and Usterts; gently or moderately sloping.
M4S Ustolls with Argids and Torriorthents; moderately sloping or steep.

Xerolls
M5a Xerolls with Argids, Cambids, Fluvents, Cryoboralfs, Cryoborolls, and Xerorthents; gently or moderately sloping.
M5S Xerolls with Cryoboralfs, Xeralfs, Xerorthents, and Xererts; moderately sloping or steep.

OXISOLS

Oxisols
O1a Ustox with Tropepts, Ustolls, Andisols, Udox, and Torrox; gently to steeply sloping.

SPODOSOLS

Aquods
S1a Aquods with Psammaquents, Aquolls, Alorthods, and Aquults; gently sloping.

Orthods
S2a Orthods with Boralfs, Aquents, Udorthents, Udipsamments, Histosols, Aquepts, Fragiochrepts, and Dystrochrepts; gently or moderately sloping.
S2S1 Orthods and Cryods with Histosols, Aquents, and Aquepts; moderately sloping or steep.

Cryods
S2S2 Cryods with Histosols; moderately sloping or steep.
S2S3 Cryods with Cryic great groups of Histosols, Cryands, and Cryaquepts; gently sloping to steep.

ULTISOLS

Aquults
U1a Aquults with Aquents, Histosols, Quartzipsamments, and Udults; gently sloping.

Humults
U2Sa Humults and Xeralfs with Xerolls, Aquepts, and Xerults; gently sloping to moderately sloping.
U2Sb Humults, Xeralfs, and Xerults with Xerolls and Rock outcrop; gently sloping to steep.

Udults
U3a Udults with Udalfs, Fluvents, Aquents, Quartzipsamments, Aquepts, Dystrochrepts, and Aquults; gently or moderately sloping.
U3S Udults with Dystrochrepts; moderately sloping or steep.

VERTISOLS

Uderts
V1a Uderts with Aqualfs, Aquerts, Eutrochrepts, Aquolls, and Ustolls; gently sloping.

Usterts
V2a Usterts with Aqualfs, Cambids, Aquerts, Fluvents, Aquolls, Ustolls, and Torrerts; gently sloping.

Miscellaneous Areas (with little soil)
X1 Salt flats and Playas
X2 Rock outcrop (plus permanent snow fields, glaciers, and lava flows).

Slope Classes
Gently sloping—Slopes mainly less than 10%, including nearly level.
Moderately sloping—Slopes mainly between 10 and 25%.
Steep—Slopes mainly steeper than 25%.

Major soil concept		Soil order

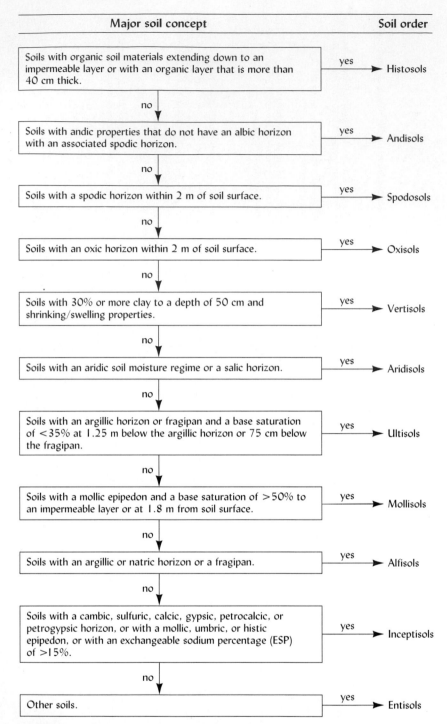

Soils with organic soil materials extending down to an impermeable layer or with an organic layer that is more than 40 cm thick.	yes →	Histosols
no ↓		
Soils with andic properties that do not have an albic horizon with an associated spodic horizon.	yes →	Andisols
no ↓		
Soils with a spodic horizon within 2 m of soil surface.	yes →	Spodosols
no ↓		
Soils with an oxic horizon within 2 m of soil surface.	yes →	Oxisols
no ↓		
Soils with 30% or more clay to a depth of 50 cm and shrinking/swelling properties.	yes →	Vertisols
no ↓		
Soils with an aridic soil moisture regime or a salic horizon.	yes →	Aridisols
no ↓		
Soils with an argillic horizon or fragipan and a base saturation of <35% at 1.25 m below the argillic horizon or 75 cm below the fragipan.	yes →	Ultisols
no ↓		
Soils with a mollic epipedon and a base saturation of >50% to an impermeable layer or at 1.8 m from soil surface.	yes →	Mollisols
no ↓		
Soils with an argillic or natric horizon or a fragipan.	yes →	Alfisols
no ↓		
Soils with a cambic, sulfuric, calcic, gypsic, petrocalcic, or petrogypsic horizon, or with a mollic, umbric, or histic epipedon, or with an exchangeable sodium percentage (ESP) of >15%.	yes →	Inceptisols
no ↓		
Other soils.	yes →	Entisols

FIGURE A.2 Simplified key showing the central concepts of the eleven orders in *Soil Taxonomy*.

Appendix B
CANADIAN AND FAO SOIL CLASSIFICATION SYSTEMS

TABLE B.1 Summary with Brief Descriptions of the Soil Orders in the Soil Classification System of Canada

Luvisolic	Soils with light-colored, eluvial horizons that have illuvial B horizons in which silicate clay has accumulated.
Solonetzic	Soils that occur on saline (often high in sodium) parent materials, which have B horizons that are very hard when dry and swell to a sticky mass of very low permeability when wet. Typically the solonetzic B horizon has prismatic or columnar macrostructure that breaks to hard to extremely hard, blocky pods with dark coatings.
Regosolic	Weakly developed soils that lack development of genetic horizons.
Organic	Peat or bog soils.
Brunisolic	Soils with sufficient development to exclude them from the Regosolic order, but lacking the degree or kind of horizon development specified for other soil orders.
Gleysolic	Gleysolic soils have features indicative of periodic or prolonged saturation (i.e., gleying, mottling) with water and reducing conditions.
Chernozemic	Well to imperfectly drained soils with high base saturation and surface horizons darkened by the accumulation of organic matter from the decomposition of plants from grassland or grassland-forest ecosystems.
Podzolic	Soils with a B horizon in which the dominant accumulation product is amorphous material composed mainly of humified organic matter combined in varying degrees with Al and Fe.
Cryosolic	Soils formed in either mineral or organic materials that have permafrost either within 1 meter of the surface or within 2 meters if more than one-third of the pedon has been strongly cryoturbated, as indicated by disrupted, mixed, or broken horizons.

TABLE B.2 Comparison of U.S. Soil Taxonomy and the Canadian Soil Classification System

U.S. Soil Taxonomy Soil Order	Equivalent in Canadian Soil Classification System

Alfisols	Luvisolic
	  Grey Brown Luvisols—Hapludalfs
	  Grey Luvisols—Cryoboralfs, Eutroboralfs, Fragiboralfs, Glossoboralfs, Paleboralf
	Solonetzic
	  Solonetz—Nartriboralfs
	  Solod—Glossic subgroups of Natraqualfs, Natriboralfs

Aridisols	Solonetzic—frigid families of Natrargids
Entisols	Regosolic—Cryic great groups and frigid families of Entisols, except Aquents Regosol—Cryic great groups and frigid families of Folists, Fluvents, Orthents, and Psamments

Histosols	Organic—Histosols except Folists
	  Fibrisol—Borofibrists, Cryofibrists, and Sphagnofibrists, some Medifibrists
	  Mesisol—Borohemists and Cryohemists, some Medihemists
	  Humisol—Borosaprists and Cryosaprists, some Medisaprists

Inceptisols	Brunisolic—Cryic great groups and frigid families of Inceptisols
	  Melanic Brunisol—Some Eutrochrepts
	  Eutric Brunisol—Typic and pergelic subgroups of Cryochrepts; and frigid and mesic families of Eutrochrepts
	  Sombric Brunisol—Frigid and mesic families of Umbrepts and of Umbric Dystrochrepts
	  Dyrtric Brunisol—Frigid families of subgroups of Dystrochrepts; and dystric subgroups of Cyrochrepts
	Gleysolic—Cryic great groups and frigid families of aquic suborders
	  Humic Gleysol—Humaquepts
	  Gleysol—Cryaquepts and frigid families of Fragiquepts and Haplaquepts

Mollisols	Chernozemic—Borolls
	  Brown (Light Chestnut)—Aridic subgroups of Argiborolls and Haploborolls
	  Dark Brown (Dark Chestnut)—Subgroups of Argiborolls and Haploborolls except aridic and udic
	  Black—Udic subgroups of Argiborolls and Haploborolls
	  Dark Gray—Boralfic subgroups of Borolls
	Solonetzic—Natric great groups of Mollisols
	  Solonetz—Natriborolls, and frigid families of Natraquolls and Natralbolls
	  Solod—Glossic subgroups of Natriborolls

Oxisols	Not relevant in Canada

Spodosols	Podzolic—Spodosols
	  Humic Podzol—Cryohumods, Placohumods, and frigid families of other Humods
	  Ferro-Humic Podzol—Humic Cryorthods, Placorrthods, and frigid families of humic subgroups of other Orthods
	  Humo-Ferric Podzol—Cryorthods, Placorthods, and frigid families of other Orthods, except humic subgroups

Ultisols	Not relevant in Canada
Andosols	Component of Brunisolic and Cryosolic
Vertisols	Grumic subgroups of Chernozemic

TABLE B.3   Soil Map Units for the FAO/UNESCO Soil Map of the World

Fluvisols	Water-deposited soils with little alteration
Regosols	Thin soil over unconsolidated material
Arenosols	Soils formed from sand
Gleysols	Soils with mottled or reduced horizons due to wetness
Rendzinas	Shallow soils over limestone
Rankers	Thin soils over siliceous material
Andosols	Soils formed in volcanic ash that have dark surfaces
Vertisols	Self-mulching, inverting soils, rich in montmorillinitic clay
Yermosols	Desert soils
Xerosols	Dry soils of semiarid regions
Solonchaks	Soils with soluble salt accumulation
Solonetz	Soils with high sodium content
Planosols	Soils with abrupt A–B horizon contact
Kastanozems	Soils with chestnut surface color, steppe vegetation
Chernozems	Soils with black surface, high humus under prairie vegetation
Phaeozems	Soils with dark surface, more leached than Kastanozem or Chernozem
Greyzems	Soils with dark surface, bleaced A2, and textural B
Cambisols	Soils with light color, structure, or consistency change due to weathering
Luvisols	Medium to high base status soils with argillic horizons
Podzoluvisols	Soils with leached horizons tonguing into argillic B horizons
Podzols	Soils with light-colored eluvial horizon and subsoil accumulation of iron aluminum, and humus
Acrisols	Low base status soils with argillic horizons
Nitosols	Soils with low CEC clay in argillic horizons
Ferralsols	Highly weathered soils with sesquioxide-rich clays
Histosols	Organic soils
Lithosols	Shallow soils over hard rock

Appendix C
FAMILY DIFFERENTIAE FOR SOIL TAXONOMY

The *Soil Taxonomy* system embodies six categories. From the highest to the lowest rank they are (1) orders, (2) suborders, (3) great groups, (4) subgroups, (5) families, and (6) series. The nomenclature for all categories except the family involves only one or two words. Since the family includes soils with common properties that affect the soil's response to management, the family name must include terms that describe these properties. This is accomplished by using eight different property classes, several of which may be used to differentiate the family, as follows:

1. Particle size classes
2. Mineralogical classes
3. Soil temperature classes
4. Calcareous and reaction classes
5. Soil depth classes
6. Soil consistence classes
7. Classes of coatings on sand
8. Permanent crack classes

Particle Size Classes

Particle size distribution is for the whole soil, not just the fine-earth fraction. Seven classes are used for most soils.

1. *Fragmental (90% or more)*—Stones, cobbles, gravel, and very coarse sand.

Classes with less than 35% rock fragments (>2 mm in diameter):
2. *Sandy*—Fine earth is sand or loamy sand.
3. *Loamy*—Fine earth is loamy, very fine sand, or finer, but less than 35% clay.
4. *Clayey*—Fine earth contains 35% or more clay.

Classes with more than 35% rock fragments:
5. *Sandy-skeletal*—Fine earth is sand or loamy sand.

6. *Loamy-skeletal*—Fine earth is loamy, very fine sand, or finer, but less than 35% clay.
7. *Clayey-skeletal*—Fine earth is clay.

In soils developed in volcanic ejecta, these particle size classes are replaced by terms that connote not only particle size but also mineralogy, such as *cindery, ashy, medial,* and *thixotropic.*

Mineralogical Classes

Mineralogical classes refer to the mineral or minerals that are most prominent in the soil. Two terms (*vermiculitic/chloritic*) may be used as well as mixed where several minerals are prominent. Common mineralogical classes include the following:

	Class	Dominant feature
	Applied to soils of any particle size class	
1.	Carbonatic	Carbonates, some gypsum
2.	Ferritic	Iron oxides
3.	Gibbsitic	Aluminum oxides (gibbsite)
4.	Oxidic	High Fe, Al oxides in clay
5.	Gypsic	Gypsum
6.	Serpentinitic	Serpentine
7.	Glauconitic	Glauconite
	Applied to coarser classes (e.g., fragmental, sandy)	
8.	Micaceous	Micas
9.	Siliceous	Silica minerals
10.	Mixed	Mixtures
	Applied to finer classes (e.g., clayey)	
11.	Halloysitic	Halloysite
12.	Kaolinitic	Kaolinite
13.	Smectitic	Smectite
14.	Illitic[a]	Illite
15.	Vermiculitic	Vermiculite
16.	Chloritic	Chlorite
17.	Mixed	Mixtures of clays

[a]Refers to fine-grained micas.

Soil Temperature Classes

These classes relate to the mean annual temperature at a standard depth (50 cm) and to differences between summer and winter temperatures, as follows.

Mean annual temperature (°C)	Differences between summer and winter temperatures	
	Greater than 5°C	*Less than 5°C*
<8	Frigid	Isofrigid
8–15	Mesic	Isomesic
15–22	Thermic	Isothermic
>22	Hyperthermic	Isohyperthermic

Calcareous and Reaction Classes

These classes refer to the presence or absence of carbonates and to the soil acidity/alkalinity. Terms such as *calcareous, noncalcareous, acid, nonacid,* and *allic* (high in Al^{3+}) are used.

Soil Depth Classes

Depth of soil classes may be used in all orders of mineral soils, terms such as *micro* and *shallow* being used.

Soil Consistence Classes

Consistence classes are used to indicate cementation of some families, especially of Spodosols.

Classes of Coatings on Sand

Coatings classes are used to indicate the coating of sand grains with silt and clay.

Permanent Crack Classes

These classes are used to identify soils with permanent cracks.

GLOSSARY OF SOIL SCIENCE TERMS[1]

A horizon   The surface horizon of a mineral soil having maximum organic matter accumulation, maximum biological activity, and/or eluviation of materials such as iron and aluminum oxides and silicate clays.

absorption, active   Movement of ions and water into the plant root as a result of metabolic processes by the root, frequently against an activity gradient.

absorption, passive   Movement of ions and water into the plant root as a result of diffusion along a gradient.

accelerated erosion   *See* erosion.

acid rain   Atmospheric precipitation with pH values less than about 5.6, the acidity being due to inorganic acids such as nitric and sulfuric that are formed when oxides of nitrogen and sulfur are emitted into the atmosphere.

acid soil   A soil with a pH value <7.0. Usually applied to surface layer or root zone, but may be used to characterize any horizon. *See also* reaction, soil.

acid sulfate soils   Soils which are potentially extremely acid (pH <3.5) because of the presence of large amounts of reduced forms of sulfur that are oxidized to sulfuric acid if the soils are exposed to oxygen when they are drained or excavated. A sulfuric horizon containing the yellow mineral jarosite is often present. *See also* cat clays.

acidity, active   The activity of hydrogen ion in the aqueous phase of a soil. It is measured and expressed as a pH value.

acidity, residual   Soil acidity that can be neutralized by lime or other alkaline materials but cannot be replaced by an unbuffered salt solution.

acidity, salt replaceable   Exchangeable hydrogen and aluminum that can be replaced from an acid soil by an unbuffered salt solution such as KCl or NaCl.

acidity, total   The total acidity in a soil. It is approximated by the sum of the salt-replaceable acidity plus the residual acidity.

---

[1]This glossary was compiled and modified from several sources including *Glossary of Soil Science Terms* (Madison, Wis.: Soil Sci. Soc. Amer., 1987), *Resource Conservation Glossary* (Anheny, Iowa: Soil Cons. Soc. Amer., 1982), and *Soil Taxonomy* (Washington, D.C.: U.S. Department of Agriculture, 1976).

actinomycetes A group of organisms intermediate between the bacteria and the true fungi that usually produce a characteristic branched mycelium. Includes many, but not all, organisms belonging to the order of Actinomycetales.

activated sludge Sludge that has been aerated and subjected to bacterial action.

active organic matter A portion of the soil organic matter that is relatively easily metabolized by microorganisms and cycles with a half-life in the soil of a few days to a few years.

adhesion Molecular attraction that holds the surfaces of two substances (e.g., water and sand particles) in contact.

adsorption The attraction of ions or compounds to the surface of a solid. Soil colloids adsorb large amounts of ions and water.

adsorption complex The group of organic and inorganic substances in soil capable of adsorbing ions and molecules.

aerate To impregnate with gas, usually air.

aeration, soil The process by which air in the soil is replaced by air from the atmosphere. In a well-aerated soil, the soil air is similar in composition to the atmosphere above the soil. Poorly aerated soils usually contain more carbon dioxide and correspondingly less oxygen than the atmosphere above the soil.

aerobic (1) Having molecular oxygen as a part of the environment. (2) Growing only in the presence of molecular oxygen, as aerobic organisms. (3) Occurring only in the presence of molecular oxygen (said of certain chemical or biochemical processes, such as aerobic decomposition).

aggregate (soil) Many soil particles held in a single mass or cluster such as a clod, crumb, block, or prism.

agric horizon *See* diagnostic subsurface horizons.

agronomy A specialization of agriculture concerned with the theory and practice of field-crop production and soil management. The scientific management of land.

air-dry (1) The state of dryness (of a soil) at equilibrium with the moisture content in the surrounding atmosphere. The actual moisture content will depend upon the relative humidity and the temperature of the surrounding atmosphere. (2) To allow to reach equilibrium in moisture content with the surrounding atmosphere.

air porosity The proportion of the bulk volume of soil that is filled with air at any given time or under a given condition, such as a specified moisture potential; usually the large pores.

albic horizon *See* diagnostic subsurface horizons.

Alfisols *See* soil classification.

algal bloom A population explosion of algae in surface waters such as lakes and streams, often resulting in high turbidity and green- or red-colored water, and commonly stimulated by nutrient enrichment with phosphorus and nitrogen.

alkali soil (Obsolete) A soil that contains sufficient alkali (sodium) to interfere with the growth of most crop plants. *See also* saline-sodic soil; sodic soil.

alkaline soil Any soil that has pH >7. Usually applied to surface layer or root zone but may be used to characterize any horizon or a sample thereof. *See also* reaction, soil.

allelopathy The process by which one plant may affect other plants by biologically active chemicals introduced into the soil either directly by leaching or exudation from the source plant, or as a result of the decay of the plant residues. The effects, though usually negative, may also be positive.

allophane An aluminosilicate mineral that has an amorphous or poorly crystalline structure and is commonly found in soils developed from volcanic ash.

alluvial soil (Obsolete) A soil developing from recently deposited alluvium and exhibiting essentially no horizon development or modification of the recently deposited materials.

alluvium A general term for all detrital material deposited or in transit by streams, including gravel, sand, silt, clay, and all variations and mixtures of these. Unless otherwise noted, alluvium is unconsolidated.

alpha particle A positively charged particle (consisting of two protons and two neutrons) that is emitted by certain radioactive compounds.

aluminosilicates Compounds containing aluminum, silicon, and oxygen as main constituents. An example is mocrocline, $KAlSi_3O_8$.

amendment, soil Any substance other than fertilizers, such as lime, sulfur, gypsum, and sawdust, used to alter the chemical or physical properties of a soil, generally to make it more productive.

amino acids Nitrogen-containing organic acids that couple together to form proteins. Each acid molecule contains one or more amino groups ($-NH_2$) and at least one carboxyl group ($-COOH$). In addition, some amino acids contain sulfur.

ammonification The biochemical process whereby ammoniacal nitrogen is released from nitrogen-containing organic compounds.

ammonium fixation The entrapment of ammonium ions by the mineral or organic fractions of the soil in forms that are insoluble in water and at least temporarily nonexchangeable.

amorphous material Noncrystalline constituents of soils.

anaerobic (1) Without molecular oxygen. (2) Living or functioning in the absence of air or free oxygen.

Andisols *See* soil classification.

anion Negatively charged ion; during electrolysis it is attracted to the positively charged anode.

anion exchange capacity The sum total of exchangeable anions that a soil can adsorb. Expressed as centimoles of charge per kilogram ($cmol_c/kg$) of soil (or of other adsorbing material such as clay).

anoxic *See* anaerobic.

anthropic epipedon *See* diagnostic surface horizons.

antibiotic A substance produced by one species of organism that, in low concentrations, will kill or inhibit growth of certain other organisms.

Ap The surface layer of a soil disturbed by cultivation or pasturing.

apatite A naturally occurring complex calcium phosphate that is the original source of most of the phosphate fertilizers. Formulas such as $[3Ca_3(PO_4)_2] \cdot CaF_2$ illustrate the complex compounds that make up apatite.

arbuscule Specialized branched structure formed within a root cortical cell by endotrophic mycorrhizal fungi.

argillic horizon *See* diagnostic subsurface horizons.

arid climate Climate in regions that lack sufficient moisture for crop production without irrigation. In cool regions annual precipitation is usually less than 25 cm. It may be as high as 50 cm in tropical regions. Natural vegetation is desert shrubs.

Aridisols *See* soil classification.

aspect (of slopes) The direction (e.g., south or north) that a slope faces with respect to the sun.

association, soil *See* soil association.

Atterberg limits Water contents of fine-grained soils at different states of consistency.

plastic limit (PL) The water content corresponding to an arbitrary limit between the plastic and semisolid states of consistency of a soil.

liquid limit (LL) The water content corresponding to the arbitrary limit between the liquid and plastic states of consistency of a soil.

autochthonous organisms Those microorganisms thought to subsist on the more resistant soil organic matter and little effected by the addition of fresh organic materials. *Contrast with* zymogenous organisms.

autotroph An organism capable of utilizing carbon dioxide or carbonates as the sole source of carbon and obtaining energy for life processes from the oxidation of inorganic elements or compounds such as iron, sulfur, hydrogen, ammonium, and nitrites, or from radiant energy. *Contrast with* heterotroph.

available nutrient That portion of any element or compound in the soil that can be readily absorbed and assimilated by growing plants. ("Available" should not be confused with "exchangeable.")

available water The portion of water in a soil that can be readily absorbed by plant roots. The amount of water released between the field capacity and the permanent wilting point.

B horizon A soil horizon usually beneath the A horizon that is characterized by one or more of the following: (1) a concentration of silicate clays, iron and aluminum oxides, and humus, alone or in combination; (2) a blocky or prismatic structure, and (3) coatings of iron and aluminum oxides that give darker, stronger, or redder color.

bar A unit of pressure equal to one million dynes per square centimeter (10^6 dynes/cm^2). It approximates the pressure of a standard atmosphere.

base-forming cations Those cations that form strong (strongly dissociated) bases by reaction with hydroxyl, e.g., K+ forms potassium hydroxide ($K^+ + OH^-$).

base saturation percentage The extent to which the adsorption complex of a soil is saturated with exchangeable cations other than hydrogen and aluminum. It is expressed as a percentage of the total cation exchange capacity.

BC soil A soil profile with B and C horizons but with little or no A horizon. Most BC soils have lost their A horizons by erosion.

bedding (Engineering) Arranging the surface of fields by plowing and grading into a series of elevated beds separated by shallow depressions or ditches for drainage.

bedrock The solid rock underlying soils and the regolith in depths ranging from zero (where exposed by erosion) to several hundred feet.

bench terrace An embankment constructed across sloping fields with a steep drop on the downslope side.

beta particle A high-speed electron emitted in radioactive decay.

biodegradable Subject to degradation by biochemical processes.

biomass The total mass of living material of a specified type (e.g., microbial biomass) in a given environment (e.g., in a cubic meter of soil).

bioremediation The decontamination or restoration of polluted or degraded soils by means of enhancing the chemical degradation or other activities of soil organisms.

biosolids *See* sewage sludge.

bleicherde (Obsolete) The light-colored, leached A2(E) horizon of Spodosols.

blocky soil structure Soil aggregates with blocklike shapes; common in B horizons of soils in humid regions.

blown-out land Areas from which all or almost all of the soil and soil material has been removed by wind erosion. Usually unfit for crop production. A miscellaneous land type.

border-strip irrigation *See* irrigation methods.

bottomland *See* floodplain.

breccia A rock composed of coarse angular fragments cemented together.

broad-base terrace A low embankment with such gentle slopes that it can be farmed, constructed across sloping fields to reduce erosion and runoff.

broadcast Scatter seed or fertilizer on the surface of the soil.

buffering capacity The ability of a soil to resist changes in pH. Commonly determined by presence of clay, humus, and other colloidal materials.

bulk blending Mixing dry individual granulated fertilizer materials to form a mixed fertilizer that is applied promptly to the soil.

bulk density, soil The mass of dry soil per unit of bulk volume, including the air space. The bulk volume is determined before drying to constant weight at 105°C.

buried soil Soil covered by an alluvial, loessal, or other deposit, usually to a depth greater than the thickness of the solum.

C horizon A mineral horizon generally beneath the solum that is relatively unaffected by biological activity and pedogenesis and is lacking properties diagnostic of an A or B horizon. It may or may not be like the material from which the A and B have formed.

calcareous soil Soil containing sufficient calcium carbonate (often with magnesium carbonate) to effervesce visibly when treated with cold 0.1 N hydrochloric acid.

calcic horizon *See* diagnostic subsurface horizons.

caliche A layer near the surface, more or less cemented by secondary carbonates of calcium or magnesium precipitated from the soil solution. It may occur as a soft, thin soil horizon; as a hard, thick bed just beneath the solum; or as a surface layer exposed by erosion.

cambic horizon *See* diagnostic subsurface horizons.

capillary conductivity (Obsolete) *See* hydraulic conductivity.

capillary water The water held in the capillary or *small* pores of a soil, usually with a tension >60 cm of water. *See also* moisture potential.

carbon cycle The sequence of transformations whereby carbon dioxide is fixed in living organisms by photosynthesis or by chemosynthesis, liberated by respiration and by the death and decomposition of the fixing organism, used by heterotrophic species, and ultimately returned to its original state.

carbon/nitrogen ratio The ratio of the weight of organic carbon (C) to the weight of total nitrogen (N) in a soil or in organic material.

carnivore An organism that feeds on animals.

casts, earthworm Rounded, water-stable aggregates of soil that have passed through the gut of an earthworm.

"cat" clays Wet clay soils high in reduced forms of sulfur that, upon being drained, become extremely acid because of the oxidation of the sulfur compounds and the formation of sulfuric acid. Usually found in tidal marshes. *See* acid sulfate soils.

catena A sequence of soils of about the same age, derived from similar parent material, and occurring under similar climatic conditions, but having different characteristics because of variation in *relief* and in *drainage*.

cation A positively charged ion; during electrolysis it is attracted to the negatively charged cathode.

cation exchange The interchange between a cation in solution and another cation on the surface of any surface-active material such as clay or organic matter.

cation exchange capacity The sum total of exchangeable cations that a soil can adsorb. Sometimes called *total-exchange capacity, base-exchange capacity,* or *cation-adsorption capacity.* Expressed in centimoles of charge per kilogram ($cmol_c/kg$) of soil (or of other adsorbing material such as clay).

cemented Indurated; having a hard, brittle consistency because the particles are held together by cementing substances such as humus, calcium carbonate, or the oxides of silicon, iron, and aluminum.

channery Thin, flat fragments of limestone, sandstone, or schist up to 15 cm (6 in.) in major diameter.

chelate (Greek, claw) A type of chemical compound in which a metallic ion is firmly combined with an organic molecule by means of multiple chemical bonds.

chert A structureless form of silica, closely related to flint, that breaks into angular fragments.

chisel, subsoil A tillage implement with one or more cultivator-type feet to which are attached strong knifelike units used to shatter or loosen hard, compact layers, usually in the subsoil, to depths below normal plow depth. *See also* subsoiling.

chlorite A 2:1:1-type layer-structured silicate mineral having 2:1 layers alternating with a magnesium-dominated octahedral sheet.

chlorosis A condition in plants relating to the failure of chlorophyll (the green coloring matter) to develop. Chlorotic leaves range from light green through yellow to almost white.

chronosequence A sequence of related soils that differ, one from the other, in certain properties primarily as a result of time as a soil-forming factor.

class, soil A group of soils having a definite range in a particular property such as acidity, degree of slope, texture, structure, land-use capability, degree of erosion, or drainage. *See also* soil structure; soil texture.

classification, soil *See* soil classification.

clastic Composed of broken fragments of rocks and minerals.

clay (1) A soil separate consisting of particles <0.002 mm in equivalent diameter. (2) A soil textural class containing >40% clay, <45% sand, and <40% silt.

clay mineral Naturally occurring inorganic material (usually crystalline) found in soils and other earthy deposits, the particles being of clay size, that is, <0.002 mm in diameter.

claypan A dense, compact, slowly permeable layer in the subsoil having a much higher clay content than the overlying material, from which it is separated by a sharply defined boundary. Claypans are usually hard when dry and plastic and sticky when wet. *See also* hardpan.

clod A compact, coherent mass of soil produced artificially, usually by such human activities as plowing and digging, especially when these operations are performed on soils that are either too wet or too dry for normal tillage operations.

coarse texture The texture exhibited by sands, loamy sands, and sandy loams (except very fine sandy loam).

cobblestone Rounded or partially rounded rock or mineral fragments 7.5–25 cm (3–10 in.) in diameter.

cohesion Holding together: force holding a solid or liquid together, owing to attraction between like molecules. Decreases with rise in temperature.

colloid, soil (Greek, gluelike) Organic and inorganic matter with very small particle size and a correspondingly large surface area per unit of mass.

colluvium A deposit of rock fragments and soil material accumulated at the base of steep slopes as a result of gravitational action.

color The property of an object that depends on the wavelength of light it reflects or emits.

columnar soil structure *See* soil structure types.

compost Organic residues, or a mixture of organic residues and soil, that have been piled, moistened, and allowed to undergo biological decomposition. Mineral fertilizers are sometimes added. Often called *artificial manure* or *synthetic manure* if produced primarily from plant residues.

concretion A local concentration of a chemical compound, such as calcium carbonate or iron oxide, in the form of grains or nodules of varying size, shape, hardness, and color.

conduction The transfer of heat by physical contact between two or more objects.

conductivity, hydraulic *See* hydraulic conductivity.

conifer A tree belonging to the order Coniferae, usually evergreen, with cones and needle-shaped or scalelike leaves and producing wood known commercially as *softwood.*

conservation tillage *See* tillage, conservation.

consistence The combination of properties of soil material that determine its resistance to crushing and its ability to be molded or changed in shape. Such terms as *loose, friable, firm, soft, plastic,* and *sticky* describe soil consistence.

constant charge The net surface charge of mineral particles, the magnitude of which depends only on the chemical and structural composition of the mineral. The charge arises from isomorphous substitution and is not affected by soil pH.

consumptive use The water used by plants in transpiration and growth, plus water vapor loss from adjacent soil or snow, or from intercepted precipitation in any specified time. Usually expressed as equivalent depth of free water per unit of time.

contour An imaginary line connecting points of equal elevation on the surface of the soil. A contour terrace is laid out on a sloping soil at right angles to the direction of the slope and nearly level throughout its course.

contour strip-cropping Layout of crops in comparatively narrow strips in which the farming operations are performed approximately on the contour. Usually strips of grass, close-growing crops, or fallow are alternated with those of cultivated crops.

convection The transfer of heat through a gas or solution because of molecular movement.

corrugated irrigation *See* irrigation methods.

cover crop A close-growing crop grown primarily for the purpose of protecting and improving soil between periods of regular crop production or between trees and vines in orchards and vineyards.

creep Slow mass movement of soil and soil material down relatively steep slopes primarily under the influence of gravity, but facilitated by saturation with water and by alternate freezing and thawing.

crop rotation A planned sequence of crops growing in a regularly recurring succession on the same area of land, as contrasted to continuous culture of one crop or growing different crops in haphazard order.

crotovina A former animal burrow in one soil horizon that has been filled with organic matter or material from another horizon (also spelled *krotovina*).

crumb A soft, porous, more or less rounded natural unit of structure from 1 to 5 mm in diameter. *See also* soil structure types.

crushing strength The force required to crush a mass of dry soil or, conversely, the resistance of the dry soil mass to crushing. Expressed in units of force per unit area (pressure).

crust A surface layer on soils, ranging in thickness from a few millimeters to perhaps as much as 3 cm, that is much more compact, hard, and brittle, when dry, than the material immediately beneath it.

crystal A homogeneous inorganic substance of definite chemical composition bounded by plane surfaces that form definite angles with each other, thus giving the substance a regular geometrical form.

crystal structure The orderly arrangement of atoms in a crystalline material.

crystalline rock A rock consisting of various minerals that have crystallized in place from magma. *See also* igneous rock; sedimentary rock.

cultivation   A tillage operation used in preparing land for seeding or transplanting or later for weed control and for loosening the soil.

deciduous plant   A plant that sheds all its leaves every year at a certain season.

decomposition   Chemical breakdown of a compound (e.g., a mineral or organic compound) into simpler compounds, often accomplished with the aid of microorganisms.

deflocculate   (1) To separate the individual components of compound particles by chemical and/or physical means. (2) To cause the particles of the *disperse phase* of a colloidal system to become suspended in the *dispersion medium*.

delta   An alluvial deposit formed where a stream or river drops its sediment load upon entering a quieter body of water.

denitrification   The biochemical reduction of nitrate or nitrite to gaseous nitrogen, either as molecular nitrogen or as an oxide of nitrogen.

density   *See* particle density; bulk density.

desalinization   Removal of salts from saline soil, usually by leaching.

desert crust   A hard layer, containing calcium carbonate, gypsum, or other binding material, exposed at the surface in desert regions.

detritus   Debris from dead plants and animals.

detritivore   An organism that subsists on *detritus*.

desorption   The removal of sorbed material from surfaces.

diagnostic horizons   (As used in *Soil Taxonomy*): Horizons having specific soil characteristics that are indicative of certain classes of soils. Horizons that occur at the soil surface are called *epipedons;* those below the surface, *diagnostic subsurface horizons.*

diagnostic subsurface horizons   The following diagnostic subsurface horizons are used in *Soil Taxonomy:*

  agric horizon   A mineral soil horizon in which clay, silt, and humus derived from an overlying cultivated and fertilized layer have accumulated. The wormholes and illuvial clay, silt, and humus occupy at least 5% of the horizon by volume.

  albic horizon   A mineral soil horizon from which clay and free iron oxides have been removed or in which the oxides have been segregated to the extent that the color of the horizon is determined primarily by the color of the primary sand and silt particles rather than by coatings on these particles.

  argillic horizon   A mineral soil horizon characterized by the illuvial accumulation of layer-lattice silicate clays.

  calcic horizon   A mineral soil horizon of secondary carbonate enrichment that is more than 15 cm thick has a calcium carbonate equivalent of more than 15%, and has at least 5% more calcium carbonate equivalent than the underlying C horizon.

  cambic horizon   A mineral soil horizon that has a texture of loamy very fine sand or finer, contains some weatherable minerals, and is characterized by the alteration or removal of mineral material. The cambic horizon lacks cementation or induration and has too few evidences of illuviation to meet the requirements of the argillic or spodic horizon.

  duripan   A mineral soil horizon that is cemented by silica, to the point that air-dry fragments will not slake in water or HCl.

  gypsic horizon   A mineral soil horizon of secondary calcium sulfate enrichment that is more than 15 cm thick.

  kandic horizon   A horizon having a sharp clay increase relative to overlying horizons and having low-activity clays.

  natric horizon   A mineral soil horizon that satisfies the requirements of an argillic horizon, but that also has prismatic, columnar, or blocky structure and a subhorizon having more than 15% saturation with exchangeable sodium.

  oxic horizon   A mineral soil horizon that is at least 30 cm thick and characterized by the virtual *absence* of weatherable primary minerals or 2:1 lattice clays and the *presence* of 1:1 lattice clays and highly insoluble minerals such as quartz sand, hydrated oxides of iron and aluminum, low cation exchange capacity, and small amounts of exchangeable bases.

petrocalcic horizon   A continuous, indurated calcic horizon that is cemented by calcium carbonate and, in some places, with magnesium carbonate. It cannot be penetrated with a spade or auger when dry; dry fragments do not slake in water; and it is impenetrable to roots.

petrogypsic horizon   A continuous, strongly cemented, massive gypsic horizon that is cemented by calcium sulfate. It can be chipped with a spade when dry. Dry fragments do not slake in water and it is impenetrable to roots.

placic horizon   A black to dark reddish mineral soil horizon that is usually thin but that may range from 1 to 25 mm in thickness. The placic horizon is commonly cemented with iron and is slowly permeable or impenetrable to water and roots.

salic horizon   A mineral soil horizon of enrichment with secondary salts more soluble in cold water than gypsum. A salic horizon is 15 cm or more in thickness.

sombric horizon   A mineral subsurface horizon that contains illuvial humus but has a low cation exchange capacity and low percentage base saturation. Mostly restricted to cool, moist soils of high plateaus and mountainous areas of tropical and subtropical regions.

spodic horizon   A mineral soil horizon characterized by the illuvial accumulation of amorphous materials composed of aluminum and organic carbon with or without iron.

sulfuric horizon   A subsurface horizon in either mineral or organic soils that has a pH <3.5, fresh straw-colored mottles (called *jarosite mottles*). Forms by oxidation of sulfide-rich materials and is highly toxic to plants.

diagnostic surface horizons   The following diagnostic surface horizons are used in *Soil Taxonomy* and are called *epipedons:*

anthropic epipedon   A surface layer of mineral soil that has the same requirements as the mollic epipedon but that has more than 250 mg/kg of P_2O_5 soluble in 1% citric acid, or is dry more than 10 months (cumulative) during the period when not irrigated. The anthropic epipedon forms under long-continued cultivation and fertilization.

histic epipedon   A thin organic soil horizon that is saturated with water at some period of the year unless artificially drained and that is at or near the surface of a mineral soil.

melanic epipedon   A surface horizon, formed in volcanic parent material, that contains more than 6% organic carbon, is dark in color, has a very low bulk density and high anion adsorption capacity.

mollic epipedon   A surface horizon of mineral soil that is dark colored and relatively thick, contains at least 0.6% organic carbon, is not massive and hard when dry, has a base saturation of more than 50%, has less than 250 mg/kg P_2O_5 soluble in 1% citric acid, and is dominantly saturated with bivalent cations.

ochric epipedon   A surface horizon of mineral soil that is too light in color, too high in chroma, too low in organic carbon, or too thin to be a plaggen, mollic, umbric, anthropic, or histic epipedon, or that is both hard and massive when dry.

plaggen epipedon   A human-made surface horizon more than 50 cm thick that is formed by long-continued manuring and mixing.

umbric epipedon   A surface layer of mineral soil that has the same requirements as the mollic epipedon with respect to color, thickness, organic carbon content, consistence, structure, and P_2O_5 content, but that has a base saturation of less than 50%.

diatoms   Algae having siliceous cell walls that persist as a skeleton after death; any of the microscopic unicellular or colonial algae constituting the class Bacillariaceae. They occur abundantly in fresh and salt waters and their remains are widely distributed in soils.

diatomaceous earth   A geologic deposit of fine, grayish, siliceous material composed chiefly or wholly of the remains of diatoms. It may occur as a powder or as a porous, rigid material.

diffusion The movement of atoms in a gaseous mixture or of ions in a solution, primarily as a result of their own random motion.

dioctahedral sheet An octahedral sheet of silicate clays in which the sites for the six-coordinated metallic atoms are mostly filled with trivalent atoms such as Al^{3+}.

disintegration Physical or mechanical breakup or separation of a substance into its component parts (e.g., a rock breaking into its mineral components).

disperse (1) To break up compound particles, such as aggregates, into the individual component particles. (2) To distribute or suspend fine particles, such as clay, in or throughout a dispersion medium, such as water.

dissolution Process by which molecules of a gas, solid, or another liquid dissolve in a liquid, thereby becoming completely and uniformly dispersed throughout the liquid's volume.

diversion dam A structure or barrier built to divert part or all of the water of a stream to a different course.

diversion terrace *See* terrace.

drain (1) To provide channels, such as open ditches or drain tile, so that excess water can be removed by surface or by internal flow. (2) To lose water (from the soil) by percolation.

drainage, soil The frequency and duration of periods when the soil is free from saturation with water.

drift Material of any sort deposited by geological processes in one place after having been removed from another. Glacial drift includes material moved by the glaciers and by the streams and lakes associated with them.

drumlin Long, smooth cigar-shaped low hills of glacial till, with their long axes parallel to the direction of ice movement.

dryland farming The practice of crop production in low-rainfall areas without irrigation.

duff The matted, partly decomposed organic surface layer of forest soils.

duripan *See* diagnostic subsurface horizons; hardpan.

dust mulch A loose, finely granular, or powdery condition on the surface of the soil, usually produced by shallow cultivation.

E horizon Horizon characterized by maximum illuviation (washing out) of silicate clays and iron and aluminum oxides; commonly occurs above the B horizon and below the A horizon.

earthworms Animals of the Lumbricidae family that burrow into and live in the soil. They mix plant residues into the soil and improve soil aeration.

ectotrophic mycorrhiza (ectomycorrhiza) A symbiotic association of the mycelium of fungi and the roots of certain plants in which the fungal hyphae form a compact mantle on the surface of the roots and extend into the surrounding soil and inward between cortical cells, but not into these cells. Associated primarily with certain trees. *See also* endotrophic mycorrhiza.

edaphology The science that deals with the influence of soils on living things, particularly plants, including human use of land for plant growth.

electrical conductivity (EC) The capacity of a substance to conduct or transmit electrical current. In soils or water, measured in siemens/meter, and related to dissolved solutes.

electrokinetic potential In colloidal systems, the difference in potential between the immovable layer attached to the surface of the dispersed phase and the dispersion medium.

eluviation The removal of soil material in suspension (or in solution) from a layer or layers of a soil. (Usually, the loss of material in *solution* is described by the term *leaching*.) *See also* leaching.

endotrophic mycorrhiza (endomycorrhiza)   A symbiotic association of the mycelium of fungi and roots of a variety of plants in which the fungal hyphae penetrate directly into root hairs, other epidermal cells, and occasionally into cortical cells. Individual hyphae also extend from the root surface outward into the surrounding soil. *See also* vesicular arbuscular mycorrhiza.

Entisols   *See* soil classification.

eolian soil material   Soil material accumulated through wind action. The most extensive areas in the United States are silty deposits (loess), but large areas of sandy deposits also occur.

epipedon   A diagnostic surface horizon that includes the upper part of the soil that is darkened by organic matter, or the upper eluvial horizons, or both. *(Soil Taxonomy.)*

equilibrium phosphorus concentration   The concentration of phosphorus in a solution in equilibrium with a soil, the EPC_0 being the concentration of phosphorus achieved by desorption of phosphorus from a soil to phosphorus-free distilled water.

erosion   (1) The wearing away of the land surface by running water, wind, ice, or other geological agents, including such processes as gravitational creep. (2) Detachment and movement of soil or rock by water, wind, ice, or gravity. The following terms are used to describe different types of water erosion:

>   **accelerated erosion**   Erosion much more rapid than normal, natural, geological erosion; primarily as a result of the activities of humans or, in some cases, of animals.

>   **gully erosion**   The erosion process whereby water accumulates in narrow channels and, over short periods, removes the soil from this narrow area to considerable depths, ranging from 1–2 feet to as much as 23–30 m (75–100 ft).

>   **natural erosion**   Wearing away of the earth's surface by water, ice, or other natural agents under natural environmental conditions of climate, vegetation, and so on, undisturbed by man. Synonymous with *geological erosion.*

>   **rill erosion**   An erosion process in which numerous small channels of only several centimeters in depth are formed; occurs mainly on recently cultivated soils. *See also* rill.

>   **sheet erosion**   The removal of a fairly uniform layer of soil from the land surface by runoff water.

>   **splash erosion**   The spattering of small soil particles caused by the impact of raindrops on very wet soils. The loosened and separated particles may or may not be subsequently removed by surface runoff.

esker   A narrow ridge of gravelly or sandy glacial material deposited by a stream in an ice-walled valley or tunnel in a receding glacier.

essential element   A chemical element required for the normal growth of plants.

eutrophic   Having concentrations of nutrients optimal (or nearly so) for plant or animal growth. (Said of nutrient solutions or of soil solutions.)

eutrophication   A process of aging of lakes whereby aquatic plants are abundant and waters are deficient in oxygen. The process is usually accelerated by enrichment of waters with surface runoff containing nitrogen and phosphorus.

evapotranspiration   The combined loss of water from a given area, and during a specified period of time, by evaporation from the soil surface and by transpiration from plants.

exchange capacity   The total ionic charge of the adsorption complex active in the adsorption of ions. *See also* anion exchange capacity; cation exchange capacity.

exchangeable sodium percentage   The extent to which the adsorption complex of a soil is occupied by sodium. It is expressed as follows:

$$ESP = \frac{\text{exchangeable sodium (cmol}_c\text{/kg soil)}}{\text{cation exchange capacity (cmol}_c\text{/kg soil)}} \times 100$$

exfoliation   Peeling away of layers of a rock from the surface inward, usually as the result of expansion and contraction that accompany changes in temperature.

external surface The area of surface exposed on the top, bottom, and sides of a clay crystal or micelle.

facultative organism An organism capable of both aerobic and anaerobic metabolism.

fallow Cropland left idle in order to restore productivity, mainly through accumulation of water, nutrients, or both. Summer fallow is a common stage before cereal grain in regions of limited rainfall. The soil is kept free of weeds and other vegetation, thereby conserving nutrients and water for the next year's crop.

family, soil In soil classification, one of the categories intermediate between the great group and the soil series. Families are defined largely on the basis of physical and mineralogical properties of importance to plant growth. *(Soil Taxonomy.)*

fauna The animal life of a region or ecosystem.

ferrihydrite, $Fe_5HO_8 \cdot 4H_2O$ A dark reddish brown poorly crystalline iron oxide that forms in wet soils.

fertility, soil The quality of a soil that enables it to provide essential chemical elements in quantities and proportions for the growth of specified plants.

fertilizer Any organic or inorganic material of natural or synthetic origin added to a soil to supply certain elements essential to the growth of plants. The major types of fertilizers include:

> **bulk blended fertilizers** Solid fertilizer materials blended together in small blending plants, delivered to the farm in bulk, and usually spread directly on the fields by truck or other special applicator.

> **granulated fertilizers** Fertilizers that are present in the form of rather stable granules of uniform size, which facilitate ease of handling the materials and reduce undesirable dusts.

> **liquid fertilizers** Fluid fertilizers that contain essential elements in liquid forms either as soluble nutrients or as liquid suspensions or both.

> **mixed fertilizers** Two or more fertilizer materials mixed together. May be as dry powders, granules, pellets, bulk blends, or liquids.

fertilizer requirement The quantity of certain plant nutrient elements needed, in addition to the amount supplied by the soil, to increase plant growth to a designated optimum.

fibric materials *See* organic soil materials.

field capacity (field moisture capacity) The percentage of water remaining in a soil two or three days after its having been saturated and after free drainage has practically ceased.

fine-grained mica A silicate clay having a 2:1-type lattice structure with much of the silicon in the tetrahedral sheet having been replaced by aluminum and with considerable interlayer potassium, which binds the layers together and prevents interlayer expansion and swelling, and limits interlayer cation exchange capacity.

fine texture Consisting of or containing large quantities of the fine fractions, particularly of silt and clay. (Includes clay loam, sandy clay loam, silty clay loam, sandy clay, silty clay, and clay textural classes.)

first bottom The normal floodplain of a stream.

fixation (1) For other than elemental nitrogen: the process or processes in a soil by which certain chemical elements essential for plant growth are converted from a soluble or exchangeable form to a much less soluble or to a nonexchangeable form; for example, potassium, ammonium, and phosphate fixation. (2) For elemental nitrogen: process by which gaseous elemental nitrogen is chemically combined with hydrogen to form ammonia.

> **biological nitrogen fixation** Occurs at ordinary temperatures and pressures. It is commonly carried out by certain bacteria, algae, and actinomycetes, which may or may not be associated with higher plants.

chemical nitrogen fixation Takes place at high temperatures and pressures in manufacturing plants; produces ammonia, which is used to manufacture most fertilizers.

flagstone A relatively thin rock or mineral fragment 15–38 cm in length commonly composed of shale, slate, limestone, or sandstone.

flocculate To aggregate or clump together individual, tiny soil particles, especially fine clay, into small clumps or floccules. Opposite of *deflocculate* or *disperse*.

floodplain The land bordering a stream, built up of sediments from overflow of the stream and subject to inundation when the stream is at flood stage. Sometimes called *bottomland*.

flora The sum total of the kinds of plants in an area at one time. The organisms loosely considered to be of the plant kingdom.

fluorapatite A member of the apatite group of minerals containing fluorine. Most common mineral in rock phosphate.

fluvial deposits Deposits of parent materials laid down by rivers or streams.

fluvioglacial *See* glaciofluvial deposits.

foliar diagnosis An estimation of mineral nutrient deficiencies (excesses) of plants based on examination of the chemical composition of selected plant parts, and the color and growth characteristics of the foliage of the plants.

fragipan Dense and brittle pan or subsurface layer in soils that owe their hardness mainly to extreme density or compactness rather than high clay content or cementation. Removed fragments are friable, but the material in place is so dense that roots penetrate and water moves through it very slowly.

friable A soil consistency term pertaining to the ease of crumbling of soils.

frigid *See* soil temperature classes.

fritted micronutrients Sintered silicates having total guaranteed analyses of micronutrients with controlled (relatively slow) release characteristics.

fulvic acid A term of varied usage but usually referring to the mixture of organic substances remaining in solution upon acidification of a dilute alkali extract from the soil.

furrow irrigation *See* irrigation methods.

furrow slice The uppermost layer of an arable soil to the depth of primary tillage; the layer of soil sliced away from the rest of the profile and inverted by a moldboard plow.

fungi Simple plants that lack a photosynthetic pigment. The individual cells have a nucleus surrounded by a membrane, and they may be linked together in long filaments called *hyphae,* which may grow together to form a visible body.

gamma ray A high-energy ray (photon) emitted during radioactive decay of certain elements.

genesis, soil The mode of origin of the soil, with special reference to the processes responsible for the development of the solum, or true soil, from the unconsolidated parent material.

geological erosion *See* erosion, natural.

gibbsite, Al(OH)$_3$ An aluminum trihydroxide mineral most common in highly weathered soils such as oxisols.

gilgai The microrelief of soils produced by expansion and contraction with changes in moisture. Found in soils that contain large amounts of clay that swells and shrinks considerably with wetting and drying. Usually a succession of microbasins and microknolls in nearly level areas or of microvalleys and microridges parallel to the direction of the slope.

glacial drift   Rock debris that has been transported by glaciers and deposited, either directly from the ice or from the meltwater. The debris may or may not be heterogeneous.

glacial till   *See* till.

glaciofluvial deposits   Material moved by glaciers and subsequently sorted and deposited by streams flowing from the melting ice. The deposits are stratified and may occur in the form of outwash plains, deltas, kames, eskers, and kame terraces.

gley soil   (Obsolete) Soil developed under conditions of poor drainage resulting in reduction of iron and other elements and in gray colors and mottles.

goethite, FeOOH   A yellow-brown iron oxide mineral that accounts for the brown color in many soils.

granular structure   Soil structure in which the individual grains are grouped into spherical aggregates with indistinct sides. Highly porous granules are commonly called *crumbs*. A well-granulated soil has the best structure for most ordinary crop plants. *See also* soil structure types.

grassed waterway   A natural or constructed waterway covered with erosion-resistant grasses that permits removal of runoff water without excessive erosion.

gravitational potential   *See* soil water potential.

gravitational water   Water that moves into, through, or out of the soil under the influence of gravity.

great group   *See* soil classification.

greenhouse effect   The entrapment of heat by upper atmosphere gases such as carbon dioxide, water vapor, and methane just as glass traps heat for a greenhouse. Increases in the quantities of these gases in the atmosphere will likely result in global warming that may have serious consequences for humankind.

green manure   Plant material incorporated with the soil while green, or soon after maturity, for improving the soil.

groundwater   Subsurface water in the zone of saturation that is free to move under the influence of gravity, often horizontally to stream channels.

gully erosion   *See* erosion.

gypsic horizon   *See* diagnostic subsurface horizon.

halophyte   A plant that requires or tolerates a saline (high salt) environment.

hardpan   A hardened soil layer, in the lower A or in the B horizon, caused by cementation of soil particles with organic matter or with materials such as silica, sesquioxides, or calcium carbonate. The hardness does not change appreciably with changes in moisture content and pieces of the hard layer do not slake in water. *See also* caliche; claypan.

harrowing   A secondary broadcast tillage operation that pulverizes, smooths, and firms the soil in seedbed preparation, controls weeds, or incorporates material spread on the surface.

heaving   The partial lifting of plants, buildings, roadways, fenceposts, etc., out of the ground, as a result of freezing and thawing of the surface soil during the winter.

heavy metals   Metals with particle densities >5.0 Mg/m^3.

heavy soil   (Obsolete in scientific use) A soil with a high content of the fine separates, particularly clay, or one with a high drawbar pull, hence difficult to cultivate.

hematite, Fe$_2$O$_3$ A red iron oxide mineral that contributes red color to many soils.

herbicide   A chemical that kills plants or inhibits their growth; intended for weed control.

herbivore   A plant-eating animal.

hemic materials   *See* organic soil materials.

heterotroph An organism capable of deriving energy for life processes only from the decomposition of organic compounds and incapable of using inorganic compounds as sole sources of energy or for organic synthesis. *Contrast with* autotroph.

histic epipedon *See* diagnostic surface horizons.

Hisotosols *See* soil classification.

horizon, soil A layer of soil, approximately parallel to the soil surface, differing in properties and characteristics from adjacent layers below or above it. *See also* diagnostic subsurface horizons; diagnostic surface horizons.

horticulture The art and science of growing fruits, vegetables, and ornamental plants.

humic acid A mixture of variable or indefinite composition of dark organic substances, precipitated upon acidification of a dilute alkali extract from soil.

humic substances A series of complex, relatively high molecular weight, brown- to black-colored organic substances that make up 60 to 80% of the soil organic matter and are generally quite resistant to ready microbial attack.

humid climate Climate in regions where moisture, when distributed normally throughout the year, should not limit crop production. In cool climate annual precipitation may be as little as 25 cm; in hot climates, 150 cm or even more. Natural vegetation in uncultivated areas is forests.

humification The processes involved in the decomposition of organic matter and leading to the formation of humus.

humin The fraction of the soil organic matter that is not dissolved upon extraction of the soil with dilute alkali.

humus That more or less stable fraction of the soil organic matter remaining after the major portions of added plant and animal residues have decomposed. Usually it is dark in color.

hydration Chemical union between an ion or compound and one or more water molecules, the reaction being stimulated by the attraction of the ion/compound for either the hydrogen or the unshared electrons of the oxygen in the water.

hydraulic conductivity An expression of the readiness with which a liquid such as water flows through a solid such as soil in response to a given potential gradient.

hydrogen bond The chemical bond between a hydrogen atom in one molecule and a highly electronegative atom such as oxygen or nitrogen in another polar molecule.

hydrologic cycle The circuit of water movement from the atmosphere to the earth and back to the atmosphere through various stages or processes, as precipitation, interception, runoff, infiltration, percolation, storage, evaporation, and transpiration.

hydrolysis The reaction between water and a compound (commonly a salt). The hydroxyl from the water combines with the anion from the compound undergoing hydrolysis to form a base; the hydrogen ion from the water combines with the cation from the compound to form an acid.

hydronium A hydrated hydrogen ion (H_3O^+), the form of the hydrogen ion usually found in an aqueous system.

hydroponics Plant production systems that use nutrient solutions and no solid medium to grow plants.

hydrous mica *See* fine-grained mica.

hydroxyapatite A member of the apatite group of minerals rich in hydroxyl groups. A nearly insoluble calcium phosphate.

hygroscopic coefficient The amount of moisture in a dry soil when it is in equilibrium with some standard relative humidity near a saturated atmosphere (about 98%), expressed in terms of percentage on the basis of oven-dry soil.

hyperthermic *See* soil temperature classes.

hypha (pl. hyphae)   Filament of fungal cells. Actinomycetes also produce similar, but thinner, filaments of cells.

igneous rock   Rock formed from the cooling and solidification of magma and that has not been changed appreciably since its formation.

illite   *See* fine-grained mica.

illuvial horizon   A soil layer or horizon in which material carried from an overlying layer has been precipitated from solution or deposited from suspension. The layer of accumulation.

immature soil   A soil with indistinct or only slightly developed horizons because of the relatively short time it has been subjected to the various soil-forming processes. A soil that has not reached equilibrium with its environment.

immobilization   The conversion of an element from the inorganic to the organic form in microbial tissues or in plant tissues, thus rendering the element not readily available to other organisms or to plants.

imogolite   A poorly crystalline aluminosilicate mineral with an approximate formula $SiO_2Al_2O_32.5H_2O$; occurs mostly in soils formed from volcanic ash.

impervious   Resistant to penetration by fluids or by roots.

Inceptisols   *See* soil classification.

indurated (soil)   Soil material cemented into a hard mass that will not soften on wetting. *See also* consistence; hardpan.

infiltration   The downward entry of water into the soil.

infiltration rate   A soil characteristic determining or describing the *maximum* rate at which water *can* enter the soil under specified conditions, including the presence of an excess of water.

inoculation   The process of introducing pure or mixed cultures of microorganisms into natural or artificial culture media.

inorganic compounds   All chemical compounds in nature except compounds of carbon other than carbon monoxide, carbon dioxide, and carbonates.

insecticide   A chemical that kills insects.

intergrade   A soil that possesses moderately well-developed distinguishing characteristics of two or more genetically related great soil groups.

interlayer   (mineralogy) Materials between layers within a given crystal, including cations, hydrated cations, organic molecules, and hydroxide groups or sheets.

internal surface   The area of surface exposed within a clay crystal or micelle between the individual crystal layers. *Compare with* external surface.

ions   Atoms, groups of atoms, or compounds that are electrically charged as a result of the loss of electrons (cations) or the gain of electrons (anions).

iron-pan   An indurated soil horizon in which iron oxide is the principal cementing agent.

irrigation efficiency   The ratio of the water actually consumed by crops on an irrigated area to the amount of water diverted from the source onto the area.

irrigation methods   Methods by which water is artificially applied to an area. The methods and the manner of applying the water are as follows:

>   **border-strip**   The water is applied at the upper end of a strip with earth borders to confine the water to the strip.
>
>   **center-pivot**   Automated sprinkler irrigation achieved by automatically rotating the sprinkler pipe or boom, supplying water to the sprinkler heads or nozzles, as a radius from the center of the field to be irrigated.
>
>   **check-basin**   The water is applied rapidly to relatively level plots surrounded by levees. The basin is a small check.

corrugation The water is applied to small, closely spaced furrows, frequently in grain and forage crops, to confine the flow of irrigation water to one direction.

drip A planned irrigation system where all necessary facilities have been installed for the efficient application of water directly to the root zone of plants by means of applicators (orifices, emitters, porous tubing, perforated pipe, etc.) operated under low pressure. The applicators may be placed on or below the surface of the ground.

flooding The water is released from field ditches and allowed to flood over the land.

furrow The water is applied to row crops in ditches made by tillage implements.

sprinkler The water is sprayed over the soil surface through nozzles from a pressure system.

subirrigation The water is applied in open ditches or tile lines until the water table is raised sufficiently to wet the soil.

wild-flooding The water is released at high points in the field and distribution is uncontrolled.

interstratification Mixing of silicate layers within the structural framework of a given silicate clay.

isomorphous substitution The replacement of one atom by another of similar size in a crystal lattice without disrupting or changing the crystal structure of the mineral.

isotopes Two or more atoms of the same element that have different atomic masses because of different numbers of neutrons in the nucleus.

joule The SI energy unit defined as a force of 1 newton applied over a distance of 1 meter; 1 joule = 0.239 calorie.

kame A conical hill or ridge of sand or gravel deposited in contact with glacial ice.

kaolinite An aluminosilicate mineral of the 1:1 crystal lattice group; that is, consisting of single silicon tetrahedral sheets alternating with single aluminium octahedral sheets.

labile A substance that is readily transformed by microorganisms or is readily available for uptake by plants.

lacustrine deposit Material deposited in lake water and later exposed either by lowering of the water level or by the elevation of the land.

land A broad term embodying the total natural environmental of the areas of the earth not covered by water. In addition to soil, its attributes include other physical conditions such as mineral deposits and water supply; location in relation to centers of commerce, populations, and other land; the size of the individual tracts or holdings; and existing plant cover, works of improvement, and the like.

land capability classification A grouping of kinds of soil into special units, subclasses, and classes according to their capability for intensive use and the treatments required for sustained use. One such system has been prepared by the USDA Natural Resources Conservation Service.

land classification The arrangement of land units into various categories based upon the properties of the land or its suitability for some particular purpose.

land-use planning The development of plans for the uses of land that, over long periods, will best serve the general welfare, together with the formulation of ways and means for achieving such uses.

laterite An iron-rich subsoil layer found in some highly weathered humid tropical soils that, when exposed and allowed to dry, becomes very hard and will not soften when rewetted. When erosion removes the overlying layers, the laterite is exposed and a virtual pavement results. *See also* plinthite.

layer (clay mineralogy) A combination in silicate clays of (tetrahedral and octahedral) sheets in a 1:1, 2:1, or 2:1:1 combination.

leaching The removal of materials in solution from the soil by percolating waters. *See also* eluviation.

leaf area index (LAI) Area of leaves per unit of land on which the plants are growing.

legume A pod-bearing member of the Leguminosae family, one of the most important and widely distributed plant families. Includes many valuable food and forage species, such as peas, beans, peanuts, clovers, alfalfas, sweet clovers, lespedezas, vetches, and kudzu. Nearly all legumes are associated with nitrogen-fixing organisms.

Liebig's law The growth and reproduction of an organism are determined by the nutrient substance (oxygen, carbon dioxide, calcium, etc.) that is available in minimum quantity with respect to organic needs, the *limiting factor*.

light soil (Obsolete in scientific use) A coarse-textured soil; a soil with a low drawbar pull and hence easy to cultivate. *See also* coarse texture; soil texture.

lignin The complex organic constituent of woody fibers in plant tissue that, along with cellulose, cements the cells together and provides strength. Lignins resist microbial attack and after some modification may become part of the soil organic matter.

lime (agricultural) In strict chemical terms, calcium oxide. In practical terms, a material containing the carbonates, oxides and/or hydroxides of calcium and/or magnesium used to neutralize soil acidity.

lime requirement The mass of agricultural limestone, or the equivalent of other specified liming material, required to raise the pH of the soil to a desired value under field conditions.

limestone A sedimentary rock composed primarily of calcite ($CaCO_3$). If dolomite ($CaCO_3 \cdot MgCO_3$) is present in appreciable quantities, it is called a *dolomitic limestone*.

limiting factor *See* Liebig's law.

liquid limit (LL) *See* Atterberg limits.

lithosequence A group of related soils that differ, one from the other, in certain properties primarily as a result of parent material as a soil-forming factor.

loam The textural-class name for soil having a moderate amount of sand, silt, and clay. Loam soils contain 7–27% clay, 28–50% silt, and 23–52% sand.

loamy Intermediate in texture and properties between fine-textured and coarse-textured soils. Includes all textural classes with the words *loam* or *loamy* as a part of the class name, such as clay loam or loamy sand. *See also* loam; soil texture.

lodging Falling over of plants, either by uprooting or stem breakage.

loess Material transported and deposited by wind and consisting of predominantly silt-sized particles.

luxury consumption The intake by a plant of an essential nutrient in amounts exceeding what it needs. For example, if potassium is abundant in the soil, alfalfa may take in more than it requires.

lysimeter A device for measuring percolation (leaching) and evapotranspiration losses from a column of soil under controlled conditions.

macronutrient A chemical element necessary in large amounts (usually 50 mg/kg in the plant) for the growth of plants. Includes C, H, O, N, P, K, Ca, Mg, and S. (*Macro* refers to quantity and not to the essentiality of the element.) *See also* micronutrient.

macropores Larger soil pores, generally having a diameter greater than 0.06 mm, from which water drains readily by gravity.

marl Soft and unconsolidated calcium carbonate, usually mixed with varying amounts of clay or other impurities.

marsh Periodically wet or continually flooded area with the surface not deeply submerged. Covered dominantly with sedges, cattails, rushes, or other hydrophytic plants. Subclasses include freshwater and saltwater marshes.

mass flow   Movement of nutrients with the flow of water to plant roots.

matric potential   *See* soil water potential.

mature soil   A soil with well-developed soil horizons produced by the natural processes of soil formation and essentially in equilibrium with its present environment.

maximum retentive capacity   The average moisture content of a disturbed sample of soil, 1 cm high, which is at equilibrium with a water table at its lower surface.

mechanical analysis   (Obsolete) *See* particle size analysis; particle size distribution.

medium texture   Intermediate between fine-textured and coarse-textured (soils). It includes the following textural classes: very fine sandy loam, loam, silt loam, and silt.

melanic epipedon   *See* diagnostic surface horizons.

mellow soil   A very soft, very friable, porous soil without any tendency toward hardness or harshness. *See also* consistence.

mesic   *See* soil temperature classes.

mesofauna   Animals of medium size, between approximately 2 and 0.2 mm in diameter.

metamorphic rock   A rock that has been greatly altered from its previous condition through the combined action of heat and pressure. For example, marble is a metamorphic rock produced from limestone, gneiss is produced from granite, and slate from shale.

methane, CH_4   An odorless, colorless gas commonly produced under anaerobic conditions. When released to the upper atmosphere, methane contributes to global warming. *See also* greenhouse effect.

micas   Primary aluminosilicate minerals in which two silica tetrahedral sheets alternate with one alumina/magnesia octahedral sheet with entrapped potassium atoms fitting between sheets. They separate readily into visible sheets or flakes.

microfauna   That part of the animal population which consists of individuals too small to be clearly distinguished without the use of a microscope. Includes protozoans and nematodes.

microflora   That part of the plant population which consists of individuals too small to be clearly distinguished without the use of a microscope. Includes actinomycetes, algae, bacteria, and fungi.

micronutrient   A chemical element necessary in only extremely small amounts (<50 mg/kg in the plant) for the growth of plants. Examples are B, Cl, Cu, Fe, Mn, and Zn. (*Micro* refers to the amount used rather than to its essentiality.) *See also* macronutrient.

micropores   Relatively small soil pores, generally found within structural aggregates, and having a diameter less than 0.06 mm. *Contrast to* macropore.

microrelief   Small-scale local differences in topography, including mounds, swales, or pits that are only a meter or so in diameter and with elevation differences of up to 2 m. *See also* gilgai.

mineralization   The conversion of an element from an organic form to an inorganic state as a result of microbial decomposition.

mineral soil   A soil consisting predominantly of, and having its properties determined predominantly by, mineral matter. Usually contains <20% organic matter, but may contain an organic surface layer up to 30 cm thick.

minimum tillage   *See* tillage, conservation.

minor element   (Obsolete) *See* micronutrient.

moderately coarse texture   Consisting predominantly of coarse particles. In soil textural classification, it includes all the sandy loams except the very fine sandy loam. *See also* coarse texture.

moderately fine texture   Consisting predominantly of intermediate-sized (soil) particles or with relatively small amounts of fine or coarse particles. In soil textural classification, it includes clay loam, sandy loam, sandy clay loam, and silty clay loam. *See also* fine texture.

moisture equivalent   (Obsolete) The weight percentage of water retained by a previously saturated sample of soil 1 cm in thickness after it has been subjected to a centrifugal force of 1000 times gravity for 30 min.

moisture potential   *See* soil water potential.

mole drain   Unlined drain formed by pulling a bullet-shaped cylinder through the soil.

mollic epipedon   *See* diagnostic surface horizons.

Mollisols   *See* soil classification.

montmorillonite   An aluminosilicate clay mineral in the smectite group with a 2:1 expanding crystal lattice, with two silicon tetrahedral sheets enclosing an aluminum octahedral sheet. Isomorphous substitution of magnesium for some of the aluminum has occurred in the octahedral sheet. Considerable expansion may be caused by water moving between silica sheets of contiguous layers.

mor   Raw humus; type of forest humus layer of unincorporated organic material, usually matted or compacted or both; distinct from the mineral soil, unless the latter has been blackened by washing in organic matter.

moraine   An accumulation of drift, with an initial topographic expression of its own, built within a glaciated region chiefly by the direct action of glacial ice. Examples are ground, lateral, recessional, and terminal moraines.

morphology, soil   The constitution of the soil including the texture, structure, consistence, color, and other physical, chemical, and biological properties of the various soil horizons that make up the soil profile.

mottling   Spots or blotches of different color or shades of color interspersed with the dominant color.

mucigel   The gelatinous material at the surface of roots grown in unsterilized soil.

muck   Highly decomposed organic material in which the original plant parts are not recognizable. Contains more mineral matter and is usually darker in color than peat. *See also* muck soil; peat.

muck soil   (1) A soil containing 20–50% organic matter. (2) An organic soil in which the organic matter is well decomposed.

mulch   Any material such as straw, sawdust, leaves, plastic film, and loose soil that is spread upon the surface of the soil to protect the soil and plant roots from the effects of raindrops, soil crusting, freezing, evaporation, etc.

mulch tillage   *See* tillage, conservation.

mull   A humus-rich layer of forested soils consisting of mixed organic and mineral matter. A mull blends into the upper mineral layers without an abrupt change in soil characteristics.

mycelium   A stringlike mass of individual fungal or actinomycetes hyphae.

mycorrhiza   The association, usually symbiotic, of fungi with the roots of seed plants. *See also* ectotrophic mycorrhiza; endotrophic mycorrhiza; vesicular-arbuscular mycorrhiza.

natric horizon   *See* diagnostic subsurface horizon.

necrosis   Death associated with discoloration and dehydration of all or parts of plant organs, such as leaves.

nematodes   Very small worms abundant in many soils and important because some of them attack and destroy plant roots.

neutral soil   A soil in which the surface layer, at least to normal plow depth, is neither acid nor alkaline in reaction. In practice this means the soil is within the pH range of 6.6–7.3. *See also* acid soil; alkaline soil; pH; reaction, soil.

nitrate depression period   A period of time, beginning shortly after the addition of fresh, highly carbonaceous organic materials to a soil, during which decomposer microorganisms have removed most of the soluble nitrate from the soil solution.

nitrification   The biochemical oxidation of ammonium to nitrate, predominantly by autotrophic bacteria.

nitrogen assimilation   The incorporation of nitrogen into organic cell substances by living organisms.

nitrogen cycle   The sequence of chemical and biological changes undergone by nitrogen as it moves from the atmosphere into water, soil, and living organisms, and upon death of these organisms (plants and animals) is recycled through a part or all of the entire process.

nitrogen fixation   The biological conversion of elemental nitrogen (N_2) to organic combinations or to forms readily utilize in biological processes.

nodule bacteria   *See* rhizobia.

nonhumic substances   The portion of soil organic matter comprised of relatively low molecular weight organic substances; mostly identifiable biomolecules.

no-tillage   *See* tillage, conservation.

nucleic acids   Complex organic acids found in the nuclei of plant and animal cells; may be combined with proteins as nucleoproteins.

O horizon   Organic horizon of mineral soils.

ochric epipedon   *See* diagnostic surface horizons.

octahedral sheet   Sheet of horizontally linked, octahedral-shaped, units that serve as the basic structural components of silicate (clay) minerals. Each unit consists of a central, six-coordinated metallic atom (e.g., Al, Mg, or Fe) surrounded by six hydroxyl groups that, in turn, are linked with other nearby metal atoms, thereby serving as interunit linkages that hold the sheet together.

order, soil   *See* soil classification.

organic soil   A soil that contains at least 20% organic matter (by weight) if the clay content is low and at least 30% if the clay content is as high as 60%.

organic fertilizer   By-product from the processing of animal or vegetable substances that contain sufficient plant nutrients to be of value as fertilizers.

organic soil materials   (As used in *Soil Taxonomy* in the United States): (1) Saturated with water for prolonged periods unless artificially drained and having 18% or more organic carbon (by weight) if the mineral fraction is more than 60% clay, more than 12% organic carbon if the mineral fraction has no clay, or between 12 and 18% carbon if the clay content of the mineral fraction is between 0 and 60%. (2) Never saturated with water for more than a few days and having more than 20% organic carbon. Histosols develop on these organic soil materials. There are three kinds of organic materials:

>   **fibric materials**   The least decomposed of all the organic soil materials, containing very high amounts of fiber that are well preserved and readily identifiable as to botanical origin.
>   **hemic materials**   Intermediate in degree of decomposition of organic materials between the less decomposed fibric and the more decomposed sapric materials.
>   **sapric materials**   The most highly decomposed of the organic materials, having the highest bulk density, least amount of plant fiber, and lowest water content at saturation.

ortstein   An indurated layer in the B horizon of Spodosols in which the cementing material consists of illuviated sesquioxides (mostly iron) and organic matter.

osmotic pressure   Pressure exerted in living bodies as a result of unequal concentrations of salts on both sides of a cell wall or membrane. Water moves from the area having the lower salt concentration through the membrane into the area having the higher salt concentration and, therefore, exerts additional pressure on the side with higher salt concentration.

osmotic potential   *See* soil water potential.

outwash plain   A deposit of coarse-textured materials (e.g., sands, gravels) left by streams of meltwater flowing from receding glaciers.

oven-dry soil   Soil that has been dried at 105°C until it reaches constant weight.

oxic horizon   *See* diagnostic subsurface horizon.

oxidation   The loss of electrons by a substance; therefore a gain in positive valence charge, and in some cases, the chemical combination with oxygen gas.

oxidation ditch   An artificial open channel for partial digestion of liquid organic wastes in which the wastes are circulated and aerated by a mechanical device.

Oxisols   *See* soil classification.

pans   Horizons or layers, in soils, that are strongly compacted, indurated, or very high in clay content. *See also* caliche; claypan; fragipan; hardpan.

parent material   The unconsolidated and more or less chemically weathered mineral or organic matter from which the solum of soils is developed by pedogenic processes.

particle density   The mass per unit volume of the soil particles. In technical work, usually expressed as metric tons per cubic meter (Mg/m^3) or grams per cubic centimeter (g/cm^3).

particle size   The effective diameter of a particle measured by sedimentation, sieving, or micrometric methods.

particle size analysis   Determination of the various amounts of the different separates in a soil sample, usually by sedimentation, sieving, micrometry, or combinations of these methods.

particle size distribution   The amounts of the various soil separates in a soil sample, usually expressed as weight percentages.

pascal   An SI unit of pressure equal to 1 newton per square meter.

peat   Unconsolidated soil material consisting largely of undecomposed, or only slightly decomposed, organic matter accumulated under conditions of excessive moisture. *See also* organic soil materials; peat soil.

peat soil   An organic soil containing more than 50% organic matter. Used in the United States to refer to the stage of decomposition of the organic matter, *peat* referring to the slightly decomposed or undecomposed deposits and *muck* to the highly decomposed materials. *See also* muck; muck soil; peat.

ped   A unit of soil structure such as an aggregate, crumb, prism, block, or granule, formed by natural processes (in contrast to a clod, which is formed artificially).

pedon   The smallest volume that can be called *a soil*. It has three dimensions. It extends downward to the depth of plant roots or to the lower limit of the genetic soil horizons. Its lateral cross section is roughly hexagonal and ranges from 1 to $10 \ m^2$ in size depending on the variability in the horizons.

peneplain   A once high, rugged area that has been reduced by erosion to a lower, gently rolling surface resembling a plain.

penetrability   The ease with which a probe can be pushed into the soil. May be expressed in units of distance, speed, force, or work depending on the type of penetrometer used.

percolation, soil water   The downward movement of water through soil. Especially, the downward flow of water in saturated or nearly saturated soil at hydraulic gradients of the order of 1.0 or less.

perforated plastic pipe   Pipe, sometimes flexible, with holes or slits in it that allow the entrance and exit of air and water. Used for soil drainage and for septic effluent spreading into soil.

permafrost   (1) Permanently frozen material underlying the solum. (2) A perennially frozen soil horizon.

permanent charge   *See* constant charge.

permanent wilting point   *See* wilting point.

permeability, soil   The ease with which gases, liquids, or plant roots penetrate or pass through a bulk mass of soil or a layer of soil.

petrocalcic horizon   *See* diagnostic subsurface horizon.

petrogypsic horizon   *See* diagnostic subsurface horizon.

pH, soil   The negative logarithm of the hydrogen ion activity (concentration) of a soil. The degree of acidity (or alkalinity) of a soil as determined by means of a glass or other suitable electrode or indicator at a specified moisture content or soil/water ratio, and expressed in terms of the pH scale.

pH-dependent charge   That portion of the total charge of the soil particles that is affected by, and varies with, changes in pH.

phase, soil   A subdivision of a soil series or other unit of classification having characteristics that affect the use and management of the soil but do not vary sufficiently to differentiate it as a separate series. Included are such characteristics as degree of slope, degree of erosion, and content of stones.

photomap   A mosaic map made from aerial photographs to which place names, marginal data, and other map information have been added.

phyllosphere   The leaf surface.

physical properties (of soils)   Those characteristics, processes, or reactions of a soil that are caused by physical forces and that can be described by, or expressed in, physical terms or equations. Examples of physical properties are bulk density, water-holding capacity, hydraulic conductivity, porosity, pore-size distribution, and so on.

phytotoxic substances   Chemicals that are toxic to plants.

placic horizon   *See* diagnostic subsurface horizons.

plaggen epipedon   *See* diagnostic surface horizons.

plant nutrients   *See* essential elements.

plastic limit (PL)   *See* Atterberg limits.

plastic soil   A soil capable of being molded or deformed continuously and permanently, by relatively moderate pressure, into various shapes. *See also* consistence.

platy   Consisting of soil aggregates that are developed predominantly along the horizontal axes; laminated; flaky.

plinthite (brick)   A highly weathered mixture of sesquioxides of iron and aluminum with quartz and other diluents that occurs as red mottles and that changes irreversibly to hardpan upon alternate wetting and drying.

plow layer   The soil ordinarily moved when land is plowed; equivalent to *surface soil.*

plow pan   A subsurface soil layer having a higher bulk density and lower total porosity than layers above or below it, as a result of pressure applied by normal plowing and other tillage operations.

plow-plant   *See* tillage, conservation.

plowing   A primary broad-base tillage operation that is performed to shatter soil uniformly with partial to complete inversion.

point of zero charge   The pH value of a solution in equilibrium with a particle whose net charge, from all sources, is zero.

polypedon (As used in *Soil Taxonomy*) Two or more contiguous pedons, all of which are within the defined limits of a single soil series; commonly referred to as a *soil individual*.

pore size distribution The volume of the various sizes of pores in a soil. Expressed as percentages of the bulk volume (soil plus pore space).

porosity, soil The volume percentage of the total soil bulk not occupied by solid particles.

primary consumer An organism that subsists on plant material.

primary mineral A mineral that has not been altered chemically since deposition and crystallization from molten lava.

primary producer An organism (usually a photosynthetic plant) that creates organic, energy-rich material from inorganic chemicals, solar energy, and water.

primary tillage *See* tillage, primary.

priming effect The increased decomposition of relatively stable soil humus under the influence of much enhanced, generally biological, activity resulting from the addition of fresh organic materials to a soil.

prismatic soil structure A soil structure type with prism-like aggregates that have a vertical axis much longer than the horizontal axes.

productivity, soil The capacity of a soil for producing a specified plant or sequence of plants under a specified system of management. Productivity emphasizes the capacity of soil to produce crops and should be expressed in terms of yields.

profile, soil A vertical section of the soil through all its horizons and extending into the parent material.

protein Any of a group of nitrogen-containing organic compounds formed by the polymerization of a large number of amino acid molecules and that, upon hydrolysis, yield these amino acids. They are essential parts of living matter and are one of the essential food substances of animals.

protozoa One-celled eucaryotic organisms such as amoeba.

puddled soil Dense, massive soil artificially compacted when wet and having no aggregated structure. The condition commonly results from the tillage of a clayey soil when it is wet.

rain, acid *See* acid rain.

reaction, soil The degree of acidity or alkalinity of a soil, usually expressed as a pH value:

Extremely acid	<4.5
Very strongly acid	4.5–5.0
Strongly acid	5.1–5.5
Medium acid	5.6–6.0
Slightly acid	6.1–6.5
Neutral	6.6–7.3
Mildly alkaline	7.4–7.8
Moderately alkaline	7.9–8.4
Strongly alkaline	8.5–9.0
Very strongly alkaline	9.1 and higher

redox potential The electrical potential (measured in volts or millivolts) of a system due to the tendency of the substances in it to give up or acquire electrons.

reduction The gain of electrons, and therefore the loss of positive valence charge by a substance. In some cases, a loss of oxygen or a gain of hydrogen is also involved.

regolith The unconsolidated mantle of weathered rock and soil material on the earth's surface; loose earth materials above solid rock. (Approximately equivalent to the term *soil* as used by many engineers.)

residual material Unconsolidated and partly weathered mineral materials accumulated by disintegration of consolidated rock in place.

rhizobia Bacteria capable of living symbiotically with higher plants, usually in nodules on the roots of legumes, from which they receive their energy, and capable of converting atmospheric nitrogen to combined organic forms; hence the term *symbiotic nitrogen-fixing bacteria*. (Derived from the generic name *Rhizobium*.)

rhizosphere That portion of the soil in the immediate vicinity of plant roots in which the abundance and composition of the microbial population are influenced by the presence of roots.

rill A small, intermittent water course with steep sides; usually only a few centimeters deep and hence no obstacle to tillage operations.

rill erosion *See* erosion.

riparian zone The area, both above and below the ground surface, that borders a river.

riprap Broken rock, cobbles, or boulders placed on earth surfaces, such as the face of a dam or the bank of a stream, for protection against the action of water (waves); also applied to brush or pole mattresses, or brush and stone, or other similar materials used for soil erosion control.

rock The material that forms the essential part of the earth's solid crust, including loose incoherent masses such as sand and gravel, as well as solid masses of granite and limestone.

root interception Acquisition of nutrients by a root as a result of the root growing into the vicinity of the nutrient source.

root nodules Swollen growths on plant roots. Often in reference to those in which symbiotic microorganisms live.

rotary tillage *See* tillage, rotary.

runoff The portion of the precipitation on an area that is discharged from the area through stream channels. That which is lost without entering the soil is called *surface runoff* and that which enters the soil before reaching the stream is called *groundwater runoff* or *seepage flow* from groundwater. (In soil science *runoff* usually refers to the water lost by surface flow; in geology and hydraulics *runoff* usually includes both surface and subsurface flow.)

salic horizon *See* diagnostic subsurface horizons.

saline-sodic soil A soil containing sufficient exchangeable sodium to interfere with the growth of most crop plants and containing appreciable quantities of soluble salts. The exchangeable sodium adsorption ratio is >13, the conductivity of the saturation extract is >4 dS/m (at 25°C), and the pH is usually 8.5 or less in the saturated soil.

saline soil A nonsodic soil containing sufficient soluble salts to impair its productivity. The conductivity of a saturated extract is >4 dS/m, the exchangeable sodium adsorption ratio is less than about 13, and the pH is <8.5.

saline seep An area of land in which saline water seeps to the surface, leaving a high salt concentration behind as the water evaporates.

salinization The process of accumulation of salts in soil.

saltation Particle movement in water or wind where particles skip or bounce along the stream bed or soil surface.

sand A soil particle between 0.05 and 2.0 mm in diameter; a soil textural class.

sapric materials *See* organic soil materials.

saprolite Bedrock that has weathered in place to the point that it is porous and can be dug with a spade.

saturation extract The solution extracted from a saturated soil paste.

saturation percentage The water content of a saturated soil paste, expressed as a dry weight percentage.

savanna (savannah) A grassland with scattered trees, either as individuals or clumps. Often a transitional type between true grassland and forest.

secondary mineral A mineral resulting from the decomposition of a primary mineral or from the reprecipitation of the products of decomposition of a primary mineral. *See also* primary mineral.

second bottom The first terrace above the normal floodplain of a stream.

sedimentary rock A rock formed from materials deposited from suspension or precipitated from solution and usually being more or less consolidated. The principal sedimentary rocks are sandstones, shales, limestones, and conglomerates.

seedbed The soil prepared to promote the germination of seed and the growth of seedlings.

self-mulching soil A soil in which the surface layer becomes so well aggregated that it does not crust and seal under the impact of rain but instead serves as a surface mulch upon drying.

semiarid Term applied to regions or climates where moisture is more plentiful than in arid regions but still definitely limits the growth of most crop plants. Natural vegetation in uncultivated areas is short grasses.

separate, soil One of the individual-sized groups of mineral soil particles—sand, silt, or clay.

septic tank An underground tank used in the deposition of domestic wastes. Organic matter decomposes in the tank, and the effluent is drained into the surrounding soil.

series, soil *See* soil classification.

sewage effluent The liquid part of sewage or wastewater, it is usually treated to remove some portion of the dissolved organic compounds and nutrients present from the original sewage.

sewage sludge Settled sewage solids combined with varying amounts of water and dissolved materials, removed from sewage by screening, sedimentation, chemical precipitation, or bacterial digestion. Also called *biosolids*.

shear Force, as of a tillage implement, acting at right angles to the direction of movement.

sheet (mineralogy) A flat array of more than one atomic thickness and composed of one or more levels of linked coordination polyhedra. A sheet is thicker than a plane and thinner than a layer. Example: tetrahedral sheet, octahedral sheet.

sheet erosion *See* erosion.

shelterbelt A wind barrier of living trees and shrubs established and maintained for protection of farm fields. Syn. windbreak.

shifting cultivation A farming system in which land is cleared, the debris burned, and crops grown for 2–3 years. When the farmer moves on to another plot, the land is then left idle for 5–15 years; then the burning and planting process is repeated.

side-dressing The application of fertilizer alongside row-crop plants, usually on the soil surface. Nitrogen materials are most commonly side-dressed.

silica/alumina ratio The molecules of silicon dioxide (SiO_2) per molecule of aluminum oxide (Al_2O_3) in clay minerals or in soils.

silica/sesquioxide ratio The molecules of silicon dioxide (SiO_2) per molecule of aluminum oxide (Al_2O_3) plus ferric oxide (Fe_2O_3) in clay minerals or in soils.

silt (1) A soil separate consisting of particles between 0.05 and 0.002 mm in equivalent diameter. (2) A soil textural class.

silting The deposition of waterborne sediments in stream channels, lakes, reservoirs, or on floodplains, usually resulting from a decrease in the velocity of the water.

site index A quantitative evaluation of the productivity of a soil for forest growth under the existing or specified environment.

slag A product of smelting, containing mostly silicates; the substances not sought to be produced as matte or metal and having a lower specific gravity.

slash and burn *See* shifting cultivation.

slick spots Small areas in a field that are slick when wet because of a high content of alkali or exchangeable sodium.

slope The degree of deviation of a surface from horizontal, measured in a numerical ratio, percent, or degrees.

slow fraction (of soil organic matter) That portion of soil organic matter that can be metabolized with great difficulty by the microorganisms in the soil and therefore has a slow turnover rate with a half-life in the soil ranging from a few years to a few decades. Often this fraction is the product of some previous decomposition.

smectite A group of silicate clays having a 2:1-type lattice structure with sufficient isomorphous substitution in either or both the tetrahedral and octahedral sheets to give a high interlayer negative charge and high cation exchange capacity and to permit significant interlayer expansion and consequent shrinking and swelling of the clay. Montmorillonite, beidellite, and saponite are in the smectite group.

sodic soil A soil that contains sufficient sodium to interfere with the growth of most crop plants, and in which the sodium adsorption ratio is 13 or greater.

sodium adsorption ratio (SAR)

$$SAR = \frac{[Na^+]}{\sqrt{\frac{1}{2}([Ca^{2+}] + [Mg^{2+}])}}$$

where the cation concentrations are in millimoles of charge per liter ($mmol_c/L$).

soil (1) A dynamic natural body composed of mineral and organic materials and living forms in which plants grow. (2) The collection of natural bodies occupying parts of the earth's surface that support plants and that have properties due to the integrated effect of climate and living matter acting upon parent material, as conditioned by relief, over periods of time.

soil air The soil atmosphere; the gaseous phase of the soil, being that volume not occupied by soil or liquid.

soil alkalinity The degree or intensity of alkalinity of a soil, expressed by a value >7.0 on the pH scale.

soil amendment Any material, such as lime, gypsum, sawdust, or synthetic conditioner, that is worked into the soil to make it more amenable to plant growth.

soil association A group of defined and named taxonomic soil units occurring together in an individual and characteristic pattern over a geographic region, comparable to plant associations in many ways.

soil classification (*Soil Taxonomy*) The systematic arrangement of soils into groups or categories on the basis of their characteristics.

  order The category at the highest level of generalization in the soil classification system. The properties selected to distinguish the orders are reflections of the degree of horizon development and the kinds of horizons present. The 11 orders are as follows:

  Andisols Soils developed from volcanic ejecta. The colloidal fraction is dominated by allophane and/or Al-humus compounds.

  Alfisols Soils with gray to brown surface horizons, medium to high supply of bases, and B horizons of illuvial clay accumulation. These soils form mostly under forest or savanna vegetation in climates with slight to pronounced seasonal moisture deficit.

Aridisols Soils of dry climates. They have pedogenic horizons, low in organic matter, that are never moist as long as 3 consecutive months. They have an ochric epipedon and one or more of the following diagnostic horizons: argillic, natric, cambic, calcic, petrocalcic, gypsic, petrogypsic, salic, or a duripan.

Entisols Soils have no diagnostic pedogenic horizons. They may be found in virtually any climate on very recent geomorphic surfaces.

Histosols Soils formed from materials high in organic matter. Histosols with essentially no clay must have at least 20% organic matter by weight (about 78% by volume). This minimum organic matter content rises with increasing clay content to 30% (85% by volume) in soils with at least 60% clay.

Inceptisols Soils that are usually moist with pedogenic horizons of alteration of parent materials but not of illuviation. Generally, the direction of soil development is not yet evident from the marks left by various soil-forming processes or the marks are too weak to classify in another order.

Mollisols Soils with nearly black, organic-rich surface horizons and high supply of bases. They have mollic epipedons and base saturation greater than 50% in any cambic or argillic horizon. They lack the characteristics of Vertisols and must not have oxic or spodic horizons.

Oxisols Soils with residual accumulations of low-activity clays, free oxides, kaolin, and quartz. They are mostly in tropical climates.

Spodosols Soils with subsurface illuvial accumulations of organic matter and compounds of aluminum and usually iron. These soils are formed in acid, mainly coarse-textured materials in humid and mostly cool or temperate climates.

Ultisols Soils that are low in bases and have subsurface horizons of illuvial clay accumulations. They are usually moist, but during the warm season of the year some are dry part of the time.

Vertisols Clayey soils with high shrink–swell potential that have wide, deep cracks when dry. Most of these soils have distinct wet and dry periods throughout the year.

suborder This category narrows the ranges in soil moisture and temperature regimes, kinds of horizons, and composition, according to which of these is most important.

great group The classes in this category contain soils that have the same kind of horizons in the same sequence and have similar moisture and temperature regimes.

subgroup The great groups are subdivided into central concept subgroups that show the central properties of the great group, intergrade subgroups that show properties of more than one great group, and other subgroups for soils with atypical properties that are not characteristic of any great group.

family Families are defined largely on the basis of physical and mineralogic properties of importance to plant growth.

series The soil series is a subdivision of a family and consists of soils that are similar in all major profile characteristics.

soil complex A mapping unit used in detailed soil surveys where two or more defined taxonomic units are so intimately intermixed geographically that it is undesirable or impractical, because of the scale being used, to separate them. A more intimate mixing of smaller areas of individual taxonomic units than that described under *soil association.*

soil conditioner Any material added to a soil for the purpose of improving its physical condition.

soil conservation A combination of all management and land-use methods that safeguard the soil against depletion or deterioration caused by nature and/or humans.

soil correlation The process of defining, mapping, naming, and classifying the kinds of soils in a specific soil survey area, the purpose being to ensure that soils are adequately defined, accurately mapped, and uniformly named.

soil erosion   *See* erosion.

soil fertility   *See* fertility, soil.

soil genesis   The mode of origin of the soil, with special reference to the processes or soil-forming factors responsible for the development of the solum, or true soil, from the unconsolidated parent material.

soil geography   A subspecialization of physical geography concerned with the areal distributions of soil types.

soil horizon   *See* horizon, soil.

soil management   The sum total of all tillage operations, cropping practices, fertilizer, lime, and other treatments conducted on or applied to a soil for the production of plants.

soil map   A map showing the distribution of soil types or other soil mapping units in relation to the prominent physical and cultural features of the earth's surface.

soil mechanics and engineering   A subspecialization of soil science concerned with the effect of forces on the soil and the application of engineering principles to problems involving the soil.

soil moisture potential   *See* soil water potential.

soil monolith   A vertical section of a soil profile removed from the soil and mounted for display or study.

soil morphology   The physical constitution, particularly the structural properties, of a soil profile as exhibited by the kinds, thicknesses, and arrangement of the horizons in the profile, and by the texture, structure, consistence, and porosity of each horizon.

soil profile   A vertical section of the soil from the surface through all its horizons, including C horizons. *See also* horizon, soil.

soil organic matter   The organic fraction of the soil that includes plant and animal residues at various stages of decomposition, cells and tissues of soil organisms, and substances synthesized by the soil population. Commonly determined as the amount of organic material contained in a soil sample passed through a 2-mm sieve.

soil porosity   *See* porosity, soil.

soil productivity   *See* productivity, soil.

soil quality   The capacity of a specific kind of soil to function, within natural or managed ecosystem boundaries, to sustain plant and animal productivity, maintain or enhance water and air quality, and support human health and habitation. Sometimes considered in relation to this capacity in the undisturbed, natural state.

soil reaction   *See* reaction, soil; pH, soil.

soil salinity   The amount of soluble salts in a soil, expressed in terms of percentage, milligrams per kilogram, parts per million (ppm), or other convenient ratios.

soil separates   *See* separate, soil.

soil series   *See* soil classification.

soil solution   The aqueous liquid phase of the soil and its solutes consisting of ions dissociated from the surfaces of the soil particles and of other soluble materials.

soil structure   The combination or arrangement of primary soil particles into secondary particles, units, or peds. These secondary units may be, but usually are not, arranged in the profile in such a manner as to give a distinctive characteristic pattern. The secondary units are characterized and classified on the basis of size, shape, and degree of distinctness into classes, types, and grades, respectively.

soil structure classes   A grouping of soil structural units or peds on the basis of size from the very fine to very coarse.

soil structure grades   A grouping or classification of soil structure on the basis of inter- and intra-aggregate adhesion, cohesion, or stability within the profile. Four grades of structure, designated from 0 to 3, are recognized:

0: *structureless*—no observable aggregation

1: *weakly* durable peds

2: *moderately* durable peds

3: *strong,* durable peds

soil structure types   A classification of soil structure based on the shape of the aggregates or peds and their arrangement in the profile, including platy, prismatic, columnar, blocky, subangular blocky, granulated, and crumb.

soil survey   The systematic examination, description, classification, and mapping of soils in an area. Soil surveys are classified according to the kind and intensity of field examination.

soil temperature classes   (*Soil Taxonomy*) Classes are based on mean annual soil temperature and on differences between summer and winter temperatures at a depth of 50 cm.

1. Soils with 5°C and greater difference between summer and winter temperatures are classed on the basis of mean annual temperatures:

    frigid: <8°C mean annual temperature

    mesic: 8–15°C mean annual temperature

    thermic: 15–22°C mean annual temperature

    hyperthermic: >22°C mean annual temperature

2. Soils with <5°C difference between summer and winter temperatures are classed on the basis of mean annual temperatures:

    isofrigid: <8°C mean annual temperature

    isomesic: 8–15°C mean annual temperature

    isothermic: 15–22°C mean annual temperature

    isohyperthermic: <22°C mean annual temperature

soil textural class   A grouping of soil textural units based on the relative proportions of the various soil separates (sand, silt, and clay). These textural classes, listed from the coarsest to the finest in texture, are sand, loamy sand, sandy loam, loam, silt loam, silt, sandy clay loam, clay loam, silty clay loam, sandy clay, silty clay, and clay. There are several subclasses of the sand, loamy sand, and sandy loam classes based on the dominant particle size of the sand fraction (e.g., loamy fine sand, coarse sandy loam).

soil texture   The relative proportions of the various soil separates in a soil.

soil water potential   (total) A measure of the difference between the free energy state of soil water and that of pure water. Technically it is defined as "that amount of work that must be done per unit quantity of pure water in order to transport reversibly and isothermically an infinitesimal quantity of water from a pool of pure water, at a specified elevation and at atmospheric pressure to the soil water (at the point under consideration)." This *total* potential consists of the following potentials:

　　matric potential   That portion of the total soil water potential due to the attractive forces between water and soil solids as represented through adsorption and capillarity. It will always be negative.

　　osmotic potential   That portion of the total soil water potential due to the presence of solutes in soil water. It will generally be negative.

　　gravitational potential   That portion of the total soil water potential due to differences in elevation of the reference pool of pure water and that of the soil water. Since the soil water elevation is usually chosen to be higher than that of the reference pool, the gravitational potential is usually positive.

solum   (plural **sola**) The upper and most weathered part of the soil profile; the A, E, and B horizons

specific heat capacity   The amount of kinetic (heat) energy required to raise the temperature of 1 gram of a substance (usually in reference to soil or soil components).

specific surface   The solid particle surface area per unit mass or volume of the solid particles.

splash erosion   *See* erosion.

spodic horizon   *See* diagnostic subsurface horizons.

Spodosols   *See* soil classification.

sprinkler irrigation   *See* irrigation methods.

stratified   Arranged in or composed of strata or layers.

strip-cropping   The practice of growing crops that require different types of tillage, such as row and sod, in alternate strips along contours or across the prevailing direction of wind.

structure, soil   *See* soil structure.

stubble mulch   The stubble of crops or crop residues left essentially in place on the land as a surface cover before and during the preparation of the seedbed and at least partly during the growing of a succeeding crop.

subirrigation   *See* irrigation methods.

subsoil   That part of the soil below the plow layer.

subsoiling   Breaking of compact subsoils, without inverting them, with a special knifelike instrument (chisel), which is pulled through the soil at depths usually of 30–60 cm and at spacings usually of 1–2 m.

summer fallow   A cropping system that involves management of uncropped land during the summer to control weeds and store moisture in the soil for the growth of a later crop.

surface runoff   *See* runoff.

surface soil   The uppermost part of the soil, ordinarily moved in tillage, or its equivalent in uncultivated soils. Ranges in depth from 7 to 25 cm. Frequently designated as the *plow layer,* the *Ap layer,* or the *Ap horizon.*

symbiosis   The living together in intimate association of two dissimilar organisms, the cohabitation being mutually beneficial.

talus   Fragments of rock and other soil material accumulated by gravity at the foot of cliffs or steep slopes.

taxonomy, soil   The science of classification of soils; laws and principles governing the classifying of soil. *See also* soil classification.

tensiometer   A device for measuring the negative pressure (or tension) of water in soil in situ; a porous, permeable ceramic cup connected through a tube to a manometer or vacuum gauge.

tension, soil-moisture   *See* soil water potential.

terrace   (1) A level, usually narrow, plain bordering a river, lake, or the sea. Rivers sometimes are bordered by terraces at different levels. (2) A raised, more or less level or horizontal strip of earth usually constructed on or nearly on a contour and designed to make the land suitable for tillage and to prevent accelerated erosion by diverting water from undesirable channels of concentration; sometimes called *diversion terrace.*

tetrahedral sheet   Sheet of horizontally linked, tetrahedral-shaped units that serve as one of the basic structural components of silicate (clay) minerals. Each unit consists of a central four-coordinated atom (e.g., Si, Al, Fe) surrounded by four oxygen atoms that, in turn, are linked with other nearby atoms (e.g., Si, Al, Fe), thereby serving as interunit linkages to hold the sheet together.

texture   *See* soil texture.

thermal analysis (differential thermal analysis)   A method of analyzing a soil sample for constituents, based on a differential rate of heating of the unknown and standard samples when a uniform source of heat is applied.

thermic   *See* soil temperature classes.

thermophilic organisms   Organisms that grow readily at temperatures above 45°C.

tile, drain Pipe made of burned clay, concrete, or ceramic material, in short lengths, usually laid with open joints to collect and carry excess water from the soil.

till (1) Unstratified glacial drift deposited directly by the ice and consisting of clay, sand, gravel, and boulders intermingled in any proportion. (2) To plow and prepare for seeding; to seed or cultivate the soil.

tillage The mechanical manipulation of soil for any purpose; but in agriculture it is usually restricted to the modifying of soil conditions for crop production.

tillage, conservation Any tillage sequence that reduces loss of soil or water relative to conventional tillage, including the following systems:

    minimum tillage The minimum soil manipulation necessary for crop production or meeting tillage requirements under the existing soil and climatic conditions.

    mulch tillage Tillage or preparation of the soil in such a way that plant residues or other materials are left to cover the surface; also called *mulch farming, trash farming, stubble mulch tillage, plowless farming.*

    no-tillage system A procedure whereby a crop is planted directly into a seedbed not tilled since harvest of the previous crop; also *zero tillage.*

    plow-planting The plowing and planting of land in a single trip over the field by drawing both plowing and planting tools with the same power source.

    ridge till Planting on ridges formed by cultivation during the previous growing period.

    sod planting A method of planting in sod with little or no tillage.

    strip till Planting is done in a narrow strip that has been tilled and mixed, leaving the remainder of the soil surface undisturbed.

    subsurface tillage Tillage with a special sweeplike plow or blade that is drawn beneath the surface, cutting plant roots and loosening the soil without inverting it or without incorporating residues of the surface cover.

    wheel-track planting A practice of planting in which the seed is planted in tracks formed by wheels rolling immediately ahead of the planter.

tillage, conventional The combined primary and secondary tillage operations normally performed in preparing a seedbed for a given crop grown in a given geographic area.

tillage, primary Tillage that contributes to the major soil manipulation, commonly with a plow.

tillage, rotary An operation using a power-driven rotary tillage tool to loosen and mix soil.

tillage, secondary Any tillage operations following primary tillage designed to prepare a satisfactory seedbed for planting.

tilth The physical condition of soil as related to its ease of tillage, fitness as a seedbed, and its impedance to seedling emergence and root penetration.

topdressing An application of fertilizer to a soil after the crop stand has been established.

toposequence A sequence of related soils that differ, one from the other, primarily because of *topography* as a soil-formation factor.

topsoil (1) The layer of soil moved in cultivation. *See also* surface soil. (2) Presumably fertile soil material used to top-dress roadbanks, gardens, and lawns.

trace element (Obsolete) *See* micronutrient.

trioctahedral An octahedral sheet of silicate clays in which the sites for the six-coordinated metallic atoms are mostly filled with divalent cations such as Mg^{2+}.

truncated Having lost all or part of the upper soil horizon or horizons.

tuff Volcanic ash usually more or less stratified and in various states of consolidation.

tundra A level or undulating treeless plain characteristic of arctic regions.

Ultisols *See* soil classification.

umbric epipedon   *See* diagnostic surface horizons.

universal soil loss equation (USLE)   An equation for predicting the average annual soil loss per unit area per year, $A = RKLSPC$, where R is the climatic erosivity factor (rainfall plus runoff), K is the soil erodibility factor, L is the length of slope, S is the percent slope, P is the soil erosion practice factor, and C is the cropping and management factor.

unsaturated flow   The movement of water in a soil that is not filled to capacity with water.

variable charge   *See* pH-dependent charge.

varnish, desert   A glossy sheen or coating on stones and gravel in arid regions.

vermiculite   A 2:1-type silicate clay usually formed from mica that has a high net negative charge stemming mostly from extensive isomorphous substitution of aluminum for silicon in the tetrahedral sheet.

Vertisols   *See* soil classification.

vesicles   (1) Unconnected voids with smooth walls. (2) Spherical structures formed inside root cortical cells by vesicular, arbuscular mycorrhizal fungi.

vesicular arbuscular mycorrhiza   A common endomycorrhizal association produced by phycomycetous fungi and characterized by the development of two types of fungal structures: (1) within root cells, small structures known as *arbuscles* and (2) between root cells, storage organs known as *vesicles*. Host range includes many agricultural and horticultural crops. *See also* endomycorrhiza.

virgin soil   A soil that has not been significantly disturbed from its natural environment.

waterlogged   Saturated with water.

water potential, soil   *See* soil water potential.

water-stable aggregate   A soil aggregate stable to the action of water such as falling drops, or agitation as in wet-sieving analysis.

water table   The upper surface of groundwater or that level below which the soil is saturated with water.

water table, perched   The surface of a local zone of saturation held above the main body of groundwater by an impermeable layer of stratum, usually clay, and separated from the main body of groundwater by an unsaturated zone.

water use efficiency   Dry matter or harvested portion of crop produced per unit of water consumed.

weathering   All physical and chemical changes produced in rocks, at or near the earth's surface, by atmospheric agents.

wetland   An area of land that has hydric soil and hydrophytic vegetation, typically flooded for part of the year, and forming a transition zone between aquatic and terrestrial systems.

wilting point (permanent wilting point)   The moisture content of soil, on an oven-dry basis, at which plants wilt and fail to recover their turgidity when placed in a dark humid atmosphere.

windbreak   Planting of trees, shrubs, or other vegetation perpendicular, or nearly so, to the principal wind direction to protect soils, crops, homesteads, etc., from wind and snow.

xerophytes   Plants that grow in or on extremely dry soils or soil materials.

zero tillage   *See* tillage, conservation.

zeta potential   *See* electrokinetic potential.

zymogenous organisms   So-called opportunist organisms found in soils in large numbers immediately following addition of readily decomposable organic materials.

Allophanes, 73, 242, 243–45, 255
Alluvial deposits, 32, 35–37, 647
Aluminum
    ions and soil acidity, 272–75
    oxides, 27, 30, 63, 80, 81, 87, 106, 131, 441, 666
        hydrous, 243, 254–55, 571
        in Oxisols and Ultisols, 470
    toxicity in organic soils, 90, 284, 288, 291, 302–303, 666
Ammonia
    absorption, 407–408
    effects of, on soil, 325
    gas, 528, 538
    level and nitrification, 408
    volatilization, 407–408
Ammonification, 408, 611
Ammonium fixation
    by clay minerals, 405–406
    sources of, 408
Andisols, 72–74, 383, 498
    phosphorus-holding capacity of, 74, 470
    volcanic ash accumulation on, 73
Animals, 29 (*see also* Biota; Fauna)
Anion exchange, 268–69
    in boron, 505
    in chlorine, 504–505
    in molybdenum, 506
Antagonism, 506–507
Anthropic surface horizon (*see* Epipedons, human-modified)
Argillic subsurface horizon, 78, 79–80, 85 (*see also* Clay; Diagnostic subsurface horizons)
Aridisols, 85–86, 383, 434, 467, 633 (*see also* Calcareous soils)
    and boron deficiencies, 505
    need for irrigation in, 85–86, 664
    overgrazing on, 86
Arsenic, 615, 619
Aspergillus, 346
    flavus, 347
Azolla/Anabaena complex, 425
Azotobacter in nitrogen fixation, 47

B

Bacillus, 413
Bacteria
    and fumigant use, 610
    growth conditions of, 353
    importance of, 352
    and manganese oxidation, 497
    methanogenic, 372
    population in soils, 352
    super-, 613
Beijerinckia, 427
Bentonite, 623
Biodiversity of tropical forests, 662
Biogas production, 531, 536
Biological nitrogen fixation, 418–21, 520, 521, 524
    enhanced by pesticides, 611
    systems, 420–21

Biomass, 2, 16, 332–33, 337, 343, 368, 376, 432, 522
    fires, 434
    of tropical vegetative cover, 666–67
Bioremediation of soils, 612–14, 620–21
Biota and soil formation, 25, 46–48 (*see also* Microorganisms; Organisms in soil)
Boron, 480, 489, 490, 494, 496, 615, 619
    anion, 505
    deficiency, 491, 505
    use in suppressing fires, 525
Bradyrhizobium, 421
Broadcasting (*see* Fertilizer, application methods)
Buffer strips, 516–17
Buffering, 279–82
    capacity of soils, 281, 298
    curves, 281
    importance of, 282
Bulk density, 114–18
    factors affecting, 115–17

C

Calcareous soils, 496, 501, 505, 506, 507, 510, 555 (*see also* Aridisols)
Calcic subsurface horizon, 85 (*see also* Diagnostic subsurface horizons)
Calcium
    depletion of, 295, 297, 303
    similarity to strontium 90, 625–26
Cambic subsurface horizon, 62
Capillarity
    fundamentals of, 145–49
    mechanism, 146–48
    rate of, 172
Carbon
    cycling, 361–64
    metabolic, 390
    and nitrification, 409
    –nitrogen ratio, 372–76
    structural, 390
Carbon dioxide
    in global carbon cycle, 362–63
    and the greenhouse effect, 17
Catena, 633
Cations (*see also* Chelates; Clay)
    adsorbed, 245–46, 252, 261–62, 272, 279
        effect of, on aggregation, 125
    base-forming, 32, 47, 65, 74, 78, 79, 255
        and nitrification, 409
    exchange, 246, 262–63, 267–68, 473
        capacity (CEC), 60, 78, 241, 249, 263–66, 278, 281, 308, 623, 626, 666
    -holding capacity of soil, 46, 47, 63
    hydration, 144
    loss in forest fires, 525
    micronutrient
        deficiency of, 507
        and silicate clays, 499
        and soil pH, 495–99
    as products of weathering, 51
    prominence, 246, 494

Land drainage, 200–11
    benefits of, 206–208
    and micronutrient solubility, 507
    and nitrate loss, 411
    subsurface systems, 202–206
    surface systems, 202
Land forming, 202
Land resources
    continental differences, 661
    potentially arable, 661–62
Land Use Capability Classification, 594–97, 655
Landfills, 621–23
Leaching, 195, 523, 620
    acid, 494
    chemical influence of earthworms on, 336
    in landfills, 623
    losses and cover crops, 518
    nitrate, 404, 410–13, 535, 584
    pesticides, 603
    potassium, 475, 476, 478–79
        management of, 484–86
    requirement, 320
    and runoff of organic chemicals, 606–607
Leaf area index (LAI), 185
Legumes
    in crop rotation, 671
    and need for potassium, 473
    and need for sulfur, 431
    nickel-deficiency in, 490
    symbiotic fixation with, 421–24
Lignin, lumber waste high in, 532
Lime
    agricultural
        carbonate forms of, 294
        hydroxide forms of, 295
        oxide forms of, 295
    application on Oxisols and Ultisols, 663–64
    fineness, 298, 299–300
    on forested watersheds, 302
    -induced boron deficiency, 505
    and molybdenum availability, 506
    overuse, 303
    reactions
        with carbon dioxide, 295
        with soil colloids, 295
    to reduce solubility of micronutrients, 496–97
    requirements, 298
    and soil fertility management, 304–305
Limiting factors
    of plant growth, 543–44
    of world food production, 666–67
Lithosequences (see Soil spatial variability, medium-scale)
Loess deposits, windblown, 39, 42, 587, 633
Lumbricus terrestris, 334, 335, 336
Luxury consumption of potassium, 479–80

M

Macronutrients, 488, 489 (see also Minerals)
    fertilization, 507
Macrotermes, 337

Magnesium
    depletion of, 295, 297
    need in liming materials, 301
Management practices (see also Agricultural practices; Fertilizer; Nutrient management)
    in Alfisols, 79, 664 (see also other orders of soil)
    of climatic zones, 193
    to control disease, 357–58
    crop–soil adaption, 225
    to encourage earthworms, 337
    and fertilizer, 508, 510
    to increase carbon in soil, 368
    and micronutrient needs, 507–10, 520
    of natural wetlands, 226–27
    and nitrate production, 412
    of phosphorus, 445–46, 450–52
    plant, 224–25
    of soil nitrogen in agriculture, 429
    and soil temperature, 236–39
    of sulfur, 442
    tree and lawn, 225–26
    on tropical soils, 665–77
    in Vertisols, 84
Manganese, 488, 490, 493
    cation, 495–97
    chelate, 501
    deficiency, 507
    oxidation, 497
    oxides in sludge, 618
    toxicity, 498
Manure
    excess and salinity damage, 527
    handling, 530–31
    management, 528, 529, 534
    and micronutrient needs, 508
    nutrient composition of, 527–28
    recycling nutrients through animal, 526–28
    solar energy captured in, 526
    storage, 528
Mapping soils
    air photographs used for, 641
    techniques and tools for, 633–39
    by units, 650
    using scales, 649
Marine sediments, 32, 38–39
Mean annual soil temperature, 64, 65
Metabolism
    of organic chemicals by microbes, 608
    of soil by microflora, 332
Metals
    in hyperaccumulating plants, 620–21
    in sewage sludge, 617–19
Methane, 214, 223
    danger in landfills, 622
    and greenhouse effect, 364–65, 372
    produced by bacteria, 372, 531
    produced by termites, 340
    used in bioremediation, 613
Micelles, 278–80
Micrococcus, 413
Micronutrients
    anions, 504–506

recycling, 513–14, 521–25, 526–28, 675–76
resources, 520
supplementation of bioremediation, 613
-supplying capabilities of Alfisols, 79
transport, 4, 23–24, 82, 184, 404

O

Orders of soil, 65–66, 67–68, 91, 686 (*see also individual orders of soils*; Soil Taxonomy)
 key to, 90–91
 potential of, 663–65
Organic matter, 16, 17, 361 (*see also* Compost; Humus; Organic soils; Soil)
 amount of, in soils, 383–84
 effect of, on phosphorus fixation, 470
 fractions, 390–92
 influence of climate on, 384–85
 influence of natural vegetation on, 385
 influence on plant growth, 380, 382
 influence on soil properties and the environment, 382–83
 management of, 390–93
 mineralization, 518
 net loss of, 364, 368
 for potting media, 394–95
 quality, 390
 sources, 363
Organic soils (*see also* Carbon–nitrogen ratio)
 decomposition of, 356
 in Histosols, 88–89, 393
 mulches and soil temperature, 236–38
Organic waste disposal sites, 623–25
Organisms in soil (*see also specific organisms*; Biota; Earthworms; Nematodes; Termites)
 abundance of, 332
 beneficial effects of, 354, 356
 biochemical degradation by, 608
 comparative metabolic activity of, 332–33
 destruction of, by organic chemicals, 601–602
 detrimental effects of, on plants, 357
 effects of agricultural practices on, 354
 effects of pesticides on, 610–12
 as a factor of soil formation, 46–48
 involved in symbiotic fixation, 421
 roots as, 342–44
 types of, 327, 328, 330–32
Outwash, glacial, 40, 42
Overgrazing of Aridisol areas, 86
Oxic subsurface horizon (*see* Diagnostic subsurface horizons)
Oxidation, 31, 365
 of organic matter, 90
 reduction
  and micronutrients, 490
  and pH, 497–99
 -reduction (redox) potential, 215, 217–19, 223
 of sulfur, 439
Oxides (*see also* Aluminum; Hydrous oxides; Iron)
 of micronutrients, 493
Oxisols, 81–82, 383, 434, 441, 458, 470, 633, 663–668

Oxygen (*see also* Gaseous oxygen)
 deficiency, 224
 depletion by soil microflora, 359
 diffusion rate (ODR), 215, 217
Ozone layer, 1, 414

P

Pans, 91 (*see also* Diagnostic subsurface horizons)
 clay-, 639
 plow, 135
Parent materials
 alluvial, 664
 classification of, 32–34
 inorganic, 32
 organic, 43–44
 reaction of, 69
 and soil variability, 633
 transportation of, 32–40, 41–42
Particle density, 114
Peat
 deposits in Histosols, 89
 distribution and nature of, 43–44
 in glaciated areas, 89
 kinds of, 43–44
 as potting media, 394
Pedon, 58, 636
Penicillium, 346, 347
Pennisetum purpureum, 480
Percentage base saturation, 276, 278
Percolation, 532
 –evaporation balance, 195–97
 and groundwaters, 197
 in landfills, 623
 losses of water, 195–97
Pesticides
 behavior of, 604–609
 benefits of, 602
 chemical reactions of, 607–608
 continued use of, on same land, 609
 costs of, 602
 degradation of, by soil organisms, 608
 effects of
  on soil fauna, 611
  on soil microorganisms, 611–12
  on soil organisms, 610
 in food supply, 608
 kinds of, 602–603
 moderation in use of, 77
 persistence of, in soils, 609
 plant absorption of, 608
pH, 19–20, 90
 and availability of phosphorus, 466–67, 470
 buffering, 279–82
 of calcareous soils, 496
 and cation exchange capacity, 265–66, 494
 changes in, 282–88
 -dependent adsorption of organic chemicals, 606
 -dependent charges on silicate clays, 258–61, 274
 determination of, 291–93
 effect of lime and fertilizers on, 90, 283–84

W

Waste
food-processing, 531, 532
lumber industry, 531, 532
municipal solid (MSW), 531–32, 534
radioactive, 626
recycling, 535–36
sewage effluents and sludge, 531, 532–34
water treatment, 532
Water (*see also* Capillarity; Conservation of moisture; Groundwater; Hydrologic cycle; Irrigation; Rain; Runoff; Soil moisture; Soil water energy; Sources of water; Water management; Water quality)
drainage, 90, 199–200
flow, 90
hydrogen bonding, 145
management on tropical soils, 624
polarity of molecules, 144
pollution, 565
and nutrient imbalance, 515
quality, 516, 525
–soil interactions, 143
soil storage, 179
structure, 144–45
supply, regulation of, 3, 5–6
surface tension, 145
table, importance of level of, 90
Water conservation in Africa, 667, 674
Water erosion, 564, 565, 567
types of, 568
gulley, 585, 587
Water-holding capacity, 105
of Histosols, 89
Water management (*see also* Irrigation)
of arid and semiarid regions, 308–309
Water pollution, 565, 625, 662

Water quality
degradation, 450–52
of salt-affected soils, 318–20
Water vapor movement (*see* Soil moisture)
Weathering
acid solution, 30–31
of rocks and minerals
chemical processes of, 29–32
physical, 28–29, 520, 521
resistance to 26, 106, 481
of soils, 25, 106, 523
high, of Oxisols, 81
Weight of Histosol soils, 89
Wetlands
drainage of coastal, 285, 287
nitrification and denitrification in, 416–17
used for agriculture, 90
volatilization from, 407
Wind erosion (*see also* Desertification)
control, 587–88, 592, 594–97
mechanics of, 588–90
and nutrient relocation, 523
predicting, 591–92
reduced by conservation practices, 564, 565
susceptibility to, 590
in Histosols, 90
World population, 660

X

Xerophitic plants, 86

Z

Zinc, 488, 490, 493, 499
cation, 495–97
chelates, 501–502, 503–504
deficiency, 494, 497, 507, 510

Periodic Table of the Elements

Based on $^{12}_{6}C$. Numbers in parentheses are the mass numbers of the most stable isotopes of radioactive elements.

Group IA	Group IIA	Group IIIB	Group IVB	Group VB	Group VIB	Group VIIB	Group VIIIB	Group VIIIB	Group VIIIB	Group IB	Group IIB	Group IIIA	Group IVA	Group VA	Group VIA	Group VIIA	Group VIIIA
1 H 1.01																	2 He 4.00
3 Li 6.94	4 Be 9.01											5 B 10.81	6 C 12.01	7 N 14.01	8 O 16.00	9 F 19.00	10 Ne 20.18
11 Na 22.99	12 Mg 24.30											13 Al 26.98	14 Si 28.09	15 P 30.97	16 S 32.07	17 Cl 35.45	18 Ar 39.95
19 K 39.10	20 Ca 40.08	21 Sc 44.96	22 Ti 47.88	23 V 50.94	24 Cr 52.00	25 Mn 54.94	26 Fe 55.85	27 Co 58.93	28 Ni 58.69	29 Cu 63.55	30 Zn 65.38	31 Ga 69.72	32 Ge 72.59	33 As 74.92	34 Se 78.96	35 Br 79.90	36 Kr 83.80
37 Rb 85.47	38 Sr 87.62	39 Y 88.91	40 Zr 91.22	41 Nb 92.91	42 Mo 95.94	43 Tc (98)	44 Ru 101.07	45 Rh 102.91	46 Pd 106.42	47 Ag 107.87	48 Cd 112.41	49 In 114.82	50 Sn 118.71	51 Sb 121.75	52 Te 127.60	53 I 126.90	54 Xe 131.29
55 Cs 132.91	56 Ba 137.33	57 La 138.91	72 Hf 178.49	73 Ta 180.95	74 W 183.85	75 Re 186.21	76 Os 190.2	77 Ir 192.22	78 Pt 195.08	79 Au 196.97	80 Hg 200.59	81 Tl 204.38	82 Pb 207.2	83 Bi 208.98	84 Po (209)	85 At (210)	86 Rn (222)
87 Fr (223)	88 Ra (226)	89 Ac (227)	104 Unq (261)	105 Unp (262)	106 Unh (263)	107 Uns (262)	108 Uno (265)	109 Une (266)									

58 Ce 140.12	59 Pr 140.91	60 Nd 144.24	61 Pm (145)	62 Sm 150.36	63 Eu 151.96	64 Gd 157.25	65 Tb 158.93	66 Dy 162.50	67 Ho 164.93	68 Er 167.26	69 Tm 168.93	70 Yb 173.04	71 Lu 174.97
90 Th (232)	91 Pa (231)	92 U (238)	93 Np (237)	94 Pu (244)	95 Am (243)	96 Cm (247)	97 Bk (247)	98 Cf (251)	99 Es (252)	100 Fm (257)	101 Md (258)	102 No (259)	103 Lr (260)

Metals ← → Nonmetals